Advanced Dynamics

Dan B. Marghitu • Mihai Dupac

Advanced Dynamics

Analytical and Numerical Calculations with MATLAB

 Springer

Dan B. Marghitu
Mechanical Engineering Department
Auburn University
Auburn, Alabama, USA

Mihai Dupac
School of Design, Engineering & Computing
Bournemouth University
Talbot Campus
Ferne Barrow, Poole, United Kingdom

Additional material to this book can be downloaded from http://extra.springer.com

ISBN 978-1-4899-9425-7   ISBN 978-1-4614-3475-7 (eBook)
DOI 10.1007/978-1-4614-3475-7
Springer New York Heidelberg Dordrecht London

Preface

Engineering mechanics involves the development of mathematical models of the physical world. Dynamics, a branch of mechanics, addresses the effects of forces on the motion of a body or system of bodies. This book deals with the understanding of the dynamic behavior of engineering structures and components. The tools of formulating the mathematical equations and also the methods of solving the equations are discussed. A knowledge of the motion for the structures and components is most important for their design.

MATLAB® is a modern tool that has transformed the mathematical calculations methods because MATLAB not only provides numerical calculations but also facilitates analytical calculations using the computer. This book uses MATLAB as a tool to prove some concepts and to solve problems. The intent is to show the convenience of MATLAB for theory and applications in dynamics. Using example problems, the MATLAB syntax will be demonstrated. MATLAB is very useful in the process of deriving solutions for any problem in dynamics. This book includes a large number of problems that are being solved using MATLAB. Specific functions dealing with dynamic topics are created.

The main distinction of this study from other projects and books is the use of symbolic MATLAB for both theory and applications. Special attention is given to the solutions of the problems that are solved analytically and numerically using MATLAB. The figures generated with MATLAB reinforce visual learning for students as they study the programs.

This book provides a thorough, rigorous presentation of kinematics and dynamics, augmented with proven learning techniques for the benefit of instructor and student. Our first objective is to present the topics thoroughly and directly, allowing fundamental principles to emerge through applications. We emphasize concepts, derivations, and interpretations of the general principles.

Modern technical advancements in areas of multibody systems, robotics, spacecraft, and design of complex mechanical devices and mechanisms in industry need knowledge of solving advanced dynamic concepts. We discussed the tools of formulating the mathematical equations and also the methods of solving them using a modern computing tool like MATLAB. Included are analytical and numerical methods for explaining the dynamics problems using computer programs.

This book is extremely useful for a number of reasons. This book presents theory, computational aspects, and applications of dynamical systems and is a straightforward introduction to the subject. It provides coverage of basic material with both mathematical and physical. This book will assist the graduate students interested in the classical principles of dynamical systems and is used primarily for a one semester course in dynamics. This book can be used for classroom instruction, and it can be used for a self-study and can also be offered as distance learning. It would be appropriate for use as a text for senior undergraduate and first year graduate students.

This text is based on MATLAB Version 7.12 (R2011a) and requires the use of the Symbolic Math Toolbox^{TM}.

The MATLAB programs for the solved examples are found on Springer Extras at http://extras.springer.com/andathttp://www.eng.auburn.edu/~marghitu/.

Contents

1 Vector Algebra 1
  1.1 Terminology and Notation 1
  1.2 Position Vector 10
  1.3 Scalar (Dot) Product of Vectors 11
  1.4 Vector (Cross) Product of Vectors 13
  1.5 Scalar Triple Product of Three Vectors 15
  1.6 Vector Triple Product of Three Vector 18
  1.7 Derivative of a Vector Function 18
  1.8 Cauchy's Inequality, Lagrange's Identity,
     and Triangle Inequality 20
  1.9 Coordinate Transformation 22
  1.10 Tensors 26
     1.10.1 Operations with Tensors 34
     1.10.2 Some Further Properties of Second-Order Tensor 35
  1.11 Examples 36
  1.12 Problems 65
  1.13 Program 69

2 Centroids and Moments of Inertia 73
  2.1 Centroids and Center of Mass 73
     2.1.1 First Moment and Centroid of a Set of Points 73
     2.1.2 Centroid of a Curve, Surface, or Solid 75
     2.1.3 Mass Center of a Set of Particles 77
     2.1.4 Mass Center of a Curve, Surface, or Solid 77
     2.1.5 First Moment of an Area 79
     2.1.6 Center of Gravity 82
     2.1.7 Theorems of Guldinus–Pappus 82
  2.2 Moments of Inertia 85
     2.2.1 Introduction 85
     2.2.2 Translation of Coordinate Axes 88

2.2.3   Principal Axes 90
2.2.4   Ellipsoid of Inertia 92
2.2.5   Moments of Inertia for Areas 93
2.3   Examples 100
2.4   Problems 130

3 Kinematics of a Particle 143
3.1   Introduction 143
3.1.1   Position, Velocity, and Acceleration 143
3.1.2   Angular Motion of a Line 144
3.1.3   Rotating Unit Vector 145
3.2   Rectilinear Motion 146
3.3   Curvilinear Motion 147
3.3.1   Cartesian Coordinates 147
3.3.2   Normal and Tangential Coordinates 148
3.3.3   Circular Motion 154
3.3.4   Polar Coordinates 155
3.3.5   Cylindrical Coordinates 157
3.4   Relative Motion 158
3.5   Frenet's Formulas 159
3.6   Examples 166
3.7   Problems 201

4 Dynamics of a Particle 209
4.1   Newton's Second Law 209
4.2   Newtonian Gravitation 210
4.3   Inertial Reference Frames 211
4.4   Cartesian Coordinates 211
4.4.1   Projectile Problem 212
4.4.2   Straight Line Motion 213
4.5   Normal and Tangential Components 213
4.6   Polar and Cylindrical Coordinates 214
4.7   Principle of Work and Energy 215
4.8   Work and Power 217
4.8.1   Work Done on a Particle by a Linear Spring 218
4.8.2   Work Done on a Particle by Weight 219
4.9   Conservation of Energy 221
4.9.1   Exercise 221
4.9.2   Exercise 222
4.10   Conservative Forces 227
4.10.1   Potential Energy of a Force Exerted by a Spring 227
4.10.2   Potential Energy of Weight 228
4.10.3   Exercise 229
4.10.4   Exercise 232
4.11   Principle of Impulse and Momentum 234
4.12   Conservation of Linear Momentum 235

4.13   Principle of Angular Impulse and Momentum 237
4.14   Examples 238
4.15   Problems 275

5   Kinematics of Rigid Bodies 281
   5.1   Introduction 281
   5.2   Velocity Analysis for a Rigid Body 282
   5.3   Acceleration Analysis for a Rigid Body 285
         5.3.1   Translation 286
         5.3.2   Rotation 289
         5.3.3   Helical Motion 294
         5.3.4   Planar Motion 298
   5.4   Angular Velocity Vector of a Rigid Body 300
   5.5   Motion of a Point that Moves Relative to a Rigid Body 304
   5.6   Planar Instantaneous Center 309
   5.7   Fixed and Moving Centrodes 311
   5.8   Closed Loop Equations 321
         5.8.1   Closed Loop Velocity Equations 323
         5.8.2   Closed Loop Acceleration Equations 325
   5.9   Independent Closed Loops Method 328
   5.10   Closed Kinematic Chains with MATLAB Functions 335
         5.10.1   Driver Link 335
         5.10.2   Position Analysis 337
         5.10.3   Complete Rotation of the Driver Link 344
         5.10.4   Velocity and Acceleration Analysis 348
   5.11   Examples 357
   5.12   Problems 400

6   Dynamics of Rigid Bodies 411
   6.1   Equation of Motion for the Mass Center 411
   6.2   Linear Momentum and Angular Momentum 414
   6.3   Spatial Angular Momentum of a Rigid Body 417
   6.4   Kinetic Energy of a Rigid Body 421
   6.5   Equations of Motion 423
   6.6   Euler's Equations of Motion 425
   6.7   Motion of a Rigid Body About a Fixed Point 426
   6.8   Rotation of a Rigid Body About a Fixed Axis 426
   6.9   Plane Motion of Rigid Body 427
         6.9.1   D'Alembert's Principle 431
         6.9.2   Free-Body Diagrams 432
         6.9.3   Force Analysis for Closed Kinematic Chains
               Using MATLAB Functions 437
   6.10   Examples 446
   6.11   Problems 510

7 Analytical Dynamics 521
   7.1   Introduction 521
   7.2   Equations of Motion 525
   7.3   Hamilton's Equations... 528
   7.4   Poisson Bracket 531
   7.5   Rotation Transformation 533
   7.6   Examples 537
   7.7   Problems 594

References 601

Index 603

Chapter 1
Vector Algebra

1.1 Terminology and Notation

Scalars are mathematics quantities that can be fully defined by specifying their magnitude in suitable units of measure. Mass is a scalar quantity and can be expressed in kilograms, time is a scalar and can be expressed in seconds, and temperature is a scalar quantity that can be expressed in degrees Celsius.

Vectors are quantities that require the specification of magnitude, orientation, and sense. The characteristics of a vector are the magnitude, the orientation, and the sense.

The *magnitude* of a vector is specified by a positive number and a unit having appropriate dimensions. No unit is stated if the dimensions are those of a pure number.

The *orientation* of a vector is specified by the relationship between the vector and given reference lines and/or planes.

The *sense* of a vector is specified by the order of two points on a line parallel to the vector.

Orientation and sense together determine the *direction* of a vector.

The *line of action* of a vector is a hypothetical infinite straight line collinear with the vector.

Displacement, velocity, and force are examples of vectors quantities.

To distinguish vectors from scalars, it is customary to denote vectors by boldface letters. Thus, the displacement vector from point A to point B could be denoted as $\mathbf{r}$ or $\mathbf{r}_{AB}$. The symbol $|\mathbf{r}| = r$ represents the magnitude (or module, norm, or absolute value) of the vector $\mathbf{r}$. In handwritten work, a distinguishing mark is used for vectors, such as an arrow over the symbol, $\vec{r}$ or $\overrightarrow{AB}$, a line over the symbol, $\bar{r}$, or an underline, $\underline{r}$.

The vectors are most frequently depicted by straight arrows. A vector represented by a straight arrow has the direction indicated by the arrow. The displacement vector from point A to point B is depicted in Fig. 1.1a as a straight arrow. In some cases, it is necessary to depict a vector whose direction is perpendicular to the surface in which

D.B. Marghitu and M. Dupac, *Advanced Dynamics: Analytical and Numerical Calculations with MATLAB*, DOI 10.1007/978-1-4614-3475-7_1,
© Springer Science+Business Media, LLC 2012

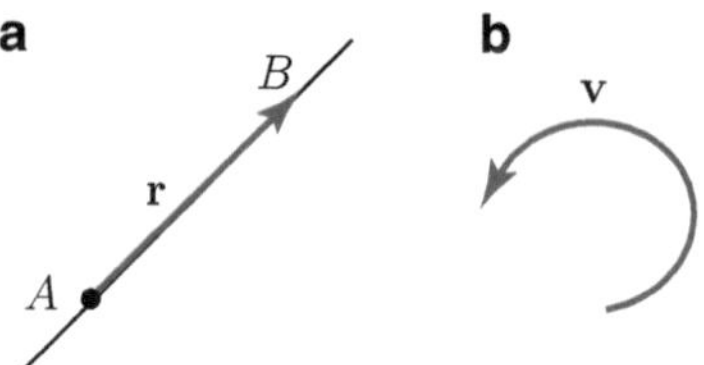

Fig. 1.1 Representations of vectors

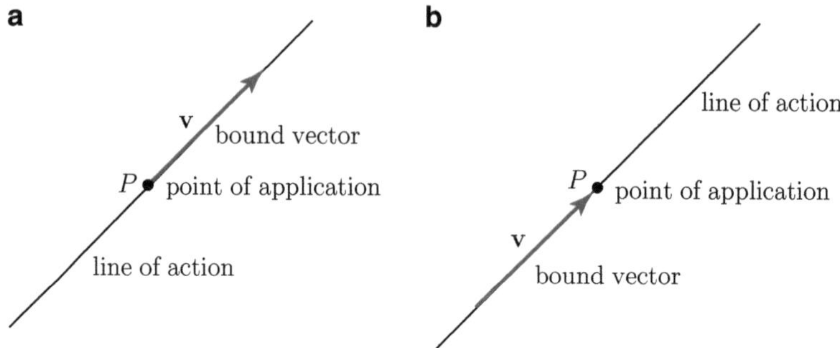

Fig. 1.2 Bound or fixed vector: (**a**) point of application represented as the tail of the vector arrow and (**b**) point of application represented as the head of the vector arrow

the representation will be drawn. Under this circumstance, the use of a portion of a circle with a direction arrow is useful. The orientation of the vector is perpendicular to the plane containing the circle, and the sense of the vector is the same as the direction in which a right-handed screw moves when the axis of the screw is normal to the plane in which the arrow is drawn and the screw is rotated as indicated by the arrow. Figure 1.1b uses this representation to depict a vector directed out of the reading surface toward the reader.

A *bound* vector is a vector associated with a particular point P in space (Fig. 1.2). The point P is the *point of application* of the vector, and the line passing through P and parallel to the vector is the line of action of the vector. The point of application may be represented as the tail, Fig. 1.2a, or the head of the vector arrow, Fig. 1.2b. A *free* vector is not associated with any particular point in space. A *transmissible* (or *sliding*) vector is a vector that can be moved along its line of action without change of meaning.

To move the rigid body in Fig. 1.3, the force vector **F** can be applied anywhere along the line Δ or may be applied at specific points A, B, and C. The force vector **F** is a transmissible vector because the resulting motion is the same in all cases.

If the body is not rigid, the force **F** applied at A will cause a different deformation of the body than **F** applied at a different point B. If one is interested in the deformation of the body, the force **F** positioned at C is a bound vector.

The operations of vector analysis deal only with the characteristics of vectors and apply, therefore, to bound, free, and transmissible vectors.

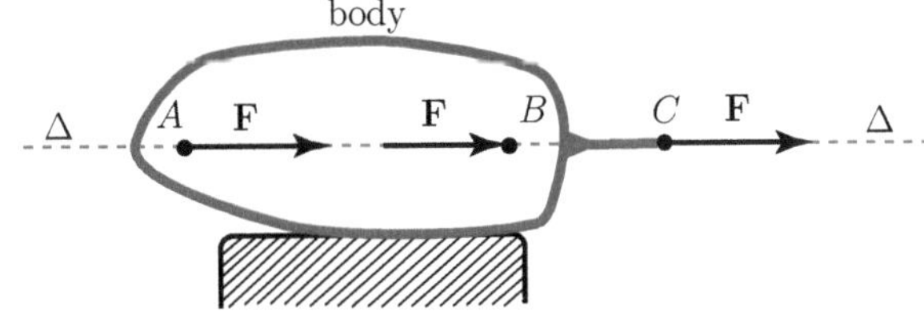

Fig. 1.3 Transmissible vector: the force vector **F** can be applied anywhere along the line Δ

Equality
Two vectors **a** and **b** are said to be equal to each other when they have the same characteristics. One then writes

$$\mathbf{a} = \mathbf{b}. \tag{1.1}$$

Equality does not imply physical equivalence. For instance, two forces represented by equal vectors do not necessarily cause identical motions of a body on which they act.

Product of a Vector and a Scalar
The product of a vector **v** and a scalar s, $s\mathbf{v}$ or $\mathbf{v}s$, is a vector having the following characteristics:

1. Magnitude. $|s\mathbf{v}| \equiv |\mathbf{v}s| = |s||\mathbf{v}|$, where $|s| = s$ denotes the absolute value (or magnitude, or module) of the scalar s.
2. Orientation. $s\mathbf{v}$ is parallel to **v**. If $s = 0$, no definite orientation is attributed to $s\mathbf{v}$.
3. Sense. If $s > 0$, the sense of $s\mathbf{v}$ is the same as that of **v**. If $s < 0$, the sense of $s\mathbf{v}$ is opposite to that of **v**. If $s = 0$, no definite sense is attributed to $s\mathbf{v}$.

Zero Vector
A *zero vector* is a vector that does not have a definite direction and whose magnitude is equal to zero. The symbol used to denote a zero vector is **0**.

Unit Vector
A *unit vector* is a vector with magnitude equal to 1. Given a vector **v**, a unit vector **u** having the same direction as **v** is obtained by forming the product of **v** with the reciprocal of the magnitude of **v**

$$\mathbf{u} = \mathbf{v}\,\frac{1}{|\mathbf{v}|} = \frac{\mathbf{v}}{|\mathbf{v}|}. \tag{1.2}$$

Vector Addition
The sum of a vector $\mathbf{v}_1$ and a vector $\mathbf{v}_2$: $\mathbf{v}_1 + \mathbf{v}_2$ or $\mathbf{v}_2 + \mathbf{v}_1$ is a vector whose characteristics can be found by either graphical or analytical processes. The vectors $\mathbf{v}_1$ and $\mathbf{v}_2$ add according to the parallelogram law: the vector $\mathbf{v}_1 + \mathbf{v}_2$ is represented by the diagonal of a parallelogram formed by the graphical representation of the vectors, see Fig. 1.4a.

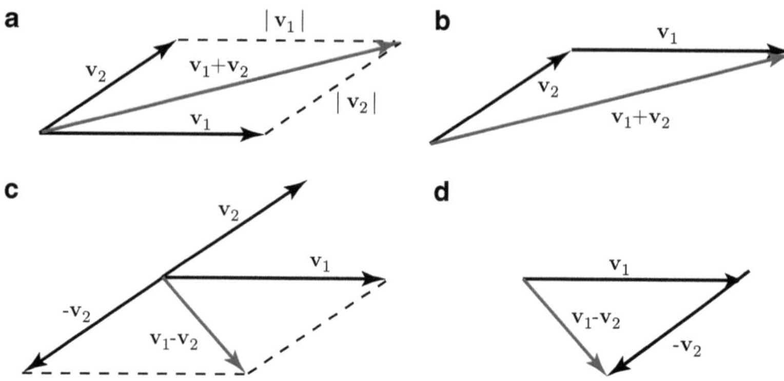

Fig. 1.4 Vector addition: (**a**) parallelogram law, (**b**) moving the vectors successively to parallel positions. Vector difference: (**c**) parallelogram law, (**d**) moving the vectors successively to parallel positions

The vector $\mathbf{v}_1 + \mathbf{v}_2$ is called the *resultant* of $\mathbf{v}_1$ and $\mathbf{v}_2$. The vectors can be added by moving them successively to parallel positions so that the head of one vector connects to the tail of the next vector. The resultant is the vector whose tail connects to the tail of the first vector, and whose head connects to the head of the last vector, see Fig. 1.4b.

The sum $\mathbf{v}_1 + (-\mathbf{v}_2)$ is called the *difference* of $\mathbf{v}_1$ and $\mathbf{v}_2$ and is denoted by $\mathbf{v}_1 - \mathbf{v}_2$, see Fig. 1.4c, d. The sum of n vectors $\mathbf{v}_i$, $i = 1,\dots,n$,

$$\sum_{i=1}^{n} \mathbf{v}_i \ \text{or} \ \mathbf{v}_1 + \mathbf{v}_2 + \cdots + \mathbf{v}_n$$

is called the *resultant* of the vectors $\mathbf{v}_i$, $i = 1,\dots,n$.

Vector addition is:

1. Commutative, that is, the characteristics of the resultant are independent of the order in which the vectors are added (commutativity law for addition)

$$\mathbf{v}_1 + \mathbf{v}_2 = \mathbf{v}_2 + \mathbf{v}_1.$$

2. Associative, that is, the characteristics of the resultant are not affected by the manner in which the vectors are grouped (associativity law for addition)

$$\mathbf{v}_1 + (\mathbf{v}_2 + \mathbf{v}_3) = (\mathbf{v}_1 + \mathbf{v}_2) + \mathbf{v}_3$$

3. Distributive, that is, the vector addition obeys the following laws of distributivity:

$$(s_1 + s_2)\mathbf{v} = s_1\mathbf{v} + s_2\mathbf{v} \ \ \text{and} \ \ s(\mathbf{v}_1 + \mathbf{v}_2) = s\mathbf{v}_1 + s\mathbf{v}_2,$$

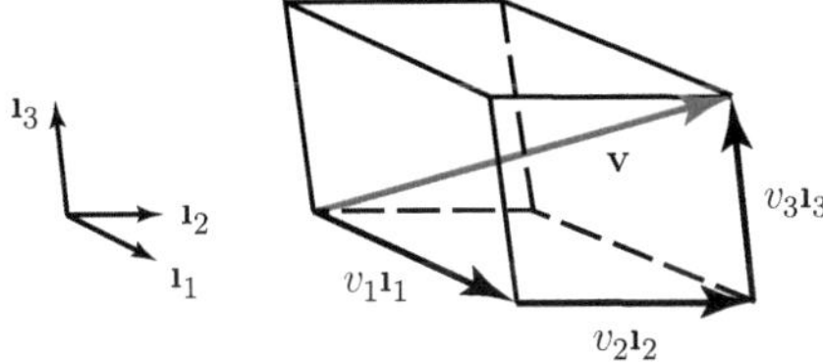

Fig. 1.5 Resolution of a vector **v** and components

or equivalent (for the general case)

$$\mathbf{v}\sum_{i=1}^{n}s_i = \sum_{i=1}^{n}(\mathbf{v}s_i) \quad \text{and} \quad s\sum_{i=1}^{n}\mathbf{v}_i = \sum_{i=1}^{n}(s\mathbf{v}_i).$$

Moreover, the characteristics of the resultant is not affected by the manner in which the vector is multiplied with scalars (associativity law for multiplication)

$$s_1\left(s_2\mathbf{v}\right) = \left(s_1 s_2\right)\mathbf{v}.$$

Every vector can be regarded as the sum of n vectors ($n = 2,3,\ldots$) of which all but one can be selected arbitrarily.

Linear Independence
If $\mathbf{v}_i$, $i = 1,\ldots,n$ are vectors and s_i, $i = 1,\ldots,n$ are scalars, then a *linear combination* of the vectors with the scalars as coefficients is defined as $\sum_{i=1}^{n}s_i\mathbf{v}_i = s_1\mathbf{v}_1 + \cdots + s_n\mathbf{v}_n$.

A collection of nonzero vectors is said to be *linearly independent* if no vector in the set can be written as a linear combination of the remaining vectors in the set. The dimension of the space is equal to the maximum number of nonzero vectors that can be included in a linearly independent set of vectors Thus, for a three-dimensional space, the maximum number of nonzero vectors in a linearly independent collection is three. Given a set of three linearly independent vectors, any other vector can be constructed as a resultant of scalar multiplication of the three vectors. Such a set of vectors is called a basis set. A set of vectors which is not linearly independent is called linearly dependent.

Resolution of Vectors and Components
Let $\mathbf{l}_1$, $\mathbf{l}_2$, $\mathbf{l}_3$ be three linearly independent unit vectors as a basis set:

$$\left|\mathbf{l}_1\right| = \left|\mathbf{l}_2\right| = \left|\mathbf{l}_3\right| = 1.$$

For a given vector **v** (Fig. 1.5), there exist three unique scalars v_1, v_1, v_3, such that **v** can be expressed as

$$\mathbf{v} = v_1\mathbf{l}_1 + v_2\mathbf{l}_2 + v_3\mathbf{l}_3. \tag{1.3}$$

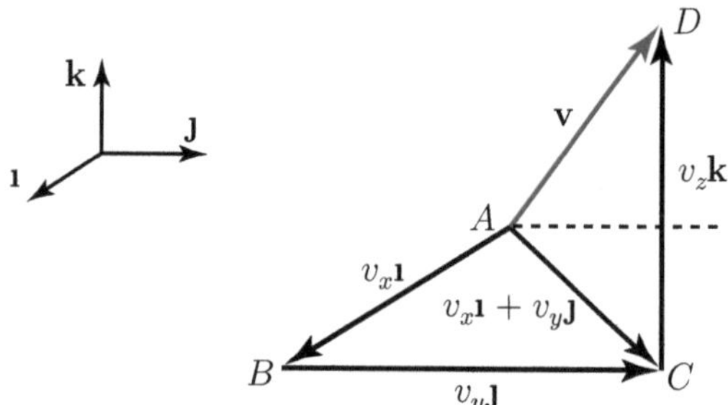

Fig. 1.6 Cartesian reference frame and the orthogonal scalar components v_x, v_y, v_z

The opposite action of addition of vectors is the *resolution* of vectors. Thus, for the given vector **v**, the vectors $v_1\mathbf{l}_1$, $v_2\mathbf{l}_2$, and $v_3\mathbf{l}_3$ sum to the original vector. The vector $v_k\mathbf{l}_k$ is called the $\mathbf{l}_k$ *component* of **v** relative to the given basis set, and v_k is called the $\mathbf{l}_k$ *scalar component* of **v** relative to the given basis set, where $k = 1, 2, 3$. A vector is often replaced by its components since the components are equivalent to the original vector.

Frequently, a vector will be given, and its components relative to a particular basis set need to be calculated. A trivial example of this situation occurs when the vector to be resolved is the zero vector. Then, each of its components is zero. Thus, under these circumstances, every vector equation $\mathbf{v} = \mathbf{0}$, where $\mathbf{v} = v_1\mathbf{l}_1 + v_2\mathbf{l}_2 + v_3\mathbf{l}_3$, is equivalent to three scalar equations $v_1 = 0$, $v_2 = 0$, $v_3 = 0$. Note that the zero vector $\mathbf{0}$ is not the number zero.

If the unit vectors $\mathbf{l}_1$, $\mathbf{l}_2$, $\mathbf{l}_3$ are mutually perpendicular, they form a *Cartesian basis* or a *Cartesian reference frame*. For a Cartesian reference frame, the following notation is used (Fig. 1.6):

$$\mathbf{l}_1 \equiv \mathbf{l}, \; \mathbf{l}_2 \equiv \mathbf{J}, \; \mathbf{l}_3 \equiv \mathbf{k} \quad \text{and} \quad \mathbf{l} \perp \mathbf{J}, \; \mathbf{l} \perp \mathbf{k}, \; \mathbf{J} \perp \mathbf{k}.$$

The symbol $\perp$ denotes perpendicular. When a vector **v** is expressed in the form $\mathbf{v} = v_x\mathbf{l} + v_y\mathbf{J} + v_z\mathbf{k}$ where $\mathbf{l}, \mathbf{J}, \mathbf{k}$ are mutually perpendicular unit vectors (Cartesian reference frame or orthogonal reference frame), the magnitude of **v** is given by

$$|\mathbf{v}| = \sqrt{v_x^2 + v_y^2 + v_z^2}. \tag{1.4}$$

The vectors $\mathbf{v}_x = v_x\mathbf{l}$, $\mathbf{v}_y = v_y\mathbf{J}$, and $\mathbf{v}_z = v_y\mathbf{k}$ are the *orthogonal* or *rectangular component vectors* of the vector **v**. The measures v_x, v_y, v_z are the *orthogonal* or *rectangular scalar components* of the vector **v**.

The resolution of a vector into components frequently facilitate the valuation of a vector equation. If $\mathbf{v}_1 = v_{1x}\mathbf{l} + v_{1y}\mathbf{J} + v_{1z}\mathbf{k}$ and $\mathbf{v}_2 = v_{2x}\mathbf{l} + v_{2y}\mathbf{J} + v_{2z}\mathbf{k}$, then the sum of the vectors is

$$\mathbf{v}_1 + \mathbf{v}_2 = (v_{1x} + v_{2x})\,\mathbf{l} + (v_{1y} + v_{2y})\,\mathbf{J} + (v_{1z} + v_{2z})\,\mathbf{k}.$$

Similarly,

$$\mathbf{v}_1 - \mathbf{v}_2 = \left(v_{1x} - v_{2x}\right)\mathbf{i} + \left(v_{1y} - v_{2y}\right)\mathbf{j} + \left(v_{1z} - v_{2z}\right)\mathbf{k}.$$

In the MATLAB® environment, a three-dimensional row vector v is written as a list of variables v = [v_x v_y v_z] or v = [v_x, v_y, v_z] where v_x, v_y, and v_z are the spatial coordinates of the vector v. The elements of a row are separated with blanks or commas. The list of elements are surrounded with square brackets, [ ]. The first component of the vector v is v_x=v(1), the second component is v_y=v(2), and the third component is v_z=v(3). The semicolon ; is used to separate the end of each row for a column vector. To create a numerical vector, the following statement is used:

```
p = [ 1 2 3 ],
```

where 1, 2, and 3 are the numerical components of the row vector p. When a variable name is assigned to data, the data is immediately displayed, along with its name. The display of the data can be suppressed by using the semicolon, ;, at the end of a statement.

The symbolic MATLAB Toolbox can perform symbolical calculation, and a vector v can be expressed in MATLAB in a symbolical fashion. In MATLAB, the sym command constructs symbolic variables and expressions. The commands

```
v_x = sym('v_x','real');
v_y = sym('v_y','real');
v_z = sym('v_z','real');
```

create a symbolic variables v_x, v_y, and v_z and also assume that the variables are real numbers. The symbolic variables can then be treated as mathematical variables. One can use the statement syms for generating a shortcut for constructing symbolic objects:

```
syms v_x v_y v_z real
v = [ v_x v_y v_z ];
```

where v is a symbolic vector. The same symbolic vector can be created with:

```
v = sym('[v_x v_y v_z]');
```

In MATLAB, a vector is defined as a matrix with either one row or one column. To make distinction between row vectors and column vectors is essential, especially when operations with vectors are required. Many errors are caused by using a row vector instead of a column vector or vice versa. The command zeros(m,n) or zeros([m n]) returns an m-by-n matrix of zeros. A zero row vector [0 0 0] is generated with zeros(1,3), and a zero column vector is generated with zeros(3,1). The command ones(m,n) or ones([m n]) returns an m-by-n matrix of ones. In MATLAB, two vectors u and v of the same size (defined either as column vectors or row vectors) can be added together using the next command:

```
u + v,
```

Vectors addition in MATLAB must follow strict rules. The vectors should be either column vectors or row vectors in order to be added and should have the same dimension. It is not possible to add a row vector to a column vector. To subtract one vector from another of the same size, use a minus $(-)$ sign. The subtraction applied to the vectors u and v can be written in MATLAB as

```
u - v
```

or

```
v - u.
```

The magnitude of the vector p can be found using the next MATLAB command:

```
norm(p).
```

The MATLAB command `norm(p)` does not work if the components of the vector p are given symbolically. Thus, a more general MATLAB function is created for the magnitude of the vector, v, with the components `v(1)`, `v(2)`, and `v(3)`. A MATLAB function is a program that performs an action and returns a result. The MATLAB function `magn` calculates the magnitude of the vector, v, in a symbolical or numerical fashion:

```
function val = magn(v)
% The symbolic magnitude function of a vector
%    v = [v(1) v(2) v(3)]
% The function accepts sym as the input argument
val = sqrt(v(1)*v(1)+v(2)*v(2)+v(3)*v(3));
```

The MATLAB statement `sqrt(x)` is the square root of the elements of x. The power of MATLAB comes into play when one can add new functions to enhance the language. The m-file function file starts with a line declaring the function, the arguments, and the outputs. Next, the statements required to produce the outputs from the inputs (arguments) are presented. It is important to note that the argument and output names used in a function file are strictly local variables that exist only within the function itself. The function returns information via the output. To calculate the magnitude of the vector `v = [v_x v_y v_z]` using the `magn` function, the following MATLAB command is used:

```
mv = magn(v),
```

and the output is

```
mv =
 (v_x^2+v_y^2+v_z^2)^(1/2).
```

To create a unit vector in the direction of the vector v the following command is used `p/norm(p)` or `v/magn(v)` where the division symbol (/) divides all the elements in the vector by the magnitude of the vector, producing a vector of the same size and direction.

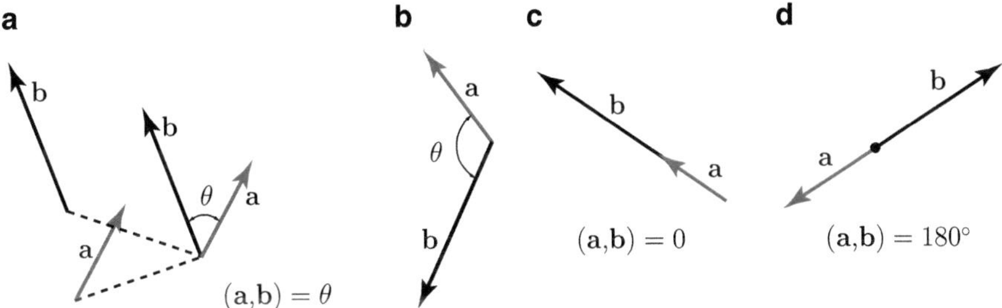

Fig. 1.7 The angle θ between the vectors **a** and **b**: (**a**) $0 < \theta < 90°$, (**b**) $90° < \theta < 180°$, and (**c**) $\theta = 0°$, and (**d**) $\theta = 180°$

Vector transposition is as easy as adding an apostrophe, $'$, (prime) to the name of the vector. Thus if v = [v_x v_y v_z], then v$'$ is:

```
v_x
v_y
v_z.
```

The mutually perpendicular unit vectors **ı**, **ȷ**, and **k** are defined in MATLAB by

```
i=[1 0 0]; j=[0 1 0]; k=[0 0 1];
```

Angle Between Two Vectors

The angle between two vectors can be determined by moving either or both vectors parallel to themselves (leaving the sense unaltered) until their initial points (tails) coincide. This angle will always be in the range between $0°$ and $180°$ inclusive. Four possible situations are shown in Fig. 1.7 where the two vectors are denoted **a** and **b**. The *angle* between **a** and **b** is the angle θ in Fig. 1.7a, b. The angle between **a** and **b** is denoted by the symbols $(\mathbf{a},\mathbf{b})$ or $(\mathbf{b},\mathbf{a})$. Figure 1.7c represents the case $(\mathbf{a},\mathbf{b}) = 0$, and Fig. 1.7d represents the case $(\mathbf{a},\mathbf{b}) = 180°$.

The direction of a vector $\mathbf{v} = v_x\mathbf{ı}+v_y\mathbf{ȷ}+v_z\mathbf{k}$ and relative to a Cartesian reference, **ı**, **ȷ**, **k**, is given by the cosines of the angles formed by the vector and the respective unit vectors. These are called *direction cosines* and are denoted as (Fig. 1.8)

$$\cos(\mathbf{v},\mathbf{ı}) = \cos\alpha = \cos\theta_x = l, \ \ \cos(\mathbf{v},\mathbf{ȷ}) = \cos\beta = \cos\theta_y = m, \text{ and}$$

$$\cos(\mathbf{v},\mathbf{k}) = \cos\gamma = \cos\theta_z = n. \tag{1.5}$$

The following relations exist: $v_x = |\mathbf{v}|\cos\alpha; \ v_y = |\mathbf{v}|\cos\beta; \ v_z = |\mathbf{v}|\cos\gamma.$

From these definitions, it follows that

$$\cos^2\alpha + \cos^2\beta + \cos^2\gamma = 1 \ \ \text{or} \ \ l^2 + m^2 + n^2 = 1. \tag{1.6}$$

Equation (1.6) is proved using the MATLAB commands:

```
syms v_x v_y v_z
v = [v_x v_y v_z];
```

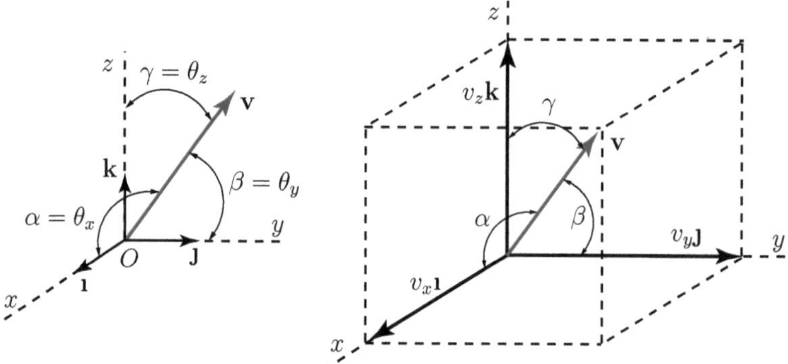

Fig. 1.8 Direction cosines

```
mv = magn(v);
l = v_x/mv;
m = v_y/mv;
n = v_z/mv;
simplify(l^2+m^2+n^2).
```

The MATLAB statement `simplify(x)` simplifies each element of the symbolic matrix x.

Recall that the formula for the unit vector of the vector **v** is

$$\mathbf{u_v} = \frac{\mathbf{v}}{|\mathbf{v}|} = \frac{\mathbf{v}}{v} = \frac{v_x}{v}\mathbf{I} + \frac{v_y}{v}\mathbf{J} + \frac{v_x}{v}\mathbf{k},$$

or written in another way

$$\mathbf{u_v} = \cos\alpha\,\mathbf{I} + \cos\beta\,\mathbf{J} + \cos\gamma\,\mathbf{k}. \tag{1.7}$$

1.2 Position Vector

The position vector of a point P relative to a point O is a vector $\mathbf{r}_{OP} = \overrightarrow{OP}$ having the following characteristics:

1. Magnitude the length of line OP.
2. Orientation parallel to line OP.
3. Sense OP (from point O to point P).

The vector $\mathbf{r}_{OP}$ is shown as an arrow connecting O to P, as depicted in Fig. 1.9a. The position of a point P relative to P is a zero vector.

Let $\mathbf{I}$, $\mathbf{J}$, $\mathbf{k}$ be mutually perpendicular unit vectors (Cartesian reference frame) with the origin at O, as shown in Fig. 1.9b. The axes of the Cartesian reference

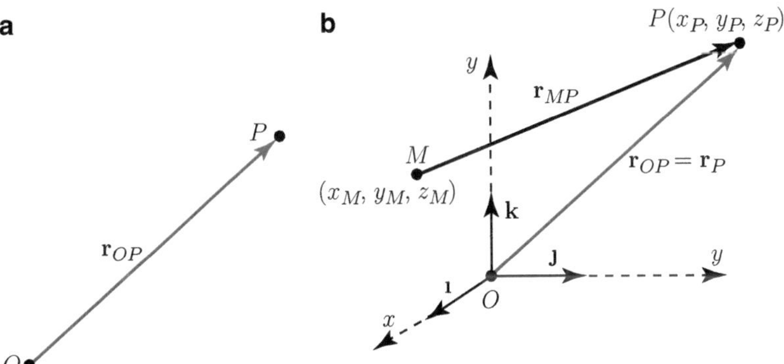

Fig. 1.9 Position vector

frame are x, y, z. The unit vectors $\mathbf{i}$, $\mathbf{j}$, $\mathbf{k}$ are parallel to x, y, z, and they have the senses of the positive x, y, z axes. The coordinates of the origin O are $x = y = z = 0$, that is, $O(0, 0, 0)$. The coordinates of a point P are $x = x_P$, $y = y_P$, $z = z_P$, that is, $P(x_P, y_P, z_P)$. The position vector of P relative to the origin O is

$$\mathbf{r}_{OP} = \mathbf{r}_P = \overrightarrow{OP} = x_P\,\mathbf{i} + y_P\,\mathbf{j} + z_P\,\mathbf{k}. \tag{1.8}$$

The position vector of the point P relative to a point M, $M \neq O$ of coordinates (x_M, y_M, z_M) is

$$\mathbf{r}_{MP} = \overrightarrow{MP} = (x_P - x_M)\,\mathbf{i} + (y_P - y_M)\,\mathbf{j} + (z_P - z_M)\,\mathbf{k}. \tag{1.9}$$

The distance d between P and M is given by

$$d = |\mathbf{r}_P - \mathbf{r}_M| = |\mathbf{r}_{MP}| = |\overrightarrow{MP}| = \sqrt{(x_P - x_M)^2 + (y_P - y_M)^2 + (z_P - z_M)^2}. \tag{1.10}$$

1.3   Scalar (Dot) Product of Vectors

Definition. The scalar (dot) product of a vector $\mathbf{a}$ and a vector $\mathbf{b}$ is

$$\mathbf{a} \cdot \mathbf{b} = |\mathbf{a}|\,|\mathbf{b}|\cos(\mathbf{a}, \mathbf{b}). \tag{1.11}$$

For the scalar (dot) product, the following rules apply:

1. For any vectors $\mathbf{a}$ and $\mathbf{b}$, one can write the commutative law for scalar product

$$\mathbf{a} \cdot \mathbf{b} = \mathbf{b} \cdot \mathbf{a}.$$

2. For any two vectors **a** and **b** and any scalar s, the following relation is written:

$$(s\mathbf{a})\cdot\mathbf{b} = s(\mathbf{a}\cdot\mathbf{b}) = \mathbf{a}\cdot(s\mathbf{b}) = s\mathbf{a}\cdot\mathbf{b}.$$

3. For any vectors **a**, **b**, and **c**, the distributive law in the first argument is

$$(\mathbf{a}+\mathbf{b})\cdot\mathbf{c} = \mathbf{a}\cdot\mathbf{c}+\mathbf{b}\cdot\mathbf{c},$$

and the distributive law in the second argument is

$$\mathbf{a}\cdot(\mathbf{b}+\mathbf{c}) = \mathbf{a}\cdot\mathbf{b}+\mathbf{a}\cdot\mathbf{c}.$$

It can be shown that the dot product is distributive, and the following relation can be written:

$$s_a\,\mathbf{a}\cdot(s_b\,\mathbf{b}+s_c\,\mathbf{c}) = s_a\,s_b\,\mathbf{a}\cdot\mathbf{b}+s_a\,s_c\,\mathbf{a}\cdot\mathbf{c}.$$

If

$$\mathbf{a} = a_x\mathbf{\imath}+a_y\mathbf{\jmath}+a_z\mathbf{k} \quad\text{and}\quad \mathbf{b} = b_x\mathbf{\imath}+b_y\mathbf{\jmath}+b_z\mathbf{k},$$

where $\mathbf{\imath}, \mathbf{\jmath}, \mathbf{k}$ are mutually perpendicular unit vectors, then

$$\mathbf{a}\cdot\mathbf{b} = a_xb_x+a_yb_y+a_zb_z. \tag{1.12}$$

The following relationships exist:

$$\mathbf{\imath}\cdot\mathbf{\imath} = \mathbf{\jmath}\cdot\mathbf{\jmath} = \mathbf{k}\cdot\mathbf{k} = 1,$$

$$\mathbf{\imath}\cdot\mathbf{\jmath} = \mathbf{\jmath}\cdot\mathbf{\imath} = \mathbf{\jmath}\cdot\mathbf{k} = \mathbf{k}\cdot\mathbf{\jmath} = \mathbf{k}\cdot\mathbf{\imath} = \mathbf{\imath}\cdot\mathbf{k} = 0.$$

Every vector **v** can be expressed in the form

$$\mathbf{v} = \mathbf{\imath}\cdot\mathbf{v}\,\mathbf{\imath}+\mathbf{\jmath}\cdot\mathbf{v}\,\mathbf{\jmath}+\mathbf{k}\cdot\mathbf{v}\,\mathbf{k}. \tag{1.13}$$

Proof. The vector **v** can always be expressed as

$$\mathbf{v} = v_x\mathbf{\imath}+v_y\mathbf{\jmath}+v_z\mathbf{k}.$$

Dot multiply both sides by $\mathbf{\imath}$

$$\mathbf{\imath}\cdot\mathbf{v} = v_x\mathbf{\imath}\cdot\mathbf{\imath}+v_y\mathbf{\imath}\cdot\mathbf{\jmath}+v_z\mathbf{\imath}\cdot\mathbf{k}.$$

But

$$\mathbf{\imath}\cdot\mathbf{\imath} = 1, \quad\text{and}\quad \mathbf{\imath}\cdot\mathbf{\jmath} = \mathbf{\imath}\cdot\mathbf{k} = 0.$$

Hence, $\mathbf{\imath}\cdot\mathbf{v} = v_x$. Similarly, $\mathbf{\jmath}\cdot\mathbf{v} = v_y$ and $\mathbf{k}\cdot\mathbf{v} = v_z$.

The MATLAB command $\mathtt{dot(v,u)}$ calculates the scalar product (or vector dot product) of the vectors $\mathtt{v}$ and $\mathtt{u}$. The dot product of two vectors $\mathtt{v}$ and $\mathtt{u}$ can be expressed as

```
sum(v.*u).
```

The command $\mathtt{sum(x)}$ with $\mathtt{x}$ defined as a vector, returns the sum of its elements. The MATLAB command $\mathtt{.*}$, named *array multiplication*, is the element-by-element product of the associated arrays, that is, $\mathtt{v.*u}$, and the arrays must have the same size, unless one of them is a scalar. To indicate an array (element-by-element) operation, the standard operator is preceded with a period (dot). Thus, $\mathtt{v.*u}$ is

```
[ v_x*u_x, v_y*u_y, v_z*u_z ].
```
$\square$

1.4   Vector (Cross) Product of Vectors

Definition. The vector (cross) product of a vector $\mathbf{a}$ and a vector $\mathbf{b}$ is the vector (Fig. 1.10)

$$\mathbf{a} \times \mathbf{b} = |\mathbf{a}|\,|\mathbf{b}|\sin(\mathbf{a},\mathbf{b})\mathbf{n}, \qquad (1.14)$$

where $\mathbf{n}$ is a unit vector whose direction is the same as the direction of advance of a right-handed screw rotated from $\mathbf{a}$ toward $\mathbf{b}$, through the angle $(\mathbf{a},\mathbf{b})$, when the axis of the screw is perpendicular to both $\mathbf{a}$ and $\mathbf{b}$. The magnitude of $\mathbf{a} \times \mathbf{b}$ is given by

$$|\mathbf{a} \times \mathbf{b}| = |\mathbf{a}|\,|\mathbf{b}|\sin(\mathbf{a},\mathbf{b}).$$

If $\mathbf{a}$ is parallel to $\mathbf{b}$, $\mathbf{a}\|\mathbf{b}$, then $\mathbf{a} \times \mathbf{b} = 0$. The symbol $\|$ denotes parallel. The relation $\mathbf{a} \times \mathbf{b} = \mathbf{0}$ implies only that the product $|\mathbf{a}|\,|\mathbf{b}|\sin(\mathbf{a},\mathbf{b})$ is equal to zero, and this is the case whenever $|\mathbf{a}| = 0$, or $|\mathbf{b}| = 0$, or $\sin(\mathbf{a},\mathbf{b}) = 0$.

For any two vectors $\mathbf{a}$ and $\mathbf{b}$ and any real scalar s, the following relation can be written:

$$(s\mathbf{a}) \times \mathbf{b} = s(\mathbf{a} \times \mathbf{b}) = \mathbf{a} \times (s\mathbf{b}) = s\mathbf{a} \times \mathbf{b}.$$

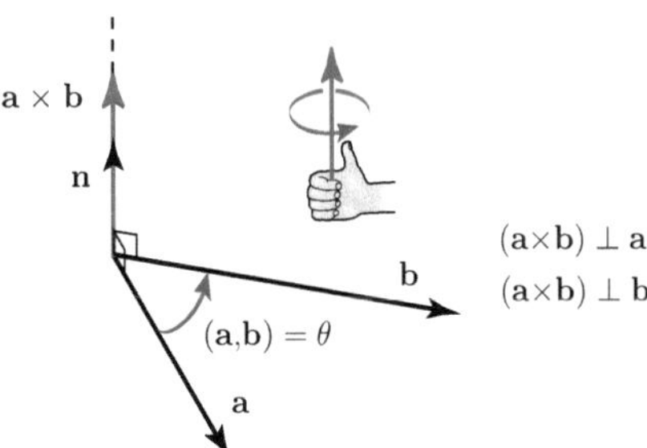

Fig. 1.10 Vector (cross) product of the vector $\mathbf{a}$ and the vector $\mathbf{b}$

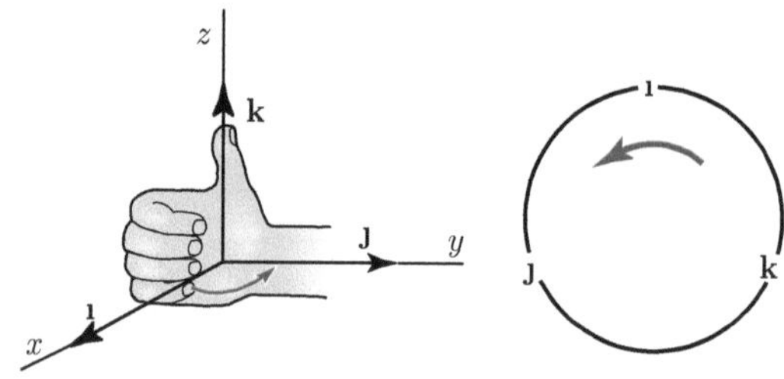

Fig. 1.11 Cartesian right-handed reference set. The cross product of two unit vectors in a counterclockwise sense around the circle yields the positive third unit vector

The sense of the unit vector **n** which appears in the definition of $\mathbf{a} \times \mathbf{b}$ depends on the order of the factors **a** and **b** in such a way that (cross product is not commutative):

$$\mathbf{b} \times \mathbf{a} = -\mathbf{a} \times \mathbf{b}. \tag{1.15}$$

The cross product distributive law for the first argument can be written as

$$(\mathbf{a} + \mathbf{b}) \times \mathbf{c} = \mathbf{a} \times \mathbf{c} + \mathbf{b} \times \mathbf{c},$$

while the distributive law for the second argument is

$$\mathbf{a} \times (\mathbf{b} + \mathbf{c}) = \mathbf{a} \times \mathbf{b} + \mathbf{a} \times \mathbf{c}.$$

Vector multiplication obeys the following law of distributivity (Varignon theorem)

$$\mathbf{a} \times \sum_{i=1}^{n} \mathbf{v}_i = \sum_{i=1}^{n} (\mathbf{a} \times \mathbf{v}_i).$$

A set of mutually perpendicular unit vectors $\mathbf{i}, \mathbf{j}, \mathbf{k}$ is called *right-handed* if $\mathbf{i} \times \mathbf{j} = \mathbf{k}$ (Fig. 1.11). A set of mutually perpendicular unit vectors $\mathbf{i}, \mathbf{j}, \mathbf{k}$ is called *left-handed* if $\mathbf{i} \times \mathbf{j} = -\mathbf{k}$.

If $\mathbf{a} = a_x \mathbf{i} + a_y \mathbf{j} + a_z \mathbf{k}$, and $\mathbf{b} = b_x \mathbf{i} + b_y \mathbf{j} + b_z \mathbf{k}$, where $\mathbf{i}, \mathbf{j}, \mathbf{k}$ are right-handed mutually perpendicular unit vectors, then $\mathbf{a} \times \mathbf{b}$ can be expressed in the following determinant form:

$$\mathbf{a} \times \mathbf{b} = \begin{vmatrix} \mathbf{i} & \mathbf{j} & \mathbf{k} \\ a_x & a_y & a_z \\ b_x & b_y & b_z \end{vmatrix}. \tag{1.16}$$

The determinant can be expanded by minors of the elements of the first row:

$$\begin{vmatrix} \mathbf{i} & \mathbf{j} & \mathbf{k} \\ a_x & a_y & a_z \\ b_x & b_y & b_z \end{vmatrix} = \mathbf{i} \begin{vmatrix} a_y & a_z \\ b_y & b_z \end{vmatrix} - \mathbf{j} \begin{vmatrix} a_x & a_z \\ b_x & b_z \end{vmatrix} + \mathbf{k} \begin{vmatrix} a_x & a_y \\ b_x & b_y \end{vmatrix}$$

$$= \mathbf{i}(a_y b_z - a_z b_y) - \mathbf{j}(a_x b_z - a_z b_x) + \mathbf{k}(a_x b_y - a_y b_x)$$

$$= (a_y b_z - a_z b_y)\mathbf{i} + (a_z b_x - a_x b_z)\mathbf{j} + (a_x b_y - a_y b_x)\mathbf{k}. \tag{1.17}$$

As a general rule, a third-order determinant can be expanded by diagonal multiplication, that is, repeating the first two columns on the right side of the determinant and adding the signed diagonal products of the diagonal elements as

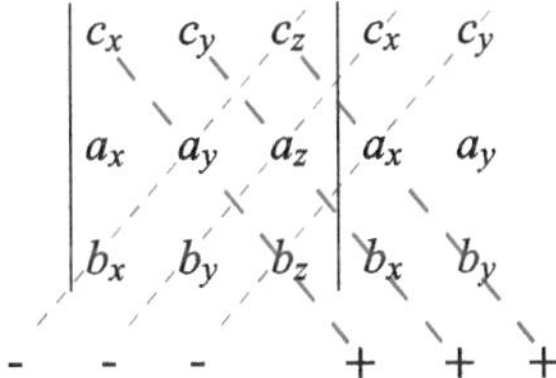

The determinant in (1.16) can be expanded using the general rule as

$$\begin{vmatrix} \mathbf{i} & \mathbf{j} & \mathbf{k} \\ a_x & a_y & a_z \\ b_x & b_y & b_z \end{vmatrix} = -\mathbf{k}a_y b_x - \mathbf{i}a_z b_y - \mathbf{j}a_x b_z + \mathbf{i}a_y b_z + \mathbf{j}a_z b_x + \mathbf{k}a_x b_y$$

$$= (a_y b_z - a_z b_y)\mathbf{i} + (a_z b_x - a_x b_z)\mathbf{j} + (a_x b_y - a_y b_x)\mathbf{k}.$$

The MATLAB command cross (a, b) calculates the cross product of the vectors a and b. Next, a MATLAB function that calculates the cross product of two vectors is presented:

```
function val = crossproduct(a,b)
% symbolic cross product function of a vector a x b
a = a(:);
% a(:) represents all elements of a,
%       regarded as a single column
b = b(:);
% b(:) represents all elements of b,
%       regarded as a single column
val = [a(2,:).*b(3,:)-a(3,:).*b(2,:) ...
       a(3,:).*b(1,:)-a(1,:).*b(3,:) ...
       a(1,:).*b(2,:)-a(2,:).*b(1,:)];
```

In the previous MATLAB function, the general MATLAB command colon (:), that is, a(:), has been used. The colon (:) is one of the most useful operators in MATLAB. It can create vectors, subscript arrays, and specify for iterations. The three full stops . . . after a MATLAB statement are used to continue the MATLAB statement to next line.

1.5   Scalar Triple Product of Three Vectors

Definition. The scalar triple product of three vectors **a**, **b**, and **c** is defines as

$$[\mathbf{a}, \mathbf{b}, \mathbf{c}] \equiv \mathbf{a} \cdot (\mathbf{b} \times \mathbf{c}) = \mathbf{a} \cdot \mathbf{b} \times \mathbf{c}. \tag{1.18}$$

The MATLAB commands for the scalar triple product of three vectors a, b, and c is

```
syms a_x a_y a_z b_x b_y b_z c_x c_y c_z real
a=[a_x a_y a_z]; b=[b_x b_y b_z]; c=[c_x c_y c_z];
% [a,b,c] = a.(b x c)
abc = dot(a, cross(b, c));
```

It does not matter whether the dot is placed between **a** and **b**, and the cross between **b** and **c**, or vice versa, that is,

$$[\mathbf{a},\mathbf{b},\mathbf{c}] = \mathbf{a}\cdot\mathbf{b}\times\mathbf{c} = \mathbf{a}\times\mathbf{b}\cdot\mathbf{c}. \tag{1.19}$$

The relation given by (1.19) is demonstrated using the MATLAB commands:

```
% [a,b,c] = a.(b x c)
abxc = simplify(dot(a, cross(b, c)));
% [a,b,c] = (a x b).c
axbc = simplify(dot(cross(a, b), c));
% a.(b x c)==(a x b).c
abxc == axbc.
```

The MATLAB relational operator == or eq is used to compare each element of array for equality. The statement LHS == RHS or eq(LHS, RHS) compares each element of the array LHS for equality with the corresponding element of the array RHS and returns an array with elements set to logical 1 (true) if LHS and RHS are equal or logical 0 (false) where they are not equal.

A change in the order of the factors appearing in a scalar triple product at most changes the sign of the product, that is,

$$[\mathbf{b},\mathbf{a},\mathbf{c}] = -[\mathbf{a},\mathbf{b},\mathbf{c}] \quad \text{and} \quad [\mathbf{b},\mathbf{c},\mathbf{a}] = [\mathbf{a},\mathbf{b},\mathbf{c}].$$

If **a**, **b**, **c** are parallel to the same plane, or if any two of the vectors **a**, **b**, **c** are parallel to each other, then $[\mathbf{a},\mathbf{b},\mathbf{c}] = 0$.

The scalar triple product $[\mathbf{a},\mathbf{b},\mathbf{c}]$ can be expressed in the following determinant form:

$$[\mathbf{a},\mathbf{b},\mathbf{c}] = \begin{vmatrix} a_x & a_y & a_z \\ b_x & b_y & b_z \\ c_x & c_y & c_z \end{vmatrix}. \tag{1.20}$$

In MATLAB, the scalar triple product of three vectors a, b, and c is expressed as

```
det([a; b; c]),
```

where det(x) is the determinant of the square matrix x. To verify (3.93), the following MATLAB command is used:

```
det([a; b; c]) == simplify(dot(a, cross(b, c))).
```

Fig. 1.12 Parallelepiped
with the sides **a**, **b**, and **c**

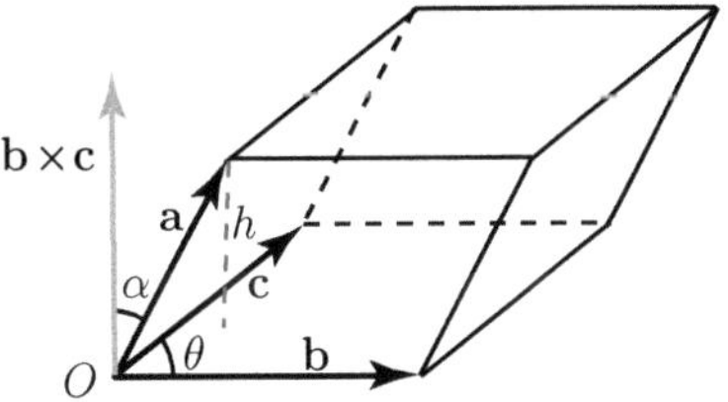

Exercise: Volume of a Parallelepiped
Figure 1.12 depicts three vectors **a**, **b**, and **c** that form a parallelepiped. Show that
the scalar $\mathbf{a} \cdot (\mathbf{b} \times \mathbf{c})$ represents the volume of the parallelepiped with the sides **a**, **b**,
and **c**.

Solution
The scalar triple product is $\mathbf{a} \cdot (\mathbf{b} \times \mathbf{c}) = |\mathbf{a}|\,|\mathbf{b}|\,|\mathbf{c}| \sin \theta \cos \alpha = hA$, where $h =$
$|\mathbf{a}| \cos \alpha$ represents the height of the parallelepiped and $A = |\mathbf{b}|\,|\mathbf{c}| \sin \theta$ represents
the area of the parallelogram with the sides **b** and **c**. The product between h and A
represents the volume of a parallelepiped, $V = hA$, so the scalar $\mathbf{a} \cdot (\mathbf{b} \times \mathbf{c})$ represents
the volume of the parallelepiped with the sides formed by the vectors **a**, **b**, and **c**.

Exercise: Vector Expressed in a Base
Let **a**, **b**, **c**, and **w** be nonzero vectors and $[\mathbf{a},\mathbf{b},\mathbf{c}] \neq 0$. The vectors **a**, **b**, **c**, and **w** are
given vectors. The vectors **a**, **b**, and **c** are free vectors and can be moved in a given
point. The vectors **a**, **b**, and **c** form the edges of a parallelepiped of nonzero volume.
Then, the scalars s_a, s_b, and s_c exist such as the vector **w** can be represented as a
linear combination of the vectors **a**, **b**, **c**: $\mathbf{w} = s_a\,\mathbf{a} + s_b\,\mathbf{b} + s_c\,\mathbf{c}$. Show that the scalars
s_a, s_b, and s_c are given by

$$s_a = \frac{[\mathbf{w},\mathbf{b},\mathbf{c}]}{[\mathbf{a},\mathbf{b},\mathbf{c}]}, \quad s_b = \frac{[\mathbf{a},\mathbf{w},\mathbf{c}]}{[\mathbf{a},\mathbf{b},\mathbf{c}]}, \quad \text{and} \quad s_c = \frac{[\mathbf{a},\mathbf{b},\mathbf{w}]}{[\mathbf{a},\mathbf{b},\mathbf{c}]}.$$

Solution
The components of the vectors **a**, **b**, **c** and the scalars s_a, s_b, and s_c are introduced
as symbolic variables using MATLAB:

```
syms a_x a_y a_z b_x b_y b_z c_x c_y c_z   real
syms s_a s_b s_c real.
```

The vectors a, b, and c are

```
a =   [ a_x a_y a_z ];
b =   [ b_x b_y b_z ];
c =   [ c_x c_y c_z ];
```

and the vector w is:

```
w = s_a*a + s_b*b + s_c*c;
```

The scalar triple products $[\mathbf{a}, \mathbf{b}, \mathbf{c}]$, $[\mathbf{w}, \mathbf{b}, \mathbf{c}]$, $[\mathbf{a}, \mathbf{w}, \mathbf{c}]$, and $[\mathbf{a}, \mathbf{b}, \mathbf{w}]$ are

```
abc = det([a; b; c]);
wbc = det([w; b; c]);
awc = det([a; w; c]);
abw = det([a; b; w]);
```

The scalars `s_a`, `s_b`, and `s_c` are obtained from $\dfrac{[\mathbf{w}, \mathbf{b}, \mathbf{c}]}{[\mathbf{a}, \mathbf{b}, \mathbf{c}]}$, $\dfrac{[\mathbf{a}, \mathbf{w}, \mathbf{c}]}{[\mathbf{a}, \mathbf{b}, \mathbf{c}]}$, and $\dfrac{[\mathbf{a}, \mathbf{b}, \mathbf{w}]}{[\mathbf{a}, \mathbf{b}, \mathbf{c}]}$
or

```
simplify(wbc/abc)
simplify(awc/abc)
simplify(abw/abc)
```

1.6   Vector Triple Product of Three Vector

Definition. The vector triple product of three vectors $\mathbf{a}$, $\mathbf{b}$, $\mathbf{c}$ is the vector $\mathbf{a} \times (\mathbf{b} \times \mathbf{c})$. The parentheses are essential because $\mathbf{a} \times (\mathbf{b} \times \mathbf{c})$ is not, in general, equal to $(\mathbf{a} \times \mathbf{b}) \times \mathbf{c}$. For any three vectors $\mathbf{a}$, $\mathbf{b}$, and $\mathbf{c}$,

$$\mathbf{a} \times (\mathbf{b} \times \mathbf{c}) = (\mathbf{a} \cdot \mathbf{c})\mathbf{b} - (\mathbf{a} \cdot \mathbf{b})\mathbf{c}. \tag{1.21}$$

The previous relation given by (1.21) can be explained using the MATLAB statements:

```
% a x (b x c)
axbxc = cross(a, cross(b, c));
% (a.c)b - (a.b)c
RHS = dot(a, c)*b - dot(a, b)*c;
% a x (b x c) - (a.c)b + (a.b)c = [0, 0, 0]
simplify(axbxc-RHS).
```

1.7   Derivative of a Vector Function

The derivative of a vector function is defined in exactly the same way as is the derivative of a scalar function. Thus

$$\frac{\mathrm{d}}{\mathrm{d}t}\mathbf{a} = \lim_{\Delta t \to 0} \frac{\mathbf{a}(t + \Delta t) - \mathbf{a}(t)}{\Delta t}.$$

The derivative of a vector has some of the properties of the derivative of a scalar function. The derivative of the sum of two vector functions $\mathbf{a}$ and $\mathbf{b}$ is

$$\frac{\mathrm{d}}{\mathrm{d}t}(\mathbf{a} + \mathbf{b}) = \frac{\mathrm{d}\mathbf{a}}{\mathrm{d}t} + \frac{\mathrm{d}\mathbf{b}}{\mathrm{d}t}. \tag{1.22}$$

The components of the vectors a and b are functions of time, t, and are introduced in MATLAB with

```
syms t real
a_x = sym('a_x(t)');
a_y = sym('a_y(t)');
a_z = sym('a_z(t)');
b_x = sym('b_x(t)');
b_y = sym('b_y(t)');
b_z = sym('b_z(t)');
a = [a_x a_y a_z];
b = [b_x b_y b_z];
```

To calculate symbolically the derivative of a vector using the MATLAB, the command diff(p,t) is used, which gives the derivative of p with respect to t. The relation given by (1.22) can be demonstrated using the MATLAB command:

```
diff(a+b, t)  ==  diff(a, t) + diff(b, t).
```

The time derivative of the product of a scalar function f and a vector function **a** is

$$\frac{d(f\mathbf{a})}{dt} = \frac{df}{dt}\mathbf{a} + f\frac{d\mathbf{a}}{dt}. \tag{1.23}$$

Equation (1.23) is verified using the MATLAB command:

```
syms f real
diff(f*a, t)  ==  diff(f, t)*a + f*diff(a, t).
```

Combining the previous results, one can conclude

$$\frac{d}{dt}(\mathbf{a}\cdot\mathbf{b}) = \frac{d\mathbf{a}}{dt}\cdot\mathbf{b} + \mathbf{a}\cdot\frac{d\mathbf{b}}{dt} \quad \text{and} \quad \frac{d}{dt}(\mathbf{a}\times\mathbf{b}) = \frac{d\mathbf{a}}{dt}\times\mathbf{b} + \mathbf{a}\times\frac{d\mathbf{b}}{dt}. \tag{1.24}$$

Equation (1.24) is demonstrated with the MATLAB commands:

```
diff(a*b.', t)  ==  diff(a, t)*b.' + a*diff(b, t).'
diff(cross(a, b), t)  ==  cross(diff(a, t), b) ...
                        + cross(a, diff(b, t)),
```

where A.' is the array transpose of A.

The general derivative of a vector v can be expressed as

$$\frac{d\mathbf{v}}{dt} = \frac{d}{dt}(v_x\mathbf{\imath} + v_y\mathbf{\jmath} + v_z\mathbf{k},) = \frac{dv_x}{dt}\mathbf{\imath} + v_x\frac{d\mathbf{\imath}}{dt} + \frac{dv_y}{dt}\mathbf{\jmath} + v_y\frac{d\mathbf{\jmath}}{dt} + \frac{dv_z}{dt}\mathbf{k} + v_z\frac{d\mathbf{k}}{dt},$$

and if the reference basis or reference frame $[\mathbf{\imath}, \mathbf{\jmath}, \mathbf{k}]$ is unchanging, then

$$\frac{d\mathbf{v}}{dt} = \frac{dv_x}{dt}\mathbf{\imath} + \frac{dv_y}{dt}\mathbf{\jmath} + \frac{dv_z}{dt}\mathbf{k}.$$

1.8   Cauchy's Inequality, Lagrange's Identity, and Triangle Inequality

The vectors **a** and **b** are nonzero vectors. The *Cauchy's inequality* can be written in vector form as

$$(\mathbf{a} \cdot \mathbf{b})^2 \le a^2\, b^2, \tag{1.25}$$

where $a^2 = |\mathbf{a}|^2 = \mathbf{a} \cdot \mathbf{a}$ and $b^2 = |\mathbf{b}|^2 = \mathbf{b} \cdot \mathbf{b}$. If **a** and **b** are parallel vectors, then

$$(\mathbf{a} \cdot \mathbf{b})^2 = a^2\, b^2.$$

The vector derivation of the inequality is

$$(\mathbf{a} \cdot \mathbf{b})^2 = a^2 b^2 \cos^2 \theta \le a^2 b^2.$$

The *Lagrange's identity* in vector form is

$$(\mathbf{a} \cdot \mathbf{b})^2 = a^2\, b^2 - (\mathbf{a} \times \mathbf{b}) \cdot (\mathbf{a} \times \mathbf{b}). \tag{1.26}$$

The vectors **a** and **b** are nonzero vectors, and the vectorial product between **a** and **b** is

$$\mathbf{a} \times \mathbf{b} = (a_y b_z - a_z b_y)\mathbf{\imath} + (a_z b_x - a_x b_z)\mathbf{\jmath} + (a_x b_y - a_y b_x)\mathbf{k},$$

where a_x, a_y, a_z and b_x, b_y, b_z are the Cartesian components of the vectors **a** and **b**. One can compute

$$(\mathbf{a} \times \mathbf{b}) \cdot (\mathbf{a} \times \mathbf{b}) = (a_y b_z - a_z b_y)^2 + (a_z b_x - a_x b_z)^2 + (a_x b_y - a_y b_x)^2. \tag{1.27}$$

The scalar product definition gives

$$(\mathbf{a} \cdot \mathbf{b})^2 = [(a_x\mathbf{\imath} + a_y\mathbf{\jmath} + a_z\mathbf{k}) \cdot (b_x\mathbf{\imath} + b_y\mathbf{\jmath} + b_z\mathbf{k})]^2 = (a_x b_x + a_y b_y + a_z b_z)^2, \tag{1.28}$$

and

$$a^2 b^2 = |\mathbf{a}|^2 |\mathbf{b}|^2 = \left|a_x\mathbf{\imath} + a_y\mathbf{\jmath} + a_z\mathbf{k}\right|^2 \left|b_x\mathbf{\imath} + b_y\mathbf{\jmath} + b_z\mathbf{k}\right|^2$$
$$= (a_x^2 + a_y^2 + a_z^2)(b_x^2 + b_y^2 + b_z^2). \tag{1.29}$$

Using (1.27), (1.28), and (1.29), it results

$$(a_x^2 + a_y^2 + a_z^2)(b_x^2 + b_y^2 + b_z^2) - (a_x b_x + a_y b_y + a_z b_z)^2$$
$$= (a_y b_z - a_z b_y)^2 + (a_z b_x - a_x b_z)^2 + (a_z b_y - a_y b_x)^2,$$

or

$$a^2\,b^2 - (\mathbf{a}\cdot\mathbf{b})^2 = (\mathbf{a}\times\mathbf{b})\cdot(\mathbf{a}\times\mathbf{b}).$$

The previous equation can be written as the identity given by (1.26).

The MATLAB proof for Lagrange's identity is

```
syms a_x a_y a_z b_x b_y b_z real
a = [ a_x a_y a_z ];
b = [ b_x b_y b_z ];
% LHS =  (a.b)^2
% RHS =  (a.a)*(b.b)  -  (a x b).(a x b)
LHS = (dot(a,b))^2;
RHS = dot(a,a)*dot(b,b)-dot(cross(a,b),cross(a,b));
expand(LHS)==expand(RHS) .
```

If $\mathbf{a}$ and $\mathbf{b}$ are nonzero vectors, the following relation can be obtained:

$$|\mathbf{a}+\mathbf{b}| \le |\mathbf{a}| + |\mathbf{b}|. \tag{1.30}$$

The inequality given by (1.30) is known as *triangle inequality*.

Proof: It is obvious that

$$(\mathbf{a}+\mathbf{b})\cdot(\mathbf{a}+\mathbf{b}) = \mathbf{a}\cdot\mathbf{a}+\mathbf{a}\cdot\mathbf{b}+\mathbf{b}\cdot\mathbf{a}+\mathbf{b}\cdot\mathbf{b} = |\mathbf{a}|^2 + \mathbf{a}\cdot\mathbf{b}+\mathbf{b}\cdot\mathbf{a}+|\mathbf{b}|^2. \tag{1.31}$$

The following relation exists;

$$\mathbf{a}\cdot\mathbf{b}+\mathbf{b}\cdot\mathbf{a} \le 2\,|\mathbf{a}\cdot\mathbf{b}| \le 2\,|\mathbf{a}|\,|\mathbf{b}|. \tag{1.32}$$

Equations (1.31) and (1.32) give

$$|\mathbf{a}+\mathbf{b}|^2 = (\mathbf{a}+\mathbf{b})\cdot(\mathbf{a}+\mathbf{b}) \le |\mathbf{a}|^2 + |\mathbf{b}|^2 + 2\,|\mathbf{a}|\,|\mathbf{b}| = (|\mathbf{a}| + |\mathbf{b}|)^2.$$

Moreover, one can prove that

$$|\mathbf{a}| - |\mathbf{b}| \le |\mathbf{a}+\mathbf{b}| \le |\mathbf{a}| + |\mathbf{b}|,\ \text{for } |\mathbf{a}| > |\mathbf{b}|,$$

$$|\mathbf{b}| - |\mathbf{a}| \le |\mathbf{a}+\mathbf{b}| \le |\mathbf{a}| + |\mathbf{b}|,\ \text{for } |\mathbf{a}| < |\mathbf{b}|.$$

For this, let $\mathbf{a} = (\mathbf{a}+\mathbf{b}) - \mathbf{b}$, and applying the inequality given by (1.30) for $\mathbf{a}+\mathbf{b}$ and $-\mathbf{b}$, it results

$$|\mathbf{a}| = |(\mathbf{a}+\mathbf{b})+(-\mathbf{b})| \le |\mathbf{a}+\mathbf{b}| + |-\mathbf{b}|,$$

or

$$|\mathbf{a}+\mathbf{b}| \ge |\mathbf{a}| - |-\mathbf{b}| = |\mathbf{a}| - |\mathbf{b}|. \tag{1.33}$$

Using (1.30) and (1.33), the following relations can be written:

$$|\mathbf{a}| - |\mathbf{b}| \leq |\mathbf{a} + \mathbf{b}| \leq |\mathbf{a}| + |\mathbf{b}|, \text{ for } |\mathbf{a}| > |\mathbf{b}|,$$

$$|\mathbf{b}| - |\mathbf{a}| \leq |\mathbf{a} + \mathbf{b}| \leq |\mathbf{a}| + |\mathbf{b}|, \text{ for } |\mathbf{a}| < |\mathbf{b}|.$$

$\square$

1.9   Coordinate Transformation

Let $\mathbf{i}, \mathbf{j}$ and $\mathbf{k}$ be the unit vectors of an orthogonal Cartesian reference frame $Oxyz$ and $\mathbf{i}', \mathbf{j}'$ and $\mathbf{k}'$ be the unit vectors of an orthogonal Cartesian reference frame $Ox'y'z'$. Then, the transformation from the reference frame $Oxyz$ to the reference frame $Ox'y'z'$ is given by the next equations:

$$\mathbf{i}' = \alpha_{11}\mathbf{i} + \alpha_{12}\mathbf{j} + \alpha_{13}\mathbf{k},$$

$$\mathbf{j}' = \alpha_{21}\mathbf{i} + \alpha_{22}\mathbf{j} + \alpha_{23}\mathbf{k},$$

$$\mathbf{k}' = \alpha_{31}\mathbf{i} + \alpha_{32}\mathbf{j} + \alpha_{33}\mathbf{k}, \tag{1.34}$$

where

$$\alpha_{11} = \mathbf{i}' \cdot \mathbf{i} = \cos(Ox', Ox), \; \alpha_{12} = \mathbf{i}' \cdot \mathbf{j} = \cos(Ox', Oy), \; \alpha_{13} = \mathbf{i}' \cdot \mathbf{k} = \cos(Ox', Oz),$$

$$\alpha_{21} = \mathbf{j}' \cdot \mathbf{i} = \cos(Oy', Ox), \; \alpha_{22} = \mathbf{j}' \cdot \mathbf{j} = \cos(Oy', Oy), \; \alpha_{23} = \mathbf{j}' \cdot \mathbf{k} = \cos(Oy', Oz),$$

$$\alpha_{31} = \mathbf{k}' \cdot \mathbf{i} = \cos(Oz', Ox), \; \alpha_{32} = \mathbf{k}' \cdot \mathbf{j} = \cos(Oz', Oy), \; \alpha_{33} = \mathbf{k}' \cdot \mathbf{k} = \cos(Oz', Oz)$$

are the direction cosines between unit vectors along the coordinate axes. The matrix

$$A = \begin{pmatrix} \alpha_{11} & \alpha_{12} & \alpha_{13} \\ \alpha_{21} & \alpha_{22} & \alpha_{23} \\ \alpha_{31} & \alpha_{32} & \alpha_{33} \end{pmatrix} \tag{1.35}$$

depends only on the Cartesian coordinates of the reference frame $Oxyz$ and $Ox'y'z'$ and represents the transforms matrix from the old reference frame to the new one. The following relations exist:

$$\mathbf{i} \cdot \mathbf{i} = 1, \, \mathbf{j} \cdot \mathbf{j} = 1, \, \mathbf{k} \cdot \mathbf{k} = 1, \tag{1.36}$$

$$\mathbf{j} \cdot \mathbf{k} = 0, \, \mathbf{k} \cdot \mathbf{i} = 0, \, \mathbf{i} \cdot \mathbf{j} = 0, \tag{1.37}$$

and

$$\mathbf{i}' \cdot \mathbf{i}' = 1, \, \mathbf{j}' \cdot \mathbf{j}' = 1, \, \mathbf{k}' \cdot \mathbf{k}' = 1, \tag{1.38}$$

$$\mathbf{j}' \cdot \mathbf{k}' = 0, \, \mathbf{k}' \cdot \mathbf{i}' = 0, \, \mathbf{i}' \cdot \mathbf{j}' = 0. \tag{1.39}$$

Using (1.34)–(1.39), the following relation is introduced:

$$\sum_{k=1}^{3} \alpha_{ik}\alpha_{jk} = \delta_{ij}, \qquad i = 1,2,3 \text{ and } j = 1,2,3, \tag{1.40}$$

where the symbol

$$\delta_{ij} = \begin{cases} 1 & \text{if } i = j \\ 0 & \text{if } i \neq j \end{cases}$$

is called the *Kronecker delta*.

The matrix

$$E = \begin{pmatrix} \delta_{11} & \delta_{12} & \delta_{13} \\ \delta_{21} & \delta_{22} & \delta_{23} \\ \delta_{31} & \delta_{32} & \delta_{33} \end{pmatrix} = \begin{pmatrix} 1 & 0 & 0 \\ 0 & 1 & 0 \\ 0 & 0 & 1 \end{pmatrix}$$

is the Kronecker matrix. Let $\mathbf{X}$ be a vector. In the Cartesian reference frame $Oxyz$, the vector $\mathbf{X}$ has the form

$$\mathbf{X} = X_1 \mathbf{i} + X_2 \mathbf{j} + X_3 \mathbf{k}, \tag{1.41}$$

and in the Cartesian reference frame $Ox'y'z'$, the vector $\mathbf{X}$ has the form

$$\mathbf{X} = X_1' \mathbf{i}' + X_2' \mathbf{j}' + X_3' \mathbf{k}'. \tag{1.42}$$

Using (1.34), (1.41), and (1.42), the relations among the components X_1, X_2, X_3 and the components X_1', X_2', X_3' are

$$\begin{aligned} X_1 &= \alpha_{11} X_1' + \alpha_{21} X_2' + \alpha_{31} X_3', \\ X_2 &= \alpha_{12} X_1' + \alpha_{22} X_2' + \alpha_{32} X_3', \\ X_3 &= \alpha_{13} X_1' + \alpha_{23} X_2' + \alpha_{33} X_3', \end{aligned} \tag{1.43}$$

or

$$X_i = \sum_{r=1}^{3} \alpha_{ri} X_r', \text{ where } i = 1,2,3. \tag{1.44}$$

For a given i, multiply (1.44) with α_{ji}, add the relation for $i = 1,2,3$, and using (1.40), the following relations are obtained:

$$\sum_{i=1}^{3} \alpha_{ji} X_i = \sum_{i=1}^{3} \alpha_{ji} \sum_{r=1}^{3} \alpha_{ri} X_r' = \sum_{r=1}^{3} X_r' \sum_{i=1}^{3} \alpha_{ji} \alpha_{ri} = \sum_{r=1}^{3} \delta_{jr} X_r' = X_j',$$

where $j = 1, 2, 3$. Changing the notation index, the previous relation gives

$$X'_i = \sum_{j=1}^{3} \alpha_{ij} X_j. \tag{1.45}$$

Equations (1.42) and (1.45) give the transformation of a vector components from a reference frame to another reference frame. The same equations can be written in a matrix form. If the matrix notation is used

$$X = \begin{pmatrix} X_1 \\ X_2 \\ X_3 \end{pmatrix},$$

and

$$X' = \begin{pmatrix} X'_1 \\ X'_2 \\ X'_3 \end{pmatrix},$$

then (1.45) can be written as

$$X' = AX. \tag{1.46}$$

Using (1.44), it is obtained

$$X = A^{\mathrm{T}} X'. \tag{1.47}$$

The matrix A^{T}

$$A^{\mathrm{T}} = \begin{pmatrix} \alpha_{11} & \alpha_{21} & \alpha_{31} \\ \alpha_{12} & \alpha_{22} & \alpha_{32} \\ \alpha_{13} & \alpha_{23} & \alpha_{33} \end{pmatrix}$$

represents the transpose of the matrix A.

Using (1.40), the following relation is written:

$$\det A \det A = \det(\delta_{ij}) = 1 \Rightarrow \det A \neq 0. \tag{1.48}$$

Equation (1.48) can be proved using the relations: $\det A \det B = \det(AB)$ (the determinant of a product of matrices equals the product of the determinants of the individual matrices), $\det A = \det A^{\mathrm{T}}$ (the determinant of the transpose of a matrix equals the determinant of the matrix), and (1.40), that is, $\sum_{k=1}^{3} \alpha_{ik}\alpha_{jk} = \delta_{ij}$, $i = 1,2,3$ and $j = 1,2,3$. The following relation can be written:

$$\det A \det A = \det A \det A^{\mathrm{T}}$$
$$= \det A A^{\mathrm{T}}$$

$$
= \det \left(\begin{pmatrix} \alpha_{11} & \alpha_{12} & \alpha_{13} \\ \alpha_{21} & \alpha_{22} & \alpha_{23} \\ \alpha_{31} & \alpha_{32} & \alpha_{33} \end{pmatrix} \begin{pmatrix} \alpha_{11} & \alpha_{21} & \alpha_{31} \\ \alpha_{12} & \alpha_{22} & \alpha_{32} \\ \alpha_{13} & \alpha_{23} & \alpha_{33} \end{pmatrix} \right)
$$

$$
= \det \begin{pmatrix} \sum\limits_{k=1}^{3} \alpha_{1k}\alpha_{1k} & \sum\limits_{k=1}^{3} \alpha_{1k}\alpha_{2k} & \sum\limits_{k=1}^{3} \alpha_{1k}\alpha_{3k} \\ \sum\limits_{k=1}^{3} \alpha_{2k}\alpha_{1k} & \sum\limits_{k=1}^{3} \alpha_{2k}\alpha_{2k} & \sum\limits_{k=1}^{3} \alpha_{2k}\alpha_{3k} \\ \sum\limits_{k=1}^{3} \alpha_{3k}\alpha_{1k} & \sum\limits_{k=1}^{3} \alpha_{3k}\alpha_{2k} & \sum\limits_{k=1}^{3} \alpha_{3k}\alpha_{3k} \end{pmatrix}
$$

$$
= \det \begin{pmatrix} \delta_{11} & \delta_{12} & \delta_{13} \\ \delta_{21} & \delta_{22} & \delta_{23} \\ \delta_{31} & \delta_{32} & \delta_{33} \end{pmatrix}
$$

$$
= \det \begin{pmatrix} 1 & 0 & 0 \\ 0 & 1 & 0 \\ 0 & 0 & 1 \end{pmatrix}
$$

$$
= \det(\delta_{ij}) = 1 \Rightarrow \det A \neq 0.
$$

Because $\det A \neq 0$, the inverse matrix A^{-1} exists, and using (1.46), it results

$$
X = A^{-1} X'. \tag{1.49}
$$

Equations (1.47) and (1.49) give

$$
(A^{\mathrm{T}} - A^{-1})X = 0.
$$

Because the previous relation is true for all the matrices X, the following relation exists:

$$
A^{\mathrm{T}} = A^{-1}. \tag{1.50}
$$

Equation (1.50) characterizes the *orthogonal matrices*. The following matrix equation can be written:

$$
AA^{\mathrm{T}} = E.
$$

It is obvious that

$$
AA^{-1} = A^{-1}A = E. \tag{1.51}
$$

Equations (1.50) and (1.51) give

$$
A^{\mathrm{T}} A = E.
$$

Equations (1.36) and (1.37) give

$$\sum_{k=1}^{3} \alpha_{ki}\alpha_{kj} = \delta_{ij}. \tag{1.52}$$

From the (1.34) and (1.44), it results

$$\mathbf{i} = \alpha_{11}\mathbf{i}' + \alpha_{21}\mathbf{j}' + \alpha_{31}\mathbf{k}',$$
$$\mathbf{j} = \alpha_{12}\mathbf{i}' + \alpha_{22}\mathbf{j}' + \alpha_{32}\mathbf{k}',$$
$$\mathbf{k} = \alpha_{13}\mathbf{i}' + \alpha_{23}\mathbf{j}' + \alpha_{33}\mathbf{k}'.$$

Then, using (1.34), the following result is obtained:

$$(\mathbf{i}',\mathbf{j}',\mathbf{k}')^{\mathrm{T}} = A(\mathbf{i},\mathbf{j},\mathbf{k})^{\mathrm{T}}.$$

The following equation can be written:

$$1 = \det E = \det(AA^{\mathrm{T}}) = \det A \det A^{\mathrm{T}} = (\det A)^2 \Leftrightarrow \det A = \pm 1.$$

Thus (because A is an orthogonal matrix),

$$\det A = 1.$$

1.10 Tensors

Let $Oxyz$ and $Ox'y'z'$ be two sets of rectangular Cartesian reference frames. The components of the vector X change according to the (1.44) and (1.45) as

$$X_i' = \sum_{j=1}^{3} \alpha_{ij} X_j \quad \text{and} \quad X_i = \sum_{j=1}^{3} \alpha_{ij} X_j', \quad \text{where } i = 1,2,3.$$

Generalizing these relations, the following scalar can be introduced:

$$X_{ij} = X_i X_j, \tag{1.53}$$

where X_k ($k = 1, 2, 3$) are the components of a vector $\mathbf{X}$. There are nine scalars X_{11}, X_{12}, X_{13}, X_{21}, X_{22}, X_{23}, X_{31}, X_{32}, and X_{33}. A vector defined in the Euclidean space has three components and is called a *tensor of rank one*. A *tensor of rank two*, in a Cartesian reference frame $Oxyz$, is a mathematical entity with nine components X_{ij} ($i = 1, 2, 3$ and $j = 1, 2, 3$). In the Cartesian reference frame $Ox'y'z'$, the new tensor of rank two will have the components X_{ij}' ($i = 1, 2, 3$ and $j = 1, 2, 3$). Equation (1.53) gives

$$X'_{ij} = X'_i X'_j$$

$$= \sum_{k=1}^{3} \alpha_{ik} X_k \sum_{l=1}^{3} \alpha_{jl} X_l$$

$$= \sum_{k=1}^{3} \sum_{j=1}^{3} \alpha_{ik} X_k \alpha_{jl} X_l$$

$$= \sum_{k=1}^{3} \sum_{j=1}^{3} \alpha_{ik} \alpha_{jl} X_k X_l$$

$$= \sum_{k=1}^{3} \sum_{l=1}^{3} \alpha_{ik} \alpha_{jl} X_{kl}, \quad \text{where } i = 1,2,3; \ j = 1,2,3. \tag{1.54}$$

Multiplying (1.54) with $\alpha_{is}\alpha_{jt}$ and adding the relations for $i = 1,2,3$ and $j = 1,2,3$, it results

$$\sum_{i=1}^{3} \sum_{j=1}^{3} \alpha_{is} \alpha_{jt} X'_{ij} = \sum_{i=1}^{3} \sum_{j=1}^{3} \sum_{k=1}^{3} \sum_{l=1}^{3} \alpha_{is} \alpha_{jt} \alpha_{ik} \alpha_{jl} X_{kl}$$

$$= \sum_{k=1}^{3} \sum_{l=1}^{3} X_{kl} \sum_{i=1}^{3} \alpha_{is} \alpha_{ik} \sum_{j=1}^{3} \alpha_{jt} \alpha_{jl},$$

where $t = 1,2,3$ and $s = 1,2,3$. The previous relation gives

$$\sum_{i=1}^{3} \sum_{j=1}^{3} \alpha_{is} \alpha_{jt} X'_{ij} = \sum_{k=1}^{3} \sum_{l=1}^{3} \delta_{sk} \delta_{tl} X_{kl} = X_{st}, \quad \text{where } t = 1,2,3; \ s = 1,2,3.$$

Changing the notation index, the following relation is obtained:

$$X_{ij} = \sum_{k=1}^{3} \sum_{l=1}^{3} \alpha_{ki} \alpha_{lj} X'_{kl}, \quad \text{where } i = 1,2,3; \ j = 1,2,3. \tag{1.55}$$

Definition. A tensor of rank two is a mathematical entity with the components $X_{ij}(i = 1,2,3$ and $j = 1,2,3)$ in the reference frame $Oxyz$. Viewed in another reference frame $Ox'y'z'$, the components X_{ij} change to $X'_{ij}(i = 1,2,3$ and $j = 1,2,3)$ according to the mathematical law given by (1.54) and (1.55). The *tensor notion* was introduce by A. L. Cauchy (1789–1857) as a part of the deformable continuous medium. The matrix attached to a tensor of rank two in a Cartesian reference frame $Oxyz$ can be written as

$$X = \begin{pmatrix} X_{11} & X_{21} & X_{31} \\ X_{12} & X_{22} & X_{32} \\ X_{13} & X_{23} & X_{33} \end{pmatrix},$$

and in the Cartesian reference frame, $Ox'y'z'$ is

$$X' = \begin{pmatrix} X'_{11} & X'_{21} & X'_{31} \\ X'_{12} & X'_{22} & X'_{32} \\ X'_{13} & X'_{23} & X'_{33} \end{pmatrix}.$$

Then, using (1.54) and (1.52), it results

$$X' = AXA^{T} = AXA^{-1}, \tag{1.56}$$

and using (1.56), it is obtained

$$X = A^{-1}X'A.$$

The previous relation represents the matrix form of (1.55).

Let $\mathbf{X} = (X_1, X_2, X_3)$ and $\mathbf{Y} = (Y_1, Y_2, Y_3)$ be two nonzero vectors written in a Cartesian reference frame $Oxyz$. The components $Z_{ij} = X_i Y_j$ ($i = 1,2,3$ and $j = 1,2,3$) define in the Cartesian reference frame $Oxyz$ a tensor $\overset{\Rightarrow}{\mathbf{Z}}$. Using a matrix form, the tensor $\overset{\Rightarrow}{\mathbf{Z}}$ can be written as

$$\overset{\Rightarrow}{\mathbf{Z}} = \begin{bmatrix} Z_{11} & Z_{12} & Z_{13} \\ Z_{21} & Z_{22} & Z_{23} \\ Z_{31} & Z_{32} & Z_{33} \end{bmatrix}. \tag{1.57}$$

The components of this tensor in the Cartesian reference frame $Ox'y'z'$ can be written $Z'_{ij} = X'_i Y'_j$. Indeed, one can write

$$Z'_{ij} = X'_i Y'_j = \sum_{k=1}^{3} \alpha_{ik} X_k \sum_{l=1}^{3} \alpha_{jl} Y_l = \sum_{k=1}^{3} \sum_{l=1}^{3} \alpha_{ik} \alpha_{jl} Z_{kl}.$$

The tensor $\overset{\Rightarrow}{\mathbf{Z}}$ is named the *tensorial product* of the vectors $\mathbf{X}$ and $\mathbf{Y}$, and it can be written symbolically as

$$\overset{\Rightarrow}{\mathbf{Z}} = \mathbf{X} \oplus \mathbf{Y}.$$

The tensor $\overset{\Rightarrow}{\mathbf{T}} = \mathbf{Y} \oplus \mathbf{X}$ given in the Cartesian reference frame $Oxyz$ has the components

$$T_{ij} = Y_i X_j, \text{ where } i = 1,2,3, \text{ and } j = 1,2,3,$$

and it is not always true that $\mathbf{X} \oplus \mathbf{Y} = \mathbf{Y} \oplus \mathbf{X}$.

The *unity tensor* $\overset{\Rightarrow}{\mathbf{E}}$ in the Cartesian reference frame $Oxyz$ has the values

$$E_{ij} = \delta_{ij}, \text{ where } i = 1,2,3 \text{ and } j = 1,2,3.$$

Using a matrix form, the unity tensor can be written as

$$\overset{\Rightarrow}{\mathbf{E}} = \begin{bmatrix} 1 & 0 & 0 \\ 0 & 1 & 0 \\ 0 & 0 & 1 \end{bmatrix}.$$

The transpose of the tensor $\overset{\Rightarrow}{\mathbf{Z}}$ given as

$$\overset{\Rightarrow}{\mathbf{Z}} = \begin{bmatrix} Z_{11} & Z_{21} & Z_{31} \\ Z_{12} & Z_{22} & Z_{32} \\ Z_{13} & Z_{23} & Z_{33} \end{bmatrix}.$$

is obtained by interchanging the subscripts of each element. If $\overset{\Rightarrow}{\mathbf{Z}} = \overset{\Rightarrow}{\mathbf{Z}}^{T}$, the tensor $\overset{\Rightarrow}{\mathbf{Z}}$ is called a symmetric tensor.

Two tensors $\overset{\Rightarrow}{\mathbf{Z}}$ and $\overset{\Rightarrow}{\mathbf{W}}$ are added:

$$\overset{\Rightarrow}{\mathbf{Z}} + \overset{\Rightarrow}{\mathbf{W}} = \begin{bmatrix} Z_{11} + W_{11} & Z_{12} + W_{12} & Z_{13} + W_{13} \\ Z_{21} + W_{21} & Z_{22} + W_{22} & Z_{23} + W_{32} \\ Z_{31} + W_{31} & Z_{32} + W_{32} & Z_{33} + W_{33} \end{bmatrix}$$

or multiplied by scalars

$$s\,\overset{\Rightarrow}{\mathbf{Z}} = \begin{bmatrix} sZ_{11} & sZ_{21} & sZ_{31} \\ sZ_{12} & sZ_{22} & sZ_{32} \\ sZ_{13} & sZ_{23} & sZ_{33} \end{bmatrix}.$$

The magnitude of a tensor is defined as

$$|\overset{\Rightarrow}{\mathbf{Z}}| = \sqrt{\frac{1}{2}\sum_{i}\sum_{j} Z_{ij}^{2}}.$$

Remark

The *permutation symbol* ϵ_{ijk}, $i = 1, 2, 3$, $j = 1, 2, 3$, and $k = 1, 2, 3$ is a three-index object sometimes called alternating tensor or Levi-Civita symbol. The permutation symbol ϵ_{ijk} is defined by the equations

$$\epsilon_{111} = \epsilon_{222} = \epsilon_{333} = \epsilon_{112} = \epsilon_{121} = \epsilon_{211} = \epsilon_{221} = \epsilon_{331} = \ldots = 0,$$

$$\epsilon_{123} = \epsilon_{231} = \epsilon_{312} = 1,$$

$$\epsilon_{213} = \epsilon_{321} = \epsilon_{132} = -1,$$

or equivalent

$$\epsilon_{ijk} = \begin{cases} +1 & \text{if } (i,j,k) \text{ is an even permutation of } (1,2,3), \\ -1 & \text{if } (i,j,k) \text{ is an odd permutation of } (1,2,3), \\ 0 & \text{otherwise,} \end{cases} \qquad (1.58)$$

that is, ϵ_{ijk} is a tensor antisymmetric in all indices (in any pair of indices) also called *alternating* tensor of third order.

A tensor T_{ij} is said to be symmetric in a pair of indices (consider the indices i and j for this case) if

$$T_{ij} = T_{ji},$$

and antisymmetric in i and j if

$$T_{ij} = -T_{ji}.$$

A tensor is called *completely antisymmetric* if changes sign under the exchange of any pair of indices.

When a permutation σ can be written as a product of r transpositions, that is, $\sigma = \tau_1 \tau_2 ... \tau_r$, the sign of the permutation is defined as $\mathrm{sgn}(\sigma) = (-1)^r$. Permutations with sign 1 are called *even* and those with sign -1 are called *odd*. A transposition is a 2-cycle, that is, a cycle of length 2. Thus, a transposition is a permutation (xy) which simply swaps round the two elements x and y.

Let $x_1, x_2, ..., x_r$, $1 \leq n \leq n$ be distinct elements of $1, 2, ..., n$. An r-cycle $(x_1 x_2 ... x_r)$ is a permutation which maps

$$x_1 \mapsto x_2, x_2 \mapsto x_3, ..., x_{r-1} \mapsto x_r, x_r \mapsto x_1,$$

and fixes all other points in $1, 2, \ldots, n$. The r-cycle could be written also as

$$(x_2 x_3 ... x_r x_1).$$

Equation (1.58) represents a very important relation for the permutation symbol (mainly used to derive identities between various vector products). The product of two permutation symbols using Kronecker's delta function is

$$\epsilon_{ijk}\,\epsilon_{lmn} = \begin{vmatrix} \delta_{il} & \delta_{im} & \delta_{in} \\ \delta_{jl} & \delta_{jm} & \delta_{jn} \\ \delta_{kl} & \delta_{km} & \delta_{kn} \end{vmatrix} \qquad (1.59)$$

or equivalent (after calculating the determinant)

$$\begin{aligned} \epsilon_{ijk}\epsilon_{lmn} =\ & \delta_{il}\delta_{jm}\delta_{kn} + \delta_{im}\delta_{jn}\delta_{kl} + \delta_{in}\delta_{jl}\delta_{km} \\ & -\delta_{in}\delta_{jm}\delta_{kl} - \delta_{jn}\delta_{km}\delta_{il} - \delta_{kn}\delta_{im}\delta_{jl}. \end{aligned} \qquad (1.60)$$

The next relation related to a Levi-Civita symbol

$$a_i \epsilon_{jkl} - a_j \epsilon_{ikl} - a_k \epsilon_{jil} - al \epsilon_{jki} = 0 \tag{1.61}$$

follows from the observation that is not possible to construct a totally antisymmetric symbol with more than three indices. The Kronecker delta δ_{ij} is a symmetric second-order tensor since $\delta_{ij} = \delta_{ji}$.

The permutation symbol can also be defined as the scalar triple product of unit vectors in a right-handed coordinate system

$$\epsilon_{ijk} = \mathbf{x}_i \cdot (\mathbf{x}_j \times \mathbf{x}_k),$$

where $\mathbf{x}_i$, $\mathbf{x}_j$, and $\mathbf{x}_k$ represent the units vectors for an arbitrary Cartesian reference frame. It follows that

$$\mathbf{x}_j \times \mathbf{x}_k = \epsilon_{ijk} \mathbf{x}_i,$$

that is, $\mathbf{x}_2 \times \mathbf{x}_3 = \mathbf{x}_1$, $\mathbf{x}_3 \times \mathbf{x}_1 = \mathbf{x}_2$, $\mathbf{x}_1 \times \mathbf{x}_3 = -\mathbf{x}_2$, and so on.

The next MATLAB program calculates the values of the two-index and three-index *permutation symbol*. The MATLAB program starts with the following statements:

```
clear all % clears all variables and functions
clc  % clears the command window and homes the cursor
close all % closes all the open figure windows
```

The statements that clear all the objects in the MATLAB workspace and reset the symbolic engine clear the command window and home the cursor and close all the open figure windows. To write the text `the values of the permutation symbol:` and `epsilon_ij:` on the computer screen, the following MATLAB commands are used:

```
disp('the values of the permutation symbol:')
fprintf('\n')
fprintf('epsilon_ij:\n')
```

The statement `fprintf(f,format,s)` writes data in the real part of array `s` to the file `f`. The data is formatted under control of the specified `format` string. The statement `disp(x)` displays the array, without printing the array name.

To calculate the two-index and three-index symbol, the program uses the MATLAB statement: `for` *var=startval:step:endval, statement* `end`. It repeatedly evaluates *statement* in a loop. The counter variable of the loop is *var*. At the start, the variable is initialized to value *startval* and is incremented (or decremented when *step* is negative) by the value *step* for each iteration. The *statement* is repeated until *var* has incremented to the value *endval*. The next Matalb command assigns the value 2 to the MATLAB variable `val`:

```
val=2;
```

Using two `for` statements and two counter variables, `i` and `j`, the two-index permutation symbol is calculated and printed. The calculation of the permutation symbol is performed using a user-defined function denoted by `epsilon([i j])`, and this function will be presented later. The values of the permutation symbol are printed inside the `for` loop using a `display` statement. The MATLAB statement `display(x)` prints the value of a variable or expression. The MATLAB software calls `display(x)` when it interprets a variable or expression that is not terminated by a semicolon. Inside the `display(x)` function, the MATLAB function `sprintf(x)` has been used. The `sprintf(x)` function formats the data under control of the specified format string and returns it in the MATLAB. The `sprintf(x)` function returns an error message string `errmsg` if an error occurred. The following commands are used to calculate ϵ_{ij}:

```
val=2;
for j=1:val
    for i=1:val
% results of the two-index permutation symbol
% function epsilon([i,j])) is used
display(sprintf(...
    '(i j) = (%d %d) -> epsilon_{%d%d}= %d',...
    i, j, i, j, epsilon([i j])));
    end
end
```

Using the same approach as before, the three-index permutation symbol is calculated using three counter variables, `i`, `j`, and `k`. The same user-defined function `epsilon([i j k])` is used for the computation of the permutation symbol:

```
val=3;
for k=1:val
    for j=1:val
        for i=1:val
% results of the three-index permutation symbol
% function epsilon([i,j,k])) is used
display(sprintf(...
 '(i j k) = (%d %d %d) -> epsilon_{%d%d%d}= %d',...
  i, j, k, i, j, k, epsilon([i j k])));
        end
    end
end
fprintf('\n')
```

The MATLAB function `epsilon` listed below calculates the value of permutation symbol. The function uses an `if` statement to calculate the value of the permutation symbol. The MATLAB statement `if expression, statements, end` evaluates expression and, if the evaluation yields logical 1 (true) or a nonzero result, executes one or more MATLAB commands denoted as `statements`.

The `expression` is a MATLAB expression, usually consisting of variables or smaller expressions joined by relational operators (e.g., `count < limit`) or logical functions. Simple expressions can be combined by logical operators into compound expressions. MATLAB evaluates compound expressions from left to right, adhering to operator precedence rules. The MATLAB function `epsilon` is listed below:

```
function [val]=epsilon(string)
if sum(sort(string)~=(1:length(string)))>0
   val=0;
else
val=0;
m=length(string);
   for i=1:m-1
       if string(i)>string(i+1)
           string_new=string(i);
           string(i)=string(i+1);
           string(i+1)=string_new;
           val=val+1;
       end
   end
string=sort(string);
val=(-1)^mod(val,2);
end
```

The MATLAB sequences

```
       for i=1:m-1
           if string(i)>string(i+1)
               string_new=string(i);
               string(i)=string(i+1);
               string(i+1)=string_new;
               val=val+1;
           end
       end
```

inside the `else` part of the MATLAB function `epsilon` interchange (permute) the string elements of `string` based on the condition `string(i)>string(i+1)`. The `then` part of the MATLAB function `epsilon` assigns the value 0 to the variable `val`, while the `else` part of the function assigns $+1$ or -1 to the same variable using the statement:

```
val=(-1)^mod(val,2);
```

MATLAB program for the calculation of the two-index and three-index permutation symbol is given in Sect. 1.13.

The permutation symbol can be generalized to an arbitrary number of elements as $\epsilon_{i_1 i_2 \ldots i_m}$, in which case it can be defined as follows. The subscripts can have any value from 1 to m. If at least two subscripts have the same value, then the generalized permutation symbol is zero. If all the subscripts are separately distinct, the generalized permutation symbol has the value -1 or 1, according to whether it requires an even or odd number of permutations to arrange the subscripts in the order $i_1 i_2 ... i_m$, that is, the permutation symbol is $-1^{n(p)}$ where $n(p)$ is the number of transpositions of pairs of elements (i.e., permutation inversions) that must be composed to build up the permutation $p = i_1 i_2 ... i_m$. For 2 elements, one can write $\epsilon_{12} = 1$, $\epsilon_{21} = -1$, $\epsilon_{11} = \epsilon_{22} = 0$. The symbol can also be interpreted as a tensor, in which case it is called the *permutation tensor*.

1.10.1 Operations with Tensors

1. The tensors $\overrightarrow{\overrightarrow{X}}$ and $\overrightarrow{\overrightarrow{Y}}$ are *equal*, and write $\overrightarrow{\overrightarrow{X}} = \overrightarrow{\overrightarrow{Y}}$ if and only if $X_{ij} = Y_{ij}$ for all $i = 1, 2, 3$ and $j = 1, 2, 3$.
2. If $\overrightarrow{\overrightarrow{X}}$ is a tensor and λ is scalar, a new tensor can be defined $\lambda \overrightarrow{\overrightarrow{X}}$, with the components λX_{ij}. This is the *product of a tensor with a scalar*.
3. If $\overrightarrow{\overrightarrow{X}}$ and $\overrightarrow{\overrightarrow{Y}}$ are second-order tensors with the components X_{ij} and Y_{ij}, one can define

$$\overrightarrow{\overrightarrow{Z}} = \overrightarrow{\overrightarrow{X}} + \overrightarrow{\overrightarrow{Y}} = \overrightarrow{\overrightarrow{Y}} + \overrightarrow{\overrightarrow{X}},$$

   as the arithmetic sum of the tensors. The previous equality is known as *commutative law of addition*.
4. The following properties can be written:
   Associative law of addition

$$\overrightarrow{\overrightarrow{X}} + \left(\overrightarrow{\overrightarrow{Y}} + \overrightarrow{\overrightarrow{Z}} \right) = \left(\overrightarrow{\overrightarrow{X}} + \overrightarrow{\overrightarrow{Y}} \right) + \overrightarrow{\overrightarrow{Z}}.$$

   For every tensor $\overrightarrow{\overrightarrow{X}}$, there is a tensor $-\overrightarrow{\overrightarrow{X}}$ such that

$$\overrightarrow{\overrightarrow{X}} + \left(-\overrightarrow{\overrightarrow{X}} \right) = \left(-\overrightarrow{\overrightarrow{X}} \right) + \overrightarrow{\overrightarrow{X}} = \mathbf{0}.$$

   Commutative law of scalar multiplication

$$\lambda \overrightarrow{\overrightarrow{X}} = \overrightarrow{\overrightarrow{X}} \lambda.$$

   Associative law of scalar multiplication

$$\lambda \left(\mu \overrightarrow{\overrightarrow{X}} \right) = \mu \left(\lambda \overrightarrow{\overrightarrow{X}} \right) = \lambda \mu \overrightarrow{\overrightarrow{X}}.$$

Distributive law

$$\lambda \left(\vec{\vec{X}} + \vec{\vec{Y}} \right) = \lambda \, \vec{\vec{X}} + \lambda \, \vec{\vec{Y}} \, .$$

For any $\lambda, \mu \in R$ *and for every tensor* $\vec{\vec{X}}$, *one can write*

$$(\lambda + \mu) \, \vec{\vec{X}} = \lambda \, \vec{\vec{X}} + \mu \, \vec{\vec{X}} \, .$$

Regarding to the previous relations, one can conclude that *the set of second-orders tensors form a linear space.*

1.10.2 Some Further Properties of Second-Order Tensor

1. *Symmetry.* Let X_{ij} be the components of the tensor $\vec{\vec{X}}$ in the Cartesian reference frame *Oxyz*.

   If $X_{ij} = X_{ji}$, the second-order tensor $\vec{\vec{X}}$ is said to be symmetric. Its matrix is diagonally symmetric.

2. *Skew-Symmetry.* Let X_{ij} be the components of the tensor $\vec{\vec{X}}$ in the Cartesian reference frame *Oxyz*.

   If $X_{ij} = -X_{ji}$, the second-order tensor $\vec{\vec{X}}$ is said to be skew-symmetric. Its matrix is skew-symmetric about the principal diagonal, having zeros along this diagonal.

   In general, a second-order tensor is neither symmetric nor skew-symmetric, but any second-order tensor may be expressed as the sum of a symmetric and skew-symmetric tensor, in the following way:

$$X_{ij} = \frac{1}{2} \left(X_{ij} + X_{ji} \right) + \frac{1}{2} \left(X_{ij} - X_{ji} \right).$$

   The quantities $T_{ij} = \frac{1}{2} \left(X_{ij} + X_{ji} \right)$ and $V_{ij} = \frac{1}{2} \left(X_{ij} - X_{ji} \right)$ are easily shown to be second-order tensors, and since $T_{ij} = T_{ji}$ and $V_{ji} = -V_{ij}$, they are, respectively, symmetric and skew-symmetric second-order tensors.

3. *Contraction.* Let X_{ij} be the components of the tensor $\vec{\vec{X}}$ in the Cartesian reference frame *Oxyz* and X'_{ij} be the components of the tensor $\vec{\vec{X}}$ in the Cartesian reference frame *Ox'y'z'* where $i = 1, 2, 3$ and $j = 1, 2, 3$. A tensor contraction exists if

$$\sum_{i=1}^{3} X_{ii} = X_{11} + X_{22} + X_{33}.$$

Theorem 1.1. *The result of a contraction of a second-order tensor* $\overset{\Rightarrow}{\mathbf{X}}$ *is a zero-order tensor, that is, a scalar, who stay invariant under orthogonal transformation.*

Proof. Using (1.54) and (1.45), one can write

$$\sum_{i=1}^{3} X'_{ii} = \sum_{i=1}^{3}\sum_{k=1}^{3}\sum_{l=1}^{3} \alpha_{ik}\alpha_{il} X_{kl} = \sum_{k=1}^{3}\sum_{l=1}^{3} X_{kl} \sum_{i=1}^{3} \alpha_{ik}\alpha_{il}$$

$$= \sum_{k=1}^{3}\sum_{l=1}^{3} X_{kl}\delta_{kl} = \sum_{l=1}^{3} X_{ll},$$

thus the scalar

$$X'_{11} + X'_{22} + X'_{33} = X_{11} + X_{22} + X_{33}$$

stays invariant if the reference frame is changed. The result says that the trace of the matrix $X_{ij}(j = 1,2,3$ and $j = 1,2,3)$, that is, the sum of its diagonal elements, *stays invariant under orthogonal transformation.* $\square$

1.11   Examples

Example 1.1. In Fig. 1.13a, the rectangular component of the vector $\mathbf{F}$ on the *OA* direction is $\mathbf{f}$, with the magnitude $|\mathbf{f}| = f$. The vector $\mathbf{F}$ acts at an angle β with the positive direction of the x-axis. Find the magnitude $|\mathbf{F}| = F$ of the vector $\mathbf{F}$. Numerical application: $f = 20$, $\alpha = 30°$, and $\beta = 60°$.

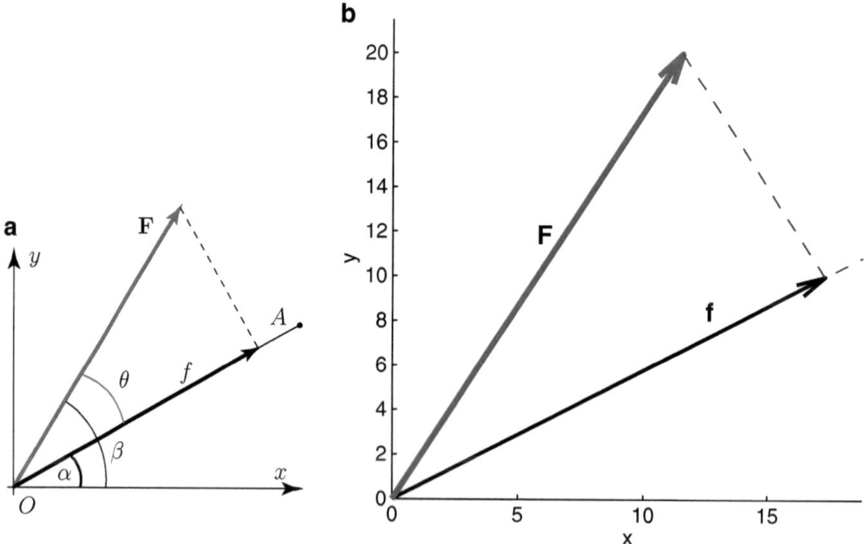

Fig. 1.13 Example 1.1

Solution
The component of $\mathbf{F}$ on the OA direction is $|\mathbf{F}|\cos\theta - f$. From Fig. 1.13, the angle θ of the vector $\mathbf{F}$ with the OA direction is $\theta = \beta - \alpha = 60° - 30° = 30°$. The magnitude F is calculated from the equation

$$|\mathbf{F}|\cos\theta = f \Leftrightarrow |\mathbf{F}|\cos 30° = 20 \Rightarrow F = |\mathbf{F}| = \frac{f}{\cos\theta} = \frac{20}{\cos 30°} \text{ or } F = 23.094.$$

The MATLAB program starts with the following statements:

```
clear all
% clears all the objects in the MATLAB workspace and
% resets the default MuPAD symbolic engine
clc  % clears the command window and homes the cursor
close all  % closes all the open figure windows
```

The MATLAB commands for the input data are

```
f = 20;
alpha = pi/6;
beta = pi/3;
```

The angle θ and the magnitude of the vector $\mathbf{F}$ are calculated with

```
theta = beta-alpha;
F = f/cos(theta);
```

The statement `cos(s)` is the cosine of the argument s in radians. The numerical solution for F is printed using the statement:

```
fprintf('F = %f',F)
```

The statement `fprintf(f,format,s)` writes data in the real part of array s to the file f. The data is formatted under control of the specified `format` string and contains ordinary characters and/or C language conversion specifications. The conversion character `%f` specifies the output as fixed-point notation. For more details about `fprintf`, see online help.

Next, the two vectors $\mathbf{f}$ and $\mathbf{F}$ will be plotted. The x and y components of the vectors $\mathbf{f}$ and $\mathbf{F}$ are

```
% components of vector f
f_x=f*cos(alpha);
f_y=f*sin(alpha);

% components of vector F
F_x=F*cos(beta);
F_y=F*sin(beta);
```

The following MATLAB commands are used to introduce the plotting of the vectors:

```
hold on
s = 1.5; % scale factor
axis([0 f_x+s 0 F_y+s])
axis square
```

The MATLAB command hold on retains the current graph and all axis properties so that succeeding plot commands add to the existing graph. The MATLAB command axis(([xMIN xMAX yMIN yMAX]) sets scaling for the x and y axes on the current plot, and the statement axis square makes the current axis box square in size. The direction of vector f is represented with

```
line([0 s*f_x],[0 s*f_y],'LineStyle','--')
```

where the command line(x,y) creates the line in vectors x and y to the current axes. The LineStyle specifies the line style '-' solid line (default), '--' dashed line, ':' dotted line, and '?.' dash-dot line.

The vector f is represented with

```
quiver(0,0,f_x,f_y,0,'Color','k','LineWidth',1.5)
```

The MATLAB command quiver(x,y,u,v,s,LineSpec) draws vectors specified by u and v at the coordinates x and y. The parameter s automatically scales the arrows to fit within the grid: s = 2 doubles the relative length, s = 0.5 halves the length, and s = 0 plots the vectors without automatic scaling. The parameter LineSpec specifies line style, marker symbol, and the 'Color' specifiers are 'r' red, 'g' green, 'b' blue, 'y' yellow, and 'k' black. The 'LineWidth' creates the width of the line in points (1 point = 1/72 inch), and by default, the line width is 0.5 point. The vector f is denoted with the MATLAB command

```
text(f_x/s+s,f_y/s+s,'f',...
     'fontsize',14,'fontweight','b')
```

where text(x,y,'text') adds the text in the quotes to location (x,y). The font size for the vector is 14, and the font is bold, 'fontweight','b'. The three full stops . . . after a MATLAB statement are used to continue the MATLAB statement to next line. The vector F is plotted and denoted with the MATLAB commands

```
quiver(0,0,F_x,F_y,0,'Color','r','LineWidth',2.5)
text(F_x/s-s,F_y/s-s,'F',...
     'fontsize',14,'fontweight','b')
```

The line that connects the end of the vector F with the end of the vector f is represented in MATLAB with

```
line([F_x f_x],[F_y f_y],'LineStyle','--')
```

The labels for the x and y axes are placed on the figure using

```
xlabel('x')
ylabel('y')
```

The MATLAB figure of the vectors is shown in Fig. 1.13b.

Example 1.2. The coordinates of two points A and B relative to the origin $O(0, 0, 0)$ are given by $A(x_A = 1, y_A = 2, z_A = 3)$ and $B(x_B = 3, y_B = 3, z_B = 3)$. Determine the unit vector of the line Δ that starts at point $A(x_A, y_A, z_A)$ and passes through the point $B(x_B, y_B, z_B)$.

Solution
The position vectors of the points $A(x_A, y_A, z_A)$ and $B(x_B, y_B, z_B)$ with respect to the origin $O(0, 0, 0)$ are

$$\overrightarrow{OA} = \mathbf{r}_A = x_A\mathbf{\imath} + y_A\mathbf{J} + z_A\mathbf{k} \quad \text{and} \quad \overrightarrow{OB} = \mathbf{r}_B = x_B\mathbf{\imath} + y_B\mathbf{J} + z_B\mathbf{k}.$$

The symbolic expressions of the vectors $\mathbf{r}_A$ and $\mathbf{r}_B$ are introduced in MATLAB as

```
syms x_A y_A z_A x_B y_B z_B real
r_A = [ x_A y_A z_A ];
r_B = [ x_B y_B z_B ];
```

The vector $\overrightarrow{AB} = \mathbf{r}_{AB}$ is defined as

$$\overrightarrow{AB} = \mathbf{r}_{AB} = \mathbf{r}_B - \mathbf{r}_A = (x_B - x_A)\mathbf{\imath} + (y_B - y_A)\mathbf{J} + (z_B - z_A)\mathbf{k},$$

or in MATLAB

```
r_AB = r_B - r_A;
```

The magnitude of the vector $\mathbf{r}_{AB}$ is

$$|\mathbf{r}_{AB}| = r_{AB} = \sqrt{(x_B - x_A)^2 + (y_B - y_A)^2 + (z_B - z_A)^2}$$

or in MATLAB

```
mr_AB = sqrt(dot(r_AB,r_AB));
```

The unit vector, $\mathbf{u}_\Delta$, of the line Δ (line AB) is

$$\mathbf{u}_\Delta = \frac{\mathbf{r}_{AB}}{|\mathbf{r}_{AB}|} = \frac{(x_B - x_A)\mathbf{\imath} + (y_B - y_A)\mathbf{J} + (z_B - z_A)\mathbf{k}}{\sqrt{(x_B - x_A)^2 + (y_B - y_A)^2 + (z_B - z_A)^2}}$$

$$= \frac{x_B - x_A}{\sqrt{(x_B - x_A)^2 + (y_B - y_A)^2 + (z_B - z_A)^2}}\mathbf{\imath}$$

$$+ \frac{y_B - y_A}{\sqrt{(x_B - x_A)^2 + (y_B - y_A)^2 + (z_B - z_A)^2}} \mathbf{J}$$

$$+ \frac{z_B - z_A}{\sqrt{(x_B - x_A)^2 + (y_B - y_A)^2 + (z_B - z_A)^2}} \mathbf{k}.$$

Using MATLAB, the unit vector is

```
u_AB = r_AB/mr_AB;
```

The numerical values of the components of the unit vector $\mathbf{u}_A = u_x\mathbf{1} + u_y\mathbf{J} + u_z\mathbf{k}$ are obtained, replacing the symbolic expressions with their numerical values

$$u_x = \frac{x_B - x_A}{r_{AB}} = \frac{3 - 1}{\sqrt{(3-1)^2 + (3-2)^2 + (3-3)^2}} = \frac{2}{2.2361} = 0.894,$$

$$u_y = \frac{y_B - y_A}{r_{AB}} = \frac{3 - 2}{\sqrt{(3-2)^2 + (3-2)^2 + (3-3)^2}} = \frac{1}{2.2361} = 0.447,$$

$$u_z = \frac{z_B - z_A}{r_{AB}} = \frac{3 - 3}{\sqrt{(3-3)^2 + (3-2)^2 + (3-3)^2}} = \frac{0}{2.2361} = 0,$$

where the magnitude of the vector $\mathbf{r}_{AB}$ is

$$r_{AB} = \sqrt{(x_B - x_A)^2 + (y_B - y_A)^2 + (z_B - z_A)^2} = \sqrt{(3-3)^2 + (3-2)^2 + (3-3)^2}$$

$$= \sqrt{5} = 2.2361.$$

To obtain the numerical values in MATLAB, x_A, y_A, z_A are replaced with 1, 2, 3 and x_B, y_B, z_B are replaced with 3, 3, 3. For this purpose, two lists are created: a list with the symbolical variables {x_A, y_A, z_A, x_B, y_B, z_B} and a list with the corresponding numeric values {1, 2, 3, 3, 3, 3}:

```
slist = {x_A, y_A, z_A, x_B, y_B, z_B};
nlist = {1, 2, 3, 3, 3, 3};
```

To obtain the numerical value for the symbolic unit vector u_AB, the following statement is introduced:

```
u_ABn = subs(u_AB, slist, nlist);
```

The statement subs(expr,lhs,rhs) replaces lhs with rhs in the symbolic expression expr. The numerical results are printed with the following command:

```
fprintf('u_AB = [%6.3f %6.3f %6.3f] \n', u_ABn)
```

The escape character \n specifies new line.

Next, the vectors $\mathbf{r}_A$, $\mathbf{r}_B$, and $\mathbf{r}_{AB}$ will be plotted using MATLAB. The numerical values of the vectors $\mathbf{r}_A$ and $\mathbf{r}_B$ are obtained with

```
rA = eval(subs(r_A, slist, nlist));
rB = eval(subs(r_B, slist, nlist));
```

The command `eval(x)`, where x is a string, executes the string as an expression. If the command `axis(([xMIN xMAX yMIN yMAX zMIN zMAX])` sets the limits of the x, y, and z axis of the current axes on the 3D plot. The statement `axis ij` positions MATLAB into its "matrix" axes mode, the coordinate system origin is at `y=z=0`, the y-axis is numbered from top to bottom, the x-axis is numbered from left to right, and the z-axis is vertical with values increasing from bottom to top. For this example, the axes are defined by

```
a=3.5;
axis ([0 a  0 a   0 a])
axis ij, grid on, hold on
```

The MATLAB command `grid on` adds major grid lines to the current axes and `hold on` locks up the current plot and all axis properties so that following graphing commands add to the existing graph. The vectors rA and rB are defined in MATLAB as

```
quiver3(0,0,0, rA(1),rA(2),rA(3),1,...
    'Color','k','LineWidth',1)
quiver3(0,0,0, rB(1),rB(2),rB(3),1,...
    'Color','k','LineWidth',1)
```

where the statement `quiver3(x,y,z,u,v,w)` represents a vector as an arrow at the point `(x,y,z)` with the components `(u,v,w)`. The dashed line `(--)` between the points A and B is plotted with the command

```
line([rA(1)  rB(1)],[rA(2)  rB(2)],[rA(3)  rB(3)],...
    'LineStyle','--')
```

and the unit vector u is represented with

```
quiver3(...
    rA(1),rA(2),rA(3), u_ABn(1),u_ABn(2),u_ABn(3),...
    1,'Color','r','LineWidth',2)
```

The labels for the vectors and the axes are printed in MATLAB with

```
text(rA(1)/2, rA(2)/2, rA(3)/2+.3,...
    'r_A','fontsize',14,'fontweight','b')
text(rB(1)/2, rB(2)/2, rB(3)/2+.3,...
    'r_B','fontsize',14,'fontweight','b')
text(...
    (rA(1)+rB(1))/2-.4,...
    (rA(2)+rB(2))/2,...
```

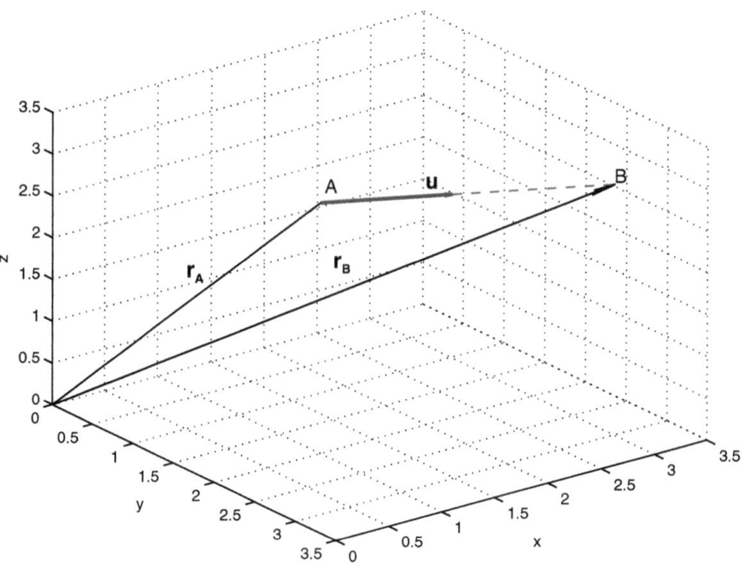

Fig. 1.14 MATLAB figure for Example 1.2

```
   (rA(3)+ rB(3))/2+.3,...
   'u','fontsize',14,'fontweight','b')
 xlabel('x'), ylabel('y'), zlabel('z')
```

The MATLAB representation of the vectors is shown in Fig. 1.14.

Example 1.3. The vectors $\mathbf{V}_1$, $\mathbf{V}_2$, $\mathbf{V}_3$, and $\mathbf{V}_4$ with the magnitude $|\mathbf{V}_1| = V_1$, $|\mathbf{V}_2| = V_2$, $|\mathbf{V}_3| = V_3$, and $|\mathbf{V}_4| = V_4$ are concurrent at the origin $O(0, 0, 0)$ and are directed through the points of coordinates $A_1(x_1, y_1, z_1)$, $A_2(x_2, y_2, z_2)$, $A_3(x_3, y_3, z_3)$, and $A_4(x_4, y_4, z_4)$, respectively. Determine the resultant vector of the system. Numerical application: $V_1 = 10$, $V_2 = 25$, $V_3 = 15$, $V_4 = 40$, $A_1(3, 1, 7)$, $A_2(5, -3, 4)$, $A_3(-4, -3, 1)$, and $A_4(4, 2, -3)$.

Solution

The magnitudes, V_i, of the vectors $\mathbf{V}_i$ and the coordinates, x_i, y_i, z_i, of the points $A_i, i = 1, 2, 3, 4$ are introduced with MATLAB as

```
V(1)=10; V(2)=25; V(3)=15; V(4)=40; % magnitudes V_i
x(1)= 3; y(1)= 1; z(1)= 7; % A_1
x(2)= 5; y(2)=-3; z(2)= 4; % A_2
x(3)=-4; y(3)=-3; z(3)= 1; % A_3
x(4)= 4; y(4)= 2; z(4)=-3; % A_4.
```

The direction cosines of the vectors $\mathbf{V}_i$ are

$$\cos\theta_{ix} = \frac{x_i}{\sqrt{x_i^2 + y_i^2 + z_i^2}}, \quad \cos\theta_{iy} = \frac{y_i}{\sqrt{x_i^2 + y_i^2 + z_i^2}}, \quad \cos\theta_{iz} = \frac{z_i}{\sqrt{x_i^2 + y_i^2 + z_i^2}},$$

and the x, y, z components of the vectors $\mathbf{V}_i$ are

$$V_{ix} = V_i \cos\theta_{ix}, \; V_{iy} = V_i \cos\theta_{iy}, \; V_{iz} = V_i \cos\theta_{iz}.$$

To calculate the direction cosines and components of the vectors for $i = 1, 2, 3, 4$ the MATLAB statement `for` *var=startval:step:endval, statement* `end` is used. It repeatedly evaluates *statement* in a loop. The counter variable of the loop is *var*. At the start, the variable is initialized to value *startval* and is incremented (or decremented when *step* is negative) by the value *step* for each iteration. The *statement* is repeated until *var* has incremented to the value *endval*. For the vectors, the following applies `for i=1:4, Program block, end` or

```
for i = 1:4
% dirction cosines of the vector v(i)
c_x(i) = x(i)/sqrt(x(i)^2+y(i)^2+z(i)^2);
c_y(i) = y(i)/sqrt(x(i)^2+y(i)^2+z(i)^2);
c_z(i) = z(i)/sqrt(x(i)^2+y(i)^2+z(i)^2);
% x, y, z components of the vector v(i)
v_x(i) = V(i)*c_x(i);
v_y(i) = V(i)*c_y(i);
v_z(i) = V(i)*c_z(i);
fprintf('vector %g: \n',i)
fprintf('direction cosines=')
fprintf(' [%6.3f,%6.3f,%6.3f]\n',c_x(i),c_y(i),c_z(i))
fprintf('vector V=')
fprintf(' [%6.3f,%6.3f,%6.3f]\n',v_x(i),v_y(i),v_z(i))
fprintf('\n')
end
```

The results in MATLAB are

```
vector 1:
direction cosines=[ 0.391,  0.130,  0.911]
vector V=[ 3.906,  1.302,  9.113]

vector 2:
direction cosines=[ 0.707,-0.424,  0.566]
vector V=[17.678,-10.607,14.142]

vector 3:
direction cosines=[-0.784,-0.588,  0.196]
vector V=[-11.767,-8.825,  2.942]

vector 4:
direction cosines=[ 0.743,  0.371,-0.557]
vector V=[29.711,14.856,-22.283]
```

or using a table form

i	V_i	A_i	$\cos\theta_{ix}$	$\cos\theta_{iy}$	$\cos\theta_{iz}$	V_{ix}	V_{iy}	V_{iz}
1	10	$(3,1,7)$	0.391	0.130	0.911	3.906	1.302	9.113
2	25	$(5,-3,4)$	0.70	-0.424	0.566	17.678	-10.607	14.142
3	15	$(-4,-3,1)$	-0.784	-0.588	0.196	-11.767	-8.825	2.942
4	40	$(4,2,-3)$	0.743	0.371	-0.557	29.711	14.856	-22.283

The vector $\mathbf{V}_i$ can be written as $\mathbf{V}_i = V_{ix}\mathbf{i} + V_{iy}\mathbf{j} + V_{iz}\mathbf{k}$, $i = 1,2,3,4$. The resultant of the system is

$$R = \sqrt{(R_x)^2 + (R_y)^2 + (R_z)^2} = \sqrt{\left(\sum V_{ix}\right)^2 + \left(\sum V_{iy}\right)^2 + \left(\sum V_{iz}\right)^2}.$$

The direction cosines of the resultant are

$$\cos\theta_x = \frac{\sum V_{ix}}{R}, \quad \cos\theta_y = \frac{\sum V_{iy}}{R}, \quad \cos\theta_z = \frac{\sum V_{iz}}{R}.$$

The resultant and the direction cosines in MATLAB are

```
Rx = sum(v_x);
Ry = sum(v_y);
Rz = sum(v_z);
R = [Rx Ry Rz];
modR = norm(R);
fprintf('R=V1+V2+V3+V4=[%6.3f,%6.3f,%6.3f]\n',R)
fprintf('|R|=%6.3g\n',modR)
fprintf('direction cosines=')
fprintf('uR=R/|R|=[%6.3f,%6.3f,%6.3f]\n',R/modR)
```

The MATLAB results are

```
R=V1+V2+V3+V4=[39.528,-3.274, 3.914]
|R|=  39.9
direction cosines=uR=R/|R|=[ 0.992,-0.082, 0.098]
```

or in table form

R	R_x	R_y	R_z	$\cos\theta_x$	$\cos\theta_y$	$\cos\theta_z$
39.9	39.528	-3.274	3.914	0.992	-0.082	0.098

The negative value of $\cos\theta_y$ signifies that the resultant has a negative component in the y direction.

Next, the vectors will be plotted using MATLAB. The axes are defined in MATLAB with

```
a = 26;
axis([-a a  -a a  -a a])
axis ij, grid on, hold on
```

```
xlabel('x'), ylabel('y'), zlabel('z')
text(0,0,0 1.5,'   O','HorizontalAlignment','right')
```
The vectors $\mathbf{V}_1$, $\mathbf{V}_2$, $\mathbf{V}_3$, $\mathbf{V}_4$, and $\mathbf{R}$ are plotted and labeled with the statements

```
quiver3(0,0,0,v_x(1),v_y(1),v_z(1),1,...
    'Color','k','LineWidth',1.5)
text(v_x(1),v_y(1),v_z(1),' V_1',...
    'fontsize',12,'fontweight','b')


quiver3(0,0,0,v_x(2),v_y(2),v_z(2),1,...
    'Color','k','LineWidth',1.5)
text(v_x(2),v_y(2),v_z(2),' V_2',...
    'fontsize',12,'fontweight','b')


quiver3(0,0,0,v_x(3),v_y(3),v_z(3),1,...
    'Color','k','LineWidth',1.5)
text(v_x(3),v_y(3),v_z(3)+1,'V_3',...
    'fontsize',12,'fontweight','b')


quiver3(0,0,0,v_x(4),v_y(4),v_z(4),1,...
    'Color','k','LineWidth',1.5)
text(v_x(4),v_y(4),v_z(4),' V_4',...
    'fontsize',12,'fontweight','b')


quiver3(0,0,0,Rx,Ry,Rz,1,...
    'Color','r','LineWidth',2.5)
text(Rx,Ry,Rz,' R','fontsize',14,'fontweight','b')
```
The MATLAB representation of the vectors is shown in Fig. 1.15.

Example 1.4. Two vectors $\mathbf{V}_1$ and $\mathbf{V}_2$ are shown in Fig. 1.16a. (a) Find the resultant of the two vectors. (b) Determine the cross product $\mathbf{V}_1 \times \mathbf{V}_1$. (c) Find the angle between the vectors $\mathbf{V}_1$ and $\mathbf{V}_2$. Numerical application: $|\mathbf{V}_1| = V_1 = 3$, $|\mathbf{V}_2| = V_2 = 3$, $a = 4$, $b = 5$, and $c = 3$.

Solution
(a) The vectors $\mathbf{V}_1$ and $\mathbf{V}_2$ are given by

$$\mathbf{V}_1 = V_{1x}\mathbf{I} + V_{1y}\mathbf{J} + V_{1z}\mathbf{k} = |\mathbf{V}_1|\frac{\mathbf{r}_{BG}}{|\mathbf{r}_{BG}|} = V_1\frac{\mathbf{r}_{BG}}{r_{BG}},$$

$$\mathbf{V}_2 = V_{2x}\mathbf{I} + V_{2y}\mathbf{J} + V_{2z}\mathbf{k} = |\mathbf{V}_2|\frac{\mathbf{r}_{BP}}{|\mathbf{r}_{BP}|} = V_2\frac{\mathbf{r}_{BP}}{r_{BP}}.$$

Next the vectors $\mathbf{r}_{BG}$ and $\mathbf{r}_{BP}$ will be calculated. From Fig. 1.16, the coordinates of the points $B, D, P,$ and Q are $B = B(x_B, y_B, z_B) = B(0, b, 0) = B(0, 5, 0)$, $G = G(x_G, y_G, z_G) = G(a, 0, c) = G(4, 0, 3)$, and $P = P(x_P, y_P, z_P) = P(a, b/2, 0) = P(4, 5/2, 0)$. The position vectors of the points $B, G,$ and P are

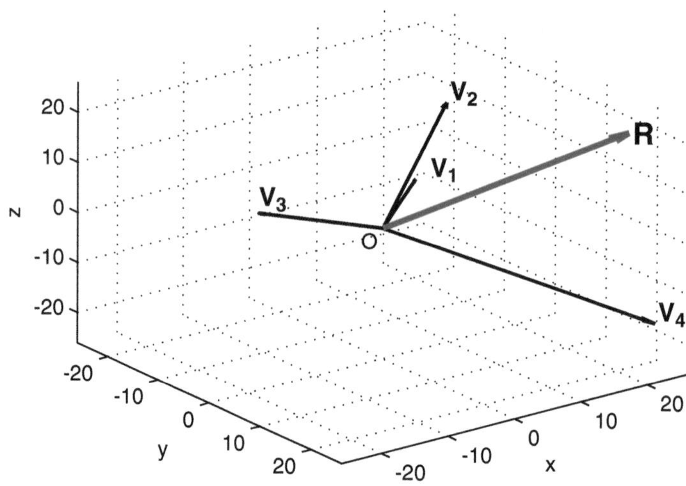

Fig. 1.15 MATLAB figure for Example 1.3

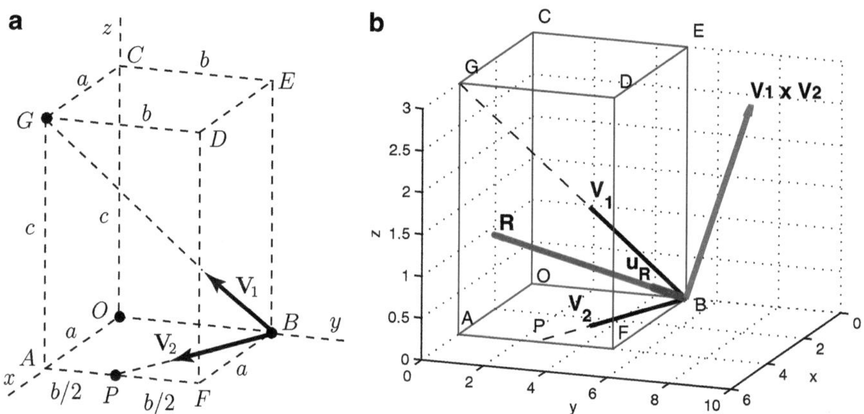

Fig. 1.16 Example 1.4

$$\mathbf{r}_B = x_B\mathbf{1} + y_B\mathbf{J} + z_B\mathbf{k} = b\mathbf{J} = 5\mathbf{J},$$

$$\mathbf{r}_G = x_G\mathbf{1} + y_G\mathbf{J} + z_G\mathbf{k} = a\mathbf{1} + c\mathbf{k} = 4\mathbf{1} + 3\mathbf{k},$$

$$\mathbf{r}_P = x_P\mathbf{1} + y_P\mathbf{J} + z_P\mathbf{k} = a\mathbf{1} + b/2\mathbf{J} = 4\mathbf{1} + 5/2\mathbf{J}.$$

The MATLAB commands for input data and for $\mathbf{r}_B$, $\mathbf{r}_G$, and $\mathbf{r}_P$ are:

```
V_1=3; V_2=3;
a=4; b=5; c=3;
```

```
x_B=0; y_B-b; z_B-0; r_B=[x_B, y_B, z_B];
x_G=a; y_G=0; z_G=c; r_G=[x_G, y_G, z_G];
x_P=a; y_P=b/2; z_P=0; r_P=[x_P, y_P, z_P];
```

The vectors $\mathbf{r}_{BG}$ and $\mathbf{r}_{BP}$ are

$$\mathbf{r}_{BG} = \mathbf{r}_G - \mathbf{r}_B = (x_G - x_B)\,\mathbf{i} + (y_G - y_B)\,\mathbf{j} + (z_G - z_B)\,\mathbf{k}$$

$$= (a-0)\,\mathbf{i} + (0-b)\,\mathbf{j} + (c-0)\,\mathbf{k}$$

$$= a\mathbf{i} - b\mathbf{j} + c\mathbf{k} = 4\mathbf{i} - 5\mathbf{j} + 3\mathbf{k},$$

$$\mathbf{r}_{BP} = \mathbf{r}_P - \mathbf{r}_B = (x_P - x_B)\,\mathbf{i} + (y_P - y_B)\,\mathbf{j} + (z_P - z_B)\,\mathbf{k}$$

$$= (a-0)\,\mathbf{i} + \left(\frac{b}{2} - b\right)\,\mathbf{j} + (0-0)\,\mathbf{k}$$

$$= a\mathbf{i} - \frac{b}{2}\mathbf{j} = 4\mathbf{i} - \frac{5}{2}\mathbf{j}.$$

The magnitudes of the vectors $\mathbf{r}_{BG}$ and $\mathbf{r}_{BP}$ are

$$|\mathbf{r}_{BG}| = r_{BG} = \sqrt{(x_G - x_B)^2 + (y_G - y_B)^2 + (z_G - z_B)^2}$$

$$= \sqrt{(a-0)^2 + (0-b)^2 + (c-0)^2} = \sqrt{a^2 + b^2 + c^2}$$

$$= \sqrt{4^2 + 5^2 + 3^2} = 7.071,$$

$$|\mathbf{r}_{BP}| = r_{BP} = \sqrt{(x_P - x_B)^2 + (y_P - y_B)^2 + (z_P - z_B)^2}$$

$$= \sqrt{(a-0)^2 + \left(\frac{b}{2} - b\right)^2 + (0-0)^2} = \sqrt{a^2 + \frac{b^2}{4}}$$

$$= \sqrt{4^2 + \frac{5^2}{4}} = 4.717.$$

The vectors $\mathbf{r}_{BG}$ and $\mathbf{r}_{BP}$ and their magnitudes in MATLAB are

```
r_BG = r_G-r_B;
r_BP = r_P-r_B;
fprintf('r_BG = [%6.3f %6.3f %6.3f]\n', r_BG)
fprintf('r_BP = [%6.3f %6.3f %6.3f]\n', r_BP)

mr_BG = sqrt(dot(r_BG, r_BG));
mr_BP = sqrt(dot(r_BP, r_BP));
fprintf('|r_BG| = %6.3f\n', mr_BG)
fprintf('|r_BP| = %6.3f\n', mr_BP)
```

The vectors $\mathbf{V}_1$ and $\mathbf{V}_2$ are

$$\mathbf{V}_1 = V_1 \frac{\mathbf{r}_{BG}}{r_{BG}} = V_1 \frac{a\mathbf{i} - b\mathbf{j} + c\mathbf{k}}{\sqrt{a^2 + b^2 + c^2}} = 3\frac{4\mathbf{i} - 5\mathbf{j} + 3\mathbf{k}}{7.071}$$

$$= 1.697\mathbf{i} - 2.121\mathbf{j} + 1.273\mathbf{k},$$

$$\mathbf{V}_2 = V_2 \frac{\mathbf{r}_{BP}}{r_{BP}} = V_2 \frac{a\mathbf{i} - (b/2)\mathbf{j}}{\sqrt{a^2 + b^2/4}} = 3\frac{4\mathbf{i} - (5/2)\mathbf{j}}{4.717}$$

$$= 2.544\mathbf{i} - 1.590\mathbf{j},$$

or with MATLAB

```
u_BD = r_BD/mr_BD;
u_PQ = r_PQ/mr_PQ;
V1  = V_1*u_BD
V2  = V_2*u_PQ
V1n = eval(subs(V1, slist, nlist));
V2n = eval(subs(V2, slist, nlist));
fprintf('V1 = [%6.3f %6.3f %6.3f]\n', V1n)
fprintf('V2 = [%6.3f %6.3f %6.3f]\n', V2n)
```

The Cartesian components of the vectors $\mathbf{V}_1$ and $\mathbf{V}_2$ are

$$V_{1x} = 1.697, V_{1y} = -2.121, V_{1z} = 1.273, V_{2x} = 2.544, V_{2y} = -1.590, V_{2z} = 0.$$

The resultant vector has the components

$$R_x = \sum V_{ix} = V_{1x} + V_{2x} = 1.697 + 2.544 = 4.241,$$

$$R_y = \sum V_{iy} = V_{1y} + V_{2y} = -2.121 - 1.590 = -3.711,$$

$$R_z = \sum V_{iz} = V_{1z} + V_{2z} = 1.273 + 0 = 1.273$$

and can be written in a vector form as

$$\mathbf{R} = R_x\mathbf{i} + R_y\mathbf{j} + R_z\mathbf{k} = 4.241\mathbf{i} - 3.711\mathbf{j} + 1.273\mathbf{k}.$$

The magnitude of $\mathbf{R}$ is

$$|\mathbf{R}| = R = \sqrt{R_x^2 + R_y^2 + R_z^2} = \sqrt{(4.241)^2 + (-3.711)^2 + (1.273)^2} = 5.778.$$

The angles of the vector $\mathbf{R}$ with the Cartesian axes are calculated from the direction cosines

$$\cos\alpha = \frac{R_x}{|\mathbf{R}|} = \frac{4.241}{5.778} = 0.734, \ \cos\beta = \frac{R_y}{|\mathbf{R}|} = \frac{-3.711}{5.778} = -0.642, \text{ and}$$

$$\cos\gamma = \frac{R_z}{|\mathbf{R}|} = \frac{1.273}{5.778} = 0.220.$$

The MATLAB commands for the resultant and direction cosines are

```
R_x = V1n(1) + V2n(1);
R_y = V1n(2) + V2n(2);
R_z = V1n(3) + V2n(3);
R = [R_x, R_y, R_z];
nR = norm(R);
u_R = R/nR; % direction cosines
fprintf('R = [%6.3f %6.3f %6.3f]\n', R)
fprintf('|R| = %6.3f\n', nR)
fprintf('u_R = [%6.3f %6.3f %6.3f]\n', u_R)
```

(b) The cross product between the vectors $\mathbf{V}_1$ and $\mathbf{V}_2$ is

$$\mathbf{V}_1 \times \mathbf{V}_2 = \begin{vmatrix} \mathbf{i} & \mathbf{j} & \mathbf{k} \\ V_{1x} & V_{1y} & V_{1z} \\ V_{2x} & V_{2y} & V_{2z} \end{vmatrix}$$

$$= \begin{vmatrix} \mathbf{i} & \mathbf{j} & \mathbf{k} \\ 1.697 & -2.121 & 1.273 \\ 2.544 & -1.590 & 0 \end{vmatrix} = 2.024\,\mathbf{i} + 3.238\,\mathbf{j} + 2.698\,\mathbf{k},$$

or with MATLAB:

```
VC = cross(V1, V2);
fprintf('V1 x V2 = [%6.3f %6.3f %6.3f]\n', VC)
```

(c) The angle θ between the vectors $\mathbf{V}_1$ and $\mathbf{V}_2$ is calculated with

$$\cos\theta = \frac{\mathbf{V}_1 \cdot \mathbf{V}_2}{V_1\,V_2} = \frac{V_{1x}\,V_{2x} + V_{1y}\,V_{2y} + V_{1z}\,V_{2z}}{V_1\,V_2}$$

$$= \frac{2.024(2.544) + (-2.121)(-1.590) + 1.273(0)}{3(3)} = 0.8545.$$

The angle is $\theta = 31.299°$. The MATLAB commands for calculating the angle between the vectors are

```
costheta = dot(V1, V2)/(V_1*V_2);
fprintf('theta = %6.3f (deg)\n', acosd(costheta))
```

The MATLAB function `acos(phi)` is the arccosine of the element `phi` and `acosd(phi)` is the inverse cosine , expressed in degrees, of the element of `phi`.

Next, the vectors $\mathbf{V}_1$, $\mathbf{V}_2$, $\mathbf{R}$, $\mathbf{u}_R$, and $\mathbf{V}_1 \times \mathbf{V}_2$ will be plotted using MATLAB. The axes are defined in MATLAB with

```
axis(1.5*[0 a 0 b 0 c])
grid on, hold on
xlabel('x'), ylabel('y'), zlabel('z')
```

For the default "Cartesian" axes mode, the coordinate system origin is at `x=y=0`. The x-axis is numbered from left to right, the y-axis is numbered from bottom to top, and the z-axis is vertical with values increasing from bottom to top. The coordinates of the points A, C, D, E, and F are

```
x_A=a; y_A=0; z_A=0;
x_C=0; y_C=0; z_C=c;
x_D=a; y_D=b; z_D=c;
x_E=0; y_E=b; z_E=c;
x_F=a; y_F=b; z_F=0;
```

The labels of the points O, A, B, C, D, E, F, G, and P are

```
text(0, 0, 0+.1,' O','fontsize',12)
text(x_A, y_A, z_A+.2,' A','fontsize',12)
text(x_B-.2, y_B, z_B-.1,' B','fontsize',12)
text(x_C, y_C, z_C+.2,' C','fontsize',12)
text(x_D, y_D, z_D+.2,' D','fontsize',12)
text(x_E, y_E, z_E+.2,' E','fontsize',12)
text(x_F, y_F, z_F+.2,' F','fontsize',12)
text(x_G, y_G, z_G+.2,' G','fontsize',12)
text(x_P, y_P, z_P+.2,' P','fontsize',12)
```

The parallelepiped $OABCDEFG$ is plotted using the MATLAB commands:

```
line([0 x_A],[0 y_A],[0 z_A])
line([0 x_B],[0 y_B],[0 z_B])
line([0 x_C],[0 y_C],[0 z_C])
line([x_B x_E],[y_B y_E],[z_B z_E])
line([x_B x_F],[y_B y_F],[z_B z_F])
line([x_A x_F],[y_A y_F],[z_A z_F])
line([x_A x_G],[y_A y_G],[z_A z_G])
line([x_C x_G],[y_C y_G],[z_C z_G])
line([x_C x_E],[y_C y_E],[z_C z_E])
line([x_D x_G],[y_D y_G],[z_D z_G])
line([x_D x_E],[y_D y_E],[z_D z_E])
line([x_D x_F],[y_D y_F],[z_D z_F])
```

Another way of drawing the parallelepiped *OABCDEFG* is

```
plot3(...
[x_G x_A x_F x_D x_G x_C x_E x_B 0 x_C],...
[y_G y_A y_F y_D y_G y_C y_E y_B 0 y_C],...
[z_G z_A z_F z_D z_G z_C z_E z_B 0 z_C])
line([0 x_A],[0 y_A],[0 z_A])
line([x_B x_F],[y_B y_F],[z_B z_F])
line([x_D x_E],[y_D y_E],[z_D z_E])
```

where the MATLAB statement `plot3(x,y,z)` plots a line in 3D through the
points whose coordinates are the elements of the vectors x, y, and z.

The lines *BG* and *BP* are plotted with

```
line([x_B x_G],[y_B y_G],[z_B z_G],...
    'Color','k','LineStyle','--')
line([x_B x_P],[y_B y_P],[z_B z_P],...
    'Color','k','LineStyle','--')
```

The vectors $\mathbf{V}_1$, $\mathbf{V}_2$, $\mathbf{R}$, $\mathbf{u}_R$, and $\mathbf{V}_1 \times \mathbf{V}_2$ and their labels are described by the
following MATLAB commands

```
quiver3(x_B,y_B,z_B, V1(1),V1(2),V1(3),1,...
    'Color','k','LineWidth',2)
quiver3(x_B,y_B,z_B, V2(1),V2(2),V2(3),1,...
    'Color','k','LineWidth',2)
quiver3(x_B,y_B,z_B, R(1),R(2),R(3),1,...
    'Color','r','LineWidth',3)
quiver3(x_B,y_B,z_B, u_R(1),u_R(2),u_R(3),1,...
    'Color','b','LineWidth',4)
quiver3(x_B,y_B,z_B, VC(1),VC(2),VC(3),1,...
    'Color','m','LineWidth',3)

text(x_B+V1(1), y_B+V1(2), z_B+V1(3)+.1,...
    'V_1','fontsize',14,'fontweight','b')
text(x_B+V2(1), y_B+V2(2), z_B+V2(3)+.1,...
    'V_2','fontsize',14,'fontweight','b')
text(x_B+R(1), y_B+R(2), z_B+R(3)+.1,...
    'R','fontsize',14,'fontweight','b')
text(x_B+u_R(1), y_B+u_R(2), z_B+u_R(3)+.1,...
    'u_R','fontsize',14,'fontweight','b')
text(x_B+VC(1), y_B+VC(2), z_B+VC(3)+.1,...
    'V_1 x V_2','fontsize',14,'fontweight','b')
```

A rotated MATLAB drawing of the vectors is shown in Fig. 1.16b.

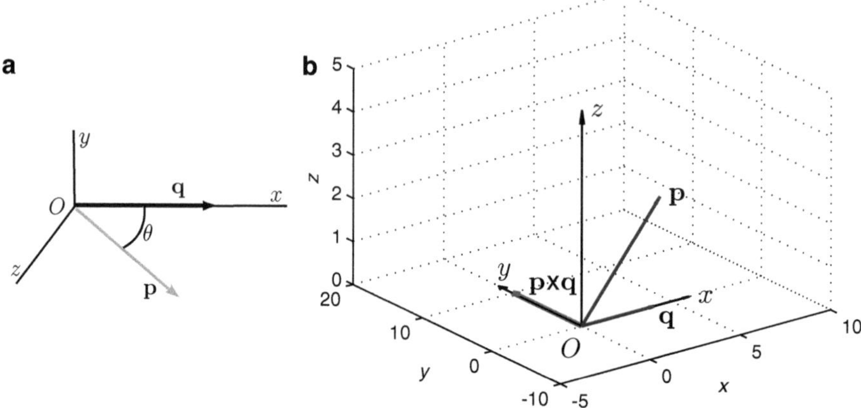

Fig. 1.17 Example 1.5

Example 1.5. The vector **p** of magnitude $|\mathbf{p}| = p$ is located in the $x - z$ plane and makes an angle θ with x-axis as shown in Fig. 1.17a. The vector **q** of magnitude $|\mathbf{q}| = q$ is situated along the x-axis. Compute the vector (cross) product $\mathbf{v} = \mathbf{p} \times \mathbf{q}$. Numerical application: $|\mathbf{p}| = p = 5$, $|\mathbf{q}| = q = 4$, and $\theta = 30°$.

Solution

The vector product **v** is perpendicular to the vectors **p** and **q** and that is why the vector **v** is along the y-axis and with has the magnitude

$$|\mathbf{v}| = |\mathbf{p}|\,|\mathbf{q}|\sin\theta = pq\sin\theta = 5(4)\sin 30° = 10.$$

From Fig. 1.17a, the direction of the vector **v** is upward.

The solution could also be obtained by expressing the vector product $\mathbf{v} = \mathbf{p} \times \mathbf{q}$ of the given vectors **p** and **q** in terms of the their rectangular components. Resolving **p** and **q** into components, one can write

$$\mathbf{v} = \mathbf{p} \times \mathbf{q} = (p_x\mathbf{\imath} + p_y\mathbf{\jmath} + p_z\mathbf{k}) \times (q_x\mathbf{\imath} + q_y\mathbf{\jmath} + q_z\mathbf{k})$$

$$= \begin{vmatrix} \mathbf{\imath} & \mathbf{\jmath} & \mathbf{k} \\ p_x & p_y & p_z \\ q_x & q_y & q_z \end{vmatrix}$$

$$= (p_yq_z - p_zq_y)\mathbf{\imath} + (p_zq_x - p_xq_z)\mathbf{\jmath} + (p_xq_y - p_yq_x)\mathbf{k}.$$

The components p_x, p_y, and p_z of the vector **p** are

$$p_x = |\mathbf{p}|\cos\theta = p\cos\theta = 5\cos 30° = 5\frac{\sqrt{3}}{2} = \frac{5\sqrt{3}}{2},\ p_y = 0,\ \text{and}$$

$$p_z = |\mathbf{p}|\sin\theta = p\sin\theta = 5\left(\frac{1}{2}\right) = \frac{5}{2}.$$

The components q_x, q_y, and q_z of the vector **q** are $q_x = q = 4$, $q_y = 0$ and $q_z = 0$.

It results

$$\mathbf{v} = \mathbf{p} \times \mathbf{q} = (p_y q_z - p_z q_y)\mathbf{i} + (p_z q_x - p_x q_z)\mathbf{j} + (p_x q_y - p_y q_x)\mathbf{k}$$

$$= \left(0(0) - \frac{5}{2}(0)\right)\mathbf{i} + \left(\frac{5}{2}(4) - \frac{5\sqrt{3}}{2}(0)\right)\mathbf{j} + \left(\frac{5\sqrt{3}}{2}(0) - 0(4)\right)\mathbf{k}$$

$$= \frac{5}{2}(4)\mathbf{j} = 10\mathbf{j}.$$

The MATLAB program for the cross product $\mathbf{v} = \mathbf{p} \times \mathbf{q}$ is

```
syms p q theta real
p_x = p*cos(theta); p_y = 0; p_z = p*sin(theta);
q_x = q; q_y = 0; q_z = 0;
v = cross([p_x p_y p_z],[q_x q_y q_z]);
slist = {p, q, theta}; nlist = {5, 4, pi/6};
vn = subs(v, slist, nlist);
fprintf('p x q = ')
fprintf('[%s %s %s]',char(v(1)),char(v(2)),char(v(3)))
fprintf(' = [%g %g %g] \n', vn)
```

and the output is

```
p x q = [0 p*sin(theta)*q 0] = [0 10 0]
```

The function `char(x)` converts the array x into MATLAB character array.

Next, the vectors $\mathbf{p}$, $\mathbf{q}$, and $\mathbf{p} \times \mathbf{q}$ will be plotted using MATLAB. The numerical values of the components of the vectors $\mathbf{p}$ and $\mathbf{q}$ are calculated with

```
p_xn=double(subs(p_x,slist,nlist));
p_yn=double(subs(p_y,slist,nlist));
p_zn=double(subs(p_z,slist,nlist));

q_xn=double(subs(q_x,slist,nlist));
q_yn=double(subs(q_y,slist,nlist));
q_zn=double(subs(q_z,slist,nlist));
```

The statement `double(x)` converts the symbolic matrix x to a matrix of double precision floating point numbers. The Cartesian axes x, y, z are ploted with

```
axis ([0 6 0 8 0 5])
axis auto,  grid on, hold on
xlabel('\it x'), ylabel('\it y'), zlabel('\it z')

quiver3(0,0,0,6,0,0,1,'Color','k','LineWidth',1)
text('Interpreter','latex','String','  $x$',...
     'Position',[6,0,0],'FontSize',14)
quiver3(0,0,0,0,12,0,1,'Color','k','LineWidth',1)
text('Interpreter','latex','String','  $y$',...
     'Position',[0,13,0],'FontSize',14)
```

```
quiver3(0,0,0,0,0,5,1,'Color','k','LineWidth',1)
text('Interpreter','latex','String','  $z$',...
    'Position',[0,0,5],'FontSize',14)
```

The statement `axis auto` returns the axis scaling to its default automatic mode. The vectors $\mathbf{p}$, $\mathbf{q}$, and $\mathbf{v} = \mathbf{p} \times \mathbf{q}$ are plotted with the MATLAB commands

```
quiver3(0,0,0,p_xn,p_yn,p_zn,1,...
    'Color','b','LineWidth',1.5)

quiver3(0,0,0,q_xn,q_yn,q_zn,1,...
    'Color','b','LineWidth',1.5)

quiver3(0,0,0,vn(1),vn(2),vn(3),1,...
    'Color','r','LineWidth',2.5)

text('Interpreter','latex',...
  'String','  \bf q',...
  'Position',[q_xn,q_yn,q_zn],...
  'FontSize',14)

text('Interpreter','latex',...
  'String','  \bf p',...
  'Position',[p_xn,p_yn,p_zn],...
  'FontSize',14)

text('Interpreter','latex',...
  'String','  {\bf p}$\times${\bf q}',...
  'Position',[vn(1)+.5,vn(2),vn(3)],...
  'FontSize',14)

text('Interpreter','latex','String','  $O$',...
  'Position',[0,0,0-.5],'FontSize',14,...
  'HorizontalAlignment','right')
```

The MATLAB drawing of the vectors is shown in Fig. 1.17b.

Example 1.6. Compute $\mathbf{a} \cdot (\mathbf{b} \times \mathbf{c})$, $(\mathbf{a} \times \mathbf{b}) \cdot \mathbf{c}$ and $(\mathbf{c} \times \mathbf{b}) \cdot \mathbf{a}$ where $\mathbf{a} = a_x \mathbf{i} + a_y \mathbf{j} + a_z \mathbf{k}$, $\mathbf{b} = b_x \mathbf{i} + b_y \mathbf{j} + b_z \mathbf{k}$, and $\mathbf{c} = c_x \mathbf{i} + c_y \mathbf{j} + c_z \mathbf{k}$. Numerical application: $a_x = 2$, $a_y = 1$, $a_z = 3$, $b_x = 2$, $b_y = 1$, $b_z = 0$, $c_x = 2$, $c_y = 0$, and $c_z = 0$.

Solution
The scalar $\mathbf{a} \cdot (\mathbf{b} \times \mathbf{c})$ is

$$\mathbf{a} \cdot (\mathbf{b} \times \mathbf{c}) = (a_x \mathbf{i} + a_y \mathbf{j} + a_z \mathbf{k}) \cdot \begin{vmatrix} \mathbf{i} & \mathbf{j} & \mathbf{k} \\ b_x & b_y & b_z \\ c_x & c_y & c_z \end{vmatrix}$$

$$= \begin{vmatrix} a_x & a_y & a_z \\ b_x & b_y & b_z \\ c_x & c_y & c_z \end{vmatrix}$$

$$= a_x \left(b_y c_z - b_z c_y \right) + a_y \left(b_z c_x - b_x c_z \right) + a_z \left(b_x c_y - b_y c_x \right)$$

$$= 2 \left(1(0) - 0(0) \right) + 1 \left(0(2) - 2(0) \right) + 3 \left(2(0) - 1(2) \right) = -6.$$

The scalar $(\mathbf{a} \times \mathbf{b}) \cdot \mathbf{c}$ is

$$(\mathbf{a} \times \mathbf{b}) \cdot \mathbf{c} = \begin{vmatrix} \mathbf{i} & \mathbf{j} & \mathbf{k} \\ a_x & a_y & a_z \\ b_x & b_y & b_z \end{vmatrix} \cdot \left(c_x \mathbf{i} + c_y \mathbf{j} + c_z \mathbf{k} \right)$$

$$= \left(c_x \mathbf{i} + c_y \mathbf{j} + c_z \mathbf{k} \right) \cdot \begin{vmatrix} \mathbf{i} & \mathbf{j} & \mathbf{k} \\ a_x & a_y & a_z \\ b_x & b_y & b_z \end{vmatrix}$$

$$= \begin{vmatrix} c_x & c_y & c_z \\ a_x & a_y & a_z \\ b_x & b_y & b_z \end{vmatrix} = - \begin{vmatrix} a_x & a_y & a_z \\ c_x & c_y & c_z \\ b_x & b_y & b_z \end{vmatrix} = \begin{vmatrix} a_x & a_y & a_z \\ b_x & b_y & b_z \\ c_x & c_y & c_z \end{vmatrix}$$

$$= a_x \left(b_y c_z - b_z c_y \right) + a_y \left(b_z c_x - b_x c_z \right) + a_z \left(b_x c_y - b_y c_x \right)$$

$$= 2 \left(1(0) - 0(0) \right) + 1 \left(0(2) - 2(0) \right) + 3 \left(2(0) - 1(2) \right) = -6.$$

The scalar $(\mathbf{c} \times \mathbf{b}) \cdot \mathbf{a}$ is

$$(\mathbf{c} \times \mathbf{b}) \cdot \mathbf{a} = \begin{vmatrix} \mathbf{i} & \mathbf{j} & \mathbf{k} \\ c_x & c_y & c_z \\ b_x & b_y & b_z \end{vmatrix} \cdot \left(a_x \mathbf{i} + a_y \mathbf{j} + a_z \mathbf{k} \right)$$

$$= \left(a_x \mathbf{i} + a_y \mathbf{j} + a_z \mathbf{k} \right) \cdot \begin{vmatrix} \mathbf{i} & \mathbf{j} & \mathbf{k} \\ c_x & c_y & c_z \\ b_x & b_y & b_z \end{vmatrix}$$

$$= \begin{vmatrix} a_x & a_y & a_z \\ c_x & c_y & c_z \\ b_x & b_y & b_z \end{vmatrix} = - \begin{vmatrix} a_x & a_y & a_z \\ b_x & b_y & b_z \\ c_x & c_y & c_z \end{vmatrix}$$

$$= - \left[a_x \left(b_y c_z - b_z c_y \right) + a_y \left(b_z c_x - b_x c_z \right) + a_z \left(b_x c_y - b_y c_x \right) \right]$$

$$= - \left[2 \left(1(0) - 0(0) \right) + 1 \left(0(2) - 2(0) \right) + 3 \left(2(0) - 1(2) \right) \right] = 6.$$

Note that $\mathbf{a} \cdot (\mathbf{b} \times \mathbf{c}) = (\mathbf{a} \times \mathbf{b}) \cdot \mathbf{c} = - (\mathbf{c} \times \mathbf{b}) \cdot \mathbf{a}$.

The MATLAB program for the example is

```
syms a_x a_y a_z b_x b_y b_z c_x c_y c_z real
a=[a_x a_y a_z]; b=[b_x b_y b_z]; c=[c_x c_y c_z];
```

```
d = dot(a,cross(b,c)); %a.(b x c)
e = dot(cross(a,b),c); % (a x b).c
f = dot(cross(c,b),a); % (c x b).a
fprintf('a.(b x c)-(a x b).c=%s\n',char(simplify(d-e)))
fprintf('a.(b x c)+(c x b).a=%s\n',char(simplify(d+f)))
slist={a_x,a_y,a_z,b_x,b_y,b_z,c_x,c_y,c_z};
nlist={2,1,3,2,1,0,2,0,0};
fprintf('a.(b x c)=%g\n',subs(d,slist,nlist))
fprintf('(a x b).c=%g\n',subs(e,slist,nlist))
fprintf('(c x b).a=%g\n',subs(f,slist,nlist))
```

Example 1.7. Find the c_z component of the vector **c** such as the vectors $\mathbf{a} = a_x\mathbf{i} + a_y\mathbf{j} + a_z\mathbf{k}$, $\mathbf{b} = b_x\mathbf{i} + b_y\mathbf{j} + b_z\mathbf{k}$, and $\mathbf{c} = c_x\mathbf{i} + c_y\mathbf{j} + c_z\mathbf{k}$ are coplanar. Numerical application: $a_x = 2$, $a_y = 3$, $a_z = 0$, $b_x = 3$, $b_y = 2$, $b_z = -2$, $c_x = 2$, and $c_y = 3$.

Solution

The three vectors are coplanar if $\mathbf{a} \cdot (\mathbf{b} \times \mathbf{c}) = 0$. The scalar $\mathbf{a} \cdot (\mathbf{b} \times \mathbf{c})$ is

$$
\mathbf{a} \cdot (\mathbf{b} \times \mathbf{c}) = (a_x\mathbf{i} + a_y\mathbf{j} + a_z\mathbf{k}) \cdot \begin{vmatrix} \mathbf{i} & \mathbf{j} & \mathbf{k} \\ b_x & b_y & b_z \\ c_x & c_y & c_z \end{vmatrix} = \begin{vmatrix} a_x & a_y & a_z \\ b_x & b_y & b_z \\ c_x & c_y & c_z \end{vmatrix}
$$

$$
= a_x(b_y c_z - b_z c_y) + a_y(b_z c_x - b_x c_z) + a_z(b_x c_y - b_y c_x)
$$

$$
= a_x b_y c_z - a_x b_z c_y + a_y b_z c_x - a_y b_x c_z + a_z b_x c_y - a_z b_y c_x
$$

$$
= a_x b_y c_z - a_y b_x c_z - a_x b_z c_y + a_y b_z c_x + a_z b_x c_y - a_z b_y c_x
$$

$$
= c_z(a_x b_y - a_y b_x) - a_x b_z c_y + a_y b_z c_x + a_z b_x c_y - a_z b_y c_x.
$$

The scalar triple product of the three vectors in MATLAB is given by

```
syms a_x a_y a_z b_x b_y b_z c_x c_y c_z real
a=[a_x a_y a_z]; b=[b_x b_y b_z]; c=[c_x c_y c_z];
d=det([a; b; c]); % a.(b x c)
```

The vectors **a**, **b**, and **c** are coplanar if

$$
\mathbf{a} \cdot (\mathbf{b} \times \mathbf{c}) = 0 \Leftrightarrow c_z(a_x b_y - a_y b_x) - a_x b_z c_y + a_y b_z c_x + a_z b_x c_y - a_z b_y c_x = 0,
$$

or

$$
c_z = \frac{a_x b_z c_y - a_y b_z c_x - a_z b_x c_y + a_z b_y c_x}{a_x b_y - a_y b_x}.
$$

Substituting with the numerical values, it results

$$
c_z = \frac{2(-2)(3) - 3(-2)(2) - 0(3)(3) + 0(2)(2)}{2(2) - 3(3)} = \frac{-12 + 12 - 0 + 0}{4 - 9} = 0.
$$

The given numerical vectors **a**, **b**, and **c** are coplanar if $c_z = 0$.

To solve the equation $\mathbf{a} \cdot (\mathbf{b} \times \mathbf{c}) = 0$, a specific MATLAB command will be used. The command `solve('eqn1', 'eqn2', ..., 'eqnN', 'var1', 'var2', ...'varN')` attempts to solve an equation or set of equations `'eqn1'`, `'eqn2'`, `...`, `'eqnN'` for the variables `'eqnN'`, `'var1'`, `'var2'`, `...`, `'varN'`. The set of equations are symbolic expressions or strings specifying equations. The MATLAB command to find the solution c_z of the equation `det([a; b; c])=0` is

```
x = solve(d, c_z);
```

and the numerical solution for c_z is displayed with

```
slist={a_x,a_y,a_z,b_x,b_y,b_z,c_x,c_y};
nlist={2,3,0,3,2,-2,2,3};
fprintf('c_z= %g\n', subs(x, slist, nlist))
```

Example 1.8. Show that the determinant $\begin{vmatrix} a_{11} & u_{12} \\ a_{21} & a_{22} \end{vmatrix}$ can be shortened as

$$\begin{vmatrix} a_{11} & a_{12} \\ a_{21} & a_{22} \end{vmatrix} = \epsilon_{ij} a_{1i} a_{2j} = \epsilon_{ij} a_{i1} a_{j2}.$$

Solution

By computation

$$\begin{vmatrix} a_{11} & a_{12} \\ a_{21} & a_{22} \end{vmatrix} = a_{11} a_{22} - a_{21} a_{12}$$

$$= \epsilon_{12} a_{11} a_{22} + \epsilon_{21} a_{21} a_{12}$$

$$= \epsilon_{11} a_{11} a_{12} + \epsilon_{12} a_{11} a_{22} + \epsilon_{21} a_{21} a_{12} + \epsilon_{22} a_{21} a_{22}$$

$$= \epsilon_{ij} a_{i1} a_{j2}.$$

The next MATLAB statement

```
syms a_11 a_12 a_21 a_22
```

creates the symbolic variables (objects) a_11, a_12, a_21, and a_22. Then, the associated matrix a is expressed in MATLAB in a symbolic fashion and printed on the computer screen as text using the `fprintf` and `pretty` MATLAB functions. The `pretty` function prints symbolic output in a format that resembles typeset mathematics:

```
a=[a_11 a_12; a_21 a_22];

fprintf('matrix a is: ')
fprintf('\n\n')
fprintf('a= '); pretty(a)
fprintf('\n')
```

The MATLAB computer screen printing gives

```
The matrix a is:

a=
   +-                -+
   |   a_11,  a_12    |
   |                  |
   |   a_21,  a_22    |
   +-                -+
```

The determinant is calculated using the MATLAB command `det(a)`. The obtained value is assigned to the MATLAB variable `determinant` and printed on the computer screen as text using the `fprintf` and `pretty` functions.

```
determinant=det(a);
fprintf...
('determinant calculated using the definition: ')
fprintf('\n\n')
fprintf(' det(a)='); pretty(determinant)
fprintf('\n')
```

The formula $\epsilon_{ij}a_{1i}a_{2j}$ is calculated using a MATLAB `for` statement. The obtained value is assigned to the MATLAB variable `formula1`. The variable initialized its value to 0 using the command `formula1=0`:

```
formula1=0;
val=2;
for j=1:val
    for i=1:val
        formula1=...
        formula1+epsilon([i j])*a(1,i)*a(2,j);
    end
end
```

The value of the variable `formula1` was printed on the computer screen using the `fprintf` and `pretty` functions below:

```
fprintf('formula eps_{i,j}*a(1,i)*a(2,j) gives:')
fprintf('\n\n')
fprintf(' eps_{i,j}*a(i,1)*a(j,2)=')
pretty(formula1); fprintf('\n')
```

The MATLAB commands

```
fprintf('remark det(a)=eps_{i,j}*a(1,i)*a(2,j)')
fprintf('\n\n')
```

print on the computer screen the message

```
remark det(a)=eps_{i,j}*a(1,i)*a(2,j).
```

On the other side

$$\begin{vmatrix} a_{11} & a_{12} \\ a_{21} & a_{22} \end{vmatrix} = a_{11}a_{22} - a_{21}a_{12}$$

$$= \epsilon_{12}a_{11}a_{22} + \epsilon_{21}a_{12}a_{21}$$

$$= \epsilon_{11}a_{11}a_{21} + \epsilon_{12}a_{11}a_{22} + \epsilon_{21}a_{12}a_{21} + \epsilon_{22}a_{12}a_{22}$$

$$= \epsilon_{ij}a_{1i}a_{2j}.$$

Using the same approach as before, formula $\epsilon_{ij}a_{i1}a_{j2}$ is calculated using two MATLAB `for` statements and the counter variables `i` and `j`. The obtained value is assigned to the MATLAB variable `formula2`. The variable `formula2` initialized its value to 0 using the command `formula2=0`:

```
formula2=0;
val=2;
for j=1:val
    for i=1:val
        formula2=...
        formula2+epsilon([i j])*a(i,1)*a(j,2);
    end
end
```

The value of the variable `formula2` was printed on the computer screen using the next MATLAB statement:

```
fprintf('formula eps_{i,j}*a(i,1)*a(j,2) gives:')
fprintf('\n\n')
fprintf(' eps_{i,j}*a(i,1)*a(j,2)=')
pretty(formula2); fprintf('\n')
```

Finally, the message

```
remark det(a)=eps_i,j*a(i,1)*a(j,2)
```

was printed on the computer screen using the next MATLAB commands

```
fprintf('remark det(a)=eps_{i,j}*a(i,1)*a(j,2)')
fprintf('\n\n')
```

Example 1.9. Show that the scalar triple product $\mathbf{a}_1 \cdot (\mathbf{a}_2 \times \mathbf{a}_3)$ of the vectors $\mathbf{a}_1 = a_{11}\mathbf{i} + a_{12}\mathbf{j} + a_{13}\mathbf{k}$, $\mathbf{a}_2 = a_{21}\mathbf{i} + a_{22}\mathbf{j} + a_{23}\mathbf{k}$ and $\mathbf{a}_3 = a_{31}\mathbf{i} + a_{32}\mathbf{j} + a_{33}\mathbf{k}$ can be expressed as

$$\mathbf{a}_1 \cdot (\mathbf{a}_2 \times \mathbf{a}_3) = \epsilon_{ijk}a_{i1}a_{j2}a_{k3} = \epsilon_{ijk}a_{1i}a_{2j}a_{3k}.$$

Solution

Using the definition of the scalar triple product, one can compute

$$
\mathbf{a}_1 \cdot (\mathbf{a}_2 \times \mathbf{a}_3) = (a_{11}\mathbf{i} + a_{12}\mathbf{j} + a_{13}\mathbf{k}) \cdot \begin{vmatrix} \mathbf{i} & \mathbf{j} & \mathbf{k} \\ a_{21} & a_{22} & a_{23} \\ a_{31} & a_{32} & a_{33} \end{vmatrix}
$$

$$
= \begin{vmatrix} a_{11} & a_{12} & a_{13} \\ a_{21} & a_{22} & a_{23} \\ a_{31} & a_{32} & a_{33} \end{vmatrix}
$$

$$
= a_{11}(a_{22}a_{33} - a_{23}a_{32}) + a_{12}(a_{23}a_{31} - a_{21}a_{33})
$$

$$
+ a_{13}(a_{21}a_{32} - a_{22}a_{31})
$$

$$
= (a_{11}a_{22}a_{33} - a_{11}a_{23}a_{32}) + (a_{12}a_{23}a_{31} - a_{12}a_{21}a_{33})
$$

$$
+ (a_{13}a_{21}a_{32} - a_{13}a_{22}u_{31})
$$

$$
= a_{11}a_{22}a_{33} - a_{11}a_{23}a_{32} + a_{12}a_{23}a_{31} - a_{12}a_{21}a_{33}
$$

$$
+ a_{13}a_{21}a_{32} - a_{13}a_{22}a_{31}
$$

$$
= \epsilon_{123}a_{11}a_{22}a_{33} + \epsilon_{132}a_{11}a_{23}a_{32} + \epsilon_{231}a_{12}a_{23}a_{31}
$$

$$
+ \epsilon_{213}a_{12}a_{21}a_{33} + \epsilon_{312}a_{13}a_{21}a_{32} + \epsilon_{321}a_{13}a_{22}a_{31}
$$

$$
= \epsilon_{123}a_{11}a_{22}a_{33} + \epsilon_{132}a_{11}a_{23}a_{32} + \epsilon_{231}a_{12}a_{23}a_{31}
$$

$$
+ \epsilon_{213}a_{12}a_{21}a_{33} + \epsilon_{312}a_{13}a_{21}a_{32} + \epsilon_{321}a_{13}a_{22}a_{31}
$$

$$
+ \epsilon_{111}a_{11}a_{21}a_{31} + \epsilon_{222}a_{12}a_{22}a_{32} + \epsilon_{333}a_{13}a_{23}a_{33}
$$

$$
+ \epsilon_{112}a_{11}a_{21}a_{32} + \epsilon_{113}a_{11}a_{21}a_{33} + \epsilon_{122}a_{11}a_{22}a_{32}
$$

$$
+ \epsilon_{133}a_{11}a_{23}a_{33} + \epsilon_{121}a_{11}a_{22}a_{31} + \epsilon_{131}a_{11}a_{23}a_{31}
$$

$$
+ \epsilon_{211}a_{12}a_{21}a_{31} + \epsilon_{212}a_{12}a_{21}a_{32} + \epsilon_{221}a_{12}a_{22}a_{31}
$$

$$
+ \epsilon_{223}a_{12}a_{22}a_{33} + \epsilon_{233}a_{12}a_{23}a_{33} + \epsilon_{232}a_{12}a_{23}a_{32}
$$

$$
+ \epsilon_{311}a_{13}a_{21}a_{31} + \epsilon_{313}a_{13}a_{21}a_{33} + \epsilon_{331}a_{13}a_{23}a_{31}
$$

$$
+ \epsilon_{322}a_{13}a_{22}a_{32} + \epsilon_{323}a_{13}a_{22}a_{33} + \epsilon_{332}a_{13}a_{23}a_{32}
$$

$$
= \epsilon_{ijk}a_{1i}a_{2j}a_{3k}.
$$

The MATLAB program starts with the following statements:

```
clear all; clc; close
```

The next MATLAB statement

```
syms a_11 a_12 a_13 a_21 a_22 a_23 a_31 a_32 a_33 real
```

creates the symbolic variables (objects) a_11, a_12, a_13, a_21, a_22, a_23, a_31, a_32, and a_33. Then, the associated vectors a_1, a_2, and a_3 are created in a symbolical fashion using the next MATLAB commands:

```
a1=[a_11 a_12 a_13];
a2=[a_21 a_22 a_23];
a3=[a_31 a_32 a_33];
```

The scalar triple product of the vectors is calculated using the dot and cross MATLAB functions inside a simplify statement, statement that simplifies the results symbolic computation of the product:

```
a1a2xa3 = simplify(dot(a1, cross(a2, a3)));
```

The scalar triple product of the vectors is printed on the computer screen as text using the fprintf and pretty MATLAB functions, as below:

```
fprintf('scalar triple product a1.a2xa3 is:')
fprintf('\n\n')
fprintf(' a1.a2xa3=');  pretty(a1a2xa3)
fprintf('\n')
```

The MATLAB computer screen printing gives

```
scalar triple product a1.a2xa3 is:


a1.a2xa3=
a_11 a_22 a_33 - a_11 a_23 a_32 - a_12 a_21 a_33
+ a_12 a_23 a_31 + a_13 a_21 a_32 - a_13 a_22 a_31
```

The associated matrix a is expressed in MATLAB in a symbolical fashion using

```
a=[a_11 a_12 a_13; a_21 a_22 a_23; a_31 a_32 a_33];
```

The associated matrix a is printed on the computer screen as text using the next commands:

```
fprintf('matrix a is: ')
fprintf('\n\n')
fprintf('a= ');  pretty(a)
fprintf('\n')
```

and the MATLAB screen gives

```
matrix a is:


a=
  +-                    -+
  |  a_11, a_12, a_13  |
  |                      |
  |  a_21, a_22, a_23  |
  |                      |
  |  a_31, a_32, a_33  |
  +-                    -+
```

The value of the determinant is calculated using the MATLAB command

```
determinant=det(a);
```

and printed using the next MATLAB statement

```
fprintf...
('determinant calculated using the definition:')
fprintf('\n\n')
fprintf(' det(a)='); pretty(determinant)
fprintf('\n')
```

resulting in the next computer screen printing:

```
determinant calculated using the definition:

det(a)=
a_11 a_22 a_33 - a_11 a_23 a_32 - a_12 a_21 a_33
+ a_12 a_23 a_31 + a_13 a_21 a_32 - a_13 a_22 a_31
```

Next, the statement remark det(a)=a1.a2xa3 is printed on the computer screen using the MATLAB commands:

```
fprintf('remark det(a)=a1.a2xa3')
fprintf('\n\n')
```

The value of $\epsilon_{ijk}a_{1i}a_{2j}a_{3k}$ is calculated using a triple MATLAB for statement and assigned to the MATLAB variable formula1. The variable initialized its value to 0 using the command formula1=0:

```
fprintf('\n')
formula1=0;
val=3;
for k=1:val
    for j=1:val
        for i=1:val
            formula1=formula1...
            +epsilon([i j k])*a(1,i)*a(2,j)*a(3,k);
        end
    end
end
fprintf('\n')
```

The value of $\epsilon_{ijk}a_{1i}a_{2j}a_{3k}$ is printed on the computer screen using the MATLAB commands

```
fprintf...
('formula eps_{i,j,k}*a(1,i)*a(2,j)*a(3,k) gives:');
fprintf('\n\n');
fprintf(' eps_{i,j,k}*a(1,i)*a(2,j)*a(3,k)=');
pretty(formula1); fprintf('\n')
```

giving the next MATLAB result

```
formula eps_{i,j,k}*a(1,i)*a(2,j)*a(3,k) gives:

eps_{i,j,k}*a(1,i)*a(2,j)*a(3,k)=
a_11 a_22 a_33 - a_11 a_23 a_32 - a_12 a_21 a_33
 - a_12 a_23 a_31 + a_13 a_21 a_32 + a_13 a_22 a_31
```

The MATLAB statement

```
remark det(a)=eps_i,j,k*a(1,i)*a(2,j)*a(3,k)
```

is then printed on the screen using

```
fprintf...
('remark det(a)=eps_{i,j,k}*a(1,i)*a(2,j)*a(3,k)')
fprintf('\n\n')
```

In a similar way, one can obtain

$$
\mathbf{a}_1 \cdot (\mathbf{a}_2 \times \mathbf{a}_3) = (a_{11}\mathbf{i} + a_{12}\mathbf{j} + a_{13}\mathbf{k}) \cdot
\begin{vmatrix}
\mathbf{i} & \mathbf{j} & \mathbf{k} \\
a_{21} & a_{22} & a_{23} \\
a_{31} & a_{32} & a_{33}
\end{vmatrix}
$$

$$
= \begin{vmatrix}
a_{11} & a_{12} & a_{13} \\
a_{21} & a_{22} & a_{23} \\
a_{31} & a_{32} & a_{33}
\end{vmatrix}
$$

$$
= a_{11}\left(a_{22}a_{33} - a_{23}a_{32}\right) + a_{12}\left(a_{23}a_{31} - a_{21}a_{33}\right)
$$

$$
+ a_{13}\left(a_{21}a_{32} - a_{22}a_{31}\right)
$$

$$
= \left(a_{11}a_{22}a_{33} - a_{11}a_{23}a_{32}\right) + \left(a_{12}a_{23}a_{31} - a_{12}a_{21}a_{33}\right)
$$

$$
+ \left(a_{13}a_{21}a_{32} - a_{13}a_{22}a_{31}\right)
$$

$$
= a_{11}a_{22}a_{33} - a_{11}a_{32}a_{23} + a_{31}a_{12}a_{23} - a_{21}a_{12}a_{33}
$$

$$
+ a_{21}a_{32}a_{13} - a_{31}a_{22}a_{13}
$$

$$
= \epsilon_{123}a_{11}a_{22}a_{33} + \epsilon_{132}a_{11}a_{32}a_{23} + \epsilon_{312}a_{31}a_{12}a_{23}
$$

$$
+ \epsilon_{213}a_{21}a_{12}a_{33} + \epsilon_{231}a_{21}a_{32}a_{13} + \epsilon_{321}a_{31}a_{22}a_{13}
$$

$$
= \epsilon_{123}a_{11}a_{22}a_{33} + \epsilon_{132}a_{11}a_{32}a_{23} + \epsilon_{312}a_{31}a_{12}a_{23}
$$

$$
+ \epsilon_{213}a_{21}a_{12}a_{33} + \epsilon_{231}a_{21}a_{32}a_{13} + \epsilon_{321}a_{31}a_{22}a_{13}
$$

$$
+ \epsilon_{111}a_{11}a_{12}a_{13} + \epsilon_{222}a_{21}a_{22}a_{23} + \epsilon_{333}a_{31}a_{32}a_{33}
$$

$$
+ \epsilon_{112}a_{11}a_{12}a_{23} + \epsilon_{113}a_{11}a_{12}a_{33} + \epsilon_{122}a_{11}a_{22}a_{23}
$$

$$
+ \epsilon_{133}a_{11}a_{32}a_{33} + \epsilon_{121}a_{11}a_{22}a_{13} + \epsilon_{131}a_{11}a_{32}a_{13}
$$

$$+\epsilon_{211}a_{21}a_{12}a_{13} + \epsilon_{212}a_{21}a_{12}a_{23} + \epsilon_{221}a_{21}a_{22}a_{13}$$

$$+\epsilon_{223}a_{21}a_{22}a_{23} + \epsilon_{233}a_{21}a_{32}a_{33} + \epsilon_{232}a_{21}a_{32}a_{23}$$

$$+\epsilon_{311}a_{31}a_{12}a_{13} + \epsilon_{313}a_{31}a_{12}a_{33} + \epsilon_{331}a_{31}a_{32}a_{13}$$

$$+\epsilon_{322}a_{31}a_{22}a_{23} + \epsilon_{323}a_{31}a_{22}a_{33} + \epsilon_{332}a_{31}a_{32}a_{23}$$

$$= \epsilon_{ijk}a_{i1}a_{j2}a_{k3}.$$

Using the same approach as before, $\epsilon_{ijk}a_{i1}a_{j2}a_{k3}$ is calculated using three MATLAB
for statements and the counter variables i, j, and k. The obtained value is assigned
to the MATLAB variable formula2. The variable formula2 initialized its value
to 0 using the command formula2=0:

```
fprintf('\n')
formula2=0;
val=3;
for k=1:val
    for j=1:val
        for i=1:val
            formula2=formula2...
            +epsilon([i j k])*a(i,1)*a(j,2)*a(k,3);
        end
    end
end
fprintf('\n')
```

The value of eps_i,j,k*a(i,1)*a(j,2)*a(k,3) was printed on the computer screen using the next MATLAB statement

```
fprintf...
('formula eps_{i,j,k}*a(i,1)*a(j,2)*a(k,3) gives:')
fprintf('\n\n')
fprintf(' eps_{i,j,k}*a(i,1)*a(j,2)*a(k,3)=')
pretty(formula2); fprintf('\n')
```

Finally, the statement:

```
remark det(a)=eps\_{i,j,k}*a(i,1)*a(j,2)*a(k,3)
```

was printed on the computer screen using the next MATLAB commands:

```
fprintf...
('remark det(a)=eps_{i,j,k}*a(i,1)*a(j,2)*a(k,3)')
fprintf('\n\n')
```

So the product $\mathbf{a}_1 \cdot (\mathbf{a}_2 \times \mathbf{a}_3)$ can be expressed as

$$\mathbf{a}_1 \cdot (\mathbf{a}_2 \times \mathbf{a}_3) = \epsilon_{ijk}a_{i1}a_{j2}a_{k3} = \epsilon_{ijk}a_{1i}a_{2j}a_{3k}.$$

Also, one can conclude that

$$\det A = \begin{vmatrix} a_{11} & a_{12} & a_{13} \\ a_{21} & a_{22} & a_{23} \\ a_{31} & a_{32} & a_{33} \end{vmatrix}$$

$$= \epsilon_{ijk} a_{1i} a_{2j} a_{3k}$$

$$= \epsilon_{ijk} a_{i1} a_{j2} a_{k3}.$$

Example 1.10. Let A and B be square matrices. Using the definition of a determinant (based on a permutation symbol) shown below

$$\det A = \epsilon_{i_1 \ldots i_n} a_{1 i_1} \ldots a_{n i_n},$$

prove that

$$\det AB = \det A \det B.$$

Solution
One can write

$$\det A \det B = \det B \det A$$

$$= \epsilon_{i_1 \ldots i_n} b_{1 i_1} \ldots b_{n i_n} \epsilon_{j_1 \ldots j_n} a_{1 j_1} \ldots a_{n j_n}$$

$$= \epsilon_{i_1 \ldots i_n} \epsilon_{j_1 \ldots j_n} b_{1 i_1} \ldots b_{n i_n} a_{1 j_1} \ldots a_{n j_n}$$

$$= \epsilon_{i_1 \ldots i_n} b_{j_1 i_1} \ldots b_{j_n i_n} a_{1 j_1} \ldots a_{n j_n}$$

$$= \epsilon_{i_1 \ldots i_n} (ab)_{1 i_1} \ldots (ab)_{n i_n}$$

$$= \det AB.$$

1.12   Problems

1.1 (a) Find the angle θ made by the vector $\mathbf{v} = -10\mathbf{i} + 5\mathbf{j}$ with the positive x-axis and determine the unit vector in the direction of $\mathbf{v}$. The angle θ is measured counterclockwise (ccw) and has the values $0 \le \theta \le 2\pi$ or $-\pi \le \theta \le \pi$.

(b) Determine the magnitude of the resultant $\mathbf{p} = \mathbf{v}_1 + \mathbf{v}_2$ and the angle that $\mathbf{p}$ makes with the positive x-axis, where the vectors $\mathbf{v}_1$ and $\mathbf{v}_2$ are shown in Fig. 1.18. The magnitudes of the vectors are $|\mathbf{v}_1| = v_1 = 10$, $|\mathbf{v}_2| = v_2 = 5$ and the angles of the vectors with the positive x-axis are $\theta_1 = 30°$ and $\theta_2 = 60°$.

1.2 The planar vectors $\mathbf{a}$, $\mathbf{b}$, and $\mathbf{c}$ are given in xOy plane as shown in Fig. 1.19. The magnitude of the vectors are $a = P$, $b = 2P$, and $c = P\sqrt{2}$. The angles in the figure are $\alpha = 45°$, $\beta = 120°$, and $\gamma = 30°$. Determine the resultant $\mathbf{v} = \mathbf{a} + \mathbf{b} + \mathbf{c}$ and the angle that $\mathbf{v}$ makes with the positive x-axis.

Fig. 1.18 Problem 1.1

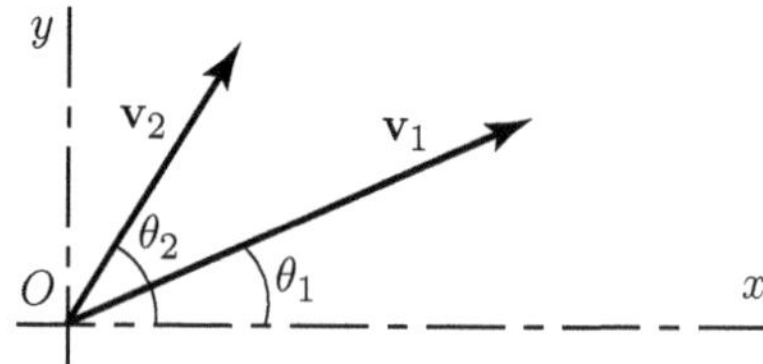

Fig. 1.19 Problem 1.2

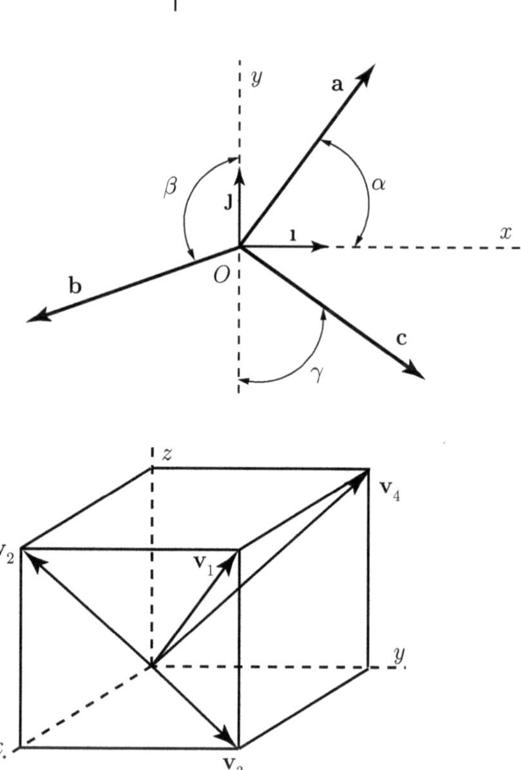

Fig. 1.20 Problem 1.3

1.3 The cube in Fig. 1.20 has the sides equal to $l = 1$. (a) Find the direction cosines
of the resultant $\mathbf{v} = \mathbf{v}_1 + \mathbf{v}_2 + \mathbf{v}_3 + \mathbf{v}_4$. (b) Determine the angle between the
vectors $\mathbf{v}_2$ and $\mathbf{v}_3$. (c) Find the projection of the vector $\mathbf{v}_2$ on the vector $\mathbf{v}_4$. (d)
Calculate $\mathbf{v}_2 \cdot \mathbf{v}_4$, $\mathbf{v}_2 \times \mathbf{v}_4$, $\mathbf{v}_1 \cdot (\mathbf{v}_2 \times \mathbf{v}_3)$, $(\mathbf{v}_2 \times \mathbf{v}_3) \times \mathbf{v}_4$, and $\mathbf{v}_2 \times (\mathbf{v}_3 \times \mathbf{v}_4)$.

1.4 The vectors $\mathbf{F}_1$, $\mathbf{F}_2$, $\mathbf{F}_3$, and $\mathbf{F}_4$, shown in Fig. 1.21, act on the sides of a cube
(the side of the cube is $l = 2$). The magnitudes of the vectors are $\mathbf{F}_1 = \mathbf{F}_2 =
F = 1$, and $\mathbf{F}_3 = \mathbf{F}_4 = F\sqrt{2}$. (a) Find the resultant $\mathbf{F}_1 + \mathbf{F}_2 + \mathbf{F}_3 + \mathbf{F}_4$. (b) Find
the direction cosines of the vector $\mathbf{F}_4$. (c) Determine the angle between the
vectors $\mathbf{F}_1$ and $\mathbf{F}_3$. (d) Find the projection of the vector $\mathbf{F}_2$ on the vector $\mathbf{F}_4$.
(e) Calculate $\mathbf{F}_1 \cdot \mathbf{F}_3$, $\mathbf{F}_2 \times \mathbf{F}_4$, and $\mathbf{F}_1 \cdot (\mathbf{F}_2 \times \mathbf{F}_3)$.

Fig. 1.21 Problem 1.4

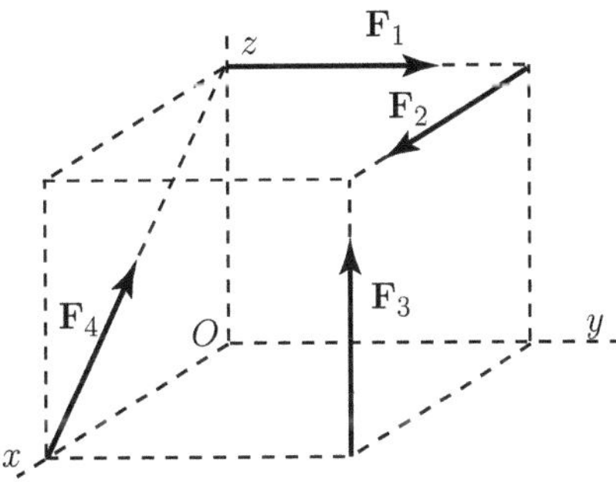

Fig. 1.22 Problem 1.5

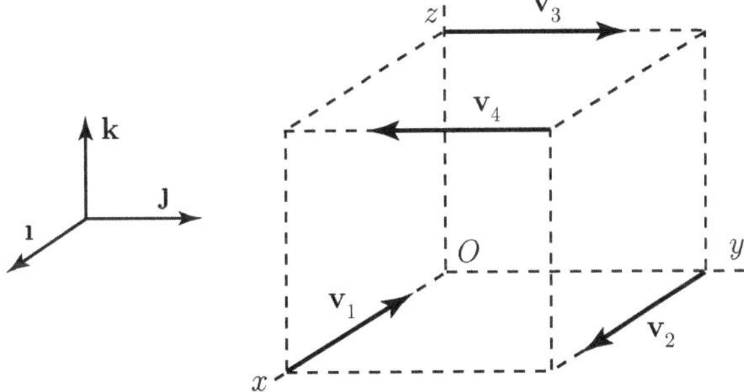

Fig. 1.23 Problem 1.6

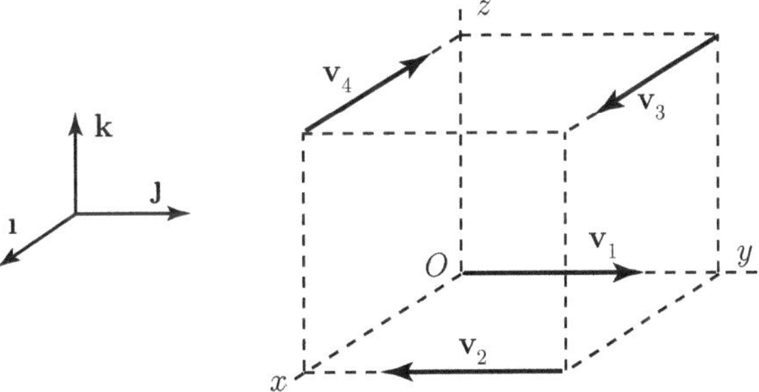

1.5 Figure 1.22 represents the vectors $\mathbf{v}_1$, $\mathbf{v}_2$, $\mathbf{v}_3$, and $\mathbf{v}_4$ acting on a cube with the side $l = 2$. The magnitude of the forces are $\mathbf{v}_1 = V = 2$ and $\mathbf{v}_2 = \mathbf{v}_3 = \mathbf{v}_4 = 2V$. (a) Find the resultant and the direction cosines of the resultant $\mathbf{v} = \mathbf{v}_1 + \mathbf{v}_2 + \mathbf{v}_3 + \mathbf{v}_4$. (b) Determine the angle between the vectors $\mathbf{v}_1$ and $\mathbf{v}_3$. (c) Find the projection of the vector $\mathbf{v}_4$ on the resultant vector $\mathbf{v}$. (d) Calculate $\mathbf{v}_2 \cdot \mathbf{v}$, $\mathbf{v}_1 \times \mathbf{v}_2$, and $\mathbf{v}_2 \times \mathbf{v}_4$.

1.6 Repeat the previous problem for Fig. 1.23.

1.7 The parallelepiped shown in Fig. 1.24 has the sides $l = 1$ m, $w = 2$ m, and $h = 3$ m. The magnitude of the vectors are $\mathbf{F}_1 = \mathbf{F}_2 = 10$ N and $\mathbf{F}_3 = \mathbf{F}_4 = 20$ N. (a) Find the resultant $\mathbf{F}_1 + \mathbf{F}_2 + \mathbf{F}_3 + \mathbf{F}_4$. (b) Find the unit vectors of the vectors $\mathbf{F}_1$ and $\mathbf{F}_4$. (c) Determine the angle between the vectors $\mathbf{F}_1$ and $\mathbf{F}_4$. (d) Find

Fig. 1.24 Problem 1.7

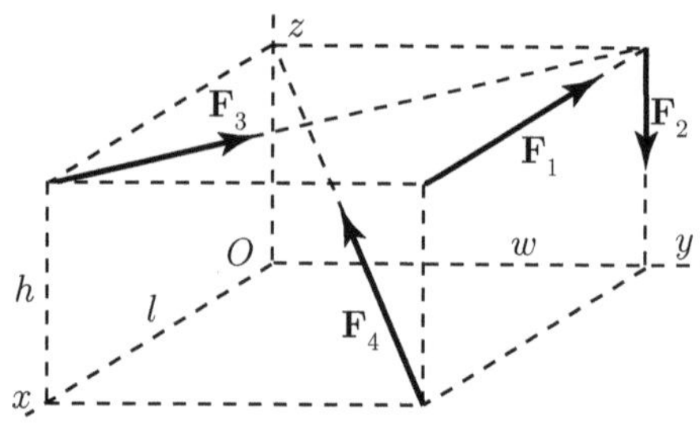

Fig. 1.25 Problem 1.8

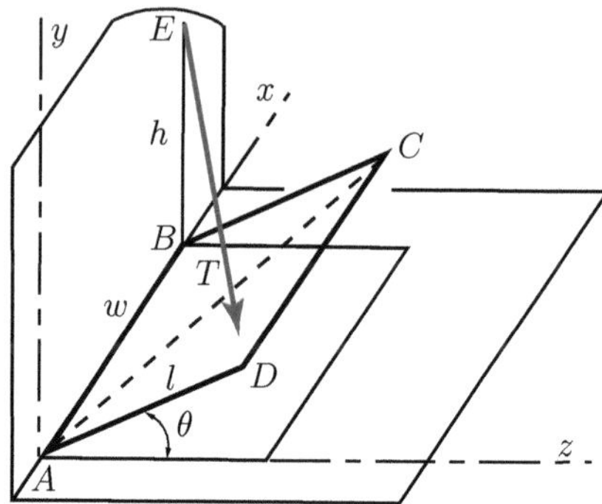

the projection of the vector $\mathbf{F}_2$ on the vector $\mathbf{F}_4$. (e) Calculate $\mathbf{F}_1 \cdot \mathbf{F}_4$, $\mathbf{F}_2 \times \mathbf{F}_3$, and $\mathbf{F}_1 \cdot (\mathbf{F}_2 \times \mathbf{F}_3)$.

1.8 A uniform rectangular plate of length l and width w is held open by a cable (Fig. 1.25). The plate is hinged about an axis parallel to the plate edge of length l. Points A and B are at the extreme ends of this hinged edge. Points D and C are at the ends of the other edge of length l and are respectively adjacent to points A and B. Points D and C move as the plate opens. In the closed position, the plate is in a horizontal plane. When held open by a cable, the plate has rotated through an angle θ relative to the closed position. The supporting cable runs from point D to point E where point E is located a height h directly above the point B on the hinged edge of the plate. The cable tension required to hold the plate open is T. Find the projection of the tension force onto the diagonal axis AC of the plate. Numerical application: $l = 1.0$ m, $w = 0.5$ m, $\theta = 45°$, $h = 1.0$ m, and $T = 100$ N.

1.9 The following spatial vectors are given: $\mathbf{v}_1 = -3\,\mathbf{i} + 4\,\mathbf{j} - 3\,\mathbf{k}$, $\mathbf{v}_2 = 3\,\mathbf{i} + 3\,\mathbf{k}$, and $\mathbf{v}_3 = 1\,\mathbf{i} + 2\,\mathbf{j} + 3\,\mathbf{k}$. Find the expressions $\mathbf{E}_1 = \mathbf{v}_1 + \mathbf{v}_2 + \mathbf{v}_3$, $\mathbf{E}_2 = \mathbf{v}_1 + \mathbf{v}_2 - \mathbf{v}_3$, $\mathbf{E}_3 = (\mathbf{v}_1 \times \mathbf{v}_2) \times \mathbf{v}_3$, and $E_4 = (\mathbf{v}_1 \times \mathbf{v}_2) \cdot \mathbf{v}_3$.

1.10 Find the angle between the vectors $\mathbf{v}_1 = 2\,\mathbf{i} - 4\,\mathbf{j} + 4\,\mathbf{k}$ and $\mathbf{v}_2 = 4\,\mathbf{i} + 2\,\mathbf{j} + 4\,\mathbf{k}$. Find the expressions $\mathbf{v}_1 \times \mathbf{v}_2$ and $\mathbf{v}_1 \cdot \mathbf{v}_2$.

1.11 The following vectors are given, $\mathbf{v}_1 = 2\,\mathbf{i} + 4\,\mathbf{j} + 6\,\mathbf{k}$, $\mathbf{v}_2 = 1\,\mathbf{i} + 3\,\mathbf{j} + 5\,\mathbf{k}$ and $\mathbf{v}_3 = -2\,\mathbf{i} + 2\,\mathbf{k}$. Find the vector triple product of $\mathbf{v}_1$, $\mathbf{v}_2$, and $\mathbf{v}_3$, and explain the result.

1.12 Solve the vectorial equation $\mathbf{x} \times \mathbf{a} = \mathbf{x} \times \mathbf{b}$, where $\mathbf{a}$ and $\mathbf{b}$ are two known given vectors.

1.13 Solve the vectorial equation $\mathbf{v} = \mathbf{a} \times \mathbf{x}$, where $\mathbf{v}$ and $\mathbf{a}$ are two known given vectors.

1.14 Solve the vectorial equation $\mathbf{a} \cdot \mathbf{x} = m$, where $\mathbf{a}$ is a known given vector and m is a known given scalar.

1.15 Show that (a) $\delta_{ii} = 3$, (b) $\delta_{ij}\delta_{ij} = 3$, (c) $\epsilon_{ijk}\epsilon_{jki} = 6$, (d) $\delta_{ij}\delta_{jk} = \delta_{ik}$, and (e) $\delta_{ij}\epsilon_{ijk} = 0$.

1.16 The vector product of two vectors $\mathbf{u} = u_1\mathbf{\imath} + u_2\mathbf{\jmath} + u_3\mathbf{k}$ and $\mathbf{v} = v_1\mathbf{\imath} + v_2\mathbf{\jmath} + v_3\mathbf{k}$ is the vector $\mathbf{w} = \mathbf{u} \times \mathbf{v}$ whose components are

$$w_1 = u_2 v_3 - u_3 v_2, \quad w_2 = u_3 v_1 - u_1 v_3, \quad w_3 = u_1 v_2 - u_2 v_1.$$

Show that this can be shortened as

$$w_i = \epsilon_{ijk}\, u_j\, v_k.$$

1.13 Program

```
% two-index and three-index permutation symbol
clear all % clears all variables and functions
clc  % clears the command window and homes the cursor
close all % closes all the open figure windows

disp('the values of the permutation symbol:')
fprintf('\n')
fprintf('epsilon_ij:\n')

val=2;
for j=1:val
    for i=1:val
% results of the two-index permutation symbol
% function epsilon([i,j])) is used
display(sprintf(...
    '(i j) = (%d %d) -> epsilon_{%d%d}= %d',...
    i, j, i, j, epsilon([i j])));
    end
end


fprintf('\n')
fprintf('epsilon_ijk:\n')
```

```
val=3;
for k=1:val
    for j=1:val
        for i=1:val
% results of the three-index permutation symbol
% function epsilon([i,j,k])) is used
display(sprintf(...
 '(i j k) = (%d %d %d) -> epsilon_{%d%d%d}= %d',...
  i, j, k, i, j, k, epsilon([i j k])));
        end
    end
end
fprintf('\n')

% end of program
```

Function epsilon

```
% permutation symbol
function [val]=epsilon(string)
if sum(sort(string)~=(1:length(string)))>0
   val=0;
else
val=0;
m=length(string);
    for i=1:m-1
        if string(i)>string(i+1)
            string_new=string(i);
            string(i)=string(i+1);
            string(i+1)=string_new;
            val=val+1;
        end
    end
string=sort(string);
val=(-1)^mod(val,2);
end
% end of program
```

Results:

```
the values of the permutation symbol:

epsilon_ij:
(i j) = (1 1) -> epsilon_{11}= 0
(i j) = (2 1) -> epsilon_{21}= -1
```

```
(i j) = (1 2) -> epsilon_{12}= 1
(i j) = (2 2) -> epsilon_{22}= 0
epsilon_ijk:
(i j k) = (1 1 1) -> epsilon_{111}= 0
(i j k) = (2 1 1) -> epsilon_{211}= 0
(i j k) = (3 1 1) -> epsilon_{311}= 0
(i j k) = (1 2 1) -> epsilon_{121}= 0
(i j k) = (2 2 1) -> epsilon_{221}= 0
(i j k) = (3 2 1) -> epsilon_{321}= 1
(i j k) = (1 3 1) -> epsilon_{131}= 0
(i j k) = (2 3 1) -> epsilon_{231}= -1
(i j k) = (3 3 1) -> epsilon_{331}= 0
(i j k) = (1 1 2) -> epsilon_{112}= 0
(i j k) = (2 1 2) -> epsilon_{212}= 0
(i j k) = (3 1 2) -> epsilon_{312}= 1
(i j k) = (1 2 2) -> epsilon_{122}= 0
(i j k) = (2 2 2) -> epsilon_{222}= 0
(i j k) = (3 2 2) -> epsilon_{322}= 0
(i j k) = (1 3 2) -> epsilon_{132}= -1
(i j k) = (2 3 2) -> epsilon_{232}= 0
(i j k) = (3 3 2) -> epsilon_{332}= 0
(i j k) = (1 1 3) -> epsilon_{113}= 0
(i j k) = (2 1 3) -> epsilon_{213}= -1
(i j k) = (3 1 3) -> epsilon_{313}= 0
(i j k) = (1 2 3) -> epsilon_{123}= 1
(i j k) = (2 2 3) -> epsilon_{223}= 0
(i j k) = (3 2 3) -> epsilon_{323}= 0
(i j k) = (1 3 3) -> epsilon_{133}= 0
(i j k) = (2 3 3) -> epsilon_{233}= 0
(i j k) = (3 3 3) -> epsilon_{333}= 0
```

Chapter 2
Centroids and Moments of Inertia

2.1 Centroids and Center of Mass

2.1.1 First Moment and Centroid of a Set of Points

The position vector of a point P relative to a point O is $\mathbf{r}_P$ and a scalar associated with P is s, for example, the mass m of a particle situated at P. The *first moment* of a point P with respect to a point O is the vector $\mathbf{M} = s\,\mathbf{r}_P$. The scalar s is called the *strength* of P. The set of n points P_i, $i = 1, 2, \ldots, n$, is $\{S\}$, Fig. 2.1a

$$\{S\} = \{P_1, P_2, \ldots, P_n\} = \{P_i\}_{i=1,2,\ldots,n}.$$

The strengths of the points P_i are s_i, $i = 1, 2, \ldots, n$, that is, n scalars, all having the same dimensions, and each associated with one of the points of $\{S\}$.

The *centroid* of the set $\{S\}$ is the point C with respect to which the sum of the first moments of the points of $\{S\}$ is equal to zero. The centroid is the point defining the geometric center of the system or of an object.

The position vector of C relative to an arbitrarily selected reference point O is $\mathbf{r}_C$, Fig. 2.1b. The position vector of P_i relative to O is $\mathbf{r}_i$. The position vector of P_i relative to C is $\mathbf{r}_i - \mathbf{r}_C$. The sum of the first moments of the points P_i with respect to C is $\sum_{i=1}^{n} s_i(\mathbf{r}_i - \mathbf{r}_C)$. If C is to be centroid of $\{S\}$, this sum is equal to zero:

$$\sum_{i=1}^{n} s_i(\mathbf{r}_i - \mathbf{r}_C) = \sum_{i=1}^{n} s_i \mathbf{r}_i - \mathbf{r}_C \sum_{i=1}^{n} s_i = 0.$$

D.B. Marghitu and M. Dupac, *Advanced Dynamics: Analytical and Numerical Calculations with MATLAB*, DOI 10.1007/978-1-4614-3475-7_2,
© Springer Science+Business Media, LLC 2012

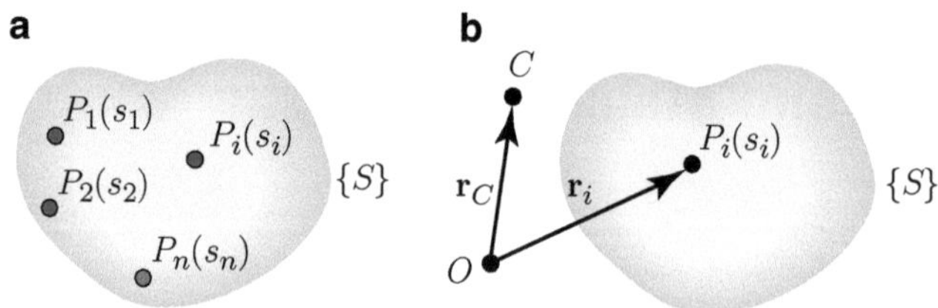

Fig. 2.1 (**a**) Set of points and (**b**) centroid of a set of points

The position vector $\mathbf{r}_C$ of the centroid C, relative to an arbitrarily selected reference point O, is given by

$$\mathbf{r}_C = \frac{\displaystyle\sum_{i=1}^{n} s_i\,\mathbf{r}_i}{\displaystyle\sum_{i=1}^{n} s_i}.$$

If $\sum_{i=1}^{n} s_i = 0$, the centroid is not defined. The centroid C of a set of points of given strength is a unique point, its location being independent of the choice of reference point O.

The Cartesian coordinates of the centroid $C(x_C, y_C, z_C)$ of a set of points P_i, $i = 1,\ldots,n$, of strengths s_i, $i = 1,\ldots,n$, are given by the expressions

$$x_C = \frac{\displaystyle\sum_{i=1}^{n} s_i x_i}{\displaystyle\sum_{i=1}^{n} s_i}, \quad y_C = \frac{\displaystyle\sum_{i=1}^{n} s_i y_i}{\displaystyle\sum_{i=1}^{n} s_i}, \quad z_C = \frac{\displaystyle\sum_{i=1}^{n} s_i z_i}{\displaystyle\sum_{i=1}^{n} s_i}.$$

The *plane of symmetry* of a set is the plane where the centroid of the set lies, the points of the set being arranged in such a way that corresponding to every point on one side of the plane of symmetry there exists a point of equal strength on the other side, the two points being equidistant from the plane.

A set $\{S'\}$ of points is called a *subset* of a set $\{S\}$ if every point of $\{S'\}$ is a point of $\{S\}$. The centroid of a set $\{S\}$ may be located using the *method of decomposition*:

- Divide the system $\{S\}$ into subsets.
- Find the centroid of each subset.
- Assign to each centroid of a subset a strength proportional to the sum of the strengths of the points of the corresponding subset.
- Determine the centroid of this set of centroids.

2.1.2   *Centroid of a Curve, Surface, or Solid*

The position vector of the centroid C of a curve, surface, or solid relative to a point O is

$$\mathbf{r}_C = \frac{\int_\tau \mathbf{r}\, d\tau}{\int_\tau d\tau}, \tag{2.1}$$

where τ is a curve, surface, or solid; $\mathbf{r}$ denotes the position vector of a typical point of τ, relative to O; and $d\tau$ is the length, area, or volume of a differential element of τ. Each of the two limits in this expression is called an "integral over the domain τ (curve, surface, or solid)." The integral $\int_\tau d\tau$ gives the total length, area, or volume of τ, that is,

$$\int_\tau d\tau = \tau.$$

The position vector of the centroid is

$$\mathbf{r}_C = \frac{1}{\tau} \int_\tau \mathbf{r}\, d\tau.$$

Let $\mathbf{\imath}$, $\mathbf{\jmath}$, $\mathbf{k}$ be mutually perpendicular unit vectors (Cartesian reference frame) with the origin at O. The coordinates of C are x_C, y_C, z_C and

$$\mathbf{r}_C = x_C \mathbf{\imath} + y_C \mathbf{\jmath} + z_C \mathbf{k}.$$

It results that

$$x_C = \frac{1}{\tau} \int_\tau x\, d\tau, \quad y_C = \frac{1}{\tau} \int_\tau y\, d\tau, \quad z_C = \frac{1}{\tau} \int_\tau z\, d\tau. \tag{2.2}$$

The coordinates for the centroid of a curve L, Fig. 2.2, is determined by using three scalar equations

$$x_C = \frac{\int_L x\, dL}{\int_L dL}, \quad y_C = \frac{\int_L y\, dL}{\int_L dL}, \quad z_C = \frac{\int_L z\, dL}{\int_L dL}. \tag{2.3}$$

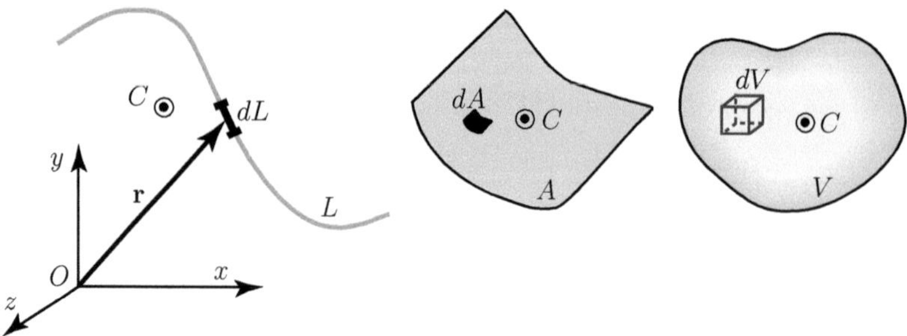

Fig. 2.2 Centroid of a curve L, an area A, and a volume V

The centroid of an area A, Fig. 2.2, is

$$x_C = \frac{\int_A x\,dA}{\int_A dA}, \quad y_C = \frac{\int_A y\,dA}{\int_A dA}, \quad z_C = \frac{\int_A z\,dA}{\int_A dA}, \tag{2.4}$$

and similarly, the centroid of a volume V, Fig. 2.2, is

$$x_C = \frac{\int_V x\,dV}{\int_V dV}, \quad y_C = \frac{\int_V y\,dV}{\int_V dV}, \quad z_C = \frac{\int_V z\,dV}{\int_V dV}. \tag{2.5}$$

For a curved line in the xy plane, the centroidal position is given by

$$x_C = \frac{\int x\,dl}{L}, \quad y_C = \frac{\int y\,dl}{L}, \tag{2.6}$$

where L is the length of the line. Note that the centroid C is not generally located along the line. A curve made up of simple curves is considered. For each simple curve, the centroid is known. The line segment, L_i, has the centroid C_i with coordinates $x_{C_i}, y_{C_i}, i = 1, \ldots, n$. For the entire curve,

$$x_C = \frac{\sum_{i=1}^{n} x_{C_i} L_i}{L}, \quad y_C = \frac{\sum_{i=1}^{n} y_{C_i} L_i}{L}, \quad \text{where } L = \sum_{i=1}^{n} L_i.$$

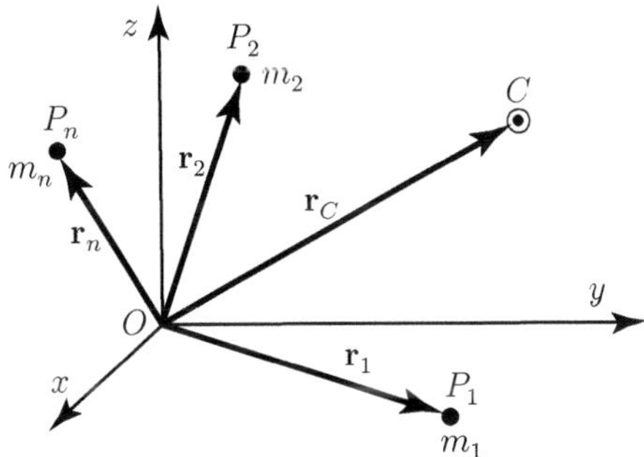

Fig. 2.3 Mass center position vector

2.1.3 Mass Center of a Set of Particles

The *mass center* of a set of particles $\{S\} = \{P_1, P_2, \ldots, P_n\} = \{P_i\}_{i=1,2,\ldots,n}$ is the centroid of the set of points at which the particles are situated with the strength of each point being taken equal to the mass of the corresponding particle, $s_i = m_i$, $i = 1, 2, \ldots, n$. For the system of n particles in Fig. 2.3, one can write

$$\left(\sum_{i=1}^{n} m_i \right) \mathbf{r}_C = \sum_{i=1}^{n} m_i \mathbf{r}_i,$$

and the mass center position vector is

$$\mathbf{r}_C = \frac{\sum_{i=1}^{n} m_i \mathbf{r}_i}{M}, \tag{2.7}$$

where M is the total mass of the system.

2.1.4 Mass Center of a Curve, Surface, or Solid

To study problems concerned with the motion of matter under the influence of forces, that is, dynamics, it is necessary to locate the mass center. The position vector of the mass center C of a continuous body τ, curve, surface, or solid, relative to a point O is

$$\mathbf{r}_C = \frac{\int_\tau \mathbf{r} \rho \, \mathrm{d}\tau}{\int_\tau \rho \, \mathrm{d}\tau} = \frac{1}{m} \int_\tau \mathbf{r} \rho \, \mathrm{d}\tau, \tag{2.8}$$

or using the orthogonal Cartesian coordinates

$$x_C = \frac{1}{m} \int_\tau x\rho \; d\tau, \; y_C = \frac{1}{m} \int_\tau y\rho \; d\tau, \; z_C = \frac{1}{m} \int_\tau z\rho \; d\tau,$$

where ρ is the mass density of the body: mass per unit of length if τ is a curve, mass per unit area if τ is a surface, and mass per unit of volume if τ is a solid; $\mathbf{r}$ is the position vector of a typical point of τ, relative to O; $d\tau$ is the length, area, or volume of a differential element of τ; $m = \int_\tau \rho \; d\tau$ is the total mass of the body; and x_C, y_C, z_C are the coordinates of C.

If the mass density ρ of a body is the same at all points of the body, $\rho = $ constant, the density, as well as the body, are said to be *uniform*. The mass center of a uniform body coincides with the centroid of the figure occupied by the body.

The density ρ of a body is its mass per unit volume. The mass of a differential element of volume dV is $dm = \rho \; dV$. If ρ is not constant throughout the body and can be expressed as a function of the coordinates of the body, then

$$x_C = \frac{\int_\tau x\rho dV}{\int_\tau \rho dV}, \quad y_C = \frac{\int_\tau y\rho dV}{\int_\tau \rho dV}, \quad z_C = \frac{\int_\tau z\rho dV}{\int_\tau \rho dV}. \tag{2.9}$$

The centroid of a volume defines the point at which the total moment of volume is zero. Similarly, the center of mass of a body is the point at which the total moment of the body's mass about that point is zero.

The *method of decomposition* can be used to locate the mass center of a continuous body:

- Divide the body into a number of simpler body shapes, which may be particles, curves, surfaces, or solids; holes are considered as pieces with negative size, mass, or weight.
- Locate the coordinates x_{C_i}, y_{C_i}, z_{C_i} of the mass center of each part of the body.
- Determine the mass center using the equations

$$x_C = \frac{\sum_{i=1}^n \int_\tau x \, d\tau}{\sum_{i=1}^n \int_\tau d\tau}, \quad y_C = \frac{\sum_{i=1}^n \int_\tau y \, d\tau}{\sum_{i=1}^n \int_\tau d\tau}, \quad z_C = \frac{\sum_{i=1}^n \int_\tau z \, d\tau}{\sum_{i=1}^n \int_\tau d\tau}, \tag{2.10}$$

where τ is a curve, area, or volume, depending on the centroid that is required. Equation (2.10) can be simplify as

$$x_C = \frac{\sum_{i=1}^n x_{C_i} \tau_i}{\sum_{i=1}^n \tau_i}, \quad y_C = \frac{\sum_{i=1}^n y_{C_i} \tau_i}{\sum_{i=1}^n \tau_i}, \quad z_C = \frac{\sum_{i=1}^n z_{C_i} \tau_i}{\sum_{i=1}^n \tau_i}, \tag{2.11}$$

where τ_i is the length, area, or volume of the ith object, depending on the type of centroid.

Fig. 2.4 Planar surface
of area A

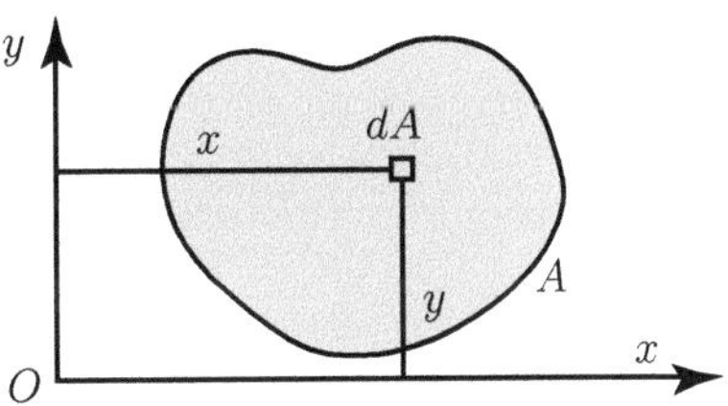

Fig. 2.5 Centroid and
centroidal coordinates
for a planar surface

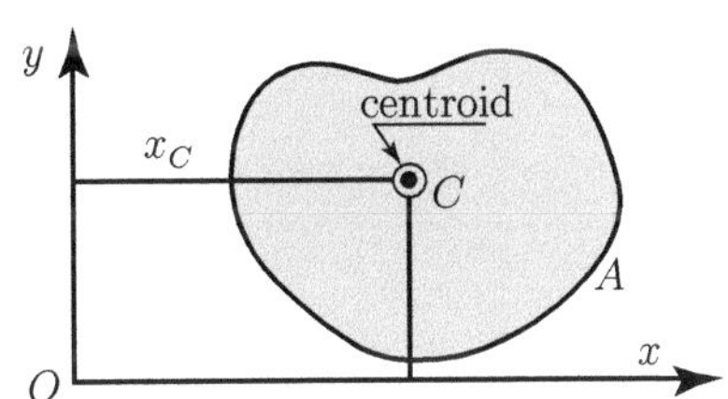

2.1.5   First Moment of an Area

A planar surface of area A and a reference frame xOy in the plane of the surface are
shown in Fig. 2.4. The first moment of area A about the x-axis is

$$M_x = \int_A y \, \mathrm{d}A, \qquad (2.12)$$

and the first moment about the y-axis is

$$M_y = \int_A x \, \mathrm{d}A. \qquad (2.13)$$

The first moment of area gives information of the shape, size, and orientation of
the area.

The entire area A can be concentrated at a position $C(x_C, y_C)$, the centroid,
Fig. 2.5. The coordinates x_C and y_C are the centroidal coordinates. To compute the
centroidal coordinates, one can equate the moments of the distributed area with that
of the concentrated area about both axes

$$A y_C = \int_A y \, \mathrm{d}A, \implies y_C = \frac{\int_A y \, \mathrm{d}A}{A} = \frac{M_x}{A}, \qquad (2.14)$$

$$A x_C = \int_A x \, \mathrm{d}A, \implies x_C = \frac{\int_A x \, \mathrm{d}A}{A} = \frac{M_y}{A}. \qquad (2.15)$$

The location of the centroid of an area is independent of the reference axes
employed, that is, the centroid is a property only of the area itself.

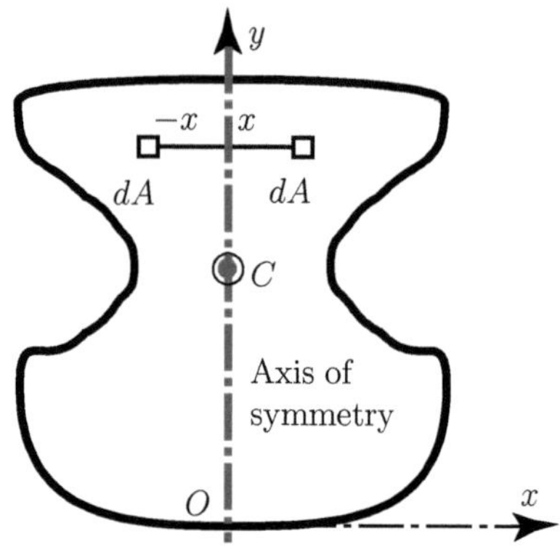

Fig. 2.6 Plane area with axis of symmetry

If the axes xy have their origin at the centroid, $O \equiv C$, then these axes are called *centroidal axes*. The first moments about centroidal axes are zero. All axes going through the centroid of an area are called centroidal axes for that area, and the first moments of an area about any of its centroidal axes are zero. The perpendicular distance from the centroid to the centroidal axis must be zero.

Finding the centroid of a body is greatly simplified when the body has axis of symmetry. Figure 2.6 shows a plane area with the axis of symmetry collinear with the axis y. The area A can be considered as composed of area elements in symmetric pairs such as shown in Fig. 2.6. The first moment of such a pair about the axis of symmetry y is zero. The entire area can be considered as composed of such symmetric pairs and the coordinate x_C is zero

$$x_C = \frac{1}{A} \int_A x \, dA = 0.$$

Thus, *the centroid of an area with one axis of symmetry must lie along the axis of symmetry*. The axis of symmetry then is a centroidal axis, which is another indication that the first moment of area must be zero about the axis of symmetry. With two orthogonal axes of symmetry, the centroid must lie at the intersection of these axes. For such areas as circles and rectangles, the centroid is easily determined by inspection. If a body has a single plane of symmetry, then the centroid is located somewhere on that plane. If a body has more than one plane of symmetry, then the centroid is located at the intersection of the planes.

In many problems, the area of interest can be considered formed by the addition or subtraction of simple areas. For simple areas, the centroids are known by inspection. The areas made up of such simple areas are *composite* areas. For composite areas

$$x_C = \frac{\sum\limits_i A_i x_{Ci}}{A} \quad \text{and} \quad y_C = \frac{\sum\limits_i A_i y_{Ci}}{A}, \tag{2.16}$$

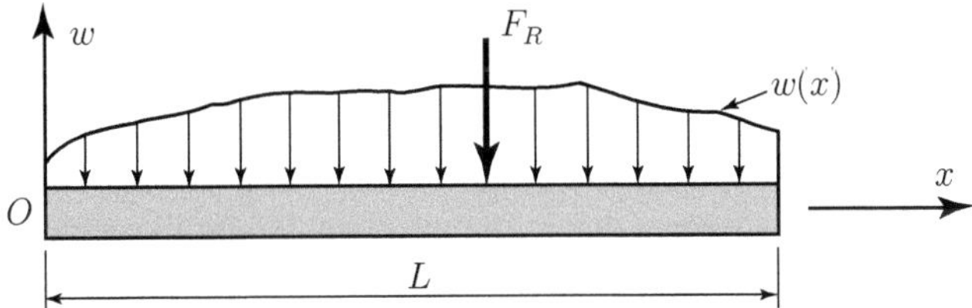

Fig. 2.7 Distributed load

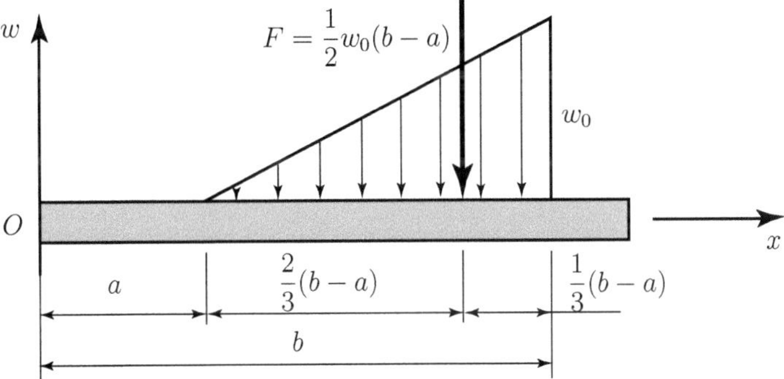

Fig. 2.8 Triangular distributed load

where x_{Ci} and y_{Ci} are the centroidal coordinates to simple area A_i (with proper signs) and A is the total area.

The centroid concept can be used to determine the simplest resultant of a distributed loading. In Fig. 2.7, the distributed load $w(x)$ is considered. The resultant force F_R of the distributed load $w(x)$ loading is given as

$$F_R = \int_0^L w(x) \, \mathrm{d}x. \tag{2.17}$$

From the equation above, the *resultant force equals the area under the loading curve*. The position, x_C, of the *simplest* resultant load can be calculated from the relation

$$F_R x_C = \int_0^L x w(x) \, \mathrm{d}x \implies x_C = \frac{\int_0^L x w(x) \, \mathrm{d}x}{F_R}. \tag{2.18}$$

The position x_C is actually the centroidal coordinate of the loading curve area. Thus, the *simplest resultant force of a distributed load acts at the centroid of the area under the loading curve*. For the triangular distributed load shown in Fig. 2.8, one can replace the distributed loading by a force F equal to $\left(\frac{1}{2}\right)(w_0)(b-a)$ at a position $\left(\frac{1}{3}\right)(b-a)$ from the right end of the distributed loading.

2.1.6   Center of Gravity

The *center of gravity* is a point which locates the resultant weight of a system of particles or body. The sum of moments due to individual particle weight about any point is the same as the moment due to the resultant weight located at the center of gravity. The sum of moments due to the individual particles weights about center of gravity is equal to zero. Similarly, the center of mass is a point which locates the resultant mass of a system of particles or body. The center of gravity of a body is the point at which the total moment of the force of gravity is zero. The coordinates for the center of gravity of a body can be determined with

$$x_C = \frac{\int_V x \rho\, g\, \mathrm{d}V}{\int_V \rho\, g\, \mathrm{d}V}, \quad y_C = \frac{\int_V y \rho\, g\, \mathrm{d}V}{\int_V \rho\, g\, \mathrm{d}V}, \quad z_C = \frac{\int_V z \rho\, g\, \mathrm{d}V}{\int_V \rho\, g\, \mathrm{d}V}. \tag{2.19}$$

The acceleration of gravity is g, $g = 9.81$ m/s^2 or $g = 32.2$ ft/s^2. If g is constant throughout the body, then the location of the center of gravity is the same as that of the center of mass.

2.1.7   Theorems of Guldinus–Pappus

The theorems of Guldinus–Pappus are concerned with the relation of a surface of revolution to its generating curve, and the relation of a volume of revolution to its generating area.

Theorem 2.1.   Consider a coplanar generating curve and an axis of revolution in the plane of this curve in Fig. 2.9. The surface of revolution A developed by rotating the generating curve about the axis of revolution equals the product of the length of the generating L curve times the circumference of the circle formed by the centroid of the generating curve y_C in the process of generating a surface of revolution

$$A = 2\pi y_C L. \tag{2.20}$$

The generating curve can touch but must not cross the axis of revolution.

Proof. An element $\mathrm{d}l$ of the generating curve is considered in Fig. 2.9. For a single revolution of the generating curve about the x-axis, the line segment $\mathrm{d}l$ traces an area

$$\mathrm{d}A = 2\pi y\, \mathrm{d}l.$$

For the entire curve, this area, $\mathrm{d}A$, becomes the surface of revolution, A, given as

$$A = 2\pi \int y\, \mathrm{d}l = 2\pi y_C L,$$

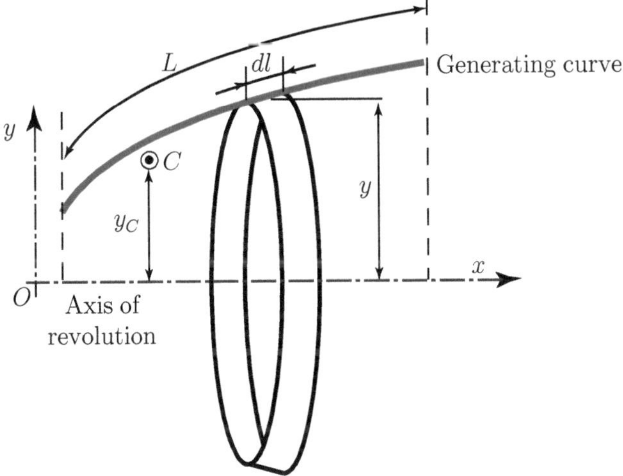

Fig. 2.9 Surface of revolution developed by rotating the generating curve about the axis of revolution

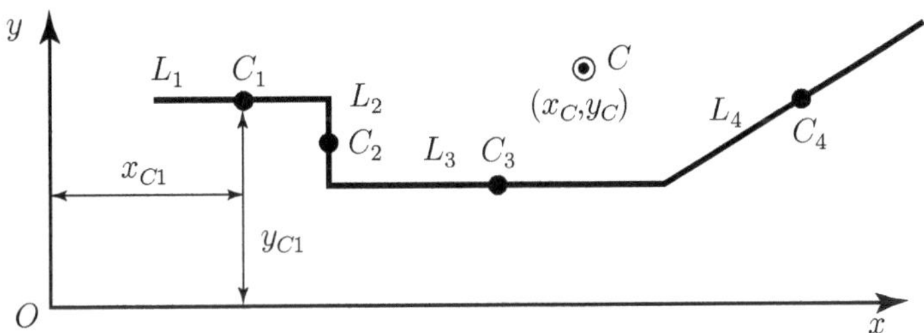

Fig. 2.10 Composed generating curve

where L is the length of the curve and y_C is the centroidal coordinate of the curve. The circumferential length of the circle formed by having the centroid of the curve rotate about the x-axis is $2\pi y_C$, q.e.d.

The surface of revolution A is equal to 2π times the first moment of the generating curve about the axis of revolution.

If the generating curve is composed of simple curves, L_i, whose centroids are known, Fig. 2.10, the surface of revolution developed by revolving the composed generating curve about the axis of revolution x is

$$A = 2\pi \left(\sum_{i=1}^{4} L_i y_{Ci} \right),$$ (2.21)

where y_{Ci} is the centroidal coordinate to the ith line segment L_i. $\square$

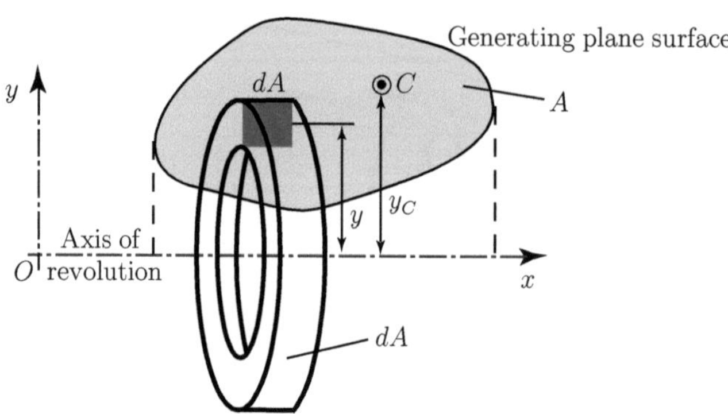

Fig. 2.11 Volume of revolution developed by rotating the generating plane surface about the axis of revolution

Theorem 2.2. Consider a generating plane surface A and an axis of revolution coplanar with the surface Fig. 2.11. The volume of revolution V developed by rotating the generating plane surface about the axis of revolution equals the product of the area of the surface times the circumference of the circle formed by the centroid of the surface y_C in the process of generating the body of revolution

$$V = 2\pi y_C A. \tag{2.22}$$

The axis of revolution can intersect the generating plane surface only as a tangent at the boundary or have no intersection at all.

Proof. The plane surface A is shown in Fig. 2.11. The volume generated by rotating an element dA of this surface about the x-axis is

$$dV = 2\pi y\, dA.$$

The volume of the body of revolution formed from A is then

$$V = 2\pi \int_A y\, dA = 2\pi y_C A.$$

Thus, the volume V equals the area of the generating surface A times the circumferential length of the circle of radius y_C, q.e.d.

The volume V equals 2π times the first moment of the generating area A about the axis of revolution.   □

2.2 Moments of Inertia

2.2.1 Introduction

A system of n particle P_i, $i = 1, 2, \ldots, n$ is considered. The mass of the particle P_i is m_i as shown in Fig. 2.12.

The position vector of the particle P_i is

$$\mathbf{r}_i = x_i \mathbf{1} + y_i \mathbf{J} + z_i \mathbf{k}.$$

The *moments of inertia* of the system about the planes xOy, yOz, and zOx are

$$I_{xOy} = \sum_i m_i z_i^2, \quad I_{yOz} = \sum_i m_i x_i^2, \quad I_{zOx} = \sum_i m_i y_i^2. \tag{2.23}$$

The *moments of inertia* of the system about x, y, and z axes are

$$I_{xx} = A = \sum_i m_i \left(y_i^2 + z_i^2 \right),$$

$$I_{yy} = B = \sum_i m_i \left(z_i^2 + x_i^2 \right),$$

$$I_{zz} = C = \sum_i m_i \left(x_i^2 + y_i^2 \right). \tag{2.24}$$

The *moment of inertia* of the system about the origin O is

$$I_O = \sum_i m_i \left(x_i^2 + y_i^2 + z_i^2 \right). \tag{2.25}$$

The *products of inertia* of the system about the axes xy, yz, and zx are

$$I_{yz} = D = \sum_i m_i y_i z_i, \quad I_{zx} = E = \sum_i m_i z_i x_i, \quad I_{xy} = F = \sum_i m_i x_i y_i. \tag{2.26}$$

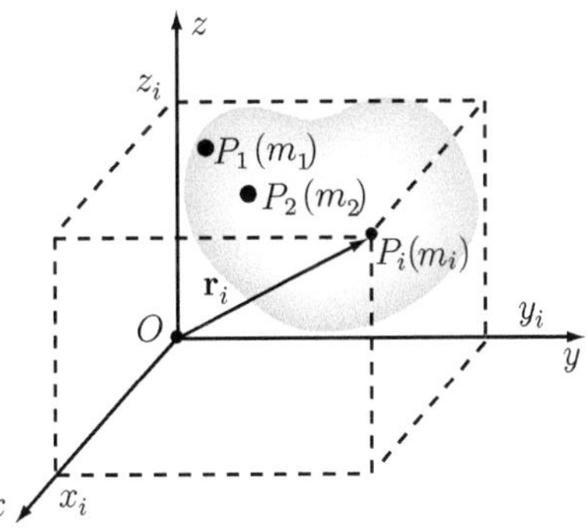

Fig. 2.12 Particle P_i with the mass m_i

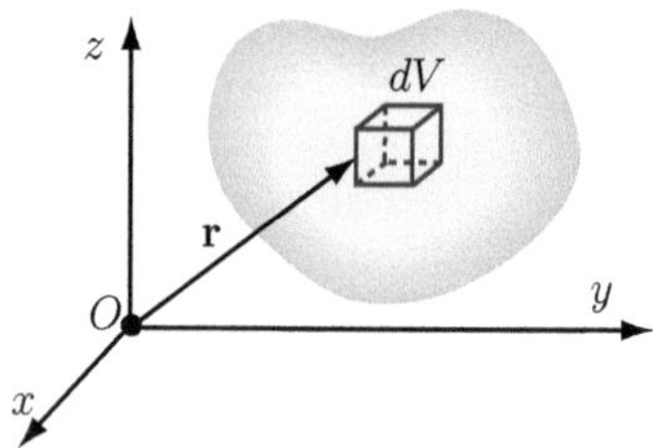

Fig. 2.13 Rigid body in space with mass m and differential volume dV

Between the different moments of inertia, one can write the relations

$$I_O = I_{xOy} + I_{yOz} + I_{zOx} = \frac{1}{2}\left(I_{xx} + I_{yy} + I_{zz}\right),$$

and

$$I_{xx} = I_{yOz} + I_{zOx}.$$

For a continuous domain D, the previous relations become

$$I_{xOy} = \int_D z^2 \mathrm{d}m, \quad I_{yOz} = \int_D x^2 \mathrm{d}m, \quad I_{zOx} = \int_D y^2 \mathrm{d}m,$$

$$I_{xx} = \int_D \left(y^2 + z^2\right)\mathrm{d}m, \quad I_{yy} = \int_D \left(x^2 + z^2\right)\mathrm{d}m, \quad I_{zz} = \int_D \left(x^2 + y^2\right)\mathrm{d}m,$$

$$I_O = \int_D \left(x^2 + y^2 + z^2\right)\mathrm{d}m,$$

$$I_{xy} = \int_D xy\,\mathrm{d}m, \quad I_{xz} = \int_D xz\,\mathrm{d}m, \quad I_{yz} = \int_D yz\,\mathrm{d}m. \tag{2.27}$$

The infinitesimal mass element $\mathrm{d}m$ can have the values

$$\mathrm{d}m = \rho_v\,\mathrm{d}V, \quad \mathrm{d}m = \rho_A\,\mathrm{d}A, \quad \mathrm{d}m = \rho_l\,\mathrm{d}l,$$

where ρ_v, ρ_A, and ρ_l are the volume density, area density, and length density.

Moment of Inertia about an Arbitrary Axis
For a rigid body with mass m, density ρ, and volume V, as shown in Fig. 2.13, the moments of inertia are defined as follows:

$$I_{xx} = \int_V \rho\left(y^2 + z^2\right)\mathrm{d}V, \quad I_{yy} = \int_V \rho\left(z^2 + x^2\right)\mathrm{d}V, \quad I_{zz} = \int_V \rho\left(x^2 + y^2\right)\mathrm{d}V, \tag{2.28}$$

Fig. 2.14 Rigid body and an
arbitrary axis Δ of unit
vector $\mathbf{u}_\Delta$

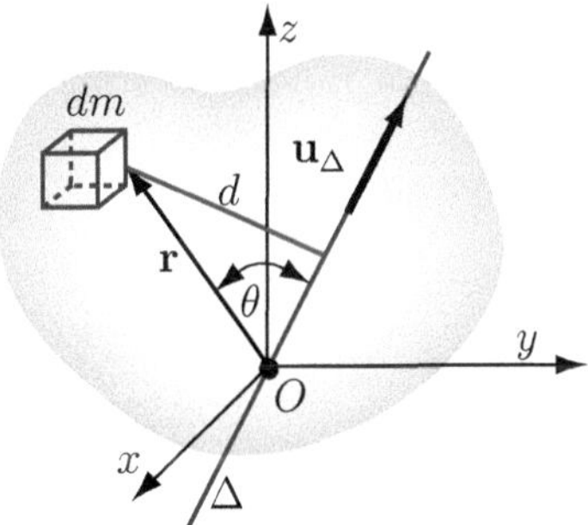

and the products of inertia

$$
I_{xy} = I_{yx} = \int_V \rho\, x y\, \mathrm{d}V, \quad I_{xz} = I_{zx} = \int_V \rho\, x z\, \mathrm{d}V, \quad I_{yz} = I_{zy} = \int_V \rho\, y z\, \mathrm{d}V. \quad (2.29)
$$

The moment of inertia given in (2.28) is just the *second moment* of the mass
distribution with respect to a Cartesian axis. For example, I_{xx} is the integral of
summation of the infinitesimal mass elements $\rho\, \mathrm{d}V$, each multiplied by the square
of its distance from the x-axis.

The effective value of this distance for a certain body is known as its *radius of
gyration* with respect to the given axis. The *radius of gyration* corresponding to I_{jj}
is defined as

$$
k_j = \sqrt{\frac{I_{jj}}{m}},
$$

where m is the total mass of the rigid body and where the symbol j can be replaced
by x, y or z. The *inertia matrix* of a rigid body is represented by the matrix

$$
[I] = \begin{bmatrix} I_{xx} & -I_{xy} & -I_{xz} \\ -I_{yx} & I_{yy} & -I_{yz} \\ -I_{zx} & -I_{zy} & I_{zz} \end{bmatrix}.
$$

Moment of Inertia about an Arbitrary Axis
Consider the rigid body shown in Fig. 2.14. The reference frame x, y, z has the origin
at O. The direction of an arbitrary axis Δ through O is defined by the unit vector $\mathbf{u}_\Delta$

$$
\mathbf{u}_\Delta = \cos\alpha\,\mathbf{i} + \cos\beta\,\mathbf{j} + \cos\gamma\,\mathbf{k},
$$

where $\cos\alpha$, $\cos\beta$, $\cos\gamma$ are the direction cosines. The moment of inertia about the
Δ axis, for a differential mass element $\mathrm{d}m$ of the body, is by definition

$$
I_\Delta = \int_D d^2\, \mathrm{d}m,
$$

where d is the perpendicular distance from dm to Δ. The position of the mass element dm is located using the position vector $\mathbf{r}$ and then $d = r\sin\theta$, which represents the magnitude of the cross product $\mathbf{u}_\Delta \times \mathbf{r}$. The moment of inertia can be expressed as

$$I_\Delta = \int_D |\mathbf{u}_\Delta \times \mathbf{r}|^2 \, dm = \int_D (\mathbf{u}_\Delta \times \mathbf{r}) \cdot (\mathbf{u}_\Delta \times \mathbf{r}) \, dm.$$

If the position vector is $\mathbf{r} = x\mathbf{i} + y\mathbf{J} + z\mathbf{k}$, then

$$\mathbf{u}_\Delta \times \mathbf{r} = (z\cos\beta - y\cos\gamma)\mathbf{i} + (x\cos\gamma - z\cos\alpha)\mathbf{J} + (y\cos\alpha - x\cos\beta)\mathbf{k}.$$

After substituting and performing the dot-product operation, one can write the moment of inertia as

$$\begin{aligned}
I_\Delta &= \int_D \left[(z\cos\beta - y\cos\gamma)^2 + (x\cos\gamma - z\cos\alpha)^2 + (y\cos\alpha - x\cos\beta)^2 \right] dm \\
&= \cos^2\alpha \int_D (y^2 + z^2)\, dm + \cos^2\beta \int_D (z^2 + x^2)\, dm + \cos^2\gamma \int_D (x^2 + y^2)\, dm \\
&\quad - 2\cos\alpha\cos\beta \int_D xy\, dm - 2\cos\beta\cos\gamma \int_D yz\, dm - 2\cos\gamma\cos\alpha \int_D zx\, dm.
\end{aligned}$$

The moment of inertia with respect to the Δ axis is

$$\begin{aligned}
I_\Delta &= I_{xx}\cos^2\alpha + I_{yy}\cos^2\beta + I_{zz}\cos^2\gamma \\
&\quad - 2I_{xy}\cos\alpha\cos\beta - 2I_{yz}\cos\beta\cos\gamma - 2I_{zx}\cos\gamma\cos\alpha. \tag{2.30}
\end{aligned}$$

2.2.2   Translation of Coordinate Axes

The defining equations for the moments and products of inertia, as given by (2.28) and (2.29), do not require that the origin of the Cartesian coordinate system be taken at the mass center. Next, one can calculate the moments and products of inertia for a given body with respect to a set of parallel axes that do not pass through the mass center. Consider the body shown in Fig. 2.15. The mass center is located at the origin $O' \equiv C$ of the primed system $x'y'z'$. The coordinate of O' with respect to the unprimed system xyz is (x_C, y_C, z_C). An infinitesimal volume element dV is located at (x, y, z) in the unprimed system and at (x', y', z') in the primed system. These coordinates are related by the equations

$$x = x' + x_C, \quad y = y' + y_C, \quad z = z' + z_C. \tag{2.31}$$

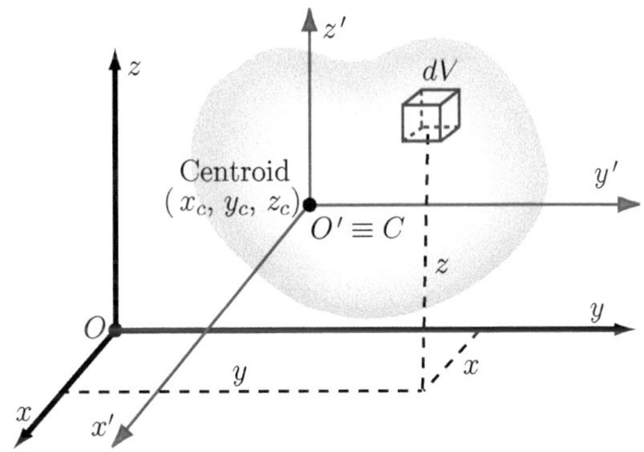

Fig. 2.15 Rigid body and centroidal axes $x'y'z'$: $x = x' + x_c$, $y = y' + y_c$, $z = z' + z_c$

The moment of inertia about the x-axis can be written in terms of primed coordinates by using (2.28) and (2.31)

$$
I_{xx} = \int_V \rho \left[(y' + y_c)^2 + (z' + z_c)^2 \right] dV
$$

$$
= I_{Cx'x'} + 2y_c \int_V \rho y' dV + 2z_c \int_V \rho z' dV + m \left(y_c^2 + z_c^2 \right), \qquad (2.32)
$$

where m is the total mass of the rigid body, and the origin of the primed coordinate system was chosen at the mass center. One can write

$$
\int_V \rho x' dV = \int_V \rho y' dV = \int_V \rho z' dV = 0, \qquad (2.33)
$$

and therefore, the two integrals on the right-hand side of (2.32) are zero. In a similar way, one can obtain I_{yy} and I. The results are summarized as follows:

$$
\begin{aligned}
I_{xx} &= I_{Cx'x'} + m \left(y_c^2 + z_c^2 \right), \\
I_{yy} &= I_{Cy'y'} + m \left(x_c^2 + z_c^2 \right), \\
I_{zz} &= I_{Cz'z'} + m \left(x_c^2 + y_c^2 \right),
\end{aligned} \qquad (2.34)
$$

or, in general,

$$
I_{kk} = I_{Ck'k'} + m d^2, \qquad (2.35)
$$

where d is the distance between a given unprimed axis and a parallel primed axis passing through the mass center C. Equation (2.35) represents the *parallel − axes*

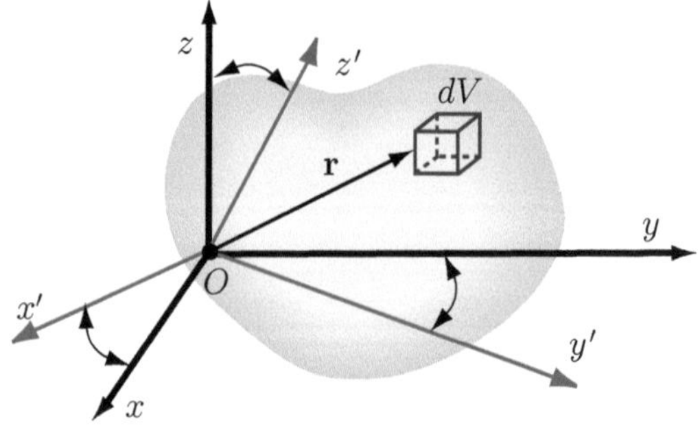

Fig. 2.16 Rotation of coordinate axes

Theorem. The products of inertia are obtained in a similar manner, using (2.29) and (2.31)

$$I_{xy} = \int_V \rho \left(x' + x_c \right) \left(y' + y_c \right) dV$$

$$= I_{Cx'y'} + x_c \int_V \rho y' dV + y_c \int_V \rho x' dV + m x_c y_c.$$

The two integrals on the previous equation are zero. The other products of inertia can be calculated in a similar manner, and the results can be written as follows:

$$I_{xy} = I_{Cx'y'} + m x_c y_c,$$

$$I_{xz} = I_{Cx'z'} + m x_c z_c,$$

$$I_{yz} = I_{Cy'z'} + m y_c z_c. \tag{2.36}$$

Equations (2.34) and (2.36) shows that a translation of axes away from the mass center results in an increase in the moments of inertia. The products of inertia may increase or decrease, depending upon the particular case.

2.2.3   Principal Axes

Next, the changes in the moments and product of inertia of a rigid body due to a rotation of coordinate axes are considered, as shown in Fig. 2.16. The origin of the coordinate axes is located at the fixed point O. In general, the origin O is not the mass center C of the rigid body. From the definitions of the moments of inertia given in (2.28), it results that the moments of inertia cannot be negative. Furthermore,

$$I_{xx} + I_{yy} + I_{zz} = 2 \int_V \rho r^2 dV, \tag{2.37}$$

where r is the square of the distance from the origin O,

$$r^2 = x^2 + y^2 + z^2.$$

The distance r corresponding to any mass element $\rho\,dV$ of the rigid body does not change with a rotation of axes from xyz to $x'y'z'$ (Fig. 2.16). Therefore, the sum of the moments of inertia is invariant with respect to a coordinate system rotation. In terms of matrix notation, the sum of the moments of inertia is just the sum of the elements on the principal diagonal of the inertia matrix and is known as the trace of that matrix. So the trace of the inertia matrix is unchanged by a coordinate rotation because the trace of any square matrix is invariant under an orthogonal transformation.

Next, the products of inertia are considered. A coordinate rotation of axes can result in a change in the signs of the products of inertia. A 180° rotation about the x-axis, for example, reverses the signs of I_{xy} and I_{xz}, while the sign of I_{yz} is unchanged. This occurs because the directions of the positive y and z axes are reversed. On the other hand, a 90° rotation about the x-axis reverses the sign of I_{yz}. It can be seen that the moments and products of inertia vary smoothly with changes in the orientation of the coordinate system because the direction cosines vary smoothly. Therefore, an orientation can always be found for which a given product of inertia is zero. It is always possible to find an orientation of the coordinate system relative to a given rigid body such that all products of inertia are zero simultaneously, that is, the inertia matrix is *diagonal*. The three mutually orthogonal coordinate axes are known as *principal axes* in this case, and the corresponding moments of inertia are the *principal moments of inertia*. The three planes formed by the principal axes are called *principal planes*.

If I is a principal moment of inertia, then I satisfies the cubic characteristic equation

$$\begin{vmatrix} I_{xx} - I & -I_{xy} & -I_{xz} \\ -I_{yx} & I_{yy} - I & -I_{yz} \\ -I_{zx} & -I_{zy} & I_{zz} - I \end{vmatrix} = 0. \tag{2.38}$$

Equation (2.38) is used to determine the associated principal moments of inertia.

Suppose that $\mathbf{u}_1$, $\mathbf{u}_2$, $\mathbf{u}_3$ are mutually perpendicular unit vectors each parallel to a principal axis of the rigid body relative to O. The principal moments of inertia associated to $\mathbf{u}_1$, $\mathbf{u}_2$, $\mathbf{u}_3$ for the rigid body relative to O are I_1, I_2, and I_3. The inertia matrix, in this case, is

$$I = \begin{bmatrix} I_1 & 0 & 0 \\ 0 & I_2 & 0 \\ 0 & 0 & I_3 \end{bmatrix}.$$

When the point O under consideration is the mass center of the rigid body, one speaks of *central principal moments of inertia*.

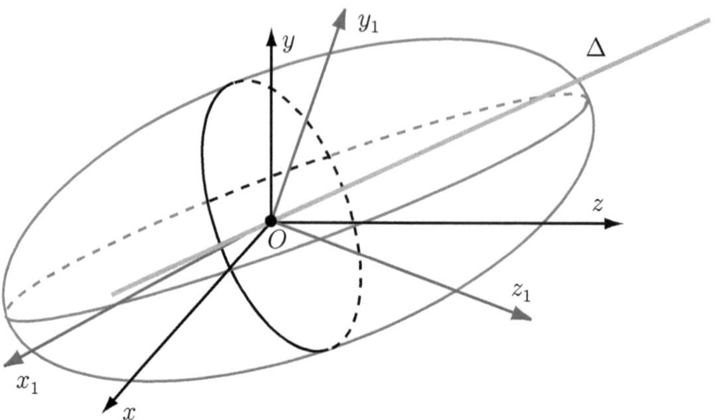

Fig. 2.17 Ellipsoid of inertia

2.2.4 *Ellipsoid of Inertia*

The ellipsoid of inertia for a given body and reference point is a plot of the moment of inertia of the body for all possible axis orientations through the reference point. This graph in space has the form of an ellipsoid surface. Consider a rigid body in rotational motion about an axis Δ. The *ellipsoid of inertia* with respect to an arbitrary point O is the geometrical locus of the points Q, where Q is the extremity of the vector $\overrightarrow{OQ}$ with the module $\left|\overrightarrow{OQ}\right| = \dfrac{1}{\sqrt{I_\Delta}}$ and where I_Δ is the moment of inertia about the instantaneous axis of rotation Δ, as shown in Fig. 2.17. The segment OQ is calculated with

$$\left|\overrightarrow{OQ}\right| = \frac{1}{\sqrt{I_\Delta}} = \frac{1}{k_0\, m},$$

where k_0 is the radius of gyration of the body about the given axis and m is the total mass. For a Cartesian system of axes, the equation of the ellipsoid surface centered at O is

$$I_{xx}\,x^2 + I_{yy}\,y^2 + I_{zz}\,z^2 + 2I_{xy}\,xy + 2I_{xz}\,xz + 2I_{yz}\,yz = 1. \tag{2.39}$$

The radius of gyration of a given rigid body depends upon the location of the axis relative to the body and is not depended upon the position of the body in space. The ellipsoid of inertia is fixed in the body and rotates with it. The x_1, y_1, and z_1 axes are assumed to be the principal axes of the ellipsoid, as shown in Fig. 2.17. For the principal axes, the equation of the inertia ellipsoid, (2.39), takes the following simple form

$$I_1\,x_1^2 + I_2\,y_1^2 + I_3\,z_3^2 = 1. \tag{2.40}$$

The previous equation is of the same form as (2.39) for the case where the $x_1, y_1,$ and z_1 axes are the principal axes of the rigid body, and all products of inertia vanish.

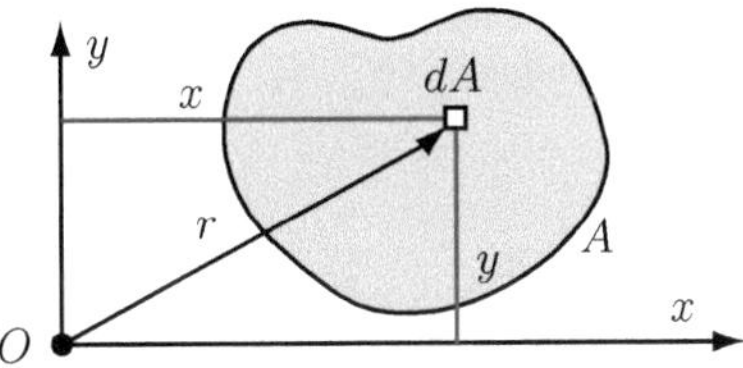

Fig. 2.18 Moments of inertia for area A about x and y axes

Therefore, I_1, I_2, and I_3 are the principal moments of inertia of the rigid body, and furthermore, the principal axes of the body coincide with those of the ellipsoid of inertia. From (2.40), the lengths of the principal semiaxes of the ellipsoid of inertia are

$$l_1 = \frac{1}{\sqrt{I_1}},$$

$$l_2 = \frac{1}{\sqrt{I_2}},$$

$$l_3 = \frac{1}{\sqrt{I_3}}.$$

From the parallel-axis theorem, (2.35), one can remark that the minimum moment of inertia about the mass center is also the smallest possible moment of inertia for the given body with respect to any reference point.

If the point O is the same as the mass center $(O \equiv C)$, the ellipsoid is named *principal ellipsoid of inertia.*

2.2.5 *Moments of Inertia for Areas*

The moment of inertia (*second moment*) of the area A about x and y axes, see Fig. 2.18, denoted as I_{xx} and I_{yy}, respectively, are

$$I_{xx} = \int_A y^2 \, dA, \tag{2.41}$$

$$I_{yy} = \int_A x^2 \, dA. \tag{2.42}$$

The second moment of area cannot be negative.

The entire area may be concentrated at a single point (k_x, k_y) to give the same second moment of area for a given reference. The distances k_x and k_y are called the radii of gyration. Thus,

$$A k_x^2 = I_{xx} = \int_A y^2 \, dA \implies k_x^2 = \frac{\int_A y^2 \, dA}{A} = \frac{I_{xx}}{A},$$

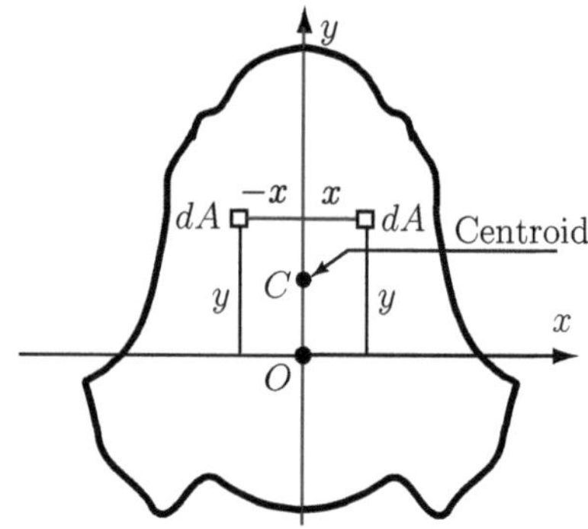

Fig. 2.19 Area, A, with an axis of symmetry Oy

$$A k_y^2 = I_{yy} = \int_A x^2 \, dA \quad \Longrightarrow \quad k_y^2 = \frac{\int_A x^2 \, dA}{A} = \frac{I_{yy}}{A}. \tag{2.43}$$

This point (k_x, k_y) depends on the shape of the area and on the position of the reference. The centroid location is independent of the reference position.

The *product of inertia* for an area A is defined as

$$I_{xy} = \int_A xy \, dA. \tag{2.44}$$

This quantity may be positive or negative and relates an area directly to a set of axes.

If the area under consideration has an axis of symmetry, the product of area for this axis is zero. Consider the area in Fig. 2.19, which is symmetrical about the vertical axis y. The planar Cartesian frame is xOy. The centroid is located somewhere along the symmetrical axis y. Two differential element of areas that are positioned as mirror images about the y-axis are shown in Fig. 2.19. The contribution to the product of area of each elemental area is $xy \, dA$, but with opposite signs, and so the result is zero. The entire area is composed of such elemental area pairs, and the product of area is zero. The product of inertia for an area I_{xy} is zero ($I_{xy} = 0$) if either the x- or y-axis is an axis of symmetry for the area.

Transfer Theorem or Parallel-Axis Theorem
The x-axis in Fig. 2.20 is parallel to an axis x', and it is at a distance b from the axis x'. The axis x' is going through the centroid C of the A area, and it is a centroidal axis. The second moment of area about the x-axis is

$$I_{xx} = \int_A y^2 \, dA = \int_A (y' + b)^2 \, dA,$$

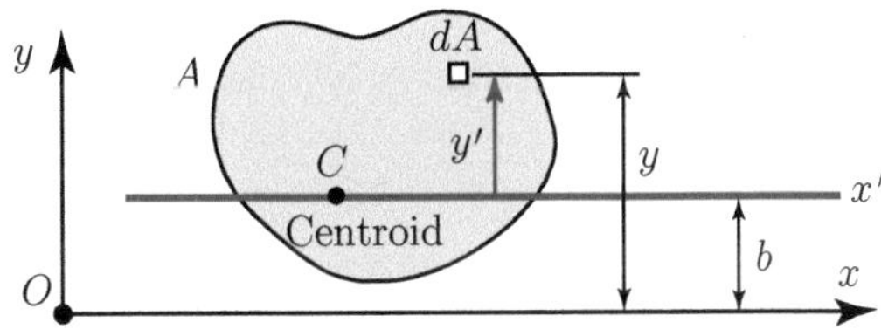

Fig. 2.20 Area and centroidal axis $Cx'x'||xx$

where the distance $y = y' + b$. Carrying out the operations

$$I_{xx} = \int_A y'^2 \, dA + 2b \int_A y' \, dA + A b^2.$$

The first term of the right-hand side is by definition $I_{x'x'}$

$$I_{Cx'x'} = \int_A y'^2 \, dA.$$

The second term involves the first moment of area about the x' axis, and it is zero because the x' axis is a centroidal axis

$$\int_A y' \, dA = 0.$$

The second moment of the area A about any axis I_{xx} is equal to the second moment of the area A about a parallel axis at centroid $I_{Cx'x'}$ plus $A b^2$, where b is the perpendicular distance between the axis for which the second moment is being computed and the parallel centroidal axis

$$I_{xx} = I_{Cx'x'} + A b^2.$$

With the transfer theorem, the second moments or products of area about any axis can be computed in terms of the second moments or products of area about a parallel set of axes going through the centroid of the area in question.

In handbooks, the areas and second moments about various centroidal axes are listed for many of the practical configurations, and using the parallel-axis theorem (or Huygens–Steiner theorem), second moments can be calculated for axes not at the centroid.

In Fig. 2.21 are shown two references, one $x'y'$ at the centroid C and the other xy arbitrary but positioned parallel relative to $x'y'$. The coordinates of the centroid $C(x_C, y_C)$ of area A measured from the reference x, y are c and b, $x_C = c, y_C = b$. The centroid coordinates must have the proper signs. The product of area about the noncentroidal axes xy is

$$I_{xy} = \int_A xy \, dA = \int_A (x' + c)(y' + b) \, dA,$$

Fig. 2.21 Centroidal axes $x'y'$ parallel to reference axes xy: $Cx'x'||xx$ and $Cy'y'||yy$

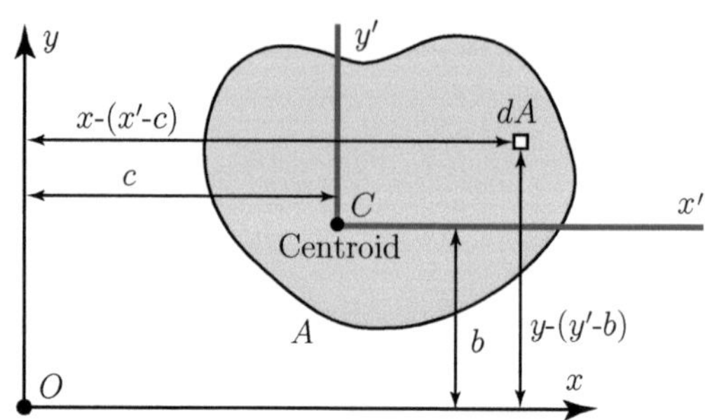

or

$$I_{xy} = \int_A x'\, y'\, \mathrm{d}A + c \int_A y'\, \mathrm{d}A + b \int_A x'\, \mathrm{d}A + A\,b\,c.$$

The first term of the right-hand side is by definition $I_{x'y'}$

$$I_{x'y'} = \int_A x'\, y'\, \mathrm{d}A.$$

The next two terms of the right-hand side are zero since x' and y' are centroidal axes

$$\int_A y'\, \mathrm{d}A = 0 \ \text{ and } \ \int_A x'\, \mathrm{d}A = 0.$$

Thus, the parallel-axis theorem for products of area is as follows.

The product of area for any set of axes I_{xy} is equal to the product of area for a parallel set of axes at centroid $I_{Cx'y'}$ plus $A\,c\,b$, where c and b are the coordinates of the centroid of area A,

$$I_{xy} = I_{Cx'y'} + A\,c\,b.$$

With the transfer theorem, the second moments or products of area can be found about any axis in terms of second moments or products of area about a parallel set of axes going through the centroid of the area.

Polar Moment of Area
In Fig. 2.18, there is a reference xy associated with the origin O. Summing I_{xx} and I_{yy},

$$I_{xx} + I_{yy} = \int_A y^2\, \mathrm{d}A + \int_A x^2\, \mathrm{d}A$$

$$= \int_A (x^2 + y^2)\, \mathrm{d}A = \int_A r^2\, \mathrm{d}A,$$

where $r^2 = x^2 + y^2$. The distance r^2 is independent of the orientation of the reference, and the sum $I_{xx} + I_{yy}$ is independent of the orientation of the coordinate system.

Fig. 2.22 Reference xy and reference $x'y'$ rotated with an angle α

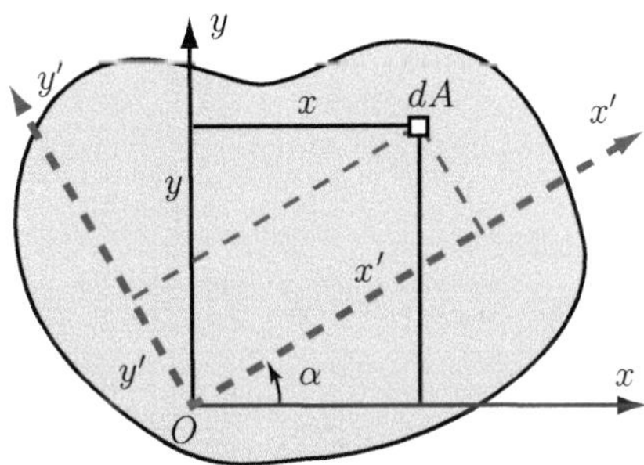

Therefore, the sum of second moments of area about orthogonal axes is a function only of the position of the origin O for the axes.

The polar moment of area about the origin O is

$$I_O = I_{xx} + I_{yy}. \tag{2.45}$$

The polar moment of area is an *invariant* of the system. The group of terms $I_{xx}\,I_{yy} - I_{xy}^2$ is also invariant under a rotation of axes. The polar radius of gyration is

$$k_O = \sqrt{\frac{I_O}{A}}. \tag{2.46}$$

Principal Axes

In Fig. 2.22, an area A is shown with a reference xy having its origin at O. Another reference $x'y'$ with the same origin O is rotated with an angle α from xy (counterclockwise as positive). The relations between the coordinates of the area elements dA for the two references are

$$x' = x\cos\alpha + y\sin\alpha, \quad y' = -x\sin\alpha + y\cos\alpha.$$

The second moment $I_{x'x'}$ can be expressed as

$$
\begin{aligned}
I_{x'x'} &= \int_A (y')^2 \, dA = \int_A (-x\sin\alpha + y\cos\alpha)^2 \, dA \\
&= \sin^2\alpha \int_A x^2 \, dA - 2\sin\alpha\cos\alpha \int_A xy \, dA + \cos^2\alpha \int_A y^2 \, dA \\
&= I_{yy}\sin^2\alpha + I_{xx}\cos^2\alpha - 2I_{xy}\sin\alpha\cos\alpha.
\end{aligned} \tag{2.47}
$$

Using the trigonometric identities

$$\cos^2\alpha = \frac{1+\cos 2\alpha}{2}, \quad \sin^2\alpha = \frac{1-\cos 2\alpha}{2}, \quad \sin 2\alpha = 2\sin\alpha\cos\alpha,$$

Fig. 2.23 Principal axis
of area

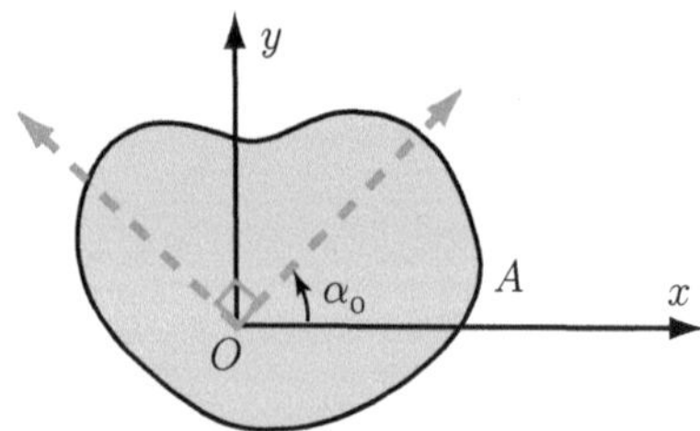

(2.47) becomes

$$I_{x'x'} = \frac{I_{xx} + I_{yy}}{2} + \frac{I_{xx} - I_{yy}}{2}\cos 2\alpha - I_{xy}\sin 2\alpha. \tag{2.48}$$

Replacing α with $\alpha + \pi/2$ in (2.48) and using the trigonometric relations

$$\cos(2\alpha + \pi) = -\cos 2\alpha, \quad \sin(2\alpha + \pi) = -\sin 2\alpha,$$

the second moment $I_{y'y'}$ is

$$I_{y'y'} = \frac{I_{xx} + I_{yy}}{2} - \frac{I_{xx} - I_{yy}}{2}\cos 2\alpha + I_{xy}\sin 2\alpha. \tag{2.49}$$

The product of area $I_{x'y'}$ is computed in a similar manner

$$I_{x'y'} = \int_A x'y'\, \mathrm{d}A = \frac{I_{xx} - I_{yy}}{2}\sin 2\alpha + I_{xy}\cos 2\alpha. \tag{2.50}$$

If I_{xx}, I_{yy}, and I_{xy} are known for a reference xy with an origin O, then the second moments and products of area for every set of axes at O can be computed. Next, it is assumed that I_{xx}, I_{yy}, and I_{xy} are known for a reference xy. The sum of the second moments of area is constant for any reference with origin at O. The *minimum* second moment of area corresponds to an axis at *right angles* to the axis having the *maximum* second moment, as shown in Fig. 2.23. This particular set of axes if called *principal axis* of area and the corresponding moments of inertia with respect to these axes are called *principal moments of inertia*.

The second moments of area can be expressed as functions of the angle variable α. The maximum second moment may be determined by setting the partial derivative of $I_{x'x'}$ with respect to α equal to zero. Thus,

$$\frac{\partial I_{x'x'}}{\partial \alpha} = (I_{xx} - I_{yy})(-\sin 2\alpha) - 2I_{xy}\cos 2\alpha = 0, \tag{2.51}$$

or

$$(I_{yy} - I_{xx})\sin 2\alpha_0 - 2I_{xy}\cos 2\alpha_0 = 0,$$

where α_0 is the value of α which defines the orientation of principal axes. Hence,

$$\tan 2\alpha_0 = \frac{2I_{xy}}{I_{yy} - I_{xx}}. \tag{2.52}$$

The angle α_0 corresponds to an extreme value of $I_{x'x'}$ (i.e., to a maximum or minimum value). There are two roots for $2\alpha_0$, which are π radians apart, that will satisfy the previous equation. Thus,

$$2\alpha_{0_1} = \tan^{-1} \frac{2I_{xy}}{I_{yy} - I_{xx}} \implies \alpha_{0_1} = \frac{1}{2} \tan^{-1} \frac{2I_{xy}}{I_{yy} - I_{xx}},$$

and

$$2\alpha_{0_2} = \tan^{-1} \frac{2I_{xy}}{I_{yy} - I_{xx}} + \pi \implies \alpha_{0_2} = \frac{1}{2} \tan^{-1} \frac{2I_{xy}}{I_{yy} - I_{xx}} + \frac{\pi}{2}.$$

This means that there are two axes orthogonal to each other having extreme values for the second moment of area at O. One of the axes is the maximum second moment of area, and the minimum second moment of area is on the other axis. These axes are the principal axes.

With $\alpha = \alpha_0$, the product of area $I_{x'y'}$ becomes

$$I_{x'y'} = \frac{I_{xx} - I_{yy}}{2} \sin 2\alpha_0 + I_{xy} \cos 2\alpha_0. \tag{2.53}$$

For $\alpha_0 = \alpha_{0_1}$, the sine and cosine expressions are

$$\sin 2\alpha_{0_1} = \frac{2I_{xy}}{\sqrt{(I_{yy} - I_{xx})^2 + 4I_{xy}^2}}, \quad \cos 2\alpha_{0_1} = \frac{-(I_{xx} - I_{yy})}{\sqrt{(I_{yy} - I_{xx})^2 + 4I_{xy}^2}}.$$

For $\alpha_0 = \alpha_{0_2}$, the sine and cosine expressions are

$$\sin 2\alpha_{0_2} = \frac{-2I_{xy}}{\sqrt{(I_{yy} - I_{xx})^2 + 4I_{xy}^2}}, \quad \cos 2\alpha_{0_2} = \frac{I_{xx} - I_{yy}}{\sqrt{(I_{yy} - I_{xx})^2 + 4I_{xy}^2}}.$$

Equation (2.53) and $\alpha_0 = \alpha_{0_1}$ give

$$I_{x'y'} = -(I_{yy} - I_{xx}) \frac{I_{xy}}{\left[(I_{yy} - I_{xx})^2 + 4I_{xy}^2\right]^{1/2}} + I_{xy} \frac{I_{yy} - I_{xx}}{\left[(I_{yy} - I_{xx})^2 + 4I_{xy}^2\right]^{1/2}} = 0.$$

In a similar way, (2.53) and $\alpha_0 = \alpha_{0_2}$ give $I_{x'y'} = 0$. The product of area corresponding to the principal axes is zero.

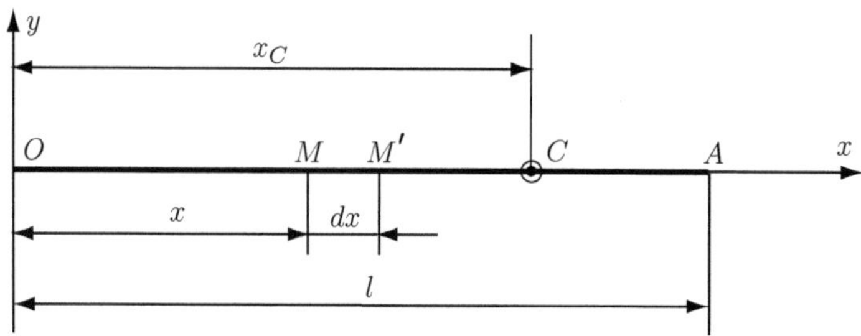

Fig. 2.24 Example 2.1

The maximum or minimum moment of inertia for the area are

$$I_{1,2} = I_{\max, \min} = \frac{I_{xx} + I_{yy}}{2} \pm \sqrt{\left(\frac{I_{xx} - I_{yy}}{2}\right)^2 + I_{xy}^2}.$$

(2.54)

If I is a principal moment of inertia, then I satisfies the quadratic characteristic equation

$$\begin{vmatrix} I_{xx} - I & I_{xy} \\ I_{yx} & I_{yy} - I \end{vmatrix} = 0.$$

(2.55)

2.3 Examples

Example 2.1. Find the position of the mass center for a nonhomogeneous straight rod, with the length $OA = l$ (Fig. 2.24). The linear density ρ of the rod is a linear function with $\rho = \rho_0$ at O and $\rho = \rho_1$ at A.

Solution
A reference frame xOy is selected with the origin at O and the x-axis along the rectilinear rod (Fig. 2.24). Let $M(x, 0)$ be an arbitrarily given point on the rod, and let MM' be an element of the rod with the length dx and the mass $dm = \rho\, dx$. The density ρ is a linear function of x given by

$$\rho = \rho(x) = \rho_0 + \frac{\rho_1 - \rho_0}{l}x.$$

The center mass of the rod, C, has x_c as abscissa. The mass center x_c, with respect to point O, is

$$x_c = \frac{\displaystyle\int_L x\, dm}{\displaystyle\int_L dm} = \frac{\displaystyle\int_0^l x\rho\, dx}{\displaystyle\int_0^l \rho\, dx} = \frac{\displaystyle\int_0^l x\left(\rho_0 + \frac{\rho_1 - \rho_0}{l}x\right) dx}{\displaystyle\int_0^l \left(\rho_0 + \frac{\rho_1 - \rho_0}{l}x\right) dx} = \frac{\dfrac{\rho_0 + 2\rho_1}{6}l^2}{\dfrac{\rho_0 + \rho_1}{2}l}.$$

Fig. 2.25 Example 2.2

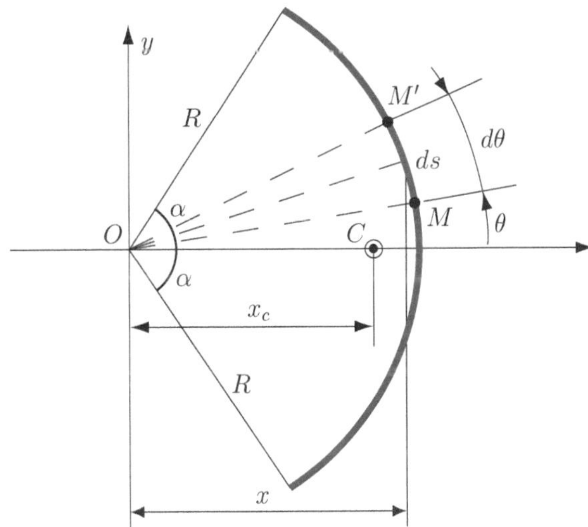

The mass center x_c is

$$x_c = \frac{\rho_0 + 2\rho_1}{3(\rho_0 + \rho_1)} l.$$

In the special case of a homogeneous rod, the density and the position of the mass center are given by

$$\rho_0 = \rho_1 \quad \text{and} \quad x_c = \frac{l}{2}.$$

The MATLAB program is

```
syms rho0 rhoL L x
rho = rho0 + (rhoL - rho0)*x/L;
m = int(rho, x, 0, L);
My = simplify(int(x*rho, x, 0, L));
xC = simplify(My/m);
fprintf('m = %s \n', char(m))
fprintf('My = %s \n', char(My))
fprintf('xC = My/m = %s \n', char(xC))
```

The MATLAB statement `int(f, x, a, b)` is the definite integral of `f` with respect to its symbolic variable `x` from `a` to `b`.

Example 2.2. Find the position of the centroid for a homogeneous circular arc. The radius of the arc is R, and the center angle is 2α radians as shown in Fig. 2.25.

Solution

The axis of symmetry is selected as the x-axis ($y_C = 0$). Let MM' be a differential element of arc with the length $ds = R\,d\theta$. The mass center of the differential element of arc MM' has the abscissa

$$x = R\cos\left(\theta + \frac{d\theta}{2}\right) \approx R\cos(\theta).$$

The abscissa x_C of the centroid for a homogeneous circular arc is calculated from (2.8)

$$x_c \int_{-\alpha}^{\alpha} R d\theta = \int_{-\alpha}^{\alpha} x R d\theta. \tag{2.56}$$

Because $x = R\cos(\theta)$ and with (2.56), one can write

$$x_c \int_{-\alpha}^{\alpha} R d\theta = \int_{-\alpha}^{\alpha} R^2 \cos(\theta) d\theta. \tag{2.57}$$

From (5.19), after integration, it results

$$2x_c \alpha R = 2R^2 \sin(\alpha),$$

or

$$x_c = \frac{R\sin(\alpha)}{\alpha}.$$

For a semicircular arc when $\alpha = \dfrac{\pi}{2}$, the position of the centroid is

$$x_c = \frac{2R}{\pi},$$

and for the quarter-circular $\alpha = \dfrac{\pi}{4}$,

$$x_c = \frac{2\sqrt{2}}{\pi}R.$$

Example 2.3. Find the position of the mass center for the area of a circular sector. The center angle is 2α radians, and the radius is R as shown in Fig. 2.26.

Solution
The origin O is the vertex of the circular sector. The x-axis is chosen as the axis of symmetry and $y_C = 0$. Let $MNPQ$ be a surface differential element with the area $dA = \rho d\rho d\theta$. The mass center of the surface differential element has the abscissa

$$x = \left(\rho + \frac{d\rho}{2}\right)\cos\left(\theta + \frac{d\theta}{2}\right) \approx \rho\cos(\theta). \tag{2.58}$$

Using the first moment of area formula with respect to Oy, (2.15), the mass center abscissa x_C is calculated as

$$x_C \iint \rho d\rho d\theta = \iint x\rho d\rho d\theta. \tag{2.59}$$

Fig. 2.26 Example 2.3

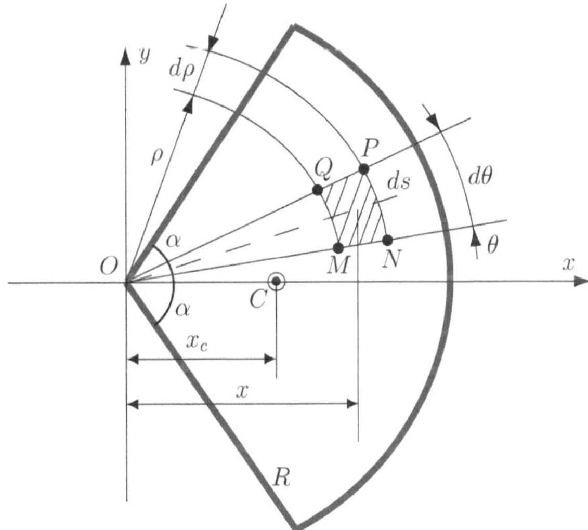

Equations (2.58) and (2.59) give

$$x_C \iint \rho \, \mathrm{d}\rho \, \mathrm{d}\theta = \iint \rho^2 \cos(\theta) \, \mathrm{d}\rho \, \mathrm{d}\theta,$$

or

$$x_C \int_0^R \rho \, \mathrm{d}\rho \int_{-\alpha}^{\alpha} \mathrm{d}\theta = \int_0^R \rho^2 \mathrm{d}\rho \int_{-\alpha}^{\alpha} \cos(\theta) \, \mathrm{d}\theta. \tag{2.60}$$

From (2.60), after integration, it results

$$x_C \left(\frac{1}{2} R^2 \right) (2\alpha) = \left(\frac{1}{3} R^3 \right) (2 \sin(\alpha)),$$

or

$$x_C = \frac{2R \sin(\alpha)}{3\alpha}. \tag{2.61}$$

For a semicircular area, $\alpha = \dfrac{\pi}{2}$, the x-coordinate to the centroid is

$$x_C = \frac{4R}{3\pi}.$$

For the quarter-circular area, $\alpha = \dfrac{\pi}{2}$, the x-coordinate to the centroid is

$$x_C = \frac{4R\sqrt{2}}{3\pi}.$$

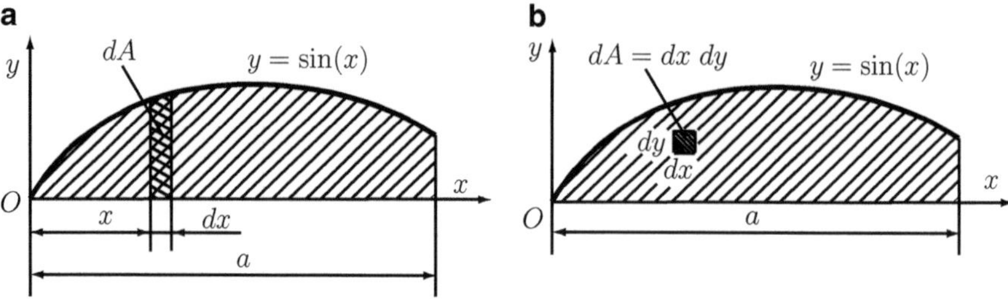

Fig. 2.27 Example 2.4

Example 2.4. Find the coordinates of the mass center for a homogeneous planar plate located under the curve of equation $y = \sin x$ from $x = 0$ to $x = a$.

Solution
A vertical differential element of area $dA = y\,dx = (\sin x)(dx)$ is chosen as shown in Fig. 2.27a. The x-coordinate of the mass center is calculated from (2.15):

$$x_c \int_0^a (\sin x)\,dx = \int_0^a x(\sin x)\,dx \quad \text{or} \quad x_c\{-\cos x\}_0^a = \int_0^a x(\sin x)\,dx, \quad \text{or}$$

$$x_c(1 - \cos a) = \int_0^a x(\sin x)\,dx. \tag{2.62}$$

The integral $\displaystyle\int_0^a x(\sin x)\,dx$ is calculated with

$$\int_0^a x(\sin x)\,dx = \{x(-\cos x)\}_0^a - \int_0^a (-\cos x)\,dx = \{x(-\cos x)\}_0^a + \{\sin x\}_0^a$$

$$= \sin a - a\cos a. \tag{2.63}$$

Using (2.62) and (2.63) after integration, it results

$$x_c = \frac{\sin a - a\cos a}{1 - \cos a}.$$

The x-coordinate of the mass center, x_C, can be calculated using the differential element of area $dA = dx\,dy$, as shown in Fig. 2.27b. The area of the figure is

$$A = \int_A dx\,dy = \int_0^a \int_0^{\sin x} dx\,dy = \int_0^a dx \int_0^{\sin x} dy$$

$$= \int_0^a dx\,\{y\}_0^{\sin x} = \int_0^a (\sin x)\,dx = \{-\cos x\}_0^a = 1 - \cos a.$$

The first moment of the area A about the y-axis is

$$M_y = \int_A x\,dA = \int_0^a \int_0^{\sin x} x\,dx\,dy = \int_0^a x\,dx \int_0^{\sin x} dy$$

$$= \int_0^a x\,dx\,\{y\}_0^{\sin x} = \int_0^a x(\sin x)\,dx = \sin a - a\cos a.$$

The x-coordinate of the mass center is $x_C = M_y/A$. The y-coordinate of the mass center is $y_C = M_x/A$, where the first moment of the area A about the x-axis is

$$M_x = \int_A y\,dA = \int_0^a \int_0^{\sin x} y\,dx\,dy = \int_0^a dx \int_0^{\sin x} y\,dy$$

$$= \int_0^a dx \left\{\frac{y^2}{2}\right\}_0^{\sin x} = \int_0^a \frac{\sin^2 x}{2}\,dx = \frac{1}{2}\int_0^a \sin^2\,dx.$$

The integral $\int_0^a \sin^2 x\,dx$ is calculated with

$$\int_0^a \sin^2 x\,dx = \int_0^a \sin x\,d(-\cos x) = \{\sin x(-\cos x)\}_0^a + \int_0^a \cos^2 x\,dx$$

$$= -\sin a(\cos a) + \int_0^a (1 - \sin^2 x)\,dx = -\sin a(\cos a) + a - \int_0^a \sin^2 x\,dx,$$

or

$$\int_0^a \sin^2 x\,dx = \frac{a - \sin a\cos a}{2}.$$

The coordinate y_C is

$$y_C = \frac{M_x}{A} = \frac{a - \sin a\cos a}{4(1 - \cos a)}.$$

The MATLAB program is given by

```
syms a x y
% dA = dx dy
% A = int dx dy ; 0<x<a 0<y<sin(x)
% Ay = int dy ; 0<y<sin(x)
Ay=int(1,y,0,sin(x))
% A = int Ay dx ; ; 0<x<a
A=int(Ay,x,0,a)

% My = int x dx dy ; 0<x<a 0<y<sin(x)
% Qyy = int dy ; 0<y<sin(x)
Qyy=int(1,y,0,sin(x))
% My = int x Qyy dx ; 0<x<a
My=int(x*Qyy,x,0,a)
```

Fig. 2.28 Example 2.5

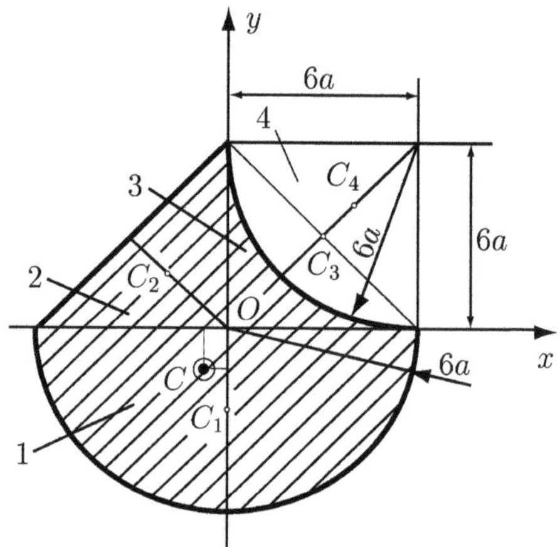

```
% Mx = int y dx dy ; 0<x<a 0<y<sin(x)
% Qxy = int y dy ; 0<y<sin(x)
% Mx = int Qxy dx ; 0<x<a
Qxy=int(y,y,0,sin(x))
Mx=int(Qxy,x,0,a)


xC=My/A;
yC=Mx/A;
pretty(xC)
pretty(yC)
```

Example 2.5. Find the position of the mass center for a homogeneous planar plate ($a = 1$ m), with the shape and dimensions given in Fig. 2.28.

Solution

The plate is composed of four elements: the circular sector area 1, the triangle 2, the square 3, and the circular area 4 to be subtracted. Using the decomposition method, the positions of the mass center x_i, the areas A_i, and the first moments with respect to the axes of the reference frame M_{y_i} and M_{x_i}, for all four elements are calculated. The results are given in the following table:

i	x_i	y_i	A_i	$M_{y_i} = x_i A_i$	$M_{x_i} = y_i A_i$
Circular sector 1	0	$-\dfrac{8}{\pi}a$	$18\pi a^2$	0	$-144a^3$
Triangle 2	$-2a$	$2a$	$18a^2$	$-36a^3$	$36a^3$
Square 3	$3a$	$3a$	$36a^2$	$108a^3$	$108a^3$
Circular sector 4	$6a - \dfrac{8a}{\pi}$	$6a - \dfrac{8a}{\pi}$	$-9\pi a^2$	$-54\pi a^3 + 72a^3$	$-54\pi a^3 + 72a^3$
Σ	$-$	$-$	$9(\pi+6)a^2$	$18(8-3\pi)a^3$	$9(8-6\pi)a^3$

The x and y coordinates of the mass center C are

$$x_c = \frac{\sum x_i A_i}{\sum A_i} = \frac{2(8-3\pi)}{\pi+6} a = -0.311\,a = -0.311 \text{ m},$$

$$y_c = \frac{\sum y_i A_i}{\sum A_i} = \frac{8-6\pi}{\pi+6} a = -1.186\,a = -1.186 \text{ m}.$$

The MATLAB program is

```
syms a

x1=0;
y1=-8*a/pi;
A1=18*pi*a^2;
x1A1=x1*A1;
y1A1=y1*A1;
fprintf('x1 A1 = %s \n', char(x1A1))
fprintf('y1 A1 = %s \n', char(y1A1))

x2=-2*a;
y2=2*a;
A2=18*a^2;
x2A2=x2*A2;
y2A2=y2*A2;
fprintf('x2 A2 = %s \n', char(x2A2))
fprintf('y2 A2 = %s \n', char(y2A2))

x3=3*a;
y3=3*a;
A3=36*a^2;
x3A3=x3*A3;
y3A3=y3*A3;
fprintf('x3 A3 = %s \n', char(x3A3))
fprintf('y3 A3 = %s \n', char(y3A3))

x4=6*a-8*a/pi;
y4=x4;
A4=-9*pi*a^2;
x4A4=simplify(x4*A4);
y4A4=simplify(y4*A4);
fprintf('x4 A4 = %s \n', char(x4A4))
fprintf('y4 A4 = %s \n', char(y4A4))

xC=(x1*A1+x2*A2+x3*A3+x4*A4)/(A1+A2+A3+A4);
xC=simplify(xC);
```

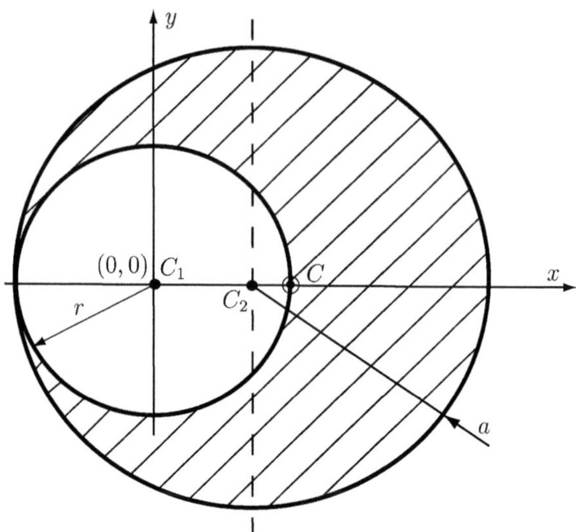

Fig. 2.29 Example 2.6


```
yC=(y1*A1+y2*A2+y3*A3+y4*A4)/(A1+A2+A3+A4);
yC=simplify(yC);

fprintf('xC = %s = %s \n', char(xC), char(vpa(xC,6)))
fprintf('yC = %s = %s \n', char(yC), char(vpa(yC,6)))
```

Example 2.6. The homogeneous plate shown in Fig. 2.29 is delimitated by the hatched area. The circle, with the center at C_1, has the unknown radius r. The circle, with the center at C_2, has the given radius $a = 2$ m, $(r < a)$. The position of the mass center of the hatched area, C, is located at the intersection of the circle, with the radius r and the center at C_1, and the positive x-axis. Find the radius r.

Solution

The x-axis is a symmetry axis for the planar plate, and the origin of the reference frame is located at $C_1 = O(0, 0)$. The y-coordinate of the mass center of the hatched area is $y_C = 0$. The coordinates of the mass center of the circle with the center at C_2 and radius a are $C_2(a - r, 0)$. The area of the of circle C_1 is $A_1 = -\pi r^2$, and the area of the circle C_2 is $A_2 = \pi a^2$. The total area is given by

$$A = A_1 + A_2 = \pi \left(a^2 - r^2 \right).$$

The coordinate x_C of the mass center is given by

$$x_C = \frac{x_{C_1} A_1 + x_{C_2} A_2}{A_1 + A_2} = \frac{x_{C_2} A_2}{A_1 + A_2} = \frac{x_{C_2} A_2}{A} = \frac{(a - r) \pi a^2}{\pi \left(a^2 - r^2 \right)}$$

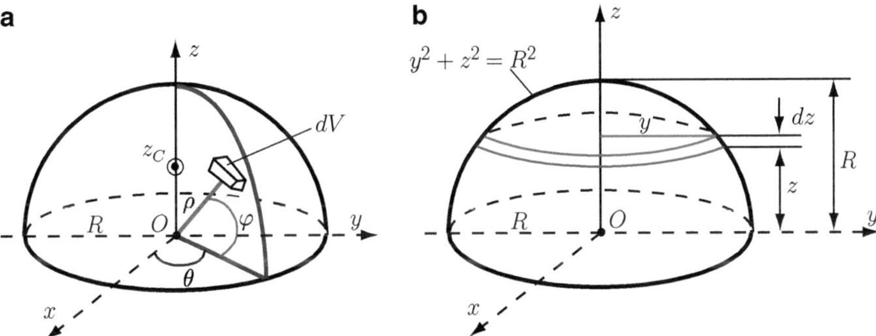

Fig. 2.30 Example 2.7

or

$$x_C\left(a^2 - r^2\right) = a^2\left(a - r\right).$$

If $x_C = r$, the previous equation gives

$$r^2 + ar - a^2 = 0,$$

with the solutions

$$r = \frac{-a \pm a\sqrt{5}}{2}.$$

Because $r > 0$, the correct solution is

$$r = \frac{a\left(\sqrt{5} - 1\right)}{2} \approx 0.62\,a = 0.62\,(2) = 1.24 \text{ m}.$$

Example 2.7. Locate the position of the mass center of the homogeneous volume of a hemisphere of radius R with respect to its base, as shown in Fig. 2.30.

Solution

The reference frame is selected as shown in Fig. 2.30a, and the z-axis is the symmetry axis for the body: $x_C = 0$ and $y_C = 0$. Using the spherical coordinates, $z = \rho \sin\varphi$, and the differential volume element is $dV = \rho^2 \cos\varphi\, d\rho\, d\theta\, d\varphi$. The z coordinate of the mass center is calculated from

$$z_C \iiint \rho^2 \cos\varphi\, d\rho\, d\theta\, d\varphi = \iiint \rho^3 \sin\varphi \cos\varphi\, d\rho\, d\theta\, d\varphi,$$

or

$$z_C \int_0^R \rho^2 d\rho \int_0^{2\pi} d\theta \int_0^{\pi/2} \cos\varphi\, d\varphi = \int_0^R \rho^3 d\rho \int_0^{2\pi} d\theta \int_0^{\pi/2} \sin\varphi \cos\varphi\, d\varphi,$$

or

$$
z_C = \frac{\displaystyle\int_0^R \rho^3 \mathrm{d}\rho \int_0^{2\pi} \mathrm{d}\theta \int_0^{\pi/2} \sin\varphi \cos\varphi \, \mathrm{d}\varphi}{\displaystyle\int_0^R \rho^2 \mathrm{d}\rho \int_0^{2\pi} \mathrm{d}\theta \int_0^{\pi/2} \cos\varphi \, \mathrm{d}\varphi}. \tag{2.64}
$$

From (2.64), after integration, it results

$$
z_C = \frac{3R}{8}.
$$

Another way of calculating the position of the mass center z_C is shown in Fig. 2.30b. The differential volume element is

$$
\mathrm{d}V = \pi y^2 \, \mathrm{d}z = \pi \left(R^2 - z^2\right) \mathrm{d}z,
$$

and the volume of the hemisphere of radius R is

$$
V = \int_V \mathrm{d}V = \int_0^R \pi \left(R^2 - z^2\right) \mathrm{d}z = \pi \left(R^2 \int_0^R \mathrm{d}z - \int_0^R z^2 \, \mathrm{d}z\right) = \pi \left(R^3 - \frac{R^3}{3}\right) = \frac{2\pi R^3}{3}.
$$

The coordinate z_C is calculated from the relation

$$
z_C = \frac{\displaystyle\int_V z \, \mathrm{d}V}{V} = \frac{\pi \left(R^2 \displaystyle\int_0^R z \, \mathrm{d}z - \int_0^R z^3 \, \mathrm{d}z\right)}{V} = \frac{\pi}{V}\left(R^2 \frac{R^2}{2} - \frac{R^4}{4}\right)
$$

$$
= \frac{\pi R^4}{4V} = \frac{\pi R^4}{4}\left(\frac{3}{2\pi R^3}\right) = \frac{3R}{8}.
$$

```
syms rho theta phi R real

% dV = rho^2 cos(phi) drho dtheta dphi
% 0<rho<R 0<theta<2pi 0<phi<pi/2
Ir = int(rho^2, rho, 0, R);
It = int(1, theta, 0, 2*pi);
Ip = int(cos(phi), phi, 0, pi/2);
V = Ir*It*Ip;
fprintf('V = %s \n', char(V))

% dMz = rho rho^2 cos(phi) drho dtheta dphi
% 0<rho<R 0<theta<2pi 0<phi<pi/2
Mr = int(rho^3, rho, 0, R);
Mt = int(1, theta, 0, 2*pi);
Mp = int(sin(phi)*cos(phi), phi, 0, pi/2);
Mz = Mr*Mt*Mp;
fprintf('Mz = %s \n', char(Mz))
```

Fig. 2.31 Example 2.8

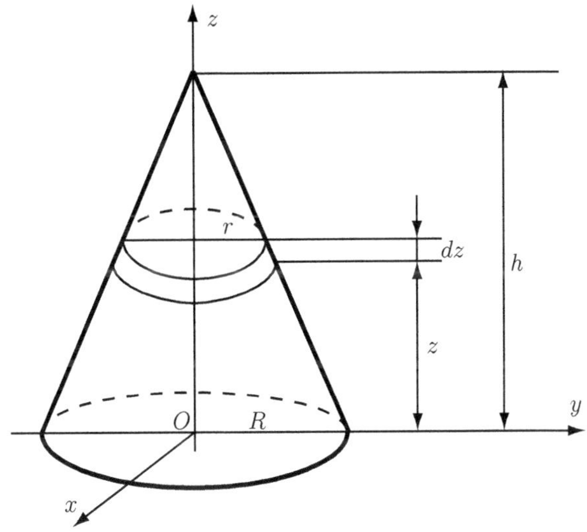


```
zC = Mz/V;
fprintf('zC = Mz/V = %s \n', char(zC))

fprintf('another method \n')
syms y z R real
% y^2 + z^2 = R^2
% dV = pi y^2 dz = pi (R^2-z^2) dz
V = int(pi*(R^2-z^2), z, 0, R);
fprintf('V = %s \n', char(V))
Mxy = int(z*pi*(R^2-z^2), z, 0, R);
fprintf('Mxy = %s \n', char(Mxy))
zC = Mxy/V;
fprintf('zC = Mxy/V = %s \n', char(zC))
```

Example 2.8. Find the position of the mass center for a homogeneous right circular cone, with the base radius R and the height h, as shown in Fig. 2.31.

Solution

The reference frame is shown in Fig. 2.31. The z-axis is the symmetry axis for the right circular cone. By symmetry, $x_C = 0$ and $y_C = 0$. The volume of the thin disk differential volume element is $dV = \pi r^2 \, dz$. From geometry

$$\frac{r}{R} = \frac{h-z}{h},$$

Fig. 2.32 Example 2.9

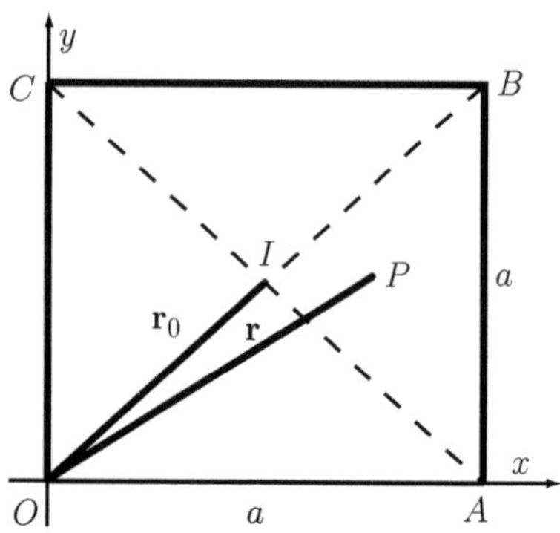

or

$$r = R\left(1 - \frac{z}{h}\right).$$

The z coordinate of the centroid is calculated from

$$z_C \pi R^2 \int_0^h \left(1 - \frac{z}{h}\right)^2 dz = \pi R^2 \int_0^h z\left(1 - \frac{z}{h}\right)^2 dz,$$

or

$$z_C = \frac{\int_0^h z\left(1 - \frac{z}{h}\right)^2 dz}{\int_0^h \left(1 - \frac{z}{h}\right)^2 dz} = \frac{h}{4}.$$

The MATLAB program is

```
syms z h real
V = int((1-z/h)^2, z, 0, h);
Mxy = int(z*(1-z/h)^2, z, 0, h);
zC=Mxy/V;

fprintf('V = %s  \n',char(V))
fprintf('Mxy = %s  \n',char(Mxy))
fprintf('zC = %s  \n',char(zC))
```

Example 2.9. The density of a square plate with the length a is given as $\rho = kr$, where $k = $ constant and r is the distance from the origin O to a current point $P(x, y)$ on the plate as shown in Fig. 2.32. Find the mass M of the plate.

Solution

The mass of the plate is given by

$$M = \iint \rho \, dx \, dy.$$

The density is

$$\rho = k\sqrt{x^2 + y^2}, \tag{2.65}$$

where

$$\rho_0 = k\,r_0 = k\,a\frac{\sqrt{2}}{2} \Rightarrow k = \frac{\sqrt{2}}{a}\rho_0. \tag{2.66}$$

Using (2.65) and (2.66), the density is

$$\rho = \frac{\sqrt{2}}{a}\rho_0\sqrt{x^2 + y^2}. \tag{2.67}$$

Using (2.67), the mass of the plate is

$$
\begin{aligned}
M &= \frac{\sqrt{2}}{a}\rho_0\int_0^a dx\int_0^a \sqrt{x^2 + y^2}\,dy \\
&= \frac{\sqrt{2}}{a}\rho_0\int_0^a \left[\frac{y}{2}\sqrt{x^2+y^2} + \frac{x^2}{2}\ln\left(y + \sqrt{x^2+y^2}\right)\right]dx \\
&= \frac{\sqrt{2}}{a}\rho_0\int_0^a \left[\frac{a}{2}\sqrt{x^2+a^2} + \frac{x^2}{2}\ln\left(\frac{a + \sqrt{x^2+a^2}}{x}\right)\right]dx \\
&= \frac{\sqrt{2}}{a}\rho_0\left\{\frac{a}{2}\int_0^a \sqrt{x^2+a^2}\,dx + \frac{1}{2}\int_0^a x^2\ln\left(\frac{a + \sqrt{x^2+a^2}}{x}\right)dx\right\} \\
&= \left\{\frac{\sqrt{2}}{a}\rho_0\frac{a}{2}\left[\frac{x}{2}\sqrt{x^2+a^2} + \frac{a^2}{2}\ln\left(a + \sqrt{x^2+a^2}\right)\right]\right\}_0^a \\
&\quad + \left\{\frac{1}{2}\frac{\sqrt{2}}{a}\rho_0\left[\ln\left(\frac{a+\sqrt{x^2+a^2}}{x}\right) + \frac{a}{3}\frac{x}{2}\sqrt{x^2+a^2} - \frac{a}{3}\frac{a^2}{2}\ln\left(a+\sqrt{x^2+a^2}\right)\right]\right\}_0^a \\
&= \frac{1}{2}\frac{\sqrt{2}}{a}\rho_0\left[\ln\left(\frac{a+\sqrt{x^2+a^2}}{x}\right) + \frac{a}{3}\frac{x}{2}\sqrt{x^2+a^2} - \frac{a}{3}\frac{a^2}{2}\ln\left(a+\sqrt{x^2+a^2}\right)\right] \\
&= \frac{\rho_0 a^2}{3}\left[2 + \sqrt{2}\ln\left(1 + \sqrt{2}\right)\right].
\end{aligned}
$$

Example 2.10. Revolving the circular area of radius R through $360°$ about the x-axis, a complete torus is generated. The distance between the center of the circle and the x-axis is d, as shown in Fig. 2.33. Find the surface area and the volume of the obtained torus.

Solution
Using the Guldinus–Pappus formulas

$$A = 2\pi y_C L,$$

Fig. 2.33 Example 2.10

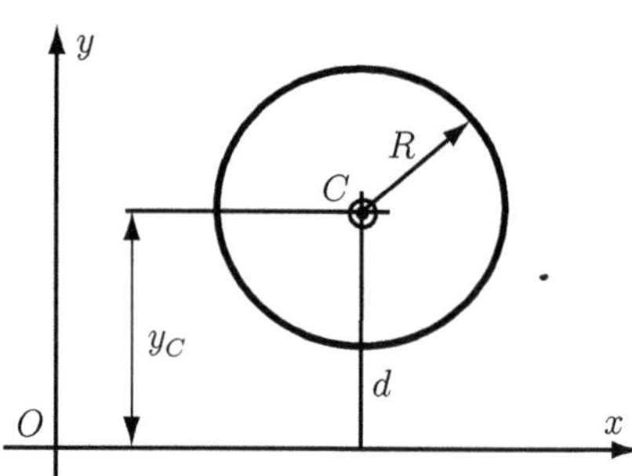

Fig. 2.34 Example 2.11

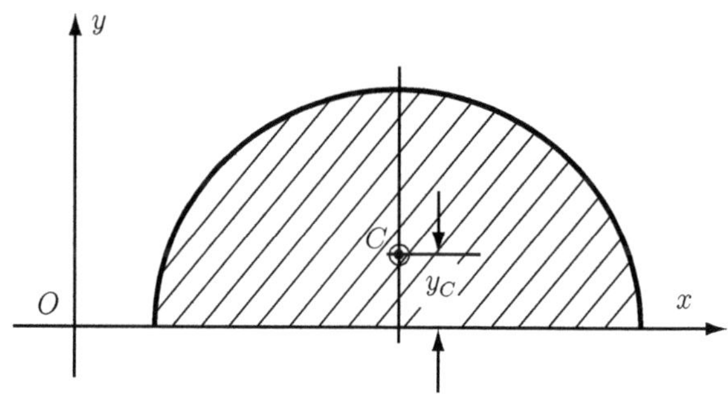

$$V = 2\pi y_C A,$$

and with $y_C = d$ the area and the volume are

$$A = (2\pi d)(2\pi R) = 4\pi^2 R d,$$

$$V = (2\pi d)(\pi R^2) = 2\pi^2 R^2 d.$$

Example 2.11. Find the position of the mass center for the semicircular area shown in Fig. 2.34.

Solution

Rotating the semicircular area with respect to x-axis, a sphere is obtained. The volume of the sphere is given by

$$V = \frac{4\pi R^3}{3}.$$

The area of the semicircular area is

$$A = \frac{\pi R^2}{2}.$$

Fig. 2.35 Example 2.12

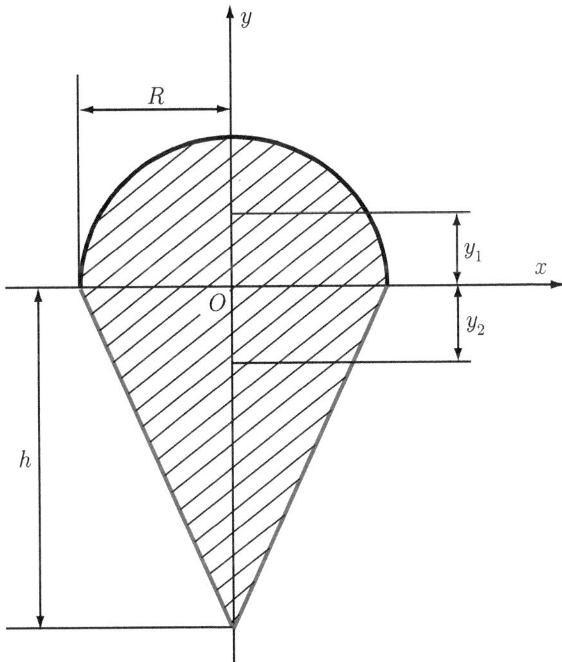

Using the second Guldinus–Pappus theorem, the position of the mass center is

$$y_C = \frac{V}{2\pi A} = \frac{\dfrac{4\pi R^3}{3}}{2\pi \dfrac{\pi R^2}{2}} = \frac{\dfrac{4\pi R^3}{3}}{\pi^2 R^2} = \frac{4R}{3\pi}.$$

Example 2.12. Find the first moment of the area with respect to the *x*-axis for the surface given in Fig. 2.35.

Solution
The composite area is considered to be composed of the semicircular area and the triangle. The mass center of the semicircular surface is given by

$$y_1 = \frac{2}{3}\frac{R\sin\dfrac{\pi}{2}}{\dfrac{\pi}{2}} = \frac{4R}{3\pi}.$$

The mass center of the triangle is given by

$$y_2 = -\frac{b}{3}.$$

Fig. 2.36 Example 2.13

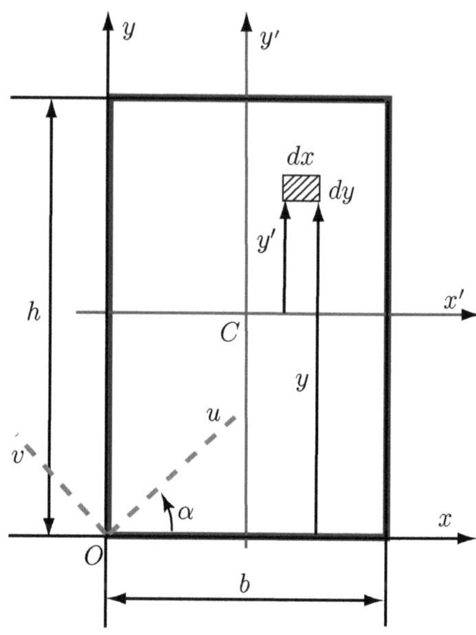

The first moment of area with respect to x-axis for the total surface is given by

$$M_x = \sum_{i=1}^{2} y_i A_i = \frac{4R}{3\pi}\frac{\pi R^2}{2} - \frac{h}{3}\frac{2Rh}{2},$$

or

$$M_x = \frac{R}{3}\left(2R^2 - h^2\right).$$

Example 2.13. A rectangular planar plate with the sides $b = 1$ *m* and $h = 2$ *m* is shown in Fig. 2.36.

(a) Find the product of inertia and the moments of inertia with respect to the axes of the reference frame xy with the origin at O.
(b) Determine the product of inertia and the moments of inertia with respect to the centroidal axes that are located at the mass center C of the rectangle and are parallel to its sides.
(c) Another reference uv with the same origin O is rotated with an angle $\alpha = 45°$ from xy (counterclockwise as positive). Find the inertia matrix of the plate with respect to uv axes.
(d) Find the principal moments and the principal directions with the reference frame xy with the origin at O.

Solution

(a) The differential element of area is $dA = dx\,dy$. The product of inertia of the rectangle about the xy axes is

$$I_{xy} = \int_A xy\,dA = \int_0^h \int_0^b xy\,dx\,dy = \int_0^b x\,dx \int_0^h y\,dy = \frac{b^2}{2}\frac{h^2}{2} = \frac{b^2 h^2}{4} = \frac{1^2\left(2^2\right)}{4} = 1 \text{ m}^4.$$

The moment of inertia of the rectangle about x-axis is

$$I_{xx} = \int_A y^2\,dA = \int_0^h \int_0^b y^2\,dx\,dy = \int_0^b dx \int_0^h y^2\,dy = b\frac{h^3}{3} = \frac{bh^3}{3} = \frac{(1)2^3}{3} = 2.666 \text{ m}^4.$$

The moment of inertia of the rectangle about y-axis is

$$I_{yy} = \int_A x^2\,dA = \int_0^h \int_0^b x^2\,dx\,dy = \int_0^b x^2\,dx \int_0^h dy = \frac{b^3}{3}h = \frac{hb^3}{3} = \frac{(2)1^3}{3} = 0.666 \text{ m}^4.$$

The moment of inertia of the rectangle about z-axis (the polar moment about O) is

$$I_O = I_{zz} = I_{xx} + I_{yy} = \frac{A}{3}\left(b^2 + h^2\right) = \frac{1(2)}{3}\left(1^2 + 2^2\right) = 3.33 \text{ m}^4.$$

The inertia matrix of the plane figure with respect to xy axes is represented by

$$[I] = \begin{bmatrix} I_{xx} & -I_{xy} & -I_{xz} \\ -I_{yx} & I_{yy} & -I_{yz} \\ -I_{zx} & -I_{zy} & I_{zz} \end{bmatrix} = \begin{bmatrix} \dfrac{bh^3}{3} & \dfrac{b^2h^2}{4} & 0 \\[2ex] \dfrac{b^2h^2}{4} & \dfrac{hb^3}{3} & 0 \\[2ex] 0 & 0 & \dfrac{A}{3}(b^2 + h^2) \end{bmatrix}$$

$$= \begin{bmatrix} 2.666 & 1 & 0 \\ 1 & 0.666 & 0 \\ 0 & 0 & 3.33 \end{bmatrix}.$$

(b) The product of inertia of the rectangle about the $x'y'$ axes is

$$I_{x'y'} = \int_A xy\,dA = \int_{-h/2}^{h/2} \int_{-b/2}^{b/2} xy\,dx\,dy = \int_{-h/2}^{h/2} x\,dx \int_{-b/2}^{b/2} y\,dy = 0.$$

The same results is obtained using the parallel-axis theorem

$$I_{xy} = I_{x'y'} + \frac{b}{2}\frac{h}{2}A,$$

or

$$I_{x'y'} = I_{xy} - \frac{bh}{4}(bh) = \frac{b^2h^2}{4} - \frac{b^2h^2}{4} = 0.$$

The moment of inertia of the rectangle about x' axis is

$$I_{x'x'} = \int_A y^2 \, dA = \int_{-h/2}^{h/2} \int_{-b/2}^{b/2} y^2 \, dx \, dy = \int_{-h/2}^{h/2} dx \int_{-b/2}^{b/2} y^2 \, dy$$

$$= \{x\}_{-h/2}^{h/2} \left\{ \frac{y^3}{3} \right\}_{-b/2}^{b/2} = \frac{bh^3}{12} = \frac{(1)\,2^3}{12} = 0.666 \text{ m}^4.$$

Using the parallel-axis theorem, the moment of inertia of the rectangle about y' axis is

$$I_{y'y'} = I_{yy} - \left(\frac{b}{2}\right)^2 A = \frac{hb^3}{3} - \frac{hb^3}{4} = \frac{hb^3}{12} = \frac{(2)\,1^3}{12} = 0.166 \text{ m}^4.$$

The moment of inertia of the rectangle about z' axis (the centroid polar moment) is

$$I_C = I_{z'z'} = I_{x'x'} + I_{y'y'} = \frac{A}{12}(b^2 + h^2) = \frac{1(2)}{12}(1^2 + 2^2) = 0.833 \text{ m}^4.$$

The inertia matrix of the plane figure with respect to centroidal axes $x'y'$ is represented by

$$[I_C] = \begin{bmatrix} I_{x'x'} & -I_{x'y'} & -I_{x'z'} \\ -I_{y'x'} & I_{y'y'} & -I_{y'z'} \\ -I_{z'x'} & -I_{z'y'} & I_{z'z'} \end{bmatrix} = \begin{bmatrix} \dfrac{bh^3}{12} & 0 & 0 \\ 0 & \dfrac{hb^3}{12} & 0 \\ 0 & 0 & \dfrac{A}{12}(b^2 + h^2) \end{bmatrix}$$

$$= \begin{bmatrix} 0.666 & 0 & 0 \\ 0 & 1.666 & 0 \\ 0 & 0 & 0.833 \end{bmatrix}.$$

(c) The moment of inertia of the rectangle about u-axis is

$$I_{uu} = \frac{I_{xx} + I_{yy}}{2} + \frac{I_{xx} - I_{yy}}{2} \cos 2\alpha - I_{xy} \sin 2\alpha$$

$$= \frac{2.666 + 0.666}{2} + \frac{2.666 - 0.666}{2} \cos 2(45°) - (1) \sin 2(45°) = 0.666 \text{ m}^4.$$

The moment of inertia of the rectangle about v-axis is

$$I_{vv} = \frac{I_{xx} + I_{yy}}{2} - \frac{I_{xx} - I_{yy}}{2} \cos 2\alpha + I_{xy} \sin 2\alpha$$

$$= \frac{2.666 + 0.666}{2} - \frac{2.666 - 0.666}{2} \cos 2(45°) + (1) \sin 2(45°) = 2.666 \text{ m}^4.$$

The product of inertia of the rectangle about uv axes is

$$
\begin{aligned}
I_{uv} &= \frac{I_{xx} - I_{yy}}{2}\sin 2\alpha + I_{xy}\cos 2\alpha \\
&= \frac{2.666 - 0.666}{2}\sin 2(45^\circ) + (1)\cos 2(45^\circ) = 1 \text{ m}^4.
\end{aligned}
$$

The polar moment of inertia of the rectangle about O is

$$
I_O = I_{zz} = I_{uu} + I_{vv} = I_{xx} + I_{yy} = 0.666 + 2.666 = 3.33 \text{ m}^4.
$$

The inertia matrix of the plane figure with respect to uv axes is

$$
[I\alpha] = \begin{bmatrix} I_{uu} & -I_{uv} & 0 \\ -I_{vu} & I_{vv} & 0 \\ 0 & 0 & I_{zz} \end{bmatrix} = \begin{bmatrix} 0.666 & 1 & 0 \\ 1 & 2.666 & 0 \\ 0 & 0 & 3.33 \end{bmatrix}.
$$

(d) The maximum or minimum moment of inertia for the area is

$$
I_{1,2} = I_{\max,\min} = \frac{I_{xx} + I_{yy}}{2} \pm \sqrt{\left(\frac{I_{xx} - I_{yy}}{2}\right)^2 + I_{xy}^2},
$$

$$
\begin{aligned}
I_1 = I_{\max} &= \frac{I_{xx} + I_{yy}}{2} + \sqrt{\left(\frac{I_{xx} - I_{yy}}{2}\right)^2 + I_{xy}^2} \\
&= \frac{2.666 + 0.666}{2} + \sqrt{\left(\frac{2.666 - 0.666}{2}\right)^2 + 1^2} = 3.080 \text{ m}^4,
\end{aligned}
$$

$$
\begin{aligned}
I_2 = I_{\min} &= \frac{I_{xx} + I_{yy}}{2} - \sqrt{\left(\frac{I_{xx} - I_{yy}}{2}\right)^2 + I_{xy}^2} \\
&= \frac{2.666 + 0.666}{2} - \sqrt{\left(\frac{2.666 - 0.666}{2}\right)^2 + 1^2} = 0.252 \text{ m}^4.
\end{aligned}
$$

The polar moment of inertia of the rectangle about O is

$$
I_O = I_{zz} = I_1 + I_2 = I_{uu} + I_{vv} = I_{xx} + I_{yy} = 3.0806 + 0.252 = 3.33 \text{ m}^4.
$$

The principal directions are obtained from

$$
\tan 2\alpha_0 = \frac{2I_{xy}}{I_{yy} - I_{xx}},
$$

or

$$\alpha_0 = \frac{1}{2}\tan^{-1}\frac{2I_{xy}}{I_{yy}-I_{xx}} = \frac{1}{2}\tan^{-1}\frac{2(1)}{0.666-2.666} = -22.5°.$$

The principal directions are

$$\alpha_1 = -22.5° \quad \text{and} \quad \alpha_2 = \alpha_1 + \pi/2 = 67.5°.$$

The MATLAB program for this example is

```
syms b h x y real

list={b, h};
listn={1, 2}; % m

% dA = dx dy
% A = int dx dy ; 0<x<b 0<y<h
% Ax = int dx ; 0<x<b
% Ay = int dy ; 0<y<h

Ax=int(1,x,0,b);
Ay=int(1,y,0,h);
A=Ax*Ay;

% Ixy = int x y dx dy ; 0<x<b 0<y<h
% Ixy1 = int x dx ; 0<x<b
% Ixy2 = int y dy ; 0<y<h
Ixy1=int(x, x, 0, b);
Ixy2=int(y, y, 0, h);
Ixy=Ixy1*Ixy2;
Ixyn=subs(Ixy, list, listn);
fprintf('Ixy = %s = %g (m^4) \n',char(Ixy), Ixyn);

% Ixx = int y^2 dx dy ; 0<x<b 0<y<h
% Ixx1 = int dx ; 0<x<b
% Ixx2 = int y^2 dy ; 0<y<h
Ixx1=int(1, x, 0, b);
Ixx2=int(y^2, y, 0, h);
Ixx=Ixx1*Ixx2;
Ixxn=subs(Ixx, list, listn);
fprintf('Ixx = %s = %g (m^4) \n',char(Ixx), Ixxn);

% Iyy = int x^2 dx dy ; 0<x<b 0<y<h
% Iyy1 = int dx ; 0<x<b
% Ixy2 = int y^2 dy ; 0<y<h
```

```
Iyy1=int(x^2, x, 0, b);
Iyy2=int(1, y, 0, h);
Iyy=Iyy1*Iyy2;
Iyyn=subs(Iyy, list, listn);
fprintf('Iyy = %s = %g (m^4) \n',char(Iyy), Iyyn);


IO=Ixx+Iyy;
IOn=Ixxn+Iyyn;
fprintf('IO = %s = %g (m^4) \n',char(IO), IOn);


% Ixxp = int y^2 dx  dy ; -b/2<x<b/2 ; -h/2<y<h/2
% Ixx1p = int dx ; -b/2<x<b/2
% Ixx2p = int y^2 dy ; -h/2<y<h/2
Ixx1p=int(1, x, -b/2, b/2);
Ixx2p=int(y^2, y, -h/2, h/2);
Ixxp=Ixx1p*Ixx2p;
Ixxpn=subs(Ixxp, list, listn);
fprintf('Ixxp = %s = %g (m^4) \n',char(Ixxp),
  Ixxpn);


Iyyp=Iyy-(b*h)*(b/2)^2;
Iyypn=subs(Iyyp, list, listn);
fprintf('Iyyp = %s = %g (m^4) \n',char(Iyyp),
  Iyypn);


Ixyp=Ixy-(b*h)*(-b/2)*(-h/2);
Ixypn=subs(Ixyp, list, listn);
fprintf('Ixyp = %s = %g (m^4) \n',char(Ixyp),
  Ixypn);


IC=Ixxp+Iyyp;
ICn=Ixxpn+Iyypn;
fprintf('IC = %s = %g (m^4) \n',char(IC), ICn);


fprintf('\n');
alpha=pi/4;
fprintf('alpha = %g (degree) \n', alpha*180/pi);
fprintf('\n');


Ixa=...
(Ixx+Iyy)/2+(Ixx-Iyy)/2*cos(2*alpha)
  -Ixy*sin(2*alpha);
Ixan=subs(Ixa, list, listn);
fprintf('Iuu45 = %g (m^4) \n', Ixan);
```

```
Iya=...
(Ixx+Iyy)/2-(Ixx-Iyy)/2*cos(2*alpha)
  +Ixy*sin(2*alpha);
Iyan=subs(Iya, list, listn);
fprintf('Ivv45 = %g (m^4) \n', Iyan);


Ixya=(Ixx-Iyy)/2*sin(2*alpha)+Ixy*cos(2*alpha);
Ixyan=subs(Ixya, list, listn);
fprintf('Iuv45 = %g (m^4) \n', Ixyan);


IOa=Ixan+Iyan;
fprintf('IO45 = Iuu45+Ivv45 = %g (m^4) \n', IOa);


fprintf('\n');
I1=(Ixx+Iyy)/2+sqrt((Ixx-Iyy)^2/4+Ixy^2);
I1n=subs(I1, list, listn);
I2=(Ixx+Iyy)/2-sqrt((Ixx-Iyy)^2/4+Ixy^2);
I1n=subs(I1, list, listn);
I2n=subs(I2, list, listn);
fprintf('I1 = Imax = %g (m^4) \n', I1n);
fprintf('I2 = Imin = %g (m^4) \n', I2n);
fprintf('I1+I2 = %g (m^4) \n', I1n+I2n);


fprintf('\n');
tanalpha0=simplify(2*Ixy/(Iyy-Ixx));
fprintf('tan (2 alpha0) = %s \n',char(tanalpha0));
alpha0n=atan(2*Ixyn/(Iyyn-Ixxn));
fprintf('alpha0 = %g (degrees) \n',
  alpha0n/2*180/pi);
```

Example 2.14. Determine the moment of inertia for the slender rod, shown in Fig. 2.37, with respect to axes of reference with the origin at the end O and with respect to centroidal axes. The length of the rod is l, the density is ρ, and the cross-sectional area ia A. Express the results in terms of the total mass, m, of the rod.

Solution

The mass of the rod is $m = \rho\, lA$, and the density will be $\rho = m/(lA)$. The differential element of mass is $dm = \rho\, A\, dx$. The moment of inertia of the slender rod about the y- or z-axes is

$$I_{yy} = I_{zz} = \int_0^l x^2 \rho\, A\, dx = \int_0^l \frac{m}{lA} A x^2\, dx = \frac{m}{l} \int_0^l x^2\, dx = \frac{m l^2}{3}.$$

The x-axis is a symmetry axis and that is why $I_{xx} = 0$. The moment of inertia of the slender rod about the centroidal axes y' or z' is calculated with the parallel-axis theorem

Fig. 2.37 Example 2.14

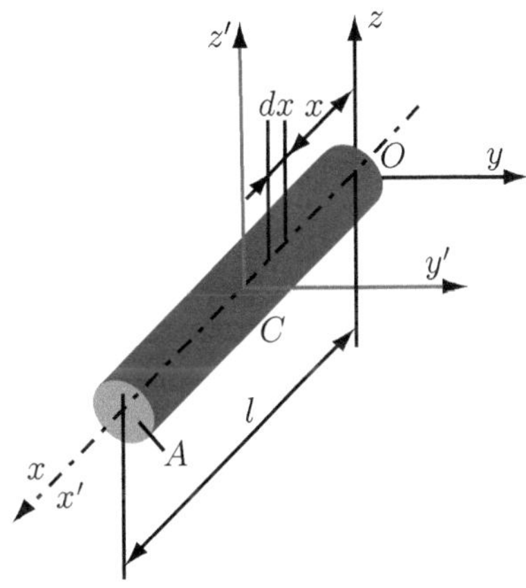

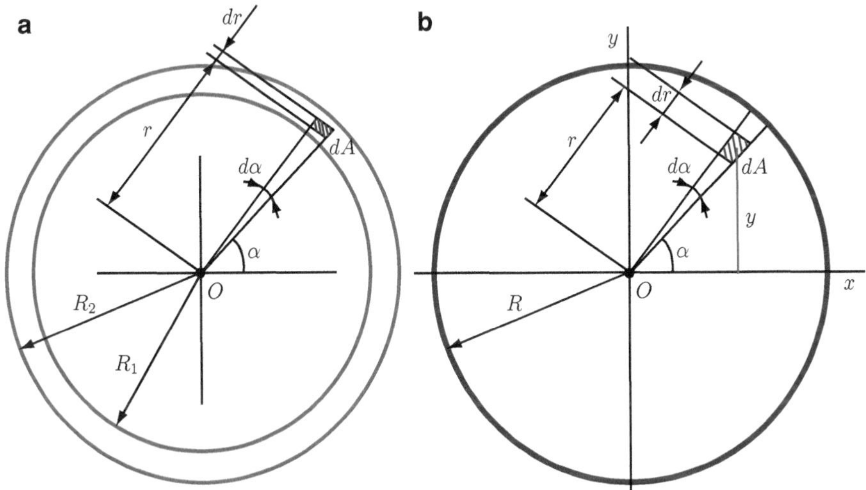

Fig. 2.38 Example 2.15

$$I_{y'y'} = I_{z'z'} = I_{yy} - \left(\frac{l}{2}\right)^2 m = \frac{ml^2}{3} - \frac{ml^2}{4} = \frac{ml^2}{12}.$$

Example 2.15. Find the polar moment of inertia of the planar flywheel shown in Fig. 2.38a. The radii of the wheel are R_1 and R_2 ($R_1 < R_2$). Calculate the moments of inertia of the area of a circle with radius R about a diametral axis and about the polar axis through the center as shown in Fig. 2.38b.

Solution

The polar moment of inertia is given by the equation

$$I_O = \int_A r^2 \, dA,$$

where r is the distance from the pole O to an arbitrary point on the wheel, and the differential element of area is

$$dA = r \, d\alpha \, dr.$$

The polar moment of inertia is

$$I_O = \int_{R_1}^{R_2} \int_0^{2\pi} r^3 \, dr \, d\alpha = \int_{R_1}^{R_2} r^3 \, dr \int_0^{2\pi} d\alpha = \frac{R_2^4 - R_1^4}{4}(2\pi) = \frac{R_2^4 - R_1^4}{2}\pi.$$

The area of the wheel is $A = \pi(R_2^2 - R_1^2)$ and

$$I_O = A\frac{R_2^2 + R_1^2}{2}. \tag{2.68}$$

If $R_1 = 0$ and $R_2 = R$, the polar moment of inertia of the circular area of radius R is, Fig. 2.38b,

$$I_O = \frac{AR^2}{2} = \frac{\pi R^4}{2}.$$

By symmetry for the circular area, shown in Fig. 2.38b, the moment of inertia about a diametral axis is $I_{xx} = I_{yy}$ and $I_O = I_{xx} + I_{yy} \implies I_{xx} = I_{yy} = I_O/2 = \pi R^4/4$. The results can be obtained using the integration

$$I_{xx} = \int_A y^2 \, dA = \int_0^R \int_0^{2\pi} (r\sin\alpha)^2 \, r \, dr \, d\alpha = \int_0^{2\pi} \frac{R^4}{4}(\sin\alpha)^2 \, d\alpha$$

$$= \frac{R^4}{4}\frac{1}{2}\left\{\alpha - \frac{\sin 2\alpha}{2}\right\}_0^{2\pi} = \frac{\pi R^4}{4}.$$

Example 2.16. Find the moments of inertia and products of inertia for the area shown in Fig. 2.39a, with respect to the xy axes and with respect to the centroidal $x'y'$ axes that pass through the mass center C. Find the principal moments of inertia for the area and the principal directions.

Solution

The plate is composed of two element area: the rectangular area 1 and the rectangular area 2, Fig. 2.39b. The x and y coordinates of the mass center C are

$$x_C = \frac{x_{C_1} A_1 + x_{C_1} A_1}{A_1 + A_2} = \frac{(a/2)(2a^2) + (2a)(2a^2)}{2a^2 + 2a^2} = \frac{5a}{4}$$

$$y_C = \frac{y_{C_1} A_1 + y_{C_1} A_1}{A_1 + A_2} = \frac{(a)(2a^2) + (a/2)(2a^2)}{2a^2 + 2a^2} = \frac{3a}{4}.$$

Fig. 2.39 Example 2.16

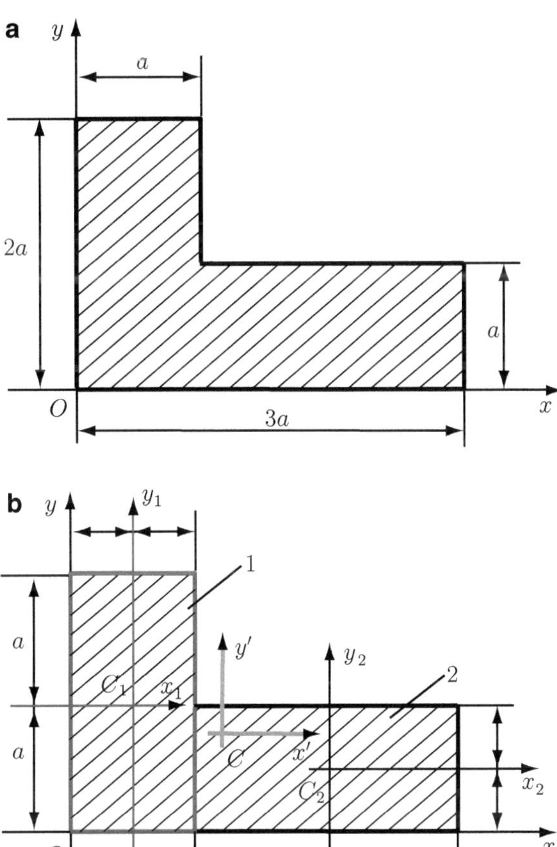

The product of inertia for the area shown in Fig. 2.39a about xy axes is given by

$$I_{xy} = \int_0^a \int_0^{2a} xy\,dxdy + \int_a^{3a} \int_0^a xy\,dxdy = \int_0^a x\,dx \int_0^{2a} y\,dy + \int_a^{3a} x\,dx \int_0^a y\,dy$$

$$= \frac{a^2\left(4a^2\right)}{4} + \frac{\left(9a^2 - a^2\right)\left(a^2\right)}{4} = 3a^4.$$

The same result is obtained if parallel-axis theorem is used:

$$I_{xy} = I_{C_1x_1y_1} + x_{C_1}y_{C_1}A_1 + I_{C_2x_2y_2} + x_{C_2}y_{C_2}A_2$$

$$= 0 + \left(\frac{a}{2}\right)(a)(2a^2) + 0 + (2a)\left(\frac{a}{2}\right)(2a^2) = 3a^4.$$

The product of inertia for the area about xz and yz axes are $I_{xz} = I_{yz} = 0$.

The moment of inertia of the figure with respect to x-axis is

$$I_{xx} = I_{C_1 x_1 x_1} + (y_{C_1})^2 A_1 + I_{C_2 x_2 x_2} + (y_{C_2})^2 A_2$$
$$= \frac{a(2a)^3}{12} + a^2(2a^2) + \frac{2a(a)^3}{12} + \left(\frac{a}{2}\right)^2 (2a^2) = \frac{10a^4}{3}.$$

The moment of inertia of the figure with respect to y-axis is

$$I_{yy} = I_{C_1 y_1 y_1} + (x_{C_1})^2 A_1 + I_{C_2 y_2 y_2} + (x_{C_2})^2 A_2$$
$$= \frac{(2a)a^3}{12} + \left(\frac{a}{2}\right)^2 (2a^2) + \frac{a(2a)^3}{12} + (2a)^2 (2a^2) = \frac{28a^4}{3}.$$

The moment of inertia of the area with respect to z-axis is

$$I_{zz} = I_{xx} + I_{yy} = \frac{10a^4}{3} + \frac{28a^4}{3} = \frac{38a^4}{3}.$$

The inertia matrix of the plane figure is represented by the matrix

$$[I] = \begin{bmatrix} I_{xx} & -I_{xy} & -I_{xz} \\ -I_{yx} & I_{yy} & -I_{yz} \\ -I_{zx} & -I_{zy} & I_{zz} \end{bmatrix} = \begin{bmatrix} \dfrac{10a^4}{3} & -3a^4 & 0 \\ -3a^4 & \dfrac{28a^4}{3} & 0 \\ 0 & 0 & \dfrac{38a^4}{3} \end{bmatrix}.$$

Using the parallel-axis theorem, the moments of inertia of the area with respect to the centroidal axes $x'y'z'$ are

$$I_{x'x'} = I_{xx} - (x_C)^2 A = \frac{10a^4}{3} - \left(\frac{3a}{4}\right)^2 (4a^2) = \frac{13a^4}{12},$$

$$I_{y'y'} = I_{yy} - (y_C)^2 A = \frac{28a^4}{3} - \left(\frac{5a}{4}\right)^2 (4a^2) = \frac{37a^4}{12},$$

$$I_{z'z'} = I_{x'x'} + I_{y'y'} = \frac{13a^4}{12} + \frac{37a^4}{12} = \frac{25a^4}{6},$$

$$I_{x'y'} = I_{xy} - (x_C)(y_C)A = 3a^4 - \left(\frac{3a}{4}\right)\left(\frac{5a}{4}\right)(4a^2) = -\frac{3a^4}{4},$$

$$I_{x'z'} = I_{y'z'} = 0.$$

The centroidal inertia matrix of the plane figure is

$$
[I'] = \begin{bmatrix} I_{x'x'} & -I_{x'y'} & -I_{x'z'} \\ -I_{y'x'} & I_{y'y'} & -I_{y'z'} \\ -I_{z'x'} & -I_{z'y'} & I_{z'z'} \end{bmatrix} = \begin{bmatrix} \dfrac{13a^4}{12} & \dfrac{3a^4}{4} & 0 \\ \dfrac{3a^4}{4} & \dfrac{37a^4}{12} & 0 \\ 0 & 0 & \dfrac{25a^4}{6} \end{bmatrix}.
$$

The principal moments of inertia for the area are

$$
I_{1,2} = I_{\max,\min} = \frac{I_{x'x'} + I_{y'y'}}{2} \pm \sqrt{\left(\frac{I_{x'x'} - I_{y'y'}}{2}\right)^2 + I_{x'y'}^2}
$$

$$
= \frac{25a^4}{12} \pm \frac{a^4}{2}\sqrt{4 + 9/4} = \frac{25a^4}{12} \pm \frac{5a^4}{4}.
$$

$$
I_1 = \frac{10a^4}{3} \quad \text{and} \quad I_2 = \frac{5a^4}{6}.
$$

The invariant of the system is $I_{x'x'} + I_{y'y'} = I_1 + I_2$. The principal directions are obtained from

$$
\tan 2\alpha = \frac{2I_{x'y'}}{I_{y'y'} - I_{x'x'}} = \frac{-2\dfrac{3a^4}{4}}{\dfrac{24a^4}{12}} = -\frac{3}{4}
$$

$$
\implies \alpha_1 = \frac{1}{2}\tan^{-1}(-3/4) = -36.869° \quad \text{and} \quad \alpha_2 = \frac{\pi}{2} + \frac{1}{2}\tan^{-1}(-3/4) = 53.131°.
$$

Example 2.17. Find the inertia matrix of the area delimited by the curve $y^2 = 2px$, from $x = 0$ to $x = a$ as shown in Fig 2.40a, about the axes of the Cartesian frame with the origin at O. Calculate the centroidal inertia matrix.

Solution
From Fig. 2.40a, when $x = a$, the value of y-coordinate is $y = b$ and $b^2 = 2pa \implies 2p = b^2/a$. The expression of the function is

$$
y^2 = 2px = \frac{b^2}{a}x.
$$

The differential element of area is $dA = dx\,dy$, and the area of the figure is

$$
A = \int_A dx\,dy = \int_0^a \int_{-\sqrt{2px}}^{\sqrt{2px}} dx\,dy = \int_0^a dx \int_{-\sqrt{2px}}^{\sqrt{2px}} dy
$$

$$
= \int_0^a dx\,\{y\}_{-\sqrt{2px}}^{\sqrt{2px}} = \int_0^a 2\sqrt{2px}\,dx = 2\sqrt{2p}\int_0^a x^{1/2}\,dx = \frac{2b}{\sqrt{a}}\left\{\frac{x^{3/2}}{3/2}\right\}_0^a = \frac{4ab}{3}.
$$

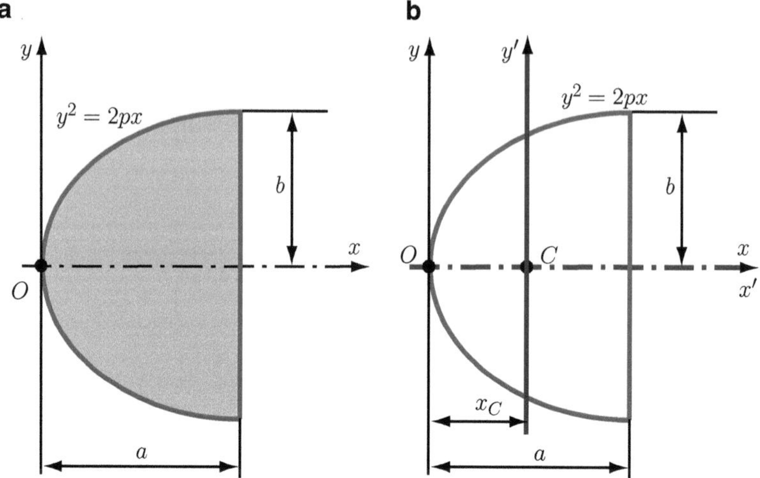

Fig. 2.40 Example 2.17

The moment of inertia of the area with respect to x-axis is

$$I_{xx} = \int_A y^2 \, \mathrm{d}x \, \mathrm{d}y = \int_0^a \int_{-\sqrt{2px}}^{\sqrt{2px}} y^2 \, \mathrm{d}x \, \mathrm{d}y = \int_0^a \mathrm{d}x \int_{-\sqrt{2px}}^{\sqrt{2px}} y^2 \mathrm{d}y$$

$$= \int_0^a \mathrm{d}x \left\{ \frac{y^3}{3} \right\}_{-\sqrt{2px}}^{\sqrt{2px}} = \frac{2}{3} \int_0^a (2px)^{3/2} \, \mathrm{d}x$$

$$= \frac{2}{3} (2p)^{3/2} \int_0^a x^{3/2} \mathrm{d}x = \frac{2}{3} (2p)^{3/2} \left\{ \frac{x^{5/2}}{5/2} \right\}_0^a = \frac{4 a b^3}{15} = \frac{4 a b}{3} \frac{b^2}{5},$$

or

$$I_{xx} = \frac{b^2 A}{5}.$$

The moment of inertia of the area with respect to y-axis is

$$I_{yy} = \int_A x^2 \, \mathrm{d}x \, \mathrm{d}y = \int_0^a \int_{-\sqrt{2px}}^{\sqrt{2px}} x^2 \, \mathrm{d}x \, \mathrm{d}y = \int_0^a x^2 \, \mathrm{d}x \int_{-\sqrt{2px}}^{\sqrt{2px}} \mathrm{d}y$$

$$= \int_0^a x^2 \, \mathrm{d}x \, \{y\}_{-\sqrt{2px}}^{\sqrt{2px}} = 2 \int_0^a x^2 \, (2px)^{1/2} \, \mathrm{d}x$$

$$= \frac{2b}{\sqrt{a}} \int_0^a x^{5/2} \mathrm{d}x = \frac{2b}{\sqrt{a}} \left\{ \frac{x^{7/2}}{7/2} \right\}_0^a = \frac{4 a^3 b}{7} = \frac{4 a b}{3} \frac{3 a^2}{7},$$

or

$$I_{yy} = \frac{3a^2 A}{7}.$$

The moment of inertia of the area with respect to z-axis is

$$I_{zz} = I_{xx} + I_{yy} = \frac{b^2 A}{5} + \frac{3a^2 A}{7} = A\left(\frac{b^2}{5} + \frac{3a^2}{7}\right).$$

The product of inertia of the area with respect to xy axes is

$$I_{xy} = \int_A xy\,dx\,dy = \int_0^a \int_{-\sqrt{2px}}^{\sqrt{2px}} xy\,dx\,dy = \int_0^a x\,dx \int_{-\sqrt{2px}}^{\sqrt{2px}} y\,dy$$

$$= \int_0^a x\,dx \left\{\frac{y^2}{2}\right\}_{-\sqrt{2px}}^{\sqrt{2px}} = 0.$$

The products of inertia of the area with respect to xz and yz axes are $I_{xz} = I_{yz} = 0$. The inertia matrix of the plane figure is

$$[I] = \begin{bmatrix} I_{xx} & -I_{xy} & -I_{xz} \\ -I_{yx} & I_{yy} & -I_{yz} \\ -I_{zx} & -I_{zy} & I_{zz} \end{bmatrix} = \begin{bmatrix} \dfrac{b^2 A}{5} & 0 & 0 \\ 0 & \dfrac{3a^2 A}{7} & 0 \\ 0 & 0 & A\left(\dfrac{b^2}{5} + \dfrac{3a^2}{7}\right) \end{bmatrix}.$$

The first moment of the area A with respect to y-axis is

$$M_y = \int_A x\,dx\,dy = \int_0^a \int_{-\sqrt{2px}}^{\sqrt{2px}} x\,dx\,dy = \int_0^a x\,dx \int_{-\sqrt{2px}}^{\sqrt{2px}} dy$$

$$= \int_0^a x\,dx\,\{y\}_{-\sqrt{2px}}^{\sqrt{2px}} = 2\int_0^a x\sqrt{2px}\,dx$$

$$= \frac{2b}{\sqrt{a}} \int_0^a x^{3/2}\,dx = \frac{2b}{\sqrt{a}}\left\{\frac{x^{5/2}}{5/2}\right\}_0^a = \frac{2b}{\sqrt{a}}\frac{a^{5/2}}{5/2} = \frac{4ba^2}{5}.$$

The x-coordinate of the mass center, Fig 2.40b, is

$$x_C = \frac{M_y}{A} = \frac{4ba^2}{5}\frac{3}{4ab} = \frac{3a}{5}.$$

The first moment of the area A with respect to x-axis is $M_x = 0$ and $y_C = M_x/A = 0$.

Using the parallel-axis theorem, the moments of inertia of the area with respect to the centroidal axes $x'y'z'$, Fig 2.40b, are

$$I_{x'x'} = I_{xx} - d^2 A = I_{xx} = \frac{b^2 A}{5},$$

$$I_{y'y'} = I_{yy} - (x_C)^2 A = I_{yy} - \left(\frac{3a}{5}\right)^2 A = \frac{3a^2 A}{7} - \frac{9a^2 A}{25} = \frac{12a^2 A}{175},$$

$$I_{z'z'} = I_{x'x'} + I_{y'y'} = \frac{b^2 A}{5} + \frac{12a^2 A}{175} = A\left(\frac{b^2}{5} + \frac{a^2}{175}\right),$$

$$I_{x'y'} = I_{xy} - (x_C)(0) A = 0,$$

$$I_{x'z'} = I_{y'z'} = 0.$$

The centroidal inertia matrix of the plane figure is

$$
[I'] = \begin{bmatrix} I_{x'x'} & -I_{x'y'} & -I_{x'z'} \\ -I_{y'x'} & I_{y'y'} & -I_{y'z'} \\ -I_{z'x'} & -I_{z'y'} & I_{z'z'} \end{bmatrix} = \begin{bmatrix} \dfrac{b^2 A}{5} & 0 & 0 \\ 0 & \dfrac{12a^2 A}{175} & 0 \\ 0 & 0 & A\left(\dfrac{b^2}{5} + \dfrac{a^2}{175}\right) \end{bmatrix}.
$$

2.4   Problems

2.1 Find the x-coordinate of the centroid of the indicated region where $A = 2$ m and $k = \pi/8$ m^{-1} (Fig. 2.41).

2.2 Find the x-coordinate of the centroid of the shaded region shown in the figure. The region is bounded by the curves $y = x^2$ and $y = \sqrt{x}$. All coordinates may be treated as dimensionless (Fig. 2.42).

2.3 The region shown is bounded by the curves $y = b$ and $y = k\,|x|^3$, where $b = ka^3$. Find the coordinates of the centroid (Fig. 2.43).

2.4 Determine the centroid of the area where $A(a, a)$. Use integration (Fig. 2.44).

2.5 Locate the center of gravity of the volume where $x^2 = a^2 y/b$, $a = b = 1$ m. The material is homogeneous (Fig. 2.45).

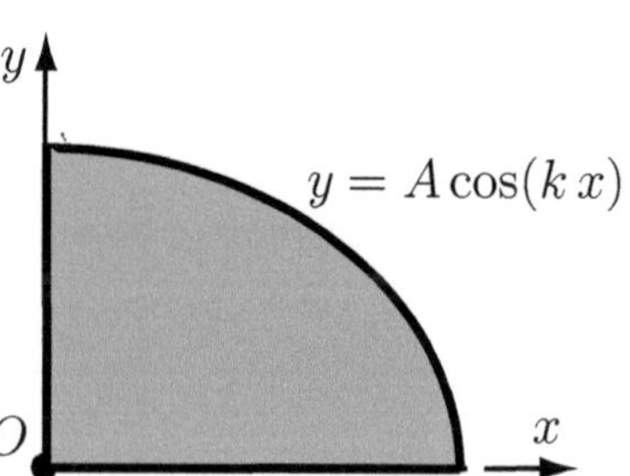

Fig. 2.41 Problem 2.1

Fig. 2.42 Problem 2.2

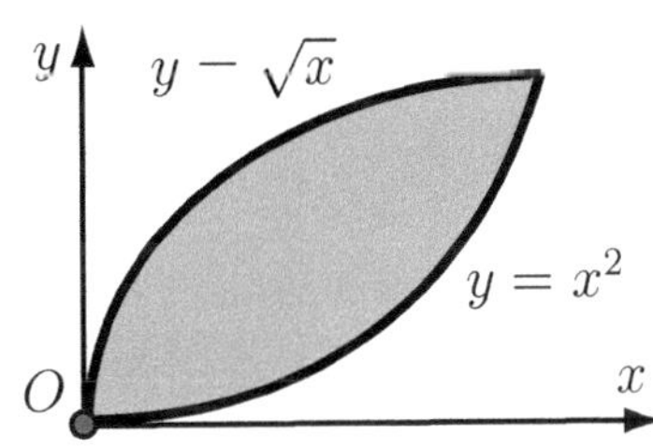

Fig. 2.43 Problem 2.3

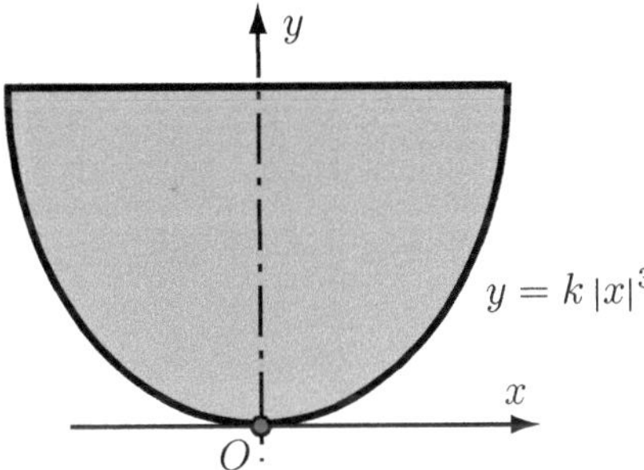

Fig. 2.44 Problem 2.4

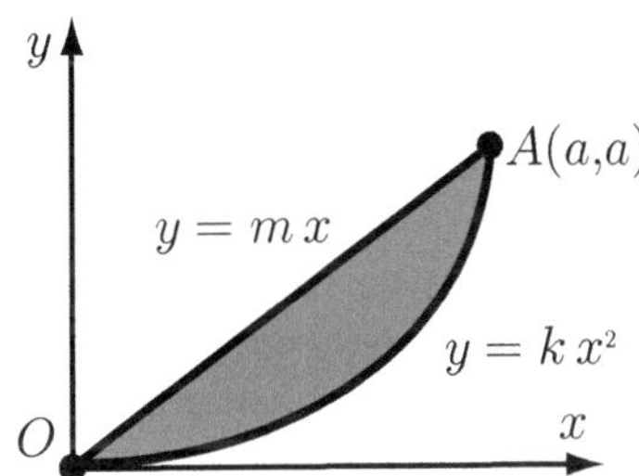

Fig. 2.45 Problem 2.5

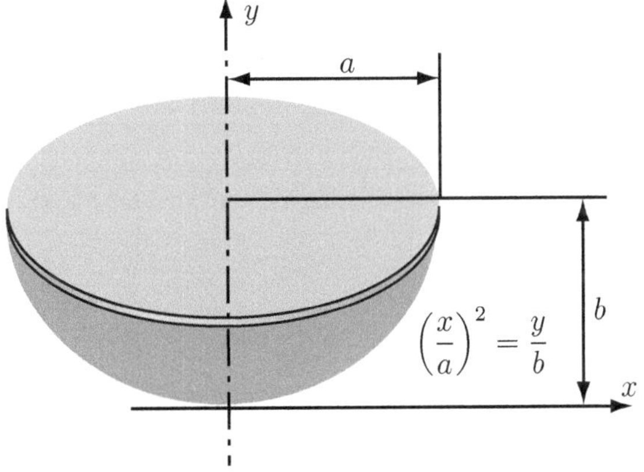

Fig. 2.46 Problem 2.6

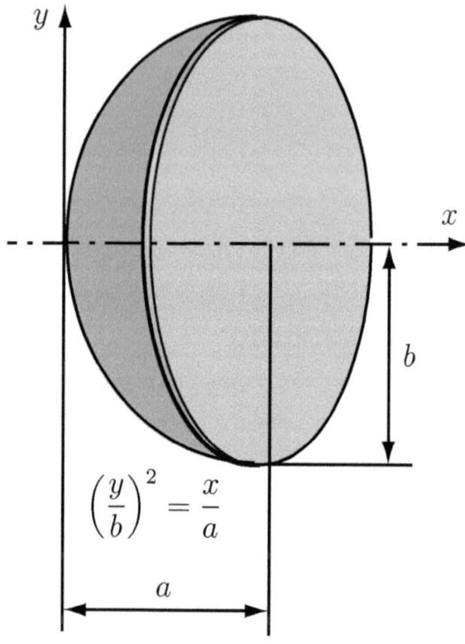

$$\left(\frac{y}{b}\right)^2 = \frac{x}{a}$$

Fig. 2.47 Problem 2.7

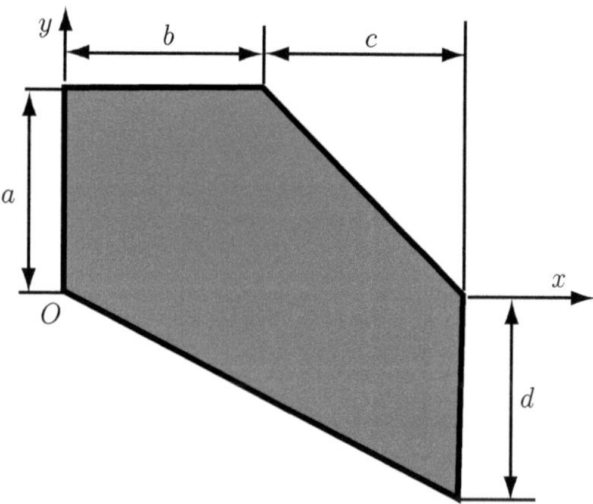

2.6 Locate the centroid of the paraboloid, shown in Fig. 2.46 defined by the equation $x/a = (y/b)^2$ where $a = b = 5$ m. The material is homogeneous.

2.7 Determine the location of the centroid C of the area where $a = 6$ m, $b = 6$ m, $c = 7$ m, and $d = 7$ m, as shown in Fig. 2.47.

2.8 The solid object, shown in Fig. 2.48 it consists of a solid half-circular base with an extruded ring. The bottom of the object is half-circular in shape with a thickness, t, equal to 10 mm. The half-circle has a radius, R, equal to 30 mm. The coordinate axes are aligned so that the origin is at the bottom of the object at the center point of the half-circle. This half-ring has an inner radius, r, of 20 mm, and an outer radius, R, of 30 mm. The extrusion height for the

Fig. 2.48 Problem 2.8

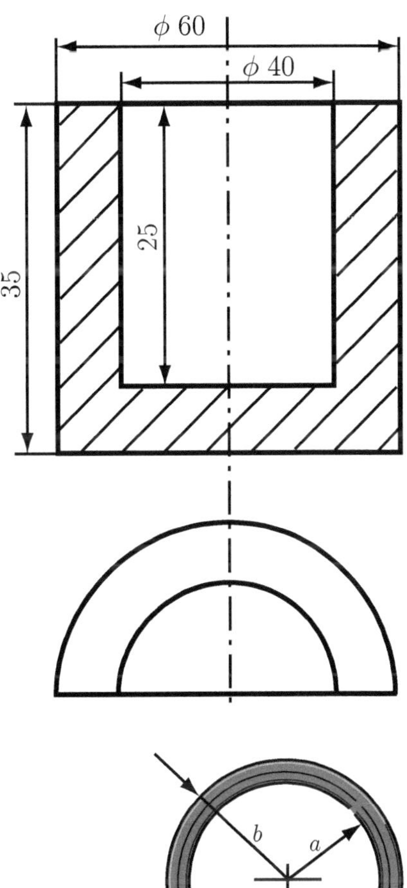

Fig. 2.49 Problem 2.9

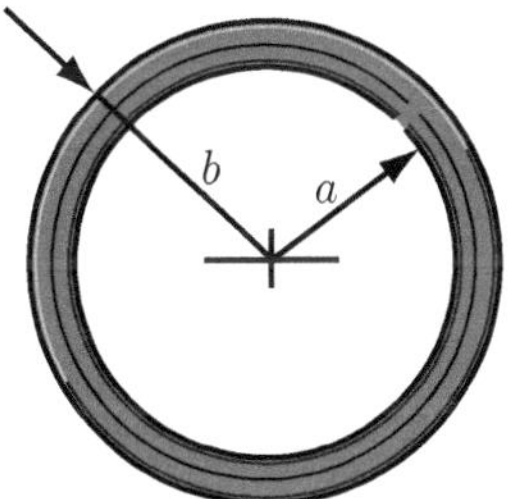

half-ring, h, is equal to 25 mm. The density of the object is uniform and will be denoted ρ. Find the coordinates of the mass center.

2.9 The ring with the circular cross section has the dimensions $a = 3$ m and $b = 5$ m. Determine the surface area of the ring (Fig. 2.49).

2.10 The belt shown in Fig. 2.50 has the dimensions of the cross area: $a = 27$ mm, $b = 66$ mm, and $c = 58$ mm. The radius of the belt is $r = 575$ mm. Find the volume of the belt (Fig. 2.50).

2.11 Determine the moment of inertia about the x-axis of the shaded area shown in Fig. 2.51 where $m = h/b$ and $b = h = 2$ m. Use integration.

2.12 Determine the moment of inertia about the y-axis of the area shown in Fig. 2.52 where $a = 2$ m and $b = 6$ m. Use integration.

Fig. 2.50 Problem 2.10

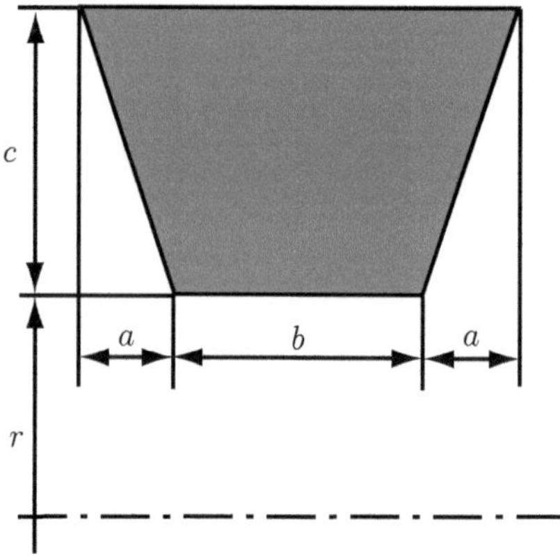

Fig. 2.51 Problem 2.11

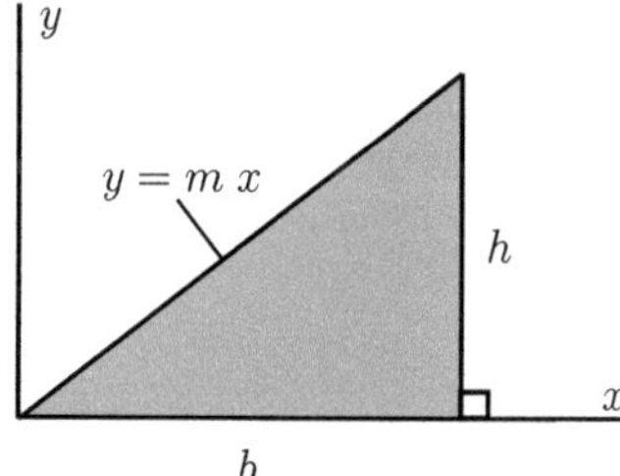

Fig. 2.52 Problem 2.12

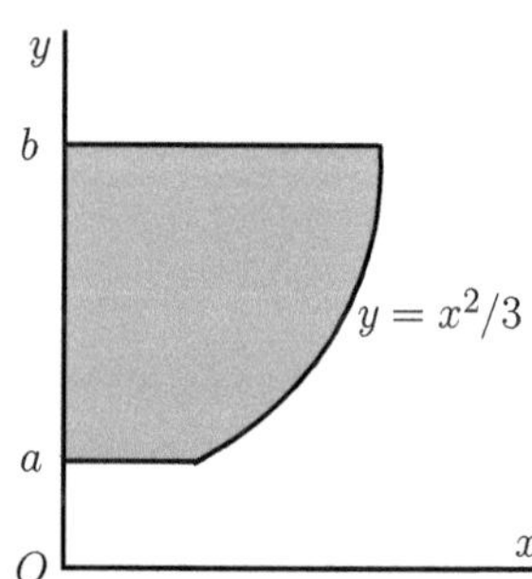

2.13 Determine the moment of inertia about the x-axis of the area shown in Fig. 2.53 where $a = 5$ m and $b = 3$ m. Use integration.

2.14 Determine the moment of inertia about the centroidal axes of the area shown in Fig. 2.54 where $r = 1$ m.

2.15 Determine the moments of inertia and the products of inertia about the centroidal axes of the shaded area shown in Fig. 2.55, where $a = 1$ in. Find the centroid polar moment of inertia. The mass center of the shaded area is at C.

Fig. 2.53 Problem 2.13

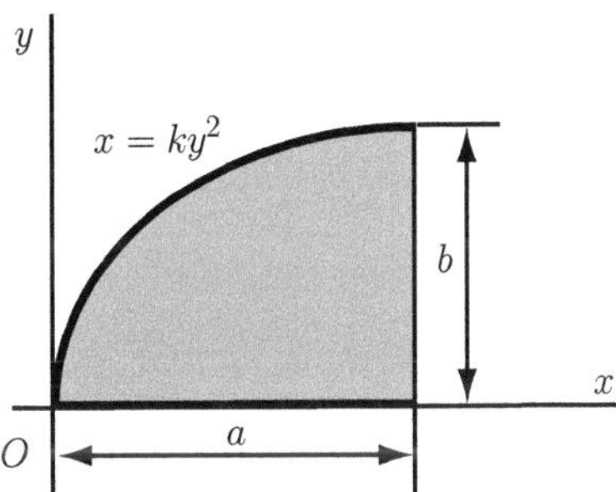

Fig. 2.54 Problem 2.14

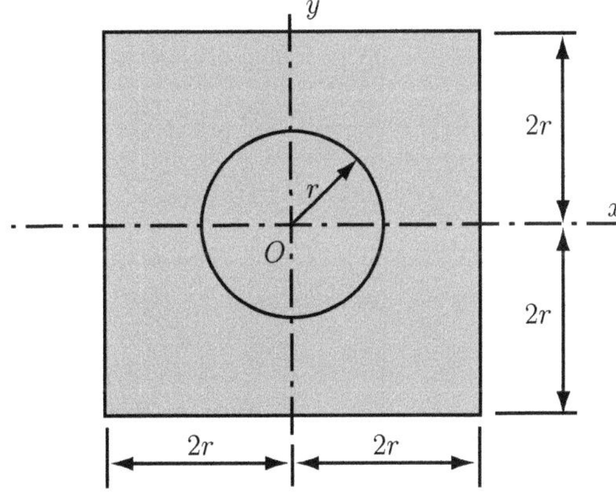

Fig. 2.55 Problem 2.15

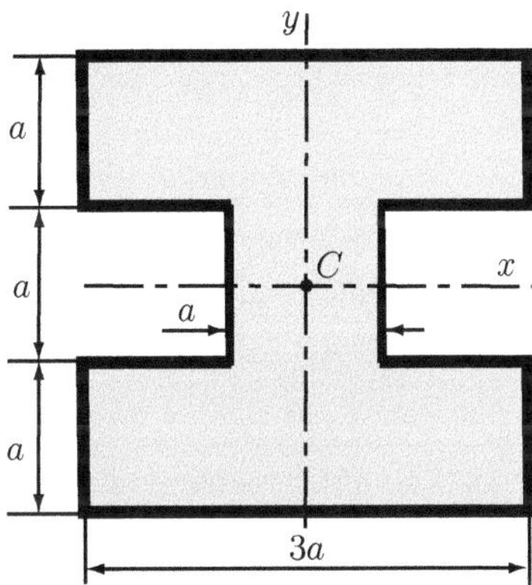

2.16 Determine the moments of inertia about the centroidal axes of the area shown in Fig. 2.56, where $a = 140$ mm, $b = 90$ mm, and the uniform thickness is $t = 18$ mm.

2.17 The region shown in Fig. 2.57 bounded by the curves $y = x$ and $y = x^2/a$, where $a = 10$ cm. Find the area moments of inertia about the x and y axes.

Fig. 2.56 Problem 2.16

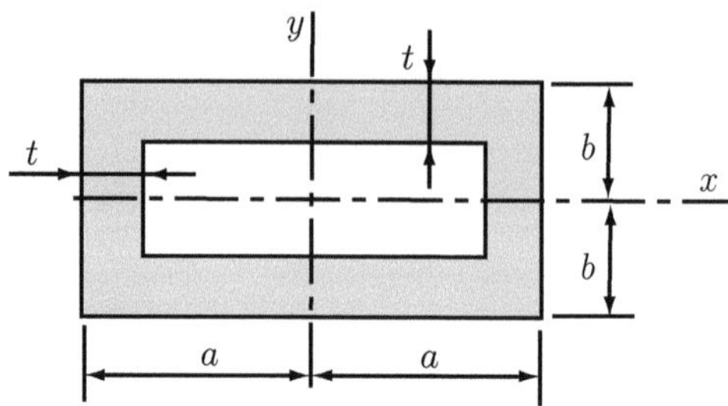

Fig. 2.57 Problem 2.17

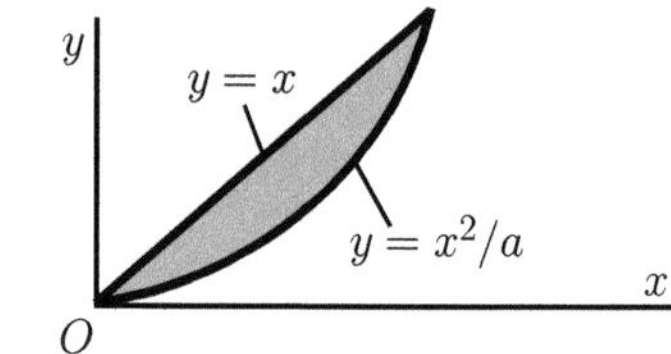

Fig. 2.58 Problem 2.18

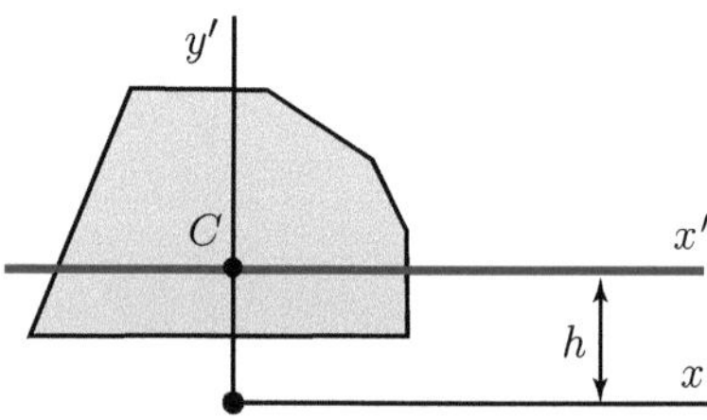

Fig. 2.59 Problem 2.19

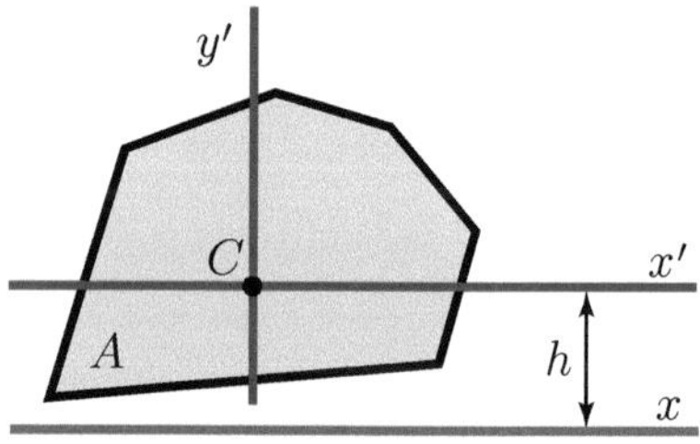

2.18 The polar moment of inertia of the area shown in Fig. 2.58 is I_{Czz} about the z-axis passing through the centroid C. The moment of inertia about the y' axis is $I_{y'y'}$ and the moment of inertia about the x-axis is I_{xx}, determine the area A. Numerical application: $I_{Czz} = 500 \times 10^6$ mm^4, $I_{y'y'} = 300 \times 10^6$ mm^4, $I_{xx} = 800 \times 10^6$ mm^4, and $h = 200$ mm.

2.19 The polar moment of inertia of the area, A, about the z-axis passing through the centroid C is I_{Czz}, see Fig. 2.59. The moment of inertia of the area about the x-axis is I_{xx}. Find the moment of inertia about the y' axis, $I_{y'y'}$. Numerical application: $I_{Czz} = 20$ in^4, $I_{xx} = 30$ in^4, $A = 5$ in^2, and $h = 2.5$ in.

Fig. 2.60 Problem 2.20

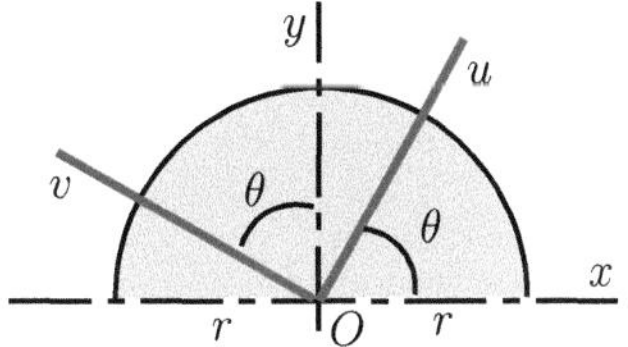

Fig. 2.61 Problem 2.21

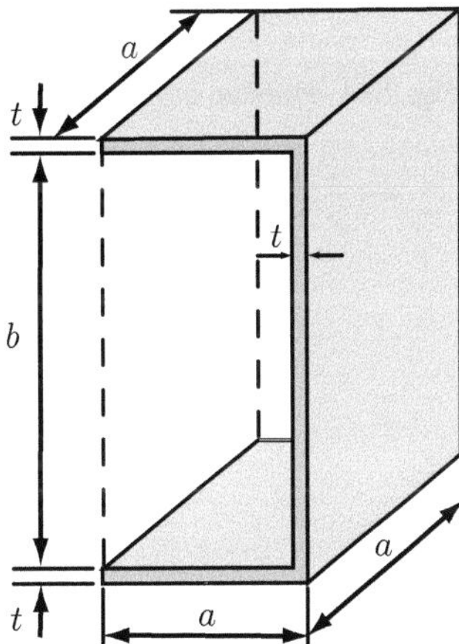

Fig. 2.62 Problem 2.22

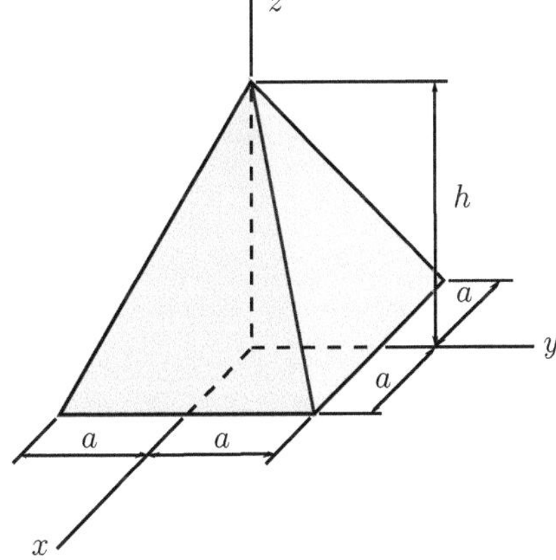

2.20 Determine the area moments of inertia I_{uu} and I_{vv} and the product of inertia I_{uv} for the semicircular area with the radius $r = 70$ mm and $\theta = 35°$, as shown in Fig. 2.60.

2.21 Determine the moments of inertia and the products of inertia about the centroidal axes of the shaded figure shown in Fig. 2.61. where $a = 1$ m, $b = 3$ m, and $t = 0.3$ m. Find the centroid polar moment of inertia.

Fig. 2.63 Problem 2.23

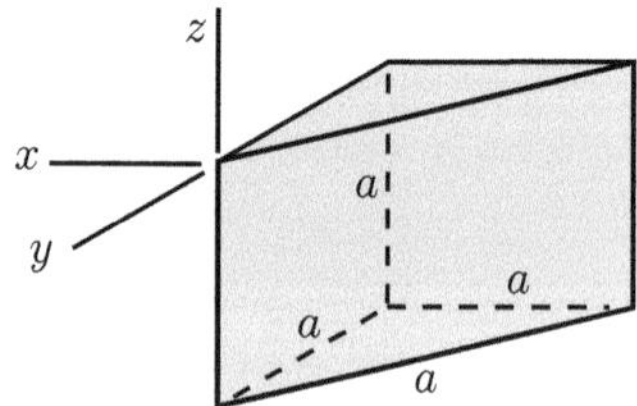

Fig. 2.64 Problem 2.24

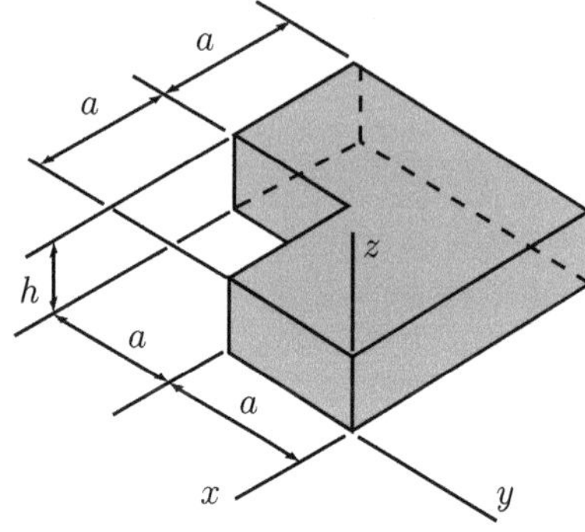

Fig. 2.65 Problem 2.25

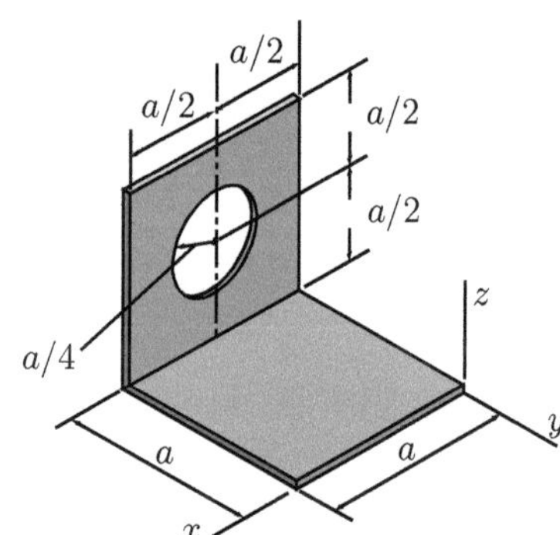

Fig. 2.66 Problem 2.26

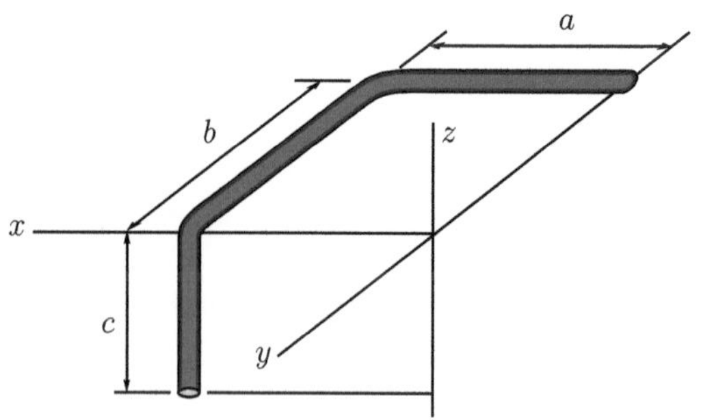

Table 2.1 Area inertia properties for some common cross sections

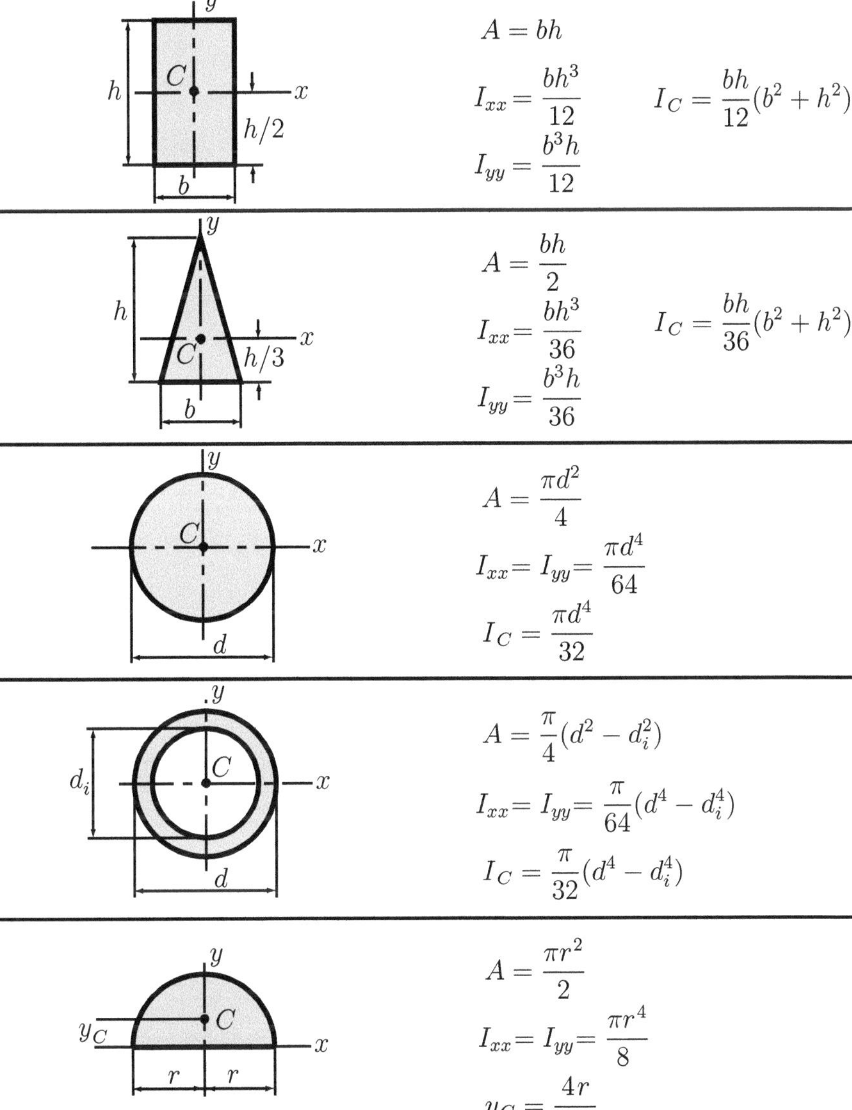

$$A = bh$$

$$I_{xx} = \frac{bh^3}{12} \qquad I_C = \frac{bh}{12}(b^2 + h^2)$$

$$I_{yy} = \frac{b^3 h}{12}$$

$$A = \frac{bh}{2}$$

$$I_{xx} = \frac{bh^3}{36} \qquad I_C = \frac{bh}{36}(b^2 + h^2)$$

$$I_{yy} = \frac{b^3 h}{36}$$

$$A = \frac{\pi d^2}{4}$$

$$I_{xx} = I_{yy} = \frac{\pi d^4}{64}$$

$$I_C = \frac{\pi d^4}{32}$$

$$A = \frac{\pi}{4}(d^2 - d_i^2)$$

$$I_{xx} = I_{yy} = \frac{\pi}{64}(d^4 - d_i^4)$$

$$I_C = \frac{\pi}{32}(d^4 - d_i^4)$$

$$A = \frac{\pi r^2}{2}$$

$$I_{xx} = I_{yy} = \frac{\pi r^4}{8}$$

$$y_C = \frac{4r}{3\pi}$$

Table 2.2 Volume inertia properties for some homogenous bodies

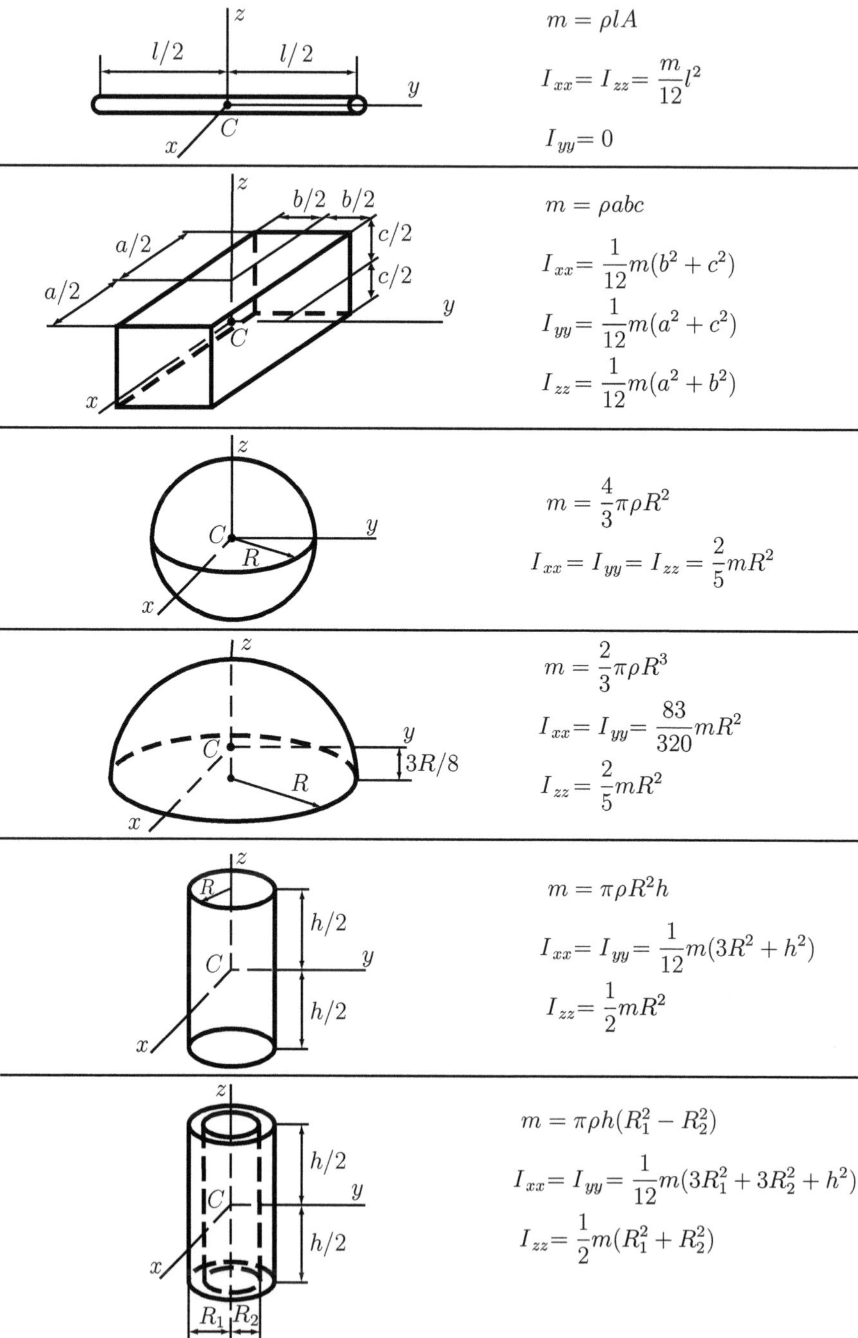

$$m = \rho l A$$

$$I_{xx} = I_{zz} = \frac{m}{12} l^2$$

$$I_{yy} = 0$$

$$m = \rho abc$$

$$I_{xx} = \frac{1}{12} m(b^2 + c^2)$$

$$I_{yy} = \frac{1}{12} m(a^2 + c^2)$$

$$I_{zz} = \frac{1}{12} m(a^2 + b^2)$$

$$m = \frac{4}{3} \pi \rho R^2$$

$$I_{xx} = I_{yy} = I_{zz} = \frac{2}{5} m R^2$$

$$m = \frac{2}{3} \pi \rho R^3$$

$$I_{xx} = I_{yy} = \frac{83}{320} m R^2$$

$$I_{zz} = \frac{2}{5} m R^2$$

$$m = \pi \rho R^2 h$$

$$I_{xx} = I_{yy} = \frac{1}{12} m(3R^2 + h^2)$$

$$I_{zz} = \frac{1}{2} m R^2$$

$$m = \pi \rho h (R_1^2 - R_2^2)$$

$$I_{xx} = I_{yy} = \frac{1}{12} m(3R_1^2 + 3R_2^2 + h^2)$$

$$I_{zz} = \frac{1}{2} m(R_1^2 + R_2^2)$$

2.22 Find the moment of inertia of the homogeneous pyramid with the mass m and density ρ about the x-axis and about the z-axis, as shown in Fig. 2.62.

2.23 Determine the products of inertia I_{xy}, I_{yz}, and I_{xz} for the homogeneous prism with the density ρ, as shown in Fig. 2.63.

2.24 Determine the products of inertia I_{xy}, I_{yz}, and I_{xz} for the homogeneous element with the material density ρ, as shown in Fig. 2.64.

2.25 Determine the products of inertia I_{xy}, I_{yz}, and I_{xz} for the homogeneous thin body with the material mass per unit area $\rho = 30$ kg/m^2 and the dimension $a = 0.8$ m, as shown in Fig. 2.65.

2.26 Determine the products of inertia I_{xy}, I_{yz}, and I_{xz} for the thin tube with the material mass per unit length $\rho = 5$ kg/m and the dimensions $a = 0.5$ m, $b = 0.8$ m, and $c = 0.4$ m, as shown in Fig. 2.66.

Table 2.1 gives area inertia properties for some homogeneous areas where A is the area C is the location of the centroid I_{xx} and I_{xx} are the moments of area about x and y axes, respectively; and I_C is the polar moment of area about C.

Table 2.2 gives volume inertia properties for some homogeneous bodies where A is the cross-sectional area, C is the location of the centroid, m is the mass, ρ is the density, I_{xx}, I_{yy}, and I_{zz} are the moments of inertia about x, y, and z axes, respectively.

Chapter 3
Kinematics of a Particle

3.1 Introduction

3.1.1 Position, Velocity, and Acceleration

The position of a particle P relative to a given reference frame with origin O is given by the position vector $\mathbf{r}$ from point O to point P, as shown in Fig. 3.1.

If the particle P is in motion relative to the reference frame, the position vector $\mathbf{r}$ is a function of time t, Fig. 3.1, and can be expressed as

$$\mathbf{r} = \mathbf{r}(t).$$

The velocity of the particle P relative to the reference frame at time t is defined by

$$\mathbf{v} = \frac{d\mathbf{r}}{dt} = \dot{\mathbf{r}} = \lim_{\Delta t \to 0} \frac{\mathbf{r}(t + \Delta t) - \mathbf{r}(t)}{\Delta t}, \tag{3.1}$$

where the vector $\mathbf{r}(t + \Delta t) - \mathbf{r}(t)$ is the change in position, or displacement of P, during the interval of time Δt, Fig. 3.1. The velocity is the rate of change of the position of the particle P. The magnitude of the velocity $\mathbf{v}$ is the speed $v = |\mathbf{v}|$.

The dimensions of $\mathbf{v}$ are (distance)/(time). The position and velocity of a particle can be specified only relative to a reference frame.

The acceleration of the particle P relative to the given reference frame at time t is defined by

$$\mathbf{a} = \frac{d\mathbf{v}}{dt} = \dot{\mathbf{v}} = \lim_{\Delta t \to 0} \frac{\mathbf{v}(t + \Delta t) - \mathbf{v}(t)}{\Delta t}, \tag{3.2}$$

where $\mathbf{v}(t + \Delta t) - \mathbf{v}(t)$ is the change in the velocity of P during the interval of time Δt, Fig. 3.1. The acceleration is the rate of change of the velocity of P at time t (the second time derivative of the displacement), and its dimensions are (distance)/(time)^2.

D.B. Marghitu and M. Dupac, *Advanced Dynamics: Analytical and Numerical Calculations with MATLAB*, DOI 10.1007/978-1-4614-3475-7_3, © Springer Science+Business Media, LLC 2012

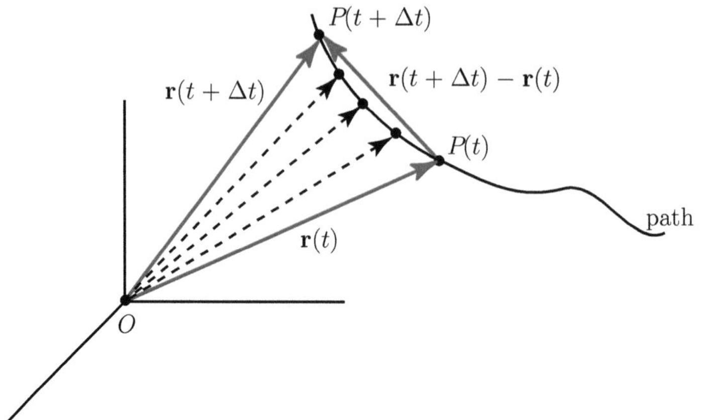

Fig. 3.1 Position of a particle P

Fig. 3.2 Angular motion
of line L relative to a
reference line L_0

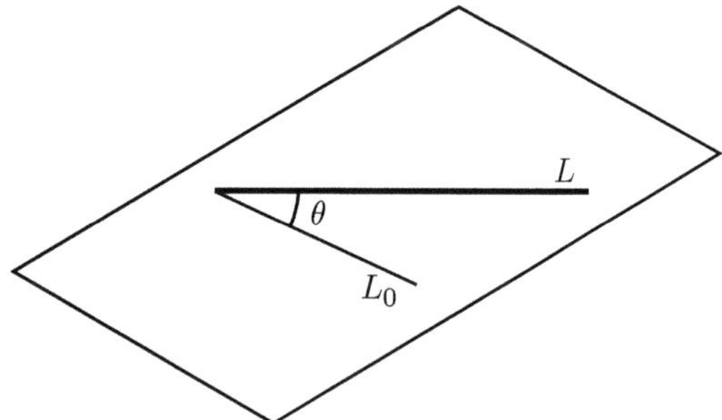

3.1.2   Angular Motion of a Line

The angular motion of the line L, in a plane, relative to a reference line L_0 in the
plane, is given by an angle θ, Fig. 3.2. The angular velocity of L relative to L_0 is
defined by

$$\omega = \frac{\mathrm{d}\theta}{\mathrm{d}t} = \dot{\theta}, \tag{3.3}$$

and the angular acceleration of L relative to L_0 is defined by

$$\alpha = \frac{\mathrm{d}\omega}{\mathrm{d}t} = \frac{\mathrm{d}^2\theta}{\mathrm{d}t^2} = \dot{\omega} = \ddot{\theta}. \tag{3.4}$$

The dimensions of the angular position, angular velocity, and angular acceleration
are [rad], [rad/s], and [rad/s^2], respectively. The scalar coordinate θ can be positive
or negative. The counterclockwise (ccw) direction is considered positive.

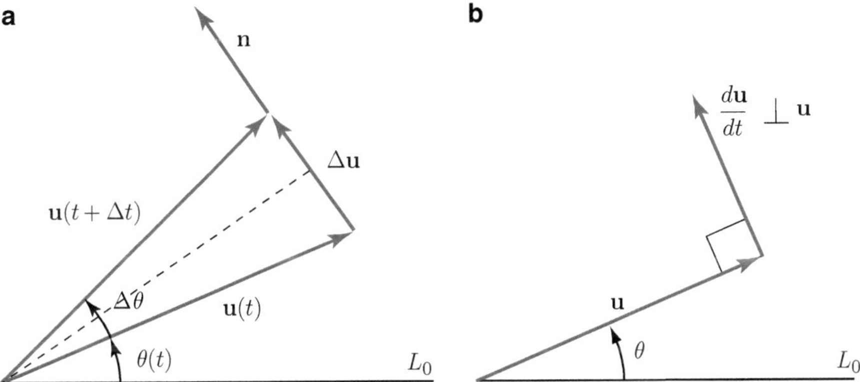

Fig. 3.3 Angular motion of a unit vector **u** in plane

3.1.3 Rotating Unit Vector

The angular motion of a unit vector **u** in a plane can be described as the angular motion of a line. The direction of **u** relative to a reference line L_0, is specified by the angle θ in Fig. 3.3a, and the rate of rotation of **u** relative to L_0 is defined by the angular velocity

$$\omega = \frac{d\theta}{dt} = \dot{\theta}.$$

The time derivative of **u** is specified by

$$\frac{d\mathbf{u}}{dt} = \lim_{\Delta t \to 0} \frac{\mathbf{u}(t+\Delta t) - \mathbf{u}(t)}{\Delta t}.$$

Figure 3.3a shows the vector **u** at time t and at time $t + \Delta t$. The change in **u** during this interval is $\Delta\mathbf{u} = \mathbf{u}(t+\Delta t)\mathbf{u}(t)$, and the angle through which **u** rotates is $\Delta\theta = \theta(t+\Delta t) - \theta(t)$. The triangle in Fig. 3.3a is isosceles, so the magnitude of $\Delta\mathbf{u}$ is

$$|\Delta\mathbf{u}| = 2|\mathbf{u}|\sin(\Delta\theta/2) = 2\sin(\Delta\theta/2).$$

The vector $\Delta\mathbf{u}$ is

$$\Delta\mathbf{u} = |\Delta\mathbf{u}|\mathbf{n} = 2\sin(\Delta\theta/2)\mathbf{n},$$

where **n** is a unit vector that points in the direction of $\Delta\mathbf{u}$, Fig. 3.3a. The time derivative of **u** is

$$\frac{d\mathbf{u}}{dt} = \lim_{\Delta t \to 0} \frac{\Delta \mathbf{u}}{\Delta t} = \lim_{\Delta t \to 0} \frac{2\sin(\Delta\theta/2)\mathbf{n}}{\Delta t} = \lim_{\Delta t \to 0} \frac{\sin(\Delta\theta/2)}{\Delta\theta/2}\frac{\Delta\theta}{\Delta t}\mathbf{n}$$

$$= \lim_{\Delta t \to 0} \frac{\sin(\Delta\theta/2)}{\Delta\theta/2}\frac{\Delta\theta}{\Delta t}\mathbf{n} = \lim_{\Delta t \to 0} \frac{\Delta\theta}{\Delta t}\mathbf{n} = \frac{d\theta}{dt}\mathbf{n},$$

where $\lim_{\Delta t \to 0} \frac{\sin(\Delta\theta/2)}{\Delta\theta/2} = 1$ and $\lim_{\Delta t \to 0} \frac{\Delta\theta}{\Delta t} = \frac{d\theta}{dt}$.

So the time derivative of the unit vector $\mathbf{u}$ is

$$\frac{d\mathbf{u}}{dt} = \frac{d\theta}{dt}\mathbf{n} = \dot{\theta}\,\mathbf{n} = \omega\,\mathbf{n},$$

where $\mathbf{n}$ is a unit vector that is perpendicular to $\mathbf{u}$, $\mathbf{n} \perp \mathbf{u}$, and points in the positive θ direction, Fig. 3.3b.

3.2   Rectilinear Motion

The position of a particle P on a straight line relative to a reference point O can be indicated by the coordinate s measured along the line from O to P, as shown in Fig. 3.4.

In this case, the reference frame is the straight line, and the origin of the reference frame is the point O. The reference frame and its origin are used to describe the position of particle P. The coordinate s is considered to be positive to the right of the origin O and is considered to be negative to the left of the origin.

Let $\mathbf{u}$ be a unit vector parallel to the straight line and pointing in the positive s, Fig. 3.4. The position vector of the point P relative to the origin O is

$$\mathbf{r} = s\mathbf{u}.$$

The velocity of the particle P relative to the origin O is

$$\mathbf{v} = \frac{d\mathbf{r}}{dt} = \frac{ds}{dt}\mathbf{u} = \dot{s}\mathbf{u}.$$

The magnitude v of the velocity vector $\mathbf{v} = v\mathbf{u}$ is the speed (velocity scalar)

$$v = \frac{ds}{dt} = \dot{s}.$$

The speed v of the particle P is equal to the slope at time t of the line tangent to the graph of s as a function of time. The acceleration of the particle P relative to O is

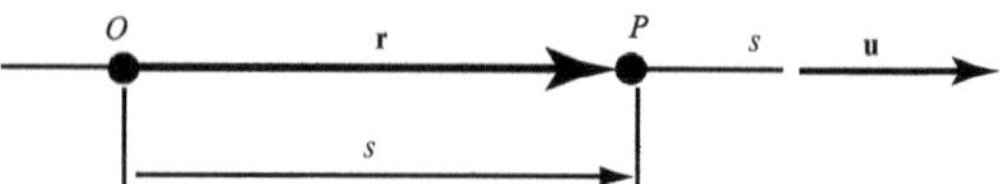

Fig. 3.4 Straight line motion of P

$$\mathbf{a} = \frac{d\mathbf{v}}{dt} = \frac{d}{dt}(v\mathbf{u}) = \frac{dv}{dt}\mathbf{u} = \dot{v}\mathbf{u} = \ddot{s}\mathbf{u}.$$

The magnitude a of the acceleration vector $\mathbf{a} = a\mathbf{u}$ is the acceleration scalar

$$a = \frac{dv}{dt} = \frac{d^2 s}{dt^2}.$$

The acceleration a is equal to the slope at time t of the line tangent to the graph of v as a function of time.

3.3   Curvilinear Motion

The motion of the particle P along a curvilinear path, relative to a reference frame, can be specified in terms of its position, velocity, and acceleration vectors. The directions and magnitudes of the position, velocity, and acceleration vectors do not depend on the particular coordinate system used to express them. The representations of the position, velocity, and acceleration vectors are different in different coordinate systems.

3.3.1   Cartesian Coordinates

Let $\mathbf{r}$ be the position vector of a particle P relative to the origin O of a Cartesian reference frame, Fig. 3.5. The components of $\mathbf{r}$ are the $x, y,$ and z coordinates of the particle P

$$\mathbf{r} = x\mathbf{i} + y\mathbf{j} + z\mathbf{k}. \tag{3.5}$$

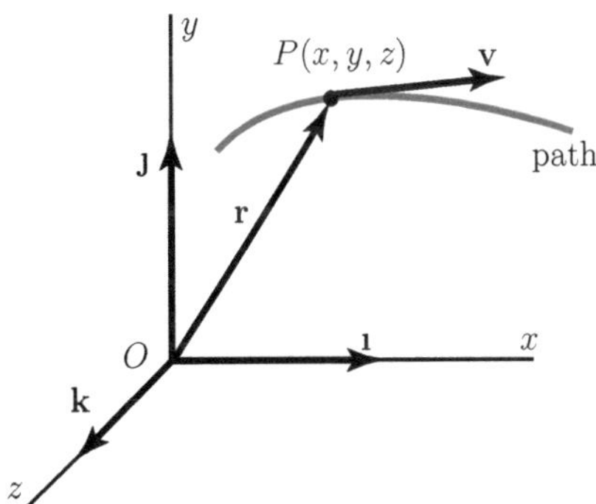

Fig. 3.5 Position vector
of a particle P in a Cartesian
reference frame

The velocity of the particle P relative to the reference frame is

$$\mathbf{v} = \frac{d\mathbf{r}}{dt} = \dot{\mathbf{r}} = \frac{dx}{dt}\mathbf{i} + \frac{dy}{dt}\mathbf{j} + \frac{dz}{dt}\mathbf{k} = \dot{x}\mathbf{i} + \dot{y}\mathbf{j} + \dot{z}\mathbf{k}. \tag{3.6}$$

The velocity in terms of scalar components is

$$\mathbf{v} = v_x\mathbf{i} + v_y\mathbf{j} + v_z\mathbf{k}. \tag{3.7}$$

Three scalar equations can be obtained:

$$v_x = \frac{dx}{dt} = \dot{x}, \quad v_y = \frac{dy}{dt} = \dot{y}, \quad v_z = \frac{dz}{dt} = \dot{z}. \tag{3.8}$$

The acceleration of the particle P relative to the reference frame is

$$\mathbf{a} = \frac{d\mathbf{v}}{dt} = \dot{\mathbf{v}} = \ddot{\mathbf{r}} = \frac{dv_x}{dt}\mathbf{i} + \frac{dv_y}{dt}\mathbf{j} + \frac{dv_z}{dt}\mathbf{k} = \dot{v}_x\mathbf{i} + \dot{v}_y\mathbf{j} + \dot{v}_z\mathbf{k} = \ddot{x}\mathbf{i} + \ddot{y}\mathbf{j} + \ddot{z}\mathbf{k}.$$

Expressing the acceleration in terms of scalar components

$$\mathbf{a} = a_x\mathbf{i} + a_y\mathbf{j} + a_z\mathbf{k}, \tag{3.9}$$

three scalar equations can be obtained

$$a_x = \frac{dv_x}{dt} = \dot{v}_x = \ddot{x}, \quad a_y = \frac{dv_y}{dt} = \dot{v}_y = \ddot{y}, \quad a_z = \frac{dv_z}{dt} = \dot{v}_z = \ddot{z}. \tag{3.10}$$

Equations (3.8) and (3.10) describe the motion of a particle relative to a Cartesian coordinate system.

3.3.2   Normal and Tangential Coordinates

The position, velocity, and acceleration of a particle will be specified in terms of their components tangential and normal (perpendicular) to the path. The particle P is moving along a plane, curvilinear path relative to a reference frame, Fig. 3.6. The position vector $\mathbf{r}$ specifies the position of the particle P relative to the reference point O. The coordinate s measures the position of the particle P along the path relative to a point O' on the path. The velocity of P relative to O is

$$\mathbf{v} = \frac{d\mathbf{r}}{dt} = \lim_{\Delta t \to 0} \frac{\mathbf{r}(t + \Delta t) - \mathbf{r}(t)}{\Delta t} = \lim_{\Delta t \to 0} \frac{\Delta \mathbf{r}}{\Delta t}, \tag{3.11}$$

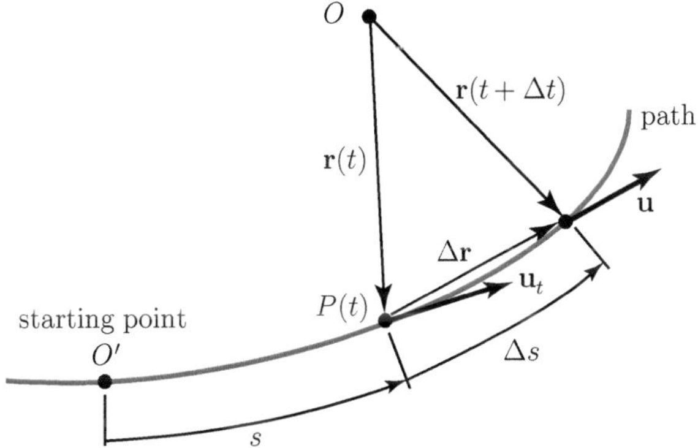

Fig. 3.6 Particle P moving along a plane, curvilinear path

where $\Delta\mathbf{r} = \mathbf{r}(t + \Delta t) - \mathbf{r}(t)$, as shown in Fig. 3.6. The distance traveled along the path from t to $t + \Delta t$ is Δs. On can write (3.11) as

$$\mathbf{v} = \lim_{\Delta t \to 0} \frac{\Delta s}{\Delta t}\mathbf{u},$$

where $\mathbf{u}$ is a unit vector in the direction of $\Delta\mathbf{r}$. In the limit, Δt approaches zero, the magnitude of $\Delta\mathbf{r}$ equals ds because a chord progressively approaches the curve. For the same reason, the direction of $\Delta\mathbf{r}$ approaches tangency to the curve, $\mathbf{u}$ becomes a unit vector, $\mathbf{u}_t$, tangent to the path at the position of P, as shown Fig. 3.6

$$\mathbf{v} = v\mathbf{u}_t = \frac{ds}{dt}\mathbf{u}_t. \tag{3.12}$$

The *tangent direction* is defined by the unit tangent vector $\mathbf{u}_t$ (or τ), which is a path variable parameter

$$\frac{d\mathbf{r}}{dt} = \frac{ds}{dt}\mathbf{u}_t,$$

or

$$\mathbf{u}_t = \frac{d\mathbf{r}}{ds}. \tag{3.13}$$

The velocity of a particle in curvilinear motion is a vector whose magnitude equals the rate of change of distance traveled along the path and whose direction is tangent to the path.

To determine the acceleration of P, the time derivative of (3.12) is taken

$$\mathbf{a} = \frac{d\mathbf{v}}{dt} = \frac{dv}{dt}\mathbf{u}_t + v\frac{d\mathbf{u}_t}{dt}. \tag{3.14}$$

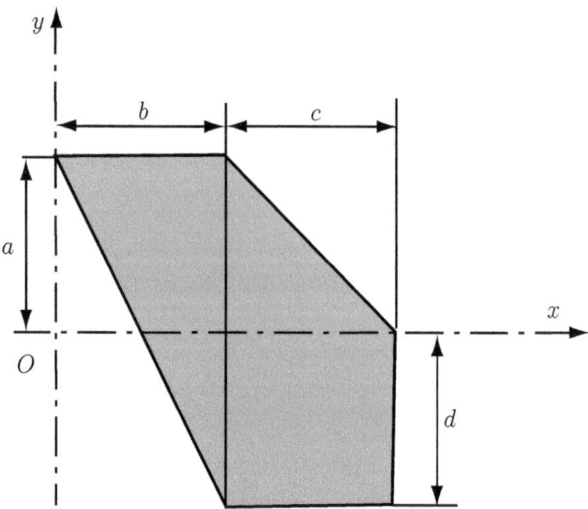

Fig. 3.7 Path angle θ and normal and tangent unit vectors to the path

If the path is not a straight line, the unit vector $\mathbf{u}_t$ rotates as P moves on the path, and the time derivative of $\mathbf{u}_t$ is not zero. The path angle θ defines the direction of $\mathbf{u}_t$ relative to a reference line shown in Fig. 3.7.

The time derivative of the rotating tangent unit vector $\mathbf{u}_t$ is

$$\frac{\mathrm{d}\mathbf{u}_t}{\mathrm{d}t} = \frac{\mathrm{d}\theta}{\mathrm{d}t}\,\mathbf{u}_n,$$

where $\mathbf{u}_n$ is a unit vector that is normal to $\mathbf{u}_t$, (or $\boldsymbol{\tau}$) and points in the positive θ direction if $\mathrm{d}\theta/\mathrm{d}t$ is positive. The normal unit vector $\mathbf{u}_n$ (or v) defines the normal direction to the path. Substituting this expression into (3.14), the acceleration of P is obtained

$$\mathbf{a} = \frac{\mathrm{d}v}{\mathrm{d}t}\,\mathbf{u}_t + v\,\frac{\mathrm{d}\theta}{\mathrm{d}t}\,\mathbf{u}_n. \tag{3.15}$$

If the path is a straight line at time t, the normal component of the acceleration equals zero because in that case, $\mathrm{d}\theta/\mathrm{d}t$ is zero.

The tangential component of the acceleration arises from the rate of change of the magnitude of the velocity. The normal component of the acceleration arises from the rate of change in the direction of the velocity vector.

Figure 3.8 shows the positions on the path reached by P at time t, $P(t)$ and at time $t + \mathrm{d}t$, $P(t + \mathrm{d}t)$.

If the path is curved, straight lines extended from these points $P(t)$ and $P(t + \mathrm{d}t)$ perpendicular to the path will intersect at C as shown in Fig. 3.8. The distance ρ from the path to the particle where these two lines intersect is called the *instantaneous radius of curvature* of the path.

If the path is circular with radius a, then the radius of curvature equals the radius of the path, $\rho = a$. The angle $\mathrm{d}\theta$ is the change in the path angle, and $\mathrm{d}s$ is the distance traveled from t to $t + \Delta t$. The radius of curvature ρ is related to $\mathrm{d}s$ by, Fig. 3.8,

$$\mathrm{d}s = \rho\,\mathrm{d}\theta.$$

Fig. 3.8 Instantaneous radius
of curvature ρ

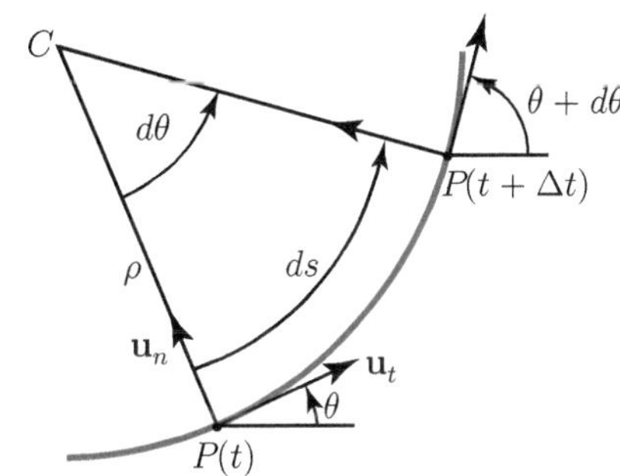

Fig. 3.9 Direction of the
normal unit vector $\mathbf{u}_n$

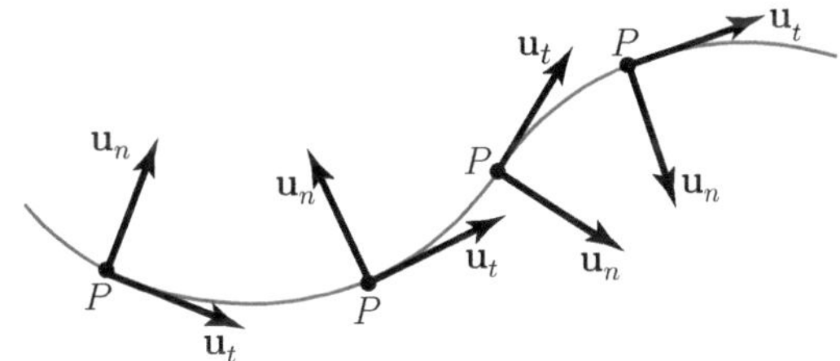

Dividing by dt, one can obtain

$$\frac{ds}{dt} = v = \rho\,\frac{d\theta}{dt}.$$

Using this relation, one can write (3.15) as

$$\mathbf{a} = \frac{dv}{dt}\,\mathbf{u_t} + \frac{v^2}{\rho}\,\mathbf{u_n}.$$

For a given value of v, the normal component of the acceleration depends on the instantaneous radius of curvature. The greater the curvature of the path, the greater the normal component of acceleration. When the acceleration is expressed in this way, the normal unit vector $\mathbf{u}_n$ must be defined to point toward the concave side of the path, Fig. 3.9.

The velocity and acceleration in terms of normal and tangential components are, Fig. 3.10,

$$\mathbf{v} = v\,\mathbf{u_t} = \frac{ds}{dt}\,\mathbf{u_t}, \tag{3.16}$$

$$\mathbf{a} = a_t\,\mathbf{u_t} + a_n\,\mathbf{u_n}, \tag{3.17}$$

where

$$a_t = \frac{dv}{dt}, \quad a_n = v\frac{d\theta}{dt} = \frac{v^2}{\rho}. \tag{3.18}$$

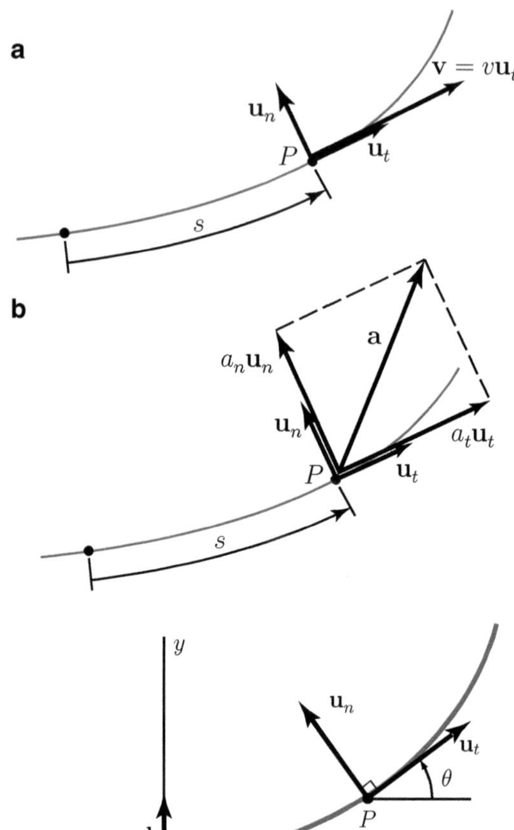

Fig. 3.10 (a) Velocity and (b) acceleration in terms of normal and tangential components

Fig. 3.11 τ and v in a Cartesian reference frame

If the motion occurs in the xy plane of a Cartesian reference frame, Fig. 3.11, and θ is the angle between the x-axis and the unit vector $\mathbf{u}_t$, the unit vectors $\mathbf{u}_t$ and $\mathbf{u}_n$ are related to the Cartesian unit vectors by

$$\mathbf{u}_t = \cos\theta \mathbf{i} + \sin\theta \mathbf{j},$$

$$\mathbf{u}_n = -\sin\theta \mathbf{i} + \cos\theta \mathbf{j}. \tag{3.19}$$

If the path in the xy plane is described by a function $y = y(x)$, it can be shown that the instantaneous radius of curvature is given by

$$\rho = \frac{\left[1 + \left(\dfrac{dy}{dx}\right)^2\right]^{3/2}}{\left|\dfrac{d^2 y}{dx^2}\right|}. \tag{3.20}$$

If the trajectory is given parametrically by

$$x = x(t) \quad \text{and} \quad y = y(t),$$

then radius of curvature is given by

$$\rho = \frac{\left(\dot{x}^2 + \dot{y}^2\right)^{3/2}}{|\dot{x}\ddot{y} - \dot{y}\ddot{x}|}. \tag{3.21}$$

Proof. The magnitude of the acceleration for an arbitrary point is

$$|\mathbf{a}| = a = \sqrt{\ddot{x}^2 + \ddot{y}^2} = \sqrt{\dot{v}^2 + \frac{v^2}{\rho^2}},$$

or

$$\ddot{x}^2 + \ddot{y}^2 = \dot{v}^2 + \frac{v^2}{\rho^2}.$$

From the previous relation, the radius of curvature is

$$\rho = \frac{v^2}{\sqrt{\ddot{x}^2 + \ddot{y}^2 - \dot{v}^2}}.$$

The velocity is

$$v = \sqrt{\dot{x}^2 + \dot{y}^2},$$

and the time derivative of the velocity is

$$\frac{dv}{dt} = \dot{v} = \frac{\dot{x}\ddot{x} + \dot{y}\ddot{y}}{\sqrt{\dot{x}^2 + \dot{y}^2}}.$$

Then, the radius of curvature is

$$\rho = \frac{\dot{x}^2 + \dot{y}^2}{\sqrt{\ddot{x}^2 + \ddot{y}^2 - \frac{(\dot{x}\ddot{x} + \dot{y}\ddot{y})^2}{\dot{x}^2 + \dot{y}^2}}} = \frac{\left(\dot{x}^2 + \dot{y}^2\right)^{3/2}}{|\dot{x}\ddot{y} - \ddot{x}\dot{y}|}.$$

The MATLAB proof for the radius of curvature formula is

```
syms t rho
x = sym('x(t)');
y = sym('y(t)');

v = sqrt(diff(x,t)^2+diff(y,t)^2);
```

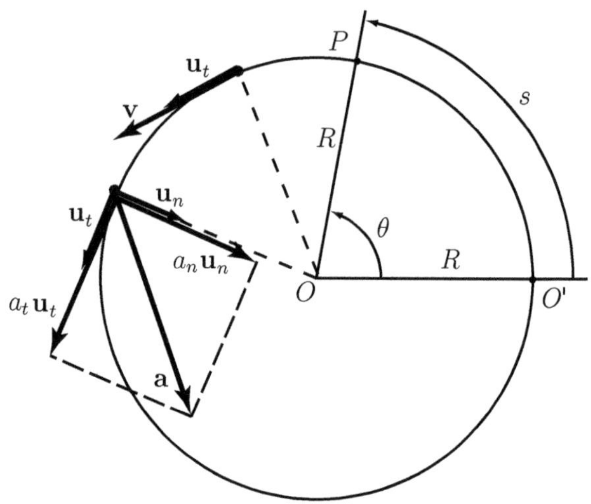

Fig. 3.12 Circular path of radius R

```
al = diff(x,t,2)^2+diff(y,t,2)^2;

a2 = diff(v,t)^2+v^4/rho^2;

rhos = simple(solve(a1-a2,rho));

pretty(rhos)
```

3.3.3 Circular Motion

The particle P moves in a plane circular path of radius R as shown in Fig. 3.12. The distance s is

$$s = R\theta, \tag{3.22}$$

where the angle θ specifies the position of the particle P along the circular path. The velocity is obtained taking the time derivative of (3.22)

$$v = \dot{s} = R\dot{\theta} = R\omega, \tag{3.23}$$

where $\omega = \dot{\theta}$ is the angular velocity of the line from the center of the path O to the particle P. The tangential component of the acceleration is $a_t = dv/dt$, and the

$$a_t = \dot{v} = R\dot{\omega} = R\alpha, \tag{3.24}$$

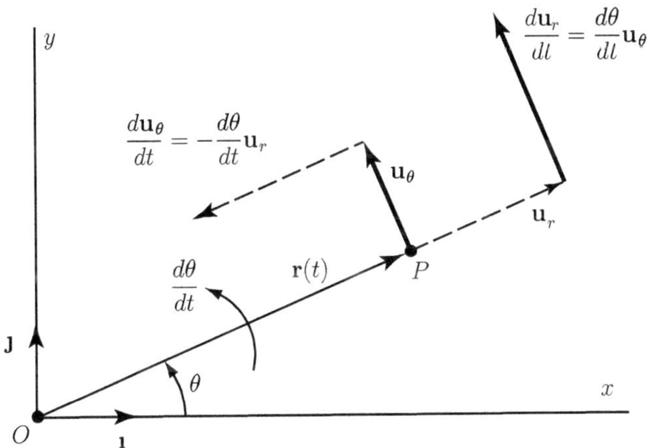

Fig. 3.13 Polar coordinates r and θ

where $\alpha = \dot{\omega}$ is the angular acceleration. The normal component of the acceleration is

$$a_n = \frac{v^2}{R} = R\omega^2. \tag{3.25}$$

For the circular path, the instantaneous radius of curvature is $\rho = R$.

3.3.4 Polar Coordinates

A particle P is considered in the $x - y$ plane of a Cartesian coordinate system. The position of the point P relative to the origin O may be specified either by its Cartesian coordinates x, y or by its polar coordinates r, θ as shown in Fig. 3.13.

The polar coordinates are defined by:

- The unit vector $\mathbf{u}_r$ that points in the direction of the radial line from the origin O to the particle P
- The unit vector $\mathbf{u}_\theta$ that is perpendicular to $\mathbf{u}_r$ and points in the direction of increasing the angle θ

The unit vectors $\mathbf{u}_r$ and $\mathbf{u}_\theta$ are related to the Cartesian unit vectors $\mathbf{1}$ and $\mathbf{J}$ by

$$\mathbf{u}_r = \cos\theta\,\mathbf{1} + \sin\theta\,\mathbf{J},$$

$$\mathbf{u}_\theta = -\sin\theta\,\mathbf{1} + \cos\theta\,\mathbf{J}. \tag{3.26}$$

The position vector $\mathbf{r}$ from O to P is

$$\mathbf{r} = r\mathbf{u}_r, \tag{3.27}$$

where r is the magnitude of the vector $\mathbf{r}$, $r = |\mathbf{r}|$. The velocity of the particle P in terms of polar coordinates is obtained by taking the time derivative of (3.27):

$$\mathbf{v} = \frac{d\mathbf{r}}{dt} = \frac{dr}{dt}\mathbf{u}_r + r\frac{d\mathbf{u}_r}{dt}. \tag{3.28}$$

The time derivative of the rotating unit vector $\mathbf{u}_r$ is

$$\frac{d\mathbf{u}_r}{dt} = \frac{d\theta}{dt}\mathbf{u}_\theta = \omega\,\mathbf{u}_\theta, \tag{3.29}$$

where $\omega = d\theta/dt$ is the angular velocity. Substituting (3.29) into (3.28), the velocity of P is

$$\mathbf{v} = \frac{dr}{dt}\mathbf{u}_r + r\frac{d\theta}{dt}\mathbf{u}_\theta = \frac{dr}{dt}\mathbf{u}_r + r\omega\,\mathbf{u}_\theta = \dot{r}\mathbf{u}_r + r\omega\,\mathbf{u}_\theta, \tag{3.30}$$

or

$$\mathbf{v} = v_r\,\mathbf{u}_r + v_\theta\,\mathbf{u}_\theta, \tag{3.31}$$

where

$$v_r = \frac{dr}{dt} = \dot{r} \text{ and } v_\theta = r\omega. \tag{3.32}$$

The acceleration of the particle P is obtained by taking the time derivative of (3.30)

$$\mathbf{a} = \frac{d\mathbf{v}}{dt} = \frac{d^2 r}{dt^2}\mathbf{u}_r + \frac{dr}{dt}\frac{d\mathbf{u}_r}{dt} + \frac{dr}{dt}\frac{d\theta}{dt}\mathbf{u}_\theta + r\frac{d^2\theta}{dt^2}\mathbf{u}_\theta + r\frac{d\theta}{dt}\frac{d\mathbf{u}_\theta}{dt}. \tag{3.33}$$

As P moves, $\mathbf{u}_\theta$ also rotates with angular velocity $d\theta/dt$. The time derivative of the unit vector $\mathbf{u}_\theta$ is in the $-\mathbf{u}_r$ direction if $d\theta/dt$ is positive:

$$\frac{d\mathbf{u}_\theta}{dt} = -\frac{d\theta}{dt}\mathbf{u}_r. \tag{3.34}$$

Substituting (3.34) and (3.29) into (3.33), the acceleration of the particle P is

$$\mathbf{a} = \left[\frac{d^2 r}{dt^2} - r\left(\frac{d\theta}{dt}\right)^2\right]\mathbf{u}_r + \left[r\frac{d^2\theta}{dt^2} + 2\frac{dr}{dt}\frac{d\theta}{dt}\right]\mathbf{u}_\theta.$$

Thus, the acceleration of P is

$$\mathbf{a} = a_r\,\mathbf{u}_r + a_\theta\,\mathbf{u}_\theta, \tag{3.35}$$

Fig. 3.14 Cylindrical coordinates r, θ, and z

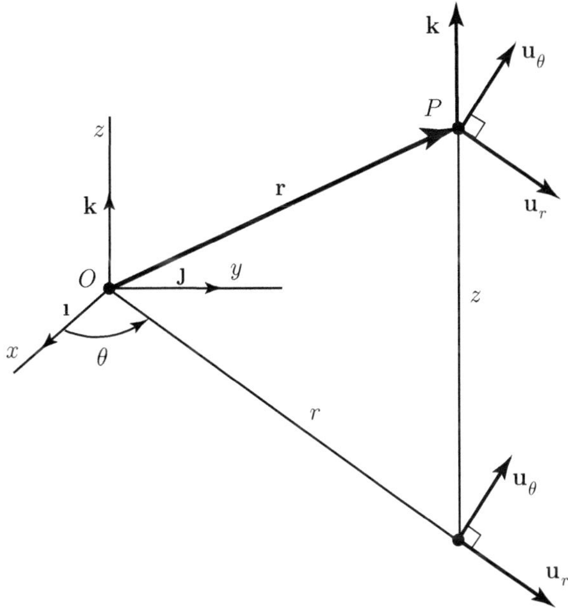

where

$$a_r = \frac{\mathrm{d}^2 r}{\mathrm{d}t^2} - r\left(\frac{\mathrm{d}\theta}{\mathrm{d}t}\right)^2 = \frac{\mathrm{d}^2 r}{\mathrm{d}t^2} - r\omega^2 = \ddot{r} - r\omega^2,$$

$$a_\theta = r\frac{\mathrm{d}^2\theta}{\mathrm{d}t^2} + 2\frac{\mathrm{d}r}{\mathrm{d}t}\frac{\mathrm{d}\theta}{\mathrm{d}t} = r\alpha + 2\omega\frac{\mathrm{d}r}{\mathrm{d}t} = r\alpha + 2\dot{r}\omega. \qquad (3.36)$$

The term

$$\alpha = \frac{\mathrm{d}^2\theta}{\mathrm{d}t^2} = \ddot{\theta}$$

is called the angular acceleration.

The radial component of the acceleration $-r\,\omega^2$ is called *the centripetal* acceleration. The transverse component of the acceleration $2\,\omega\,(\mathrm{d}r/\mathrm{d}t)$ is called the *Coriolis* acceleration.

3.3.5   Cylindrical Coordinates

The cylindrical coordinates r, θ, and z describe the motion of a particle P in the xyz space as shown in Fig. 3.14.

The cylindrical coordinates r and θ are the polar coordinates of P measured in the plane parallel to the $x - y$ plane, and the unit vectors $\mathbf{u}_r$, and $\mathbf{u}_\theta$ are the same.

The coordinate z measures the position of the particle P perpendicular to the $x-y$ plane. The unit vector $\mathbf{k}$ attached to the coordinate z points in the positive z-axis direction. The position vector $\mathbf{r}$ of the particle P in terms of cylindrical coordinates is

$$\mathbf{r} = r\mathbf{u}_r + z\mathbf{k}. \tag{3.37}$$

The coordinate r in (3.37) is not equal to the magnitude of $\mathbf{r}$ except when the particle P moves along a path in the $x-y$ plane.

The velocity of the particle P is

$$\begin{aligned}
\mathbf{v} = \frac{d\mathbf{r}}{dt} &= v_r\mathbf{u}_r + v_\theta\mathbf{u}_\theta + v_z\mathbf{k} \\
&= \frac{dr}{dt}\mathbf{u}_r + r\omega\mathbf{u}_\theta + \frac{dz}{dt}\mathbf{k} \\
&= \dot{r}\mathbf{u}_r + r\omega\mathbf{u}_\theta + \dot{z}\mathbf{k},
\end{aligned} \tag{3.38}$$

and the acceleration of the particle P is

$$\mathbf{a} = \frac{d\mathbf{v}}{dt} = a_r\mathbf{u}_r + a_\theta\mathbf{u}_\theta + a_z\mathbf{k}, \tag{3.39}$$

where

$$a_r = \frac{d^2 r}{dt^2} - r\omega^2 = \ddot{r} - r\omega^2,$$

$$a_\theta = r\alpha + 2\frac{dr}{dt}\omega = r\alpha + 2\dot{r}\omega,$$

$$a_z = \frac{d^2 z}{dt^2} = \ddot{z}. \tag{3.40}$$

3.4 Relative Motion

Suppose that A and B are two particles that move relative to a reference frame with origin at point O, Fig. 3.15.

Let $\mathbf{r}_A$ and $\mathbf{r}_B$ be the position vectors of points A and B relative to O. The vector $\mathbf{r}_{BA}$ is the position vector of point A relative to point B. These vectors are related by

$$\mathbf{r}_A = \mathbf{r}_B + \mathbf{r}_{BA}. \tag{3.41}$$

The time derivative of (3.41) is

$$\mathbf{v}_A = \mathbf{v}_B + \mathbf{v}_{AB}, \tag{3.42}$$

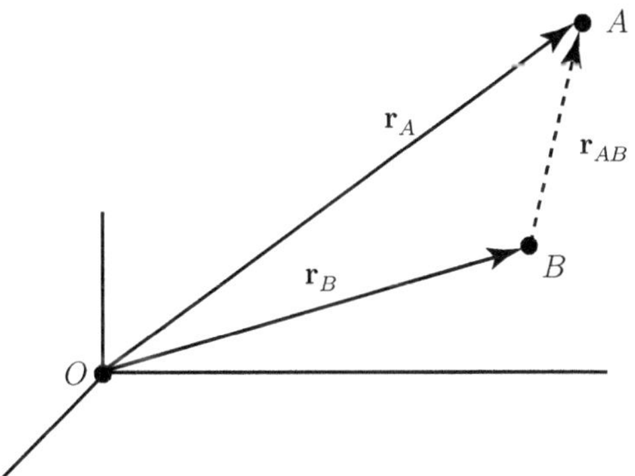

Fig. 3.15 Relative motion of two particles A and B

where $\mathbf{v}_A$ is the velocity of A relative to O, $\mathbf{v}_B$ is the velocity of B relative to O, and $\mathbf{v}_{AB} = \mathrm{d}\mathbf{r}_{AB}/\mathrm{d}t = \dot{\mathbf{r}}_{AB}$ is the velocity of A relative to B. The time derivative of (3.42) is

$$\mathbf{a}_A = \mathbf{a}_B + \mathbf{a}_{AB}, \tag{3.43}$$

where $\mathbf{a}_A$ and $\mathbf{a}_B$ are the accelerations of A and B relative to O and $\mathbf{a}_{AB} = \mathrm{d}\mathbf{v}_{AB}/\mathrm{d}t = \ddot{\mathbf{r}}_{AB}$ is the acceleration of A relative to B.

3.5 Frenet's Formulas

The motion of a particle P along a three-dimensional path is considered, Fig. 3.16a. The tangent direction is defined by the unit tangent vector $\boldsymbol{\tau}$ $(|\boldsymbol{\tau}| = 1)$:

$$\boldsymbol{\tau} = \frac{\mathrm{d}\mathbf{r}}{\mathrm{d}s}. \tag{3.44}$$

The second unit vector is derived by considering the dependence of $\boldsymbol{\tau}$ on s, $\boldsymbol{\tau} = \boldsymbol{\tau}(s)$. The dot product $\boldsymbol{\tau} \cdot \boldsymbol{\tau}$ gives the magnitude of the unit vector $\boldsymbol{\tau}$, that is,

$$\boldsymbol{\tau} \cdot \boldsymbol{\tau} = 1. \tag{3.45}$$

Equation (3.45) can be differentiated with respect to the path variable s:

$$\frac{\mathrm{d}\boldsymbol{\tau}}{\mathrm{d}s} \cdot \boldsymbol{\tau} + \boldsymbol{\tau} \cdot \frac{\mathrm{d}\boldsymbol{\tau}}{\mathrm{d}s} = 0 \implies \boldsymbol{\tau} \cdot \frac{\mathrm{d}\boldsymbol{\tau}}{\mathrm{d}s} = 0. \tag{3.46}$$

Equation (3.46) means that the vector $\mathrm{d}\boldsymbol{\tau}/\mathrm{d}s$ is always perpendicular to the vector $\boldsymbol{\tau}$. The normal direction, with the unit vector is $\boldsymbol{v}$, is defined to be parallel to

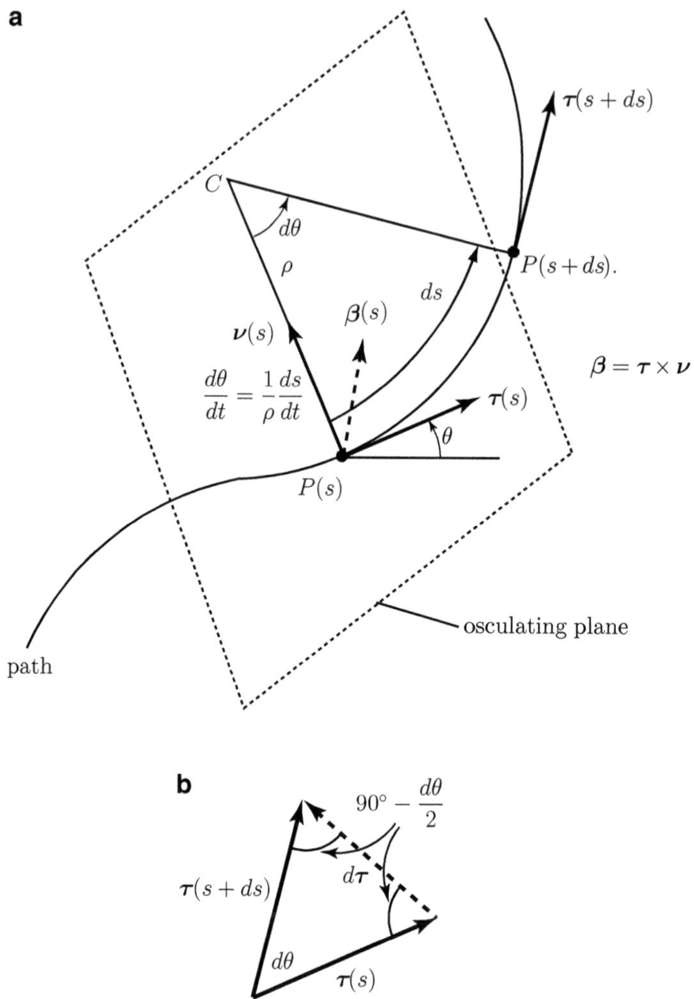

Fig. 3.16 Frenet's reference frame

the derivative $d\tau/ds$. Because parallelism of two vectors corresponds to their proportionality, the normal unit vector may be written as

$$v = \rho \frac{d\tau}{ds},\tag{3.47}$$

or

$$\frac{d\tau}{ds} = \frac{1}{\rho}v,\tag{3.48}$$

where ρ is the radius of curvature. Figure 3.16a depicts the tangent and normal vectors associated with two points, $P(s)$ and $P(s+ds)$. The two points are separated by an infinitesimal distance ds measured along an arbitrary planar path. The point C is the intersection of the normal vectors at the two positions along the curve, and it is the center of curvature. Because ds is infinitesimal, the arc $P(s)P(s+ds)$ seems to be circular. The radius ρ of this arc is the radius of curvature. The formula for the arc of a circle is

$$d\theta = ds/\rho.$$

The angle $d\theta$ between the normal vectors in Fig. 3.16a is also the angle between the tangent vectors $\tau(s+ds)$ and $\tau(s)$. The vector triangle $\tau(s+ds)$, $\tau(s)$, $d\tau = \tau(s+ds) - \tau(s)$ in Fig. 3.16b is isosceles because $|\tau(s+ds)| = |\tau(s)| = 1$. Hence, the angle between $d\tau$ and either tangent vector is $90° - d\theta/2$. Since $d\theta$ is infinitesimal, the vector $d\tau$ is perpendicular to the vector τ in the direction of v. A unit vector has a length of one, so

$$|d\tau| = d\theta|\tau| = \frac{ds}{\rho}.$$

Any vector may be expressed as the product of its magnitude and a unit vector defining the sense of the vector

$$d\tau = |d\tau|v = \frac{ds}{\rho}v. \tag{3.49}$$

Note that the radius of curvature ρ is generally not a constant.

The tangent (τ) and normal (v) unit vectors at a selected position form a plane, the *osculating plane*, that is tangent to the curve. Any plane containing τ is tangent to the curve. When the path is not planar, the orientation of the osculating plane containing the τ, v pair will depend on the position along the curve. The direction perpendicular to the osculating plane is called the *binormal*, and the corresponding unit vector is β. The cross product of two unit vectors is a unit vector perpendicular to the original two, so the binormal direction may be defined such that

$$\beta = \tau \times v. \tag{3.50}$$

Next, the derivative of the v unit vector with respect to s in terms of its tangent, normal, and binormal components will be calculated. The component of any vector in a specific direction may be obtained from a dot product with a unit vector in that direction

$$\frac{dv}{ds} = \left(\tau \cdot \frac{dv}{ds}\right)\tau + \left(v \cdot \frac{dv}{ds}\right)v + \left(\beta \cdot \frac{dv}{ds}\right)\beta. \tag{3.51}$$

The orthogonality of the unit vectors τ and v, $\tau \perp v$, requires that

$$\tau \cdot v = 0. \tag{3.52}$$

Equation (3.52) can be differentiated with respect to the path variable: s

$$\tau \cdot \frac{dv}{ds} + v \cdot \frac{d\tau}{ds} = 0,$$

or

$$\tau \cdot \frac{dv}{ds} = -v \cdot \frac{d\tau}{ds} = -v \cdot \left(\frac{1}{\rho}v\right) = -\frac{1}{\rho}. \tag{3.53}$$

Because $v \cdot v = 1$, the following relation can be written:

$$v \cdot \frac{dv}{ds} = 0. \tag{3.54}$$

The derivative of the binormal component is

$$\frac{1}{T} = \beta \cdot \frac{dv}{ds}, \tag{3.55}$$

or

$$\frac{dv}{ds} = \frac{1}{T}\beta, \tag{3.56}$$

where T is the *torsion*. The reciprocal is used for consistency with (3.48). The torsion T has the dimension of length. Substitution of (3.53)–(3.55) into (3.51) results in

$$\frac{dv}{ds} = -\frac{1}{\rho}\tau + \frac{1}{T}\beta. \tag{3.57}$$

The derivative of β

$$\frac{d\beta}{ds} = \left(\tau \cdot \frac{d\beta}{ds}\right)\tau + \left(v \cdot \frac{dv}{ds}\right)v + \left(\beta \cdot \frac{d\beta}{ds}\right)\beta, \tag{3.58}$$

may be obtained by a similar approach. Using the fact that τ, v, and β are mutually orthogonal, and (3.48), (3.57), yields

$$\tau \cdot \beta = 0 \Longrightarrow \tau \cdot \frac{d\beta}{ds} = -\frac{d\tau}{ds} \cdot \beta = -\frac{1}{\rho}v \cdot \beta = 0,$$

$$v \cdot \beta = 0 \Longrightarrow v \cdot \frac{d\beta}{ds} = -\frac{dv}{ds} \cdot \beta = -\frac{1}{T},$$

$$\beta \cdot \beta = 1 \Longrightarrow \beta \cdot \frac{d\beta}{ds} = 0. \tag{3.59}$$

The result is

$$\frac{d\boldsymbol{\beta}}{ds} = -\frac{1}{T}\boldsymbol{v}. \tag{3.60}$$

Because $\boldsymbol{v}$ is a unit vector, this relation provides an alternative to (3.56) for the torsion

$$\frac{1}{T} = -\left|\frac{d\boldsymbol{\beta}}{ds}\right|. \tag{3.61}$$

Equations (3.48), (3.57), and (3.60) are the Frenet's formulas for a spatial curve.

Next, the path is given in parametric form, the x, y, and z coordinates are given in terms of a parameter α. The position vector is written as

$$\mathbf{r} = x(\alpha)\mathbf{i} + y(\alpha)\mathbf{j} + z(\alpha)\mathbf{k}. \tag{3.62}$$

The unit tangent vector is

$$\boldsymbol{\tau} = \frac{d\mathbf{r}}{d\alpha}\frac{d\alpha}{ds} = \frac{\mathbf{r}'(\alpha)}{s'(\alpha)}, \tag{3.63}$$

where a prime denotes differentiation with respect to α and

$$\mathbf{r}' = x'\mathbf{i} + y'\mathbf{j} + z'\mathbf{k}.$$

Using the fact that $|\boldsymbol{\tau}| = 1$, the derivative of the arc length s is

$$s' = (\mathbf{r}' \cdot \mathbf{r}')^{1/2} = \left[(x')^2 + (y')^2 + (z')^2\right]^{1/2}. \tag{3.64}$$

The arc length s is computed with the relation

$$s = \int_{\alpha_o}^{\alpha} \left[(x')^2 + (y')^2 + (z')^2\right]^{1/2} d\alpha, \tag{3.65}$$

where α_o is the value at the starting position. The value of s' found from (3.64) may be substituted into (3.63) to calculate the tangent vector

$$\boldsymbol{\tau} = \frac{x'\mathbf{i} + y'\mathbf{j} + z'\mathbf{k}}{\sqrt{(x')^2 + (y')^2 + (z')^2}}. \tag{3.66}$$

From (3.63) and (3.48), the normal vector is

$$\boldsymbol{v} = \rho\frac{d\boldsymbol{\tau}}{ds} = \rho\frac{d\boldsymbol{\tau}}{d\alpha}\frac{d\alpha}{ds} = \frac{\rho}{s'}\left(\frac{\mathbf{r}''}{s'} - \frac{\mathbf{r}'s''}{(s')^2}\right)$$

$$= \frac{\rho}{(s')^3}(\mathbf{r}''s' - \mathbf{r}'s''). \tag{3.67}$$

The value of s' is given by (3.64), and the value of s'' is obtained differentiating (3.64):

$$s'' = \frac{\mathbf{r}' \cdot \mathbf{r}''}{(\mathbf{r}' \cdot \mathbf{r}')^{1/2}} = \frac{\mathbf{r}' \cdot \mathbf{r}''}{s'}. \tag{3.68}$$

The expression for the normal vector is obtained by substituting (3.68) into (3.67):

$$v = \frac{\rho}{(s')^4} \left[\mathbf{r}''(s')^2 - \mathbf{r}'(\mathbf{r}' \cdot \mathbf{r}'') \right]. \tag{3.69}$$

Because $v \cdot v = 1$, the radius of curvature is

$$\frac{1}{\rho} = \frac{\rho}{(s')^4} \left| \left[\mathbf{r}''(s')^2 - \mathbf{r}'(\mathbf{r}' \cdot \mathbf{r}'') \right] \right|$$

$$= \frac{\rho}{(s')^4} \left[\mathbf{r}'' \cdot \mathbf{r}''(s')^4 - 2(\mathbf{r}' \cdot \mathbf{r}'')^2(s')^2 + \mathbf{r}'(\mathbf{r}' \cdot \mathbf{r}'') \right]^{1/2}.$$

which simplifies to

$$\frac{1}{\rho} = \frac{1}{(s')^3} [\mathbf{r}'' \cdot \mathbf{r}''(s')^2 - (\mathbf{r}' \cdot \mathbf{r}'')^2]^{1/2}. \tag{3.70}$$

In the case of a planar curve $y = y(x)$ ($\alpha = x$), (3.70) reduces to (3.20).

The binormal vector may be calculated with the relation

$$\beta = \tau \times v = \frac{\mathbf{r}'}{s'} \times \frac{\rho}{(s')^4} \left[\mathbf{r}''(s')^2 - \mathbf{r}'(\mathbf{r}' \cdot \mathbf{r}'') \right] \tag{3.71}$$

$$= \frac{\rho}{(s')^3} \mathbf{r}' \times \mathbf{r}''.$$

The result of differentiating (3.71) is written as

$$\frac{d\beta}{ds} = \frac{1}{s'} \frac{d\beta}{d\alpha} = \frac{1}{s'} \frac{d}{d\alpha} \left[\frac{\rho}{(s')^3} \right] (\mathbf{r}' \times \mathbf{r}'') + \frac{\rho}{(s')^4} (\mathbf{r}' \times \mathbf{r}'''). \tag{3.72}$$

The torsion T may be obtained by applying the formula

$$\frac{1}{T} = -v \cdot \frac{d\beta}{ds}$$

$$= -\frac{\rho}{(s')^4} [\mathbf{r}''(s')^2 - \mathbf{r}'(\mathbf{r}' \cdot \mathbf{r}'')] \cdot \left[\frac{1}{s'} \frac{d}{d\alpha} \left(\frac{\rho}{(s')^3} \right) (\mathbf{r}' \times \mathbf{r}'') + \frac{\rho}{(s')^4} (\mathbf{r}' \times \mathbf{r}''') \right].$$

The above equation may be simplified and T can be calculated from

$$\frac{1}{T} = -\frac{\rho^2}{(s')^6}[\mathbf{r}'' \cdot (\mathbf{r}' \times \mathbf{r}''')]. \qquad (3.73)$$

The expressions for the velocity and acceleration in normal and tangential directions for three-dimensional motion are identical in form to the expressions for planar motion. The velocity is a vector whose magnitude equals the rate of change of distance and whose direction is tangent to the path. The acceleration has a component tangential to the path equal to the rate of change of the magnitude of the velocity and a component perpendicular to the path that depends on the magnitude of the velocity and the instantaneous radius of curvature of the path. In three-dimensional motion, v is parallel to the osculating plane, whose orientation depends on the nature of the path. The binormal vector β is a unit vector that is perpendicular to the osculating plane and therefore defines its orientation.

To calculate the vectors of the Frenet's reference frame, the following MATLAB function is introduced (TrFrenet.m):

```
function [vTangent,vNormal,vBinormal]=TrFrenet(x,y,z,t)

[v_r,a_r,magn_v,magn_a] = part_der(x,y,z,t);
% Tangent
vTangent = v_r./magn_v
dvTangent=diff(vTangent,t)
magn_dvTangent = ...
sqrt(dvTangent(1)^2+dvTangent(2)^2+dvTangent(3)^2);
% Normal
vNormal = dvTangent/magn_dvTangent
% Binormal
vBinormal = cross(vTangent,vNormal)
```

The function TrFrenet.m calls the function par_der:

```
function [v_r,a_r,magn_v,magn_a] = part_der(x,y,z,t)

% 1st Order Partial Derivative

% velocity components
v_x=diff(x,t);
v_y=diff(y,t);
v_z=diff(z,t);
v_r=[v_x v_y v_z];
% magnitude of the velocity
magn_v=sqrt(v_r(1)^2+v_r(2)^2+v_r(3)^2);

% 2nd Order Partial Derivative
a_x=diff(x,t,2);
```

```
a_y=diff(y,t,2);
a_z=diff(z,t,2);
a_r=[a_x a_y a_z];
% magnitude of the acceleration
magn_a=sqrt(a_r(1)^2+a_r(2)^2+a_r(3)^2);
```

3.6   Examples

Example 3.1. A particle M is moving with respect to a Cartesian reference frame
xOy. The parametric equations of the motion of the particle are given by $x = R\cos\theta$,
$y = R\sin\theta$, $\theta = \omega t$, where R, ω, are constants and t is the time. Find the velocity
and the acceleration of the particle M. Plot the trajectory, the velocity, and the
acceleration of the particle in MATLAB for $\omega = 1.5\,\text{rad/s}$ and $R = 5\,\text{m}$.

Solution
The trajectory of the particle is obtained eliminating the time t from the parametric
equations. From $\frac{x}{R} = \cos\theta$, $\frac{y}{R} = \sin\theta$, and $\cos^2\theta + \sin^2\theta = 1$ give

$$x^2 + y^2 = R^2.$$

The trajectory of the moving particle is given by the equation of the circle with
radius R and the origin at O, Fig. 3.17a.
   The components of the velocity are

$$v_x = \dot{x} = -R\dot{\theta}\sin\theta = -R\omega\sin\theta,$$

$$v_y = \dot{y} = R\dot{\theta}\cos\theta = R\omega\cos\theta,$$

and the components of the acceleration are

$$a_x = \ddot{x} = \dot{v}_x = -R\dot{\theta}^2\cos\theta = -R\omega^2\cos\theta,$$

$$a_y = \ddot{y} = \dot{v}_y = -R\dot{\theta}^2\sin\theta = -R\omega^2\sin\theta.$$

The magnitude of the acceleration is

$$a = \sqrt{\ddot{x}^2 + \ddot{y}^2} = R\omega^2.$$

The projections of the particle on the x and y axes are the points M' and M''.
The displacements of the points M' and M'' represents oscillatory motions with the
period $T = \frac{2\pi}{\omega}$. When the particle M has achieved a complete rotation $t = T$, the
projections M' and M'' have complete oscillations on x and y axes.
   The physical motion of the particle is obtained using the practical mechanism
shown in Fig. 3.17b. The crank OM (link 1) is rotating with respect to the fixed
point O, and the slider M (link 2) describes a circle of radius R. The slider M has a
rectilinear motion with respect to the slider 3.

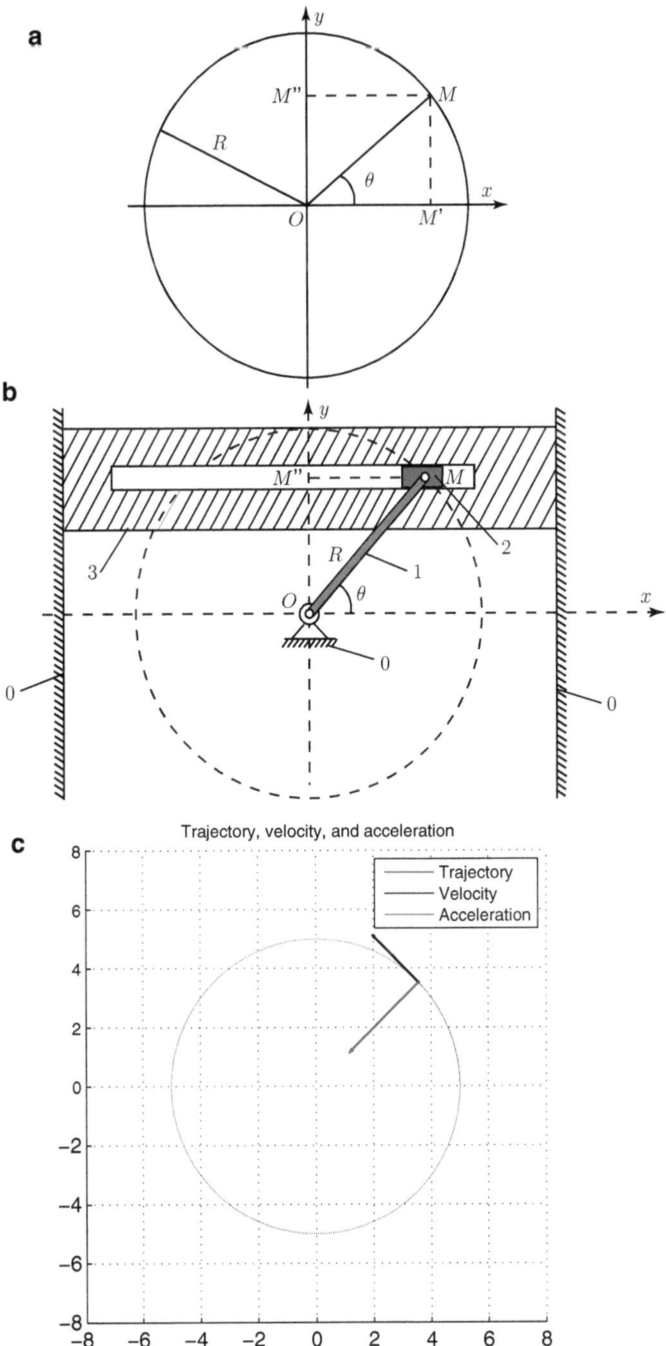

Fig. 3.17 Example 3.1

The equation of M on x-axis is

$$MM'' = x = R \cos \omega t.$$

The slider 3 has a rectilinear oscillatory motion with respect to the fixed point O and the fixed link 0. The equation of M on y-axis is

$$OM'' = y = R \sin \omega t.$$

The MATLAB program for the symbolical calculation is:

```
syms R omega t;

% parametric equation of motion
theta=omega*t;
x=R*cos(theta);
y=R*sin(theta);

% velocity
v_x=diff(x,t);
v_y=diff(y,t);
v=[v_x v_y];
% magnitude of velocity
magn_v=sqrt(v(1)^2+v(2)^2);

% acceleration
a_x=diff(x,t,2);
a_y=diff(y,t,2);
a=[a_x a_y];
% magnitude of acceleration
magn_a=sqrt(a(1)^2+a(2)^2);
```

To print the results for the velocities, the following commands are used:

```
fprintf('velocity components v_x and v_y \n')
pretty(v); fprintf('\n\n')
fprintf('magnitude of the velocity \n')
pretty(simplify(magn_v))
fprintf('\n\n')
```

and the results are

```
velocity components v_x and v_y


   +-                                                      -+
   | -R omega sin(omega t), R omega cos(omega t) |
   +-                                                      -+


magnitude of the velocity
```

```
   2           2  1/2
  (R    omega  )
```

MATLAB gives the result $a^\wedge (1/2)$, which means $a^{1/2}$, using symbolic notation for the square root operation, without actually calculating the simplified value.

Next, the trajectory, the velocity, and the acceleration of the particle will be represented in MATLAB. For the numerical application

```
omegan=1.5;
Rn= 5.0;
```

a symbolic list is introduced as

```
slist={R,omega,t};
```

A system of axis is defined as

```
axis manual
axis equal
hold on
grid on
axis([-8 8 -8 8])
title('Trajectory, velocity, and acceleration')
```

The trajectory of the particle is plotted at each x and y location defined by the parametric equation of motion for different time `tn=0, 0.05, 0.1, 0.15,` ... A for loop (statement) is used to repeatedly calculates the position, velocity, and acceleration. The numerical values are calculated replacing the symbolic variables R, omega, and t in `slist` with the numerical values from `nlist`:

```
nlist={Rn,omegan,tn};
```

The numerical values for the position, xn, yn, velocity, vxn, vyn, and acceleration, axn, ayn, will be calculated inside the for loop:

```
scalefactor = 3;
for tn = 0 : 0.01 : 3*pi/2
   nlist = {Rn, omegan, tn};
   xn = subs(x, slist, nlist);
   yn = subs(y, slist, nlist);
   vxn = subs(v_x, slist, nlist)/scalefactor;
   vyn = subs(v_y, slist, nlist)/scalefactor;
   axn = subs(a_x, slist, nlist)/scalefactor;
   ayn = subs(a_y, slist, nlist)/scalefactor;
```

The `scalefactor` was introduced for a proper representation of the velocity and acceleration vectors. First, the position of the particle (current position) is represented by a red dot and is updated at each `tm` value of the loop:

```
hm = plot(xn, yn, 'k.','Color', 'red');
```

Next, the trajectory is plotted by small blue dots (retain the particle trajectory):

```
ht = plot(xn, yn);
```

The velocity of the particle is represented by a black vector, and the acceleration of the particle is represented by a red vector and is updated at each `tm` value of the loop:

```
pv=quiver(xn,yn,vxn,vyn,'Color','k','LineWidth',1);
pa=quiver(xn,yn,axn,ayn,'Color','r','LineWidth',1);
```

The simulation is slowed down with the command (brief pause to make the animation looks good):

```
pause(0.001)
```

Next, the particle represented by the red dot is deleted from the old location (delete the particle situated at the old position before plotting the particle at the new location):

```
delete(hm);
```

The velocity vector and the acceleration vector are also deleted from the old location (before plotting the new ones at the new locations) using the commands

```
delete(pv);
delete(pa);
```

The for loop depicts at each `tm` value of the loop the trajectory, the velocity, and the acceleration, and then using a delay, the particle, the velocity, and the acceleration are deleted. For the last value of `tm`, the particle, the velocity, and the acceleration are plotted with

```
hl = plot(xn,yn,'k.','Color','r');
pause(0.5)
quiver(xn,yn,vxn,vyn,'color','k','LineWidth',1.3);
pause(0.5)
quiver(xn,yn,axn,ayn,'color','r','LineWidth',1.3);
```

The MATLAB legend for the plot is obtained with

```
[legend_h, object_h, plot_h,text_strings]=...
 legend('Trajectory','Velocity','Acceleration');
set(plot_h(1),'color','b');
set(object_h(1),'color','b');
set(plot_h(2),'color','k');
set(object_h(2),'color','k');
set(plot_h(3),'color','r');
set(object_h(3),'color','r');
```

Here, `legend_h` handles the legend axes `object_h` handles the line, patch, and text graphics objects, `plot_h` handles the lines and other objects used in the plot, and `text_strings` is a cell array of the text strings used in the legend. The command

```
set(plot_h(1),'color','b');
```

assigns a blue color to the written text `Trajectory`. The command

```
set(object_h(1),'color','b');
```

assigns blue color to the line used in legend (same color with the line used to represent the particle trajectory), line that corresponds to the written text `Trajectory`. For the `Velocity`, the color black is assigned, and the `Acceleration` is plotted red. The final MATLAB plot is depicted in Fig. 3.17c.

Example 3.2. A particle M is moving with respect to the reference frame xOy. The parametric equations of the motion of the particle are given by $x = l_1 \sin\theta$, $y = l_2 \cos\theta$, $\theta = \omega t$, where l_1, l_2, and ω are constants and t is the time. Find the velocity and the acceleration of the particle. Plot the trajectory, the velocity, and the acceleration of the particle in MATLAB for $l_1 = 5\,\text{m}$, $l_2 = 2.5\,\text{m}$, and $\omega = 1.5\,\text{rad/s}$.

Solution
From $\frac{x}{l_1} = \cos\theta$, $\frac{x}{l_2} = \sin\theta$ and $\cos^2\theta + \sin^2\theta = 1$, the trajectory of the particle is given by the equation

$$\frac{x^2}{l_1^2} + \frac{y^2}{l_2^2} = 1. \tag{3.74}$$

The trajectory of the particle is given by the equation of an ellipse with the semiaxes l_1 and l_2, Fig. 3.18a. The magnitude of the velocity is given by

$$v = \sqrt{\dot{x}^2 + \dot{y}^2} = \omega\sqrt{l_1^2 \sin^2 \omega t + l_2^2 \cos^2 \omega t}, \tag{3.75}$$

and the magnitude of the acceleration is given by

$$a = \sqrt{\ddot{x}^2 + \ddot{y}^2} = \omega^2\sqrt{l_1^2 \cos^2 \omega t + l_2^2 \sin^2 \omega t}. \tag{3.76}$$

The projections of the particle on the x and y axes are the points M' and M''. The displacements of M' and M'' represent oscillatory motions with the period $T = \frac{2\pi}{\omega}$. When the particle M has reached a complete rotation $t = T$, the projections M' and M'' have complete oscillations on x and y axes.

The physical motion of the particle using the mechanism shown in Fig. 3.18b. The crank OA (link 1) has a fixed point at O and is rotating with respect to O. The sliders 3 and 5 are moving on the fixed axes y and x at B and C.

Fig. 3.18 Example 3.2

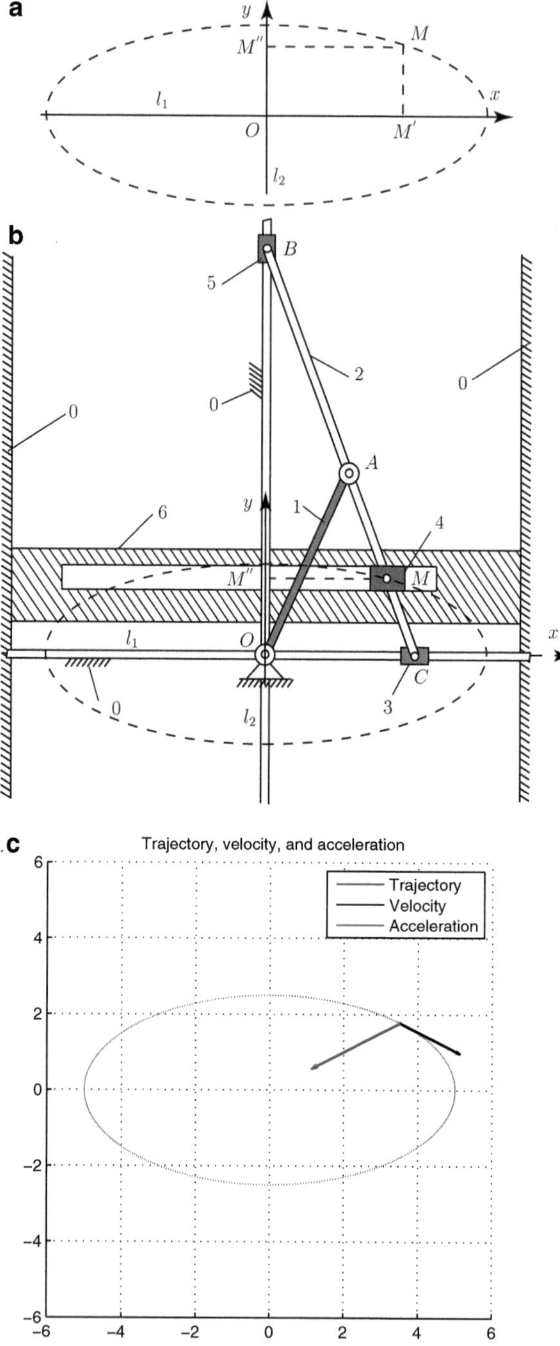

The trajectory of the slider 4 is the ellipse where $BM = l_1$ and $CM = l_2$. The slider 4 has a rectilinear oscillatory motion with respect to the link 6. The equation of M on the x-axis is

$$MM'' = x = l_1 \cos \omega t.$$

The slider 6 has a rectilinear oscillatory motion with respect to the fixed link 0. The equation of M on the y-axis is

$$OM'' = y = l_2 \sin \omega t.$$

The final MATLAB plot is depicted in Fig. 3.18c.

Example 3.3. A particle M is moving in the xy plane as seen in Fig. 3.19a. The trajectory of the particle is a cycloid defined by the parametric equations $x = r(\theta - \sin \theta)$, $y = r(1 - \cos \theta)$, $\theta = \omega t$, where r, ω are constants and t is the time. The cycloid is the locus of a point on the rim of the circle of radius r and center C rolling along a straight line. Analyze the motion of the particle.

Solution
Suppose that a circle of radius r is tangent to the x-axis at the time $t = 0$, at O. After a time t, the circle rotates with an angle θ, and the initial point O is at the position M, Fig. 3.19b. The geometrical condition of rolling without slipping is

$$OC' = \text{arc length } C'M = r\theta,$$

where C' is the new tangent point to the x-axis. The coordinates of the point M are

$$x = OC' - pr_x\,CM = r\theta - r \cos\left(\frac{\pi}{2} - \theta\right) = r(\theta - \sin \theta),$$

$$y = C'C - pr_y\,CM = r - r \cos \theta = r(1 - \cos \theta), \tag{3.77}$$

where pr_x is the projection on x-axis and pr_y is the projection on y-axis. The derivatives of (4.4) give

$$\dot{x} = r\omega(1 - \cos \theta),$$

$$\dot{y} = r\omega \sin \theta,$$

$$\dot{\theta} = \omega.$$

The velocity of point M is

$$\mathbf{v} = r\omega(1 - \cos \omega t)\,\mathbf{\imath} + r\omega \sin \omega t\,\mathbf{\jmath},$$

and the speed of M is

$$|\mathbf{v}| = v = \sqrt{\dot{x}^2 + \dot{y}^2} = r\omega \sqrt{2(1 - \cos \omega t)} = 2r\omega \sin \frac{\omega t}{2}.$$

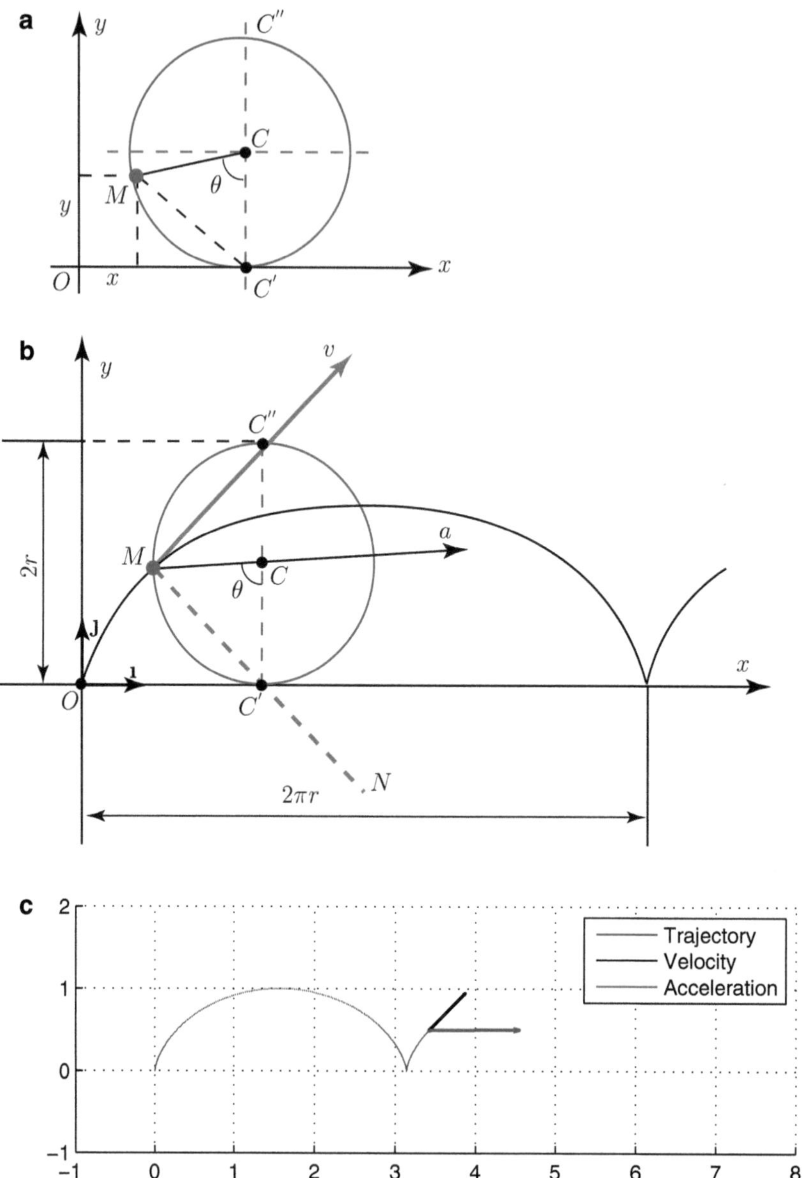

Fig. 3.19 Example 3.3

The MATLAB program for the calculation of the velocity of M is

```
syms R omega t r real

theta=omega*t;
```

```
x=r*(theta-sin(theta));
y=r*(1-cos(theta));

v_x=diff(x,t);
v_y=diff(y,t);
v=[v_x v_y];
magn_v=simplify(sqrt(v(1)^2+v(2)^2));

fprintf('velocity \n')
pretty(v); fprintf('\n\n')
fprintf('magnitude of velocity \n')
pretty(magn_v); fprintf('\n\n')
```

The position vectors of M and C' are

$$\mathbf{r}_M = r\left(\theta - \sin\theta\right)\mathbf{\imath} + r\left(1 - \cos\theta\right)\mathbf{\jmath},$$

$$\mathbf{r}_{C'} = r\theta\,\mathbf{\imath}.$$

The position vector $\mathbf{r}_{C'M}$ is

$$\mathbf{r}_{C'M} = -r\sin\omega t\,\mathbf{\imath} + r\left(1 - \cos\omega t\right)\mathbf{\jmath},$$

The scalar product of the vectors $\mathbf{r}_{C'M}$ and $\mathbf{v}$ is

$$\mathbf{r}_{C'M}\cdot\mathbf{v} = \left[-r\sin\omega t\,\mathbf{\imath} + r\left(1 - \cos\omega t\right)\mathbf{\jmath}\right]\cdot\left[r\omega\left(1 - \cos\omega t\right)\mathbf{\imath} + r\omega\sin\omega t\,\mathbf{\jmath}\right]$$
$$= -r^2\,\omega\sin\omega t\,\left(1 - \cos\omega t\right) + r^2\,\omega\left(1 - \cos\omega t\right)\sin\omega t = 0.$$

That is the velocity $\mathbf{v}$ is perpendicular to $C'M$ and tangent to cycloid curve. The following MATLAB program shows that the vectors $\mathbf{v}$ and $C'M$ are perpendicular:

```
% position vector of M
rM=[x y];
% position vector of C'
rCp=[r*theta 0];
% position vector C'M
rCpM=simplify(rM-rCp);
rCpMv=simplify(dot(rCpM,v));
fprintf('rCpM.v = %s \n',char(rCpMv))
% rC'M.v = 0 => v perpendicular to rC'M
```

The point C'' is the end of the diameter $C'CC''$, see Fig. 3.19b. The MATLAB commands that show the point C'' is located on the velocity vector $\mathbf{v}$ are

```
% position vector of C''
rCpp=[r*theta 2*r];
% position vector C''M
```

```
rCppM=simplify(rM-rCpp);
rCppM=simplify(dot(rCpM,rCppM));
fprintf('rCpM.rCppM = %s \n',char(rCppM))
% rC'M.rC''M = 0 => C'M perpendicular to C''M
```

The magnitude of the vector $\mathbf{r}_{C'M}$ is

$$C'M = |\mathbf{r}_{C'M}| = \sqrt{r^2 \sin^2 \omega t + r^2 (1 - \cos \omega t)^2} = r\sqrt{\sin^2 \omega t + (1 - \cos \omega t)^2},$$

and the magnitude of velocity is

$$v = |\mathbf{v}| = \sqrt{r^2 \omega^2 (1 - \cos \omega t)^2 + r^2 \omega^2 \sin^2 \omega t} = r\omega \sqrt{(1 - \cos \omega t)^2 + \sin^2 \omega t}.$$

It results

$$|\mathbf{v}| = \omega \, |\mathbf{r}_{C'M}| \quad \text{or} \quad v = \omega C'M.$$

In MATLAB

```
magn_CpM=sqrt(rCpM(1)^2+rCpM(2)^2);
om=simplify(magn_v/magn_CpM);
fprintf('|v|/|rCpM| = %s \n',char(om))
```

The acceleration of M is given by

$$\ddot{x} = r\omega^2 \sin \theta,$$
$$\ddot{y} = r\omega^2 \cos \theta,$$

and

$$a = \sqrt{\ddot{x}^2 + \ddot{y}^2} = r\omega^2,$$

or using MATLAB

```
a_x=diff(x,t,2);
a_y=diff(y,t,2);
a=[a_x a_y];
magn_a=sqrt(a(1)^2+a(2)^2);

fprintf('acceleration \n')
pretty(a); fprintf('\n\n')
fprintf('magnitude of the acceleration \n')
pretty(simplify(magn_a)); fprintf('\n\n')
```

The acceleration vector is

$$\mathbf{a} = r\omega^2 \sin \omega t \mathbf{i} + r\omega^2 \cos \omega t \mathbf{j},$$

and the vector $\mathbf{r}_{MC}$ is

$$\mathbf{r}_{MC} = r \sin \omega t \mathbf{i} + r \cos \omega t \mathbf{j}.$$

It results the direction of the acceleration is along the radius line MC. In MATLAB, it is shown that the acceleration vector a is equal with the vector rMC multiplied with omega squared:

```
% position vector of C
rC=[r*theta r];
rMC=simplify(rC-rM)
a==omega^2*rMC
```

The tangential acceleration is

$$a_\tau = \frac{dv}{dt} = \frac{d}{dt}\left(2r\omega\sin\frac{\omega t}{2}\right) = r\omega^2\cos\frac{\omega t}{2}.$$

The normal acceleration, a_v, is calculated from

$$a^2 = a_\tau^2 + a_v^2,$$

and

$$a_v^2 = a^2 - a_\tau^2 = r^2\omega^4 - \left(r\omega^2\cos\frac{\omega t}{2}\right)^2 = r^2\omega^4\left(\sin\frac{\omega t}{2}\right)^2.$$

The radius of curvature, ρ, of the cycloid is calculated from the relation

$$a_v = \frac{v^2}{\rho} = \frac{4r^2\omega^2\sin^2\frac{\omega t}{2}}{\rho} = r\omega^2\sin\frac{\omega t}{2}.$$

The radius of curvature is

$$\rho = 4r\sin\frac{\omega t}{2} = 4r\sin\frac{\theta}{2}.$$

Because $MC' = 2r\sin\frac{\theta}{2}$ ($\Delta C'CM$ is isoscel), it results that the center of curvature of the cycloid is on the normal MC' to the tangent MC'' at the distance $MN = 2MC' = \rho$. The MATLAB plot for $\omega = 2.5\,\text{rad/s}$ and $r = 0.5\,\text{m}$ is shown in Fig. 3.19c.

Example 3.4. A particle is moving along a straight line $OABC$ starting from the origin O when $t = 0$, as shown in Fig. 3.20a, and has no initial velocity. For the first segment OA, the motion has a constant acceleration. The velocity of the particle after $t_1 = 1$ s at the point A is $V = 1$ m/s. For the segment AB, the velocity of the particle remains constant for the next $t_2 = 1$ s . For the last segment BC, the particle is decelerating with a constant acceleration until it comes to a complete stop at C. It takes the particle $t_3 = 3$ s to go from point B to point C.

Determine and plot the acceleration, velocity, and position versus time of the particle for the segment OC.

Solution

1. *Segment OA*

For the segment OA, the acceleration is constant $a = $ constant:

$$\ddot{x}(t) = a_1.$$

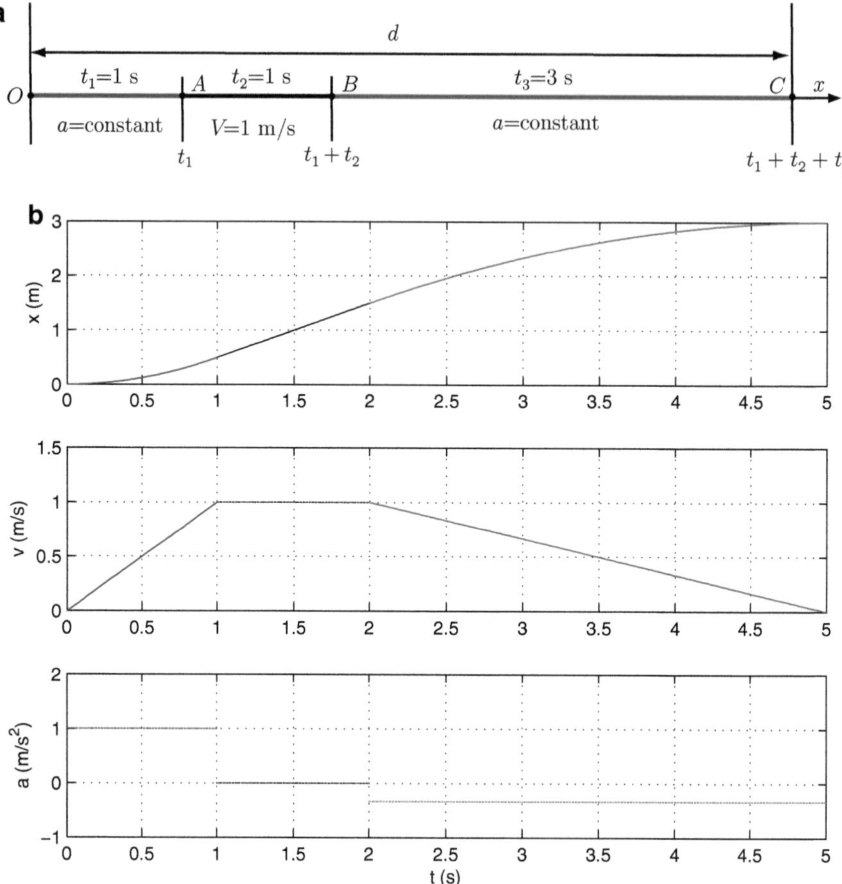

Fig. 3.20 Example 3.4

The velocity equation is obtained taking the integral of the acceleration:

$$\dot{x}(t) = \int a_1 \, dt = a_1 t + c_1.$$

The constant c_1 is obtained from the initial condition for the velocity at O:

$$t = 0 \Rightarrow \dot{x}(0) = 0 \text{ or } c_1 = 0.$$

The constant acceleration a_1 is calculated from the fact that the velocity is V at A when $t = t_1$:

$$t = t_1 \Rightarrow \dot{x}(t_1) = 0 \text{ or } a_1 t_1 = V.$$

The acceleration is

$$a_1 = \frac{V}{t_1} = \frac{1 \text{ m/s}}{1 \text{ s}} = 1 \text{ m/s}^2.$$

The velocity equation is

$$\dot{x}(t) = \frac{V}{t_1} t = t \qquad \text{for } t \in [0; t_1].$$

The position is obtained taking the integral of the velocity:

$$x(t) = \frac{V}{t_1} \int t \, dt = \frac{V}{t_1} \frac{t^2}{2} + c_2.$$

From the initial condition for displacement at O, the constant c_2 is obtained:

$$t = 0 \Rightarrow x(0) = 0 \text{ or } c_2 = 0.$$

The distance equation for the first segment OA is

$$x(t) = \frac{V}{2 t_1} t^2 = \frac{t^2}{2} \qquad \text{for } t \in [0; t_1].$$

The distance $d_1 = OA$ for this segment is

$$d_1 = x(t_1) = \frac{V t_1}{2} = \frac{1\,(1)}{2} = \frac{1}{2} \text{ m.}$$

In Fig. 3.20b, the position, velocity, and acceleration are represented as a function of time for the first segment.

2. *Segment AB*

   For the segment AB, the velocity of the particle is constant V:

$$\dot{x}(t) = V \qquad \text{for } t \in [t_1; t_1 + t_2],$$

where $t_2 = 1$ s is the time interval for the particle to travel the segment AB. The acceleration is obtained differentiating the velocity:

$$\ddot{x}(t) = \frac{d}{dt} V = \dot{V} = 0 \text{ or } \ddot{x}(t) = 0 \qquad \text{for } t \in [t_1; t_1 + t_2].$$

The position equation for the segment AB is obtained integrating the velocity:

$$x(t) = \int V \, dt = V t + c_3.$$

At the moment $t = t_1$, the displacement is $d_1 = \frac{V\,t_1}{2}$, and the constant c_3 is calculated from

$$t = t_1 \Rightarrow x(t_1) = V\,t_1 + c_3 = \frac{V\,t_1}{2},$$

or

$$c_3 = -\frac{V\,t_1}{2} = -\frac{1}{2}.$$

The position function of time is

$$x(t) = V\,t - \frac{V\,t_1}{2} = t - \frac{1}{2} \qquad \text{for } t \in [t_1; t_1 + t_2].$$

For the second segment, the distance traveled by the particle is $d_2 = AB$:

$$t = t_1 + t_2 \Rightarrow x(t_1 + t_2) = d_1 + d_2 \quad \text{or} \quad V(t_1 + t_2) - \frac{V\,t_1}{2} = \frac{V\,t_1}{2} + d_2,$$

and $d_2 = V\,t_2 = 1(1) = 1$ m. The distance $s_2 = OB = d_1 + d_2 = V\,t_1/2 + V\,t_2$ is $s_2 = 0.5 + 1.5 = 1.5$ m.

3. *Segment BC*

   For the segment BC, the acceleration is constant and negative because the particle stops at C:

$$\ddot{x}(t) = -a_3 \qquad \text{for } t \in [t_1 + t_2; t_1 + t_2 + t_3],$$

   where a_3 is the constant acceleration of the particle for the last segment. The velocity equation is given by

$$\dot{x}(t) = -\int a_3 \, dt = -a_3 t + c_4.$$

   At the moment $t = t_1 + t_2$, the velocity is V:

$$t = t_1 + t_2 \Rightarrow \dot{x}(t_1 + t_2) = V \ \text{or} \ -a_3(t_1 + t_2) + c_4 = V \ \text{or} \ c_4 = V + a_3(t_1 + t_2).$$

   The velocity equation is

$$\dot{x}(t) = -a_3 t + V + a_3(t_1 + t_2),$$

or

$$\dot{x}(t) = V - a_3[t - (t_1 + t_2)].$$

   At the moment $t = t_1 + t_2 + t_3$ at C, the velocity is zero:

$$\dot{x}(t_1 + t_2 + t_3) = V - a_3[t_1 + t_2 + t_3 - (t_1 + t_2)] = V - a_3 t_3 = 0.$$

The magnitude of the acceleration for the last segment is

$$a_3 = \frac{V}{t_3} = \frac{1}{3}\ \text{m/s}^2.$$

The velocity equation for the last segment will be

$$\dot{x}(t) = V - \frac{V}{t_3}[t - (t_1 + t_2)] = 1 - \frac{1}{3}[t - (1+1)] = 1 - \frac{1}{3}[t - 2] = -\frac{1}{3}t + \frac{5}{3}.$$

The position equation is

$$x(t) = \int \dot{x}(t)\,dt = -\frac{V}{2t_3}t^2 + V\frac{(t_3 + t_1 + t_2)}{t_3}t + c_5.$$

At the moment $t = t_1 + t_2$ at B, the displacement is $d_1 + d_2$

$$t = t_1 + t_2 \Rightarrow x(t_1 + t_2) = d_1 + d_2 \ \text{ or}$$

$$-\frac{V}{2t_3}(t_1 + t_2)^2 + V\frac{(t_3 + t_1 + t_2)}{t_3}(t_1 + t_2) + c_5 = \frac{V t_1}{2} + V t_2.$$

The integration constant c_5 is

$$c_5 = -\frac{V}{2t_3}\left[(t_1 + t_2)^2 + t_1 t_3\right] = -\frac{1}{2\,(3)}\left[2^2 + 1\,(3)\right] = -\frac{7}{6}.$$

The position equation is

$$x(t) = -\frac{V}{2t_3}t^2 + V\frac{(t_3 + t_1 + t_2)}{t_3}t - \frac{V}{2t_3}\left[(t_1 + t_2)^2 + t_1 t_3\right] = -\frac{1}{6}t^2 + \frac{5}{3}t - \frac{7}{6}.$$

At the end of the motion $t = t_1 + t_2 + t_3$, the total displacement is d:

$$t = t_1 + t_2 + t_3 \Rightarrow d = x(t_1 + t_2 + t_3)$$

or

$$d = -\frac{1}{6}5^2 + \frac{5}{3}5 - \frac{7}{6} = 3\,\text{m}.$$

Figure 3.20b shows the position, velocity, and acceleration as a function of time for the segment OC.

Example 3.5. The parametrical equations for the path of a particle are $x = l_1 \cos\theta$, $y = l_2 \sin\theta$, where $l_1 = 3$ m, $l_2 = 2$ m, $\theta = \omega t$ with $\omega = pi/2$. Find the velocity, the acceleration, and the radius of curvature at the time $t_0 = 0$ s, $t_1 = 0.5$ s, and $t_2 = 1$ s.

Solution

1. The trajectory is obtained eliminating the time t from the parametric equations $x = 3\cos\left(\frac{\pi}{2}t\right)$ and $y = 2\sin\left(\frac{\pi}{2}t\right)$ or $\frac{x}{3} = \cos\left(\frac{\pi}{2}t\right)$ and $\frac{y}{2} = \sin\left(\frac{\pi}{2}t\right)$. It results

$$\left(\frac{x}{3}\right)^2 + \left(\frac{y}{2}\right)^2 = \cos^2\left(\frac{\pi}{2}t\right) + \sin^2\left(\frac{\pi}{2}t\right) = 1.$$

The equation

$$\left(\frac{x}{3}\right)^2 + \left(\frac{y}{2}\right)^2 = 1. \tag{3.78}$$

represents an ellipse. The parametrical equations of the particle are periodic function. The period of motion, T, is calculated with the relation $\frac{\pi}{2}T = 2\pi \Rightarrow T = 4$ s. The velocity components are obtained taking the derivative of the position

$$v_x = \dot{x} = -\frac{3\pi}{2}\sin\left(\frac{\pi}{2}t\right), \ v_y = \dot{y} = \pi\cos\left(\frac{\pi}{2}t\right).$$

The magnitude of the velocity is

$$v = |\mathbf{v}| = \sqrt{\dot{x}^2 + \dot{y}^2} = \frac{\pi}{2}\sqrt{9\sin^2\left(\frac{\pi}{2}t\right) + 4\cos^2\left(\frac{\pi}{2}t\right)} = \frac{\pi}{2}\sqrt{5\sin^2\left(\frac{\pi}{2}t\right) + 4}.$$

The acceleration components are

$$a_x = \ddot{x} = -\frac{3\pi^2}{4}\cos\left(\frac{\pi}{2}t\right), \ a_y = \ddot{y} = -\frac{\pi^2}{2}\sin\left(\frac{\pi}{2}t\right).$$

The magnitude of the acceleration is

$$a = |\mathbf{a}| = \sqrt{\ddot{x}^2 + \ddot{y}^2} = \frac{\pi^2}{4}\sqrt{9\cos^2\left(\frac{\pi}{2}t\right) + 4\sin^2\left(\frac{\pi}{2}t\right)} = \frac{\pi^2}{4}\sqrt{5\cos^2\left(\frac{\pi}{2}t\right) + 4}.$$

The radius of curvature is calculated with

$$\rho = \frac{\left(\dot{x}^2 + \dot{y}^2\right)^{3/2}}{|\dot{x}\ddot{y} - \ddot{x}\dot{y}|}.$$

The MATLAB program for calculating v, a, and ρ is

```
syms R omega t theta l_1 l_2

theta = omega*t;
x = l_1*cos(theta);
y = l_2*sin(theta);

v_x = diff(x,t);
v_y = diff(y,t);
v = [v_x v_y];
magn_v=sqrt(v(1)^2+v(2)^2);

a_x = diff(x,t,2);
a_y = diff(y,t,2);
```

```matlab
a = [a_x a_y];
magn_a = sqrt(a(1)^2+a(2)^2);

rho = (v_x^2+v_y^2)^(3/2)...
    /abs(v_x*a_y-a_x*v_y);

% numerical values
omegan = pi/2;
l_1n = 3;
l_2n = 2;

lists = {l_1, l_2, omega};
listt = {l_1n, l_2n, omegan};

xt = subs(x, lists, listt);
yt = subs(y, lists, listt);

vxt = subs(v_x, lists, listt);
vyt = subs(v_y, lists, listt);
vt = [vxt vyt];
magn_vt=sqrt(vt(1)^2+vt(2)^2);

axt = subs(a_x, lists, listt);
ayt = subs(a_y, lists, listt);
at = [axt ayt];
magn_at=sqrt(at(1)^2+at(2)^2);

fprintf('v = [v_x, v_y] = \n')
pretty(v); fprintf('\n\n')
pretty(vt); fprintf('\n\n')
fprintf('|v| = \n')
pretty(magn_v); fprintf('\n\n')
pretty(magn_vt); fprintf('\n\n')

fprintf('a = [a_x, a_y] = \n')
pretty(a); fprintf('\n\n')
pretty(at); fprintf('\n\n')
fprintf('|a| = \n')
pretty(simplify(magn_a)); fprintf('\n\n')
pretty(magn_at); fprintf('\n\n')

fprintf('rho = \n')
pretty(simplify(rho))
rhot = subs(rho, lists, listt);
pretty(simplify(rhot))
```

To plot the trajectory, the vectors, and to calculate the numerical values for different times, the following MATLAB commands are used:

```
scale = 3;
axis manual
axis equal
hold on
grid on
axis([-5 5 -5 5])

syms data
data=[];
i=0;

for tn = 0 : 0.01 : 3*pi/2

   xn = subs(xt, t, tn);
   yn = subs(yt, t, tn);

   hm=plot(xn, yn,'k.','Color','red');
   ht=plot(xn, yn);
   title('Trajectory, velocity and acceleration')
   pause(0.001)
   delete(hm);

   if tn==0 | tn==0.5 | tn==1 | tn==2 | tn==3
        pause(0.2)
        vxn = subs(vxt, t, tn);
        vyn = subs(vyt, t, tn);
quiver(xn,yn,vxn/scale,vyn/scale, ...
'color','k','LineWidth',1.3);
        pause(0.7)
        axn = subs(axt, t, tn);
        ayn = subs(ayt, t, tn);
quiver(xn,yn,axn/scale,ayn/scale, ...
'color','r','LineWidth',1.3);

rhon = subs(rhot, t, tn);

        i=i+1;
data{i} = ...
{xn,yn,vxn,vyn,sqrt(vxn^2+vyn^2), ...
 axn,ayn,sqrt(axn^2+ayn^2),rhon};
    end
end
```

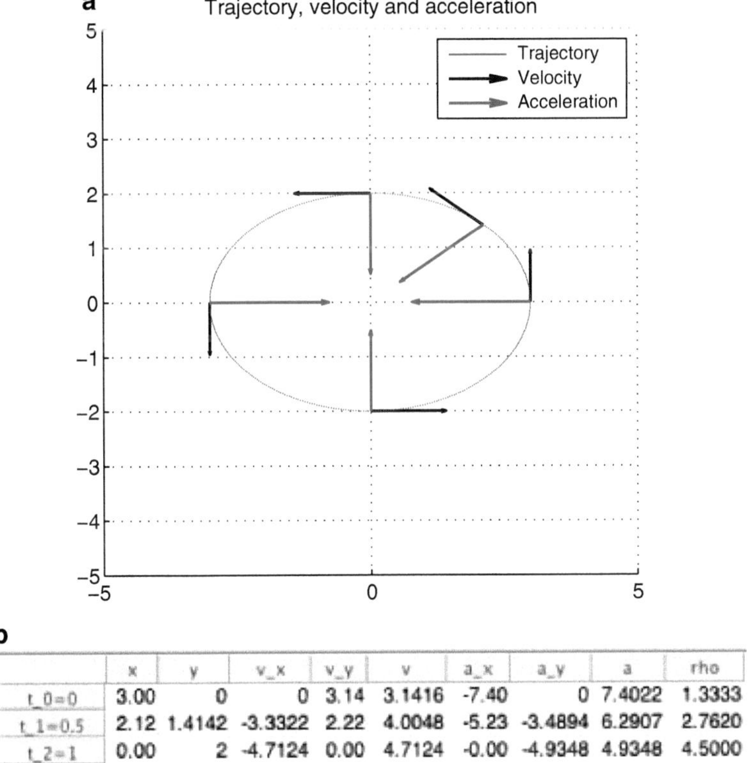

b

	x	y	v_x	v_y	v	a_x	a_y	a	rho
t_0=0	3.00	0	0	3.14	3.1416	-7.40	0	7.4022	1.3333
t_1=0.5	2.12	1.4142	-3.3322	2.22	4.0048	-5.23	-3.4894	6.2907	2.7620
t_2=1	0.00	2	-4.7124	0.00	4.7124	-0.00	-4.9348	4.9348	4.5000

Fig. 3.21   Example 3.5

```
[legend_h, object_h, plot_h,text_strings]=...
 legend('Trajectory','Velocity','Acceleration');

set(plot_h(1),'color','b');
set(object_h(1),'color','b');
set(plot_h(2),'color','k');
set(object_h(2),'color','k');
set(plot_h(3),'color','r');
set(object_h(3),'color','r');
```

The results are depicted in Fig. 3.21a.

To display the numerical results the following MATLAB table is created:

```
f = figure('Position',[280 300 720 150]);
dat = [data{1};data{2};data{3}];
cnames = ...
{'x','y','v_x','v_y',' v ','a_x','a_y',' a ','rho'};
```

```
cformat = {'bank','numeric','numeric','bank',...
    'numeric','bank','numeric','numeric'};
rnames = ...
{'t_0=0','t_1=0.5','t_2=1'};
t = uitable('Parent',f,'Data',dat,'ColumnName',...
cnames,'ColumnFormat',cformat,'RowName',rnames,...
'Position',[20 20 674 100]);
```

The numerical results are shown in Fig. 3.21b.

Example 3.6. A carousel has a radius R_c and rotates at a constant rate of $\dot{\theta} = \omega_0$ as shown in Fig. 3.36. A particle moves outward along a slot in the carousel with a speed of $\dot{r} = v_r(t) = at$. The particle starts from rest at the initial distance R_0 from the center of the carousel. Find the velocity and acceleration of the particle when is at the radius R_1 from the center of the carousel. The numerical values are $R_c = 15\,\text{m}$, $R_0 = 4\,\text{m}$, $R_1 = 14\,\text{m}$, $\omega_0 = 0.05\,\text{rad/s}$, and $a = 0.0889\,\text{m/s}^2$. Plot the particle trajectory using polar coordinates.

Solution

The motion of the particle can be described using polar (radial and transverse) coordinates as shown in Fig. 3.22b. The displacement vector $\mathbf{r}$ is the radial vector from the origin to the particle location

$$\mathbf{r} = r\mathbf{u}_r, \tag{3.79}$$

where r is the magnitude of the vector and $\mathbf{u}_r$ is the unit vector parallel to the radius vector. The velocity is the time derivative of the displacement:

$$\mathbf{v} = \frac{d}{dt}\mathbf{r} = \frac{dr}{dt}\mathbf{u}_r + r\frac{d\mathbf{u}_r}{dt} = \frac{dr}{dt}\mathbf{u}_r + r\frac{d\theta}{dt}\mathbf{u}_\theta$$

$$= \dot{r}\mathbf{u}_r + r\dot{\theta}\mathbf{u}_\theta = \dot{r}\mathbf{u}_r + r\omega\mathbf{u}_\theta, \tag{3.80}$$

where $\mathbf{u}_\theta$ is a unit vector perpendicular to $\mathbf{u}_r$ and pointing in the direction of travel along the orbit. The velocity of the particle is computed as

$$\mathbf{v} = \dot{r}\mathbf{u}_r + r\dot{\theta}\mathbf{u}_\theta = at\mathbf{u}_r + R_1\omega_0\mathbf{u}_\theta$$

$$= 0.0889t\mathbf{u}_r + 14(0.05)\mathbf{u}_\theta = 0.0889t\mathbf{u}_r + 0.7\mathbf{u}_\theta, \tag{3.81}$$

where $r = R_1 = 14\,\text{m}$, $\dot{r} = v_R(t) = at = 0.0889t\,\text{m/s}$, and $\dot{\theta} = \omega_0 = 0.05\,\text{rad/s}$. Equation (3.81) is a function of the time, t. Integrating the radial component of the velocity, one can find the required time t for the particle to travel from $r = R_0 = 4\,\text{m}$ to $r = R_1 = 14\,\text{m}$ as

$$v_r = \frac{dr}{dt} \iff \int_0^t v_r dt = \int_{R_0}^{R_1} dr \iff \int_0^t at\,dt = \int_{R_0}^{R_1} dr,$$

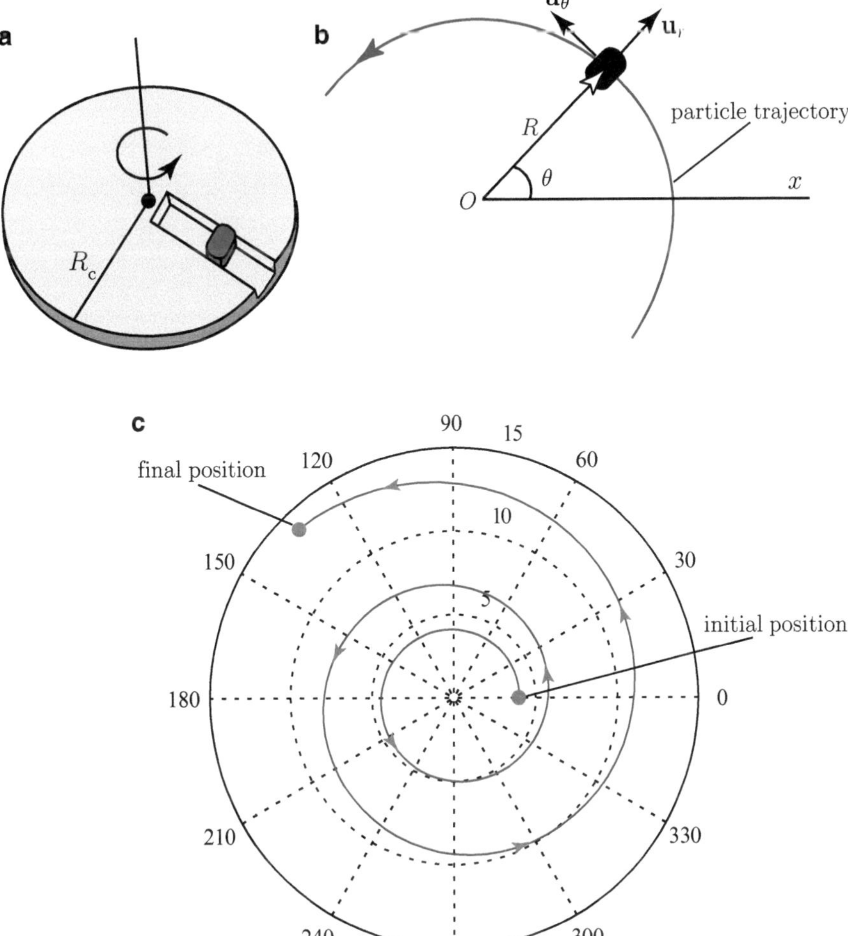

Fig. 3.22 Example 3.6

or

$$a\frac{t^2}{2} = R_1 - R_0 \iff \frac{0.0889t^2}{2} = 14 - 4 \iff t = 14.99906 \cong 15.$$

Substituting t into (3.81), the final velocity v is

$$\mathbf{v} = 1.333\,\mathbf{u}_r + 0.700\,\mathbf{u}_\theta. \tag{3.82}$$

The acceleration is the time derivative of the velocity and can be written as a vectorial sum of radial and tangential components as

$$\mathbf{a} = \frac{d}{dt}\mathbf{v} = \frac{d}{dt}\left(\dot{r}\mathbf{u}_r + r\dot{\theta}\mathbf{u}_\theta\right)$$

$$= \frac{d}{dt}\left(\dot{r}\mathbf{u}_r\right) + \frac{d}{dt}\left(r\dot{\theta}\mathbf{u}_\theta\right)$$

$$= \frac{d\dot{r}}{dt}\mathbf{u}_r + \frac{d\mathbf{u}_r}{dt}\dot{r} + \frac{dr}{dt}\dot{\theta}\mathbf{u}_\theta + \frac{d\dot{\theta}}{dt}r\mathbf{u}_\theta + \frac{d\mathbf{u}_\theta}{dt}r\dot{\theta}$$

$$= \ddot{r}\mathbf{u}_r + \left(\frac{d\theta}{dt}\mathbf{u}_\theta\right)\dot{r} + \dot{r}\dot{\theta}\mathbf{u}_\theta + \ddot{\theta}r\mathbf{u}_\theta + \left(-\frac{d\theta}{dt}\mathbf{u}_r\right)r\dot{\theta}$$

$$= \ddot{r}\mathbf{u}_r + \dot{\theta}\dot{r}\mathbf{u}_\theta + \dot{r}\dot{\theta}\mathbf{u}_\theta + \ddot{\theta}r\mathbf{u}_\theta - \dot{\theta}^2 r\mathbf{u}_r$$

$$= \left(\ddot{r} - r\dot{\theta}^2\right)\mathbf{u}_r + \left(r\ddot{\theta} + 2\dot{r}\dot{\theta}\right)\mathbf{u}_\theta. \tag{3.83}$$

Equation (3.83) is used to calculate the acceleration of the particle when $r = R_1 = 14\,\mathrm{m}$:

$$\mathbf{a} = \left(\ddot{r} - r\dot{\theta}^2\right)\mathbf{u}_r + \left(r\ddot{\theta} + 2\dot{r}\dot{\theta}\right)\mathbf{u}_\theta$$

$$= [0.0889 - 14(0.0500)^2]\mathbf{u}_r + [14(0) + 2(0.0889)(15)(0.0500)]\mathbf{u}_\theta$$

$$= 0.539\,\mathbf{u}_r + 0.1334\,\mathbf{u}_\theta, \tag{3.84}$$

where $\dot{r} = v_r(t) = at = 0.0889t\,\mathrm{m/s}$, $\ddot{r} = \dot{v}_r(t) = a = 0.0889\,\mathrm{m/s}^2$, $\dot{\theta} = \omega_0 = 0.05\,\mathrm{rad/s}$, $\ddot{\theta} = \dot{\omega}_0 = 0\,\mathrm{rad/s}^2$, and $t = 15$ s.

The MATLAB program starts with the input data:

```
R0n = 4.0;   %  (m)
R1n = 14.0;  %  (m)
an =0.0889;  %  (m/s^2)
```

The velocity and displacement (distance from the origin to the particle location) of the particle are expressed in MATLAB using the next statement:

```
v=a*t;
r=int(v);
```

To find the required time t for the particle to travel from $r = R_0 = 4\,\mathrm{m}$ to $r = R_1 = 14\,\mathrm{m}$, the radial component of the velocity is integrated, and the resulting equation is solved. The integration of the radial components is solved using the MATLAB commands `lpe=int(v,t, 0, t)` `rpe=int(1,dr, R0, R1)`. The resulting equation `eq=lpe-rpe` is solved using the MATLAB command `soleq=simplify(solve(eq,t))`. The resulting MATLAB statement is:

```
lpe=int(v,t, 0, t);
rpe=int(1,dr, R0, R1);
eq=lpe-rpe;
soleq=simplify(solve(eq,t));
tf=simplify(soleq);
```

Fig. 3.23 Problem 3.26

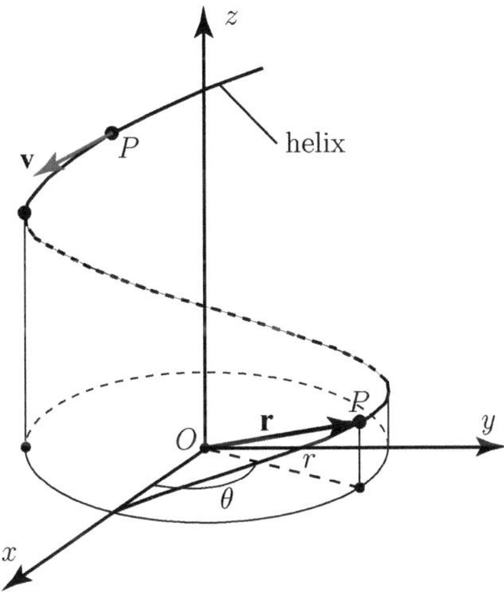

To obtain the numerical value of the final time `tf`, two lists have been created: a list with the symbolical variables `slist={R0,R1,a}` and a list with the corresponding numeric values `nlist={R0n,R1n,an}`:

```
slist={R0,R1,a};
nlist={R0n,R1n,an};
tfn=abs(subs(tf,slist,nlist))
```

The statement `tfn=abs(subs(tf,slist,nlist))` replaces `slist` with `nlist` in the symbolic expression `tf`. To obtain the particle position at each time step `t_part`, the symbolic expression `r` is replaced by its numerical value:

```
tn = 0:0.1:tfn;
rn = subs(R0n+r,{a,t},{an,tn});
```

The starting time is 0, and the final time is `tfn`. The statement `rn = subs(R0n+r, {a,t}, {an,tn})` replaces `{a,t}` with `{an,tn}` in the symbolic expression `R0n+r`. The particle trajectory shown in Fig. 3.23b is plotted using the MATLAB commands:

```
polar(tn,rn)
```

The particle trajectory, Fig. 3.22c, was plotted using polar coordinates. The polar coordinates locate each point of the particle trajectory using the particle distance (radius measured from the origin) and angle (measured from some agreed starting point). The used MATLAB function `polar(θ,ρ)` creates a polar coordinate plot of the angle θ versus the radius ρ, where θ is the angle (specified in radians) from the x-axis to the radius vector and ρ is the length of the radius vector.

Example 3.7. A particle P is moving with respect to the *xyz* reference frame as shown in Fig. 3.24a. The motion of the particle is given by the position vector

$$\mathbf{r} = a\cos(\omega t)\mathbf{i} + a\sin(\omega t)\mathbf{j} + bt\,\mathbf{k}. \tag{3.85}$$

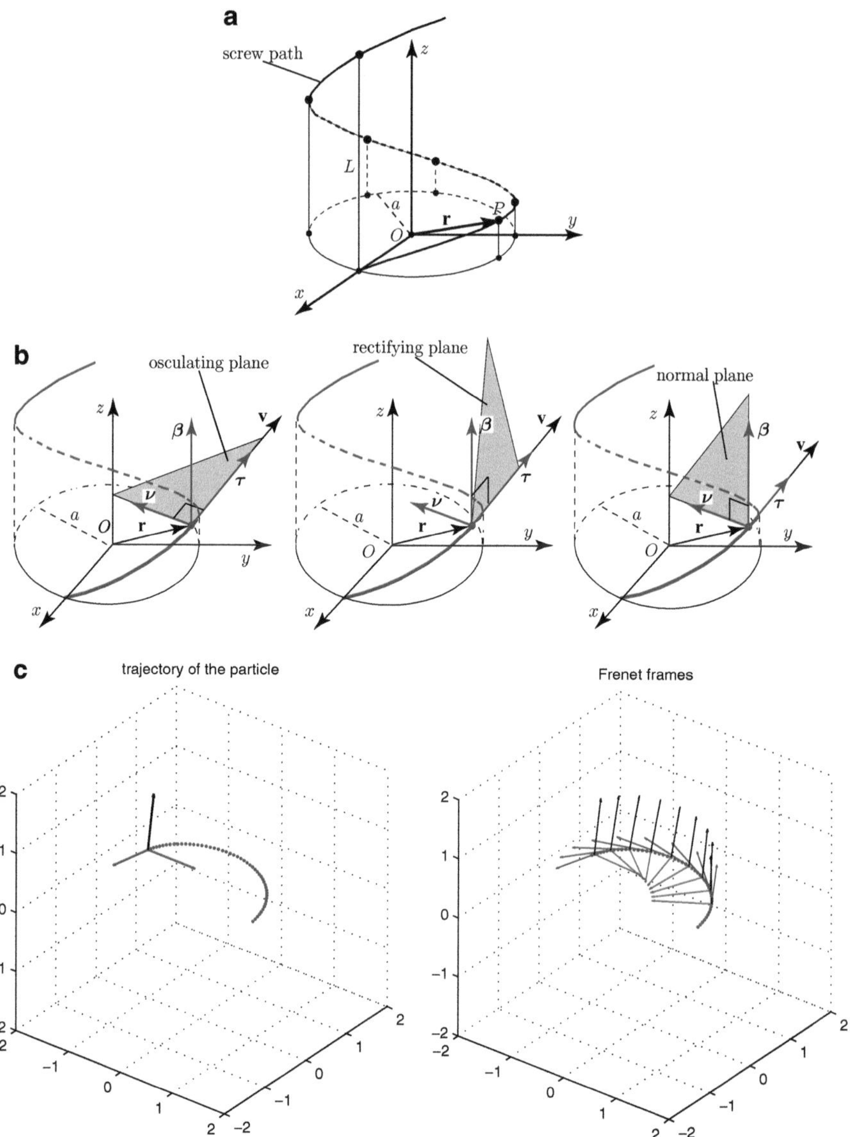

Fig. 3.24 Example 3.7

Determine the tangential unit vector, $\boldsymbol{\tau}$; the normal unit vector, $\boldsymbol{\nu}$; and the binormal unit vector, $\boldsymbol{\beta}$, to the trajectory of the motion of the particle. Find the velocity of the particle and the components of the acceleration on the vectors $\boldsymbol{\tau}$, $\boldsymbol{\nu}$, and $\boldsymbol{\beta}$. Numerical application: $a = 1\,\mathrm{m}$, $\omega = 1\,\mathrm{rad/s}$, and $b = \omega/(2\pi)$.

Solution

Equation (3.88) represents the motion along a screw line wrapped on the cylinder of radius a as shown in Fig. 3.24a. The parametric equations of the particle are given by

$$x = a \cos \omega t,$$

$$y = a \sin \omega t,$$

$$z = bt, \tag{3.86}$$

where a, b, and ω are constants and t is the time. When the point P describes the screw line, the projection on the xy plane is a circle of radius a: $x = a \cos \omega t$, $y = a \sin \omega t$. The amount of time needed for a complete revolution (a full circle on the xy plane) is $T = \frac{2\pi}{\omega}$. The vertical distance L that corresponds to a complete revolution ($t = T$) is called the lead of the screw. From the equation $z = bt$, the lead is

$$L = bT = \frac{2\pi b}{\omega}. \tag{3.87}$$

The velocity vector, $\mathbf{v}$, of the particle is

$$\mathbf{v} = \dot{\mathbf{r}} = \dot{x}\mathbf{i} + \dot{y}\mathbf{j} + \dot{z}\mathbf{k} = -a\omega \sin \omega t\, \mathbf{i} + a\omega \cos \omega t\, \mathbf{j} + b\mathbf{k}.$$

The magnitude of the velocity is

$$|\mathbf{v}| = v = \sqrt{(\dot{x})^2 + (\dot{y})^2 + (\dot{z})^2} = \sqrt{(-a\omega \sin \omega t)^2 + (a\omega \cos \omega t)^2 + b^2}$$

$$= \sqrt{a^2 \omega^2 + b^2}.$$

In MATLAB, the velocity and its magnitude are calculated with

```
syms a b omega t

% parametric equations
theta = omega*t;
x = a*cos(theta);
y = a*sin(theta);
z = b*t;

% velocity components on x,y,z
v_x=diff(x,t);
v_y=diff(y,t);
v_z=diff(z,t);
vP=[v_x v_y v_z];

% magnitude of the velocity
magn_v=sqrt(vP(1)^2+vP(2)^2+vP(3)^2);
```

The tangential unit vector is computed using

$$
\boldsymbol{\tau} = \frac{\mathbf{v}}{v} = \frac{\mathbf{v}}{|\mathbf{v}|} = \frac{\dot{\mathbf{r}}}{|\dot{\mathbf{r}}|}
$$

$$
= \frac{\dot{x}\mathbf{\imath} + \dot{y}\mathbf{\jmath} + \dot{z}\mathbf{k}}{\sqrt{(\dot{x})^2 + (\dot{y})^2 + (\dot{z})^2}}
$$

$$
= \frac{-a\omega \sin \omega t\,\mathbf{\imath} + a\omega \cos \omega t\,\mathbf{\jmath} + b\mathbf{k}}{\sqrt{(-a\omega \sin \omega t)^2 + (a\omega \cos \omega t)^2 + b^2}}
$$

$$
= \frac{-a\omega \sin \omega t\,\mathbf{\imath} + a\omega \cos \omega t\,\mathbf{\jmath} + b\mathbf{k}}{\sqrt{a^2 \omega^2 + b^2}}. \tag{3.88}
$$

The normal vector is by definition

$$
\boldsymbol{\nu} = \frac{\dfrac{d\boldsymbol{\tau}}{dt}}{\left|\dfrac{d\boldsymbol{\tau}}{dt}\right|}. \tag{3.89}
$$

Differentiating (3.88) with respect to time, it is obtained

$$
\frac{d\boldsymbol{\tau}}{dt} = \frac{1}{a^2 \omega^2 + b^2}\left(-a\omega^2 \cos \omega t\,\mathbf{\imath} - a\omega^2 \sin \omega t\,\mathbf{\jmath}\right). \tag{3.90}
$$

Hence, from (3.90), it results

$$
\left|\frac{d\boldsymbol{\tau}}{dt}\right| = \sqrt{\frac{1}{a^2 \omega^2 + b^2}\left((-a\omega^2 \cos \omega t)^2 + (-a\omega^2 \sin \omega t)^2\right)}
$$

$$
= \frac{a\omega^2}{\sqrt{a^2 \omega^2 + b^2}}. \tag{3.91}
$$

Introducing (3.91) and (3.90) into (3.89), the normal unit vector is

$$
\boldsymbol{\nu} = \frac{1}{a\omega^2}\left(-a\omega^2 \cos \omega t\,\mathbf{\imath} - a\omega^2 \sin \omega t\,\mathbf{\jmath}\right)
$$

$$
= -\cos \omega t\,\mathbf{\imath} - \sin \omega t\,\mathbf{\jmath} + (0)\mathbf{k}. \tag{3.92}
$$

The binormal vector is defined as the vector product of $\boldsymbol{\tau}$ and $\boldsymbol{\nu}$ is

$$
\boldsymbol{\beta} = \boldsymbol{\tau} \times \boldsymbol{\nu} = \frac{1}{\sqrt{a^2 \omega^2 + b^2}}\begin{vmatrix} \mathbf{\imath} & \mathbf{\jmath} & \mathbf{k} \\ -a\omega \sin \omega t & -a\omega \cos \omega t & b \\ -\cos \omega t & -\sin \omega t & 0 \end{vmatrix}
$$

$$= \frac{1}{\sqrt{a^2\,\omega^2 + b^2}}\,(b\,\sin\omega t\,\mathbf{i} - b\,\cos\omega t\,\mathbf{j} + a\,\omega\,\mathbf{k}). \tag{3.93}$$

The unit tangent vector $\boldsymbol{\tau}$, the unit normal vector $\boldsymbol{\nu}$, and the unit binormal vector $\boldsymbol{\beta}$ form a moving reference frame associated with the position of the particle on the curve. The three mutually perpendicular unit vectors, Fig. 3.24b, define:

- Normal plane—plane that is perpendicular to the tangent vector and contain the normal and the binormal.
- Osculating plane—plane that contains the unit tangent vector and the normal vector.
- Rectifying plane—plane that contains the tangent and binormal.

The unit tangent vector $\boldsymbol{\tau}$, the unit normal vector $\boldsymbol{\nu}$, and the unit binormal vector $\boldsymbol{\beta}$ are computed symbolically in MATLAB with

```
tau = vP/magn_v;
dtau = diff(tau,t);
magn_dtau = sqrt(dtau(1)^2+dtau(2)^2+dtau(3)^2);
nu = dtau/magn_dtau;
beta = cross(tau,nu);
```

Another way for calculating the unit vectors of Frenet's frame is using the function `TrFrenet` that was defined previously:

```
[Tangent,Normal,Binormal]=TrFrenet(x,y,z,t);

tau = Tangent;
nu = Normal;
beta = Binormal;
```

The components of the velocity along the reference frame defined by $\boldsymbol{\tau}$, $\boldsymbol{\nu}$, and $\boldsymbol{\beta}$ are

$$v_\tau = \boldsymbol{\tau}\cdot\mathbf{v},$$

$$v_\nu = \boldsymbol{\nu}\cdot\mathbf{v},$$

$$v_\beta = \boldsymbol{\beta}\cdot\mathbf{v},$$

and with MATLAB commands

```
fprintf('velocity components on tau, nu, beta')
v_tau = simplify(vP*tau.');
v_nu = simplify(vP*nu.');
v_beta = simplify(vP*beta.');

fprintf('v_tau = ')
pretty(v_tau);
fprintf('\n');
fprintf('v_nu = ')
```

```
pretty(v_nu);
fprintf('\n');
fprintf('v_beta = ')
pretty(v_beta);
fprintf('\n\n');
```

the following results are obtained:

```
v_tau =
      2        2      2 1/2
   (a   omega   + b  )


v_nu =
   0


v_beta =
   0
```

Differentiation of the velocity vector yields the acceleration of the particle:

$$\mathbf{a} = \dot{\mathbf{v}} = \dot{v}_x\mathbf{i} + \dot{v}_y\mathbf{j} + \dot{v}_z\mathbf{k},$$

$$= -\cos\omega t\,\mathbf{i} - \sin\omega t\,\mathbf{j} + (0)\mathbf{k}. \tag{3.94}$$

Using MATLAB, the acceleration of the particle and its magnitude are

```
% acceleration components
a_x=diff(x,t,2);
a_y=diff(y,t,2);
a_z=diff(z,t,2);
aP=[a_x a_y a_z];
% magnitude of the acceleration
magn_a=sqrt(aP(1)^2+aP(2)^2+aP(3)^2);
```

The components of the acceleration along the reference frame defined by $\boldsymbol{\tau}$, $\boldsymbol{\nu}$, and $\boldsymbol{\beta}$ are

$$a_\tau = \boldsymbol{\tau}\cdot\mathbf{a},$$

$$a_\nu = \boldsymbol{\nu}\cdot\mathbf{a},$$

$$a_\beta = \boldsymbol{\beta}\cdot\mathbf{a}.$$

The components of the acceleration along the Frenet's reference with MATLAB are

```
a_tau =
   0


a_nu =
```

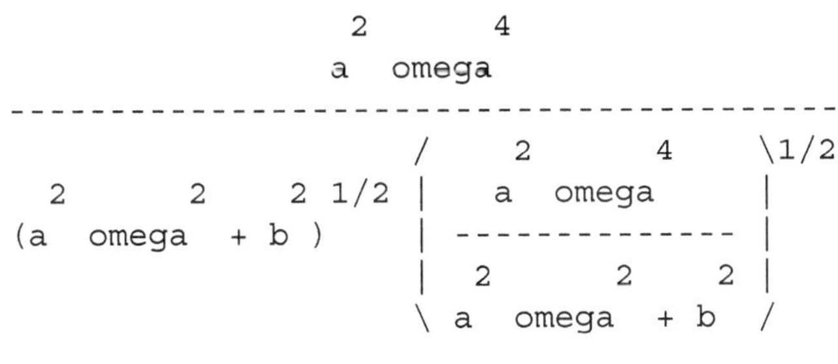

```
                           2      4
                          a   omega
        ---------------------------------------------
                               /     2        4      \1/2
     2        2      2 1/2    |     a   omega        |
    (a   omega   + b )        |     -------------    |
                              |       2        2    2 |
                              \      a   omega  + b  /
```

```
    a_beta =
       0
```

To plot the particle trajectory and the Frenet's unit vectors, the following numerical
data are input:

```
omegan=1.0;
an= 1;
bn = omegan/(2*pi);
```

The MATLAB command `axis` was used for the three-dimensional plot scaling and
appearance:

```
axis manual
axis equal
axis([-2 2 -2 2 -2 2])
grid on
```

The MATLAB command `view(az,el)` was used in the next statement to set
the viewing angle of the three-dimensional plot. The defined values `az=37.5` and
`el=30` represent the azimuth (horizontal rotation about the z-axis measured in
degrees) and the vertical elevation (of the viewpoint in degrees—the angle between
the observer viewpoint line to the center of the plot box and a horizontal line).

```
az = 37.5;
el = 30;
view(az, el);
hold
```

A scale factors `scale_factor=1` can be introduced for a proper representation
of the particle trajectory using the next MATLAB statement. For the simulation, the
initial, final, and the increment values of the time t are

```
start_value=0;
end_value=pi;
step=pi/50;
```

The trajectory of the particle will be plotted at each `xn`, `yn`, and `zn` location defined
by the parametric equation of motion:

```
i=0;
for tn = start_value : step : end_value
```

```
    i=i+1;
    slist1={a,b,omega,t};
    nlist1={an,bn,omegan,tn};

    xn = subs(x,slist1,nlist1);
    yn = subs(y,slist1,nlist1);
    zn = subs(z,slist1,nlist1);

    taun = subs(tau,slist1,nlist1)/scale_factor;
    nun = subs(nu,slist1,nlist1)/scale_factor;
    betan = subs(beta,slist1,nlist1)/scale_factor;

    ht = plot3(xn,yn,zn,'k.','Color','r');
    hm = plot3(xn,yn,zn,'k.','Color','b');

    title('trajectory of the particle')
    % Frenet frame represented along the trajectory
    ptau=quiver3(xn,yn,zn,taun(1),taun(2),taun(3),...
        'Color','b','LineWidth',1);
    pnu=quiver3(xn,yn,zn,nun(1),nun(2),nun(3),...
        'Color','r','LineWidth',1);
pbeta=quiver3(xn,yn,zn,betan(1),betan(2),betan(3),...
        'Color','k','LineWidth',1);

    pause(0.05)

    delete(ht);
    delete(ptau);
    delete(pnu);
    delete(pbeta);

    x_nn(i) = xn;
    y_nn(i) = yn;
    z_nn(i) = zn;
    for j=1:3
    taunn(i,j)=taun(j);
    nunn(i,j)=nun(j);
    betann(i,j)=betan(j);
    end
end
```

A `for` loop was used to calculate the successive positions of particle location and simulate the particle dynamics using small blue dots as shown in Fig. 3.24c.

The actual location of the particle is represented by a red dot and is updated at each time step tn of the loop. Inside the `for` loop, the simulation is slowed down with the command `pause` to better observe the motion.

The next MATLAB commands plot the Frenet's frame consisting of unit tangential, normal, and binormal vectors (vectors that collectively forms an orthonormal basis of 3-space) at some successive positions equally spaced along the particle helix trajectory. The equally spaced particle successive conditions are obtaining using the MATLAB condition `mod(i,5)==0`. The `quiver3(...)` commands inside the `for` loop plots the tangent, the normal, and the binormal, respectively. In the plot, the unit tangent vectors are represented using the color blue, the unit normal vectors are represented by the red color, and the unit binormal vectors by black color:

```
i=0;
% plot Frenet frame
for tn = start_value : step : end_value
i=i+1;
    if (i>5) && (mod(i,5)==0)
    quiver3(x_nn(i),y_nn(i),z_nn(i),...
        taunn(i,1),...
        taunn(i,2),...
        taunn(i,3),'color','b');
    quiver3(x_nn(i),y_nn(i),z_nn(i),...
        nunn(i,1),...
        nunn(i,2),...
        nunn(i,3),'color','r');
    quiver3(x_nn(i),y_nn(i),z_nn(i),...
        betann(i,1),...
        betann(i,2),...
        betann(i,3),'color','k');
    end
    title('Frenet frames')
end
```

The plot generated in MATLAB with Frenet's reference frames is shown in Fig. 3.24c.

Example 3.8. The link 1 of radius R rotates about the vertical axis z of the fixed reference frame xyz with the constant angular velocity ω as seen in Fig. 3.25. The reference frame $x_1 y_1 z_1$ is attached to the link 1. The particle 2 slides along the link 1 with a constant angular velocity Ω. Find the components of the linear velocity and linear acceleration of the particle 2 along the fixed reference frame xyz and the rotating reference frame $x_1 y_1 z_1$.

Solution
The relative motion of the link 1 with respect to the fixed reference frame is determined by the angular displacement $\alpha = \omega t$, as shown in Fig. 3.25b, and the relative motion of the particle 2 with respect to the link 1 is determined by the

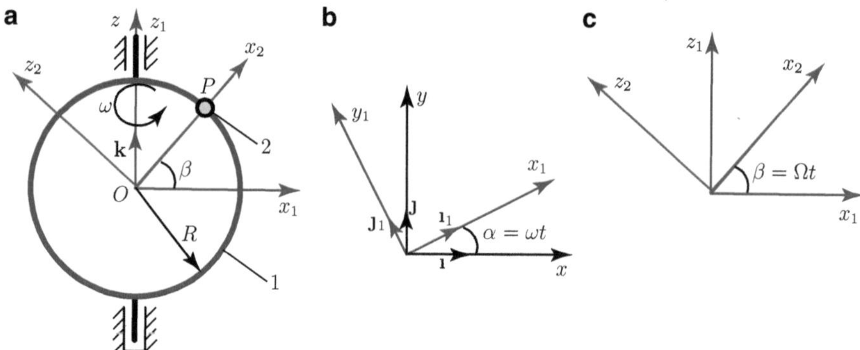

Fig. 3.25 Example 3.8

angular displacement $\beta = \Omega t$, as shown in Fig. 3.25c. The position vector of this particle 2 is

$$\mathbf{r} = R\cos\Omega t\,\mathbf{l}_1 + R\cos\Omega t\,\mathbf{k}_1. \tag{3.95}$$

The velocity of the particle is the time derivative of the position vector. The components of the position vector along the axes of the fixes reference frame, *xyz*, are

$$r_x = \mathbf{l}\cdot\mathbf{r} = R\cos\Omega t\,\mathbf{l}\cdot\mathbf{l}_1 + R\cos\Omega t\,\mathbf{l}\cdot\mathbf{k}_1 = R\cos\omega t\cos\Omega t,$$

$$r_y = \mathbf{J}\cdot\mathbf{r} = R\cos\Omega t\,\mathbf{J}\cdot\mathbf{l}_1 + R\cos\Omega t\,\mathbf{J}\cdot\mathbf{k}_1 = R\sin\omega t\cos\Omega t,$$

$$r_z = \mathbf{k}\cdot\mathbf{r} = R\cos\Omega t\,\mathbf{k}\cdot\mathbf{l}_1 + R\cos\Omega t\,\mathbf{k}\cdot\mathbf{k}_1 = R\sin\Omega t. \tag{3.96}$$

Hence, the components of the velocity along the axes of the fixes reference frame are

$$v_x = -R\omega\sin\omega t\cos\Omega t - R\Omega\cos\omega t\sin\Omega t,$$

$$v_y = R\omega\cos\omega t\cos\Omega t - R\Omega\sin\omega t\sin\Omega t,$$

$$v_z = R\Omega\cos\Omega t. \tag{3.97}$$

The second derivative yields the components of the acceleration along the axes of the fixes reference frame *xyz*:

$$a_x = -R\omega^2\cos\omega t\cos\Omega t + 2R\omega\Omega\sin\omega t\sin\Omega t - R\Omega^2\cos\omega t\cos\Omega t,$$

$$a_y = -R\omega^2\sin\omega t\cos\Omega t - 2R\omega\Omega\cos\omega t\sin\Omega t - R\Omega^2\sin\omega t\cos\Omega t,$$

$$a_z = -R\Omega^2\sin\Omega t. \tag{3.98}$$

The components of the velocity and the components of the acceleration along the axes of the rotating reference frame $x_1y_1z_1$ are

$$\begin{aligned}
v_{x1} &= \mathbf{I}_1 \cdot \mathbf{v} \\
&= v_x \mathbf{I}_1 \cdot \mathbf{I} + v_y \mathbf{I}_1 \cdot \mathbf{J} + v_z \mathbf{I}_1 \cdot \mathbf{k} \\
&= (-R\omega \sin \omega t \cos \Omega t - R\Omega \cos \omega t \sin \Omega t)\, \mathbf{I}_1 \cdot \mathbf{I} \\
&\quad + (R\omega \cos \omega t \cos \Omega t - R\Omega \sin \omega t \sin \Omega t)\, \mathbf{I}_1 \cdot \mathbf{J} \\
&\quad + (R\Omega \cos \Omega t)\, \mathbf{I}_1 \cdot \mathbf{k} \\
&= \cos \omega t\,(-R\omega \sin \omega t \cos \Omega t - R\Omega \cos \omega t \sin \Omega t) \\
&\quad + \sin \omega t\,(R\omega \cos \omega t \cos \Omega t - R\Omega \sin \omega t \sin \Omega t) \\
&\quad + 0 \cdot (R\omega \cos \Omega t) \\
&= -R\Omega \sin \Omega t,
\end{aligned} \tag{3.99}$$

$$\begin{aligned}
v_{y1} &= \mathbf{J}_1 \cdot \mathbf{v} \\
&= v_x \mathbf{J}_1 \cdot \mathbf{I} + v_y \mathbf{J}_1 \cdot \mathbf{J} + v_z \mathbf{J}_1 \cdot \mathbf{k} \\
&= (-R\omega \sin \omega t \cos \Omega t - R\Omega \cos \omega t \sin \Omega t)\, \mathbf{I}_1 \cdot \mathbf{I} \\
&\quad + (R\omega \cos \omega t \cos \Omega t - R\Omega \sin \omega t \sin \Omega t)\, \mathbf{I}_1 \cdot \mathbf{J} \\
&\quad + (R\Omega \cos \Omega t)\, \mathbf{I}_1 \cdot \mathbf{k} \\
&= -\sin \omega t\,(-R\omega \sin \omega t \cos \Omega t - R\Omega \cos \omega t \sin \Omega t) \\
&\quad + \cos \omega t\,(R\omega \cos \omega t \cos \Omega t - R\Omega \sin \omega t \sin \Omega t) \\
&\quad + 0 \cdot (R\omega \cos \Omega t) \\
&= R\omega \cos \Omega t,
\end{aligned} \tag{3.100}$$

$$\begin{aligned}
v_{z1} &= \mathbf{k}_1 \cdot \mathbf{v} \\
&= v_x \mathbf{k}_1 \cdot \mathbf{I} + v_y \mathbf{k}_1 \cdot \mathbf{J} + v_z \mathbf{k}_1 \cdot \mathbf{k} \\
&= (-R\omega \sin \omega t \cos \Omega t - R\Omega \cos \omega t \sin \Omega t)\, \mathbf{k}_1 \cdot \mathbf{I} \\
&\quad + (R\omega \cos \omega t \cos \Omega t - R\Omega \sin \omega t \sin \Omega t)\, \mathbf{k}_1 \cdot \mathbf{J} \\
&\quad + (R\Omega \cos \Omega t)\, \mathbf{k}_1 \cdot \mathbf{k} \\
&= 0 \cdot (-R\omega \sin \omega t \cos \Omega t - R\Omega \cos \omega t \sin \Omega t) \\
&\quad + 0 \cdot (R\omega \cos \omega t \cos \Omega t - R\Omega \sin \omega t \sin \Omega t) \\
&\quad + 1 \cdot (R\omega \cos \Omega t) \\
&= R\Omega \cos \Omega t
\end{aligned} \tag{3.101}$$

or

$$\begin{aligned}
v_{x1} &= -R\Omega \sin \Omega t, \\
v_{y1} &= R\omega \cos \Omega t, \\
v_{z1} &= R\Omega \cos \Omega t.
\end{aligned} \tag{3.102}$$

For the acceleration, the following results are obtained:

$$a_{x1} = \mathbf{i}_1 \cdot \mathbf{a} = -R\Omega^2 \cos\Omega t - R\omega^2 \cos\Omega t,$$

$$a_{y1} = \mathbf{j}_1 \cdot \mathbf{a} = -2R\omega\Omega \sin\Omega t,$$

$$a_{z1} = \mathbf{k}_1 \cdot \mathbf{a} = -R\Omega^2 \sin\Omega t. \tag{3.103}$$

The MATLAB program starts with the following statements:

```
clear all; clc; close all
syms R t omega Omega alpha beta real
```

The angles $\alpha = \omega t$ and $\beta = \Omega t$ are introduced in MATLAB with:

```
alpha=omega*t;
beta=Omega*t;
```

The MATLAB commands for the rotating orthogonal unit vectors i1, j1, k1 calculated in the fixed reference frame *xyz* are

```
i1=[cos(alpha) sin(alpha) 0];
j1=[-sin(alpha) cos(alpha) 0];
k1=[0 0 1];
```

The position vector r of the particle in the fixed reference frame is defined in MATLAB using:

```
r = R*cos(beta)*i1+R*sin(beta)*k1;
```

The position vector components r_x, r_y, and r_x are calculated using the next MATLAB statement:

```
r_x = r(1);
r_y = r(2);
r_z = r(3);
```

The MATLAB results for the position components with respect to the fixed reference frame *xyz* are

```
r_x = R*cos(Omega*t)*cos(omega*t)
r_y = R*cos(Omega*t)*sin(omega*t)
r_z = R*sin(Omega*t)
```

The v_x, v_y, and v_z components of the velocity (in the *xyz* reference frame) are calculated in MATLAB using

```
v = diff(r,t);
v_x = v(1);
v_y = v(2);
v_z = v(3);
```

The a_x, a_y, and a_z components of the acceleration (in the *xyz* reference frame) are in MATLAB:

```
a = diff(v,t);
a_x = simplify(a(1));
a_y = simplify(a(2));
a_z = simplify(a(3));
```

The `v_x1`, `v_y1`, and `v_z1` components of the velocity `v` expressed in terms of the rotating reference frame $x_1 y_1 z_1$ are

```
v_x1=simplify(i1*v');
v_y1=simplify(j1*v');
v_z1=simplify(k1*v');
fprintf('v_x1 = %s \n',char(v_x1))
fprintf('v_y1 = %s \n',char(v_y1))
fprintf('v_z1 = %s \n',char(v_z1))
fprintf('\n')
```

and the MATLAB results are

```
v_x1 = -Omega*R*sin(Omega*t)
v_y1 = R*omega*cos(Omega*t)
v_z1 = Omega*R*cos(Omega*t)
```

The `a_x1`, `a_y1`, and `a_z1` components of the acceleration (expressed in the $x_1 y_1 z_1$ reference frame) are calculated in MATLAB using

```
a_x1=simplify(i1*a');
a_y1=simplify(j1*a');
a_z1=simplify(k1*a');
fprintf('a_x1 = %s \n',char(a_x1))
fprintf('a_y1 = %s \n',char(a_y1))
fprintf('a_z1 = %s \n',char(a_z1))
fprintf('\n')
```

and the MATLAB results are

```
a_x1 = -R*cos(Omega*t)*(Omega^2 + omega^2)
a_y1 = -2*Omega*R*omega*sin(Omega*t)
a_z1 = -Omega^2*R*sin(Omega*t)
```

3.7 Problems

3.1 A particle moves along a straight line with the acceleration $a = bt + c$ where $b = 3\,\text{m/s}^3$ and $c = -2\,\text{m/s}^2$. At $t = 0$, the position of the particle is $d_0 = d(0) = 0.5\,\text{m}$, and the velocity is $v_0 = v(0) = 1\,\text{m/s}$. Determine the velocity of the particle and the position when $t = 5\,\text{s}$. Find the total distance the particle travels during this time period.

Fig. 3.26 Problem 3.2

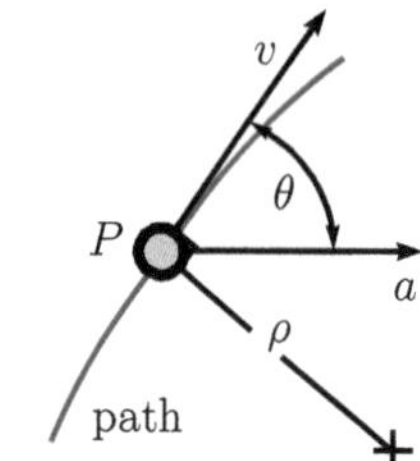

Fig. 3.27 Problem 3.3

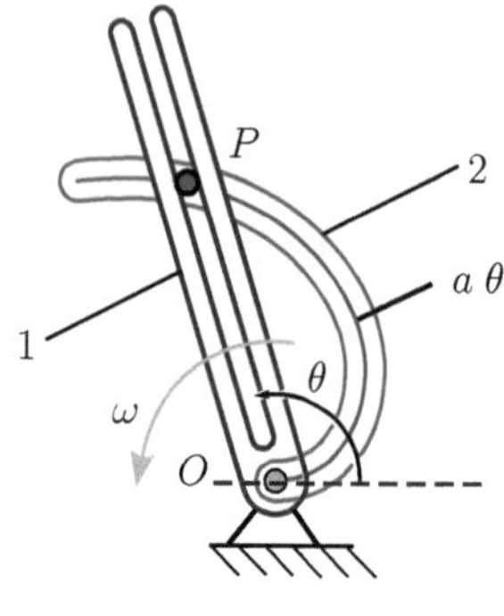

Fig. 3.28 Problem 3.4

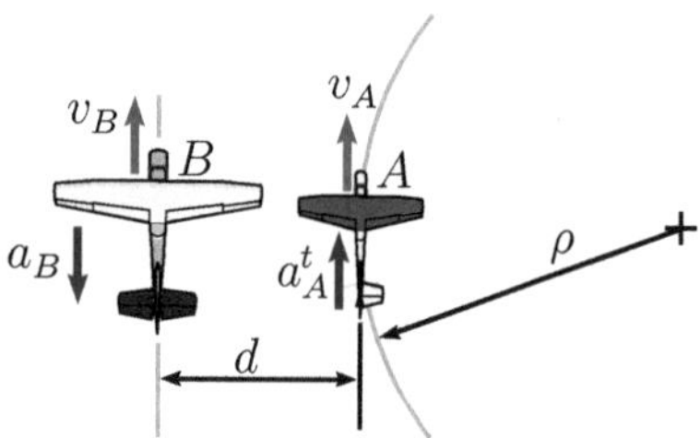

3.2 The particle P shown in the Fig. 3.26 is moving along a curvilinear path. The particle has the speed $v = 350$ ft/s and the acceleration $a = 65$ ft/s² acting in the direction shown in Fig. 3.26. The angle between the velocity and the acceleration is $\theta = 55°$. Find the rate of increase in the particle speed and the radius of curvature of the pass.

3.3 The link 1, shown in Fig. 3.27, is connected to the ground at O and has a constant angular velocity $\omega = \dot{\theta} = 4$ rad/s. The rotation slotted link carries the particle P along the link 2. The link 2 represents a curving guide of equation $r(\theta) = a\,\theta = 0.5\,\theta$ (m), where θ is in radians. Find the velocity and the acceleration of the particle P for the instant when $\theta = 3\pi/4$ rad.

3.4 Plane A is flying along a circular path with the speed $v_A = 550$ km/h, and the tangential acceleration is $a_A^t = 80$ km/h², as shown in the Fig. 3.28. The radius of curvature of the circular path of plane A is $\rho = 400$ km. Plane B is flying along a straight line path with the speed $v_B = 600$ km/h and the acceleration $a_B = 60$ km/h² acting in the direction shown in the Fig. 3.28. The distance between the planes is $d = 3$ km. Find the velocity and the acceleration of plane A measured by plane B.

3.5 The parametric equations of a particle are given in a planar Cartesian frame: $x(t), y(t)$. The coordinates x and y are in meters (m), and t is the time in seconds (s). Find and depict the trajectory of the particle. Find and depict the velocity, the acceleration, and the radius of curvature for the time $t_0 = 0$ and t_i.

(a) $x(t) = \cos t,\, y(t) = 2 \cos t$, where $t_1 = \pi/4,\, t_2 = \pi/2,\, t_3 = \pi$.
(b) $x(t) = \sin^2 t,\, y(t) = 2 \cos 2t$, where $t_1 = \pi/4,\, t_2 = \pi/2,\, t_3 = \pi$.
(c) $x(t) = \cos t,\, y(t) = 1 + \cos 2t$, where $t_1 = \pi/4,\, t_2 = \pi/2,\, t_3 = \pi$.
(d) $x(t) = t^3 + 1,\, y(t) = 1 - t^3$, where $t_1 = 1,\, t_2 = 2,\, t_3 = \sqrt{3}$.
(e) $x(t) = a + b \cos \omega t,\, y(t) = b \sin \omega t$, where $a = 1$ m, $b = 2$ m, $\omega = \pi$ rad/s, $t_1 = \pi/4,\, t_2 = \pi/2,\, t_3 = \pi$.
(f) $x(t) = 8 \cos 4t,\, y(t) = t$, where $t_1 = \pi/4,\, t_2 = \pi/2,\, t_3 = 3\pi/2$.

3.6 The parametric equations of a particle are given in planar polar coordinates: $r(t),\, \theta(t)$, where r is in meters, θ is in radians, and t is the time in seconds. Find and plot the trajectory of the particle. Find and plot the velocity, the acceleration, and the radius of curvature for $t_0 = 0$ and $t_i(s)$.

(a) $r(t) = at,\, \theta(t) = bt$, where $a = 1$ m, $b = 2$ m, $t_1 = 1,\, t_2 = 2,\, t_3 = 3$.
(b) $r(t) = a,\, \theta(t) = \pi(b - \sin t)$, where $a = 2$ m, $b = 4$ m, $t_1 = \pi/4,\, t_2 = \pi/2,\, t_3 = \pi$.

3.7 The parametric equations of a particle are given in cylindrical coordinates:

(a) $r(t) = 3,\, \theta(t) = t,\, z(t) = 3t/\pi$.
(b) $r(t) = 4,\, \theta(t) = 3 \sin t,\, z(t) = 4t$.
(c) $r(t) = (1 + \sin^2 t)^{0.5},\, \theta(t) = t,\, z(t) = \sin 2t$.
(d) $r(t) = 2t,\, \theta(t) = t,\, z(t) = t^2$.

Find and depict the trajectory of the particle. Find and depict the velocity, the acceleration, and the radius of curvature for $t_0 = 0,\, t_1 = 1,\, t_2 = 2,\, t_3 = 3$, and $t_4 = 4$ s.

3.8 The position equations of a particle P in a spatial Cartesian frame are given by: $x(t) = 2a \cos t,\, y(t) = 3a \sin t,\, z(t) = 4a \cos t$, where $a = 2$ m. Find and depict the trajectory of the particle. Find and depict the velocity, the acceleration, and the radius of curvature for $t_0 = 0,\, t_1 = \pi/4,\, t_2 = \pi/2,\, t_3 = \pi$ s.

3.9 A particle P is moving on a trajectory defined by the parabola $y = x^2$ with $x > 0$. The particle starts from the origin with an initial velocity $v_0 = 1$ m/s and with a constant tangential acceleration $a_t = $ constant. After $t_1 = 2$ s, the particle has the coordinate $x_1 = 2$ m. Find the instantaneously velocity of the particle and the radius of curvature at the origin and at t_1.

3.10 A particle P is moving on a trajectory defined by $y = x^2 + x + 9$. The x-component of the velocity is constant $v_x = 1$ m/s. The particle starts from the origin at $t = 0,\, x = 0$. Find the parametric equations of the particle, the velocity, and the acceleration of the particle.

3.11 The trajectory of a particle is defined by $y^2 = ax$, and the component of velocity along the y-axis is $v_y = bt$, where a and b are constants in meters. Determine the velocity and the acceleration of the particle.

Fig. 3.29 Problem 3.13

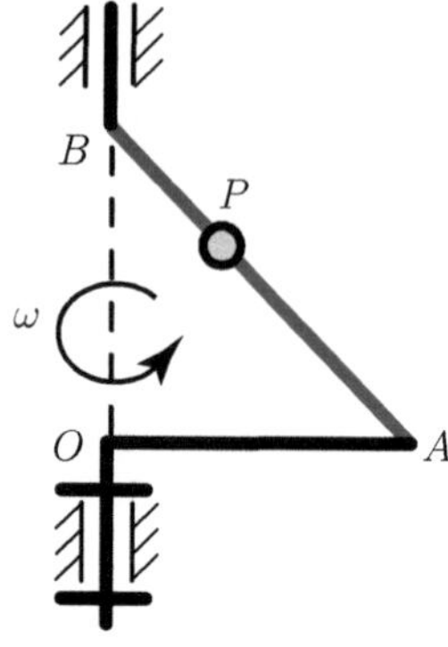

Fig. 3.30 Problem 3.14

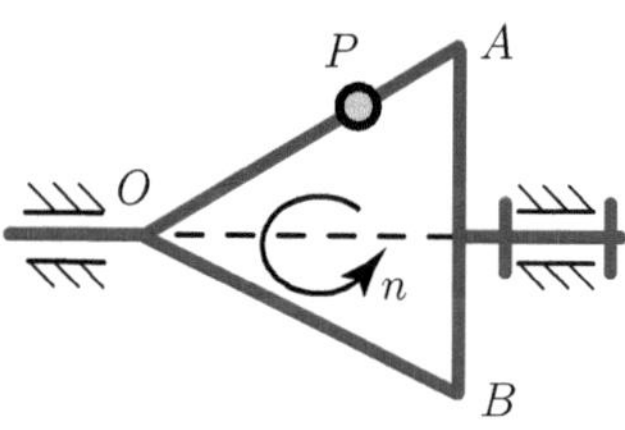

3.12 The rectangular coordinates of a particle are (m): $x(t) = 10\cos 4t$, $y(t) = 20\sin 2t$, and $z(t) = t^2 + 10t$. Find the angle between the position vector and the velocity and the angle between the position vector and the acceleration a, both at time $t = 5$ s.

3.13 The isosceles triangle $\triangle OAB$, with $OA = OB = a$ and $\angle(OA, OB) = \pi/2$, is rotating with a constant angular velocity ω along the OB axis, as shown in Fig. 3.29. A particle P is moving along the hypotenuse AB according to the function $AP = bt$ (cm), where t is the time (s). Find the velocity and the acceleration of the particle. Numerical application: $a = 3$ cm, $b = 1$ cm, and $\omega = 10$ rad/s.

3.14 The isosceles triangle $\triangle OAB$, with $OA = OB = a$ and $AB = b$, rotates about the horizontal with a constant angular rate n, as shown in Fig. 3.30. In the same time, the particle P oscillates along OA with its distance from O given by $OP = 1 + 2\cos \pi t$ (cm), where t is the time in seconds. Find the velocity of the particle P. Calculate the acceleration of P for the instant when the velocity along OA from O to A is a maximum. Numerical application: $a = 3\sqrt{2}$ cm, $b = 6$ cm and $n = 100$ rpm.

3.15 The disk and the rod OA rotate about a vertical z-axis with a constant speed $\omega = d\theta/dt = 1$ rad/s, as shown in Fig. 3.31. The angle θ is measured from the fixed reference x-axis. The rod OA rotates upward with a constant angular speed $\Omega = d\phi/dt = 2$ rad/s. The particle P moves along the rod according to the time function $OP = 0.1t^2 + 0.01$ (m). At time $t = 0$, the angles θ and ϕ are zero. Find the velocity and acceleration of the particle P for $t = 0.5$ s and $t = 1$ s.

Fig. 3.31 Problem 3.15

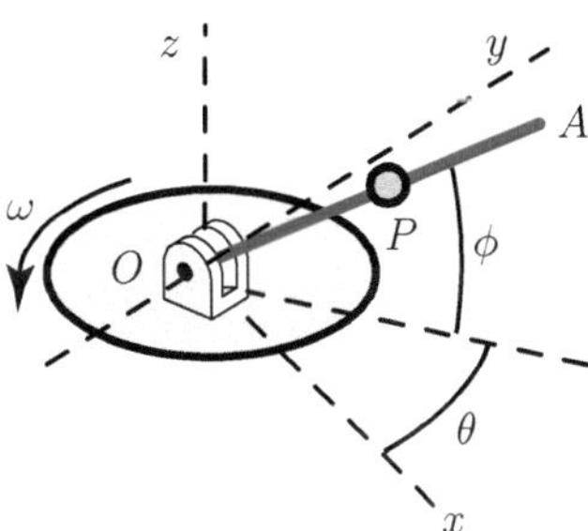

Fig. 3.32 Problem 3.20

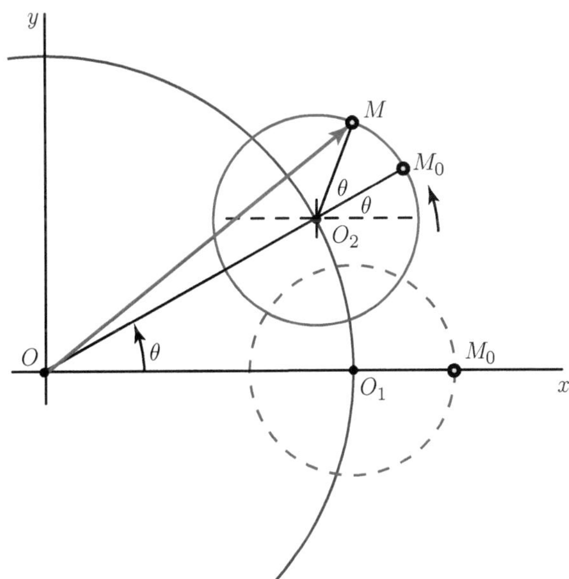

3.16 If the radius of curvature of a curve is infinity ($\rho = \infty$), show that the curve is a straight line.

3.17 If the torsion of a curve is infinity ($T = \infty$), show that the curve is planar.

3.18 Find the radius of curvature and the torsion the curve given by $x(t) = 7\cos t, y(t) = 7\sin t, z(t) = 5t$.

3.19 A particle is moving on the surface of a sphere of radius R with a constant velocity $\mathbf{v}$. The velocity $\mathbf{v}$ makes a constant angle α with the meridian (line of longitude) of the sphere. Find the path of the particle.

3.20 A planetary circle with the radius r and the center at O_1 is rotating with a uniform angular rotation $\omega = d\theta/dt$ with respect to the origin O in such a way that $OO_1 = 2r$, as shown in Fig. 3.32. Simultaneously, the circle is rotating about its center O_1 with the same angular velocity ω. Find the velocity and the acceleration of a point M on the circumference of the circle. The initial position of the point is at $M_0 \in Ox$.

Fig. 3.33 Problem 3.21

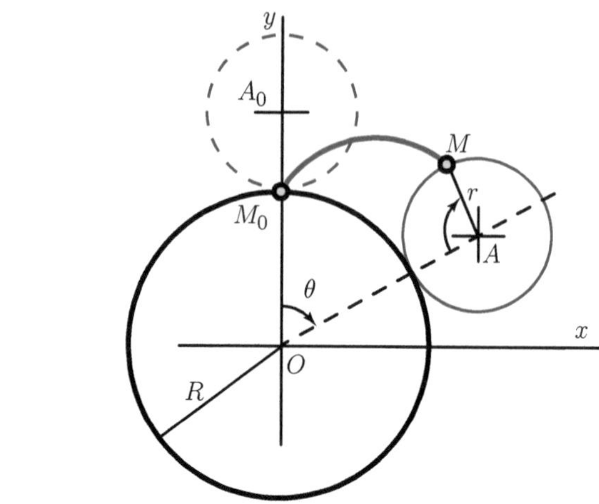

Fig. 3.34 Problem 3.22

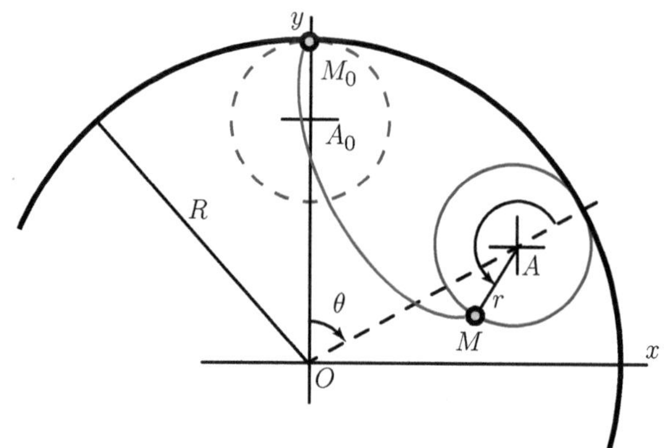

3.21 A fixed circle with the radius $R = 2r$ and the center at O is shown in Fig. 3.33.
A planetary circle with the radius r and the center at A rolls without slipping
around the fixed circle. The path of a chosen point of the planetary circle M is a
plane curve called epicycloid. Find the parametric equations for the epicycloid
and the velocity and the acceleration of the point M if at the initial moment,
$t = 0$, $\theta = 0$, and the velocity of A is $v_A = u$, where u is constant.

3.22 A fixed circle with the radius $R = 4r$ has the center at O, as shown in Fig. 3.34.
A planetary circle with the radius r and the center at A rolls without slipping
around the interior of the fixed circle. Find the path, the velocity, and the
acceleration of a point M on the rolling circle. At the initial moment $t = 0$,
$\theta = 0$, and the velocity of A is $v_A = u = $ constant.

3.23 A particle P is moving with a constant velocity $v_P = u = $ on the circle with
the center at C and radius $CA = CB = r$, as shown in Fig. 3.35. The circle is

Fig. 3.35 Problem 3.23

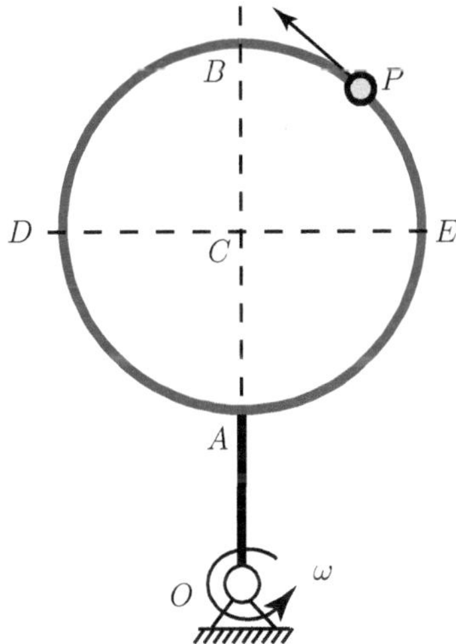

Fig. 3.36 Problem 3.25

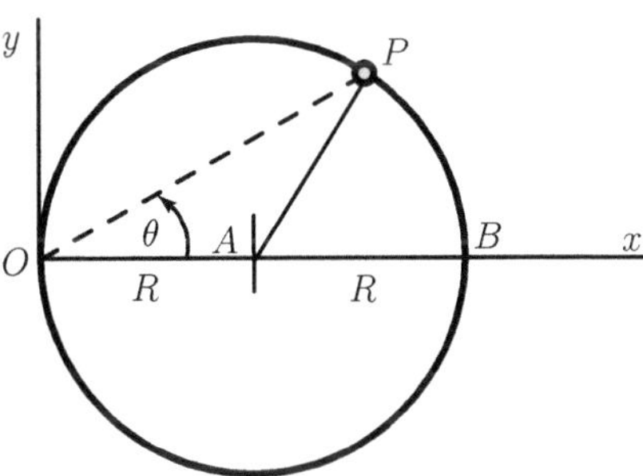

rotating with a constant speed ω about the pin joint O, where $OA = r$. Find the velocity and the acceleration of the particle at the points A, B, D, and E.

3.24 The path of a particle P is given by the parabola $y^2 - 2px = 0$, where p is a constant. The hodograph (velocity diagram) of the motion of the particle is given by the same parabola, that is, $\dot{y}^2 - 2p\dot{x} = 0$. The hodograph is a representation that gives a vectorial visual representation of the movement of the particle. At the initial time $t = 0\,\text{s}$, the particle is at the point $P_0(p/2, p)$. Find the parametric equations of motion, the velocity, and the acceleration of the particle. Find the locus of the extremities of the velocity and acceleration vectors.

3.25 The particle P is moving on a circle of equation $x^2 + y^2 - 2Rx = 0$, where R is the radius of the circle, as shown in Fig. 3.36. The angular speed of the line OP is given by

$$\frac{\mathrm{d}\theta}{\mathrm{d}t} = \frac{1}{1+t^2},$$

and at the initial moment $t = 0$ s, the initial angle is $\theta(0) = 0$. Find the velocity, the acceleration, and the hodograph of the particle P.

3.26 A particle P is moving down a helix at a constant speed of $v = 1$ m/s. The helix is defined by the equations $r = 1$ m and $z = -\theta/\pi$. Determine the angular velocity about the z-axis and the magnitude of the acceleration of the particle.

Chapter 4
Dynamics of a Particle

4.1 Newton's Second Law

Classical mechanics was established by Isaac Newton with the publication of *Philosophiae naturalis principia mathematica* in 1687. Newton stated three "laws" of motion:

1. When the sum of the forces acting on a particle is zero, its velocity is constant. In particular, if the particle is initially stationary, it will remain stationary.
2. When the sum of the forces acting on a particle is not zero, the sum of the forces is equal to the rate of change of the *linear momentum* of the particle.
3. The forces exerted by two particles on each other are equal in magnitude and opposite in direction:

$$\mathbf{F}_{ij} + \mathbf{F}_{ji} = \mathbf{0},$$

where $\mathbf{F}_{ij}$ is the force exerted by particle i on particle j and $\mathbf{F}_{ji}$ is the force exerted by particle j on particle i.

The *linear momentum* of a particle is the product of the mass of the particle, m, and the velocity of the particle, $\mathbf{v}$:

$$\mathbf{L} = m\mathbf{v}.$$

Newton's second law may be written as

$$\mathbf{F} = \frac{\mathrm{d}}{\mathrm{d}t}(m\mathbf{v}), \tag{4.1}$$

where $\mathbf{F}$ is the total force on the particle. If the mass of the particle is constant, $m =$ constant, the total force equals the product of its mass and acceleration, $\mathbf{a}$:

$$\mathbf{F} = m\frac{\mathrm{d}\mathbf{v}}{\mathrm{d}t} = m\mathbf{a}. \tag{4.2}$$

D.B. Marghitu and M. Dupac, *Advanced Dynamics: Analytical and Numerical Calculations with MATLAB*, DOI 10.1007/978-1-4614-3475-7_4,
© Springer Science+Business Media, LLC 2012

Newton's second law gives interpretation to the terms *mass* and *force*. In SI units, the unit of mass is the kilogram (kg). The unit of force is the newton (N), which is the force required to give a mass of one kilogram and an acceleration of one meter per second squared 1 N $=$ (1 kg) (1 m/s^2)$=$1 kg m/s^2. In US customary units, the unit of force is the pound [lb]. The unit of mass is the slug, which is the amount of mass accelerated at one foot per second squared by a force of one pound 1 lb $=$ (1 slug) (1 ft/s^2), or 1 slug $=$ 1 lb s^2/lb.

4.2   Newtonian Gravitation

Newton's postulate for the magnitude of gravitational force F between two particles in terms of their masses m_1 and m_2 and the distance r between them, Fig. 4.1, may be expressed as

$$F = \frac{G\, m_1\, m_2}{r^2}, \tag{4.3}$$

where G is called the *universal gravitational constant*. Equation (5.110) may be used to approximate the weight of a particle of mass m due to the gravitational attraction of the earth,

$$W = \frac{G\, m\, m_E}{r^2}, \tag{4.4}$$

where m_E is the mass of the earth and r is the distance from the center of the earth to the particle. When the weight of the particle is the only force acting on it, the resulting acceleration is called the acceleration due to gravity. In this case, Newton's second law states that $W = ma$, and from (4.4), the acceleration due to gravity is

$$a = \frac{G\, m_E}{r^2}. \tag{4.5}$$

The acceleration due to gravity at sea level is denoted by g. From (4.5), one may write $G\, m_E = g\, R_E^2$, where R_E is the radius of the earth. The expression for the

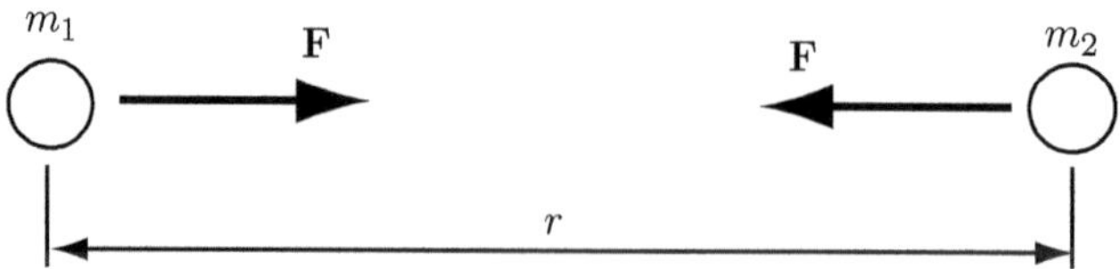

Fig. 4.1 Gravitational force between two particles

acceleration due to gravity at a distance r from the center of the earth in terms of the acceleration due to gravity at sea level is

$$a = g\frac{R_{\mathrm{E}}^2}{r^2}.$$ (4.6)

At sea level, the weight of a particle is given by

$$W = m\,g.$$ (4.7)

The value of g varies on the surface of the earth from a location to another. The values of g used in examples and problems are $g = 9.81$ m/s^2 in SI units and $g = 32.2$ ft/s^2 in US customary units.

4.3 Inertial Reference Frames

Newton's laws do not give accurate results if a problem involves velocities that are not small compared to the velocity of light (3 10^8 m/s). Einstein's theory of relativity may be applied to such problems. Newtonian mechanics also fails in problems involving atomic dimensions. Quantum mechanics may be used to describe phenomena on the atomic scale.

The position, velocity, and acceleration of a point are specified, in general, relative to an arbitrary reference frame. The Newton's second law cannot be expressed in terms of just any reference frame. Newton stated that the second law should be expressed in terms of a reference frame at rest with respect to the "fixed stars." Newton's second law, (4.2), may be expressed in terms of a reference frame that is fixed relative to the earth. Equation (4.2) may be applied using a reference that translates at constant velocity relative to the earth. If a reference frame may be used to apply (4.2), it is said to be *Newtonian* or *inertial* reference frame. Newton's second law may be applied with good results using reference frames that accelerate and rotate by properly accounting for the acceleration and rotation.

4.4 Cartesian Coordinates

To apply Newton's second law in a particular situation, one may choose a coordinate system. Newton's second law in a Cartesian reference frame, Fig. 4.2, may be expressed as

$$\sum \mathbf{F} = m\,\mathbf{a},$$ (4.8)

where $\sum \mathbf{F} = \sum F_x \mathbf{i} + \sum F_y \mathbf{j} + \sum F_z \mathbf{k}$ is the sum of the forces acting on a particle P of mass m and

$$\mathbf{a} = a_x \mathbf{i} + a_y \mathbf{j} + a_z \mathbf{k} = \ddot{x}\mathbf{i} + \ddot{y}\mathbf{j} + \ddot{z}\mathbf{k}$$

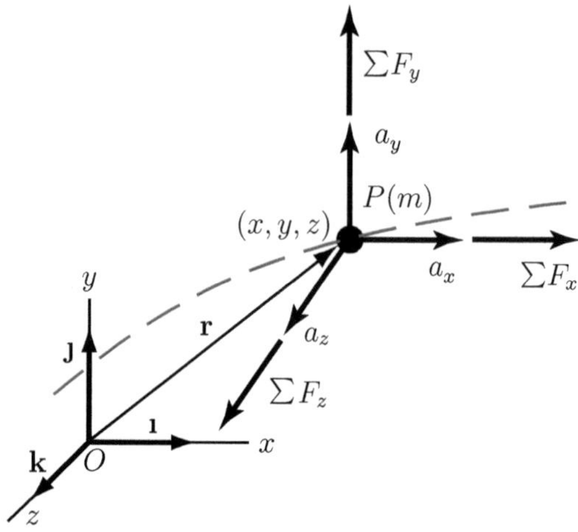

Fig. 4.2 Newton's second law in a Cartesian reference frame

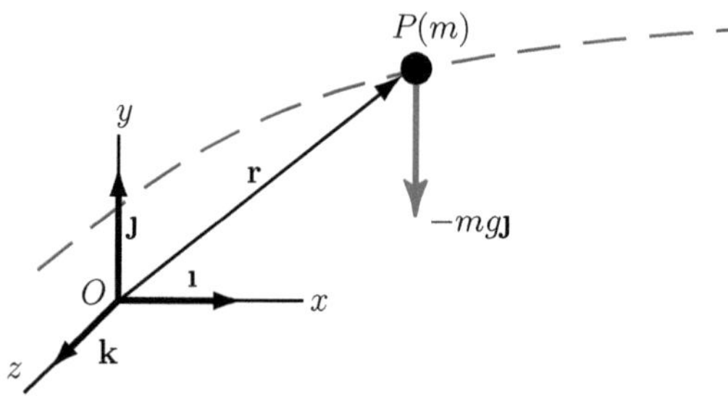

Fig. 4.3 Projectile motion

is the acceleration of the particle. Equating x, y, and z components, three scalar equations of motion are obtained:

$$\sum F_x = m a_x = m\ddot{x}, \ \sum F_y = m a_y = m\ddot{y}, \ \sum F_z = m a_z = m\ddot{z}, \qquad (4.9)$$

or the total force in each coordinate direction equals the product of the mass and component of the acceleration in that direction.

4.4.1 Projectile Problem

An object P, of mass m, is launched through the air, Fig. 4.3. The force on the object is just the weight of the object (the aerodynamic forces are neglected). The sum of the forces is $\sum \mathbf{F} = -m\,g\,\mathbf{J}$. From (4.9), one may obtain

$$a_x = \ddot{x} = 0, \ a_y = \ddot{y} = -g, \ a_z = \ddot{z} = 0.$$

The projectile accelerates downward with the acceleration due to gravity.

4.4.2 Straight Line Motion

For straight line motion along the *x*-axis, (4.9) is

$$\sum F_x = m\ddot{x}, \quad \sum F_y = 0, \quad \sum F_z = 0.$$

4.5 Normal and Tangential Components

A particle *P* of mass *m* moves on a curved path Fig. 4.4. One may resolve the sum of the forces $\sum \mathbf{F}$ acting on the particle into normal F_n and tangential F_t components:

$$\sum \mathbf{F} = F_t\,\mathbf{u}_t + F_n\,\mathbf{u}_n.$$

The acceleration of the particle in terms of normal and tangential components is

$$\mathbf{a} = a_t\,\mathbf{u}_t + a_n\,\mathbf{u}_n.$$

Newton's second law is

$$\sum \mathbf{F} = m\mathbf{a},$$

$$F_t\,\mathbf{u}_t + F_n\,\mathbf{u}_n = m\,(a_t\,\mathbf{u}_t + a_n\,\mathbf{u}_n), \tag{4.10}$$

where

$$a_t = \frac{dv}{dt} = \dot{v} \quad \text{and} \quad a_n = \frac{v^2}{\rho}.$$

Equating the normal and tangential components in (4.10), two scalar equations of motion are obtained:

$$F_t = m\dot{v} \quad \text{and} \quad F_n = m\frac{v^2}{\rho}. \tag{4.11}$$

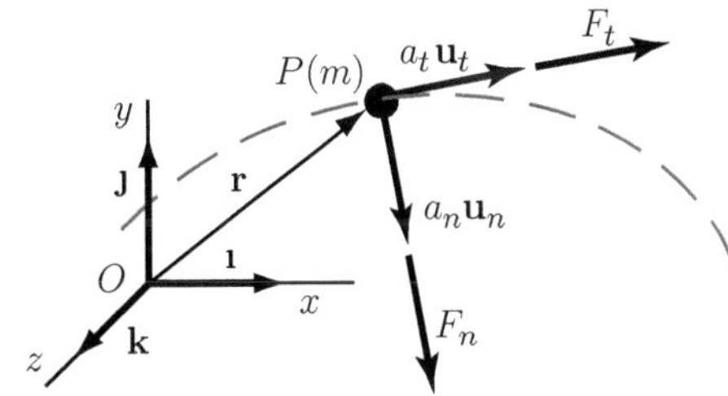

Fig. 4.4 Newton's second law in terms of normal and tangential components

The sum of the forces in the tangential direction equals the product of the mass and the rate of change of the magnitude of the velocity, and the sum of the forces in the normal direction equals the product of the mass and the normal component of acceleration. If the path of the particle lies in a plane, the acceleration of the particle perpendicular to the plane is zero, and so the sum of the forces perpendicular to the plane is zero.

4.6   Polar and Cylindrical Coordinates

The particle P with the mass m moves in a plane curved path, Fig. 4.5. The motion of the particle may be described in terms of the polar coordinates. Resolving the sum of the forces parallel to the plane into radial and transverse components

$$\sum \mathbf{F} = F_r\,\mathbf{u}_r + F_\theta\,\mathbf{u}_\theta$$

and expressing the acceleration of the particle in terms of radial and transverse components, Newton's second law may be written the form

$$F_r\,\mathbf{u}_r + F_\theta\mathbf{u}_\theta = m\left(a_r\,\mathbf{u}_r + a_\theta\,\mathbf{u}_\theta\right), \tag{4.12}$$

where

$$a_r = \frac{d^2 r}{dt^2} - r\left(\frac{d\theta}{dt}\right)^2 = \ddot{r} - r\omega^2,$$

$$a_\theta = r\frac{d^2\theta}{dt^2} + 2\frac{dr}{dt}\frac{d\theta}{dt} = r\alpha + 2\dot{r}\omega.$$

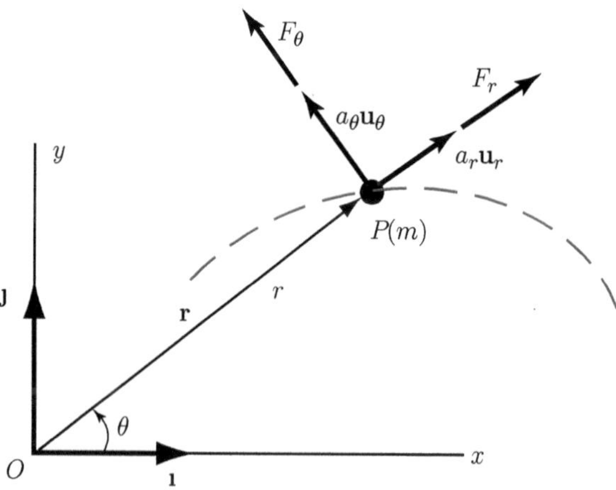

Fig. 4.5 Newton's second law in terms of polar components

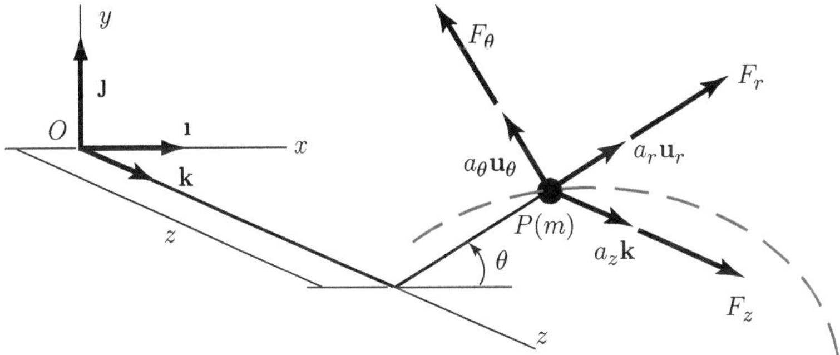

Fig. 4.6 Newton's second law in terms of cylindrical components

Two scalar equations are obtained:

$$F_r = m\left(\ddot{r} - r\omega^2\right),$$
$$F_\theta = m\left(r\alpha + 2\dot{r}\omega\right). \tag{4.13}$$

The sum of the forces in the radial direction equals the product of the mass and the radial component of the acceleration, and the sum of the forces in the transverse direction equals the product of the mass and the transverse component of the acceleration.

The three-dimensional motion of the particle P may be obtained using the cylindrical coordinates, Fig. 4.6. The position of P perpendicular to the xy plane is measured by the coordinate z and the unit vector $\mathbf{k}$. The sum of the forces is resolved into radial, transverse, and z components:

$$\sum \mathbf{F} = F_r \mathbf{u}_r + F_\theta \mathbf{u}_\theta + F_z \mathbf{k}.$$

The three scalar equations of motion are the radial and transverse relations, (4.13), and the equation of motion in the z-direction:

$$F_r = m\left(\ddot{r} - r\omega^2\right),$$
$$F_\theta = m\left(r\alpha + 2\dot{r}\omega\right),$$
$$F_z = m\ddot{z}. \tag{4.14}$$

4.7 Principle of Work and Energy

The Newton's second law for a particle of mass m can be written in the form

$$\mathbf{F} = m\frac{d\mathbf{v}}{dt} = m\dot{\mathbf{v}}. \tag{4.15}$$

The dot product of both sides of (4.15) with the velocity $\mathbf{v} = d\mathbf{r}/dt$ gives

$$\mathbf{F} \cdot \mathbf{v} = m\dot{\mathbf{v}} \cdot \mathbf{v}, \tag{4.16}$$

or

$$\mathbf{F} \cdot \frac{d\mathbf{r}}{dt} = m\dot{\mathbf{v}} \cdot \mathbf{v}. \tag{4.17}$$

But

$$\frac{d}{dt}(\mathbf{v} \cdot \mathbf{v}) = \dot{\mathbf{v}} \cdot \mathbf{v} + \mathbf{v} \cdot \dot{\mathbf{v}} = 2\dot{\mathbf{v}} \cdot \mathbf{v},$$

and

$$\dot{\mathbf{v}} \cdot \mathbf{v} = \frac{1}{2}\frac{d}{dt}(\mathbf{v} \cdot \mathbf{v}). \tag{4.18}$$

Using the previous relation, (4.17) is written as

$$\mathbf{F} \cdot d\mathbf{r} = \frac{1}{2}m\,d(\mathbf{v} \cdot \mathbf{v}). \tag{4.19}$$

The term

$$dU = \mathbf{F} \cdot d\mathbf{r}$$

is the *work* where $\mathbf{F}$ is the total external force acting on the particle of mass m and $d\mathbf{r}$ is the infinitesimal displacement of the particle. Integrating (4.19), the following result is obtained:

$$\int_{\mathbf{r}_1}^{\mathbf{r}_2} \mathbf{F} \cdot d\mathbf{r} = \int_{v_1^2}^{v_2^2} \frac{1}{2}m\,d(v^2) = \frac{1}{2}mv_2^2 - \frac{1}{2}mv_1^2, \tag{4.20}$$

where v_1 and v_2 are the magnitudes of the velocity at the positions $\mathbf{r}_1$ and $\mathbf{r}_2$. The *kinetic energy* of a particle of mass m with the velocity $\mathbf{v}$ is the term

$$T = \frac{1}{2}m\mathbf{v} \cdot \mathbf{v} = \frac{1}{2}mv^2, \tag{4.21}$$

where $|\mathbf{v}| = v$. The work done as the particle moves from position $\mathbf{r}_1$ to position $\mathbf{r}_2$ is

$$U_{12} = \int_{\mathbf{r}_1}^{\mathbf{r}_2} \mathbf{F} \cdot d\mathbf{r}. \tag{4.22}$$

The *principle of work and energy* may be expressed as

$$U_{12} = \frac{1}{2}mv_2^2 - \frac{1}{2}mv_1^2. \tag{4.23}$$

The work done on a particle as it moves between two positions equals the change in its kinetic energy. The dimensions of work, and therefore the dimensions of kinetic energy, are (force) $\times$ (length). In US customary units, work is expressed in ft lb. In SI units, work is expressed in N m, or joules [J].

One may use the principle of work and energy on a system if no net work is done by internal forces. The internal friction forces may do net work on a system.

4.8 Work and Power

The position of a particle P of mass m in curvilinear motion is specified by the coordinate s measured along its path from a reference point O, Fig. 4.7a. The velocity of the particle is

$$\mathbf{v} = \frac{ds}{dt}\mathbf{u}_t = \dot{s}\,\mathbf{u}_t,$$

where $\mathbf{u}_t$ is the tangential unit vector. Using the relation $\mathbf{v} = d\mathbf{r}/dt$, the infinitesimal displacement $d\mathbf{r}$ along the path is

$$d\mathbf{r} = \mathbf{v}\,dt = \frac{ds}{dt}\,\mathbf{u}_t\,dt = ds\,\mathbf{u}_t.$$

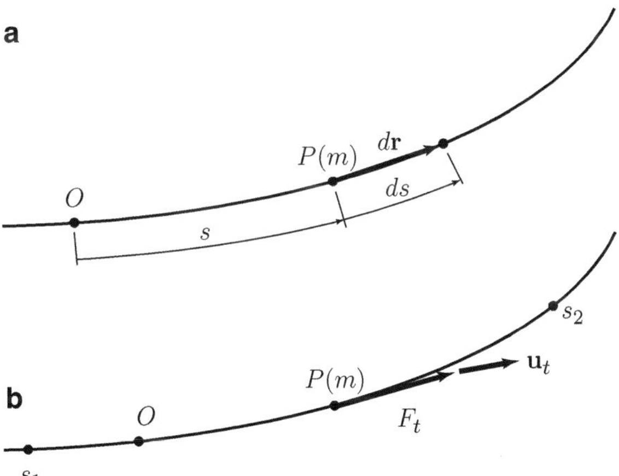

Fig. 4.7 Position of a particle in curvilinear motion

The work done by the external forces acting on the particle as result of the displacement $d\mathbf{r}$ is

$$\mathbf{F} \cdot d\mathbf{r} = \mathbf{F} \cdot ds\, \mathbf{u}_t = \mathbf{F} \cdot \mathbf{u}_t\, ds = F_t\, ds,$$

where $F_t = \mathbf{F} \cdot \mathbf{u}_t$ is the tangential component of the total force. The work as the particle moves from a position s_1 to a position s_2 is, Fig. 4.7b,

$$U_{12} = \int_{s_1}^{s_2} F_t\, ds. \tag{4.24}$$

The work is equal to the integral of the tangential component of the total force with respect to distance along the path. Components of force perpendicular to the path do not do any work.

The work done by the external forces acting on a particle during an infinitesimal displacement $d\mathbf{r}$ is

$$dU = \mathbf{F} \cdot d\mathbf{r}.$$

The *power*, P, is the rate at which work is done. The power P is obtained by dividing the expression of the work by the interval of time dt during which the displacement takes place:

$$P = \mathbf{F} \cdot \frac{d\mathbf{r}}{dt} = \mathbf{F} \cdot \mathbf{v}.$$

In SI units, the power is expressed in newton meters per second, which is joules per second [J/s] or watts [W]. In US customary units, power is expressed in foot pounds per second or in horsepower [hp], which is 746 W or approximately 550 ft lb/s. The power is also the rate of change of the kinetic energy of the object:

$$P = \frac{d}{dt}\left(\frac{1}{2}mv^2\right).$$

4.8.1 Work Done on a Particle by a Linear Spring

A linear spring connects a particle P of mass m to a fixed support, Fig. 4.8. The force exerted on the particle is

$$\mathbf{F} = -k\,(r - r_0)\mathbf{u}_r,$$

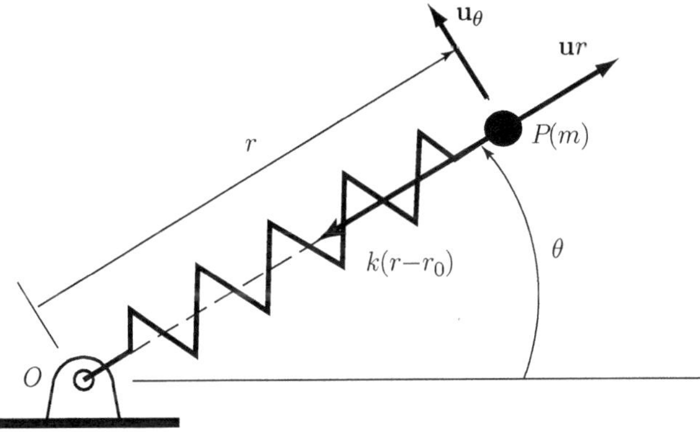

Fig. 4.8 Linear spring

where k is the spring constant, r_0 is the un-stretched length of the spring, and $\mathbf{u}_r$ is the polar unit vector. Using the expression for the velocity in polar coordinates, the vector $d\mathbf{r} = \mathbf{v}\,dt$ is

$$d\mathbf{r} = \left(\frac{dr}{dt}\mathbf{u}_r + r\frac{d\theta}{dt}\mathbf{u}_\theta \right) dt = dr\,\mathbf{u}_r + r\,d\theta\,\mathbf{u}_\theta, \tag{4.25}$$

and

$$\mathbf{F}\cdot d\mathbf{r} = \left[-k(r - r_0)\,\mathbf{u}_r \right]\cdot(dr\,\mathbf{u}_r + r\,d\theta\,\mathbf{u}_\theta) = -k(r - r_0)\,dr.$$

The stretch of a spring is defined as $\delta = r - r_0$. In terms of this variable, $\mathbf{F}\cdot d\mathbf{r} = -k\,\delta\,d\delta$, and the work is expressed as

$$U_{12} = \int_{\mathbf{r}_1}^{\mathbf{r}_2} \mathbf{F}\cdot d\mathbf{r} = \int_{\delta_1}^{\delta_2} -k\,\delta\,d\delta = -\frac{1}{2}k\left(\delta_2^2 - \delta_1^2\right),$$

where δ_1 and δ_2 are the values of the stretch at the initial and final positions.

4.8.2   Work Done on a Particle by Weight

The particle P of mass m, Fig. 4.9, moves from position 1 with coordinates (x_1, y_1, z_1) to position 2 with coordinates (x_2, y_2, z_2) in a Cartesian reference frame with the y-axis upward. The force exerted by the weight is

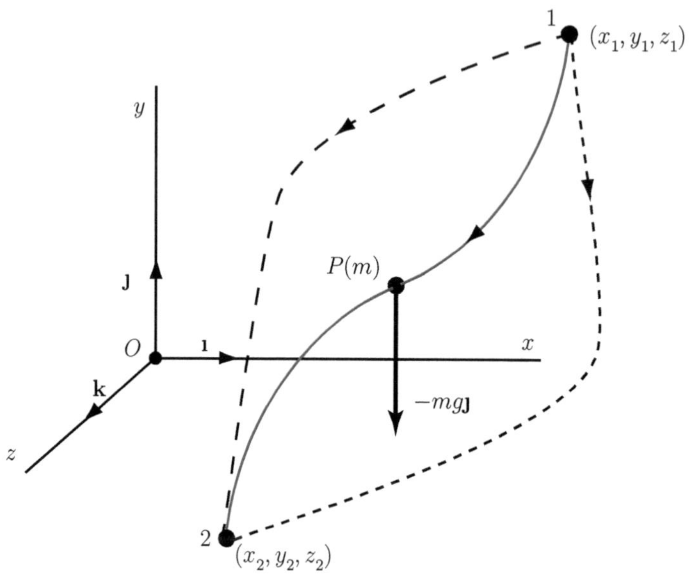

Fig. 4.9   Particle moving from position 1 to position 2

$$\mathbf{F} = -m\,g\,\mathbf{J}.$$

Because $\mathbf{v} = d\mathbf{r}/dt$, the expression for the vector $d\mathbf{r}$ is

$$d\mathbf{r} = \left(\frac{dx}{dt}\mathbf{1} + \frac{dy}{dt}\mathbf{J} + \frac{dz}{dt}\mathbf{k}\right)\,dt = dx\,\mathbf{1} + dy\,\mathbf{J} + dz\,\mathbf{k}.$$

The dot product of $\mathbf{F}$ and $d\mathbf{r}$ is

$$\mathbf{F} \cdot d\mathbf{r} = (-mg\mathbf{J}) \cdot (dx\,\mathbf{1} + dy\,\mathbf{J} + dz\,\mathbf{k}) = -m\,g\,dy.$$

The work done as P moves from position 1 to position 2 is

$$U_{12} = \int_{\mathbf{r}_1}^{\mathbf{r}_2} \mathbf{F} \cdot d\mathbf{r} = \int_{y_1}^{y_2} -m\,g\,dy = -m\,g\,(y_2 - y_1).$$

The work is the product of the weight and the change in the height of the particle. The work done is negative if the height increases and positive if it decreases. The work done is the same no matter what path the particle follows from position 1 to position 2. To determine the work done by the weight of the particle, only the relative heights of the initial and final positions must be known.

4.9 Conservation of Energy

The change in the kinetic energy is

$$U_{12} = \int_{r_1}^{r_2} \mathbf{F} \cdot d\mathbf{r} = \frac{1}{2}mv_2^2 - \frac{1}{2}mv_1^2. \tag{4.26}$$

A scalar function of position is V called *potential energy* and is determined as

$$dV = -\mathbf{F} \cdot d\mathbf{r}. \tag{4.27}$$

Using the function V, the integral defining the work is

$$U_{12} = \int_{r_1}^{r_2} \mathbf{F} \cdot d\mathbf{r} = \int_{V_1}^{V_2} -dV = -(V_2 - V_1), \tag{4.28}$$

where V_1 and V_2 are the values of V at the positions r_1 and r_2. The principle of work and energy would then have the form

$$\frac{1}{2}mv_1^2 + V_1 = \frac{1}{2}mv_2^2 + V_2, \tag{4.29}$$

which means that the sum of the kinetic energy and the potential energy V is constant

$$\frac{1}{2}mv^2 + V = \text{constant}, \tag{4.30}$$

or

$$E = T + V = \text{constant}. \tag{4.31}$$

If a potential energy V exists for a given force $\mathbf{F}$, that is, a function of position V exists such that $dV = -\mathbf{F} \cdot d\mathbf{r}$, then $\mathbf{F}$ is said to be *conservative*.

If all the forces that do work on a system are conservative, the total energy—the sum of the kinetic energy and the potential energies of the forces—is constant, or conserved. The system is said to be conservative.

4.9.1 Exercise

A heavy body of mass m falls on the ground without an initial velocity. The reference frame xy and the initial and final position of the body are shown in Fig. 4.10. The initial position of the body is at $A(h,0)$, where $AB = h$ represents the distance

Fig. 4.10 Fall of a heavy body

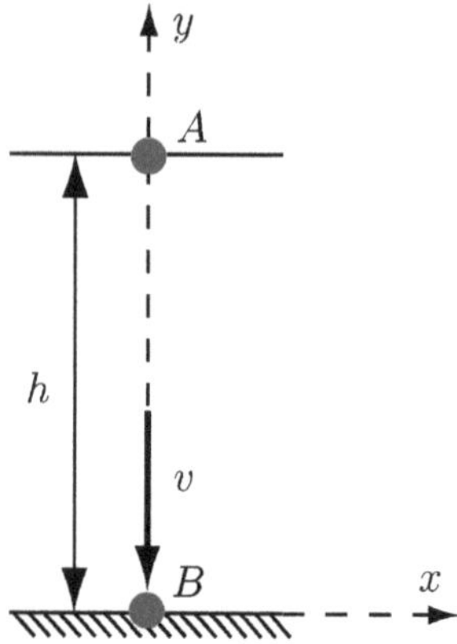

between the body and the ground. The coordinates of the final position B of the body are $B(0,0)$. Find the velocity of the body, when it hits the ground (the body hits the ground at point B).

Solution
The kinetic energy at A ($v = 0$) is

$$T_A = 0,$$

and the kinetic energy at B is

$$T_B = \frac{1}{2}mv^2,$$

where v is the velocity of the body at the point B. The work between A and B is

$$U_{AB} = mgh.$$

It results

$$\frac{1}{2}mv^2 = mgh,$$

or

$$v = \sqrt{2gh}.$$

4.9.2 Exercise

The system shown in the Fig. 4.11a is initially at rest. The system consists of two massless pulley wheels, one with the radius r and the other with radius R ($r < R$), and two bodies 1 and 2. The weight of body 1 is P, and the weight of body 2 is Q. Determine the motion of the system when it is released from rest.

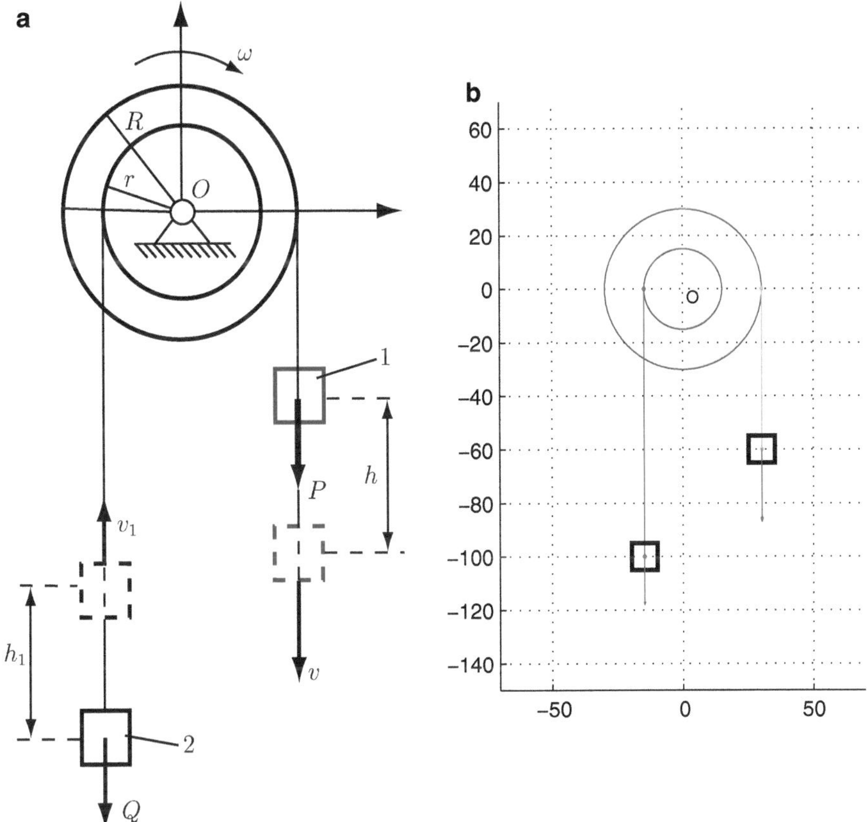

Fig. 4.11 (**a**) System of pulley wheels and particles and (**b**) MATLAB figure

Solution

Suppose that after a period of time t, the body (particle) 1 moves downward with a distance h. The pulley wheels will rotate with an angle $\theta = \frac{h}{R}$. The body (particle) 2 moves upward with a distance $h_1 = \theta r = h\frac{r}{R}$. If the velocity of the body 1 is v, then the velocity of the body 2 is $v_1 = v\frac{r}{R}$. The kinetic energy of the system is the sum of the kinetic energies of the two particle:

$$T = T(t) = \frac{1}{2}\frac{P}{g}v^2 + \frac{1}{2}\frac{Q}{g}v_1^2,$$

or

$$T = \frac{1}{2}\frac{P}{g}v^2 + \frac{1}{2}\frac{Q}{g}v^2\frac{r^2}{R^2} = \frac{v^2}{2}\left(\frac{P}{g} + \frac{Qr^2}{gR^2}\right).$$

where g is the gravitational acceleration. The total work is

$$U_{AB} = Ph - Qh_1 = h\left(P - Q\frac{r}{R}\right).$$

where A is the initial position of the system and B is the position at the time t. Using the relation $U_{AB} = T_B - T_A$, it results

$$\frac{v^2}{2}\left(\frac{P}{g} + \frac{Qr^2}{gR^2}\right) = h\left(P - Q\frac{r}{R}\right),$$

or

$$v^2 = \frac{2h\left(P - Q\dfrac{r}{R}\right)}{\dfrac{P}{g} + \dfrac{Qr^2}{gR^2}}.$$

The derivative with respect to time of the previous relation is

$$2v\dot{v} = \frac{2\dot{h}\left(P - Q\dfrac{r}{R}\right)}{\dfrac{P}{g} + \dfrac{Qr^2}{gR^2}}.$$

Using the relations

$$\dot{h} = v \text{ and } \dot{v} = a,$$

the acceleration, a, of the body 1 is determined as

$$a = \frac{P - Q\dfrac{r}{R}}{\dfrac{P}{g} + \dfrac{Qr^2}{gR^2}} = \text{const.}$$

Thus,

$$h = \frac{1}{2}at^2, \quad \theta = \frac{a}{2R}t^2, \quad h_1 = \frac{ar}{2R}t^2,$$

and

$$v = at, \quad v_1 = \frac{ar}{R}t.$$

The MATLAB program for the exercise is

```
syms theta R r v P Q g t
h = sym('h(t)');

theta=h/R;
h1=theta*r;
omega=v/R;
v1=omega*r;
```

```
T=(1/2)*(P/g)*v^2+(1/2)*(Q/g)*v1^2;

fprintf('kinetic energy of the system \n')
fprintf('T = %s  \n',char(T))

U=P*h-Q*h1;
fprintf('total work \n')
fprintf('U = %s  \n',char(U))

dTU=simplify(T-U);

v=simple(solve(dTU,v));

fprintf('velocity v^2 = \n')
pretty(v(1)^2)

% v = dh/dt
% a = dv/ dt
% d(v^2)/dt = 2 v a = 2 (dh/dt) a

a=simple(diff(v(1)^2,t)/(2*diff(h,t)));

fprintf('acceleration a = \n')
pretty(a)
```

The results of the program are

```
kinetic energy of the system

T = (P*v^2)/(2*g) + (Q*r^2*v^2)/(2*R^2*g)

total work

U = P*h(t) - (Q*r*h(t))/R

velocity v^2 =

                    2 3           2            2    2 3
   2 R g h(t) (P   R   - P Q R   r + P Q R r   - Q   r )
   --------------------------------------------------------
                       2       2 2
                  (P R   + Q r )
```

```
acceleration a =

  R g (P R - Q r)
  ---------------
       2      2
    P R   + Q r
```

The MATLAB program for Fig. 4.11b is

```
axis equal
hold on
axis([-70 70 -150 70])
R=30;
x_R=0;
y_R=0;
circle_m(x_R,y_R,R);
r=15;
x_r=0;
y_r=0;
circle_m(x_r,y_r,r);

x_O=0; y_O=0; z_O=0;

disp=1.5;
t1=text(x_O+disp, y_O-2*disp, z_O,'O','fontsize',8);

len_2=60;
len_1=100;
line([R R],[-len_2 0],'Marker','.',...
    'LineStyle','-','Color',[.8 .6 .8])
line([-r -r],[-len_1 0],'Marker','.',...
    'LineStyle','-','Color',[.8 .2 .8])

l=10;
rectangle('Position',[-r-l/2,-len_1-l/2,l,l],...
        'LineWidth',2,'LineStyle','-');
rectangle('Position',[R-l/2,-len_2-l/2,l,l],...
        'LineWidth',2,'LineStyle','-');
Q=20;
quiver(-r,-len_1,0,-Q)
P=30;
quiver(R,-len_2,0,-P)
```

where the function circle_m is

```
function H=circle_m(x_c,y_c,radius)
grid on
```

```
nr_points=1000;
theta = linspace(0,2*pi,nr_points);
x = x_c + radius*cos(theta);
y = y_c + radius*sin(theta);
plot(x,y);
```

4.10 Conservative Forces

A particle moves from a position 1 to a position 2. Equation (4.28) states that the work depends only on the values of the potential energy at positions 1 and 2. The work done by a conservative force as a particle moves from position 1 to position 2 is independent of the path of the particle.

A particle P of mass m slides with friction along a path of length L. The magnitude of the friction force is $\mu m g$ and is opposite to the direction of the motion of the particle. The coefficient of friction is μ. The work done by the friction force is

$$U_{12} = \int_0^L -\mu m g \, ds = -\mu m g L.$$

The work is proportional to the length L of the path and therefore is not independent of the path of the particle. Friction forces are not conservative.

4.10.1 *Potential Energy of a Force Exerted by a Spring*

The force exerted by a linear spring attached to a fixed support is a conservative force. In terms of polar coordinates, the force exerted on a particle, Fig. 4.8, by a linear spring is $\mathbf{F} = -k\,(r - r_0)\,\mathbf{u_r}$. The potential energy must satisfy

$$dV = -\mathbf{F} \cdot d\mathbf{r} = k\,(r - r_0)\,dr,$$

or

$$dV = k\,\delta\,d\delta,$$

where $\delta = r - r_0$ is the stretch of the spring. Integrating this equation, the potential energy of a linear spring is

$$V = \frac{1}{2}k\,\delta^2. \tag{4.32}$$

4.10.2   Potential Energy of Weight

The weight of a particle is a conservative force. The weight of the particle P of mass m, Fig. 4.9, is $F = -mg\mathbf{j}$. The potential energy V must satisfy the relation

$$dV = -\mathbf{F} \cdot d\mathbf{r} = (mg\mathbf{j}) \cdot (dx\mathbf{i} + dy\mathbf{j} + dz\mathbf{k}) = mg\,dy, \tag{4.33}$$

or

$$\frac{dV}{dy} = mg.$$

Integrating this equation, the potential energy is

$$V = mgy + C,$$

where C is an integration constant. The constant C is arbitrary because this expression satisfies (4.33) for any value of C. For $C = 0$, the potential energy of the weight of a particle is

$$V = mgy. \tag{4.34}$$

The potential energy V is a function of position and may be expressed in terms of a Cartesian reference frame as $V = V(x, y, z)$. The differential of dV is

$$dV = \frac{\partial V}{\partial x}\,dx + \frac{\partial V}{\partial y}\,dy + \frac{\partial V}{\partial z}\,dz. \tag{4.35}$$

The potential energy V satisfies the relation

$$\begin{aligned} dV &= -\mathbf{F} \cdot d\mathbf{r} = -(F_x\mathbf{i} + F_y\mathbf{j} + F_z\mathbf{k}) \cdot (dx\mathbf{i} + dy\mathbf{j} + dz\mathbf{k}) \\ &= -(F_x\,dx + F_y\,dy + F_z\,dz), \end{aligned} \tag{4.36}$$

where $\mathbf{F} = F_x\mathbf{i} + F_y\mathbf{j} + F_z\mathbf{k}$. Using (4.35) and (4.36), one may obtain

$$\frac{\partial V}{\partial x}\,dx + \frac{\partial V}{\partial y}\,dy + \frac{\partial V}{\partial z}\,dz = -(F_x\,dx + F_y\,dy + F_z\,dz),$$

which implies that

$$F_x = -\frac{\partial V}{\partial x},\, F_y = -\frac{\partial V}{\partial y},\, F_z = -\frac{\partial V}{\partial z}. \tag{4.37}$$

Given the potential energy $V = V(x, y, z)$ expressed in Cartesian coordinates, the force **F** is

$$\mathbf{F} = -\left(\frac{\partial V}{\partial x}\mathbf{i} + \frac{\partial V}{\partial y}\mathbf{j} + \frac{\partial V}{\partial z}\mathbf{k}\right) = -\nabla V, \tag{4.38}$$

where ∇V is the *gradient* of V. The gradient expressed in Cartesian coordinates is

$$\nabla = \frac{\partial}{\partial x}\mathbf{i} + \frac{\partial}{\partial y}\mathbf{j} + \frac{\partial}{\partial z}\mathbf{k}. \tag{4.39}$$

The *curl* of a vector force **F** in Cartesian coordinates is

$$\nabla \times \mathbf{F} = \begin{vmatrix} \mathbf{i} & \mathbf{j} & \mathbf{k} \\ \dfrac{\partial}{\partial x} & \dfrac{\partial}{\partial y} & \dfrac{\partial}{\partial z} \\ F_x & F_y & F_z \end{vmatrix}. \tag{4.40}$$

If a force F is conservative, its curl $\nabla \times \mathbf{F}$ is zero. The converse is also true: A force **F** is conservative if its curl is zero. In terms of cylindrical coordinates, the force **F** is

$$\mathbf{F} = -\nabla V = -\left(\frac{\partial V}{\partial r}\mathbf{u}_\mathrm{r} + \frac{1}{r}\frac{\partial V}{\partial \theta}\mathbf{u}_\theta + \frac{\partial V}{\partial z}\mathbf{k}\right). \tag{4.41}$$

In terms of cylindrical coordinates, the curl of the force **F** is

$$\nabla \times \mathbf{F} = \frac{1}{r}\begin{vmatrix} \mathbf{u}_\mathrm{r} & r\mathbf{u}_\theta & \mathbf{k} \\ \dfrac{\partial}{\partial r} & \dfrac{\partial}{\partial \theta} & \dfrac{\partial}{\partial z} \\ F_\mathrm{r} & rF_\theta & F_z \end{vmatrix}. \tag{4.42}$$

4.10.3   *Exercise*

The spring shown in Fig. 4.12a has the initial length OO_1. The spring can be stretched with the distance $OM = x$, and the spring force will be $F = kOM$, where k is the elastic spring constant. Find the potential energy and the total work done when the particle M moves from point A ($OA = x_A$) to point B ($OB = x_B$).

Solution

For the spring shown in Fig. 4.12, the force acting on the particle M is

$$\mathbf{F} = F_x\mathbf{i} + F_y\mathbf{j} + F_z\mathbf{k}$$

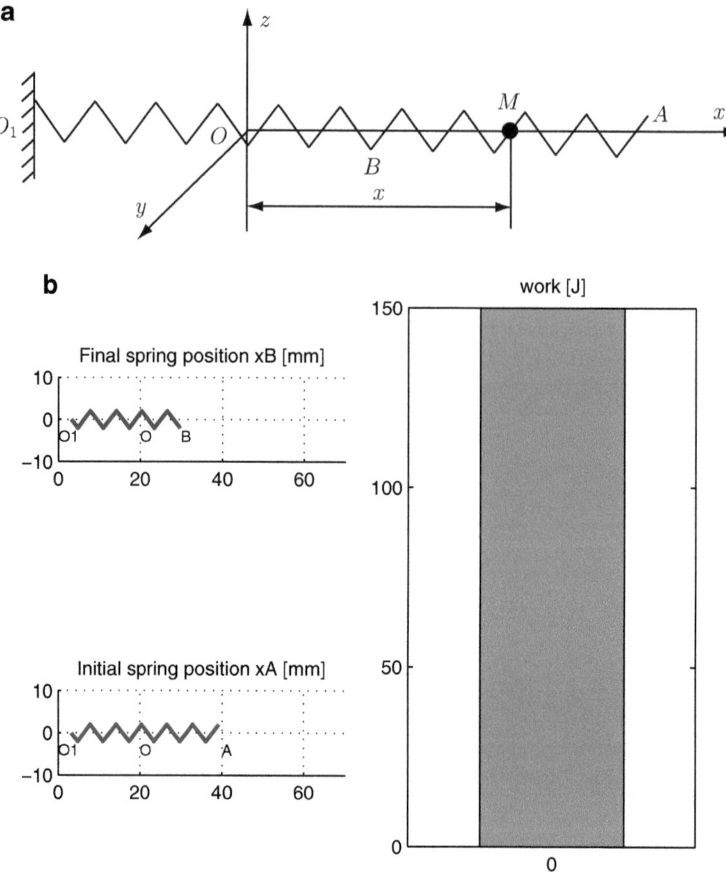

Fig. 4.12 (a) Spring and (b) MATLAB output

where

$$F_x = -kx, \quad F_y = 0, \quad F_z = 0.$$

Using (4.36), the potential energy is calculated as

$$V = \frac{1}{2} k x^2 + C,$$

and the work done as M moves from position A to position B is

$$U_{AB} = -(V_B - V_A) = \frac{1}{2} k \left(x_A^2 - x_B^2 \right).$$

The symbolic MATLAB program for the above calculations is

```
VA=dsolve('DF = - k*xA','F(0) = 0','xA');
VB=dsolve('DF = - k*xB','F(0) = 0','xB');
fprintf('Potential energy \n')
fprintf('VA = %s  \n',char(VA))
fprintf('VB = %s  \n',char(VB))

U=abs(-(VB-VA));
fprintf('Work done by the particle \n')
fprintf('UAB=-(VB-VA) = %s  \n',char(U))
```

If $O_1O = 0.02\,\text{m}$, $x_A = 0.02\,\text{m}$, $x_B = 0.01\,\text{m}$, and $k = 1{,}000\,\text{N/m}$, then the MATLAB commands for the exercise are

```
% numerical application

disp=4;
x_O1=0; % mm
x_O=20; % mm
x_A=40; % mm
x_B=30; % mm

subplot(2,2,1);
axis manual
axis equal
hold on
grid on
axis([0 70 -10 10])

x_ii = pi:pi/2:x_B;
x_if =  2*sin(x_ii);

plot(x_ii,x_if,'-','LineWidth',2)
title 'Final spring position xB [mm]'
t1=text(x_O1,-disp,'O1','fontsize',8);
t2=text(x_O,-disp, 'O','fontsize',8);
t3=text(x_B,-disp,'B','fontsize',8);

subplot(2,2,3);
axis manual
axis equal
hold on
grid on
axis([0 70 -10 10])
```

Fig. 4.13 Potential energy
and the total work done by
particle P

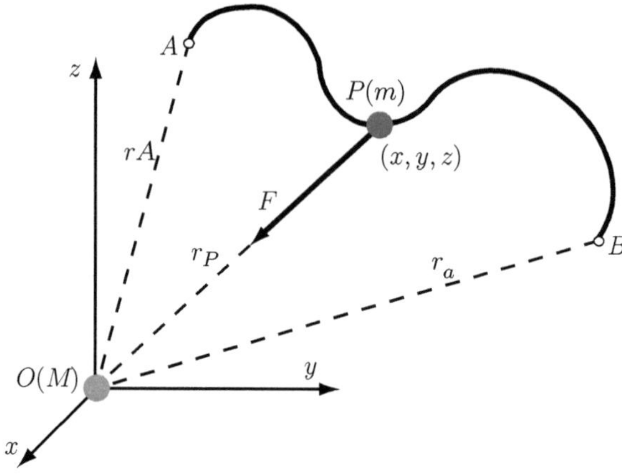

```
u_fi = pi:pi/2:x_A;
u_ff =   2*sin(u_fi);
plot(u_fi,u_ff,'-','Color','r','LineWidth',2)
title 'Initial spring position xA [mm]'
t1=text(x_O1,-disp,'O1','fontsize',8);
t2=text(x_O,-disp, 'O','fontsize',8);
t3=text(x_A,-disp,'A','fontsize',8);

subplot(2,2,[2,4]);
kn=1000; % N/m
slist={k,'xA','xB'};
nlist={kn,x_A-x_O,x_B-x_O};
Un = subs(U,slist,nlist);

bar(0,Un*10^(-3),0.5,'r');
title 'work [J]'
```

The output of the numerical program is shown in Fig. 4.12b.

4.10.4   Exercise

A particle P of mass m is located at the distance $\mathbf{r} = \mathbf{r}_{OP} = x\mathbf{i} + y\mathbf{j} + z\mathbf{k}$ from the
origin of a fixed reference frame xyz as shown in Fig. 4.13. The origin O of mass M
attracts the particle P. The attraction force between O and P is $F = G\frac{mM}{r^2}$, as shown
in Fig. 4.13, where G is called the universal gravitational constant. Find the potential
energy and the total work done when the particle P moves from point A $(OA = r_A)$
to point B $(OB = r_B)$.

Solution

With respect to the reference frame xyz, the attraction force $\mathbf{F}$ is

$$\mathbf{F} = F_x\mathbf{i} + F_y\mathbf{j} + F_z\mathbf{k},$$

where

$$F_x = -G\frac{mM}{r^2}\frac{x}{r}, \quad F_y = -G\frac{mM}{r^2}\frac{y}{r}, \quad F_z = -G\frac{mM}{r^2}\frac{z}{r}.$$

The distance $r = |\mathbf{r}|$ is computed as

$$r = \sqrt{x^2 + y^2 + z^2}.$$

The potential energy V is calculated as

$$V = -G\frac{mM}{r} = -G\frac{mM}{\sqrt{x^2 + y^2 + z^2}}.$$

The total work done when the particle M moves from point A ($OA = r_A$) to point B ($OB = r_B$) is given by

$$U_{AB} = -(V_B - V_A) = GmM\left(\frac{1}{r_B} - \frac{1}{r_A}\right).$$

The MATLAB program for the exercise is

```
syms x y z G m M xA yA zA xB yB zB

r_v = [x y z];
fprintf('r = [%s %s %s]\n',...
    char(r_v(1)),char(r_v(2)),char(r_v(3)))

r = sqrt(x^2+y^2+z^2);

F = -G*m*M*r_v/r^3;
fprintf('Fx = %s \n',char(F(1)))
fprintf('Fy = %s \n',char(F(2)))
fprintf('Fz = %s \n',char(F(3)))

% potential energy
V = -int(F(1),x);
V = -int(F(2),y);
V = -int(F(3),z);
fprintf('V = %s \n',char(V))
```

```
slist={x,y,z};
nlistA={xA,yA,zA};
nlistB={xB,yB,zB};

VA=subs(V,slist,nlistA);
VB=subs(V,slist,nlistB);

% work
UAB = simplify(VA-VB);
fprintf('UAB = VA-VB = %s \n',char(UAB))
```

4.11 Principle of Impulse and Momentum

Newton's second law

$$\mathbf{F} = m\,\frac{d\mathbf{v}}{dt}$$

is integrated with respect to time to obtain

$$\int_{t_1}^{t_2} \mathbf{F}\, dt = m\mathbf{v}_2 - m\mathbf{v}_1, \tag{4.43}$$

where $\mathbf{v}_1$ and $\mathbf{v}_2$ are the velocities of the particle P at the times t_1 and t_2. The term $\int_{t_1}^{t_2} \mathbf{F}\, dt$ is called the *linear impulse*, and the term $m\mathbf{v}$ is called the *linear momentum*. *The principle of impulse and momentum*: The impulse applied to a particle during an interval of time is equal to the change in its linear momentum, Fig. 4.14. The dimensions of the linear impulse and linear momentum are (mass) $\times$ (length)/(time). The average with respect to time of the total force acting on a particle from t_1 to t_2 is

$$\mathbf{F}_{av} = \frac{1}{t_2 - t_1} \int_{t_1}^{t_2} \mathbf{F}\, dt,$$

so (4.43) can be written as

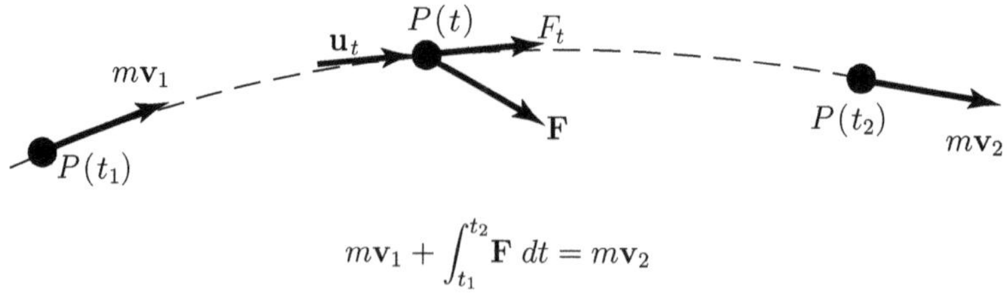

$$m\mathbf{v}_1 + \int_{t_1}^{t_2} \mathbf{F}\, dt = m\mathbf{v}_2$$

Fig. 4.14 Principle of impulse and momentum

Fig. 4.15 Impulsive force

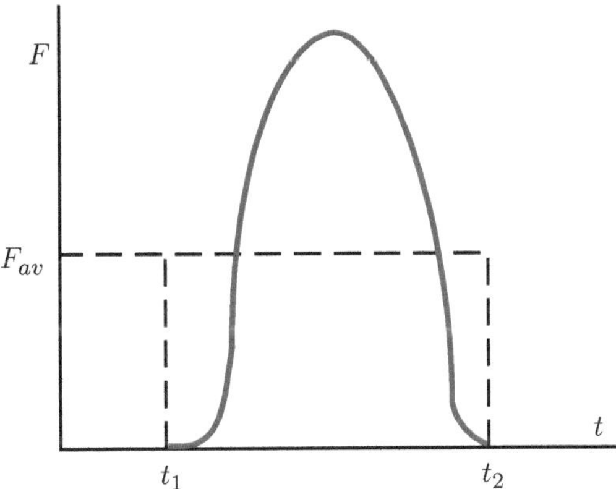

$$\mathbf{F}_{av}\left(t_2 - t_1\right) = m\mathbf{v}_2 - m\mathbf{v}_1.\tag{4.44}$$

An *impulsive force* is a force of relatively large magnitude that acts over a small interval of time, Fig. 4.15. Equations (4.43) and (4.44) may be expressed in scalar forms. The sum of the forces in the tangent direction $\mathbf{u}_t$ to the path of the particle equals the product of its mass m and the rate of change of its velocity along the path

$$F_t = m a_t = m \frac{dv}{dt}.$$

Integrating this equation with respect to time gives

$$\int_{t_1}^{t_2} F_t\, dt = m v_2 - m v_1,\tag{4.45}$$

where v_1 and v_2 are the velocities along the path at the times t_1 and t_2. The impulse applied to an object by the sum of the forces tangent to its path during an interval of time is equal to the change in its linear momentum along the path.

4.12   Conservation of Linear Momentum

Consider two particles P_1 of mass m_1 and P_2 of mass m_2 in Fig. 4.16. The vector $\mathbf{F}_{12}$ is the force exerted by P_1 on P_2, and $\mathbf{F}_{21}$ is the force exerted by P_2 on P_1. These forces could be contact forces or could be exerted by a spring connecting the particles. As a consequence of Newton's third law, these forces are equal and opposite:

$$\mathbf{F}_{12} + \mathbf{F}_{21} = \mathbf{0}.\tag{4.46}$$

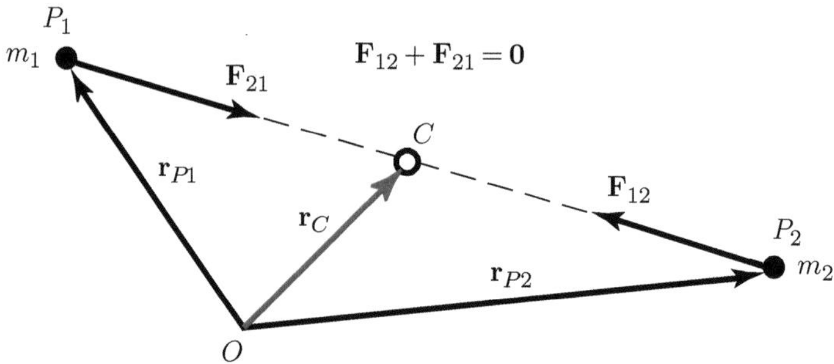

Fig. 4.16 Position of the center of mass of P_1 and P_2

Consider that no external forces act on P_1 and P_2, or the external forces are negligible. The principle of impulse and momentum to each particle for arbitrary times t_1 and t_2 gives

$$\int_{t_1}^{t_2} \mathbf{F}_{21}\, dt = m_1 \mathbf{v}_{P_1}(t_2) - m_1 \mathbf{v}_{P_1}(t_1),$$

$$\int_{t_1}^{t_2} \mathbf{F}_{12}\, dt = m_2 \mathbf{v}_{P_2}(t_2) - m_1 \mathbf{v}_{P_2}(t_1),$$

where $\mathbf{v}_{P_1}(t_1)$, $\mathbf{v}_{P_1}(t_2)$ are the velocities of P_1 at the times t_1, t_2, and $\mathbf{v}_{P_2}(t_1)$, $\mathbf{v}_{P2}(t_2)$ are the velocities of P_2 at the times t_1, t_2. The sum of these equations is

$$m_1 \mathbf{v}_{P_1}(t_1) + m_2 \mathbf{v}_{P_2}(t_1) = m_1 \mathbf{v}_{P_1}(t_2) + m_2 \mathbf{v}_{P_2}(t_2),$$

or the total linear momentum of P_1 and P_2 is conserved:

$$m_1 \mathbf{v}_{P_1} + m_2 \mathbf{v}_{P_2} = \text{constant.} \tag{4.47}$$

The position of the center of mass of P_1 and P_2 is, Fig. 4.12,

$$\mathbf{r}_C = \frac{m_1 \mathbf{r}_{P_1} + m_2 \mathbf{r}_{P_2}}{m_1 + m_2},$$

where $\mathbf{r}_{P_1}$ and $\mathbf{r}_{P_1}$ are the position vectors of P_1 and P_2. Taking the time derivative of this equation and using (4.47), one may obtain

$$(m_1 + m_2)\mathbf{v}_C = m_1 \mathbf{v}_{P_1} + m_2 \mathbf{v}_{P_2} = \text{constant,} \tag{4.48}$$

where $\mathbf{v}_C = d\mathbf{r}_C/dt$ is the velocity of the combined center of mass. The total linear momentum of the particles is conserved, and the velocity of the combined center of mass of the particles P_1 and P_2 is constant.

4.13 Principle of Angular Impulse and Momentum

The position of a particle P of mass m relative to an inertial reference frame with origin O is given by the position vector $\mathbf{r} = \mathbf{r}_{OP}$, Fig. 4.17. The cross product of Newton's second law with the position vector $\mathbf{r}$ is

$$\mathbf{r} \times \mathbf{F} = \mathbf{r} \times m\mathbf{a} = \mathbf{r} \times m\frac{d\mathbf{v}}{dt}. \tag{4.49}$$

The time derivative of the quantity $\mathbf{r} \times m\mathbf{v}$ is

$$\frac{d}{dt}(\mathbf{r} \times m\mathbf{v}) = \left(\frac{d\mathbf{r}}{dt} \times m\mathbf{v}\right) + \left(\mathbf{r} \times m\frac{d\mathbf{v}}{dt}\right) = \mathbf{r} \times m\frac{d\mathbf{v}}{dt},$$

because $\frac{d\mathbf{r}}{dt} = \mathbf{v}$, and the cross product of parallel vectors is zero. Equation (4.49) may be written as

$$\mathbf{r} \times \mathbf{F} = \frac{d\mathbf{H}_O}{dt}, \tag{4.50}$$

where the vector

$$\mathbf{H}_O = \mathbf{r} \times m\mathbf{v} \tag{4.51}$$

is called the *angular momentum* about O Fig. 4.13. The angular momentum may be interpreted as the moment of the linear momentum of the particle about point O. The moment $\mathbf{r} \times \mathbf{F}$ equals the rate of change of the moment of momentum about point O.

Integrating (4.50) with respect to time, one may obtain

$$\int_{t_1}^{t_2} (\mathbf{r} \times \mathbf{F})\, dt = (\mathbf{H}_O)_2 - (\mathbf{H}_O)_1. \tag{4.52}$$

The integral on the left-hand side is called the *angular impulse*.

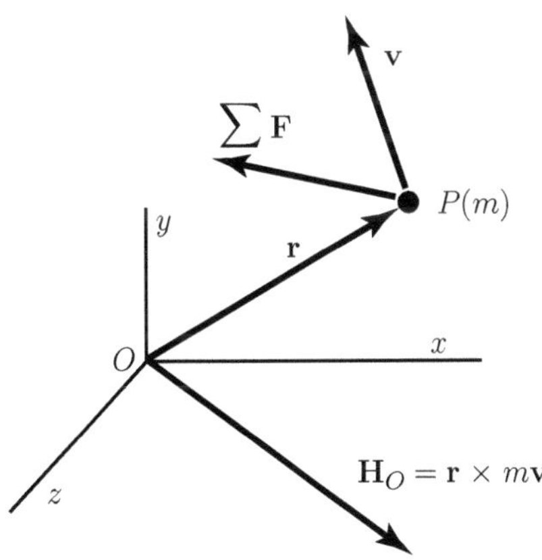

Fig. 4.17 Angular momentum

The principle of angular impulse and momentum: The angular impulse applied to a particle during an interval of time is equal to the change in its angular momentum.

The dimensions of the angular impulse and angular momentum are $(\text{mass}) \times (\text{length})^2/(\text{time})$.

4.14   Examples

Example 4.1. A particle M of mass m starts moving from the origin O on the inclined plane as shown in Fig. 4.18a. The angle of the inclined plane with the horizontal line BD is α, $\alpha = \angle(BC, BD)$. The reference frame at O has the x-axis parallel to BC ($Ox \| BC$) and the vertical z-axis perpendicular to the inclined plane. The y-axis is and perpendicular to Ox, and it is situated in the inclined plane. The initial velocity of the particle is $\mathbf{v}_0 = v_0 \mathbf{J}$ where $\mathbf{J}$ is the unit vector of the y-axis. Find the equations of motion of the particle.

Solution

The equations of the motion with respect to the reference frame xyz are

$$m\ddot{x} = mg \sin\alpha, \tag{4.53}$$

$$m\ddot{y} = 0, \tag{4.54}$$

$$m\ddot{z} = N - mg \cos\alpha. \tag{4.55}$$

Because the motion of the particle is planar and z-axis is perpendicular to the plane of the motion, it results

$$z = 0 \text{ and } \ddot{z} = 0.$$

From (4.55), the reaction is

$$N = mg \cos\alpha.$$

Integrating the (4.53) and (4.54), one can write

$$x = C_1 + C_2 t + \frac{t^2}{2} g \sin\alpha,$$

$$y = C_3 + C_4 t. \tag{4.56}$$

Using the initial conditions at $t = 0$

$$x = 0, \ y = 0, \ \dot{x} = 0, \ \dot{y} = v_0,$$

the constants are

$$C_1 = C_2 = C_3 = 0 \text{ and } C_4 = v_0. \tag{4.57}$$

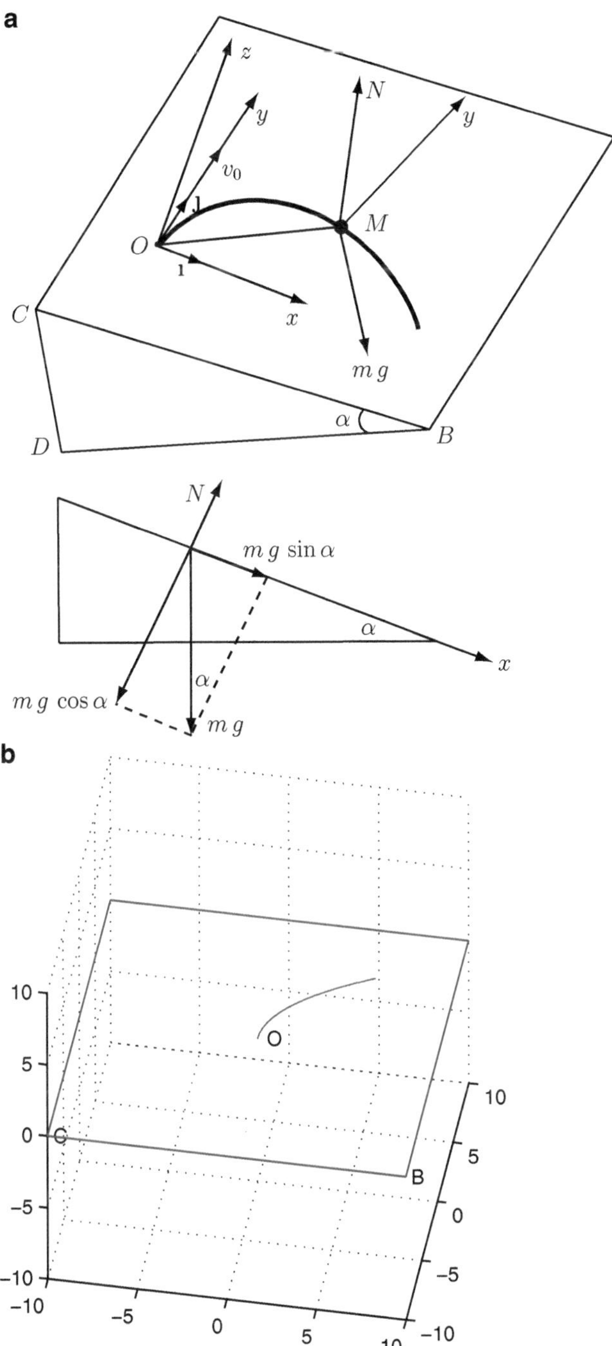

Fig. 4.18   Example 4.1

Thus, from (4.56) and (4.57), it results

$$x = \frac{t^2}{2} g \sin \alpha,$$

$$y = v_0 t. \tag{4.58}$$

The MATLAB commands for solving the equations of motion are

```
syms m g alpha v0 t

N = m*g*cos(alpha);

sx=dsolve('D2x - g*sin(alpha) = 0');
sy=dsolve('D2y = 0');

% dsolve - symbolic solution of ODE
% by default, the independent variable is 't'
% the letter 'D' denotes differentiation d/dt
% 'D' followed by a digit denotes repeated
%      differentiation
% 'D2' is d^2/dt^2

fprintf('equations of motion: \n')
fprintf('x = %s \n',char(sx))
fprintf('y = %s \n',char(sy))

fprintf ...
('equations of motion with initial conditions\n')
x=dsolve('D2x-g*sin(alpha)=0','Dx(0)=0','x(0)=0');
y=dsolve('D2y = 0','Dy(0) = v0','y(0) = 0');
fprintf('x = %s   \n',char(x))
fprintf('y = %s   \n\n',char(y))
```

The symbolic solution of ordinary differential equations is obtained with the MATLAB function dsolve.

Using (4.58), the analytical expression of the trajectory of the particle is obtained as

$$x = \frac{1}{2} \frac{g \sin \alpha}{v_0^2} y^2.$$

For $\alpha = 0$, it results

$$N = mg, \quad x = 0, \quad y = v_0 t.$$

The velocity and acceleration of the particle are obtained symbolically with the
MATLAB commands:

```
% velocity
v_x=diff(x,t);
v_y=diff(y,t);
v=[v_x v_y 0];
fprintf('velocity v = \n')
pretty(v); fprintf('\n\n')

% magnitude of velocity
magn_v=sqrt(v(1)^2+v(2)^2)+v(3)^2;
fprintf('|v| = \n')
pretty(magn_v); fprintf('\n\n')

% acceleration
a_x=diff(x,t,2);
a_y=diff(y,t,2);
a_z=0;

a=[a_x a_y a_z];
fprintf('a = \n')
pretty(a); fprintf('\n\n');

% magnitude of acceleration
magn_a=sqrt(a(1)^2+a(2)^2+a(3)^2);
fprintf('|a| = \n')
pretty(simplify(magn_a)); fprintf('\n\n')
```

For the numerical data $\alpha = \pi/6$, $v_0 = 4$ m/s, and $g = 9.81$ m/s^2, the trajectory of
the particle is visualized with the following MATLAB commands:

```
% numerical application
gn = 9.81; % m/s^2
alphan = pi/6;
v0n= 4; % m/s

axis manual
axis equal
hold on
grid on

a = 10;
axis([-a a -a a -a a])

x_O=0; y_O=0; z_O=0;
x_B=10; y_B=-10; z_B=0;
```

```
x_C=-10; y_C=-10; z_C=0;
x_E=-10; y_E=10; z_E=0;
x_F=10; y_F=10; z_F=0;

t1=text(x_O+0.2, y_O, z_O,' O');
t3=text(x_B, y_B, z_B,' B');
t4=text(x_C, y_C, z_C,' C');

line([x_C,x_B],[y_C,y_B],[z_C,z_B]);
line([x_B,x_F],[y_B,y_F],[z_B,z_F]);
line([x_F,x_E],[y_F,y_E],[z_F,z_E]);
line([x_E,x_C],[y_E,y_C],[z_E,z_C]);

view(10,40);
light('Position',[1 3 2]);

% trajectory of the particle
for tn = 0 : 0.005 : 1.5
    slist={g,alpha,v0,t};
    nlist={gn,alphan,v0n,tn};
    xn = subs(x,slist,nlist);
    yn = subs(y,slist,nlist);
    hm=plot(xn,yn,'k.','Color','red');
    ht=plot(xn,yn);
    pause(0.001)
    delete(hm);
end
```

The output represents the trajectory of the particle and is shown in Fig. 4.18b.

Example 4.2. A person of mass m_p stands at the center of a stationary wagon of mass m_w. The length of the wagon is L, as shown in Fig. 4.19a. (a) The person starts moving with the velocity v_p relative to the ground. Determine the velocity of the wagon relative to the ground. (b) The person stops when he reaches the end of the wagon. Determine the positions of the person and wagon relative to their original position.

Solution

(a) Let v_w be the velocity of the wagon. If the person moves to the right, then $v_p > 0$ and $v_w < 0$, Fig. 4.19b. The origin of the coordinate system is at the center of the stationary wagon as shown in Fig. 4.19a.

   Before the person starts moving, the total linear momentum of the person and the wagon is zero because the person and wagon are stationary. When the person is moving, the total linear momentum of the person and the wagon in

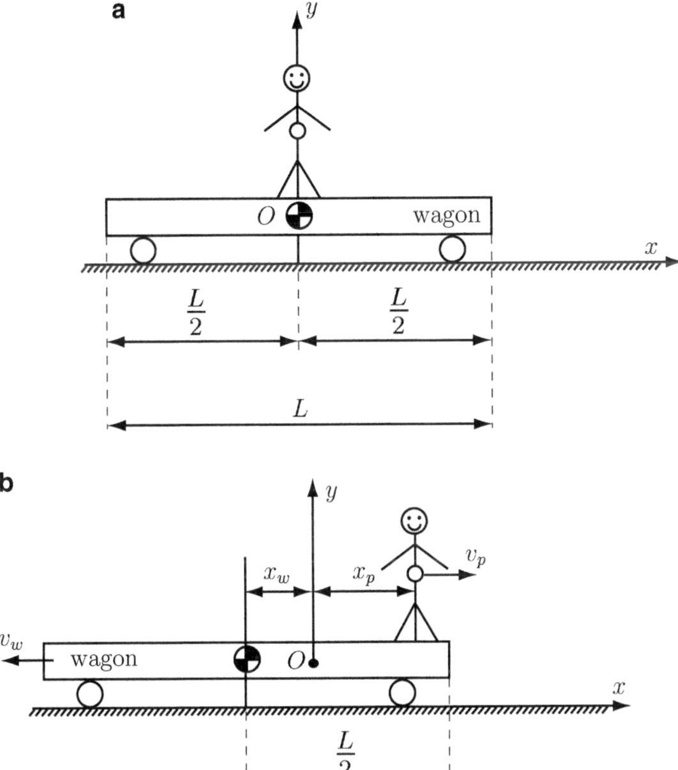

Fig. 4.19 Example 4.2

the horizontal direction is $m_{\mathrm{p}} v_{\mathrm{p}} - m_{\mathrm{w}} v_{\mathrm{w}}$. The total linear momentum in the horizontal direction is conserved, and one can obtain

$$m_{\mathrm{p}} v_{\mathrm{p}} - m_{\mathrm{w}} v_{\mathrm{w}} = 0.$$

The velocity of the wagon is

$$v_{\mathrm{w}} = \frac{m_{\mathrm{p}} v_{\mathrm{p}}}{m_{\mathrm{w}}}.$$

(b) Let x_{p} be the position of the person relative to the origin O as shown in Fig. 4.24b. The position of the mass center of the wagon with respect to the origin is x_{w}. The position of the combined center of mass of the person and the wagon is

$$x_c = \frac{m_{\mathrm{p}} x_{\mathrm{p}} - m_{\mathrm{w}} x_{\mathrm{w}}}{m_{\mathrm{p}} + m_{\mathrm{w}}}.$$

Fig. 4.20 Example 4.3

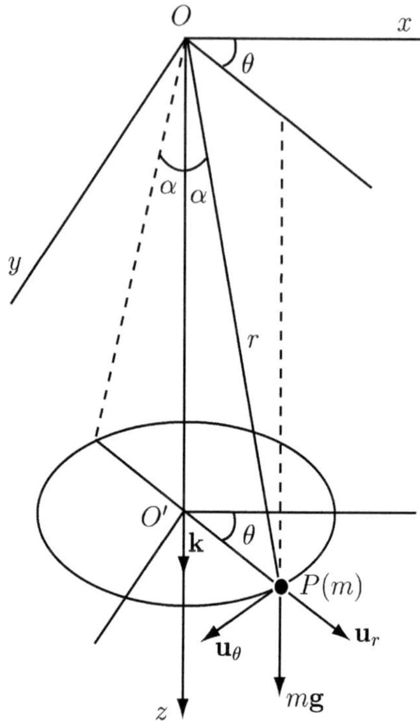

The combined center of mass is initially stationary, and it must remain
stationary. When the person has stopped at the end of the wagon, the combined
center of mass must still be at $x = 0$ (the original stationary position). Thus,

$$\frac{m_{\mathrm{p}} x_{\mathrm{p}} - m_{\mathrm{w}} x_{\mathrm{w}}}{m_{\mathrm{p}} + m_{\mathrm{w}}} = 0. \tag{4.59}$$

Another equation is the relation

$$x_{\mathrm{p}} + x_{\mathrm{w}} = \frac{L}{2}. \tag{4.60}$$

Solving together (4.59) and (4.60), the following results are obtained:

$$x_{\mathrm{p}} = \frac{m_{\mathrm{w}} L}{2 \left(m_{\mathrm{p}} + m_{\mathrm{w}}\right)}, \quad x_{\mathrm{w}} = \frac{m_{\mathrm{p}} L}{2 \left(m_{\mathrm{p}} + m_{\mathrm{w}}\right)}.$$

Example 4.3. A particle of mass m attached by a string of length r to a fixed point
O is rotating in a horizontal circle about a vertical axis Oz. This pendulum (conical
pendulum) describes a cone of constant angle 2α as shown in Fig. 4.20. Determine
the tangential velocity $\dot{\theta}$.

Solution

The position vector of the particle is

$$\mathbf{r}_{OP} = \mathbf{r} = r\sin\alpha\,\mathbf{u}_r + r\cos\alpha\,\mathbf{k},$$

where $\alpha = $ constant and $r = $ constant.

The velocity of the particle P is

$$\mathbf{v} = \dot{\mathbf{r}} = r\sin\alpha\,\dot{\mathbf{u}}_r + r\cos\alpha\,\dot{\mathbf{k}}.$$

The derivative of the polar unit vector is $\dot{\mathbf{u}}_r = \dot{\theta}\mathbf{u}_\theta$ and $\dot{\mathbf{k}} = 0$.

Thus,

$$\mathbf{v} = r\dot{\theta}\sin\alpha\,\mathbf{u}_\theta.$$

The moment of external forces about the origin O is

$$\mathbf{M}_O = \mathbf{r}\times m\mathbf{g} = \mathbf{r}\times mg\mathbf{k} = (r\sin\alpha\,\mathbf{u}_r + r\cos\alpha\,\mathbf{k})\times mg\mathbf{k}$$

$$= rmg\sin\alpha\,(\mathbf{u}_r\times\mathbf{k}) = -rmg\sin\alpha\,\mathbf{u}_\theta.$$

The angular momentum of the particle about O is the momentum of the linear momentum

$$\mathbf{H}_O = \mathbf{r}\times m\mathbf{v} = (r\sin\alpha\,\mathbf{u}_r + r\cos\alpha\,\mathbf{k})\times mr\dot{\theta}\sin\alpha\,\mathbf{u}_\theta$$

$$= -mr^2\dot{\theta}\sin\alpha\cos\alpha\,\mathbf{u}_r + mr^2\dot{\theta}\sin^2\alpha\,\mathbf{k}.$$

The derivative of the angular momentum is

$$\dot{\mathbf{H}}_O = \frac{\mathrm{d}\mathbf{H}_O}{\mathrm{d}t} = -mr^2\ddot{\theta}\sin\alpha\cos\alpha\,\mathbf{u}_r - mr^2\dot{\theta}^2\sin\alpha\cos\alpha\,\mathbf{u}_\theta + mr^2\ddot{\theta}\sin^2\alpha\,\mathbf{k}.$$

The moment of all the external forces acting on a particle about the fixed origin O is equal to the time rate of change of the angular momentum of the particle

$$\mathbf{M}_O = \dot{\mathbf{H}}_O,$$

or

$$\begin{cases} mr^2\ddot{\theta}\sin\alpha\cos\alpha = 0, \\ mr^2\dot{\theta}^2\sin\alpha\cos\alpha = mgr\sin\alpha, \\ mr^2\ddot{\theta}\sin^2\alpha = 0. \end{cases}$$

The first equation of the previous system gives $\ddot{\theta} = 0$ and $\dot{\theta} = $ constant. The tangential angular velocity $\dot{\theta}$ has a constant magnitude. Consider now $\dot{\theta} = \omega = $ constant. With $\dot{\theta} = \omega$, the second equation of the system gives $mr^2\omega^2\sin\alpha\cos\alpha = mgr\sin\alpha$ or $\omega = \sqrt{\dfrac{g}{r\cos\alpha}}$.

The MATLAB program for the example is

```
syms t r alpha m g dtheta

theta = sym('theta(t)');

c = cos(alpha);
s = sin(alpha);

rP = [r*s, 0, r*c];
omega = [0, 0, diff(theta,t)];
v = diff(rP,t) + cross(omega, rP);
fprintf('v = \n')
pretty(v); fprintf('\n')

G = [0 0 m*g];
MO = cross(rP, G);
fprintf('MO = r x G = \n')
pretty(MO); fprintf('\n')

HO = cross(rP, m*v);
fprintf('HO = r x (m v) = \n')
pretty(HO); fprintf('\n')

dHO = diff(HO, t)+cross(omega, HO);
fprintf('d(HO)/dt - MO =>\n')
eq = dHO-MO;
pretty(eq(1)); fprintf('\n')
pretty(eq(2)); fprintf('\n')
pretty(eq(3)); fprintf('\n\n')

eqf = subs(eq(2), diff(theta, t), dtheta);
sol = solve(eqf, 'dtheta');

fprintf('for d2(theta)/dt2 = 0 => d(theta)/dt =\n')
pretty(sol(1))
```

Example 4.4. A particle P of mass m is traveling down an inclined surface as shown in Fig. 4.21. The particle P is released from rest at the point A. The angle between the inclined surface and the horizontal is α, and the point A is located at the vertical distance h. The coefficient of friction between the particle and the surface is μ. (a) Find the velocity of the particle at the point B where the inclined surface intersects the horizontal. (b) The particle will stop, because of the friction, at the point C located at the distance d from the point B. Find the distance d. For the numerical application, use $\alpha = 30°$, $h = 2$ m, $\mu = 0.2$, and $g = 9.8$ m/s^2.

Fig. 4.21 Example 4.4

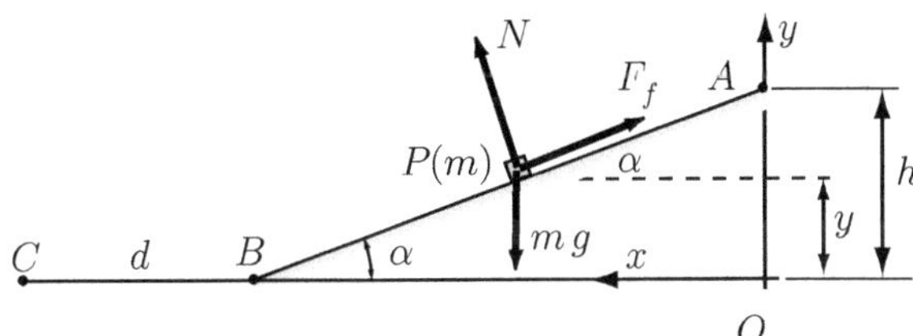

Solution

(a) A system of coordinates axes is chosen as shown in Fig. 4.21. The gravity force acts on the particle and is given by

$$\mathbf{G} = m\mathbf{g} = -mg\,\mathbf{j}.$$

The friction force on the particle is

$$\mathbf{F}_f = -\mu\,mg\cos^2\alpha\,\mathbf{i} + \mu\,mg\sin\alpha\cos\alpha\,\mathbf{j}.$$

The position vector of the particle P is

$$\mathbf{r} = \mathbf{r}_{OP} = x\mathbf{i} + y\mathbf{j}.$$

The work done on the particle as it moves between the points A and B is

$$
U_{AB} = \int_{\mathbf{r}_A}^{\mathbf{r}_B} \mathbf{F}\cdot d\mathbf{r} = \int_{\mathbf{r}_A}^{\mathbf{r}_B} (\mathbf{G} + \mathbf{F}_f)\cdot(dx\mathbf{i} + dy\mathbf{j})
$$

$$
= \int_{\mathbf{r}_A}^{\mathbf{r}_B} \left[-\mu\,mg\cos^2\alpha\,\mathbf{i} + mg\,(\mu\sin\alpha\cos\alpha - 1)\,\mathbf{j} \right]\cdot(dx\mathbf{i} + dy\mathbf{j})
$$

$$
= \int_{\mathbf{r}_A}^{\mathbf{r}_B} -\mu\,mg\cos^2\alpha\,dx + mg\,(\mu\sin\alpha\cos\alpha - 1)\,dy
$$

$$
= \int_{0}^{OB = h/\tan\alpha} \left[-\mu\,mg\cos^2\alpha \right]dx + \int_{h}^{O} mg\,(\mu\sin\alpha\cos\alpha - 1)\,dy
$$

$$
= -\mu\,mg\cos^2\alpha\left(\frac{h}{\tan\alpha}\right) - mg\,(\mu\sin\alpha\cos\alpha - 1)\,h
$$

$$
= -\mu\,mgh\frac{\cos^3\alpha}{\sin\alpha} - mg\mu h\sin\alpha\cos\alpha + mgh
$$

$$
= mgh\left(1 - \frac{\mu\cos\alpha}{\sin\alpha}\right).
$$

The change in kinetic energy between the two positions A and B is

$$\Delta T_{AB} = T_B - T_A = \frac{1}{2}mv_B^2 - \frac{1}{2}mv_A^2 = \frac{1}{2}mv_B^2.$$

The particle starts from rest, that is, $v_A = 0$. The principle of work and energy may be expressed as

$$U_{AB} = T_B - T_A,$$

or

$$mgh\left(1 - \frac{\mu\cos\alpha}{\sin\alpha}\right) = \frac{1}{2}mv_B^2.$$

The velocity of the particle at B is

$$v_B = \sqrt{2gh\left(1 - \frac{\mu}{\tan\alpha}\right)}.$$

(b) The work done on the particle as it moves between B and C is expressed as

$$U_{BC} = \int_{\mathbf{r}_B}^{\mathbf{r}_C} \mathbf{F}\cdot d\mathbf{r} = \int_{\mathbf{r}_B}^{\mathbf{r}_C} (\mathbf{G} + \mathbf{F}_f)\cdot(dx\mathbf{i} + dy\mathbf{j})$$

$$= \int_{x_B}^{y_B} (-\mu mg\mathbf{i} - mg\mathbf{j})\cdot dx\mathbf{i} = \int_{\frac{h}{\tan\alpha}}^{\frac{h}{\tan\alpha}+d} (-\mu mg)\,dx = -\mu mgd.$$

The change in the kinetic energy of the particle as it moves from B to C is

$$\Delta T_{BC} = T_C - T_B = \frac{1}{2}mv_C^2 - \frac{1}{2}mv_B^2 = -\frac{1}{2}mv_B^2 = -mgh\left(1 - \frac{\mu}{\tan\alpha}\right).$$

The principle of work and energy is expressed as

$$U_{BC} = T_C - T_B,$$

or

$$-\mu mgd = -mgh\left(1 - \frac{\mu}{\tan\alpha}\right).$$

The distance d until the particle will stop is

$$d = \frac{h}{\mu}\left(1 - \frac{\mu}{\tan\alpha}\right).$$

Fig. 4.22 Example 4.5

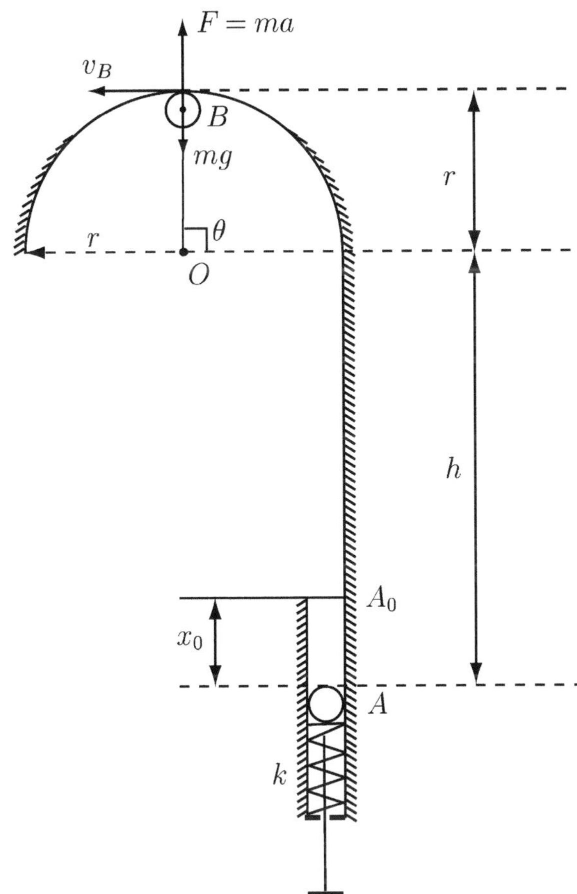

Numerical application:

$$v_B = \sqrt{2gh\left(1 - \frac{\mu}{\tan\alpha}\right)} = \sqrt{2(9.8)(2)\left(1 - \frac{0.2}{\tan 30°}\right)} = 5.062 \text{ m/s},$$

$$d = \frac{h}{\mu}\left(1 - \frac{\mu}{\tan\alpha}\right) = \frac{2}{0.2}\left(1 - \frac{0.2}{\tan 30°}\right) = 6.536 \text{ m}.$$

Example 4.5. The ball of mass m is fired up the smooth vertical and circular track using the spring plunger (Fig. 4.22). The ball is of negligible size. The uncompressed position of the spring is at A_0. The compression in the spring is $x_0 = A_0 A$. The ball will begin to leave the track when $\theta = 90°$ at the highest point B. The radius of the circular track is r, and the height of the vertical track is h. Determine the spring constant k. Numerical application: $m = 1$ kg, $h = 0.4$ m, $r = 0.2$ m, $x_0 = 0.08$ m, and $g = 9.8$ m/s^2.

Solution

The total energy (the sum of the kinetic energy T and the potential energy V) is constant or conserved at the positions A and B

$$T_A + V_A = T_B + V_B. \tag{4.61}$$

At the position, A the kinetic energy is zero $T_A = 0$ ($v_A = 0$), and the potential energy is $V_A = \frac{1}{2}kx_0^2$. At the position B, the kinetic energy is $T_B = \frac{1}{2}mv_B^2$, and the potential energy is $V_B = mg(h+r)$. At B, the free-body diagram of the ball is shown in Fig. 4.22. The equation of motion in the normal direction gives

$$mg = m\frac{v_B^2}{r},$$

or

$$v_B^2 = rg.$$

Equation (4.61) becomes

$$\frac{1}{2}kx_0^2 = \frac{1}{2}mrg + mg(h+r).$$

Solving one can have

$$k = \frac{mg}{x_0^2}(2h+3r).$$

For the numerical application, the spring constant is $k = 2143.750$ N/m.

Example 4.6. A central force $\mathbf{F}$ attracts a particle P of mass m as shown in Fig. 4.23a. The force $\mathbf{F} = -k^2 m\mathbf{r}$ is on the xy plan of motion. The force $\mathbf{F}$ is proportional to the length of the position vector $\mathbf{r}$ relative to a fixed point O (directed toward the point O), to the mass m of the particle, and to the constant k. The coordinates of the initial position of the particle are $P_0(a, -\frac{g}{k^2})$, and the initial velocity of the particle is $\mathbf{v} = v_0\mathbf{J}$. Study the motion of the particle. Numerical application: $a = 2$ m, $g = 9.8$ m/s^2, $v_0 = 10$ m/s, and $k = 3$ s^{-1}.

Solution

The Newton's second law of motion of the particle can be written as

$$m\ddot{\mathbf{r}} = m\mathbf{g} + \mathbf{F}.$$

Replacing the expression of the central force $\mathbf{F} = -k^2 m\mathbf{r}$ in the equation of motion, it is obtained

$$m\ddot{\mathbf{r}} = m\mathbf{g} - k^2 m\mathbf{r}.$$

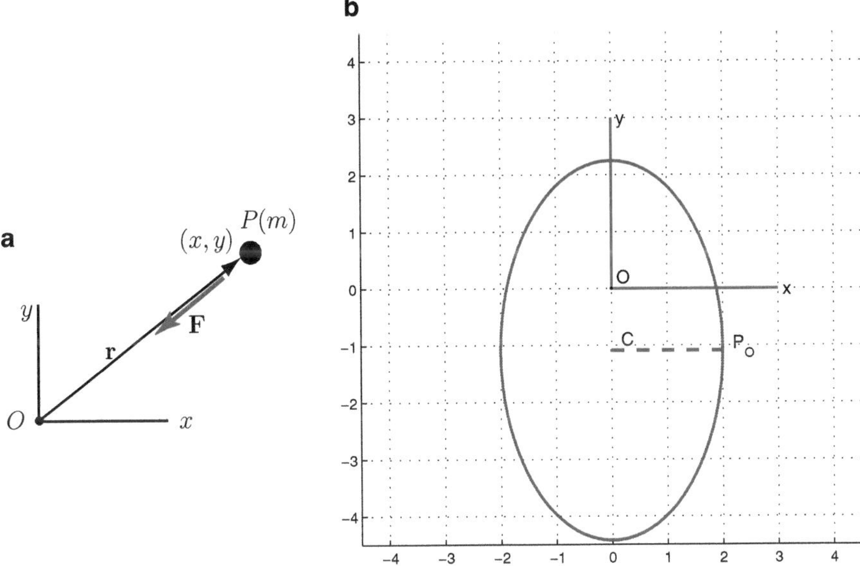

Fig. 4.23 Example 4.6

Or the two equations of motion are

$$\ddot{x} + k^2 x = 0,$$
$$\ddot{y} + k^2 y = -g. \tag{4.62}$$

Solving (4.62), it results

$$x = c_1 \cos(kt) + c_2 \sin(kt),$$
$$y = c_3 \cos(kt) + c_4 \sin(kt) - \frac{g}{k^2},$$

and

$$\dot{x} = -k c_1 \sin(kt) + k c_2 \cos(kt),$$
$$\dot{y} = -k c_3 \sin(kt) + k c_4 \cos(kt).$$

In MATLAB, the previous relations are written as

```
syms x y k g t a v0 real

% equations of motion
xs = dsolve('D2x = - k^2*x');
ys = dsolve('D2y = - k^2*y-g');
```

```
fprintf('x = %s   \n',char(xs))
fprintf('y = %s   \n',char(ys))
fprintf('\n')

dxs = diff(xs,t);
dys = diff(ys,t);
fprintf('dx/dt = %s   \n',char(dxs))
fprintf('dy/dt = %s   \n',char(dys))
fprintf('\n\n')
```

and the output is

```
x = C2*cos(k*t)  + C3*sin(k*t)
y = C5*cos(k*t)  + C6*sin(k*t)  - g/k^2

dx/dt = C3*k*cos(k*t)  - C2*k*sin(k*t)
dy/dt = C6*k*cos(k*t)  - C5*k*sin(k*t)
```

Using the initial conditions at $t = 0$

$$x(0) = a, \ y(0) = -\frac{g}{k^2}, \ \dot{x}(0) = 0, \ \dot{y}(0) = v_0,$$

the integration constants c_1, c_2, c_3, and c_4 are calculated as

$$c_1 = a,$$
$$c_2 = 0,$$
$$c_3 = 0,$$
$$c_4 = \frac{v_0}{k}.$$

The parametric equations for the trajectory of the particle are

$$x = a\cos(kt),$$
$$y = \frac{v_0}{k}\sin(kt) - \frac{g}{k^2}.$$

In MATLAB, the integration constants and the parametric equations are calculated
with

```
fprintf ...
('t=0,x(0)=a,y(0)=-g/k^2,dx(0)=0,dy(0)=v0=>\n')

eqx  = a-subs(xs, t, 0);
eqy  = -g/k^2-subs(ys, t, 0);
eqdx = 0-subs(dxs, t, 0);
eqdy = v0-subs(dys, t, 0);
```

```
fprintf(' %s = 0 \n',char(eqx))
fprintf(' %s = 0 \n',char(eqy))
fprintf(' %s = 0 \n',char(eqdx))
fprintf(' %s = 0 \n',char(eqdy))
fprintf('\n')

sol=solve(eqx,eqy,eqdx,eqdy,'C2,C3,C5,C6');
C2s = sol.C2;
C3s = sol.C3;
C5s = sol.C5;
C6s = sol.C6;
fprintf('C2 = %s \n',char(C2s))
fprintf('C3 = %s \n',char(C3s))
fprintf('C5 = %s \n',char(C5s))
fprintf('C6 = %s \n',char(C6s))
fprintf('\n\n')

xn = subs(xs,{'C2','C3'},{C2s, C3s});
yn = subs(ys,{'C5','C6'},{C5s, C6s});
fprintf('x = %s \n',char(xn))
fprintf('y = %s \n',char(yn))
fprintf('\n')
```

The set of parametric equations is converted to a single equation, eliminating the variable t from the simultaneous equations

$$\cos^2(kt) = \frac{x^2}{a^2},$$

$$\sin^2(kt) = \frac{\left(y+\frac{g}{k^2}\right)^2}{\left(\frac{v_0}{k}\right)^2},$$

and using the Pythagorean trigonometric identity

$$\cos^2(kt) + \sin^2(kt) = \frac{x^2}{a^2} + \frac{\left(y+\frac{g}{k^2}\right)^2}{\left(\frac{v_0}{k}\right)^2} = 1.$$

The trajectory of the particle

$$\frac{x^2}{a^2} + \frac{\left(y+\frac{g}{k^2}\right)^2}{\left(\frac{v_0}{k}\right)^2} = 1$$

is an elliptical curve with the center of the ellipse located at $\left(0, -\frac{g}{k^2}\right)$ and the semi-axes a and v_0/k. Since the trigonometric functions $\sin(kt)$ and $\cos(kt)$ are periodic functions, the motion of the particle is also periodic. The period for the motion T of the particle is calculated from

$$\sin(kt) = \sin[k(t+T)] = \sin(kt + 2n\pi),$$

or

$$kt + kT = kt + 2n\pi,$$

and it results

$$T = \frac{2n\pi}{k} \quad \left(\text{for } n = 1 \implies T = \frac{2\pi}{k}\right).$$

The trajectory of the particle shown in Fig. 4.23b is simulated in MATLAB for the numerical data with the commands:

```
axis manual
axis equal
hold on
grid on
axis([-4.5 4.5 -4.5 4.5])

slist={a, g, v0, k};
nlist={2, 9.8, 10, 3};

xnt=subs(xn,slist,nlist);
ynt=subs(yn,slist,nlist);

x0 = subs(xnt,t,0);
y0 = subs(ynt,t,0);

xC = 0;
yC = subs(-g/k^2,slist,nlist);

text(0,0+.2,' O','fontsize',12)
text(x0,y0,'  P_O','fontsize',12)
text(xC,yC+.2,'  C','fontsize',12)
plot(xC,yC,'.','Color','black')

line ...
([0 0],[0 3],'LineWidth',1.2,'Color','red')
line ...
([0 3],[0 0],'LineWidth',1.2,'Color','red')
```

Fig. 4.24 Example 4.7

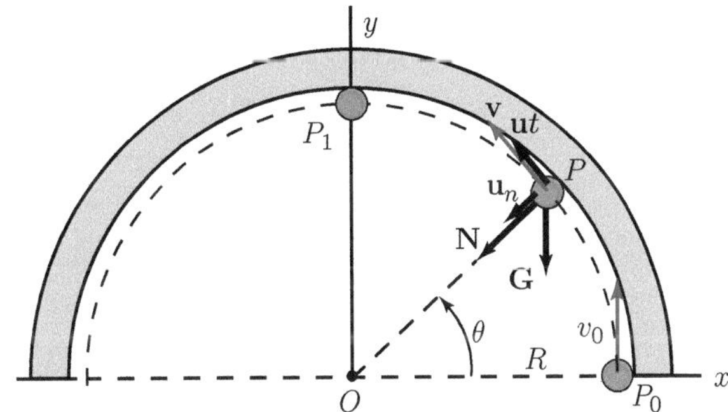


```
text(0,3,' y','fontsize',12)
text(3,0,' x','fontsize',12)
plot(0,0,'.','Color','black')
line ...
([xC x0],[yC y0],'LineStyle','--','LineWidth',1.5)
plot(xC,yC,'.','Color','red')

for t_n = 0 : 0.005 : 2.1
    x_t=subs(xnt,t,t_n);
    y_t=subs(ynt,t,t_n);
    hm=plot(x_t,y_t,'.','Color','red');
    pause(0.001)
    ht=plot(x_t,y_t,'.');
    delete(hm);
end
plot(x0,y0,'.','Color','red')
```

Example 4.7. A particle P of mass m is propelled inside a circular track of radius R as shown in Fig. 4.24. The initial velocity of the particle is $\mathbf{v}_0$ at the point $P_0(R, 0)$. (a) Determine the equation of motion of the particle. (b) Find the expression of the normal contact force N between the circular track and the particle. (c) What is be the minimum initial velocity $\mathbf{v}_0$ of the particle to prevent the particle from falling off the circular track before reaching the point P_1 with $\theta = pi/2$. (d) Find the initial acceleration of the particle and the initial contact force N. For the numerical application, use $R = 1$ m, $m = 1$ kg, $g = 10$ m/s^2.

Solution

(a) The equation of motion along the tangential and normal axes is

$$ma_\text{t} = F_\text{t},$$

$$ma_\text{n} = F_\text{n} + N,$$

or

$$m\frac{dv}{dt} = F_t,$$

$$m\frac{v^2}{R} = F_n + N.$$

For the circular motion, $a_t = R\ddot{\theta}$, $a_n = R\dot{\theta}^2$, $F_t = -mg\cos\theta$, $F_n = mg\sin\theta$, and the equations of motion are

$$mR\ddot{\theta} = -mg\cos\theta, \tag{4.63}$$

$$mR\dot{\theta}^2 = mg\sin\theta + N. \tag{4.64}$$

Multiplying by $\dot{\theta}$ both terms of (4.63), it results

$$R\dot{\theta}\ddot{\theta} = -g\dot{\theta}\cos\theta,$$

and after integration, the following relation is obtained:

$$R\dot{\theta}^2 = -2g\sin\theta + C.$$

Using the initial conditions at $t = 0$, $\theta = 0$, and $R\dot{\theta} = v_0$, the constant of integration C is

$$C = \frac{v_0^2}{R}.$$

The equation of motion of the particle is

$$R\dot{\theta}^2 = -2g\sin\theta + \frac{v_0^2}{R}. \tag{4.65}$$

(b) The expression of the normal contact force N between the circular path and the
particle is obtained from (4.64) and (4.65):

$$N = mR\dot{\theta}^2 - mg\sin\theta$$

$$= m\left(-2g\sin\theta + \frac{v_0^2}{R}\right) - mg\sin\theta$$

$$= -3mg\sin\theta + m\frac{v_0^2}{R}.$$

(c) The condition for the particle to be on the circular track is $N > 0$:

$$N = -3mg\sin\theta + m\frac{v_0^2}{R} > 0,$$

and the minimal initial velocity $\mathbf{v}_0$ of the particle to prevent it from falling off the circular track is

$$v > \sqrt{3Rg\sin\theta}.$$

The condition for the particle to reach point $P_1(\theta = \pi/2)$ is $v > \sqrt{3Rg}$.

(d) The initial conditions are at $t = 0$ s, $\theta(0) = 0$ rad and $\dot{\theta}(0) = \omega_0 = v_0/R = 1$ rad/s. The angular acceleration at $t = 0$ is obtained from (4.63):

$$\alpha_0 = \ddot{\theta}(0) = -\frac{g}{R}\cos\theta(0) = -\frac{g}{R} = -1 \text{ rad/s}^2.$$

The initial tangential and normal accelerations are

$$a_t(0) = R\ddot{\theta}(0) = -g = -100 \text{ m/s}^2,$$
$$a_n(0) = -R\dot{\theta}^2(0) = -R\omega_0^2 = -v_0^2/R = -1 \text{ m/s}^2,$$

and the total acceleration is $a = \sqrt{a_n^2 + a_t^2} = \sqrt{g^2 + 1} = \sqrt{101}$ m/s^2. The expression of the normal contact force N between the circular path and the particle at $t = 0$ is obtained from (4.64):

$$N(0) = mR\dot{\theta}^2(0) - mg\sin\theta(0) = mR\omega_0^2 = mv_0^2/R = 1 \text{ N}.$$

Example 4.8. A particle P of mass m is released from rest at the initial position $P_0(x_0, y_0, z_0)$. The particle is acted upon by the force system $\mathbf{F}_1 = mk_1\,\mathbf{I}$, $\mathbf{F}_2 = mk_2\,\mathbf{J}$, and $\mathbf{F}_3 = mk_3\,\mathbf{k}$, as shown in Fig. 4.25a. The effect of gravity on the particle is neglected. For numerical application, use $x_0 = 3$ m, $y_0 = 3$ m, $z_0 = 7$ m, $k_1 = 4$ m/s^2, $k_2 = -1.5$ m/s^2, $k_3 = -5$ m/s^2. Find the equation of motion of the particle, the acceleration of the particle, and the distance the particle is from the origin a time $t = 1$ s after it has been released.

Solution
The equations of motion of the particle are

$$m\ddot{x} = mk_1,$$
$$m\ddot{y} = mk_2,$$
$$m\ddot{z} = mk_3,$$

or

$$\ddot{x} = k_1, \ \ddot{y} = k_2, \ddot{z} = k_3.$$

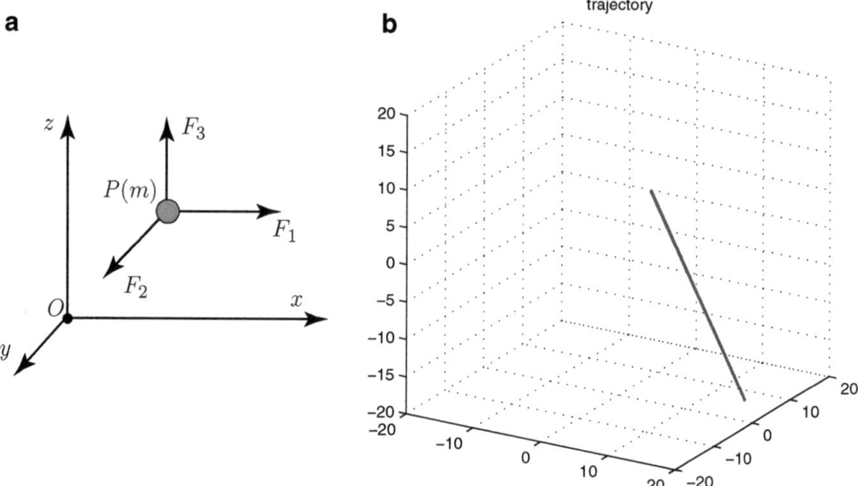

Fig. 4.25 Example 4.8

After integration, it results

$$\dot{x} = k_1 t + c_1,$$

$$\dot{y} = k_2 t + c_2,$$

$$\dot{z} = k_3 t + c_3.$$

Integrating the previous expression, the following relations are obtained:

$$x = k_1 \frac{t^2}{2} + c_1 t + c_4,$$

$$y = k_2 \frac{t^2}{2} + c_2 t + c_5,$$

$$z = k_3 \frac{t^2}{2} + c_3 t + c_6.$$

With the initial conditions at $t = 0$, $\dot{x} = \dot{y} = \dot{z} = 0$ and $x = x_0$, $y = y_0$, $z = z_0$, the constants of integration are

$$c_1 = 0, \quad c_2 = 0, \quad c_3 = 0,$$

$$c_4 = x_0, \quad c_5 = y_0, \quad c_6 = z_0.$$

Replacing the constants of integration, the parametric equations of the particle are

$$x = x_0 + k_1 \frac{t^2}{2},$$

$$y = y_0 + k_2 \frac{t^2}{2},$$

$$z = z_0 + k_3 \frac{t^2}{2},$$

or

$$\frac{x - x_0}{k_1} = \frac{t^2}{2},$$

$$\frac{y - y_0}{k_2} = \frac{t^2}{2},$$

$$\frac{z - z_0}{k_3} = \frac{t^2}{2}.$$

The implicit equation of the trajectory is

$$\frac{x - x_0}{k_1} = \frac{y - y_0}{k_2} = \frac{z - z_0}{k_3}, \tag{4.66}$$

and that is the equation of a straight line. The acceleration of the particle can be computed as

$$a = \sqrt{\ddot{x}^2 + \ddot{y}^2 + \ddot{z}^2} = \sqrt{k_1^2 + k_2^2 + k_3^2}.$$

The distance the particle is from the origin O at time $t = t_1 = 1$ s after it has been released is calculated as

$$d = \sqrt{x^2(t_1) + y^2(t_1) + z^2(t_1)}.$$

The particle is moving on a line with the equation given by (4.66) with a constant acceleration. The trajectory of the particle obtained with MATLAB is shown in Fig. 4.25b.

Example 4.9. A particle P of mass m is thrown up on an inclined plane AB with an initial velocity v_0 as shown in Fig. 4.26. The length of the plane is $AB = v_0^2/g$, and the angle between the plane and the horizontal axis is α. The particle is leaving the inclined plane at the highest point B, starting a free fall motion. The landing position of the particle is the point C as shown in Fig. 4.26. Find (a) the particle velocity at point B, (b) the expression of the normal contact force N between the particle and the inclined plane, and (c) the horizontal distance AC between the initial position and the landing position of the particle. For the numerical application, use $m = 1$ kg, $\alpha = \pi/10$, $v_0 = 10$ m/s, and $g = 9.8$ m/s^2.

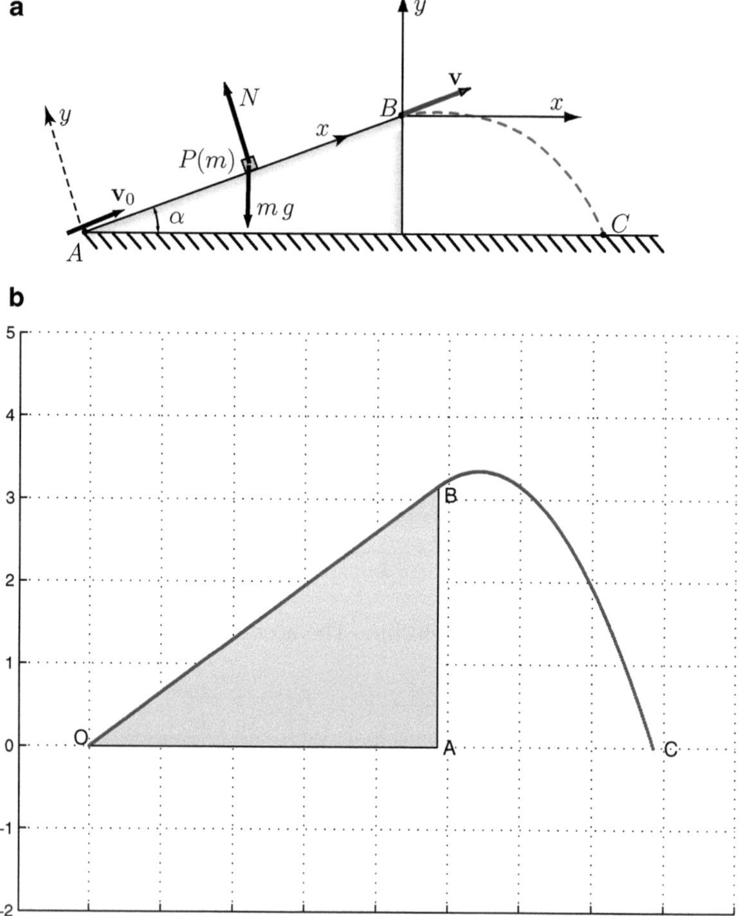

Fig. 4.26 Example 4.9

Solution

(a) The motion of the particle on the inclined plane can be described using
Newton's second law of motion. The equations of motion are

$$m\ddot{x} = -mg\sin\alpha,$$

$$m\ddot{y} = -mg\cos\alpha + N.$$

Since $\ddot{y} = 0$, the previous equations can be rewritten as

$$\ddot{x} = -m\sin\alpha,$$

$$N = mg\cos\alpha.$$

Integrating twice the first equation, the position of the particle is obtained:

$$\dot{x} = -gt \sin \alpha + c_1,$$

and

$$x = -g\frac{t^2}{2} \sin \alpha + c_1 t + c_2.$$

Using the initial condition, at $t = 0$, $x = 0$, and $\dot{x} = v_0$, it results

$$c_2 = 0 \quad \text{and} \quad c_1 = v_0.$$

Replacing c_1 and c_2 into the equation of motion, one obtains

$$\dot{x} = -gt \sin \alpha + v_0, \tag{4.67}$$

and

$$x = -g\frac{t^2}{2} \sin \alpha + v_0 t. \tag{4.68}$$

The MATLAB commands for the equation of motion along the inclined plane are

```
xs=dsolve...
('m*D2x+m*g*sin(alpha)=0','x(0)=0','Dx(0)=v_0');
fprintf('eom IC: t=0=> x(0)=0, Dx(0)=v_0 \n')
fprintf('x = %s \n',char(xs))

vxs=dsolve...
('m*Dv+m*g*sin(alpha)=0','v(0)=v_0');
fprintf('vx = %s \n\n',char(vxs))
```

Next, the time needed for the particle to travel from point A to point B is calculated. Replacing x by $\frac{v_0^2}{g}$ in (4.68), one obtains

$$\frac{v_0^2}{g} = -g\frac{t^2}{2} \sin \alpha + v_0 t,$$

or

$$gt^2 \sin \alpha - 2gv_0 t + 2v_0^2 = 0.$$

The time needed for the particle to travel on the inclined plane is

$$t_B = \frac{gv_0 \pm \sqrt{g^2 v_0^2 - 2g^2 v_0^2 \sin \alpha}}{g^2 \sin^2 \alpha} = \frac{v_0}{g \sin \alpha}\left(1 \pm \sqrt{1 - 2\sin \alpha}\right). \tag{4.69}$$

The velocity at B is calculated from (4.69) and (4.67):

$$v_B = \dot{x}_B = v_0 - g\sin\alpha\,\frac{v_0}{g\sin\alpha}\left(1 \pm \sqrt{1 - 2\sin\alpha}\right) = v_0\sqrt{1 - 2\sin\alpha}. \quad (4.70)$$

The velocity at B is computed in MATLAB with

```
ts=solve(subs(x-xs,x,v_0^2/g),t);
ts=simplify(ts);
fprintf('t_AB1 = %s   \n',char(ts(1)))
fprintf('t_AB2 = %s   \n',char(ts(2)))
fprintf('\n')

v_B1=simplify(subs(vxs,t,ts(1)));
v_B2=simplify(subs(vxs,t,ts(2)));

fprintf('v_B1 = %s   \n',char(v_B1));
fprintf('v_B2 = %s   \n',char(v_B2));
fprintf('\n\n');
```

The numerical results are obtained with MATLAB:

```
% numerical data
slist={m, g, v_0, alpha};
nlist={1,9.8,10,pi/10};

tn1=subs(ts(1),slist,nlist);
tn2=subs(ts(2),slist,nlist);
fprintf('t_AB1 = %6.3f (s)  \n',tn1)
fprintf('t_AB2 = %6.3f (s)  \n',tn2)
fprintf('\n')

v_B1n=subs(v_B1,slist,nlist);
v_B2n=subs(v_B2,slist,nlist);
fprintf('v_B1 = %6.3f (m/s)  \n',v_B1n)
fprintf('v_B2 = %6.3f (m/s)  \n',v_B2n)
fprintf('\n\n')
 if v_B1n > 0
    v_Bn = v_B1n; tn=tn1; vB=v_B1;
  else
    v_Bn = v_B2n; tn=tn2; vB=v_B2;
  end
fprintf('velocity at B \n')
fprintf('v_B = %s   \n',char(vB))
fprintf('v_B = %6.3f (m/s)  \n',v_Bn)
fprintf('\n')
```

and the results are

```
eom IC: t=0=> x(0)=0, Dx(0)=v_0
x = t*v_0 - (g*t^2*sin(alpha))/2
vx = v_0 - g*t*sin(alpha)


t_AB1 =

                                   1/2
   v_0 ((1 - 2 sin(alpha))    + 1)
   -------------------------------
                g sin(alpha)


t_AB2 =

                                    1/2
    v_0 ((1 - 2 sin(alpha))    - 1)
  - -------------------------------
                 g sin(alpha)


v_B1 = -v_0*(1 - 2*sin(alpha))^(1/2)
v_B2 = v_0*(1 - 2*sin(alpha))^(1/2)


t_AB1 =   5.343 (s)
t_AB2 =   1.261 (s)


v_B1 = -6.180 (m/s)
v_B2 =  6.180 (m/s)


velocity at B
v_B = v_0*(1 - 2*sin(alpha))^(1/2)
v_B =   6.180 (m/s)
```

(b) The expression of the normal contact force N between the particle and the inclined plane is given by $N = mg \cos \alpha$, and in MATLAB, result is

```
N = g*m*cos(alpha)  =   9.320 (N).
```

(c) The motion of the particle in free flight (after the particle is leaving the inclined plane at the point B) can be described using the following equations of motion:

$$m\ddot{x} = 0$$

$$m\ddot{y} = -mg.$$

Integrating twice the previous equation, it results

$$\dot{x} = c_3,$$

$$\dot{y} = -gt + c_4,$$

and

$$x = c_3 t + c_5,$$

$$y = -g\frac{t^2}{2} + c_4 t + c_6.$$

Using the initial conditions at $t = 0$, $x = 0$, $y = 0$, $\dot{x} = v_B \cos\alpha$, and $\dot{y} = v_B \sin\alpha$, the constants of integration are

$$c_5 = c_6 = 0, \quad c_3 = v_B \cos\alpha, \quad c_4 = v_B \sin\alpha.$$

The equations for the trajectory of the particle are

$$x = v_B t \cos\alpha,$$

$$y = -g\frac{t^2}{2} + v_B t \sin\alpha.$$

The MATLAB program for the equations of motion for the free fall is

```
fprintf('free fall motion \n\n');
% IC: t=0 => x(0)=x_0, y(0)=y_0
% Dx(0)=v_B*cos(alpha), Dy(0)=v_B*sin(alpha)

vx_p=dsolve('m*Dv=0','v(0)=v_B*cos(alpha)');
vy_p=dsolve('m*Dv+m*g=0','v(0)=v_B*sin(alpha)');
fprintf('v_x = %s   \n',char(vx_p))
fprintf('v_y = %s   \n',char(vy_p))
fprintf('\n\n')

x_p=dsolve...
('m*D2x=0','x(0)=0','Dx(0)=v_B*cos(alpha)');
y_p=dsolve...
('m*D2y+m*g=0','y(0)=0','Dy(0)=v_B*sin(alpha)');
fprintf('x = %s   \n',char(x_p))
fprintf('y = %s   \n',char(y_p))
fprintf('\n\n')

x_pn=simplify(subs(x_p,v_B,vB));
y_pn=simplify(subs(y_p,v_B,vB));
```

```
fprintf('x = %s   \n',char(x_pn))
fprintf('y = %s   \n',char(y_pn))
fprintf('\n\n')
```

Eliminating the time t from the previous parametric equations, the trajectory of the particle is

$$y = -\frac{g x^2}{2 v_0^2 (1 - 2\sin\alpha)\cos^2\alpha} + x\tan\alpha. \qquad (4.71)$$

The abscissa of the point C relative to xBy reference frame is obtained from (4.71) for $y = -\frac{v_0^2}{g}\sin\alpha$:

$$x_C = \frac{v_0^2(1 - 2\sin\alpha)}{g}\cos\alpha\left[\sin\alpha + \sqrt{\sin^2\alpha + \frac{2\sin\alpha}{1 - 2\sin\alpha}}\right].$$

The horizontal distance x_{AC} between the initial position and the landing position of the particle can now be calculated as

$$x_{AC} = \frac{v_0^2}{g}\cos\alpha + \frac{v_0^2(1 - 2\sin\alpha)}{g}\cos\alpha\left[\sin\alpha + \sqrt{\sin^2\alpha + \frac{2\sin\alpha}{1 - 2\sin\alpha}}\right]$$

$$= \frac{v_0^2\cos\alpha}{g}\left[1 + (1 - 2\sin\alpha)\left(\sin\alpha + \sqrt{\sin^2\alpha + \frac{2\sin\alpha}{1 - 2\sin\alpha}}\right)\right].$$

The MATLAB commands for the distance AC are

```
t_x=solve(x-x_pn,t);
y_eq=subs(y_pn,t,t_x);
fprintf('trajectory of the particle\n')
fprintf('y = \n')
pretty(y_eq)
fprintf('\n\n')

% yC=-v0^2*sin(alpha)/g
eq=subs(y-y_eq,y,-v_0^2*sin(alpha)/g);

x_C=simplify(solve(eq,x));
xCn=subs(x_C,slist,nlist);
fprintf('x_C1 = %6.3f  (m)  \n',xCn(1))
fprintf('x_C2 = %6.3f  (m)  \n',xCn(2))
fprintf('\n\n');

  if xCn(1) > 0
     xC = xCn(1); xCs = x_C(1);
```

```
  else
    xC = xCn(2); xCs = x_C(2);
  end
fprintf('xC = \n')
pretty(xCs)
fprintf('xC = %6.3f (m) \n',xC)
fprintf('\n\n')
x_AC=v_0^2*cos(alpha)/g+xCs;
x_ACval=subs(x_AC,slist,nlist);
fprintf('x_AC = %6.3f (m) \n',x_ACval)
fprintf('\n\n')
```

The numerical results obtained with MATLAB are

```
trajectory of the particle
y =


                                                        2
    x sin(alpha)                            g x
   ------------ + --------------------------------------------
     cos(alpha)              2            2
                        2 v_0  cos(alpha)   (2 sin(alpha) - 1)


x_C1 =   5.998  (m)
x_C2 = -3.707  (m)

xC =   5.998  (m)
x_AC = 15.702  (m)
```

The MATLAB commands for the graphical simulation of the motion are

```
axis manual
hold on
grid on
axis([-2 18 -2 5])

l=v_0^2/g;
xB=l*cos(alpha);
yB=l*sin(alpha);
xBn=subs(xB,slist,nlist);
yBn=subs(yB,slist,nlist);

a=xBn; b=yBn;
x_O=0; y_O=0;
```

```
x_A=a; y_A=0;
x_B=a; y_B=b;

t1=text...
(x_O-0.6, y_O+0.1, 0,' O','fontsize',12);
t2=text(x_A, y_A, 0,' A','fontsize',12);
t3=text(x_B, y_B-0.1, 0,' B','fontsize',12);
t3=text(x_ACval, 0, 0,'  C','fontsize',12);

% inclined plane vertices
vert = [x_O y_O 0; x_A y_A 0; x_B y_B 0;];
% inclined plane faces
fac = [ 2 1 3];
% draw the prism
prism=patch...
('Faces',fac,'Vertices',vert,'FaceColor','y');

for t_n = 0 : 0.005 : tn
    slist={m,g,v_0, alpha,'t'};
    nlist={2,9.8,10,pi/10,t_n};
    x_t=subs(xs*cos(alpha),slist,nlist);
    y_t=subs(xs*sin(alpha),slist,nlist);
    hm=plot(x_t,y_t,'.','Color','red');
    ht=plot(x_t,y_t,'.');
    pause(0.001)
    delete(hm);
end

for t_n = 0 : 0.005 : 1.02
    slist1={g,v_0, alpha,'v_B','t'};
    nlist1={9.8,10,pi/10,v_Bn,t_n};
    x_tn = xBn+subs(x_pn,slist1,nlist1);
    y_tn = yBn+subs(y_pn,slist1,nlist1);
    hm=plot(x_tn,y_tn,'.','Color','red');
    ht=plot(x_tn,y_tn,'.');
    pause(0.001)
    delete(hm);
end
```

The trajectory of the particle obtained with MATLAB is shown in Fig. 4.26b.

Example 4.10. A thin-walled cylindrical tube of length rotates about the vertical y-axis with a constant angular velocity $\dot{\theta} = \omega$ as shown in Fig. 4.27. The angle between the cylindrical tube and the y-axis is φ. A particle P with mass m moves freely and without friction inside the tube. Determine the equation of motion for the particle inside the cylindrical tube and the force acting on the particle.

Fig. 4.27 Example 4.10

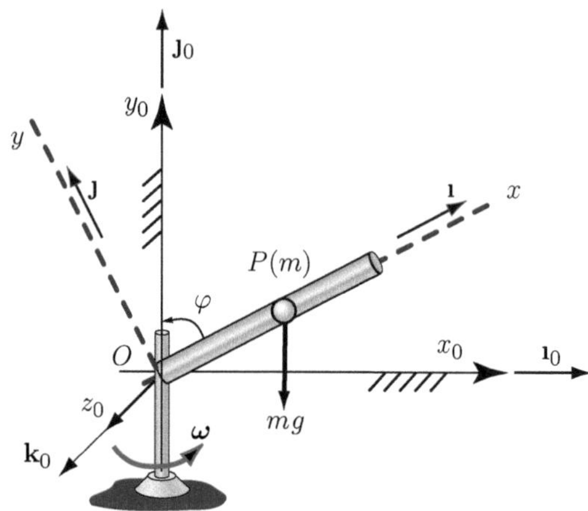

Solution

The motion of the particle is defined when the position vector is defined as function of time with respect to a fixed reference frame with the origin at O. Let $\mathbf{1}_0, \mathbf{J}_0$, and $\mathbf{k}_0$ be the constant unit vectors of a fixed orthogonal Cartesian reference frame $x_0 y_0 z_0$ and $\mathbf{1}, \mathbf{J}$, and $\mathbf{k}$ be the unit vectors of a mobile or rotating orthogonal Cartesian reference frame xyz (Fig. 4.27). The unit vectors $\mathbf{1}_0, \mathbf{J}_0$, and $\mathbf{k}_0$ of the primary reference frame are constant with respect to time. A reference frame that moves with the tube is a rotating reference frame. The unit vectors $\mathbf{1}, \mathbf{J}$, and $\mathbf{k}$ are not constant because they rotate. The position vector of the point P with respect to the fixed reference frame $x_0 y_0 z_0$ is denoted by $\mathbf{r} = \mathbf{r}_{OP}$ and is given by

$$\mathbf{r} = x\mathbf{1} + y\mathbf{J} + z\mathbf{k},$$

where x, y, and z represent the projections of the vector $\mathbf{r} = \mathbf{r}_{OP}$ on the rotating reference frame. The velocity of P with respect to the fixed reference frame $x_0 y_0 z_0$ is the derivative with respect to time of the position vector $\mathbf{r}$

$$\mathbf{v} = \frac{d\mathbf{r}}{dt} = \frac{dx}{dt}\mathbf{1} + \frac{dy}{dt}\mathbf{J} + \frac{dz}{dt}\mathbf{k} + x\frac{d\mathbf{1}}{dt} + y\frac{d\mathbf{J}}{dt} + z\frac{d\mathbf{k}}{dt}$$

$$= \dot{x}\mathbf{1} + \dot{y}\mathbf{J} + \dot{z}\mathbf{k} + \boldsymbol{\omega} \times \mathbf{r},$$

where $\boldsymbol{\omega} = \omega_x\mathbf{1} + \omega_y\mathbf{J} + \omega_z\mathbf{k}$ is the angular velocity of the rotating reference frame and

$$\frac{d\mathbf{1}}{dt} = \boldsymbol{\omega} \times \mathbf{1}, \quad \frac{d\mathbf{J}}{dt} = \boldsymbol{\omega} \times \mathbf{J}, \quad \frac{d\mathbf{k}}{dt} = \boldsymbol{\omega} \times \mathbf{k}.$$

The velocity of P relative to the rotating reference frame is a derivative in the rotating reference frame

$$\mathbf{v}^{\text{rel}}_{P(xyz)} = \frac{^{(xyz)}\mathrm{d}\,\mathbf{r}}{\mathrm{d}t} = \frac{\mathrm{d}x}{\mathrm{d}t}\mathbf{\imath} + \frac{\mathrm{d}y}{\mathrm{d}t}\mathbf{J} + \frac{\mathrm{d}z}{\mathrm{d}t}\mathbf{k} = \dot{x}\mathbf{\imath} + \dot{y}\mathbf{J} + \dot{z}\mathbf{k}.$$

The velocity of the point P relative to the primary reference frame is

$$\mathbf{v}_P = \mathbf{v}^{\text{rel}}_{P(xyz)} + \boldsymbol{\omega} \times \mathbf{r}.$$

The acceleration of the point P relative to the primary reference frame is obtained by taking the time derivative of the velocity:

$$\mathbf{a}_P = \mathbf{a}^{\text{rel}}_{P(xyz)} + 2\boldsymbol{\omega} \times \mathbf{v}^{\text{rel}}_{P(xyz)} + \boldsymbol{\alpha} \times \mathbf{r} + \boldsymbol{\omega} \times (\boldsymbol{\omega} \times \mathbf{r}),$$

where

$$\boldsymbol{\alpha} = \frac{\mathrm{d}\boldsymbol{\omega}}{\mathrm{d}t} = \frac{\mathrm{d}\omega_x}{\mathrm{d}t}\mathbf{\imath} + \frac{\mathrm{d}\omega_y}{\mathrm{d}t}\mathbf{J} + \frac{\mathrm{d}\omega_z}{\mathrm{d}t}\mathbf{k}$$

is the angular acceleration of rotating reference frame and

$$\mathbf{a}^{\text{rel}}_{P(xyz)} = \frac{^{(xyz)}\mathrm{d}^2\,\mathbf{r}}{\mathrm{d}t^2} = \frac{\mathrm{d}^2x}{\mathrm{d}t^2}\mathbf{\imath} + \frac{\mathrm{d}^2y}{\mathrm{d}t^2}\mathbf{J} + \frac{\mathrm{d}^2z}{\mathrm{d}t^2}\mathbf{k}$$

is the acceleration of P relative to the fixed reference frame. The term

$$\mathbf{a}^{\text{cor}}_{P(xyz)} = 2\boldsymbol{\omega} \times \mathbf{v}^{\text{rel}}_{P(xyz)}$$

is called the Coriolis acceleration.

For this example, $\mathbf{r} = \mathbf{OP} = x\mathbf{\imath}$ and

$$\boldsymbol{\omega} = \omega\cos\varphi\,\mathbf{\imath} + \omega\sin\varphi\,\mathbf{J}.$$

For the acceleration of the particle, the following expressions are calculated:

$$\boldsymbol{\omega} \times \mathbf{r} = \begin{vmatrix} \mathbf{\imath} & \mathbf{J} & \mathbf{k} \\ \omega\cos\varphi & \omega\sin\varphi & 0 \\ x & 0 & 0 \end{vmatrix} = -x\omega\sin\varphi\,\mathbf{k},$$

$$\boldsymbol{\omega} \times (\boldsymbol{\omega} \times \mathbf{r}) = \begin{vmatrix} \mathbf{\imath} & \mathbf{J} & \mathbf{k} \\ \omega\cos\varphi & \omega\sin\varphi & 0 \\ 0 & 0 & -x\omega\sin\varphi \end{vmatrix}$$
$$= -x\omega^2\sin^2\varphi\,\mathbf{\imath} + x\omega^2\sin\varphi\cos\varphi\,\mathbf{J},$$

and

$$2\boldsymbol{\omega} \times \mathbf{v}^{rel}_{P(xyz)} = 2 \begin{vmatrix} \mathbf{1} & \mathbf{J} & \mathbf{k} \\ \omega\cos\varphi & \omega\sin\varphi & 0 \\ \dot{x} & 0 & 0 \end{vmatrix} = -2\dot{x}\omega\sin\varphi\,\mathbf{k}.$$

The acceleration of P is

$$\begin{aligned} \mathbf{a}_P &= \mathbf{a}^{rel}_{P(xyz)} + \boldsymbol{\alpha}\times\mathbf{r} + \boldsymbol{\omega}\times(\boldsymbol{\omega}\times\mathbf{r}) + 2\boldsymbol{\omega}\times\mathbf{v}^{rel}_{P(xyz)} \\ &= \ddot{x}\mathbf{1} + (-x\omega^2\sin^2\varphi\,\mathbf{1} + x\omega^2\sin\varphi\cos\varphi\,\mathbf{J}) - 2\dot{x}\omega\sin\varphi\,\mathbf{k} \\ &= (\ddot{x} - x\omega^2\sin^2\varphi)\mathbf{1} + x\omega^2\sin\varphi\cos\varphi\,\mathbf{J}) - 2\dot{x}\omega\sin\varphi\,\mathbf{k}, \end{aligned}$$

where $\boldsymbol{\alpha} = \mathbf{0}$. The equation of the trajectory of the particle is $f = y = 0$ and $g = z = 0$. The reaction of the tube on the particle is given by

$$\begin{aligned} \mathbf{R}_l &= \lambda\,\nabla f + \mu\,\nabla g \\ &= \lambda\left(\frac{\partial f}{\partial x}\mathbf{1} + \frac{\partial f}{\partial y}\mathbf{J} + \frac{\partial f}{\partial z}\mathbf{k}\right) + \mu\left(\frac{\partial g}{\partial x}\mathbf{1} + \frac{\partial g}{\partial y}\mathbf{J} + \frac{\partial g}{\partial z}\mathbf{k}\right) \\ &= \lambda\mathbf{J} + \mu\mathbf{k} = R_y\mathbf{J} + R_z\mathbf{k}. \end{aligned}$$

Newton's second law of motion of the particle inside the cylindrical tube can be written as

$$m\mathbf{a}_P = \mathbf{G} + \mathbf{R}_l,$$

where the gravitational force acting on the particle and is calculated as

$$\mathbf{G} = m\mathbf{g} = -mg\cos\varphi\,\mathbf{1} - mg\sin\varphi\,\mathbf{J}.$$

Projecting the previous vectorial equation of motion along the axis of the rotating reference frame, it results

$$\begin{aligned} m(\ddot{x} - x\omega^2\sin^2\varphi) &= -mg\cos\varphi, \\ mx\omega^2\sin\varphi\cos\varphi &= -mg\sin\varphi + R_y, \\ -2m\dot{x}\omega\sin\varphi &= R_z, \end{aligned}$$

or

$$\begin{aligned} \ddot{x} - x\omega^2\sin^2\varphi &= -mg\cos\varphi, \\ R_y &= mx\omega^2\sin\varphi\cos\varphi + mg\sin\varphi, \\ R_z &= -2m\dot{x}\omega\sin\varphi. \end{aligned}$$

The first equation represents the equation of motion for the particle, and the reaction forces are given by the next two equations.

Example 4.11.   A particle of mass m is thrown with an initial velocity $\mathbf{v}_0$. This initial velocity makes an angle α with the horizontal axis. Find the equations of motion, the horizontal distance between the launching point and the particle landing location, and the height when the particle is at its peak (maximal height). Given: $\alpha = \pi/3$, $v_0 = 10$ m/s, and $g = 9.8$ m/s^2.

Solution
The Newton's second law of motion of the particle can be written as $m\mathbf{a} = \sum \mathbf{F}$ or using the projections on the Cartesian reference frame

$$m\ddot{x} = 0,$$
$$m\ddot{y} = -mg,$$
$$m\ddot{z} = 0.$$

Integrating the previous relatins with respect to time, it results

$$\dot{x} = c_1,$$
$$\dot{y} = -gt + c_2,$$
$$\dot{z} = c_3,$$

and

$$x = c_1 t + c_4,$$
$$y = -g\frac{t^2}{2} + c_2 t + c_5,$$
$$z = c_3 t + c_6.$$

Using the initial conditions at $t = t_0 = 0$, $x(0) = 0$, $y(0) = 0$, $z(0) = 0$, $v_x(0) = \dot{x}(0) = v_0 \cos\alpha$, $v_y(0) = \dot{y}(0) = v_0 \sin\alpha$, $v_z(0) = \dot{z}(0) = 0$ the constants of integration are

$$c_4 = 0, \quad c_5 = 0, \quad c_6 = 0,$$
$$c_1 = v_0 \cos\alpha, \quad c_2 = v_0 \sin\alpha, \quad c_3 = 0.$$

The path of the particle is given by

$$x = v_0 t \cos\alpha,$$
$$y = -g\frac{t^2}{2} + v_0 t \sin\alpha,$$
$$z = 0.$$

Differentiating the previous relations with respect to time, it results

$$\dot{x} = v_0 \cos\alpha,$$
$$\dot{y} = -gt + v_0 \sin\alpha,$$
$$\dot{z} = 0.$$

The MATLAB commands for the trajectory and velocity are

```
syms m g v_0 alpha t real

% IC t=0 =>
% x(0)=0, y(0)=0
% Dx(0)=v_0*cos(alpha), Dy(0)=v_0*sin(alpha)
xs=dsolve('m*D2x=0','x(0)=0','Dx(0)=v_0*cos(alpha)');
ys=dsolve('m*D2y=-m*g','y(0)=0',
    'Dy(0)=v_0*sin(alpha)');
zs=dsolve('m*D2z=0','z(0)=0','Dz(0)=0');

fprintf('position \n')
fprintf('x = %s   \n',char(xs))
fprintf('y = %s   \n',char(ys))
fprintf('z = %s   \n',char(zs))

Dx=diff(xs,t);
Dy=diff(ys,t);
Dz=diff(zs,t);

fprintf('velocity \n');
fprintf('vx = %s   \n',char(Dx))
fprintf('vy = %s   \n',char(Dy))
fprintf('vz = %s   \n',char(Dz))
fprintf('\n')
```

From $z = 0$, it results that the particle trajectory is situated in the xy plane. The equation for the trajectory of the particle is the equation of a parabola given by

$$y = -\frac{g}{2v_0^2\cos^2\alpha}x^2 + x\tan\alpha.$$

The magnitude of the velocity can be calculated as

$$|\mathbf{v}| = \sqrt{\dot{x}^2 + \dot{y}^2 + \dot{z}^2} = \sqrt{v_0^2 - 2gy}.$$

The MATLAB commands for the path and the magnitude of velocity are

```
syms x y real
t_x = solve(x-xs,t);
t_y = solve(y-ys,t);

yn=subs(ys,t,t_x);
fprintf('trajectory'); fprintf('\n')
fprintf('y =  %s',char(yn))
fprintf('\n\n')
```

```
magn_v=subs(sqrt(Dx^2+Dy^2),t,t_y);
magn_v1=simplify(magn_v(1));
fprintf('|v|= %s \n',char(magn_v1))
fprintf('\n')
```

The horizontal distance $d = OA$ between the launching point O and the landing location A is computed from the condition $y_A = 0$ as

$$d = \frac{v_0^2 \sin(2\alpha)}{g}.$$

The horizontal distance d is maximal when the fraction $\frac{v_0^2 \sin(2\alpha)}{g}$ is maximal. Since both g and v_0 are constants, the distance $d = \frac{v_0^2 \sin(2\alpha)}{g}$ will be maximal when the trigonometric function $\sin(2\alpha)$ its maximal, that is, $\sin(2\alpha) = 1$ or $\alpha = 45°$. Therefore, the maximal horizontal distance the particle can travel is

$$d_{\max} = \frac{v_0^2}{g}.$$

The maximum vertical displacement attained by a particle is known as the maximum height. The maximum height of the particle trajectory can be computed from $\frac{dy}{dx} = 0$

$$\frac{dy}{dx} = -2\frac{g}{2v_0^2 \cos^2 \alpha} x_B + \tan \alpha = 0,$$

or

$$x_B = \frac{v_0^2 \sin(2\alpha)}{2g}.$$

The maximum vertical displacement is then computed as

$$
\begin{aligned}
y_B &= -\frac{g}{2v_0^2 \cos^2 \alpha} x_B^2 + x_B \tan \alpha \\[2mm]
&= -\frac{g}{2v_0^2 \cos^2 \alpha}\left(\frac{v_0^2 \sin(2\alpha)}{2g}\right)^2 + \frac{v_0^2 \sin(2\alpha)}{2g}\tan \alpha \\[2mm]
&= -\frac{g}{2v_0^2 \cos^2 \alpha}\frac{4v_0^4 \sin^2 \alpha \cos^2 \alpha}{4g^2} + \frac{v_0^2 2 \sin \alpha \cos \alpha}{2g}\frac{\sin \alpha}{\cos \alpha} \\[2mm]
&= -\frac{v_0^2 \sin^2 \alpha}{2g} + \frac{v_0^2 \sin^2 \alpha}{g} \\[2mm]
&= \frac{v_0^2 \sin^2 \alpha}{2g}.
\end{aligned}
$$

The maximal maximum height of the particle is attained when the trigonometric function $\sin(2\alpha)$ in is maximal, that is, $\sin(2\alpha) = 1$ or $\alpha = 45°$. Therefore, the maximal maximum height can be written as

$$y_{\max} = \frac{v_0^2}{2g}.$$

The MATLAB program for x_B, $x_{\max}$, y_B, and $y_{\max}$ is

```
x_As=solve(yn,x);
x_A = simplify(x_As(2));
fprintf('x_A = %s \n',char(x_A))
fprintf('\n')


x_max=subs(x_A,sin(2*alpha),1);
fprintf('x_max = %s \n',char(x_max))
fprintf('\n')


x_B=simplify(solve(diff(yn,x),x));
fprintf('x_B = %s \n',char(x_B))
fprintf('\n')


y_B=simplify(subs(yn,x,x_B));
fprintf('y_B = %s \n',char(y_B))
fprintf('\n')


y_max=simplify(subs(y_B,sin(alpha),1));
fprintf('y_max = %s \n',char(y_max))
fprintf('\n')
```

The motion of the particle is simulated for the numerical data with the following MATLAB commands:

```
ynn=subs(yn,slist,nlist);
fprintf('y =   %s',char(ynn))
fprintf('\n')


axis manual
axis equal
hold on
grid on
axis([-1 10 -1 4])


for x_t = 0 : 0.01 : 8.83
    slist={g,v_0, alpha,x};
    nlist={9.8,10,pi/6,x_t};
    y_t = subs(ynn,slist,nlist);
```

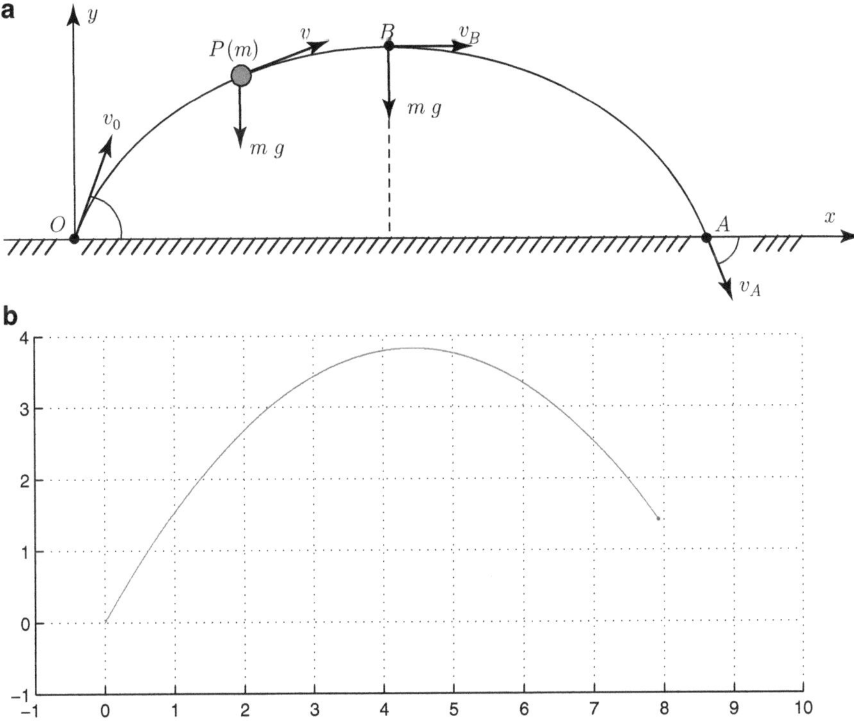

Fig. 4.28 Example 4.11

```
    hm=plot(x_t,y_t,'k.','Color','red');
    ht=plot(x_t,y_t);
    pause(0.001)
    delete(hm);
  end
```

The MATLAB output is shown in Fig. 4.28b.

4.15 Problems

4.1 A bullet with the mass m was fired horizontally into a spherical object with
the mass M suspended on a wire with the length l as shown in Fig. 4.29. The
spherical object with the bullet embedded in it moves to a height equal to h.
Find the speed of the bullet as it entered the spherical object. For the numerical
application, use $m = 40$ g, $M = 30$ kg, $l = 700$ mm, $h = 20$ mm.

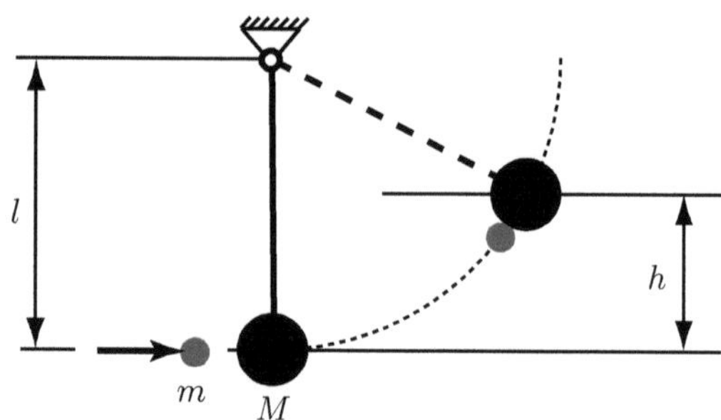

Fig. 4.29 Problem 4.1

Fig. 4.30 Problem 4.2

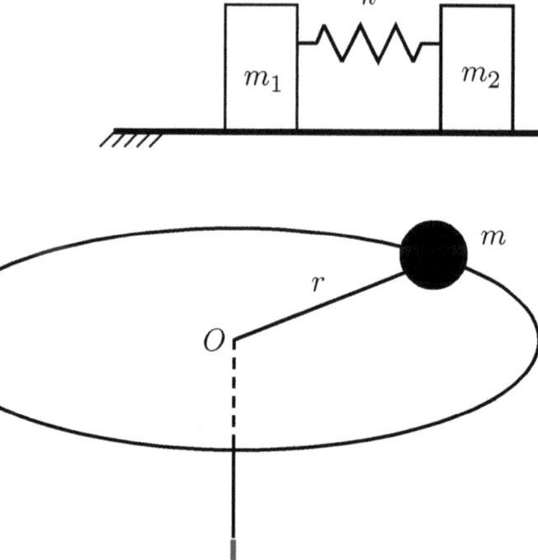

Fig. 4.31 Problem 4.3

4.2 Figure 4.30 shows two masses m_1 and m_2, on a smooth horizontal plane, connected by a spring with the normal length l_0. The spring constant is k. The spring is compressed to a length l ($l < l_0$), and the system is released. Find the speed of each mass when the spring is again its normal length l_0. For the numerical application, use $m_1 = 1.5$ kg, $m_2 = 2.5$ kg, $k = 15$ lb/in., $l_0 = 12$ cm, $l = 8$ cm.

4.3 A particle of mass m moves in a circular path of radius r on a smooth horizontal plane as shown in Fig. 4.31. The particle is connected to a string which passes through a the center O of the pane. The angular velocity of the string and the particle is ω when the radius is r. The string is pulled from underneath until the radius of the path is $r/2$. Find the final angular velocity and the final tension in the string.

4.4 Two particles, each weighing W, are connected by a string with negligible mass as shown in Fig. 4.32. The platform disk is rotating with the angular

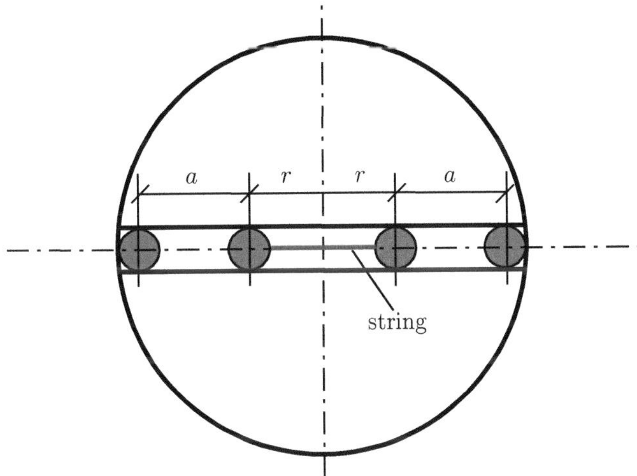

Fig. 4.32 Problem 4.4

velocity ω when the string breaks. There is no friction between the particles and the groove in which they ride. Find the angular speed of the system when the particles hit the outer stops. For the numerical application, use $W = 4$ lb, $\omega = 20$ rad/s, $r = 3$ in., $a = 10$ in.

4.5 The mass of a gun is m_g, and the mass of the projectile is m_p. The speed of the projectile immediately after the explosion is v_g. Find the speed of recoil (the speed of the gun immediately after the explosion).

4.6 A person with the mass m sitting in a boat with the mass M fires horizontally a shotgun releasing a bullet with the mass m_b. The muzzle speed is v_b, and the friction is neglected. Find the speed of the boat after the shot is fired.

4.7 A bullet with the mass m_b is moving with a speed of v_b strikes an object of mass M moving in the same direction with a speed of V. After the impact, the bullet is embedded in the block. Find the resultant speed of the bullet and the block immediately after the impact. For the numerical application, use $m_b = 50$ g, $v_b = 400$ m/s, $M = 12$ kg, $V = 30$ m/s.

4.8 A sphere of mass m falls from a height h on a vertical spring as shown in Fig. 4.33. The spring has an elastic constant k. Find (a) the total time of contact of the sphere with the spring, (b) the relative displacement of the spring, (c) the velocity jump of the sphere (before the contact and after the contact with the spring), and (d) the maximum elastic force. For the numerical application, use $m = 10$ kg, $h = 1$ m, and $k = 294 \times 10^3$ N/m.

4.9 A particle P of mass m is moving without friction on a plane astroid as shown in Fig. 4.34. The astroid curve is specified by the equation $x^{2/3} + y^{2/3} = a^{2/3}$ where $a > 1$. At the initial moment $t = 0$, the particle is at $P_0(0, a)$ and is moving toward the positive x-axis. The initial velocity of the particle is

Fig. 4.33 Problem 4.8

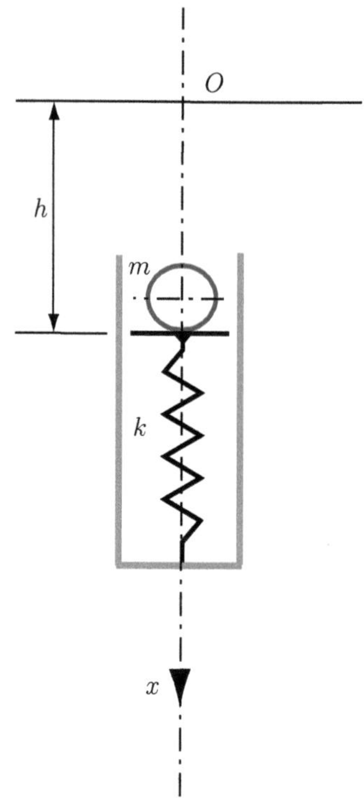

Fig. 4.34 Problem 4.9

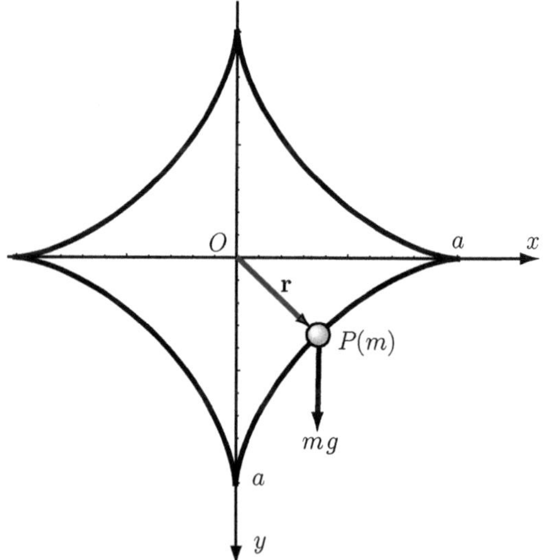

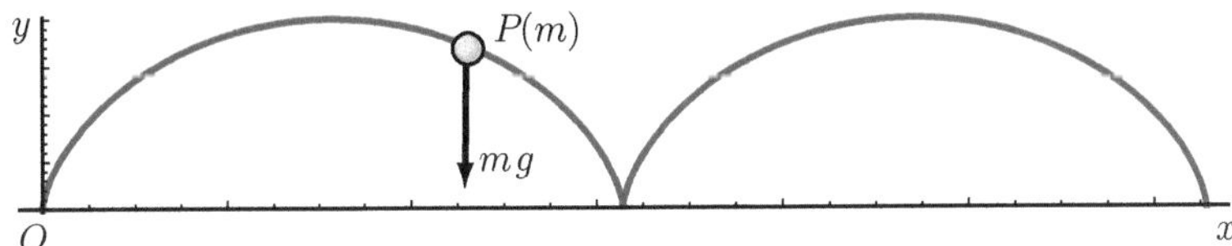

Fig. 4.35 Problem 4.10

Fig. 4.36 Problem 4.11

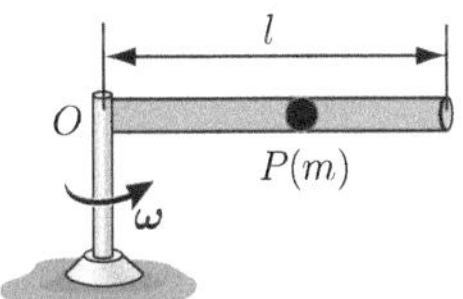

$v_0 = \sqrt{2ga}$. Find and solve the equation of motion of the particle and determine the reaction force of the astroid on the particle. For the numerical application, use $m = 1$ kg, $a = 2$ m, and $g = 9.81$ m/s^2.

4.10 A particle P of mass m is moving on a cycloid through the origin, generated by a circle of radius r and defined by the parametric equations $x = r\,(q - \sin q)$ and $y = r\,(1 - \cos q)$ where q is a real parameter. For the initial moment $t = 0$ s, the initial conditions are $q(0) = \pi$ and $\dot{q}(0) = v_0/(2r)$ where v_0 is the initial velocity of the particle. Find the motion of the particle and the reaction of the curve on the particle for the smooth curve and rough curve with the coefficient of friction μ. For the numerical application, use $m = 1$ kg, $r = 1$ m, $v_0 = 10$ m/s, and $g = 9.81$ m/s^2 (Fig. 4.35).

4.11 A particle P of mass m moves outward through a tube of length i. The tube rotates in the horizontal plane at a constant rate ω. The particle starts at point O, as shown in Fig. 4.36, with an initial radial velocity v_0. Find and solve the equation of motion. For the numerical application, use $m = 1$ kg, $L = 1$ m, $v_0 = 1.8$ m/s, and $g = 9.81$ m/s^2.

4.12 A particle of mass $m = 1$ kg moves along a horizontal path defined by the equations $r = (3t + 5)$ m and $\theta = (2t^2 - 3t)$ rad where t is in seconds. Determine the resultant force at the instant $t = 5$ s.

4.13 A particle P of mass m is sliding down the spiral at a constant speed such that its position has components $r = r_0$, $\theta = wt$, and $z = at$. Determine the components of force which the spiral exerts on the particle at the instant t_i. Given: $m = 10$ kg, $r_0 = 1$ m, $w = 1.5$ rad/s, $a = -2$ m/s, $g = 9.81$ m/s^2, $t_1 = 1$ s, and $t_2 = 2$ s.

4.14 The link 1, shown in Fig. 4.37, rotates with a constant angular velocity ω with respect to the vertical plane xy. The slider 2 that can be treated as a particle of mass m can move along the link 1 without friction. Find the equations of motion of the particle 2.

Fig. 4.37 Problem 4.14

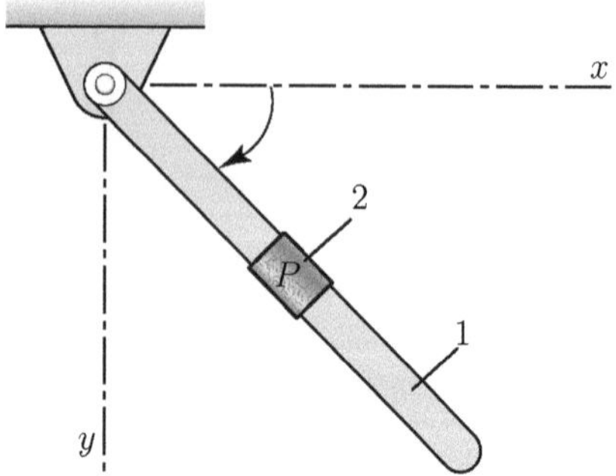

Fig. 4.38 Problem 4.13

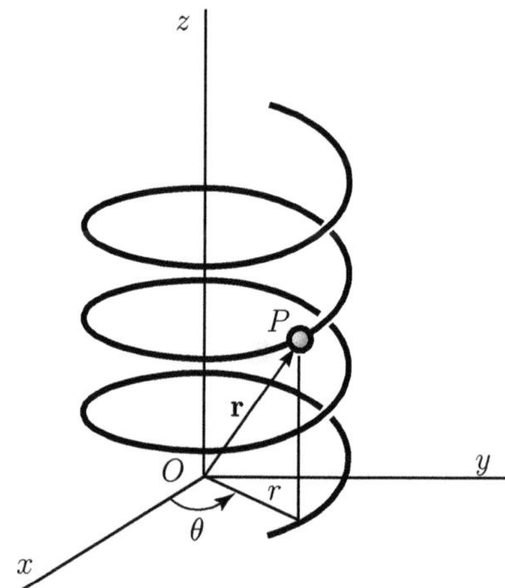

4.15 The particle of mass m moves along the screw line shown in Fig. 4.38. Its motion with respect to the inertial space xyz is given by $r = a\cos\omega t\,\mathbf{i} + a\sin\omega t\,\mathbf{j} + bt\,\mathbf{k}$. Find the reaction force between the spiral and the particle and the kinetic energy of the particle.

Chapter 5
Kinematics of Rigid Bodies

5.1 Introduction

A rigid body is an idealized model of an object that does not deform or change shape. A rigid body, (RB), is by definition an object with the property that the distance between every pair of points of the body is constant:

$$\forall\, A, B \in (RB) \implies \text{dist}(AB) = AB = \text{constant}.$$

Although any real object deforms as it moves, if its deformation is small, the motion may be approximated as a rigid body motion. The motion of a rigid body (RB) is defined when the position vector, velocity, and acceleration of all points of the rigid body are defined as functions of time with respect to a fixed reference frame with the origin at O_0. Let $\mathbf{\imath}_0$, $\mathbf{\jmath}_0$, and $\mathbf{k}_0$ be the constant unit vectors of a fixed orthogonal Cartesian reference frame $O_0 x_0 y_0 z_0$, as shown in Fig. 5.1. The vectors $\mathbf{\imath}$, $\mathbf{\jmath}$, and $\mathbf{k}$ are the unit vectors of a rotating orthogonal Cartesian reference frame $Oxyz$. The unit vectors $\mathbf{\imath}_0$, $\mathbf{\jmath}_0$, and $\mathbf{k}_0$ of the primary reference frame are constant with respect to time.

A reference frame that moves with the rigid body is a *body-fixed* (or rotating or mobile) reference frame. The unit vectors $\mathbf{\imath}$, $\mathbf{\jmath}$, and $\mathbf{k}$ of the body-fixed reference frame are not constant because they rotate with the body-fixed reference frame. The location of the point O is arbitrary.

The position vector of a point M, $M \in (RB)$ with respect to the fixed reference frame $O_0 x_0 y_0 z_0$ is denoted by $\mathbf{r}_1 = \mathbf{r}_{O_0 M}$ and with respect to the mobile reference frame $Oxyz$ is denoted by $\mathbf{r} = \mathbf{r}_{OM}$. The location of the origin O of the mobile reference frame with respect to the fixed point O_0 is defined by the position vector $\mathbf{r}_O = \mathbf{r}_{O_0 O}$.

The relation between the vectors $\mathbf{r}_1$, $\mathbf{r}$, and $\mathbf{r}_0$ is given by

$$\mathbf{r}_1 = \mathbf{r}_O + \mathbf{r} = \mathbf{r}_O + x\mathbf{\imath} + y\mathbf{\jmath} + z\mathbf{k}, \tag{5.1}$$

D.B. Marghitu and M. Dupac, *Advanced Dynamics: Analytical and Numerical Calculations with MATLAB*, DOI 10.1007/978-1-4614-3475-7_5, © Springer Science+Business Media, LLC 2012

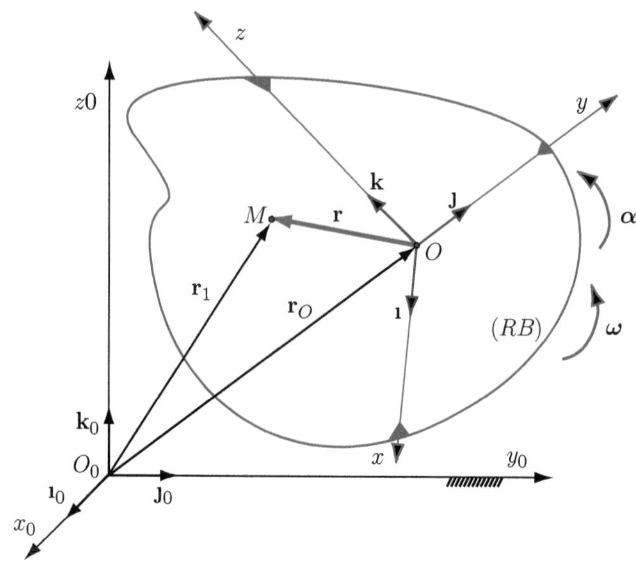

Fig. 5.1 Fixed orthogonal Cartesian reference frame $[\mathbf{1}_0, \mathbf{J}_0, \mathbf{k}_0]$ and body-fixed (or rotating) reference frame $[\mathbf{1}, \mathbf{J}, \mathbf{k}]$

where x, y, and z represent the projections of the vector $\mathbf{r}$ on the mobile reference frame. The magnitude of the vector $\mathbf{r} = \mathbf{r}_{OM}$ is constant as the distance between the points O and M is constant, $O \in (RB)$ and $M \in (RB)$. Thus, the x, y, and z components of the vector $\mathbf{r}$ with respect to the rotating reference frame are constant. The unit vectors $\mathbf{1}, \mathbf{J}$, and $\mathbf{k}$ are time-dependent vector functions.

The vectors $\mathbf{1}, \mathbf{J}$, and $\mathbf{k}$ are the unit vectors of an orthogonal Cartesian reference frame, and the following relations can be written:

$$\mathbf{1} \cdot \mathbf{1} = 1, \quad \mathbf{J} \cdot \mathbf{J} = 1, \quad \mathbf{k} \cdot \mathbf{k} = 1, \tag{5.2}$$

$$\mathbf{1} \cdot \mathbf{J} = 0, \quad \mathbf{J} \cdot \mathbf{k} = 0, \quad \mathbf{k} \cdot \mathbf{1} = 0. \tag{5.3}$$

5.2   Velocity Analysis for a Rigid Body

The velocity of an arbitrary point M of the rigid body with respect to the fixed reference frame $O_0 x_0 y_0 z_0$ is the derivative with respect to time of the position vector $\mathbf{r}_1$:

$$\mathbf{v} = \frac{d\mathbf{r}_1}{dt} = \dot{\mathbf{r}}_1 = \dot{\mathbf{r}}_O + \dot{\mathbf{r}} = \mathbf{v}_O + x\mathbf{i} + y\mathbf{j} + z\dot{\mathbf{k}} + \dot{x}\mathbf{1} + \dot{y}\mathbf{J} + \dot{z}\mathbf{k}, \tag{5.4}$$

where $\mathbf{v}_O = \dot{\mathbf{r}}_O$ represents the velocity of the origin of the rotating reference frame $Oxyz$ with respect to the fixed reference frame $O_0x_0y_0z_0$. Because all the points in the rigid body maintain their relative position, their velocity relative to the mobile reference frame $Oxyz$ is zero, that is, $\dot{x} = \dot{y} = \dot{z} = 0$.

The velocity of point M is

$$\mathbf{v} = \mathbf{v}_O + x\mathbf{i} + y\mathbf{j} + z\dot{\mathbf{k}}.$$

The derivative of the (5.2) and (5.3) with respect to time gives

$$\mathbf{i} \cdot \mathbf{i} = 0, \quad \mathbf{j} \cdot \mathbf{j} = 0, \quad \dot{\mathbf{k}} \cdot \mathbf{k} = 0, \tag{5.5}$$

and

$$\mathbf{i} \cdot \mathbf{j} + \mathbf{j} \cdot \mathbf{i} = 0, \quad \mathbf{j} \cdot \mathbf{k} + \dot{\mathbf{k}} \cdot \mathbf{j} = 0, \quad \dot{\mathbf{k}} \cdot \mathbf{i} + \mathbf{i} \cdot \mathbf{k} = 0. \tag{5.6}$$

For (5.6), one can introduce the convention:

$$\mathbf{i} \cdot \mathbf{j} = -\mathbf{j} \cdot \mathbf{i} = \omega_z,$$
$$\mathbf{j} \cdot \mathbf{k} = -\dot{\mathbf{k}} \cdot \mathbf{j} = \omega_x,$$
$$\dot{\mathbf{k}} \cdot \mathbf{i} = -\mathbf{i} \cdot \mathbf{k} = \omega_y, \tag{5.7}$$

where ω_x, ω_y, and ω_z are considered as the projections of a vector $\boldsymbol{\omega}$:

$$\boldsymbol{\omega} = \omega_x \mathbf{i} + \omega_y \mathbf{j} + \omega_z \mathbf{k}.$$

To calculate $\mathbf{i}$, $\mathbf{j}$, $\dot{\mathbf{k}}$, the following relation for an arbitrary vector $\mathbf{v}$ can be used:

$$\mathbf{p} = p_x \mathbf{i} + p_y \mathbf{j} + p_z \mathbf{k} = (\mathbf{p} \cdot \mathbf{i})\mathbf{i} + (\mathbf{p} \cdot \mathbf{j})\mathbf{j} + (\mathbf{p} \cdot \mathbf{k})\mathbf{k}. \tag{5.8}$$

Using (5.8) and the results from (5.5) and (5.6), one can write

$$\begin{aligned}
\mathbf{i} &= (\mathbf{i} \cdot \mathbf{i})\mathbf{i} + (\mathbf{i} \cdot \mathbf{j})\mathbf{j} + (\mathbf{i} \cdot \mathbf{k})\mathbf{k} \\
&= (0)\mathbf{i} + (\omega_z)\mathbf{j} - (\omega_y)\mathbf{k} \\
&= \begin{vmatrix} \mathbf{i} & \mathbf{j} & \mathbf{k} \\ \omega_x & \omega_y & \omega_z \\ 1 & 0 & 0 \end{vmatrix} = \boldsymbol{\omega} \times \mathbf{i},
\end{aligned}$$

$$\begin{aligned}
\mathbf{j} &= (\mathbf{j} \cdot \mathbf{i})\mathbf{i} + (\mathbf{j} \cdot \mathbf{j})\mathbf{j} + (\mathbf{j} \cdot \mathbf{k})\mathbf{k} \\
&= (-\omega_z)\mathbf{i} + (0)\mathbf{j} + (\omega_x)\mathbf{k} \\
&= \begin{vmatrix} \mathbf{i} & \mathbf{j} & \mathbf{k} \\ \omega_x & \omega_y & \omega_z \\ 0 & 1 & 0 \end{vmatrix} = \boldsymbol{\omega} \times \mathbf{j},
\end{aligned}$$

$$\dot{\mathbf{k}} = \left(\dot{\mathbf{k}}\cdot\mathbf{i}\right)\mathbf{i} + \left(\dot{\mathbf{k}}\cdot\mathbf{j}\right)\mathbf{j} + \left(\dot{\mathbf{k}}\cdot\mathbf{k}\right)\mathbf{k}$$

$$= (\omega_y)\mathbf{i} - (\omega_x)\mathbf{j} + (0)\mathbf{k}$$

$$= \begin{vmatrix} \mathbf{i} & \mathbf{j} & \mathbf{k} \\ \omega_x & \omega_y & \omega_z \\ 0 & 0 & 1 \end{vmatrix} = \omega \times \mathbf{k}. \tag{5.9}$$

The relations

$$\dot{\mathbf{i}} = \omega \times \mathbf{i}, \quad \dot{\mathbf{j}} = \omega \times \mathbf{j}, \quad \dot{\mathbf{k}} = \omega \times \mathbf{k} \tag{5.10}$$

are known as *Poisson formulas*, and ω is the angular velocity vector. Using (5.4) and (5.10), one can obtain

$$\mathbf{v} = \mathbf{v}_O + x\omega \times \mathbf{i} + y\omega \times \mathbf{j} + z\omega \times \mathbf{k} = \mathbf{v}_O + \omega \times (x\mathbf{i} + y\mathbf{j} + z\mathbf{k}),$$

or

$$\mathbf{v} = \mathbf{v}_O + \omega \times \mathbf{r}. \tag{5.11}$$

Combining (5.4) and (5.11), it results

$$\dot{\mathbf{r}} = \omega \times \mathbf{r}. \tag{5.12}$$

Using (5.11), the components of the velocity are

$$v_x = v_{Ox} + z\omega_y - y\omega_z,$$

$$v_y = v_{Oy} + x\omega_z - z\omega_x,$$

$$v_z = v_{Oz} + y\omega_x - x\omega_y.$$

The relation between the velocities $\mathbf{v}_M$ and $\mathbf{v}_O$ of two points M and O on the rigid body is

$$\mathbf{v}_M = \mathbf{v}_O + \omega \times \mathbf{r}_{OM}, \tag{5.13}$$

or

$$\mathbf{v}_M = \mathbf{v}_O + \mathbf{v}_{MO}^{\mathrm{rel}}, \tag{5.14}$$

where $\mathbf{v}_{MO}^{\mathrm{rel}}$ is the relative velocity, for rotational motion, of M with respect to O and is given by

$$\mathbf{v}_{MO}^{\mathrm{rel}} = \mathbf{v}_{MO} = \omega \times \mathbf{r}_{OM}. \tag{5.15}$$

The relative velocity $\mathbf{v}_{MO}$ is perpendicular to the position vector $\mathbf{r}_{OM}$, $\mathbf{v}_{MO} \perp \mathbf{r}_{OM}$ and has the direction given by the angular velocity vector ω. The magnitude of the relative velocity is $|\mathbf{v}_{MO}| = v_{MO} = \omega\, r_{OM}$.

5.3   Acceleration Analysis for a Rigid Body

The acceleration of an arbitrary point $M \in (RB)$ with respect to a fixed reference frame $O_0 x_0 y_0 z_0$ represents the double derivative with respect to time of the position vector $\mathbf{r}_1$:

$$\mathbf{a} = \ddot{\mathbf{r}}_1 = \dot{\mathbf{v}} = \frac{d}{dt}(\mathbf{v}_O + \boldsymbol{\omega} \times \mathbf{r}) = \dot{\mathbf{v}}_O + \dot{\boldsymbol{\omega}} \times \mathbf{r} + \boldsymbol{\omega} \times \dot{\mathbf{r}}. \tag{5.16}$$

The acceleration of the point O with respect to the fixed reference frame is

$$\mathbf{a}_O = \dot{\mathbf{v}}_O = \ddot{\mathbf{r}}_O. \tag{5.17}$$

The derivative of the vector $\boldsymbol{\omega}$ with respect to the time is the angular acceleration vector $\boldsymbol{\alpha}$ given by

$$\begin{aligned}
\boldsymbol{\alpha} = \dot{\boldsymbol{\omega}} &= \dot{\omega}_x \mathbf{\imath} + \dot{\omega}_y \mathbf{\jmath} + \dot{\omega}_z \mathbf{k} + \omega_x \dot{\mathbf{\imath}} + \omega_y \dot{\mathbf{\jmath}} + \omega_z \dot{\mathbf{k}} \\
&= \alpha_x \mathbf{\imath} + \alpha_y \mathbf{\jmath} + \alpha_z \mathbf{k} + \omega_x \boldsymbol{\omega} \times \mathbf{\imath} + \omega_y \boldsymbol{\omega} \times \mathbf{\jmath} + \omega_z \boldsymbol{\omega} \times \mathbf{k} \\
&= \alpha_x \mathbf{\imath} + \alpha_y \mathbf{\jmath} + \alpha_z \mathbf{k} + \boldsymbol{\omega} \times \boldsymbol{\omega} = \alpha_x \mathbf{\imath} + \alpha_y \mathbf{\jmath} + \alpha_z \mathbf{k},
\end{aligned} \tag{5.18}$$

where $\alpha_x = \dot{\omega}_x$, $\alpha_y = \dot{\omega}_y$, and $\alpha_z = \dot{\omega}_z$. In the previous expressions, the Poisson formulas $\dot{\mathbf{\imath}} = \boldsymbol{\omega} \times \mathbf{\imath}$, $\dot{\mathbf{\jmath}} = \boldsymbol{\omega} \times \mathbf{\jmath}$, and $\dot{\mathbf{k}} = \boldsymbol{\omega} \times \mathbf{k}$ have been used.

Using (5.16)–(5.18), the acceleration of the point M is

$$\mathbf{a} = \mathbf{a}_O + \boldsymbol{\alpha} \times \mathbf{r} + \boldsymbol{\omega} \times (\boldsymbol{\omega} \times \mathbf{r}). \tag{5.19}$$

Using (5.19), the components of the acceleration are

$$\begin{aligned}
a_x &= a_{Ox} + (z\alpha_y - y\alpha_z) + \omega_y (y\omega_x - x\omega_y) + \omega_z (x\omega_x - x\omega_z), \\
a_y &= a_{Oy} + (x\alpha_z - z\alpha_x) + \omega_z (z\omega_y - y\omega_z) + \omega_x (x\omega_y - y\omega_z), \\
a_z &= a_{Oz} + (y\alpha_x - x\alpha_y) + \omega_x (x\omega_z - z\omega_x) + \omega_y (y\omega_z - z\omega_y).
\end{aligned}$$

The relation between the accelerations $\mathbf{a}_M$ and $\mathbf{a}_O$ of two points M and O on the rigid body is

$$\mathbf{a}_M = \mathbf{a}_O + \boldsymbol{\alpha} \times \mathbf{r}_{OM} + \boldsymbol{\omega} \times (\boldsymbol{\omega} \times \mathbf{r}_{OM}). \tag{5.20}$$

In the case of planar motion,

$$\boldsymbol{\omega} \times (\boldsymbol{\omega} \times \mathbf{r}_{OM}) = -\omega^2 \mathbf{r}_{OM},$$

and (5.20) becomes

$$\mathbf{a}_M = \mathbf{a}_O + \boldsymbol{\alpha} \times \mathbf{r}_{OM} - \omega^2 \mathbf{r}_{OM}. \tag{5.21}$$

Equation (5.21) can be written as

$$\mathbf{a}_M = \mathbf{a}_O + \mathbf{a}_{MO}^{\text{rel}},$$
(5.22)

where $\mathbf{a}_{MO}^{\text{rel}}$ is the relative acceleration, for rotational motion, of M with respect to O and is given by

$$\mathbf{a}_{MO}^{\text{rel}} = \mathbf{a}_{MO} = \mathbf{a}_{MO}^{\text{n}} + \mathbf{a}_{MO}^{\text{t}}.$$
(5.23)

The normal relative acceleration of M with respect to O is

$$\mathbf{a}_{MO}^{\text{n}} = \boldsymbol{\omega} \times (\boldsymbol{\omega} \times \mathbf{r}_{OM}),$$
(5.24)

is parallel to the position vector $\mathbf{r}_{OM}$, $\mathbf{a}_{MO}^{\text{n}} \| \mathbf{r}_{OM}$, and has the direction toward the center of rotation, from M to O. The magnitude of the normal relative acceleration is

$$|\mathbf{a}_{MO}^{\text{n}}| = a_{MO}^{\text{n}} = \omega^2 \, r_{OM} = \frac{v_{MO}^2}{r_{OM}}.$$

The tangential relative acceleration of M with respect to O

$$\mathbf{a}_{MO}^{\text{t}} = \boldsymbol{\alpha} \times \mathbf{r}_{OM}$$
(5.25)

is perpendicular to the position vector $\mathbf{r}_{OM}$, $\mathbf{a}_{MO}^{\text{t}} \perp \mathbf{r}_{OM}$ and has the direction given by the angular acceleration $\boldsymbol{\alpha}$. The magnitude of the normal relative acceleration is

$$|\mathbf{a}_{MO}^{\text{t}}| = a_{MO}^{\text{t}} = \alpha \, r_{OM}.$$

5.3.1 Translation

A rigid body is in *pure translation* if all points on the body describe parallel paths, as shown in Fig. 5.2. Every point of a rigid body in translation has the same velocity and acceleration. The motion of the rigid body may be described by the motion of a single point. Some examples of bodies in translational motions are an elevator and the piston of a slider-crank mechanism. From the definition, it results that the axes of the body-fixed reference frame $Oxyz$ are parallel to fixed directions. Consequently, the fixed, $O_0x_0y_0z_0$, and mobile, $Oxyz$, reference frames have their axes parallel, as shown in Fig. 5.3. The position of the rigid body at a certain moment can be specified only using the position vector of the origin O with respect to the fixed reference frame $O_0x_0y_0z_0$:

$$\mathbf{r}_O = x_O(t)\,\mathbf{\imath}_0 + y_O(t)\,\mathbf{\jmath}_0 + z_O(t)\,\mathbf{k}_0.$$

Fig. 5.2 Rigid body in pure translation

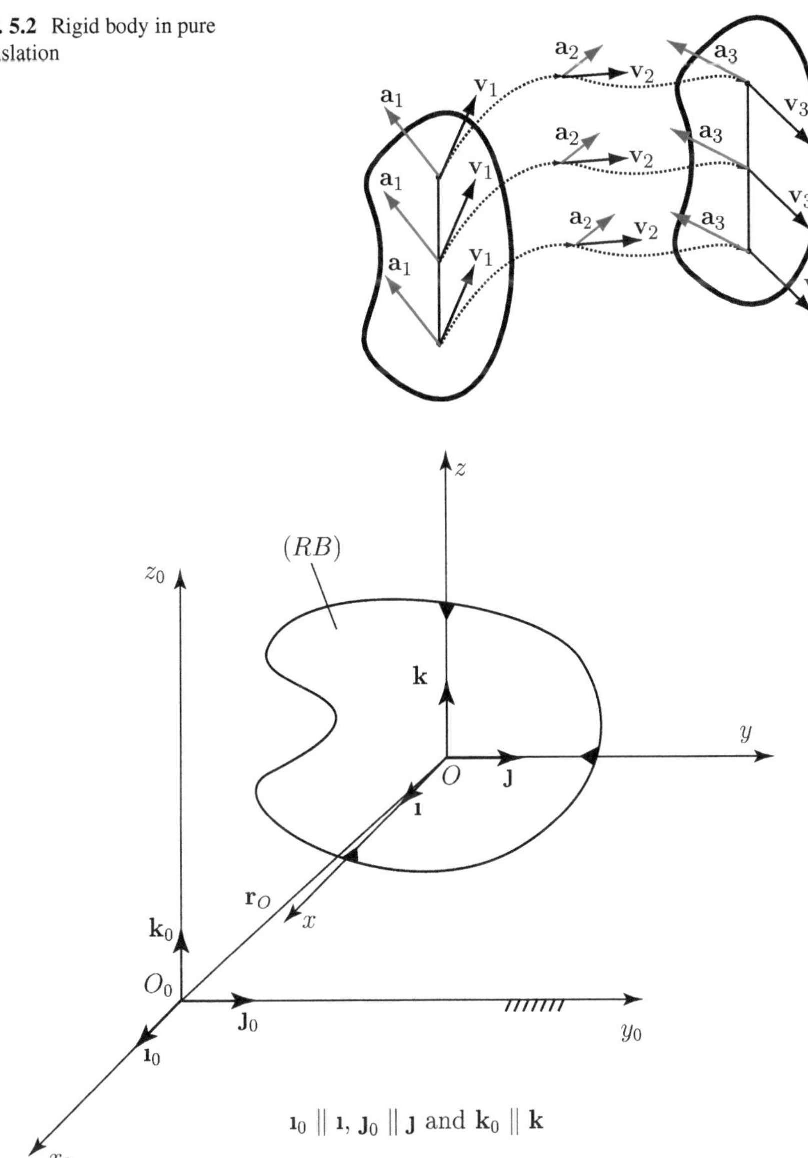

Fig. 5.3 Translation

It results that a spatial rigid body in the translational motion has three degrees of freedom, as its position at a certain moment is determined by the three independent scalar functions: $x_O(t)$, $y_O(t)$, and $z_O(t)$.

Because the fixed and mobile reference frame (body-fixed reference frame) have parallel axes, it results that the unit vectors $\mathbf{i}$, $\mathbf{j}$, and $\mathbf{k}$ have fixed directions:

$$\dot{\mathbf{i}} = \dot{\mathbf{j}} = \dot{\mathbf{k}} = \mathbf{0}. \tag{5.26}$$

Using (5.26) and Poisson formulas, it results

$$\dot{\mathbf{i}} = \boldsymbol{\omega} \times \mathbf{i} = \mathbf{0}, \ \dot{\mathbf{j}} = \boldsymbol{\omega} \times \mathbf{j} = \mathbf{0}, \ \dot{\mathbf{k}} = \boldsymbol{\omega} \times \mathbf{k} = \mathbf{0}.$$

These equations are simultaneously verified if and only if

$$\boldsymbol{\omega} = \mathbf{0}. \tag{5.27}$$

From (5.27), the derivative of the vector $\boldsymbol{\omega}$ is

$$\boldsymbol{\alpha} = \dot{\boldsymbol{\omega}} = \mathbf{0}. \tag{5.28}$$

Equations (5.27) and (5.28) characterize the translational motion of the rigid body.

Velocity
The general equation for the velocity of a point $M \in (RB)$ is

$$\mathbf{v} = \mathbf{v}_0 + \boldsymbol{\omega} \times \mathbf{r},$$

and using the (5.27), the velocity field for the translational motion of a rigid body is

$$\mathbf{v} = \mathbf{v}_O.$$

At a certain moment, all the points of the rigid body have the same velocity (direction, sense, magnitude). Thus, for this type of motion, the velocity is a free vector.

Acceleration
The general equation for the acceleration of a point $M \in (RB)$ is

$$\mathbf{a} = \mathbf{a}_O + \boldsymbol{\alpha} \times \mathbf{r} + \boldsymbol{\omega} \times (\boldsymbol{\omega} \times \mathbf{r}),$$

and using (5.27) and (5.28), the acceleration for the translational motion is

$$\mathbf{a} = \mathbf{a}_O.$$

At a certain moment of time, all the points of the rigid body have the same acceleration (direction, sense, magnitude). Thus, for this type of motion, the acceleration is a free vector.

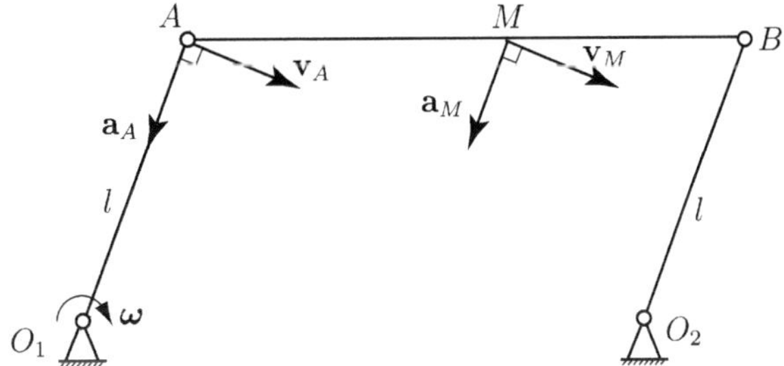

Fig. 5.4 Four-bar mechanism

Exercise

Compute the velocity and acceleration of a point M on a connecting rod AB of a four-bar mechanism, with $O_1A = O_2B = l$, $O_1O_2 = AB$, and $\omega =$ constant (Fig. 5.4).

Solution

The rod AB performs a translational motion. The motion of this rod will be parallel to the fixed line O_1O_2. Therefore, any point on this rod at a certain moment has the same velocity and acceleration as the point A. The point A is also a point of the crank O_1A. Thus, the velocity and the acceleration of M are

$$|\mathbf{v}_M| = |\mathbf{v}_A| = l\omega,$$

and

$$|\mathbf{a}_M| = |\mathbf{a}_A| = l\omega^2.$$

5.3.2 Rotation

A rigid body performs a rotational motion if two of its points, O and O_1 that form an axis, are fixed during the motion. The fixed axis is called the axis of rotation. Some examples of rigid bodies in rotational motion are the rotor of a turbine, a pendulum, etc. From the definition of the rotational motion about an axis and the rigidity hypothesis of the rigid body (*RB*), it results that every point of the body in the rotational motion describes a circular trajectory in a plane perpendicular to the axis of rotation.

In order to simplify this study, the origins of the two reference frames (fixed and mobile) are chosen at the same point ($O_0 = O$), and the axes O_0z_0 and Oz coincide with the rotation axis as shown in Fig. 5.5. The position of the rigid at a certain moment can be completely specified using the angle $\theta = \theta(t)$. Consequently, in this motion, the rigid body has only one degree of freedom. Because the origins of the two reference frames are chosen at the same point ($O_0 \equiv O$), and the axes O_0z_0 and Oz are identical, one can write

$$\mathbf{v}_0 = \mathbf{0}, \quad \mathbf{a}_0 = \mathbf{0}, \quad \mathbf{k} \equiv \mathbf{k}_0, \quad \dot{\mathbf{k}} = \mathbf{0}.$$

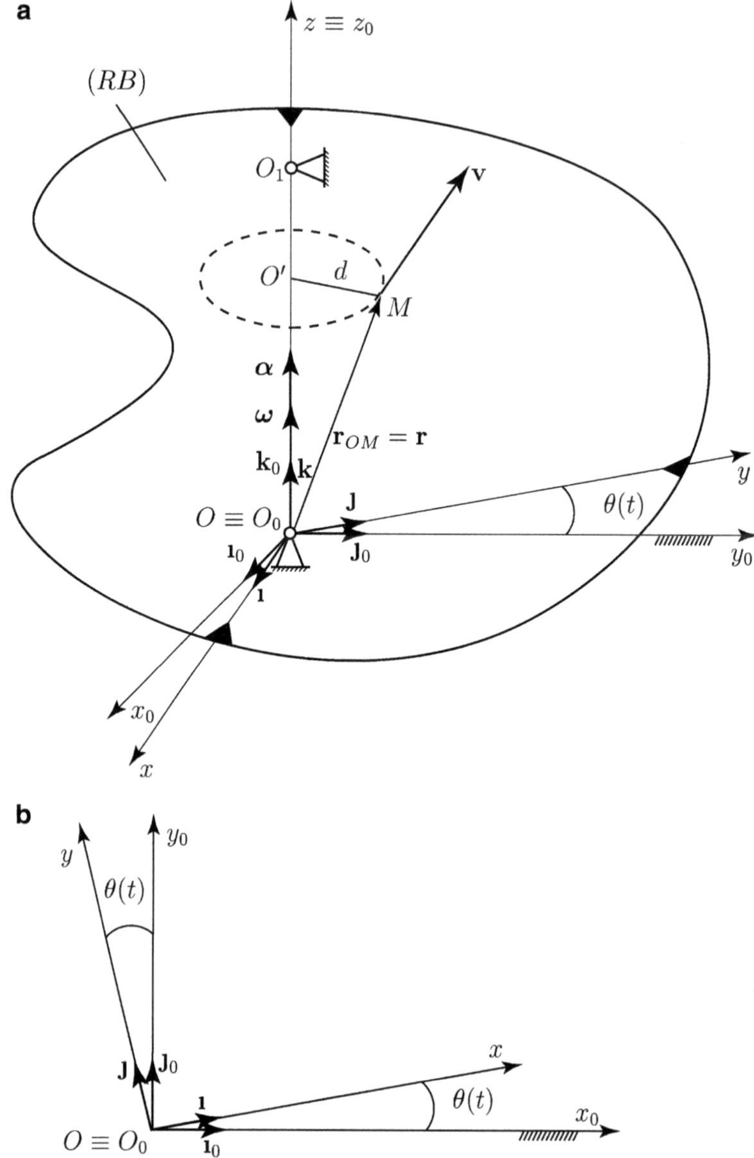

Fig. 5.5 Rigid body in rotational motion about a fixed axis

It results

$$\dot{\mathbf{k}} \cdot \mathbf{1} = \omega_y = 0 \quad \text{and} \quad \dot{\mathbf{k}} \cdot \mathbf{J} = -\omega_x = 0.$$

Because $\omega_z = \mathbf{i} \cdot \mathbf{J}$, it is necessary to compute $\mathbf{i}$. From Fig. 5.5b, the moving unit vectors $\mathbf{1}$ and $\mathbf{J}$ are

$$\mathbf{i} = \cos\theta\,\mathbf{i}_0 + \sin\theta\,\mathbf{j}_0,$$

$$\mathbf{j} = -\sin\theta\,\mathbf{i}_0 + \cos\theta\,\mathbf{j}_0.$$

It results

$$\dot{\mathbf{i}} = -\dot{\theta}\sin\theta\,\mathbf{i}_0 + \dot{\theta}\cos\theta\,\mathbf{j}_0 = \dot{\theta}\mathbf{j},$$

and

$$\omega = \omega_z = \dot{\mathbf{i}}\cdot\mathbf{j} = \dot{\theta}\mathbf{j}\cdot\mathbf{j} = \dot{\theta}.$$

Consequently

$$\boldsymbol{\omega} = \omega\mathbf{k}, \quad\text{and}\quad \boldsymbol{\alpha} = \alpha\mathbf{k}. \tag{5.29}$$

The vectors $\boldsymbol{\omega}$ and $\boldsymbol{\alpha}$ are collinear. The line of action of the vector $\boldsymbol{\omega}$ is the axis of rotation. The magnitude of the vector $\boldsymbol{\omega}$ is the angular speed, $\dot{\theta}, |\boldsymbol{\omega}| = \dot{\theta}$.

Velocity
Using (5.11), the velocity field for the rotational motion is

$$\mathbf{v} = \boldsymbol{\omega}\times\mathbf{r}. \tag{5.30}$$

Equation (5.30) shows that the velocity vector is perpendicular to the plane generated by the vectors $\boldsymbol{\omega}$ and $\mathbf{r}$ and has the magnitude

$$|\mathbf{v}| = |\boldsymbol{\omega}||\mathbf{r}|\sin\alpha = |\boldsymbol{\omega}|\,d = \omega d = \dot{\theta}d,$$

where $d = O'M$ is in the plane defined by the vectors $\boldsymbol{\omega}$ and $\mathbf{r}$. The vector $\mathbf{v}$ is perpendicular to $O'M$. A physical interpretation of the vector $\boldsymbol{\omega}$ now. The vector $\boldsymbol{\omega}$ characterizes the rotational motion of the rigid body and is called angular velocity. The line of action of the angular velocity is the fixed axis, and the sense of $\boldsymbol{\omega}$ is given by the right-hand rule: If you point the thumb of your right hand in the direction of $\boldsymbol{\omega}$, the fingers curl around $\boldsymbol{\omega}$ in the direction of rotation.

All the points of a rigid body have the same angular velocity magnitude $\omega = \dot{\theta}$. Similarly, $\boldsymbol{\alpha}$ is called angular acceleration and has the direction of the fixed axis. The magnitude of the angular acceleration is $\ddot{\theta}$.

Using (5.30), the analytical equations for the velocity is

$$\mathbf{v} = v_x\mathbf{i} + v_y\mathbf{j} + v_z\mathbf{k} = \boldsymbol{\omega}\times\mathbf{r} = \begin{vmatrix} \mathbf{i} & \mathbf{j} & \mathbf{k} \\ 0 & 0 & \omega \\ x & y & z \end{vmatrix} = -y\omega\mathbf{i} + x\omega\mathbf{j} + (0)\mathbf{k},$$

and

$$v_x = -y\omega,$$

$$v_y = x\omega,$$

$$v_z = 0.$$

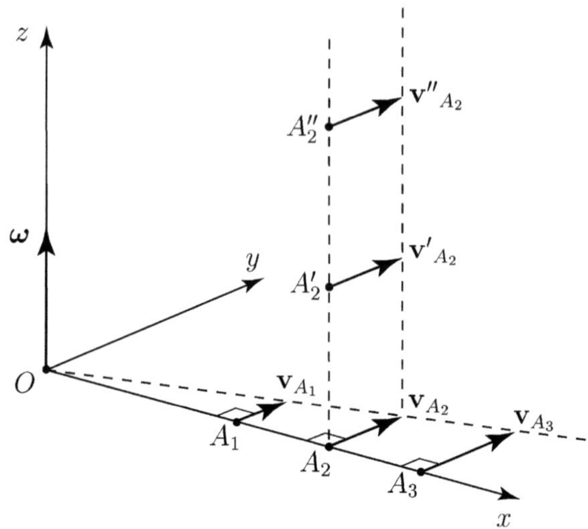

Fig. 5.6 Velocity field for rigid body in rotational motion

The properties of the velocity field are better explained if three points $A_1(x_1,0,0)$, $A_2(x_2,0,0)$, and $A_3(x_3,0,0)$ are arbitrarily placed on a perpendicular line to the rotation axis Ox (Fig. 5.6). The velocities of these points are

$$\mathbf{v}_{A1} = \boldsymbol{\omega} \times \mathbf{r}_1 = \omega\mathbf{k} \times x_1\mathbf{i} = \omega x_1\mathbf{j},$$

$$\mathbf{v}_{A2} = \boldsymbol{\omega} \times \mathbf{r}_2 = \omega\mathbf{k} \times x_2\mathbf{i} = \omega x_2\mathbf{j},$$

$$\mathbf{v}_{A3} = \boldsymbol{\omega} \times \mathbf{r}_3 = \omega\mathbf{k} \times x_3\mathbf{i} = \omega x_3\mathbf{j}. \tag{5.31}$$

From (5.31), the following remarks can be made:

(a) The velocities of the points situated on a perpendicular line to the axis of rotation are perpendicular to the axis. Their magnitude is directly proportional to the distance from the point to the axis of rotation.
(b) The points on the axis of rotation have the velocity equal to zero.
(c) The velocities are in perpendicular planes to the axis of rotation because $v_z = 0$.

Acceleration
Starting from (5.19), the acceleration field is

$$\mathbf{a} = \boldsymbol{\alpha} \times \mathbf{r} + \boldsymbol{\omega} \times (\boldsymbol{\omega} \times \mathbf{r}). \tag{5.32}$$

The angular velocity $\boldsymbol{\omega}$ and the angular acceleration $\boldsymbol{\alpha}$ are along the rotation axis, as shown in (5.29). It results that the term $\boldsymbol{\alpha} \times \mathbf{r}$ represents the tangential acceleration, and the term $\boldsymbol{\omega} \times (\boldsymbol{\omega} \times \mathbf{r})$ represents the normal acceleration. Using (5.32), the acceleration is

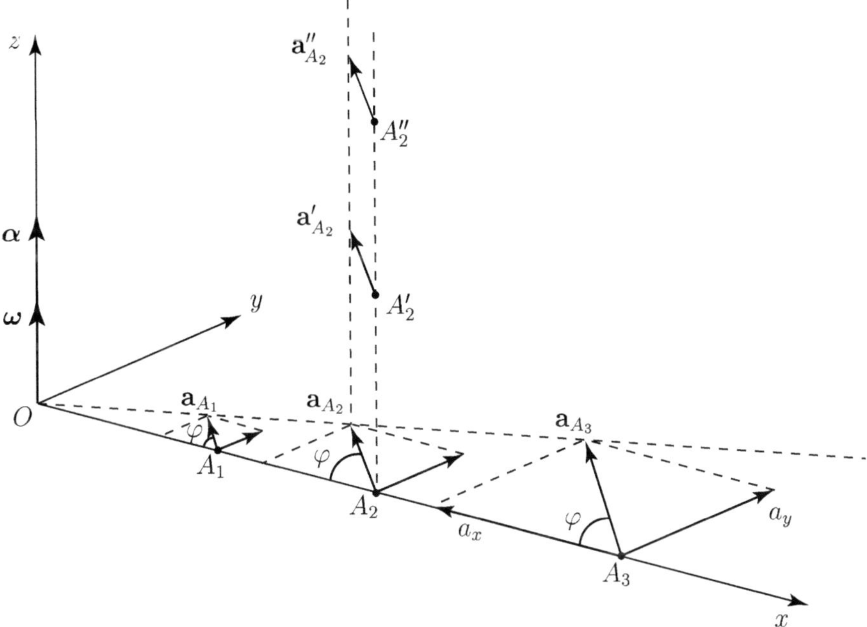

Fig. 5.7 Acceleration field for rigid body in rotational motion

$$\mathbf{a} = \begin{vmatrix} \mathbf{i} & \mathbf{j} & \mathbf{k} \\ 0 & 0 & \alpha \\ x & y & z \end{vmatrix} + \begin{vmatrix} \mathbf{i} & \mathbf{j} & \mathbf{k} \\ 0 & 0 & \omega \\ -y\omega & x\omega & 0 \end{vmatrix},$$

and

$$a_x = -y\alpha - x\omega^2,$$
$$a_y = x\alpha - y\omega^2,$$
$$a_z = 0.$$

The properties of the acceleration field are better explained if three points $A_1(x_1,0,0)$, $A_2(x_2,0,0)$, and $A_3(x_3,0,0)$ are arbitrarily placed on a perpendicular line to the rotation axis Ox (Fig. 5.7). Using (5.32), the accelerations of these points can be written as

$$\mathbf{a}_{A1} = \boldsymbol{\alpha} \times \mathbf{r}_1 + \boldsymbol{\omega} \times (\boldsymbol{\omega} \times \mathbf{r}_1) = \alpha\mathbf{k} \times x_1\mathbf{i} + \omega\mathbf{k} \times \omega x_1\mathbf{j} = \alpha x_1\mathbf{j} - \omega^2 x_1\mathbf{i},$$
$$\mathbf{a}_{A2} = \boldsymbol{\alpha} \times \mathbf{r}_2 + \boldsymbol{\omega} \times (\boldsymbol{\omega} \times \mathbf{r}_2) = \alpha\mathbf{k} \times x_2\mathbf{i} + \omega\mathbf{k} \times \omega x_2\mathbf{j} = \alpha x_2\mathbf{j} - \omega^2 x_2\mathbf{i},$$
$$\mathbf{a}_{A3} = \boldsymbol{\alpha} \times \mathbf{r}_3 + \boldsymbol{\omega} \times (\boldsymbol{\omega} \times \mathbf{r}_3) = \alpha\mathbf{k} \times x_3\mathbf{i} + \omega\mathbf{k} \times \omega x_3\mathbf{j} = \alpha x_3\mathbf{j} - \omega^2 x_3\mathbf{i}.$$

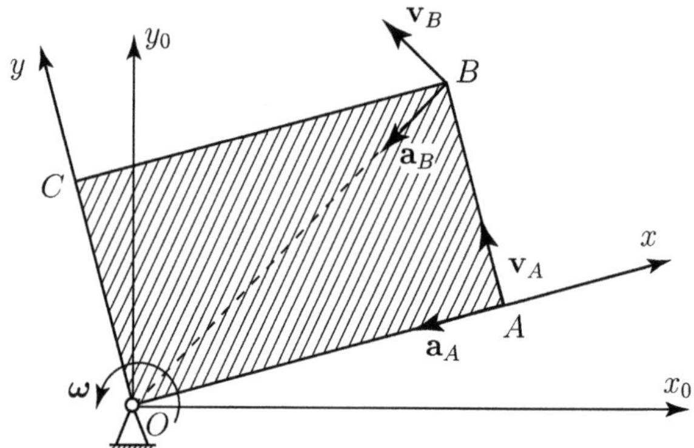

Fig. 5.8 Rectangular plate in rotational motion

Exercise

A rectangular plate $OABC$ with $|\mathbf{r}_{OA}| = OA = b$, $|\mathbf{r}_{OC}| = OC = h$ is rotating with angular velocity $\omega = $ constant around a fixed axis. This axis passes through O and is perpendicular to the plate (Fig. 5.8). Compute the velocities and acceleration for the points A and B.

Solution

Using (5.30), the velocities of A and B are

$$\mathbf{v}_A = \boldsymbol{\omega} \times \mathbf{r}_{OA} = \omega \mathbf{k} \times b\mathbf{i} = \omega b \mathbf{j}, \quad \mathbf{v}_A \perp \mathbf{r}_{OA}, \quad |\mathbf{v}_A| = \omega b,$$

and

$$\mathbf{v}_B = \boldsymbol{\omega} \times \mathbf{r}_{OB} = \begin{vmatrix} \mathbf{i} & \mathbf{j} & \mathbf{k} \\ 0 & 0 & \omega \\ -h & b & 0 \end{vmatrix} = -h\omega \mathbf{i} + b\omega \mathbf{j}, \quad \mathbf{v}_B \perp \mathbf{r}_{OB}, \quad |\mathbf{v}_B| = \omega\sqrt{b^2 + h^2}.$$

Using (5.32), the accelerations are

$$\mathbf{a}_A = \boldsymbol{\omega} \times (\boldsymbol{\omega} \times \mathbf{r}_{OA}) = \omega \mathbf{k} \times \omega b \mathbf{j} = -b\omega^2 \mathbf{i}, \quad |\mathbf{a}_A| = b\omega^2,$$

and

$$\mathbf{a}_B = \boldsymbol{\omega} \times (\boldsymbol{\omega} \times \mathbf{r}_{OB}) = \begin{vmatrix} \mathbf{i} & \mathbf{j} & \mathbf{k} \\ 0 & 0 & \omega \\ -h\omega & b\omega & 0 \end{vmatrix} = -b\omega^2 \mathbf{i} - h\omega^2 \mathbf{j}, \quad |\mathbf{a}_B| = \omega^2\sqrt{b^2 + h^2}.$$

5.3.3 Helical Motion

A rigid body performs a helical motion if two of its points remain on fixed line during the motion. This fixed line is called the helical axis of motion. From the definition, it results that the body rotates about this axis and advances

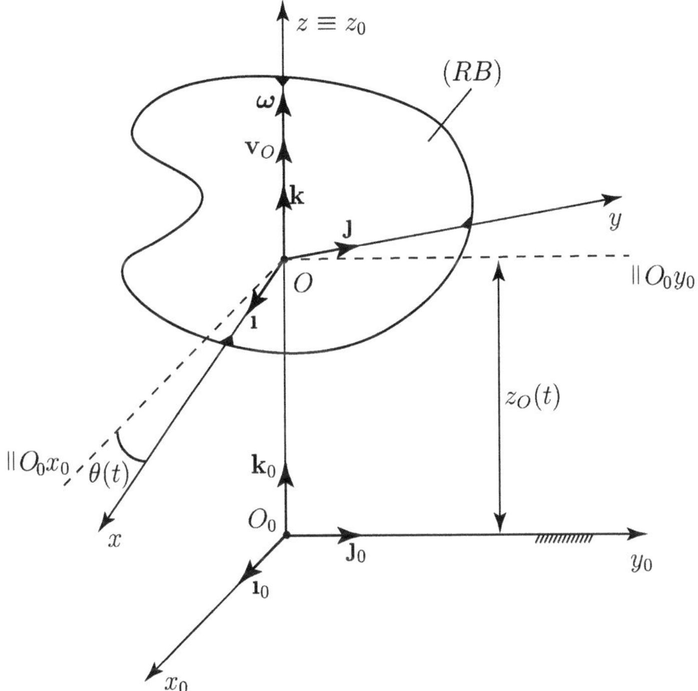

Fig. 5.9 Helical motion

simultaneously. One can consider this motion as composed of a translation and a rotation with respect to the helical axis of motion.

In order to study this motion, it is considered $O_0 z_0 \equiv Oz$ as the helical motion axis (Fig. 5.9). The origin of the mobile reference system O is moving on the z-axis with $z_O = z_O(t)$. The position of the rigid body at a certain moment is defined by the independent scalar parameters $z_O = z_O(t)$ and $\theta = \theta(t)$. Consequently, in the helical motion, the rigid has two degrees of freedom. As in the rotational motion case, it results

$$\mathbf{k}_0 = \mathbf{k},$$

$$\dot{\mathbf{k}} = \mathbf{0},$$

$$\dot{\mathbf{i}} = \dot{\theta}\,\mathbf{j}.$$

Thus,

$$\omega_x = -\mathbf{j} \cdot \dot{\mathbf{k}} = 0,$$

$$\omega_y = \dot{\mathbf{k}} \cdot \mathbf{i} = 0,$$

$$\omega_z = \dot{\mathbf{i}} \cdot \mathbf{j} = \dot{\theta}.$$

The angular velocity and acceleration are

$$\boldsymbol{\omega} = \omega\mathbf{k} = \dot{\theta}\,\mathbf{k},$$

$$\boldsymbol{\alpha} = \alpha\mathbf{k} = \ddot{\theta}\,\mathbf{k}. \tag{5.33}$$

The velocity and acceleration of the origin of the mobile reference frame O are

$$\mathbf{v}_O = \dot{z}_O\,\mathbf{k} = v_O\,\mathbf{k},$$

$$\mathbf{a}_O = \ddot{z}_O\,\mathbf{k} = a_O\,\mathbf{k}. \tag{5.34}$$

As z_O and θ are independent scalars, it results that the pairs $|\mathbf{v}_O|$ and $|\boldsymbol{\omega}|$ and $|\mathbf{a}_O|$ and $|\boldsymbol{\alpha}|$ are also independent. Thus, (5.33) and (5.34) imply that the vectors $\mathbf{v}_O$ and $\boldsymbol{\omega}$ are collinear during the motion.

Velocity
In the case of helical motion, the velocity field is

$$\mathbf{v} = \mathbf{v}_O + \boldsymbol{\omega} \times \mathbf{r} = v_O\mathbf{k} + \begin{vmatrix} \mathbf{i} & \mathbf{j} & \mathbf{k} \\ 0 & 0 & \omega \\ x & y & z \end{vmatrix} = -y\,\omega\mathbf{i} + x\,\omega\mathbf{j} + v_O\mathbf{k}. \tag{5.35}$$

The velocity components with respect to the body-fixed reference frame are

$$v_x = -y\,\omega,$$

$$v_y = x\,\omega,$$

$$v_z = v_O.$$

From the previous relations, the velocity distribution is obtained by a superposition of two velocity fields: the first one specific to a rotation about the z-axis and the second one specific to a translation along the z-axis.

The velocity of five arbitrary points $A_1(x_1,0,0)$, $A_2(x_2,0,0)$, $A_3(x_3,0,0)$, $A_2'(x_2,0,z')$, and $A_2''(x_2,0,z'')$ of the rigid body in helical motion are

$$v_{A_1} = v_O\mathbf{k} + \begin{vmatrix} \mathbf{i} & \mathbf{j} & \mathbf{k} \\ 0 & 0 & \omega \\ x_1 & 0 & 0 \end{vmatrix} = v_O\mathbf{k} + x_1\,\omega\mathbf{j},$$

$$v_{A_2} = v_O\mathbf{k} + \begin{vmatrix} \mathbf{i} & \mathbf{j} & \mathbf{k} \\ 0 & 0 & \omega \\ x_2 & 0 & 0 \end{vmatrix} = v_O\mathbf{k} + x_2\,\omega\mathbf{j},$$

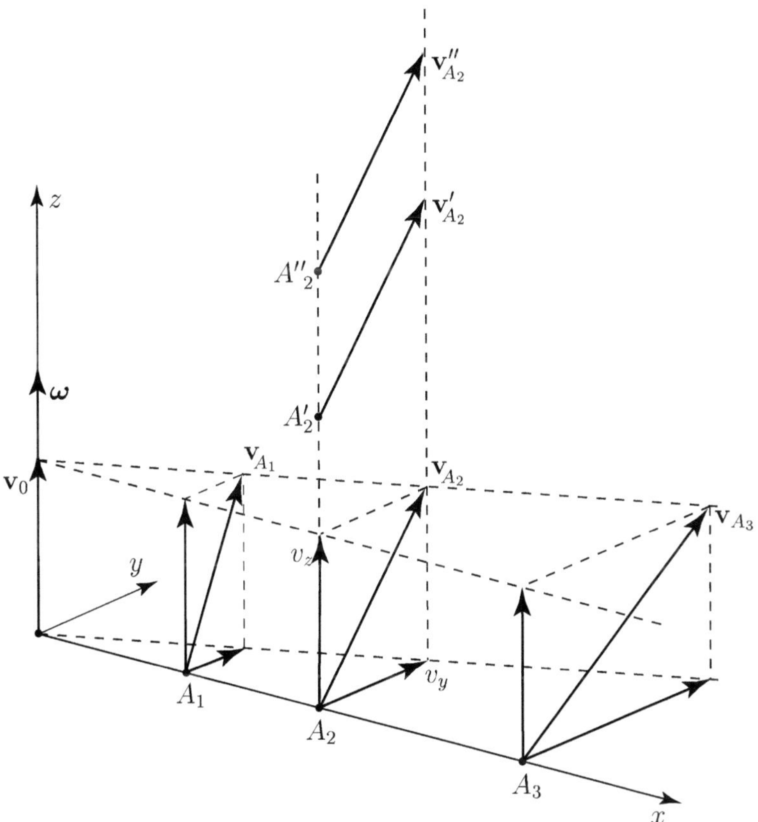

Fig. 5.10 Velocity field for helical motion

$$v_{A_3} = v_O\,\mathbf{k} + \begin{vmatrix} \mathbf{\imath} & \mathbf{J} & \mathbf{k} \\ 0 & 0 & \omega \\ x_3 & 0 & 0 \end{vmatrix} = v_O\,\mathbf{k} + x_3\,\omega\,\mathbf{J},$$

$$v_{A_2'} = v_O\,\mathbf{k} + \begin{vmatrix} \mathbf{\imath} & \mathbf{J} & \mathbf{k} \\ 0 & 0 & \omega \\ x_2 & 0 & z' \end{vmatrix} = v_O\,\mathbf{k} + x_2\,\omega\,\mathbf{J},$$

$$v_{A_2''} = v_O\,\mathbf{k} + \begin{vmatrix} \mathbf{\imath} & \mathbf{J} & \mathbf{k} \\ 0 & 0 & \omega \\ x_2 & 0 & z'' \end{vmatrix} = v_O\,\mathbf{k} + x_2\,\omega\,\mathbf{J}.$$

The velocity field is shown in Fig. 5.10. From (5.35) and Fig. 5.10, it results that there are no points with zero velocity. The points located on the helical axis of rotation (z-axis) have a minimum speed. The points located on a parallel to the helical axis ($A_2 A_2'' \parallel Oz$) have the same velocity, $v_{A_2} = v_{A_2'} = v_{A_2''}$.

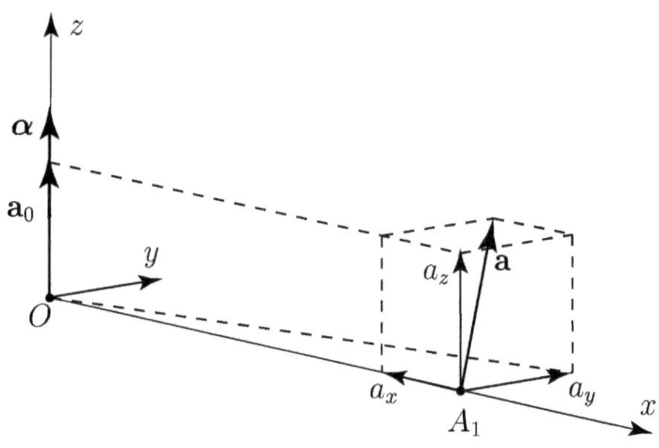

Fig. 5.11 Acceleration field for helical motion

Acceleration

Starting from the (5.19), the acceleration field is

$$\mathbf{a} = \mathbf{a}_O + \boldsymbol{\alpha} \times \mathbf{r} + \boldsymbol{\omega} \times (\boldsymbol{\omega} \times \mathbf{r}).$$

or

$$\mathbf{a} = a_O\,\mathbf{k} + \begin{vmatrix} \mathbf{i} & \mathbf{j} & \mathbf{k} \\ 0 & 0 & \alpha \\ x & y & z \end{vmatrix} + \begin{vmatrix} \mathbf{i} & \mathbf{j} & \mathbf{k} \\ 0 & 0 & \omega \\ -y\omega & x\omega & 0 \end{vmatrix}.$$

The acceleration components with respect to the mobile reference frame are

$$a_x = -y\,\alpha - x\,\omega^2,$$

$$a_y = x\,\alpha - y\,\omega^2,$$

$$a_z = a_O.$$

Similar to the velocity distribution, the acceleration distribution is obtained by the superposition of two acceleration fields: the first one specific to a rotation about the z-axis and the second one specific to a translation along the z-axis (Fig. 5.11).

5.3.4 Planar Motion

Consider a rigid body intersected by a plane fixed relative to a given reference frame as shown in Fig. 5.12. The points of the rigid body intersected by the plane remain in the plane for two-dimensional, or planar, motion. The fixed plane is the plane of the motion. Planar motion or complex motion exhibits a simultaneous combination of rotation and translation. Points on the rigid body will travel nonparallel paths, and

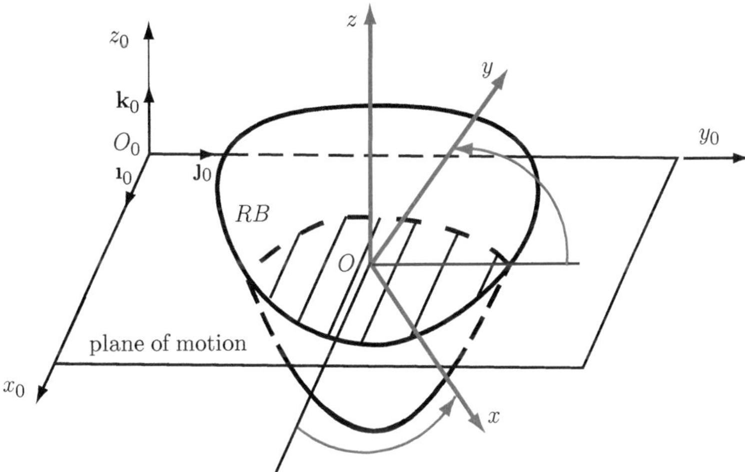

Fig. 5.12 Planar motion

there will be, at every instant, a center of rotation, which will continuously change location. The rotation of a rigid body about a fixed axis is a special case of planar motion.

The motion of a rigid body can be classified as a function of the vectors $\mathbf{v}_O$ and $\boldsymbol{\omega}$. This classification is given in the following table:

	Motion	$\mathbf{v}_O$	$\boldsymbol{\omega}$	Remarks
1	Translational motion	$\mathbf{v}_O \neq \mathbf{0}$	$\boldsymbol{\omega} = \mathbf{0}$	–
2	Rotational motion (rigid body with a fixed axis)	$\mathbf{v}_O = \mathbf{0}$	$\boldsymbol{\omega} \neq \mathbf{0}$	$\boldsymbol{\omega}$ is collinear with a fixed axis Δ
3	Spherical motion (rigid body with a fixed point)	$\mathbf{v}_O = \mathbf{0}$	$\boldsymbol{\omega} \neq \mathbf{0}$	$\boldsymbol{\omega}$ has an arbitrarily direction variable in time. The vectors $\boldsymbol{\omega}$ and $\boldsymbol{\alpha}$ have different directions
4	Helicoidal motion	$\mathbf{v}_O \neq \mathbf{0}$	$\boldsymbol{\omega} \neq \mathbf{0}$	$\mathbf{v}_O$ and $\boldsymbol{\omega}$ are collinear with a fixed axis Δ and $\mathbf{v}_O \parallel \boldsymbol{\omega}$
5	Planar motion	$\mathbf{v}_O \neq \mathbf{0}$	$\boldsymbol{\omega} \neq \mathbf{0}$	$\mathbf{v}_O \perp \boldsymbol{\omega}$ $\mathbf{v}_O$ included in a fixed plane
6	Spatial motion	$\mathbf{v}_O \neq \mathbf{0}$	$\boldsymbol{\omega} \neq \mathbf{0}$	$\mathbf{v}_O$ and $\boldsymbol{\omega}$ have arbitrary directions

5.4 Angular Velocity Vector of a Rigid Body

The angular velocity can be introduced also in another way. If the orientation of a rigid body RB in a reference frame RF_0 depends on only a single scalar variable ζ, there exists for each value of ζ a vector $\boldsymbol{\omega}$ such that the derivative with respect to ζ in RF_0 of every vector $\mathbf{c}$ fixed in rigid body RB is given by

$$\frac{d\mathbf{c}}{d\zeta} = \boldsymbol{\omega} \times \mathbf{c}, \tag{5.36}$$

where the vector $\boldsymbol{\omega}$ is the rate of change of orientation of the rigid body RB in the reference frame RF_0 with respect to ζ. The vector $\boldsymbol{\omega}$ is given by

$$\boldsymbol{\omega} = \frac{\dfrac{d\mathbf{a}}{d\zeta} \times \dfrac{d\mathbf{b}}{d\zeta}}{\dfrac{d\mathbf{a}}{d\zeta} \cdot \mathbf{b}}, \tag{5.37}$$

where $\mathbf{a}$ and $\mathbf{b}$ are any two nonparallel vectors fixed in the rigid body RB.

Proof. The vectors $\mathbf{a}$ and $\mathbf{b}$ are fixed in the rigid body. The magnitudes $\mathbf{a} \cdot \mathbf{a}$, $\mathbf{b} \cdot \mathbf{b}$ and the angle between $\mathbf{a}$ and $\mathbf{b}$ are independent of ζ:

$$\frac{d(\mathbf{a} \cdot \mathbf{a})}{d\zeta} = 0, \quad \frac{d(\mathbf{b} \cdot \mathbf{b})}{d\zeta} = 0, \quad \frac{d(\mathbf{a} \cdot \mathbf{b})}{d\zeta} = 0,$$

or

$$\frac{d\mathbf{a}}{d\zeta} \cdot \mathbf{a} = 0, \quad \frac{d\mathbf{b}}{d\zeta} \cdot \mathbf{b} = 0, \quad \frac{d\mathbf{a}}{d\zeta} \cdot \mathbf{b} + \mathbf{a} \cdot \frac{d\mathbf{b}}{d\zeta} = 0.$$

Using the vector triple product of three vectors $\mathbf{p}$, $\mathbf{q}$, and $\mathbf{r}$, the following expressions exists:

$$\mathbf{p} \times (\mathbf{q} \times \mathbf{r}) = \mathbf{p} \cdot \mathbf{r}\,\mathbf{q} - \mathbf{p} \cdot \mathbf{q}\,\mathbf{r}, \quad (\mathbf{p} \times \mathbf{q}) \times \mathbf{r} = \mathbf{r} \cdot \mathbf{p}\,\mathbf{q} - \mathbf{r} \cdot \mathbf{q}\,\mathbf{r}.$$

From these expression, it follows that

$$\frac{\dfrac{d\mathbf{a}}{d\zeta} \times \dfrac{d\mathbf{b}}{d\zeta}}{\dfrac{d\mathbf{a}}{d\zeta} \cdot \mathbf{b}} \times \mathbf{a} = \frac{\left(\dfrac{d\mathbf{a}}{d\zeta} \times \dfrac{d\mathbf{b}}{d\zeta}\right) \times \mathbf{a}}{\dfrac{d\mathbf{a}}{d\zeta} \cdot \mathbf{b}} = \frac{\mathbf{a} \cdot \dfrac{d\mathbf{a}}{d\zeta}\dfrac{d\mathbf{b}}{d\zeta} - \mathbf{a} \cdot \dfrac{d\mathbf{b}}{d\zeta}\dfrac{d\mathbf{a}}{d\zeta}}{\dfrac{d\mathbf{a}}{d\zeta} \cdot \mathbf{b}}$$

$$= \frac{-\mathbf{a} \cdot \dfrac{d\mathbf{b}}{d\zeta}\dfrac{d\mathbf{a}}{d\zeta}}{\dfrac{d\mathbf{a}}{d\zeta} \cdot \mathbf{b}} = \frac{\dfrac{d\mathbf{a}}{d\zeta} \cdot \mathbf{b}\dfrac{d\mathbf{a}}{d\zeta}}{\dfrac{d\mathbf{a}}{d\zeta} \cdot \mathbf{b}} = \frac{d\mathbf{a}}{d\zeta}, \tag{5.38}$$

and

$$\frac{\dfrac{d\mathbf{a}}{d\zeta}\times\dfrac{d\mathbf{b}}{d\zeta}}{\dfrac{d\mathbf{a}}{d\zeta}\cdot\mathbf{b}}\times\mathbf{b}=\frac{\left(\dfrac{d\mathbf{a}}{d\zeta}\times\dfrac{d\mathbf{b}}{d\zeta}\right)\times\mathbf{b}}{\dfrac{d\mathbf{a}}{d\zeta}\cdot\mathbf{b}}=\frac{\mathbf{b}\cdot\dfrac{d\mathbf{a}}{d\zeta}\dfrac{d\mathbf{b}}{d\zeta}-\mathbf{b}\cdot\dfrac{d\mathbf{b}}{d\zeta}\dfrac{d\mathbf{a}}{d\zeta}}{\dfrac{d\mathbf{a}}{d\zeta}\cdot\mathbf{b}}$$

$$=\frac{\mathbf{b}\cdot\dfrac{d\mathbf{a}}{d\zeta}\dfrac{d\mathbf{b}}{d\zeta}}{\dfrac{d\mathbf{a}}{d\zeta}\cdot\mathbf{b}}=\frac{d\mathbf{b}}{d\zeta}. \tag{5.39}$$

The following vector is defined:

$$\boldsymbol{\omega}=\frac{\dfrac{d\mathbf{a}}{d\zeta}\times\dfrac{d\mathbf{b}}{d\zeta}}{\dfrac{d\mathbf{a}}{d\zeta}\cdot\mathbf{b}},$$

and the (5.38) and (5.39) can be written as

$$\frac{d\mathbf{a}}{d\zeta}=\boldsymbol{\omega}\times\mathbf{a},\quad\frac{d\mathbf{b}}{d\zeta}=\boldsymbol{\omega}\times\mathbf{b}.$$

In general, a given vector $\mathbf{v}$ can be expressed as

$$\mathbf{v}=v_1\mathbf{n}_1+v_2\mathbf{n}_2+v_3\mathbf{n}_3,$$

where $\mathbf{n}_1,\mathbf{n}_2,\mathbf{n}_3$ are three units vectors not parallel to the same plane, and v_1,v_2,v_3 are three scalars.

Any vector $\mathbf{c}$ fixed in the rigid body RB can be expressed as

$$\mathbf{c}=c_1\mathbf{a}+c_2\mathbf{b}+c_3\mathbf{a}\times\mathbf{b}, \tag{5.40}$$

where c_1, c_2, and c_3 are constant and independent of ζ. Differentiating (5.40) with respect to ζ one can obtain

$$\frac{d\mathbf{c}}{d\zeta}=c_1\frac{d\mathbf{a}}{d\zeta}+c_2\frac{d\mathbf{b}}{d\zeta}+c_3\frac{d\mathbf{a}}{d\zeta}\times\mathbf{b}+c_3\mathbf{a}\times\frac{d\mathbf{b}}{d\zeta}$$

$$=c_1\boldsymbol{\omega}\times\mathbf{a}+c_2\boldsymbol{\omega}\times\mathbf{b}+c_3\left[(\boldsymbol{\omega}\times\mathbf{a})\times\mathbf{b}+\mathbf{a}\times(\boldsymbol{\omega}\times\mathbf{b})\right]$$

$$=c_1\boldsymbol{\omega}\times\mathbf{a}+c_2\boldsymbol{\omega}\times\mathbf{b}+c_3\left[\mathbf{b}\cdot\boldsymbol{\omega}\,\mathbf{a}-\mathbf{b}\cdot\mathbf{a}\,\boldsymbol{\omega}+\mathbf{a}\cdot\mathbf{b}\,\boldsymbol{\omega}-\mathbf{a}\cdot\boldsymbol{\omega}\,\mathbf{b}\right]$$

$$=c_1\boldsymbol{\omega}\times\mathbf{a}+c_2\boldsymbol{\omega}\times\mathbf{b}+c_3\left[\boldsymbol{\omega}\cdot\mathbf{b}\,\mathbf{a}-\mathbf{a}\cdot\mathbf{b}\,\boldsymbol{\omega}+\mathbf{a}\cdot\mathbf{b}\,\boldsymbol{\omega}-\mathbf{a}\cdot\boldsymbol{\omega}\,\mathbf{b}\right]$$

$$=c_1\boldsymbol{\omega}\times\mathbf{a}+c_2\boldsymbol{\omega}\times\mathbf{b}+c_3\left[\boldsymbol{\omega}\cdot\mathbf{b}\,\mathbf{a}-\boldsymbol{\omega}\cdot\mathbf{a}\,\mathbf{b}\right]$$

$$= c_1 \boldsymbol{\omega} \times \mathbf{a} + c_2 \boldsymbol{\omega} \times \mathbf{b} + c_3 \boldsymbol{\omega} \times (\mathbf{a} \times \mathbf{b})$$

$$= \boldsymbol{\omega} \times (c_1 \mathbf{a} + c_2 \mathbf{b} + c_3 \mathbf{a} \times \mathbf{b})$$

$$= \boldsymbol{\omega} \times \mathbf{c}. \tag{5.41}$$

The vector $\boldsymbol{\omega}$ is a free vector, that is, it is not associated with any particular point. With the help of $\boldsymbol{\omega}$, one can replace the process of differentiation with that of cross multiplication.

The vector $\boldsymbol{\omega}$ may be expressed in a symmetrical relation in $\mathbf{a}$ and $\mathbf{b}$:

$$\boldsymbol{\omega} = \frac{1}{2} \left(\frac{\dfrac{d\mathbf{a}}{d\zeta} \times \dfrac{d\mathbf{b}}{d\zeta}}{\dfrac{d\mathbf{a}}{d\zeta} \cdot \mathbf{b}} + \frac{\dfrac{d\mathbf{b}}{d\zeta} \times \dfrac{d\mathbf{a}}{d\zeta}}{\dfrac{d\mathbf{b}}{d\zeta} \cdot \mathbf{a}} \right). \tag{5.42}$$

The first derivatives of a vector $\mathbf{p}$ with respect to a scalar variable ζ in two reference frames RF_i and RF_j are related as follows:

$$\frac{^{(j)}d\mathbf{p}}{d\zeta} = \frac{^{(i)}d\mathbf{p}}{d\zeta} + \boldsymbol{\omega}_{ij} \times \mathbf{p}, \tag{5.43}$$

where $\boldsymbol{\omega}_{ij}$ is the rate of change of orientation of RF_i in RF_j with respect to ζ and $\frac{^{(j)}d\mathbf{p}}{d\zeta}$ is the total derivative of $\mathbf{p}$ with respect to ζ in RF_j.

Proof. The vector $\mathbf{p}$ can be expressed as

$$\mathbf{p} = p_1 \mathbf{l}_1 + p_2 \mathbf{l}_2 + p_3 \mathbf{l}_3,$$

where $\mathbf{l}_1, \mathbf{l}_2, \mathbf{l}_3$ are three units vectors not parallel to the same plane fixed in RF_i, and p_x, p_y, p_z are the scalar measure numbers of $\mathbf{p}$. Differentiating in RF_j,

$$\frac{^{(j)}d\mathbf{p}}{d\zeta} = \frac{^{(j)}d}{d\zeta} (p_1 \mathbf{l}_1 + p_2 \mathbf{l}_2 + p_3 \mathbf{l}_3)$$

$$= \frac{^{(j)}d p_2}{d\zeta} \mathbf{l}_1 + \frac{^{(j)}d p_2}{d\zeta} \mathbf{l}_2 + \frac{^{(j)}d p_3}{d\zeta} \mathbf{l}_3 + p_1 \frac{^{(j)}d \mathbf{l}_1}{d\zeta} + p_2 \frac{^{(j)}d \mathbf{l}_2}{d\zeta} + p_3 \frac{^{(j)}d \mathbf{l}_3}{d\zeta}$$

$$= \frac{d p_2}{d\zeta} \mathbf{l}_1 + \frac{d p_2}{d\zeta} \mathbf{l}_2 + \frac{d p_3}{d\zeta} \mathbf{l}_3 + p_1 \boldsymbol{\omega}_{ij} \times \mathbf{l}_1 + p_2 \boldsymbol{\omega}_{ij} \times \mathbf{l}_2 + p_3 \boldsymbol{\omega}_{ij} \times \mathbf{l}_3$$

$$= \frac{^{(i)}d p_2}{d\zeta} \mathbf{l}_1 + \frac{^{(i)}d p_2}{d\zeta} \mathbf{l}_2 + \frac{^{(i)}d p_3}{d\zeta} \mathbf{l}_3 + \boldsymbol{\omega}_{ij} \times (p_1 \mathbf{l}_1 + p_2 \mathbf{l}_2 + p_3 \mathbf{l}_3)$$

$$= \frac{^{(i)}d\mathbf{p}}{d\zeta} + \boldsymbol{\omega}_{ij} \times \mathbf{p}. \tag{5.44}$$

The *angular velocity* of a rigid body RB in a reference frame RF_0 is the rate of change of orientation with respect to the time t:

$$\omega = \frac{1}{2}\left(\frac{\dfrac{d\mathbf{a}}{dt}\times\dfrac{d\mathbf{b}}{dt}}{\dfrac{d\mathbf{a}}{dt}\cdot\mathbf{b}} + \frac{\dfrac{d\mathbf{b}}{dt}\times\dfrac{d\mathbf{a}}{dt}}{\dfrac{d\mathbf{b}}{dt}\cdot\mathbf{a}}\right) = \frac{1}{2}\left(\frac{\dot{\mathbf{a}}\times\dot{\mathbf{b}}}{\dot{\mathbf{a}}\cdot\mathbf{b}} + \frac{\dot{\mathbf{b}}\times\dot{\mathbf{a}}}{\dot{\mathbf{b}}\cdot\mathbf{a}}\right).$$

The direction of ω is related to the direction of the rotation of the rigid body through a right-hand rule.

Let RF_i, $i = 1, 2, \ldots, n$ be n reference frames. The angular velocity of a rigid body r in the reference frame RF_n can be expressed as

$$\omega_{rn} = \omega_{r1} + \omega_{12} + \omega_{23} + \cdots + \omega_{r,n-1}.$$

Proof. Let $\mathbf{p}$ be any vector fixed in the rigid body. Then,

$$\frac{{}^{(i)}d\mathbf{p}}{dt} = \omega_{ri}\times\mathbf{p}$$

$$\frac{{}^{(i-1)}d\mathbf{p}}{dt} = \omega_{r,i-1}\times\mathbf{p}.$$

On the other hand,

$$\frac{{}^{(i)}d\mathbf{p}}{dt} = \frac{{}^{(i-1)}d\mathbf{p}}{dt} + \omega_{i,i-1}\times\mathbf{p}.$$

Hence,

$$\omega_{ri}\times\mathbf{p} = \omega_{r,i-1}\times\mathbf{p} + \omega_{i,i-1}\times\mathbf{p},$$

as this equation is satisfied for all $\mathbf{p}$ fixed in the rigid body:

$$\omega_{ri} = \omega_{r,i-1} + \omega_{i,i-1}. \tag{5.45}$$

With $i = n$, (5.45) gives

$$\omega_{rn} = \omega_{r,n-1} + \omega_{n,n-1}. \tag{5.46}$$

With $i = n - 1$, (5.45) gives

$$\omega_{r,n-1} = \omega_{r,n-2} + \omega_{n-1,n-2}. \tag{5.47}$$

Substitute (5.47) into (5.46):

$$\omega_{rn} = \omega_{r,n-2} + \omega_{n-1,n-2} + \omega_{n,n-1}.$$

Next, use (5.45) with $i = n - 2$, then with $i = n - 3$, and so forth.

5.5   Motion of a Point that Moves Relative to a Rigid Body

A reference frame that moves with the rigid body is a body-fixed reference frame. Figure 5.13 shows a rigid body (RB) in motion relative to a primary reference frame with its origin at point O_0, $x_0y_0z_0$. The primary reference frame is a fixed reference frame or an earth-fixed reference frame. The unit vectors $\mathbf{i}_0, \mathbf{J}_0$, and $\mathbf{k}_0$ of the primary reference frame are constant. The body-fixed reference frame, xyz, has its origin at a point O of the rigid body ($O \in (RB)$) and is a moving reference frame relative to the primary reference. The unit vectors $\mathbf{i}, \mathbf{J}$, and $\mathbf{k}$ of the body-fixed reference frame are not constant because they rotate with the body-fixed reference frame.

The position vector of a point P of the rigid body ($P \in (RB)$) relative to the origin, O, of the body-fixed reference frame is the vector $\mathbf{r}_{OP}$. The velocity of P relative to O is

$$\frac{\mathrm{d}\mathbf{r}_{OP}}{\mathrm{d}t} = \mathbf{v}_{PO}^{\mathrm{rel}} = \boldsymbol{\omega} \times \mathbf{r}_{OP},$$

where $\boldsymbol{\omega}$ is the angular velocity vector of the rigid body.

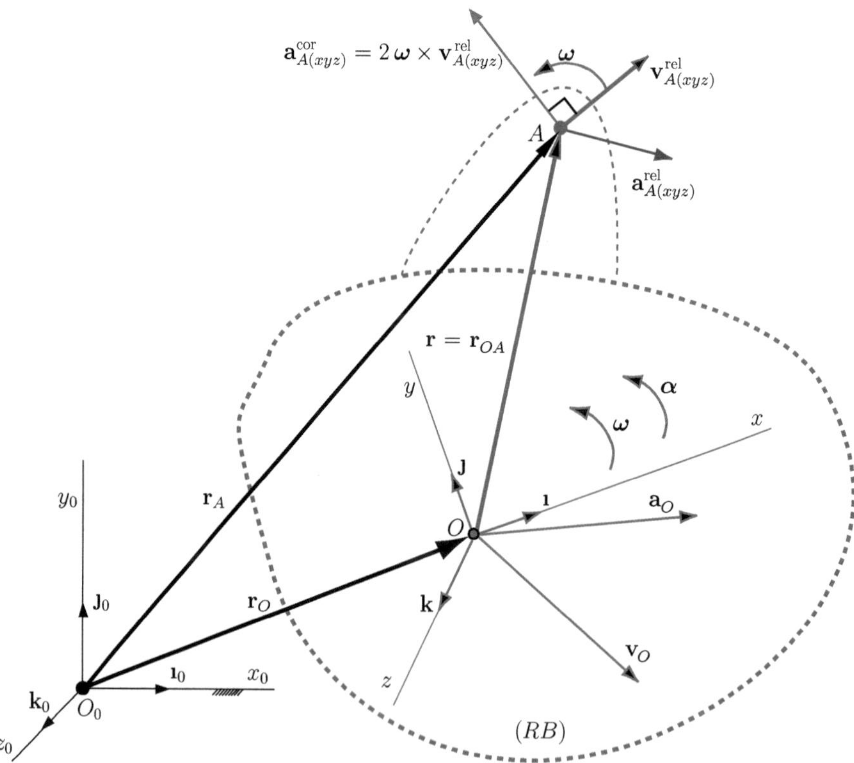

Fig. 5.13 Rigid body in motion; the point A is not assumed to be a point of the rigid body, $A \notin (RB)$

The position vector of a point A (the point A is not assumed to be a point of the rigid body, $A \notin (RB)$), relative to the origin O_0 of the primary reference frame is, Fig. 5.13,

$$\mathbf{r}_A = \mathbf{r}_O + \mathbf{r},$$

where

$$\mathbf{r} = \mathbf{r}_{OA} = x\mathbf{\imath} + y\mathbf{\jmath} + z\mathbf{k}$$

is the position vector of A relative to the origin O, of the body-fixed reference frame, and x, y, and z are the coordinates of A in terms of the body-fixed reference frame. The velocity of the point A is the time derivative of the position vector $\mathbf{r}_A$:

$$\mathbf{v}_A = \frac{\mathrm{d}\mathbf{r}_O}{\mathrm{d}t} + \frac{\mathrm{d}\mathbf{r}}{\mathrm{d}t} = \mathbf{v}_O + \mathbf{v}_{AO}^{\mathrm{rel}}$$

$$= \mathbf{v}_O + \frac{\mathrm{d}x}{\mathrm{d}t}\mathbf{\imath} + x\frac{\mathrm{d}\mathbf{\imath}}{\mathrm{d}t} + \frac{\mathrm{d}y}{\mathrm{d}t}\mathbf{\jmath} + y\frac{\mathrm{d}\mathbf{\jmath}}{\mathrm{d}t} + \frac{\mathrm{d}z}{\mathrm{d}t}\mathbf{k} + z\frac{\mathrm{d}\mathbf{k}}{\mathrm{d}t}.$$

Using Poisson formulas, the total derivative of the position vector $\mathbf{r}$ is

$$\frac{\mathrm{d}\mathbf{r}}{\mathrm{d}t} = \dot{\mathbf{r}} = \dot{x}\mathbf{\imath} + \dot{y}\mathbf{\jmath} + \dot{z}\mathbf{k} + \boldsymbol{\omega} \times \mathbf{r}.$$

The velocity of A relative to the body-fixed reference frame is a derivative in the body-fixed reference frame:

$$\mathbf{v}_{A(xyz)}^{\mathrm{rel}} = \frac{^{(xyz)}\mathrm{d}\mathbf{r}}{\mathrm{d}t} = \frac{\mathrm{d}x}{\mathrm{d}t}\mathbf{\imath} + \frac{\mathrm{d}y}{\mathrm{d}t}\mathbf{\jmath} + \frac{\mathrm{d}z}{\mathrm{d}t}\mathbf{k} = \dot{x}\mathbf{\imath} + \dot{y}\mathbf{\jmath} + \dot{z}\mathbf{k}. \tag{5.48}$$

A general formula for the total derivative of a moving vector $\mathbf{r}$ may be written as

$$\frac{\mathrm{d}\mathbf{r}}{\mathrm{d}t} = \frac{^{(xyz)}\mathrm{d}\mathbf{r}}{\mathrm{d}t} + \boldsymbol{\omega} \times \mathbf{r}, \tag{5.49}$$

where $\frac{\mathrm{d}\mathbf{r}}{\mathrm{d}t} = \frac{^{(0)}\mathrm{d}\mathbf{r}}{\mathrm{d}t}$ is the derivative in the fixed (primary) reference frame (0) $(x_0 y_0 z_0)$ and $\frac{^{(xyz)}\mathrm{d}\mathbf{r}}{\mathrm{d}t}$ is the derivative in the rotating (mobile or body-fixed) reference frame (xyz).

The velocity of the point A relative to the primary reference frame is

$$\mathbf{v}_A = \mathbf{v}_O + \mathbf{v}_{A(xyz)}^{\mathrm{rel}} + \boldsymbol{\omega} \times \mathbf{r}. \tag{5.50}$$

Equation (5.50) expresses the velocity of a point A as the sum of three terms:

- The velocity of a point O of the rigid body
- The velocity $\mathbf{v}^{\text{rel}}_{A(xyz)}$ of A relative to the rigid body
- The velocity $\boldsymbol{\omega} \times \mathbf{r}$ of A relative to O due to the rotation of the rigid body

The acceleration of the point A relative to the primary reference frame is obtained by taking the time derivative of (5.50):

$$\mathbf{a}_A = \mathbf{a}_O + \mathbf{a}_{AO}$$

$$= \mathbf{a}_O + \mathbf{a}^{\text{rel}}_{A(xyz)} + 2\boldsymbol{\omega} \times \mathbf{v}^{\text{rel}}_{A(xyz)} + \boldsymbol{\alpha} \times \mathbf{r} + \boldsymbol{\omega} \times (\boldsymbol{\omega} \times \mathbf{r}), \qquad (5.51)$$

where

$$\mathbf{a}^{\text{rel}}_{A(xyz)} = \frac{^{(xyz)}\mathrm{d}^2\,\mathbf{r}}{\mathrm{d}t^2} = \frac{\mathrm{d}^2 x}{\mathrm{d}t^2}\mathbf{i} + \frac{\mathrm{d}^2 y}{\mathrm{d}t^2}\mathbf{j} + \frac{\mathrm{d}^2 z}{\mathrm{d}t^2}\mathbf{k} \qquad (5.52)$$

is the acceleration of A relative to the body-fixed reference frame or relative to the rigid body. The term

$$\mathbf{a}^{\text{cor}}_{A(xyz)} = 2\boldsymbol{\omega} \times \mathbf{v}^{\text{rel}}_{A(xyz)}$$

is called the Coriolis acceleration. The direction of the Coriolis acceleration is obtained by rotating the linear relative velocity $\mathbf{v}^{\text{rel}}_{A(xyz)}$ through $90°$ in the direction of rotation given by $\boldsymbol{\omega}$.

In the case of planar motion, (5.51) becomes

$$\mathbf{a}_A = \mathbf{a}_O + \mathbf{a}_{OA}$$

$$= \mathbf{a}_O + \mathbf{a}^{\text{rel}}_{A(xyz)} + 2\boldsymbol{\omega} \times \mathbf{v}^{\text{rel}}_{A(xyz)} + \boldsymbol{\alpha} \times \mathbf{r} - \omega^2 \mathbf{r}. \qquad (5.53)$$

The motion of the rigid body (RB) is described relative to the primary reference frame. The velocity $\mathbf{v}_A$ and the acceleration $\mathbf{a}_A$ of a point A are relative to the primary reference frame. The terms $\mathbf{v}^{\text{rel}}_{A(xyz)}$ and $\mathbf{a}^{\text{rel}}_{A(xyz)}$ are the velocity and acceleration of point A relative to the body-fixed reference frame, that is, they are the velocity and acceleration measured by an observer moving with the rigid body, Fig. 5.13. If A is a point of the rigid body, $A \in (RB)$, $\mathbf{v}^{\text{rel}}_{A(xyz)} = \mathbf{0}$ and $\mathbf{a}^{\text{rel}}_{A(xyz)} = \mathbf{0}$.

Motion of a Point Relative to a Moving Reference Frame
The velocity and acceleration of an arbitrary point A relative to a point O of a rigid body, in terms of the body-fixed reference frame, are given by (5.50) and (5.51):

$$\mathbf{v}_A = \mathbf{v}_O + \mathbf{v}^{\text{rel}}_{AO} + \boldsymbol{\omega} \times \mathbf{r}_{OA}, \qquad (5.54)$$

$$\mathbf{a}_A = \mathbf{a}_O + \mathbf{a}^{\text{rel}}_{AO} + 2\boldsymbol{\omega} \times \mathbf{v}^{\text{rel}}_{AO} + \boldsymbol{\alpha} \times \mathbf{r}_{OA} + \boldsymbol{\omega} \times (\boldsymbol{\omega} \times \mathbf{r}_{OA}). \qquad (5.55)$$

These results apply to any reference frame having a moving origin O and rotating with angular velocity $\boldsymbol{\omega}$ and angular acceleration $\boldsymbol{\alpha}$ relative to a primary reference

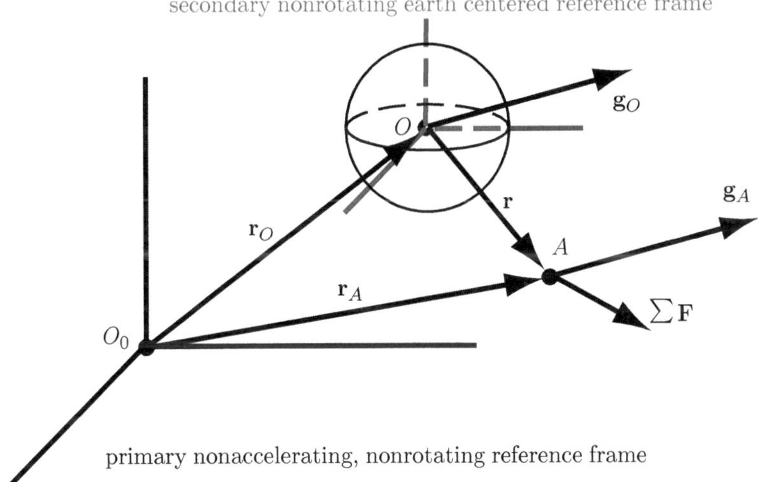

Fig. 5.14 Secondary earth-centered reference frame with the origin at O

frame (Fig. 5.13). The terms $\mathbf{v}_A$ and $\mathbf{a}_A$ are the velocity and acceleration of an arbitrary point A relative to the primary reference frame. The terms $\mathbf{v}_{AO}^{\text{rel}}$ and $\mathbf{a}_{AO}^{\text{rel}}$ are the velocity and acceleration of A relative to the secondary moving reference frame, that is, they are the velocity and acceleration measured by an observer moving with the secondary reference frame. The Coriolis acceleration is $\mathbf{a}_{AO}^{\text{cor}} = 2\boldsymbol{\omega} \times \mathbf{v}_{AO}^{\text{rel}}$.

Inertial Reference Frames
A reference frame is inertial if Newton's second law ca be applied in the form $\sum \mathbf{F} = m\mathbf{a}$.

Figure 5.14 shows a nonaccelerating, nonrotating reference frame with the origin at O_0, and a secondary nonrotating, earth-centered reference frame with the origin at O. The nonaccelerating, nonrotating reference frame with the origin at O_0 is assumed to be an inertial reference. The acceleration of the earth, due to the gravitational attractions of the sun and moon, is $\mathbf{g}_O$. The earth-centered reference frame has also the acceleration $\mathbf{g}_O$. Newton's second law for an object A of mass m, using the hypothetical nonaccelerating, nonrotating reference frame with the origin at O_0 is written as

$$m\mathbf{a}_A = \mathbf{G}_A + \sum \mathbf{F}, \tag{5.56}$$

where $\mathbf{a}_A$ is the acceleration of A relative to O_0, $\mathbf{G}_A = m\mathbf{g}_A$, $\mathbf{g}_A$ is the resulting gravitational acceleration (attraction of the sun, moon), and $\sum \mathbf{F}$ is the sum of other external forces acting on A. The acceleration of A relative to O_0 is

$$\mathbf{a}_A = \mathbf{a}_O + \mathbf{a}_A^{\text{rel}},$$

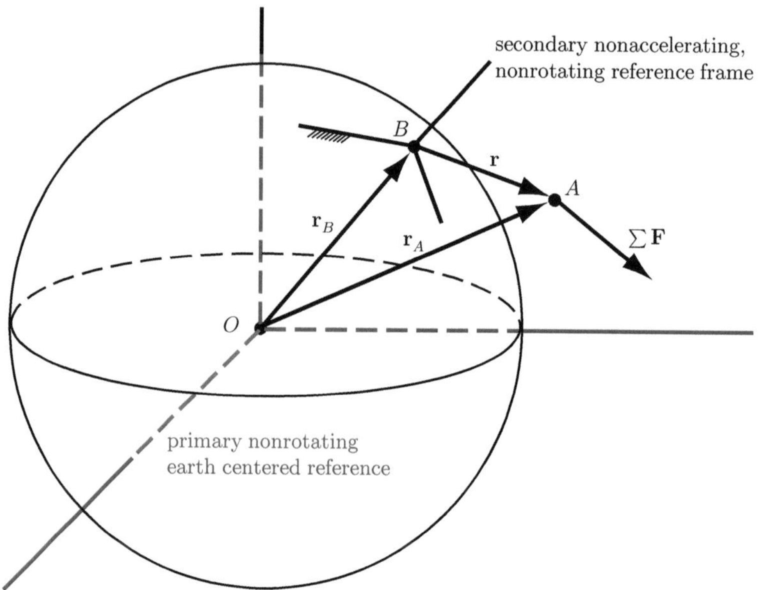

Fig. 5.15 Secondary earth-fixed reference frame with the origin at B

where $\mathbf{a}_A^{\mathrm{rel}}$ is the acceleration of A relative to the nonrotationg earth-centered reference frame. The acceleration of the origin O is equal to the gravitational acceleration of the earth $\mathbf{a}_O = \mathbf{g}_O$. The earth-centered reference frame does not rotate ($\boldsymbol{\omega} = \mathbf{0}$). If the object A is near the earth or on the earth, then $\mathbf{g}_A \approx \mathbf{g}_O$, and (5.56) becomes

$$\sum \mathbf{F} = m\,\mathbf{a}_A^{\mathrm{rel}}. \tag{5.57}$$

It results that Newton's second law can be applied using a nonrotating, earth-centered reference frame for an object near to the earth.

In most applications, Newton's second law may be applied using an earth-fixed reference frame. Figure 5.15 shows a nonrotating reference frame with its origin at the center of the earth O and a secondary earth-fixed reference frame with its origin at a point B. The earth-fixed reference frame with the origin at B may be assumed to be an inertial reference and

$$\sum \mathbf{F} = m\,\mathbf{a}_A^{\mathrm{rel}}, \tag{5.58}$$

where $\mathbf{a}_A^{\mathrm{rel}}$ is the acceleration of A relative to the earth-fixed reference frame. The motion of an object A may be analyzed using a primary inertial reference frame with its origin at the point O, Fig. 5.16. A secondary reference frame with its origin at B

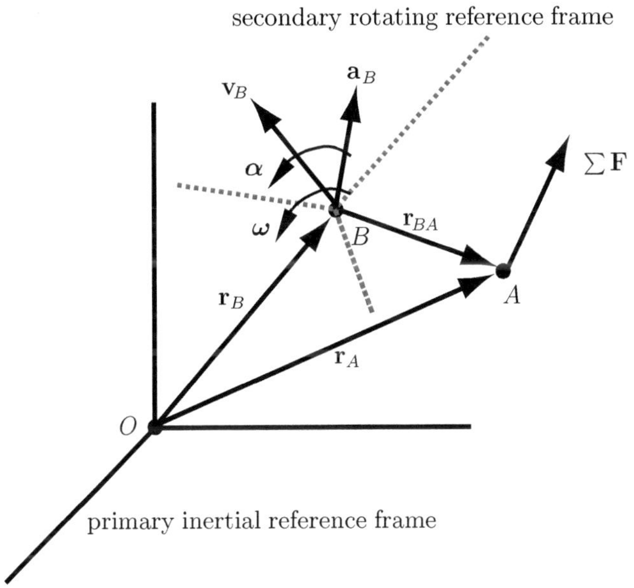

Fig. 5.16 Primary inertial reference frame and secondary reference frame

undergoes an arbitrary motion with angular velocity ω and angular acceleration α. The Newton's second law for the object A of mass m is

$$\sum\mathbf{F} = m\,\mathbf{a}_A, \tag{5.59}$$

where $\mathbf{a}_A$ is the acceleration of A acceleration relative to O. Equation (5.59) may be written in the form

$$\sum\mathbf{F} = m\left[\mathbf{a}_B + \mathbf{a}_A^{\mathrm{rel}} + 2\,\omega \times \mathbf{v}_A^{\mathrm{rel}} + \alpha \times \mathbf{r}_{BA} + \omega \times (\omega \times \mathbf{r}_{BA})\right], \tag{5.60}$$

where $\mathbf{a}_A^{\mathrm{rel}}$ is the acceleration of A relative to the secondary reference frame. The term $\mathbf{a}_B$ is the acceleration of the origin B of the secondary reference frame relative to the primary inertial reference. The term $2\,\omega \times \mathbf{v}_A^{\mathrm{rel}}$ is the Coriolis acceleration, and the term $-2\,m\,\omega \times \mathbf{v}_A^{\mathrm{rel}}$ is the Coriolis force. This is Newton's second law expressed in terms of a secondary reference frame undergoing an arbitrary motion relative to an inertial primary reference frame.

5.6 Planar Instantaneous Center

The *instantaneous center* of a rigid body is a point whose velocity is zero at the instant under consideration. Every point of the rigid body rotates about the instantaneous center at the instant under consideration.

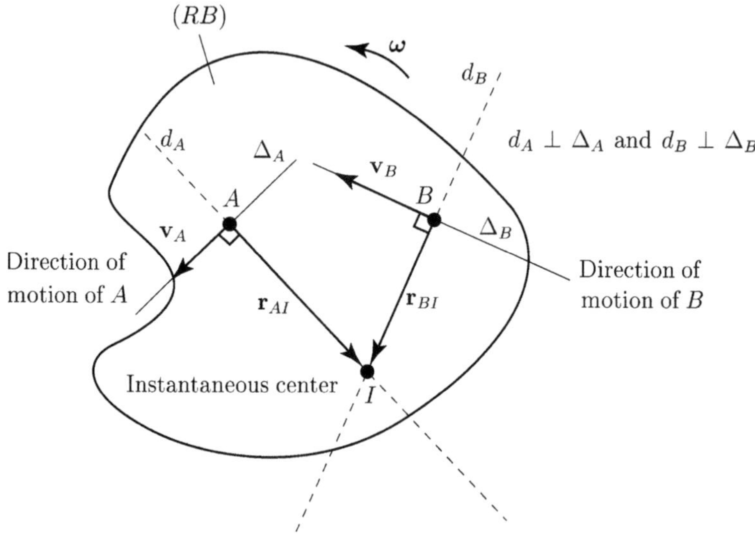

Fig. 5.17 Instantaneous center

The instantaneous center may be or may not be a point of the rigid body. When the instantaneous center is not a point of the rigid body, the rigid body is rotating about an external point at that instant. Figure 5.17 shows two points A and B of a rigid body and their directions of the motion Δ_A and Δ_B:

$$\mathbf{v}_A \| \Delta_A \text{ and } \mathbf{v}_B \| \Delta_B,$$

where $\mathbf{v}_A$ is the velocity of point A and $\mathbf{v}_B$ is the velocity of point B. Through the points A and B, perpendicular lines are drawn to their directions of motion:

$$d_A \perp \Delta_A \text{ and } d_B \perp \Delta_B.$$

The perpendicular lines intersect at the point I:

$$d_A \cap d_B = I.$$

The velocity of point I in terms of the velocity of point A is

$$\mathbf{v}_I = \mathbf{v}_A + \boldsymbol{\omega} \times \mathbf{r}_{AI},$$

where $\boldsymbol{\omega}$ is the angular velocity vector of the rigid body. Since the vector $\boldsymbol{\omega} \times \mathbf{r}_{AI}$ is perpendicular to $\mathbf{r}_{AI}$,

$$(\boldsymbol{\omega} \times \mathbf{r}_{AI}) \perp \mathbf{r}_{AI},$$

this equation states that the direction of motion of I is parallel to the direction of motion of A:

$$\mathbf{v}_I \| \mathbf{v}_A. \tag{5.61}$$

The velocity of point I in terms of the velocity of point B is

$$\mathbf{v}_I = \mathbf{v}_B + \boldsymbol{\omega} \times \mathbf{r}_{BI}.$$

The vector $\boldsymbol{\omega} \times \mathbf{r}_{BI}$ is perpendicular to $\mathbf{r}_{BI}$,

$$(\boldsymbol{\omega} \times \mathbf{r}_{BI}) \perp \mathbf{r}_{BI},$$

so this equation states that the direction of motion of I is parallel to the direction of motion of B:

$$\mathbf{v}_I \| \mathbf{v}_B. \tag{5.62}$$

But C cannot be moving parallel to A and parallel to B, so (5.61) and (5.62) are contradictory unless $\mathbf{v}_I = \mathbf{0}$. So the point I, where the perpendicular lines through A and B to their directions of motion intersect, is the instantaneous center. This is a simple method to locate the instantaneous center of a rigid body in planar motion.

   If the rigid body is in translation (the angular velocity of the rigid body is zero), the instantaneous center of the rigid body I moves to infinity.

5.7 Fixed and Moving Centrodes

In general, the instantaneous center does not retain the same position. The locus of the instantaneous centers forms a curve in space known as the *fixed centrode* (*space centrode*). Similarly, the locus of the instantaneous centers relative to the moving rigid body is known as the *moving centrode* (*body centrode*). Both curves lie in the same plane, of course, but neither curve need to be closed. Now suppose that, at a given moment, the fixed centrode C_f and the moving centrode C_m corresponding to a given motion are as shown in Fig. 5.18. The instantaneous center C is a point common to both curves at this moment. The curve C_f is fixed in space, and curve C_m is rotating with angular velocity ω. In general, the curves C_m and C_s are tangent at C but have different curvatures at this point.

Instant Centers of Velocity for Mechanisms
The instant center of velocities is a point, common to two bodies in plane motion, which point has the same instantaneous velocity in each body. The instant centers are also called centros or poles. Two bodies are needed to create an instant center.

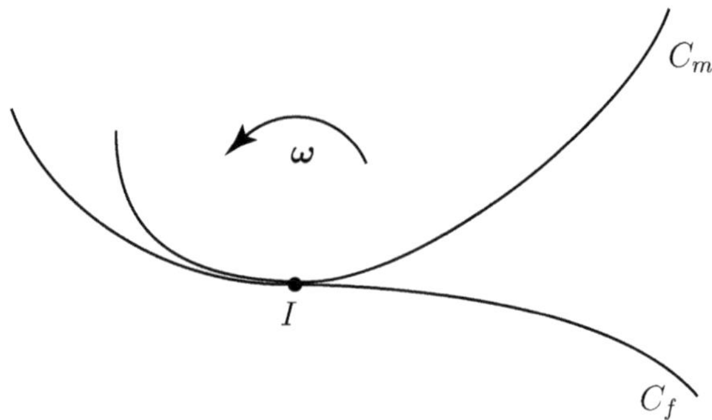

Fig. 5.18 Fixed centrode C_f and moving centrode C_m

The number of instant centers is determined from a system of n bodies and represents the combination for n bodies taken 2 at a time:

$$C_n^2 = \frac{n(n-1)}{2}. \tag{5.63}$$

The combination formula for n things taken r at time is

$$
\begin{aligned}
C_n^r &= \frac{n!}{r!(n-r)!}\\
&= \frac{1(2)\ldots(n)}{[1(2)\ldots(r)]\,[1(2)\cdots\cdots(n-r)]}\\
&= \frac{(n-r+1)\cdots\cdots(n-2)(n-1)n}{r!}.
\end{aligned}
$$

A four-bar mechanism ($n = 4$), has 6 instant centers, a six-bar mechanism ($n = 6$) has 15, and an eight-bar mechanism ($n = 8$) has 28. Figure 5.19a shows a four-bar mechanism in an arbitrary position. Figure 5.19b is a graph which is useful for keeping track of the instant centers of velocities. The graph is created by drawing a diagram on which the links 0, 1, 2 and 3 are represented with circles. The line between the circles represents the joint each time an instant center is found. This graph is a geometric solution to (5.63) since connecting all the circles in pairs gives all the possible combinations of the circles (links) taken two at a time.

The four-bar mechanism has four pin joints, and they have the same velocity in both links at all the time. These instant centers have been labeled I_{01}, I_{12}, I_{23}, and I_{03}. The order of the subscripts is unimportant. Instant center I_{21} is the same as I_{12}. The center of velocity at the pin joint I_{12} has the same velocity whether it is considered to be part of link 1 or link 2. Same for I_{01}, I_{23}, and I_{03}. The instant centers move to new locations as the mechanism changes position. The remaining instant centers for the four-bar example will be found with the help of Aronhold–Kennedy theorem.

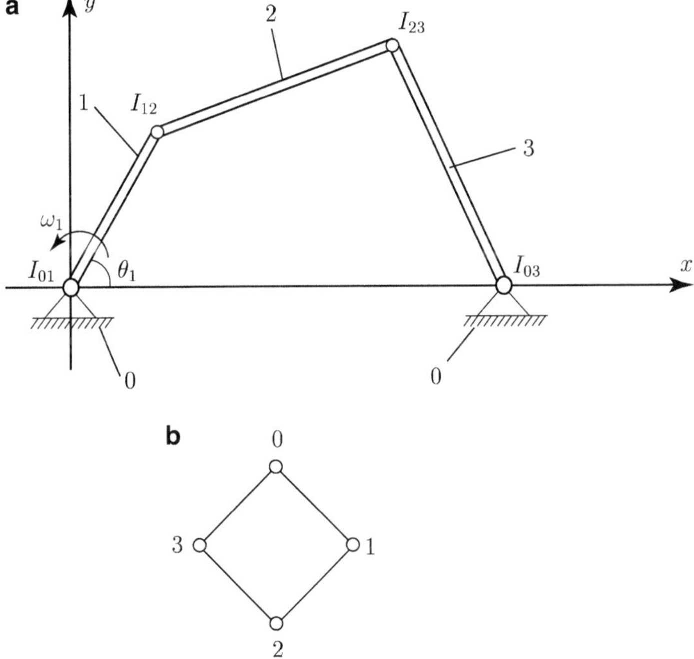

Fig. 5.19 Four-bar mechanism and the corresponding graph

Aronhold–Kennedy Theorem

Any three bodies in plane motion will have exactly three instant centers, and their center will lie on the same straight line.

For $n = 3$ with (5.63), three bodies there will have 3 center of rotation.

The second clause is that the instant centers for three bodies in plane will lie on the same straight line. This clause not require that the three bodies be connected in any way.

With this rule the remaining instant centers for the four-bar mechanism, which are not obvious from inspection, will be found. Figure 5.20b, shows the construction necessary to find instant center I_{02}. On the graph in Fig. 5.20b a dotted line connects the links 0 and 2. Two triangles on the graph which each contain dotted line and whose other two sides are solid lines representing the instant centers already found are identified. The triangles $\Delta 023$ and $\Delta 021$ define the three instant centers. The instant centers I_{01}, I_{12}, and I_{02} must lie on the same straight line (links 0, 1, 2, and $\Delta 021$). The instant centers I_{03}, I_{23}, and I_{02} must lie on the same straight line (bodies 0, 3, 2, and $\Delta 032$). In Fig. 5.20a, a line has been drawn through I_{01} and I_{12} and extended. The instant center I_{02} must lie on this line. Another line has been drawn through I_{03} and I_{23} and extended. The instant center I_{02} must lie on this line. The instant center I_{02} represents the intersection of the extended lines $I_{01}I_{12}$ and $I_{03}I_{23}$ $(I_{02} = I_{01}I_{12} \cap I_{03}I_{23})$.

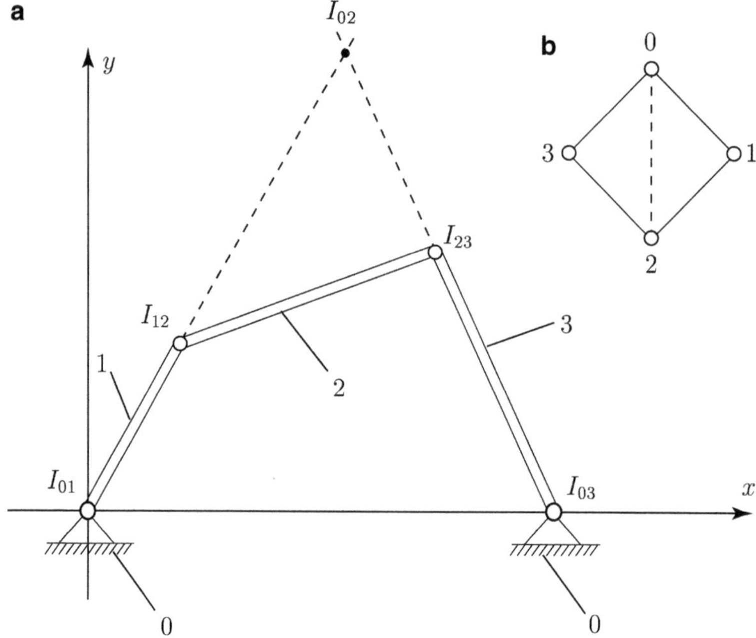

Fig. 5.20 Centrode I_{02}

Figure 5.21b shows the construction necessary to find the instant center I_{13}. On the graph in Fig. 5.21b, connect links 1 and 3 with a dotted line. The line forms the triangles $\Delta 013$, and $\Delta 123$. For the triangle $\Delta 013$ the instant centers I_{01}, I_{03}, and I_{13} (the unknown) must lie on the same straight line (links 0, 1, 3 and $\Delta 021$). For the triangle $\Delta 123$, the instant centers I_{12}, I_{23}, and I_{13} (the unknown) must lie on the same straight line (links 0, 1, 2 and $\Delta 123$). In Fig. 5.21a, a line has been drawn through I_{01} and I_{03} and extended. The instant center I_{13} must lie on this line. Another line has been drawn through I_{12} and I_{23} and extended. The instant center I_{13} must lie on this line. The instant center I_{13} represents the intersection of the extended lines $I_{01}I_{03}$ and $I_{12}I_{23}$ ($I_{13} = I_{01}I_{03} \cap I_{12}I_{23}$).

Instant Center of a Curved Slider
If link 1 is a slider moving on a circular arc on link 2 as shown in Fig. 5.22, then the center of the arc is a stationary location common to both links 1 and 2. This point is the instant center. If the curve is not circular at the location of interest, the center of curvature of the path at the given point would be the instant center.

Instant Center of a Slider Joint
If the radius of curvature, ρ, in the case of the curved slider is allowed to become very large, the arc will approach a straight line (Fig. 5.23). The location of the instant center will be toward infinity. The velocity of a point P of the slider, relative to the link 2, will remain perpendicular to the line from P to the instant center.

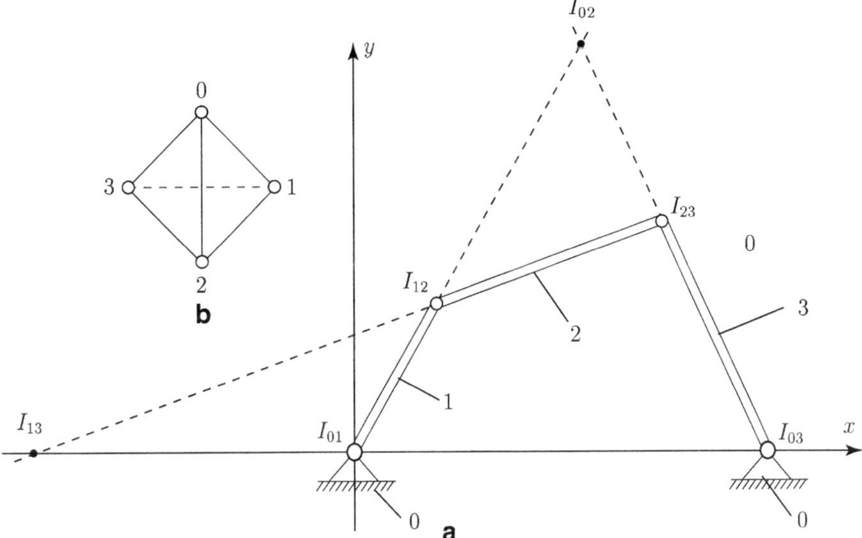

Fig. 5.21 Centrode I_{13}

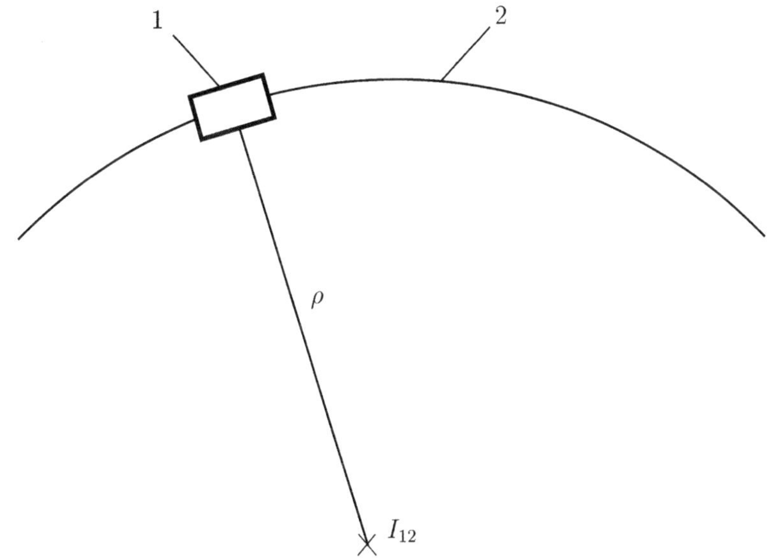

Fig. 5.22 Instant center of a curved slider

Therefore, if one knows the direction of the velocity of any point P relative to link 2, one can find the locus of the instant center, that is, it must lie on a line perpendicular to the velocity vector as shown in Fig. 5.23. Note that the location of the line to infinity is unimportant, only the direction is defined by the velocity direction.

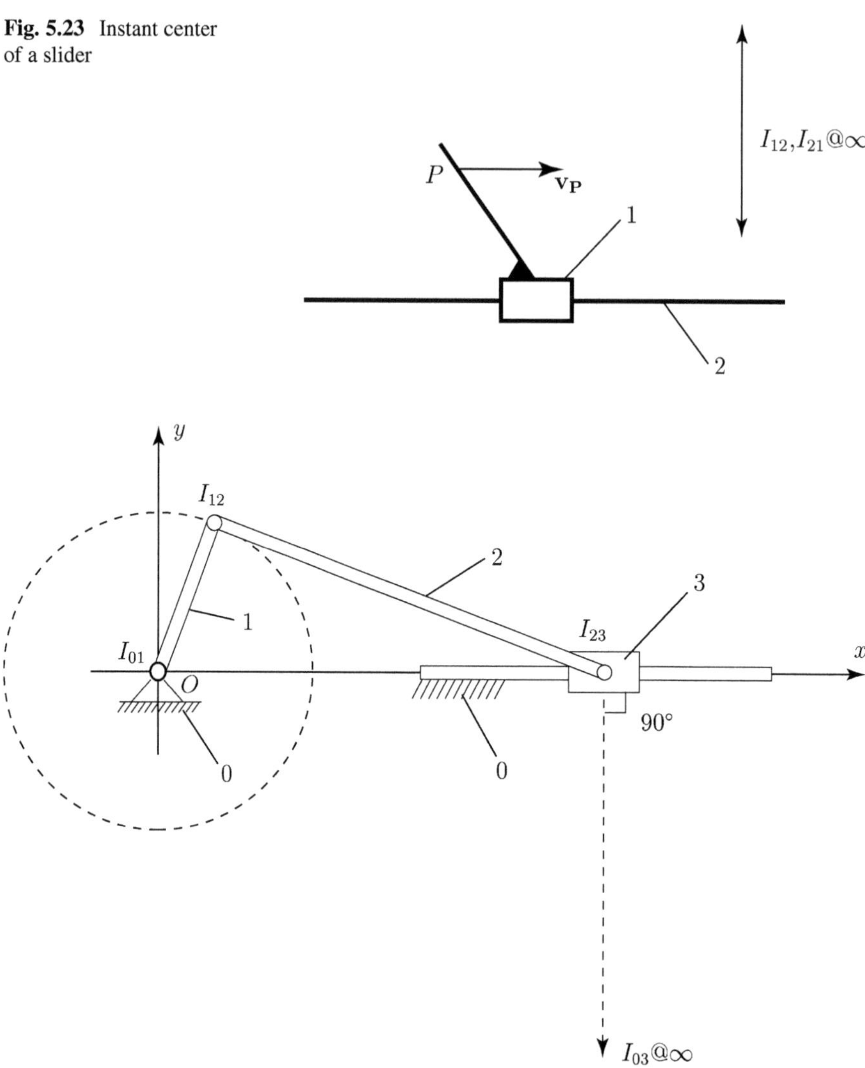

Fig. 5.23 Instant center
of a slider

Fig. 5.24 Instant centers for slider-crank mechanism

Figure 5.24 shows a slider-crank mechanism. There are only three pin joints at
the instant centers I_{01}, I_{12}, and I_{23}. The joint between links 0 and 3 is a sliding joint.
A sliding joint will have its instant center at infinity along a line perpendicular to
the direction of sliding as shown in Fig. 5.24 ($I_{03}\,@\infty$).

Instant Center of a Rolling Contact Joint
The instant center of a pure rolling contact between two rigid bodies 1 and 2 is
located at the point of contact of the two links as shown in Fig. 5.25. The rolling
condition is that the points in contact be at rest relative to one another. Figure 5.26a
shows a mechanism in an arbitrary position, and Fig. 5.26b is the graph which is

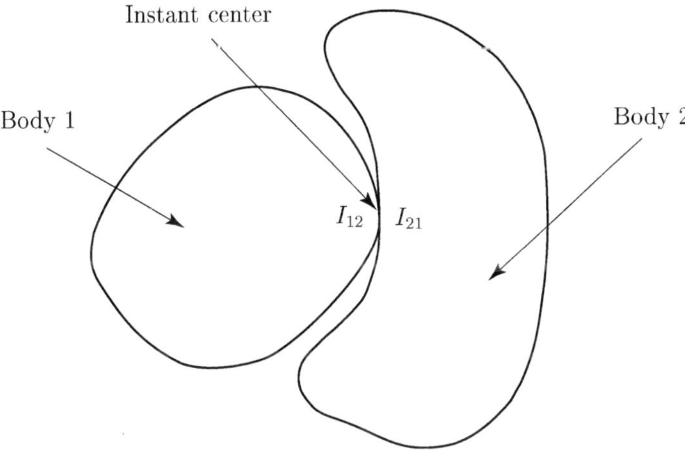

Fig. 5.25 Instant center of a rolling contact joint

used to find the instant centers of velocities. The joint between links 0 and 3 is a sliding joint and will have its instant center I_{03} at infinity along a line perpendicular to the direction of sliding as shown in Fig. 5.26a. The instant centers I_{01}, I_{12}, and I_{23} are found by inspection.

The instant centers I_{01}, I_{12}, and I_{02} are on the same straight line (bodies 0, 1, 2 and $\Delta 021$). The instant centers I_{03}, I_{23}, and I_{02} are on the same straight line (bodies 0, 3, 2 and $\Delta 032$). In Fig. 5.26a, a line has been drawn through I_{01} and I_{12} and extended. The instant center I_{02} must lie on this line. Another line perpendicular to the direction of sliding as shown in Fig. 5.26a has been drawn through I_{23} and extended. The instant center I_{02} must lie on this line. The instant center I_{02} represents the intersection of the extended lines $I_{01}I_{12}$ and the line that passes through I_{23} and is perpendicular to the direction of sliding.

The instant centers I_{01}, I_{03}, and I_{13} must lie on the same straight line (bodies 0, 1, 3 and $\Delta 013$). The instant centers I_{12}, I_{23}, and I_{13} must lie on the same straight line (bodies 1, 2, 3 and $\Delta 123$). In Fig. 5.26a, a line has been drawn through I_{12} and I_{23} and extended. The instant center I_{13} must lie on this line. Another line (a line parallel to the line that pass through I_{23} and is perpendicular to the direction of sliding) has been drawn through I_{01}. The instant center I_{13} must lie on this line. The instant center I_{13} represents the intersection of the extended lines $I_{12}I_{23}$ and the line that passes through I_{01} as shown in Fig. 5.26a.

Velocity Analysis Using Instant Centers for Mechanisms
Once the instant centers have been found, they can be used for a graphical velocity analysis. Figure 5.27 shows a four-bar mechanism with the center I_{02} located and labeled. From the definition of the instant center, both links sharing the instant center will have identical velocity at that point. Instant center I_{02} involves the link 2 which is in complex plane motion and the ground link 0 which is stationary. All points on link 0 have zero velocity in the fixed reference frame Oxy, which represents link 0.

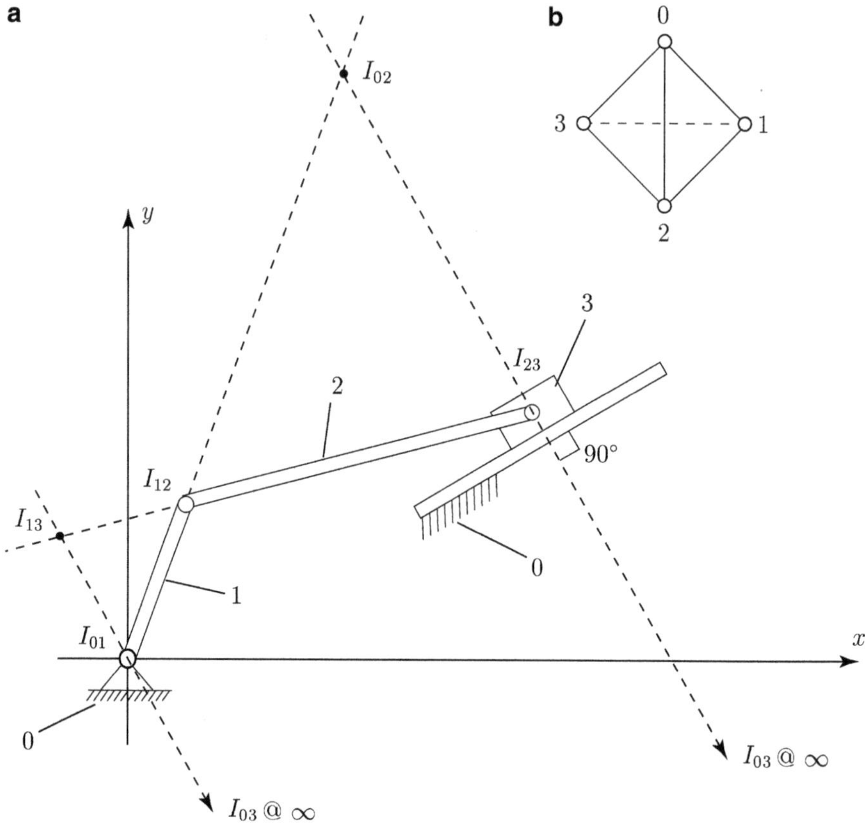

Fig. 5.26 Instant centers for a mechanism

Therefore, I_{02} must have zero velocity at this instant. If I_{02} has zero velocity, then it can be considered to be an instantaneous "fixed pivot" about which link 2 is in pure rotation with respect to link 0. A moment later, I_{02} will move to a new location, and link 2 will be "pivoting" about a new instant center. The velocity $\mathbf{v}_A$ of point A is shown on Fig. 5.27.

The magnitude of $\mathbf{v}_A$ can be computed. Its direction is perpendicular to line OA ($\mathbf{v}_A \perp OA$), and its sense is given by ω_1 (the velocity $\mathbf{v}_A$ will rotate link 1 in the same sense as ω_1). The joint A is also instant center I_{12}. It has the same velocity as part of link 1 and as part of link 2. Since link 2 is effectively pivoting about I_{02} at this instant, the angular velocity ω_2 can be found by rearranging equation

$$\mathbf{v}_A = \boldsymbol{\omega}_2 \times \mathbf{r}_{I_{02}A}, \text{ and } (\mathbf{v}_A \perp \mathbf{r}_{I_{02}A}).$$

Thus,

$$v_A = \omega_2 \left(I_{02}A \right),$$

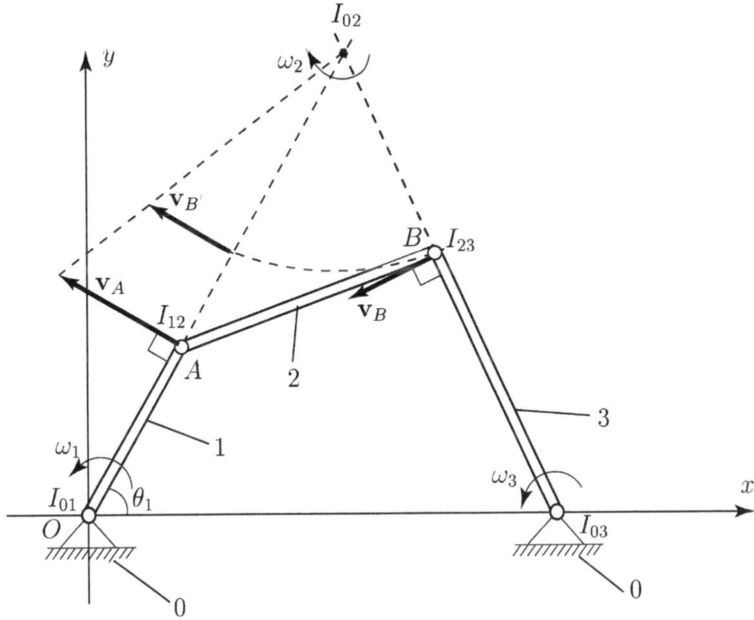

Fig. 5.27 Instant center analysis for a four-bar mechanism

or

$$\omega_2 = \frac{v_A}{I_{02}A}.$$

Once ω_2 is known, the magnitude of $\mathbf{v}_B$ can also be found from

$$\mathbf{v}_B = \boldsymbol{\omega}_2 \times \mathbf{r}_{I_{02}B}.$$

Thus,

$$v_B = \omega_2 \left(I_{02}B\right).$$

Once $\mathbf{v}_B$ is known, ω_3 can also be found from

$$\mathbf{v}_B = \boldsymbol{\omega}_3 \times \mathbf{r}_{I_{03}B}.$$

Thus,

$$v_B = \omega_3 \left(I_{03}B\right),$$

or

$$\omega_3 = \frac{v_B}{I_{03}B}.$$

Only the scalar magnitude of the velocities vectors has been determined. The center I_{02} represents an instant "fixed" point for link 2, and the location of I_{02} is known.

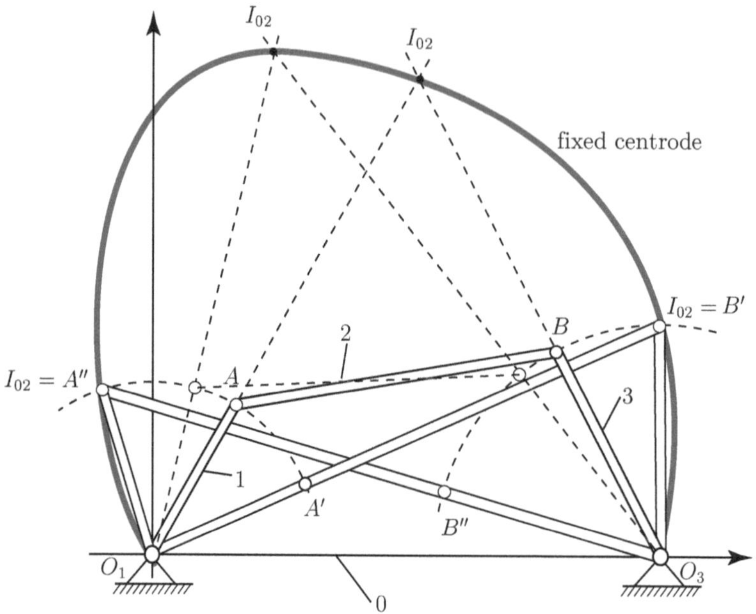

Fig. 5.28 Fixed centrode for the four-bar mechanism

The velocity of the point B on link 2 is $\mathbf{v}_B$ and is perpendicular to the radius $\mathbf{r}_{I_{02}B}$ ($\mathbf{v}_B = \omega_2 \times \mathbf{r}_{I_{02}B}$ and $\mathbf{v}_B \perp \mathbf{r}_{I_{02}B}$). The velocity of the point B on link 3 is also $\mathbf{v}_B$ and is perpendicular to the radius $\mathbf{r}_{I_{03}B}$ ($\mathbf{v}_B = \omega_3 \times \mathbf{r}_{I_{03}B}$ and $\mathbf{v}_B \perp \mathbf{r}_{I_{03}B}$). A graphical solution is shown in the figure in Fig. 5.27. Arcs centered at I_{02} with radius $I_{02}B$ are drawn from points B to intersect line AI_{02}. The magnitude of velocity $\mathbf{v}_B$ is found from the vectors drawn perpendicular to the line at the intersections of the arcs and line AI_{02}. The length of the vector is defined by the line from the tip of $\mathbf{v}_A$ to the instant center I_{02}. This vector $\mathbf{v}_B$ can then be moved along the arc back to point B, keeping it tangent to the arc.

Centrodes for Mechanism

Figure 5.28 shows the successive positions of an instant center I_{02} that form a path. The locus of the instant center (the path) is the centrode. There are two links needed to create an instant center, and there will be two centrodes associated with any one instant center. These are formed by projecting the path of the instant center first on one link and then on the other. Figure 5.28 shows the locus of instant center I_{02} as projected onto link 0. Because link 0 is fixed, this is the fixed centrode. Next, fixing link 2 as the ground and moving link 0, the projection of the locus of I_{02} onto link 2 is shown in Fig. 5.29. In the original mechanism, link 2 was a moving element, so this locus is called the moving centrode. Figure 5.29 shows the mechanism with both fixed and moving centrodes superposed. Using the definition of the instant center, both links have the same velocity at the point, at this instant. Link 0 is fixed and has zero velocity as does the fixed centrode. As the mechanism moves, the moving

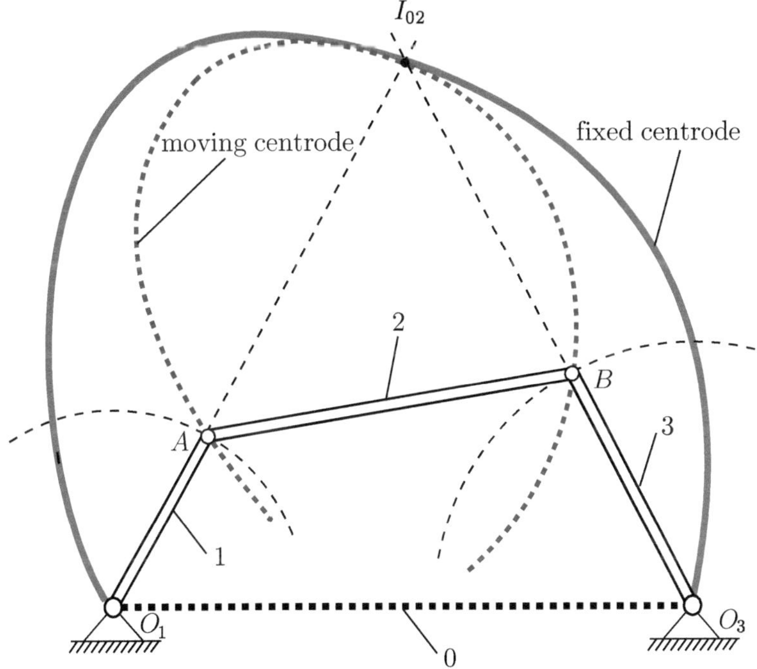

Fig. 5.29 Fixed and moving centrodes of the mechanism

centrode must roll against the fixed centrode without slipping. All instant centers of a mechanism have centrodes. If the links are directly connected by a joint, such as I_{12}, I_{23}, I_{01}, and I_{03}, their fixed and moving centrodes will degenerate to a point at that location on each link. The most interesting centrodes are those involving links not directly connected to one another such as I_{02} and I_{13}.

5.8   Closed Loop Equations

Two rigid bodies (j) and (k) are connected by a joint or kinematic pair at A as shown in Fig. 5.30. The point A_j of the rigid body (j) is guided along a path prescribed in the body (k). The points A_j belonging to body (j) and the A_k belonging to body (k) are coincident at the instant of motion under consideration. The following relation exists between the velocity $\mathbf{v}_{A_j}$ of the point A_j and the velocity $\mathbf{v}_{A_k}$ of the point A_k:

$$\mathbf{v}_{A_j} = \mathbf{v}_{A_k} + \mathbf{v}^r_{A_{jk}}, \tag{5.64}$$

where $\mathbf{v}^r_{A_{jk}} = \mathbf{v}^r_{A_j A_k}$ indicates the velocity of A_j as seen by an observer at A_k attached to body (k) or the relative velocity of A_j with respect to A_k, allowed at the joint A.

$$\text{Point } A \quad \begin{cases} A_j \text{ on link } (j) \\ A_k \text{ on link } (k) \end{cases}$$

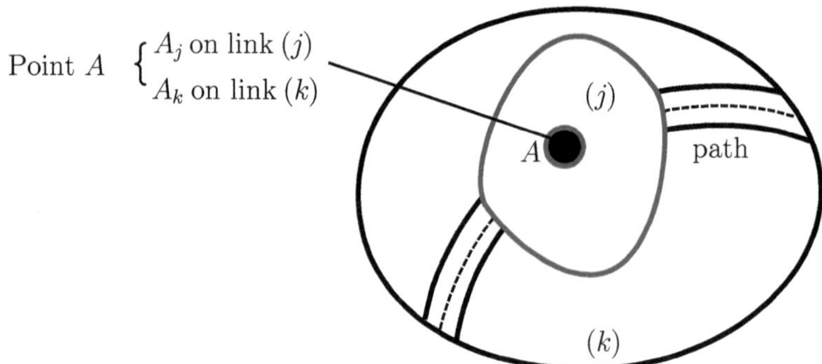

Fig. 5.30 Relative motion and coincident points

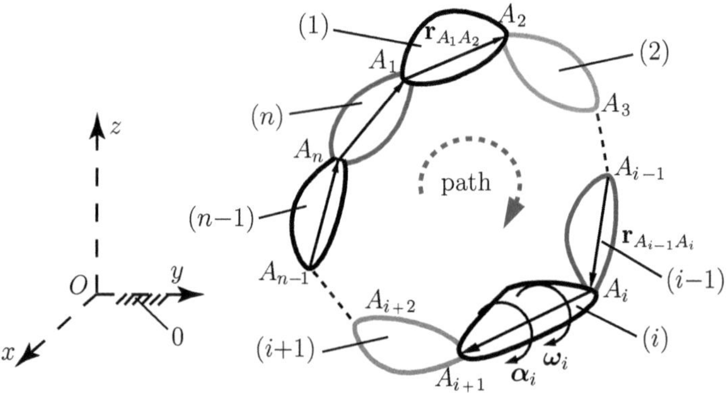

Fig. 5.31 Monoloop of a closed kinematic chain

The direction of $\mathbf{v}^{\mathrm{r}}_{A_{jk}}$ is obviously tangent to the path prescribed in the body (k). From (5.64), the accelerations of A_j and A_k are expressed as

$$\mathbf{a}_{A_j} = \mathbf{a}_{A_k} + \mathbf{a}^{\mathrm{r}}_{A_{jk}} + \mathbf{a}^{\mathrm{c}}_{A_{jk}}, \tag{5.65}$$

where $\mathbf{a}^{\mathrm{r}}_{A_{jk}} = \mathbf{a}^{\mathrm{r}}_{A_j A_k}$ is the relative velocity of A_j with respect to A_k and $\mathbf{a}^{\mathrm{c}}_{A_{jk}} = \mathbf{a}^{\mathrm{c}}_{A_j A_k}$ is the Coriolis acceleration given by

$$\mathbf{a}^{\mathrm{c}}_{A_{jk}} = 2\boldsymbol{\omega}_k \times \mathbf{v}^{\mathrm{r}}_{A_{jk}}, \tag{5.66}$$

where $\boldsymbol{\omega}_k$ is the angular velocity of the body (k). Equations (5.64) and (5.65) are useful even for coincident points belonging to two rigid bodies that may not be directly connected.

Figure 5.31 shows a monoloop closed kinematic chain with n rigid links. The joint A_i where $i = 1, 2, \ldots, n$ is the connection between the links (i) and $(i-1)$.

The last link n is connected with the first link 1 of the chain. For the closed kinematic chain, a path is chosen from link 1 to link n. At the joint A_i, there are two instantaneously coincident points: (1) the point $A_{i,i}$ belonging to link (i), $A_{i,i} \in (i)$ and (2) the point $A_{i,i-1}$ belonging to body $(i-1)$, $A_{i,i-1} \in (i-1)$.

5.8.1   Closed Loop Velocity Equations

The absolute angular velocity, $\omega_i = \omega_{i0}$, of the rigid body (i), or the angular velocity of the rigid body (i) with respect to the "fixed" reference frame $Oxyz$ is

$$\omega_i = \omega_{i-1} + \omega_{i,i-1}, \tag{5.67}$$

where ω_{i-1} is the absolute angular velocity of the rigid body $(i-1)$ (or the angular velocity of the rigid body $(i-1)$ with respect to the "fixed" reference frame $Oxyz$) and $\omega_{i,i-1}$ is the relative angular velocity of the rigid body (i) with respect to the rigid body $(i-1)$. For the n link closed kinematic chain, the following expressions are obtained for the angular velocities:

$$\omega_1 = \omega_n + \omega_{1,n}$$
$$\omega_2 = \omega_1 + \omega_{2,1}$$
$$\dotsb$$
$$\omega_i = \omega_{i-1} + \omega_{i,i-1}$$
$$\dotsb$$
$$\omega_n = \omega_{n-1} + \omega_{n,n-1}. \tag{5.68}$$

Summing the expressions given in (5.68), the following relation is obtained:

$$\omega_{1,n} + \omega_{2,1} + \cdots + \omega_{n,n-1} = \mathbf{0}, \tag{5.69}$$

which is rewritten as

$$\sum_{(i)} \omega_{i,i-1} = \mathbf{0}. \tag{5.70}$$

Equation (5.70) represents the first vectorial equation for the angular velocities of a simple closed kinematic chain.

The following relation exists between the velocity $\mathbf{v}_{A_{i,i}}$ of the point $A_{i,i}$ and the velocity $\mathbf{v}_{A_{i,i-1}}$ of the point $A_{i,i-1}$ (see Fig. 5.32):

$$\mathbf{v}_{A_{i,i}} = \mathbf{v}_{A_{i,i-1}} + \mathbf{v}^{\mathrm{r}}_{A_{i,i-1}}, \tag{5.71}$$

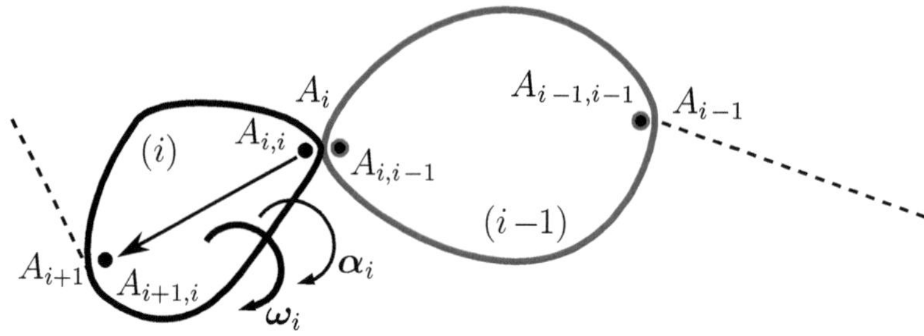

Fig. 5.32 Rigid body $(i-1)$ and rigid body (i)

where $\mathbf{v}^r_{A_{i,i-1}} = \mathbf{v}^r_{A_{i,i}A_{i,i-1}}$ is the relative velocity of $A_{i,i}$ on link (i) with respect to $A_{i,i-1}$ on link $(i-1)$. Using the velocity relation for two particles on the same rigid body (i), the following relation can be written:

$$\mathbf{v}_{A_{i+1,i}} = \mathbf{v}_{A_{i,i}} + \boldsymbol{\omega}_i \times \mathbf{r}_{A_iA_{i+1}}, \tag{5.72}$$

where $\boldsymbol{\omega}_i$ is the absolute angular velocity of the link (i) in the reference frame $Oxyz$ and $\mathbf{r}_{A_iA_{i+1}}$ is the distance vector from A_i to A_{i+1}. Using (5.71) and (5.72), the velocity of the point $A_{i+1,i} \in (i+1)$ is written as

$$\mathbf{v}_{A_{i+1,i}} = \mathbf{v}_{A_{i,i-1}} + \boldsymbol{\omega}_i \times \mathbf{r}_{A_iA_{i+1}} + \mathbf{v}^r_{A_{i,i-1}}. \tag{5.73}$$

For the n link closed kinematic chain, the following expressions are obtained:

$$\mathbf{v}_{A_{3,2}} = \mathbf{v}_{A_{2,1}} + \boldsymbol{\omega}_2 \times \mathbf{r}_{A_2A_3} + \mathbf{v}^r_{A_{2,1}}$$

$$\mathbf{v}_{A_{4,3}} = \mathbf{v}_{A_{3,2}} + \boldsymbol{\omega}_3 \times \mathbf{r}_{A_3A_4} + \mathbf{v}^r_{A_{3,2}}$$

$$\cdots\cdots\cdots\cdots\cdots\cdots\cdots\cdots\cdots\cdots\cdots\cdots\cdots$$

$$\mathbf{v}_{A_{i+1,i}} = \mathbf{v}_{A_{i,i-1}} + \boldsymbol{\omega}_i \times \mathbf{r}_{A_iA_{i+1}} + \mathbf{v}^r_{A_{i,i-1}}$$

$$\cdots\cdots\cdots\cdots\cdots\cdots\cdots\cdots\cdots\cdots\cdots\cdots\cdots$$

$$\mathbf{v}_{A_{1,n}} = \mathbf{v}_{A_{n,n-1}} + \boldsymbol{\omega}_n \times \mathbf{r}_{A_nA_1} + \mathbf{v}^r_{A_{n,n-1}}$$

$$\mathbf{v}_{A_{2,1}} = \mathbf{v}_{A_{1,n}} + \boldsymbol{\omega}_1 \times \mathbf{r}_{A_1A_2} + \mathbf{v}^r_{A_{1,n}}. \tag{5.74}$$

Summing the relations in (5.74),

$$\left(\boldsymbol{\omega}_1 \times \mathbf{r}_{A_1A_2} + \boldsymbol{\omega}_2 \times \mathbf{r}_{A_2A_3} + \cdots + \boldsymbol{\omega}_i \times \mathbf{r}_{A_iA_{i+1}} + \cdots + \boldsymbol{\omega}_n \times \mathbf{r}_{A_nA_1} \right)$$

$$+ \left(\mathbf{v}^r_{A_{2,1}} + \mathbf{v}^r_{A_{3,2}} + \cdots + \mathbf{v}^r_{A_{i,i-1}} + \cdots + \mathbf{v}^r_{A_{n,n-1}} + \mathbf{v}^r_{A_{1,n}} \right) = \mathbf{0}. \tag{5.75}$$

Because the reference system $Oxyz$ is considered "fixed," the vector $\mathbf{r}_{A_{i-1}A_i}$ is written in terms of the position vectors of the points A_{i-1} and A_i:

$$\mathbf{r}_{A_{i-1}A_i} = \mathbf{r}_{A_i} - \mathbf{r}_{A_{i-1}},$$

where $\mathbf{r}_{A_i} = \mathbf{r}_{OA_i}$ and $\mathbf{r}_{A_{i-1}} = \mathbf{r}_{OA_{i-1}}$. Equation (5.75) becomes

$$\left[\mathbf{r}_{A_1} \times (\boldsymbol{\omega}_1 - \boldsymbol{\omega}_n) + \mathbf{r}_{A_2} \times (\boldsymbol{\omega}_2 - \boldsymbol{\omega}_1) + \cdots + \mathbf{r}_{A_n} \times (\boldsymbol{\omega}_n - \boldsymbol{\omega}_{n-1})\right]$$
$$+ \left(\mathbf{v}^{\mathrm{r}}_{A_{1,n}} + \mathbf{v}^{\mathrm{r}}_{A_{2,1}} + \cdots + \mathbf{v}^{\mathrm{r}}_{A_{i,i-1}} + \cdots + \mathbf{v}^{\mathrm{r}}_{A_{n,n-1}}\right) = \mathbf{0},$$

or

$$\left(\mathbf{r}_{A_1} \times \boldsymbol{\omega}_{1,n} + \mathbf{r}_{A_2} \times \boldsymbol{\omega}_{2,1} + \cdots + \mathbf{r}_{A_n} \times \boldsymbol{\omega}_{n,n-1}\right) + \left(\mathbf{v}^{\mathrm{r}}_{A_{1,n}} + \mathbf{v}^{\mathrm{r}}_{A_{2,1}} + \cdots + \mathbf{v}^{\mathrm{r}}_{A_{n,n-1}}\right) = \mathbf{0}.$$
$$\tag{5.76}$$

The previous equation is written as

$$\sum_{(i)} \mathbf{r}_{A_i} \times \boldsymbol{\omega}_{i,i-1} + \sum_{(i)} \mathbf{v}^{\mathrm{r}}_{A_{i,i-1}} = \mathbf{0}. \tag{5.77}$$

Equation (5.77) represents the second vectorial equation for the angular velocities of a simple closed kinematic chain.

Equations such as

$$\sum_{(i)} \boldsymbol{\omega}_{i,i-1} = \mathbf{0} \quad \text{and} \quad \sum_{(i)} \mathbf{r}_{A_i} \times \boldsymbol{\omega}_{i,i-1} + \sum_{(i)} \mathbf{v}^{\mathrm{r}}_{A_{i,i-1}} = \mathbf{0} \tag{5.78}$$

represent the velocity equations for a simple closed kinematic chain.

5.8.2 Closed Loop Acceleration Equations

The absolute angular acceleration, $\boldsymbol{\alpha}_i = \boldsymbol{\alpha}_{i0}$, of the rigid body (i) (or the angular acceleration of the rigid body (i) with respect to the "fixed" reference frame $Oxyz$) is

$$\boldsymbol{\alpha}_i = \boldsymbol{\alpha}_{i-1} + \boldsymbol{\alpha}_{i,i-1} + \boldsymbol{\omega}_i \times \boldsymbol{\omega}_{i,i-1}, \tag{5.79}$$

where $\boldsymbol{\alpha}_{i-1}$ is the absolute angular acceleration of the rigid body $(i-1)$ (or the angular acceleration of the rigid body $(i-1)$ with respect to the "fixed" reference frame $Oxyz$) and $\boldsymbol{\alpha}_{i,i-1}$ is the relative angular acceleration of the rigid body (i) with respect to the rigid body $(i-1)$. For the n link closed kinematic chain, the following expressions are obtained for the angular accelerations:

$$\alpha_2 = \alpha_1 + \alpha_{2,1} + \omega_2 \times \omega_{2,1}$$

$$\alpha_3 = \alpha_2 + \alpha_{3,2} + \omega_3 \times \omega_{3,2}$$

$$\dots\dots\dots\dots\dots\dots\dots\dots\dots$$

$$\alpha_i = \alpha_{i-1} + \alpha_{i,i-1} + \omega_i \times \omega_{i,i-1}$$

$$\dots\dots\dots\dots\dots\dots\dots\dots\dots$$

$$\alpha_1 = \alpha_n + \alpha_{1,n} + \omega_1 \times \omega_{1,n}. \tag{5.80}$$

Summing all the expressions in (5.80),

$$\alpha_{2,1} + \alpha_{3,2} + \cdots + \alpha_{1,n} + \omega_2 \times \omega_{2,1} + \cdots + \omega_1 \times \omega_{1,n} = \mathbf{0}. \tag{5.81}$$

Equation (5.81) is rewritten as

$$\sum_{(i)} \alpha_{i,i-1} + \sum_{(i)} \omega_i \times \omega_{i,i-1} = \mathbf{0}. \tag{5.82}$$

Equation (5.82) represents the first vectorial equation for the angular accelerations of a simple closed kinematic chain.

Using the acceleration distributions of the relative motion of two rigid bodies (i) and $(i-1)$,

$$\mathbf{a}_{A_{i,i}} = \mathbf{a}_{A_{i,i-1}} + \mathbf{a}^{\mathrm{r}}_{A_{i,i-1}} + \mathbf{a}^{\mathrm{c}}_{A_{i,i-1}}, \tag{5.83}$$

where $\mathbf{a}_{A_{i,i}}$ and $\mathbf{a}_{A_{i,i-1}}$ are the linear accelerations of the points $A_{i,i}$ and $A_{i,i-1}$ and $\mathbf{a}^{\mathrm{r}}_{A_{i,i-1}} = \mathbf{a}^{\mathrm{r}}_{A_{i,i}A_{i,i-1}}$ is the relative acceleration between $A_{i,i}$ on link (i) and $A_{i,i-1}$ on link $(i-1)$. Finally, $\mathbf{a}^{\mathrm{c}}_{A_{i,i-1}}$ is the Coriolis acceleration defined as

$$\mathbf{a}^{\mathrm{c}}_{A_{i,i-1}} = 2\omega_{i-1} \times \mathbf{v}^{\mathrm{r}}_{A_{i,i-1}}. \tag{5.84}$$

Using the acceleration distribution relations for two particles on a rigid body,

$$\mathbf{a}_{A_{i+1,i}} = \mathbf{a}_{A_{i,i}} + \alpha_i \times \mathbf{r}_{A_iA_{i+1}} + \omega_i \times (\omega_i \times \mathbf{r}_{A_iA_{i+1}}), \tag{5.85}$$

where α_i is the angular acceleration of the link (i). From (5.83) and (5.85), it results

$$\mathbf{a}_{A_{i+1,i}} = \mathbf{a}_{A_{i,i-1}} + \mathbf{a}^{\mathrm{r}}_{A_{i,i-1}} + \mathbf{a}^{\mathrm{c}}_{A_{i,i-1}} + \alpha_i \times \mathbf{r}_{A_iA_{i+1}} + \omega_i \times (\omega_i \times \mathbf{r}_{A_iA_{i+1}}). \tag{5.86}$$

Writing similar equations for all the links of the kinematic chain, the following relations are obtained:

$$\mathbf{a}_{A_{3,2}} = \mathbf{a}_{A_{2,1}} + \mathbf{a}^{\mathrm{r}}_{A_{2,1}} + \mathbf{a}^{\mathrm{c}}_{A_{2,1}} + \boldsymbol{\alpha}_2 \times \mathbf{r}_{A_2 A_3} + \boldsymbol{\omega}_2 \times (\boldsymbol{\omega}_2 \times \mathbf{r}_{A_2 A_3}),$$

$$\mathbf{a}_{A_{4,3}} = \mathbf{a}_{A_{3,2}} + \mathbf{a}^{\mathrm{r}}_{A_{3,2}} + \mathbf{a}^{\mathrm{c}}_{A_{3,2}} + \boldsymbol{\alpha}_3 \times \mathbf{r}_{A_3 A_4} + \boldsymbol{\omega}_3 \times (\boldsymbol{\omega}_3 \times \mathbf{r}_{A_3 A_4}),$$

$$\cdots\cdots\cdots\cdots\cdots\cdots\cdots\cdots\cdots\cdots\cdots\cdots\cdots\cdots\cdots\cdots\cdots\cdots\cdots$$

$$\mathbf{a}_{A_{1,n}} = \mathbf{a}_{A_{n,n-1}} + \mathbf{a}^{\mathrm{r}}_{A_{n,n-1}} + \mathbf{a}^{\mathrm{c}}_{A_{n,n-1}} + \boldsymbol{\alpha}_n \times \mathbf{r}_{A_n A_1} + \boldsymbol{\omega}_n \times (\boldsymbol{\omega}_n \times \mathbf{r}_{A_n A_1}),$$

$$\mathbf{a}_{A_{2,1}} = \mathbf{a}_{A_{1,n}} + \mathbf{a}^{\mathrm{r}}_{A_{1,n}} + \mathbf{a}^{\mathrm{c}}_{A_{1,n}} + \boldsymbol{\alpha}_1 \times \mathbf{r}_{A_1 A_2} + \boldsymbol{\omega}_1 \times (\boldsymbol{\omega}_1 \times \mathbf{r}_{A_1 A_2}). \tag{5.87}$$

Summing the expressions in (5.87),

$$\left(\mathbf{a}^{\mathrm{r}}_{A_{1,n}} + \mathbf{a}^{\mathrm{r}}_{A_{2,1}} + \cdots + \mathbf{a}^{\mathrm{r}}_{A_{n,n-1}}\right) + \left(\mathbf{a}^{\mathrm{c}}_{A_{1,n}} + \mathbf{a}^{\mathrm{c}}_{A_{2,1}} + \cdots + \mathbf{a}^{\mathrm{c}}_{A_{n,n-1}}\right)$$

$$+ \left(\boldsymbol{\alpha}_1 \times \mathbf{r}_{A_1 A_2} + \boldsymbol{\alpha}_2 \times \mathbf{r}_{A_2 A_3} + \cdots + \boldsymbol{\alpha}_n \times \mathbf{r}_{A_n A_1}\right) + \boldsymbol{\omega}_1 \times (\boldsymbol{\omega}_1 \times \mathbf{r}_{A_1 A_2})$$

$$+ \boldsymbol{\omega}_2 \times (\boldsymbol{\omega}_2 \times \mathbf{r}_{A_2 A_3}) + \cdots + \boldsymbol{\omega}_n \times (\boldsymbol{\omega}_n \times \mathbf{r}_{A_n A_1}) = \mathbf{0}. \tag{5.88}$$

Using the relation $\mathbf{r}_{A_{i-1} A_i} = \mathbf{r}_{A_i} - \mathbf{r}_{A_{i-1}}$ in (5.88),

$$\left(\mathbf{a}^{\mathrm{r}}_{A_{1,n}} + \mathbf{a}^{\mathrm{r}}_{A_{2,1}} + \cdots + \mathbf{a}^{\mathrm{r}}_{A_{n,n-1}}\right) + \left(\mathbf{a}^{\mathrm{c}}_{A_{1,n}} + \mathbf{a}^{\mathrm{c}}_{A_{2,1}} + \cdots + \mathbf{a}^{\mathrm{c}}_{A_{n,n-1}}\right)$$

$$+ \left[\mathbf{r}_{A_1} \times (\boldsymbol{\alpha}_{1,n} + \boldsymbol{\omega}_1 \times \boldsymbol{\omega}_{1,n}) + \cdots + \mathbf{r}_{A_n} \times (\boldsymbol{\alpha}_{n,n-1} + \boldsymbol{\omega}_n \times \boldsymbol{\omega}_{n,n-1})\right]$$

$$+ \boldsymbol{\omega}_1 \times (\boldsymbol{\omega}_1 \times \mathbf{r}_{A_1 A_2}) + \boldsymbol{\omega}_2 \times (\boldsymbol{\omega}_2 \times \mathbf{r}_{A_2 A_3}) + \cdots + \boldsymbol{\omega}_n \times (\boldsymbol{\omega}_n \times \mathbf{r}_{A_n A_1}) = \mathbf{0}.$$

$$\tag{5.89}$$

Equation (5.89) is rewritten as

$$\sum_{(i)} \mathbf{a}^{\mathrm{r}}_{A_{i,i-1}} + \sum_{(i)} \mathbf{a}^{\mathrm{c}}_{A_{i,i-1}} + \sum_{(i)} \mathbf{r}_{A_i} \times (\boldsymbol{\alpha}_{i,i-1} + \boldsymbol{\omega}_i \times \boldsymbol{\omega}_{i,i-1}) + \sum_{(i)} \boldsymbol{\omega}_i \times (\boldsymbol{\omega}_i \times \mathbf{r}_{A_i A_{i+1}}) = \mathbf{0}.$$

$$\tag{5.90}$$

Equation (5.90) represents the second vectorial equation for the angular accelerations of a simple closed kinematic chain. Thus,

$$\sum_{(i)} \boldsymbol{\alpha}_{i,i-1} + \sum_{(i)} \boldsymbol{\omega}_i \times \boldsymbol{\omega}_{i,i-1} = \mathbf{0} \text{ and}$$

$$\sum_{(i)} \mathbf{r}_{A_i} \times (\boldsymbol{\alpha}_{i,i-1} + \boldsymbol{\omega}_i \times \boldsymbol{\omega}_{i,i-1}) + \sum_{(i)} \mathbf{a}^{\mathrm{r}}_{A_{i,i-1}} + \sum_{(i)} \mathbf{a}^{\mathrm{c}}_{A_{i,i-1}} + \sum_{(i)} \boldsymbol{\omega}_i \times (\boldsymbol{\omega}_i \times \mathbf{r}_{A_i A_{i+1}}) = \mathbf{0}$$

$$\tag{5.91}$$

are the acceleration equations for the case of a simple closed kinematic chain.

Remarks

1. For a closed kinematic chain in planar motion, simplified relations are obtained
   because

$$\boldsymbol{\omega}_i \times (\boldsymbol{\omega}_i \times \mathbf{r}_{A_iA_{i+1}}) = -\omega_i^2\, \mathbf{r}_{A_iA_{i+1}} \text{ and } \boldsymbol{\omega}_i \times \boldsymbol{\omega}_{i,i-1} = \mathbf{0}. \tag{5.92}$$

   Equations

$$\sum_{(i)} \boldsymbol{\alpha}_{i,i-1} = \mathbf{0} \text{ and}$$

$$\sum_{(i)} \mathbf{r}_{A_i} \times \boldsymbol{\alpha}_{i,i-1} + \sum_{(i)} \mathbf{a}^{\mathrm{r}}_{A_{i,i-1}} + \sum_{(i)} \mathbf{a}^{\mathrm{c}}_{A_{i,i-1}} - \omega_i^2\, \mathbf{r}_{A_iA_{i+1}} = \mathbf{0} \tag{5.93}$$

   represent the acceleration equations for a simple closed kinematic chain in planar
   motion.
2. The Coriolis acceleration, given by the expression

$$\mathbf{a}^{\mathrm{c}}_{A_{i,i-1}} = 2\,\boldsymbol{\omega}_{i-1} \times \mathbf{v}^{\mathrm{r}}_{A_{i,i-1}},$$

   vanishes when $\boldsymbol{\omega}_{i-1} = \mathbf{0}$, or $\mathbf{v}^{\mathrm{r}}_{A_{i,i-1}} = \mathbf{0}$, or when $\boldsymbol{\omega}_{i-1}$ is parallel to $\mathbf{v}^{\mathrm{r}}_{A_{i,i-1}}$.

5.9 Independent Closed Loops Method

A diagram is used to represent a mechanism in the following way: The numbered
links are the nodes of the diagram and are represented by circles, and the joints
are represented by lines which connect the nodes. The maximum number of
independent closed loops is given by

$$N = c - n \quad \text{or} \quad n_c = N = c - p + 1, \tag{5.94}$$

where c is the number of joints, n is the number of moving links, and p is the number
of links.

The equations for velocities and accelerations are written for any closed loop of
the mechanism. However, it is best to write the closed loops equations only for the
independent loops of the diagram representing the mechanism.

Step 1. Determine the position analysis of the mechanism.
Step 2. Draw a diagram representing the mechanism and select the independent
closed loops. Determine a path for each closed loop.
Step 3. For each closed loop, write the closed loop velocity relations and the closed
loops acceleration relations. For a closed kinematic chain in planar motion,
the following equations will be used:

$$\sum_{(i)} \boldsymbol{\omega}_{i,i-1} = \mathbf{0},$$

$$\sum_{(i)} \mathbf{r}_{A_i} \times \boldsymbol{\omega}_{i,i-1} + \sum_{(i)} \mathbf{v}^{\mathrm{r}}_{A_{i,i-1}} = \mathbf{0}. \tag{5.95}$$

$$\sum_{(i)} \boldsymbol{\alpha}_{i,i-1} = \mathbf{0},$$

$$\sum_{(i)} \mathbf{r}_{A_i} \times \boldsymbol{\alpha}_{i,i-1} + \sum_{(i)} \mathbf{a}^{\mathrm{r}}_{A_{i,i-1}} + \sum_{(i)} \mathbf{a}^{\mathrm{c}}_{A_{i,i-1}} - \omega_i^2 \mathbf{r}_{A_i A_{i+1}} = \mathbf{0}. \tag{5.96}$$

Step 4. Project on a Cartesian reference system the velocity and acceleration equations. Linear algebraic equations are obtained where the unknowns are

- The components of the relative angular velocities $\boldsymbol{\omega}_{j,j-1}$
- The components of the relative angular accelerations $\boldsymbol{\alpha}_{j,j-1}$
- The components of the relative linear velocities $\mathbf{v}^{\mathrm{r}}_{Aj,j-1}$
- The components of the relative linear accelerations $\mathbf{a}^{\mathrm{r}}_{Aj,j-1}$

Solve the algebraic system of equations and determine the unknown kinematic parameters.

Step 5. Determine the absolute angular velocities $\boldsymbol{\omega}_j$ and the absolute angular accelerations $\boldsymbol{\alpha}_j$. Compute the velocities and accelerations of the characteristic points and joints.

Exercise

A planetary gear train is shown in Fig. 5.33. The ring gear 1 has $N_1 = 66$ internal gear teeth, the planet gear 2 has $N_2 = 26$ external gear teeth, and the planet gear $2'$ has $N_{2'} = 20$ external gear teeth. Gears 2 and $2'$ are fixed on the same shaft. The planet gear 3 has $N_3 = 24$ teeth, and gears 3 and $3'$ are fixed on the same shaft. The ring gear 4 has $N_4 = 108$ internal gear teeth. Gear 1 rotates with the input angular speed $n_1 = 300$ rpm, and arm 5 rotates at $n_5 = -150$ rpm (n_1 is opposite to n_5). The module of the gears is $m = 18$, and the pressure angle of the gears is $20°$. Find the angular velocity of the output ring gear 4, ω_4.

Solution

There are five moving links (1, 2, 3, 4, and 5), $n = 5$, connected by:

- Five one degree of freedom joints ($c_5 = 5$): one at A, between the frame 0 and the gear 1; one at C, between the link 2 and the arm 5; one at E, between the link 3 and the arm 5; one at G, between the arm 5 and the frame 0; and one at G, between the gear 4 and the frame 0.
- Three two degrees of freedom joints ($c_4 = 3$): one at B, between the gear 1 and the gear 2; one at D, between the gear 2' and the gear 3; and one at F, between the gear 3' and the gear 4.

Using Chebyshev's formula, the system has two degrees of freedom:

$$M = 3n - 2c_5 - c_4 = 3(5) - 2(5) - 3 = 2.$$

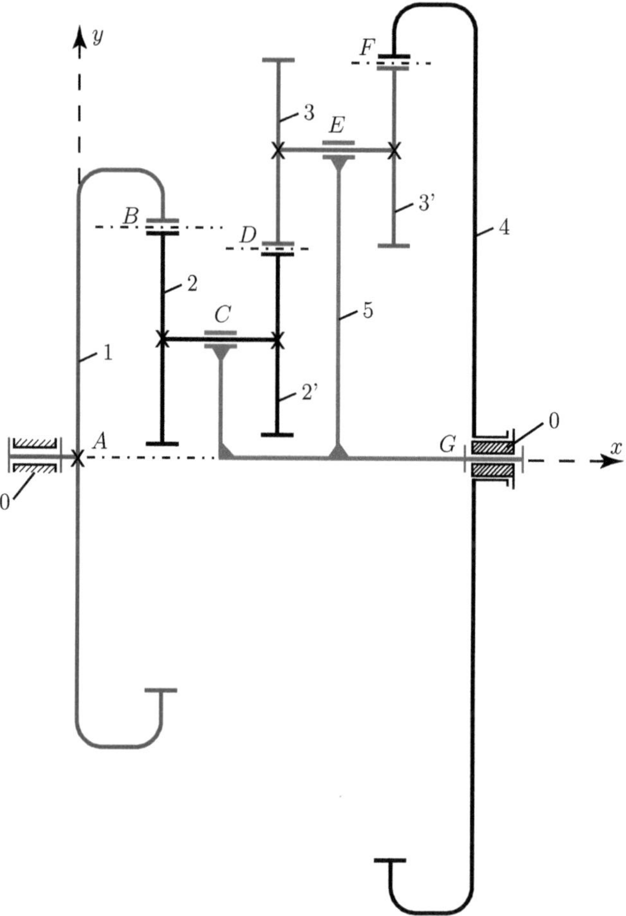

Fig. 5.33 Planetary gear train

The system shown in Fig. 5.33 has a total of six links $(0, 1, 2, 3, 4, 5)$, $p = 6$, and eight joints, $c = 8$. The number of independent closed loops is given by

$$n_c = c - p + 1 = 8 - 6 + 1 = 3.$$

This gear system has three independent closed loops. The graph representing the kinematic chain and the independent contours are shown in Fig. 5.34.

The gear 1 has a radius of the pitch circle equal to r_1, the planet gear 2 has a radius of the pitch circle equal to r_2, the planet gear 2' has a radius of the pitch circle equal to $r_{2'}$, the planet gear 3 has a radius of the pitch circle equal to r_3, the planet gear 3' has a radius of the pitch circle equal to $r_{3'}$, and the gear 4 has a radius

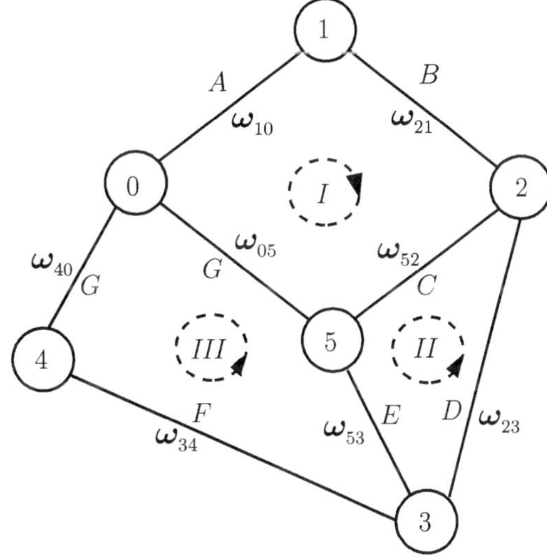

Fig. 5.34 Independent closed loops for planetary gear train

of the pitch circle equal to r_4. The radius of the pitch circles are calculated with the following relations:

$$r_1 = mN_1/2, \ r_2 = mN_2/2, \ r_{2'} = mN_{2'}/2, \ r_3 = mN_3/2, \ r_4 = mN_4/2,$$

$$r_{3'} = r_4 - (r_1 - r_2 + r_{2'} + r_3).$$

The position vectors of the representative points are defined as follows:

$$\mathbf{r}_A = [0,0,0],$$

$$\mathbf{r}_B = [x_B, y_B, 0] = [x_B, r_1, 0],$$

$$\mathbf{r}_C = [x_C, y_C, 0] = [x_C, r_1 - r_2, 0],$$

$$\mathbf{r}_D = [x_D, y_D, 0] = [x_D, r_1 - r_2 + r_{2'}, 0],$$

$$\mathbf{r}_E = [x_E, y_E, 0] = [x_E, r_1 - r_2 + r_{2'} + r_3, 0],$$

$$\mathbf{r}_F = [x_F, y_F, 0] = [x_F, r_1 - r_2 + r_{2'} + r_3 + r_{3'}, 0],$$

$$\mathbf{r}_G = [x_G, y_G, 0] = [x_G, 0, 0].$$

The MATLAB commands are

```
r1  = m*N1/2;
r2  = m*N2/2;
r2p= m*N2p/2;
r3  = m*N3/2;
r4  = m*N4/2;
r3p= r4 - (r1 - r2 + r2p + r3);
```

```
syms xB xC xD xE xF xG

rA = [0, 0, 0];
rB = [xB, r1, 0];
rC = [xC, r1 - r2, 0];
rD = [xD, r1 - r2 + r2p, 0];
rE = [xE, r1 - r2 + r2p + r3, 0];
rF = [xF, r1 - r2 + r2p + r3 + r3p, 0];
rG = [xG, 0, 0];
```

First Independent Closed Loop
The first independent closed loop (following the clockwise path) contains

$$0 - \frac{A}{-} \rightarrow 1 - \frac{B}{-} \rightarrow 2 - \frac{C}{-} \rightarrow 5 - \frac{G}{-} \rightarrow 0.$$

For the velocity analysis, the following vectorial equations can be written:

$$\boldsymbol{\omega}_{10} + \boldsymbol{\omega}_{21} + \boldsymbol{\omega}_{52} + \boldsymbol{\omega}_{05} = \mathbf{0},$$

$$\mathbf{r}_A \times \boldsymbol{\omega}_{10} + \mathbf{r}_B \times \boldsymbol{\omega}_{21} + \mathbf{r}_C \times \boldsymbol{\omega}_{52} + \mathbf{r}_G \times \boldsymbol{\omega}_{05} = \mathbf{0}, \tag{5.97}$$

where the input angular velocities are

$$\boldsymbol{\omega}_{10} = [\omega_{10}, 0, 0] = [n_1, 0, 0],$$
$$\boldsymbol{\omega}_{05} = [\omega_{05}, 0, 0] = [-n_5, 0, 0],$$

and the unknown angular velocities are

$$\boldsymbol{\omega}_{21} = [n_{21}, 0, 0],$$
$$\boldsymbol{\omega}_{52} = [n_{52}, 0, 0].$$

The MATLAB statements for the first loop are

```
% Loop I: 0-A-1-B-2-C-5-G-0
fprintf('\n')
fprintf('Loop I \n')
syms n21s n52s
omega10  = [n10 , 0, 0];
omega21s = [n21s, 0, 0];
omega52s = [n52s, 0, 0];
omega05  = [n05 , 0, 0];
eqIomega = omega10+omega21s+omega52s+omega05;
eqIx = eqIomega(1);
digits(3);
eIx = vpa(eqIx);
```

```
fprintf('%s = 0 (1)\n', char(eIx))
eqIv = ...
 cross(rA,omega10)+cross(rB,omega21s)...
+cross(rC,omega52s)+cross(rG,omega05);
eqIz = eqIv(3);
fprintf('%s = 0 (2)\n\n', char(vpa(eqIz)))
solIv = solve(eqIx, eqIz);
n21n = eval(solIv.n21s);
n52n = eval(solIv.n52s);
fprintf('n21 = %6.3f rpm\n', n21n)
fprintf('n52 = %6.3f rpm\n', n52n)
omega21n = [n21n, 0, 0];
omega52n = [n52n, 0, 0];
```

Solving the algebraic equations, the following values are obtained:

```
n21 = 692.308 rpm
n52 = -1142.308 rpm
```

Second Independent Closed Loop
The second independent closed loop contains

$$5 - \frac{E}{-} \to 3 - \frac{D}{-} \to 2 - \frac{C}{-} \to 5.$$

For the velocity analysis, the following vectorial equations can be written:

$$\omega_{35} + \omega_{23} + \omega_{52} = \mathbf{0},$$

$$\mathbf{r}_E \times \omega_{35} + \mathbf{r}_D \times \omega_{23} + \mathbf{r}_C \times \omega_{52} = \mathbf{0}, \tag{5.98}$$

where the angular velocity $\omega_{52} = [n_{52}, 0, 0]$ is known and

$$\omega_{35} = [n_{35}, 0, 0],$$

$$\omega_{23} = [n_{23}, 0, 0],$$

are the unknown angular velocities. The MATLAB statements for this loop are

```
% Loop II: 5-E-3-D-2-C-5
fprintf('\n')
fprintf('Loop II \n')
syms n35s n23s
omega35s = [n35s, 0, 0];
omega23s = [n23s, 0, 0];
eqIIomega = omega35s+omega23s+omega52n;
eqIIx = eqIIomega(1);
fprintf('%s = 0 (3)\n', char(vpa(eqIIx)))
```

```
eqIIv = ...
 cross(rE,omega35s)+cross(rD,omega23s)...
+cross(rC,omega52n);
eqIIz = eqIIv(3);
fprintf('%s = 0  (4)\n\n', char(vpa(eqIIz)))
solIIv = solve(eqIIx, eqIIz);
n35n = eval(solIIv.n35s);
n23n = eval(solIIv.n23s);
fprintf('n35 = %6.3f rpm\n', n35n)
fprintf('n23 = %6.3f rpm\n', n23n)
omega35n = [n35n, 0, 0];
omega23n = [n23n, 0, 0];
```

Solving the algebraic equations the following values are obtained:

```
n35 = -951.923 rpm
n23 = 2094.231 rpm
```

Third Independent Closed Loop
The third independent closed loop contains

$$0 - \frac{G}{-} \to 4 - \frac{F}{-} \to 3 - \frac{E}{-} \to 5 - \frac{G}{-} \to 0.$$

For the velocity analysis, the following vectorial equations can be written:

$$\boldsymbol{\omega}_{40} + \boldsymbol{\omega}_{34} + \boldsymbol{\omega}_{53} + \boldsymbol{\omega}_{05} = \mathbf{0},$$

$$\mathbf{r}_G \times \boldsymbol{\omega}_{40} + \mathbf{r}_F \times \boldsymbol{\omega}_{34} + \mathbf{r}_E \times \boldsymbol{\omega}_{53} + \mathbf{r}_G \times \boldsymbol{\omega}_{05} = \mathbf{0}, \qquad (5.99)$$

where the angular velocity $\boldsymbol{\omega}_{53} = -\boldsymbol{\omega}_{35}$ is known and

$$\boldsymbol{\omega}_{40} = [n_{40}, \, 0, \, 0],$$

$$\boldsymbol{\omega}_{34} = [n_{34}, \, 0, \, 0],$$

are the unknown angular velocities. The MATLAB statements for the loop are

```
% Loop III: 0-G-4-F-3-E-5-G-0
fprintf('\n')
fprintf('Loop III \n')
omega53 = -omega35n;
syms n40s n34s
omega40s = [n40s, 0, 0];
omega34s = [n34s, 0, 0];
eqIIIomega = omega40s+omega34s+omega53+omega05;
eqIIIx = eqIIIomega(1);
fprintf('%s = 0 (5)\n', char(vpa(eqIIIx)))
```

```
eqIIIv = ...
 cross(rG,omega40s)+cross(rF,omega34s)...
+cross(rE,omega53)+cross(rG,omega05);
eqIIIz = eqIIIv(3);
fprintf('%s = 0 (6)\n\n', char(vpa(eqIIIz)))
solIIIv = solve(eqIIIx, eqIIIz);
n34n = eval(solIIIv.n34s);
n40n = eval(solIIIv.n40s);
fprintf('n34 = %6.3f rpm\n', n34n)
fprintf('n40 = %6.3f rpm = %6.3f rad/s\n',...
    n40n, n40n*pi/30)
omega34n = [n34n, 0, 0];
omega40n = [n40n, 0, 0];
```

Solving the algebraic equations, the following values are obtained:

```
n34 = -740.385 rpm
n40 = -361.538 rpm = -37.860 rad/s
```

5.10   Closed Kinematic Chains with MATLAB Functions

The analysis of planar mechanisms with dyads is studied with the help of m functions written in MATLAB. The solution of the whole system can then be obtained by composing partial solutions. This approach will eliminate the need of storing complete mechanism information in a large database.

The advantage of the system group classification lies in the fact that the solution can then be obtained by composing the partial solutions. Using MATLAB packages for the system groups, the planar mechanisms can be analyzed in a systematic way.

5.10.1   Driver Link

Functions can be used a to calculate the position, velocity, and acceleration of a driver link in rotational motion Fig. 5.35. For the position analysis, the input data are the coordinates (x_A, y_A) of the joint A (the joint between the driver link and the ground) with respect to the reference frame $xOyz$, the length of the link AB, and the angle ϕ with the horizontal axis. For the velocity and the acceleration analysis, the angular velocity $\omega = \dot{\phi}$ and the angular acceleration $\alpha = \ddot{\phi}$ are considered. The output data are the position, velocity, and acceleration of the end point B.

The position equations for the driver link are

$$x_B = x_A + AB\cos\phi,$$

$$y_B = y_A + AB\sin\phi,$$

where x_B and y_B are the coordinates of the point B.

Fig. 5.35 Driver link

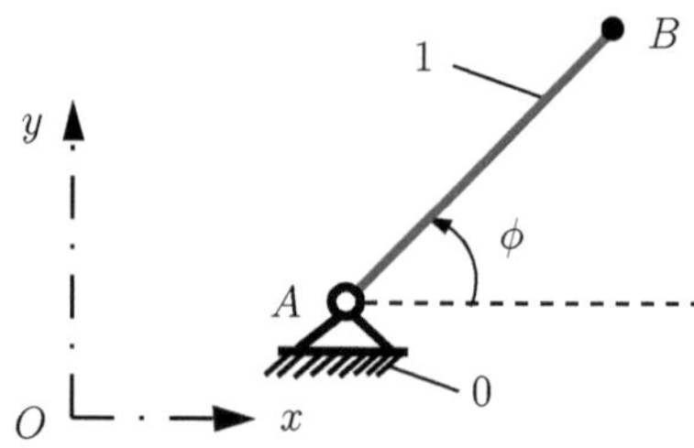

The velocity equations for the driver link are

$$v_{Bx} = -AB\omega \sin\phi,$$

$$v_{By} = AB\omega \cos\phi,$$

where v_{Bx} and v_{By} are the velocity components of the point B on the x- and y-axes.
The acceleration equations for the driver link are

$$a_{Bx} = -AB\omega^2 \cos\phi - AB\alpha \sin\phi,$$

$$a_{By} = -AB\omega^2 \sin\phi + AB\alpha \cos\phi,$$

where a_{Bx} and a_{By} are the acceleration components of the point B on the x- and y-axes.

In order to compute the position, velocity, and acceleration of the joint B, using MATLAB, the necessary commands can be can collected in a function. The name of the function is `driver`, and the m-file is `driver.m`. The input data, the variable parts of the computation, and the output data are defined as parameters to this function. All the variables local to `driver` are declared in the `function` statement to isolate them from any values they might have globally. The MATLAB command `function out=driver(xA, yA, AB, phi, omega, alpha);` specifies that `driver` is to be evaluated with the local values `xA, yA, AB, phi, omega, alpha`.

The MATLAB command `out` returns the value of the `driver` function. For the driver link , `driver` represents the output data, and it is a vector that contains the elements `xB, yB, vBx, vBy, aBx,` and `aBy`. The following MATLAB function is introduced:

```
% driver.m
% driver link
% position, velocity, and acceleration

function out = driver(xA, yA, AB, phi, omega, alpha);
xB = xA + AB*cos(phi);
yB = yA + AB*sin(phi);
```

Fig. 5.36 Dyad RRR

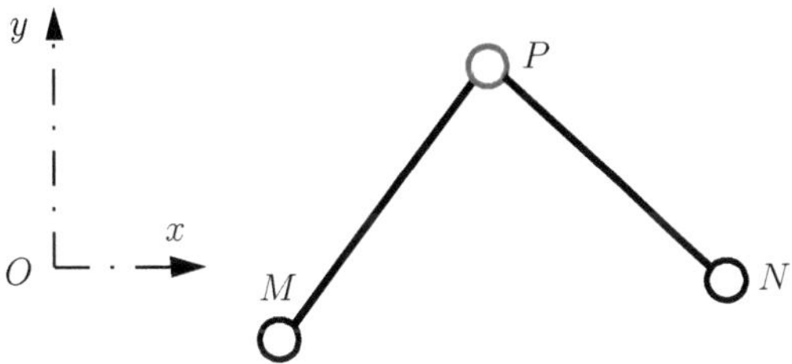

```
vBx  = -  AB*omega*sin(phi);
vBy  =    AB*omega*cos(phi);

aBx  = -  AB*omega^2*cos(phi)  -  AB*alpha*sin(phi);
aBy  = -  AB*omega^2*sin(phi)  +  AB*alpha*cos(phi);

out  =  [xB, yB, vBx, vBy, aBx, aBy];
```

5.10.2   Position Analysis

RRR Dyad

The RRR dyad is shown in Fig. 5.36. The input data for position analysis are the coordinates of the joint $M(x_M, y_M)$, the coordinates of the joint $N(x_N, y_N)$, and the lengths of the segments MP and NP.

The output data are the coordinates of the joint $P(x_P, y_P)$. The position equations for the RRR dyad are

$$(x_M - x_P)^2 + (y_M - y_P)^2 = MP^2,$$
$$(x_N - x_P)^2 + (y_N - y_P)^2 = NP^2,$$

where the unknowns are the coordinates x_P and y_P of the joint P. There are two solutions for the position of the joint P: (x_{P1}, y_{P1}) and (x_{P2}, y_{P2}). The MATLAB function for the positions of $x_{P_1}, y_{P_1}, x_{P_2}, y_{P_2}$ is

```
% pRRR.m
% position RRR dyad

function out = pRRR(xM, yM, xN, yN, MP, PN);
xP=sym('xP','real');
yP=sym('yP','real');

eqRRR1  =  (xM-xP)^2+(yM-yP)^2-MP^2;
eqRRR2  =  (xN-xP)^2+(yN-yP)^2-PN^2;
```

Fig. 5.37 Dyad RRT

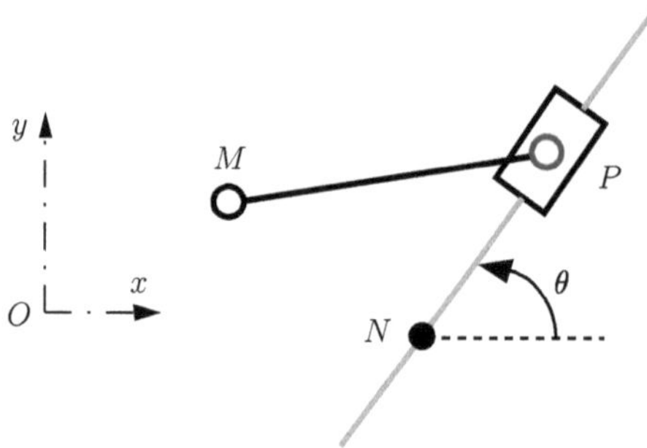

```
solRRR = solve(eqRRR1, eqRRR2);
xPpos = eval(solRRR.xP);
yPpos = eval(solRRR.yP);
xP1 = xPpos(1);  xP2 = xPpos(2);
yP1 = yPpos(1);  yP2 = yPpos(2);

out = [xP1 yP1 xP2 yP2];
end
```

RRT Dyad

The RRT dyad is shown in Fig. 5.37. The input data are the coordinates of the
joint $M(x_M, y_M)$, the coordinates of the point $N(x_N, y_N)$ on the sliding direction,
the length of the segment MP, and the value of the angle θ. The output data are the
coordinates of the joint $P(x_P, y_P)$. The position equations for the RRT dyad are

$$(x_M - x_P)^2 + (y_M - y_P)^2 = MP^2,$$

$$\tan \theta = \frac{y_P - y_N}{x_P - x_N},$$

where the unknowns are the coordinates x_P and y_P of the joint P. There are two
solutions for the position of the joint P; those are (x_{P1}, y_{P1}) and (x_{P2}, y_{P2}).

If the value of the angle θ is $0°$ or $180°$, then $y_P = y_N$ and the following equation
is used to find the coordinate x_P of the point P:

$$(x_M - x_P)^2 + (y_M - y_N)^2 = MP^2. \tag{5.100}$$

If the value of the angle θ is $90°$ or $270°$, then $x_P = x_N$ and the following equation
is used to find the coordinate y_P of the point P:

$$(x_M - x_N)^2 + (y_M - y_P)^2 = MP^2. \tag{5.101}$$

The MATLAB function for the position analysis is

```matlab
% pRRT.m
% position RRT dyad

function out = pRRT(xM, yM, xN, yN, MP, Theta);
xP=sym('xP','real');
yP=sym('yP','real');
if Theta==pi/2 || Theta==3*pi/2
   xP1 = xN; xP2 = xN;
   eqRRT = (xM-xN)^2+(yM-yP)^2-MP^2;
   solRRT = solve(eqRRT);
   yP1 = eval(solRRT(1));
   yP2 = eval(solRRT(2));
elseif Theta==0 || Theta==pi
    yP1 = yN; yP2 = yN;
    eqRRT = (xM-xP)^2+(yM-yN)^2-MP^2;
    solRRT = solve(eqRRT);
    xP1 = eval(solRRT(1));
    xP2 = eval(solRRT(2));
else
    eqRRT1 = (xM-xP)^2 + (yM-yP)^2 - MP^2';
    eqRRT2 = tan(Theta) - (yP-yN)/(xP-xN)';
    solRRT = solve(eqRRT1, eqRRT2);
    xPpos = eval(solRRT.xP);
    yPpos = eval(solRRT.yP);
    xP1 = xPpos(1); xP2 = xPpos(2);
    yP1 = yPpos(1); yP2 = yPpos(2);
   end
out = [xP1 yP1 xP2 yP2];
end
```

R-RTR-RRT Mechanism

The planar R-RTR-RRT mechanism considered is shown in Fig. 5.38a. Given the input data $AB = 0.20$ m, $AD = 0.40$ m, $CD = 0.70$ m, $CE = 0.30$ m, the angle of the driver link AB with the horizontal axis, $\phi = 45°$, calculate the positions of the joints. The distance from the slider 5 to the horizontal x-axis is $y_E = 0.35$ m.

A Cartesian reference frame xyz is chosen with the origin at $A \equiv O$. The coordinates of the joint D are $x_D = 0$, $y_D = -AD = -0.400$ m. The MATLAB commands of the file R_RTR_RRT_p.m for the input data are

```matlab
% R_RTR_RRT_p.m
% R-RTR-RRT mechanism
% position analysis

clear all; clc; close all
```

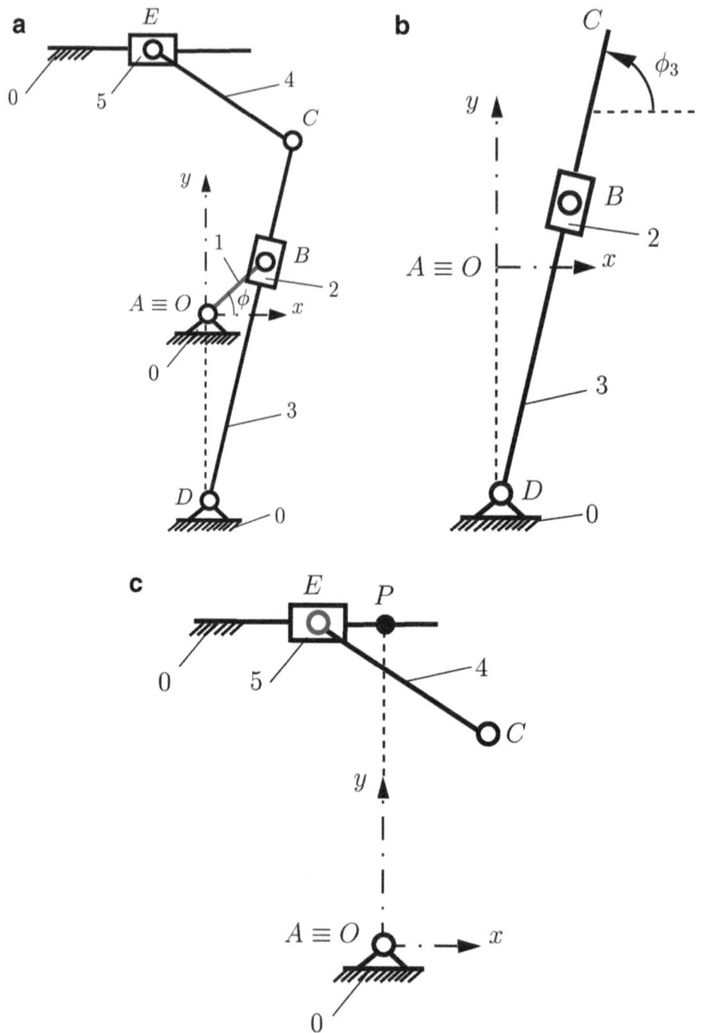

Fig. 5.38 R-RTR-RRT mechanism

```
AB  = 0.20 ;          % m
AD  = 0.40 ;          % m
CD  = 0.70 ;          % m
CE  = 0.30 ;          % m
yE  = 0.35 ;          % m
phi = pi/4;           % rad
```

```
% position of joint A
xA = 0; yA = 0;

% position of joint D
xD = 0; yD = -AD;
rD = [xD yD 0]; % position vector of D
```

Position of Joint B

The coordinates of the joint B are obtained using the MATLAB function `driver`:

```
% position of joint B
% call function driver
jointB = driver(xA,yA,AB,phi,200*pi/30,0);
xB = jointB(1);
yB = jointB(2);
rB = [xB yB 0]; % position vector of B
fprintf('rB = [%6.3f, %6.3f, %g] (m)\n', rB)
```

The function `driver` calculates also the velocity and acceleration of point B. For this example, the constant angular speed of the driver link is 200 rpm. For the R-RTR-RRT mechanism, the segment CD is given, and the points B, C, and D are on the same straight line DBC. The following equations can be written:

$$(x_C - x_D)^2 + (y_C - y_D)^2 = CD^2,$$

$$\tan\phi_2 = \tan\phi_3 = \frac{y_C - y_D}{x_C - x_D} = \frac{y_B - y_D}{x_D - x_D}.$$

The coordinates of point C are calculated from the previous equations. The function, `p3p.m` is introduced as

```
% p3P.m
% position 3 points on a line

function out = p3P(xM, yM, xN, yN, MP);
xP=sym('xP','real');
yP=sym('yP','real');
  eqP1 = (xM-xP)^2 + (yM-yP)^2 - MP^2;
  eqP2 = (yP-yM)/(xP-xM)-(yM-yN)/(xM-xN);
  solP = solve(eqP1, eqP2);
  xPpos = eval(solP.xP);
  yPpos = eval(solP.yP);
  xP1 = xPpos(1); xP2 = xPpos(2);
  yP1 = yPpos(1); yP2 = yPpos(2);
out = [xP1 yP1 xP2 yP2];
end
```

The positions of C is determined with the commands:

```
posC = p3P(xD,yD,xB,yB,CD);
xC1 = posC(1); yC1 = posC(2);
xC2 = posC(3); yC2 = posC(3);
```

The correct solution for this mechanism is selected from the condition $y_C > y_B$ and in MATLAB:

```
if yC1 > yB xC = xC1; yC = yC1;
else xC = xC2; yC = yC2; end
rC = [xC yC 0]; % position vector of C
fprintf('rC = [%6.3f, %6.3f, %g] (m)\n', rC)
```

The angle ϕ_3 of the link 3 with the horizontal is

$$\phi_3 = \arctan\frac{y_B - y_D}{x_B - x_D} = 75.36°.$$

or with MATLAB

```
phi3 = atan((yB-yD)/(xB-xD));
fprintf('phi3 = %6.3f (degrees) \n', phi3*180/pi)
```

The next dyad RRT (CEE) is represented in Fig. 5.38c.

Position of Joint E
In this mechanism, the coordinate y_E of the joint E is constant: $y_E = 0.350$ m. The numerical solution for x_E is obtained using the MATLAB function pRRT. The input data are the coordinates of the joint $C(x_C, y_C)$, the coordinates of the point $P(0, y_E)$ located on the sliding direction, the length of the link CE, and the angle between the sliding direction and the horizontal x-axis, phi5=180°:

```
% position of joint E
phi5 = pi;
% xP=0; yP=yE
% call function pRRT
posE = pRRT(xC,yC,0,yE,CE,phi5);


xE1 = posE(1); xE2 = posE(3);
```

The output data are the first and the third element of the vector returned by the function pRRT, which are the x-coordinates of the joint E. The second and the fourth elements are constant and equal to the y-coordinate of the joint E. The correct solution for x_E for the input driver angle is selected using the condition $x_E < x_C$:

```
if xE1 < xC xE = xE1; else xE = xE2; end
rE = [xE yE 0]; % position vector of E
fprintf('rE = [%6.3f, %6.3f, %g] (m)\n', rE)
```

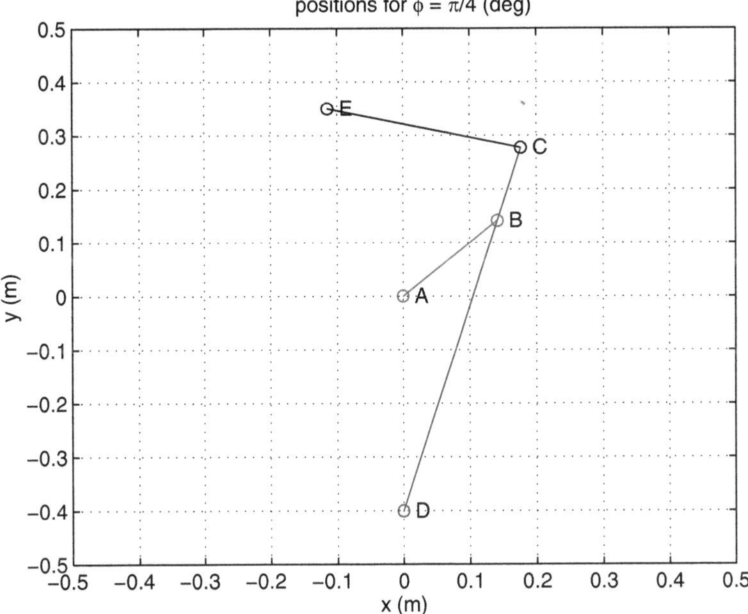

Fig. 5.39 MATLAB representation of R-RTR-RRT mechanism

The angle ϕ_4 of the link 4 with the horizontal is in MATLAB:

```
phi4 = atan2(yE-yC, xE-xC);
fprintf('phi4 = %6.3f (degrees) \n', phi4*180/pi)
```

The following results are obtained:

```
phi = phi1 = 45 (degrees)
rA = [0, 0, 0] (m)
rD = [ 0.000, -0.400, 0] (m)
rB = [ 0.141,  0.141, 0] (m)
rC = [ 0.177,  0.277, 0] (m)
phi2 = phi3 = 75.361 (degrees)
rE = [-0.114,  0.350, 0] (m)
phi4 = 165.971 (degrees)
```

The MATLAB figure of the mechanism is shown in Fig. 5.39. The plot of the
mechanism is obtained with

```
axis manual
axis equal
hold on
grid on
```

```
axis([-0.5 0.5 -0.5 0.5])
xlabel('x (m)'),ylabel('y (m)')

plot([xA,xB],[yA,yB],'r-o',...
     [xD,xC],[yD,yC],'b-o',...
     [xC,xE],[yC,yE],'k-o'),...
xlabel('x (m)'),ylabel('y (m)')

title('positions for \phi = \pi/4 (deg)')
text(xA,yA,'  A'), text(xB,yB,'  B')
text(xC,yC,'  C'), text(xD,yD,'  D')
text(xE,yE,'  E')
```

5.10.3   *Complete Rotation of the Driver Link*

To calculate the position analysis for a complete a complete rotation of the driver link AB, $0 \leq \phi \leq 360°$, the MATLAB statement `for` *var=startval:step:endval, statement* `end` is used. It repeatedly evaluates *statement* in a loop. The counter variable of the loop is *var*. At the start, the variable is initialized to value *startval* and is incremented (or decremented when *step* is negative) by the value *step* for each iteration. The *statement* is repeated until *var* has incremented to the value *endval*. For the considered mechanism, the following applies:

```
step=pi/18;
for phi=pi/4:step:2*pi+pi/4,
Program block;
end;
```

A method for the position analysis for a complete rotation of the driver link uses constraint conditions only for the initial value of the angle ϕ. Next, for the mechanism, the correct position of the joint C is calculated using a simple function, the Euclidian distance between two points P and Q:

$$d = \sqrt{(x_P - x_Q)^2 + (y_P - y_Q)^2}. \tag{5.102}$$

In MATLAB, the following function is introduced with a m-file (`Dist.m`):

```
% Dist.m
function d=Dist(xP,yP,xQ,yQ);
d=sqrt((xP-xQ)^2+(yP-yQ)^2);
end
```

For the R-RTR-RRT mechanism, the initial angle $\phi = \pi/4$, the constraint is $y_B \leq x_C$, so the first position of the joint C, that is, C_0, is calculated for the first step

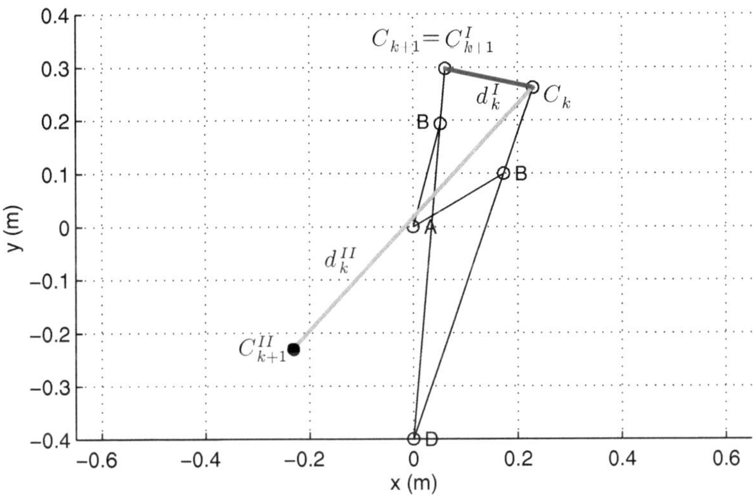

Fig. 5.40 Selection of the correct position: $d^I_k < d^{II}_k \Rightarrow C_{k+1} = C^I_{k+1}$

$C = C_0 = C_k$. For the next position of the joint, C_{k+1}, there are two solutions C^I_{k+1} and C^{II}_{k+1}, $k = 0,\ 1,\ 2,\ldots$. In order to choose the correct solution of the joint, C_{k+1}, the distances between the old position, C_k, and each new calculated positions C^I_{k+1} and C^{II}_{k+1}. The distances between the known solution C_k and the new solutions C^I_{k+1} and C^{II}_{k+1} are d^I_k and d^{II}_k are compared. If the distance to the first solution is less than the distance to the second solution, $d^I_k < d^{II}_k$, then the correct answer is $C_{k+1} = C^I_{k+1}$ or else $C_{k+1} = C^{II}_{k+1}$ (Fig. 5.40). The following MATLAB statements are used to determine the correct position of the joints C and E using single conditions for all four quadrants:

```
% R_RTR_RRRT_p360
% R-RTR-RRT Mechanism
% complete rotation of the driver

clear all; clc; close all

AB = 0.20 ;        % m
AD = 0.40 ;        % m
CD = 0.70 ;        % m
CE = 0.30 ;        % m
yE = 0.35 ;        % m

n = 200;           % rpm
omega=pi*n/30.;    % rad/s
alpha = 0.;        % rad/s^2
```

```matlab
xA = 0; yA = 0;
xD = 0; yD = -AD;
rD = [xD yD 0];

% initial moment phi=pi/4 => increment = 0
increment = 0 ;

% the step has to be small for this method
step=pi/18;
for phi=pi/4:step:2*pi+pi/4,

% position of joint B
jointB = driver(xA,yA,AB,phi,omega,alpha);
xB = jointB(1);
yB = jointB(2);

% position of joint C
posC = p3P(xD,yD,xB,yB,CD);
xC1 = posC(1); yC1 = posC(2);
xC2 = posC(3); yC2 = posC(3);

if increment == 0
  if yC1 > yB xC = xC1; yC = yC1;
  else xC = xC2; yC = yC2; end
else
  dist1 = Dist(xC1,yC1,xCold,yCold);
  dist2 = Dist(xC2,yC2,xCold,yCold);
  if dist1<dist2 xC=xC1; yC=yC1;
  else xC=xC2; yC=yC2; end
end
xCold=xC;
yCold=yC;

% position of joint E
phi5 = pi;
posE = pRRT(xC,yC,0,yE,CE,phi5);
xE1 = posE(1); xE2 = posE(3);

if increment == 0
  if xE1 < xC xE = xE1;
  else xE = xE2; end
else
  dist1 = Dist(xE1,yE,xEold,yE);
  dist2 = Dist(xE2,yE,xEold,yE);
```

```matlab
    if dist1<dist2 xE=xE1;
    else xE=xE2; end
end
xEold=xE;

increment=increment+1;

% centroid of link 4
xC4(increment)=(xC+xE)/2;
yC4(increment)=(yC+yE)/2;

% graphic of the mechanism
axis manual
axis equal
hold on
grid on
axis([-0.65 0.65 -0.4 0.4])
xlabel('x (m)'),ylabel('y (m)')

pM=plot(...
[xA,xB],[yA,yB],'r-o',...
[xD,xC],[yD,yC],'b-o',...
[xC,xE],[yC,yE],'k-o');

plot(xC4, yC4, 'k.','Color', 'red')

pause(1)
delete(pM)

end

text(xA,yA,'  A')
text(xB,yB,'  B')
text(xC,yC,'  C')
text(xD,yD,'  D')
text(xE,yE,'  E')

plot(...
[xA,xB],[yA,yB],'r-o',...
[xD,xC],[yD,yC],'b-o',...
[xC,xE],[yC,yE],'k-o')

title('Path described by C_4')

% end of program
```

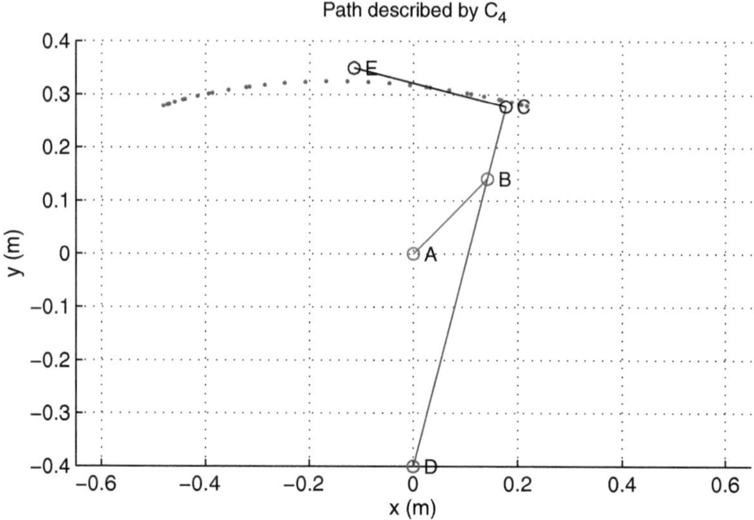

Fig. 5.41 Trajectory of the centroid C_4 of link 4 in general motion

At the beginning of the rotation, the driver link makes an angle `phi=pi/4` with the horizontal, and the value of counter `increment` is 0. The MATLAB statement

```
increment=increment+1;
```

specifies that 1 is to be added to the value in `increment`, and the result stored back in `increment`. The value `increment` should be incremented by 1.

With this algorithm, the correct solution is selected using just one constraint relation for the initial step, and then, automatically, the problem is solved. In this way, it is not necessary to have different constraints for different quadrants.

For the Euclidian distance method, the selection of the step of the angle ϕ is very important. If the step of the angle has a large value, the method might give wrong answers and that is why it is important to check the graphic of the mechanism.

The previous MATLAB program for a complete rotation of the driver link simulates the motion of the mechanism and plots the trajectory of the centroid of link 4 as seen in Fig. 5.41.

5.10.4 Velocity and Acceleration Analysis

RRR Dyad

The input data are the coordinates $x_M, y_M, x_N, y_N, x_P, y_P$ of the joints M, N, and P; the velocities $\dot{x}_M, \dot{y}_M, \dot{x}_N, \dot{y}_N$; and the accelerations $\ddot{x}_M, \ddot{y}_M, \ddot{x}_N, \ddot{y}_N$ of the joints M and N. The output data are the velocities $\dot{x}_P, \dot{y}_P$ and acceleration components $\ddot{x}_P$ and $\ddot{y}_P$ of the joint P.

The velocity equations for the RRR dyad are obtained, taking the derivative of the position equations

$$(x_M - x_P)(\dot{x}_M - \dot{x}_P) + (y_M - y_P)(\dot{y}_M - \dot{y}_P) = 0,$$

$$(x_N - x_P)(\dot{x}_N - \dot{x}_P) + (y_N - y_P)(\dot{y}_N - \dot{y}_P) = 0,$$

where the unknowns are the velocity components $\dot{x}_P$ and $\dot{y}_P$ of the joint P.

The acceleration equations for the RRR dyad are obtained taking the derivative of the velocity equations:

$$(x_M - x_P)(\ddot{x}_M - \ddot{x}_P) + (\dot{x}_M - \dot{x}_P)^2 + (y_M - y_P)(\ddot{y}_M - \ddot{y}_P) + (\dot{y}_M - \dot{y}_P)^2 = 0,$$

$$(x_N - x_P)(\ddot{x}_N - \ddot{x}_P) + (\dot{x}_N - \dot{x}_P)^2 + (y_N - y_P)(\ddot{y}_N - \ddot{y}_P) + (\dot{y}_N - \dot{y}_P)^2 = 0,$$

where the unknowns are the acceleration components $\ddot{x}_P$ and $\ddot{y}_P$ of the joint P.

The MATLAB function for the velocity and acceleration analysis is

```
% vaRRR.m
% velocity and acceleration RRR dyad
function out=vaRRR...
(xM,yM,xN,yN,xP,yP,vMx,vMy,vNx,vNy,aMx,aMy,aNx,aNy);

% velocity
vPxs=sym('vPxs','real');
vPys=sym('vPys','real');
eqRRR1v =...
(xM-xP)*(vMx-vPxs)+(yM-yP)*(vMy-vPys);
eqRRR2v =...
(xN-xP)*(vNx-vPxs)+(yN-yP)*(vNy-vPys);

solRRRv=solve(eqRRR1v, eqRRR2v,'vPxs','vPys');
vPx = eval(solRRRv.vPxs);
vPy = eval(solRRRv.vPys);

% acceleration
aPxs=sym('aPxs','real');
aPys=sym('aPys','real');
eqRRR1a =...
(xM-xP)*(aMx-aPxs)+(vMx-vPx)^2+...
(yM-yP)*(aMy-aPys)+(vMy-vPy)^2;
eqRRR2a =...
(xN-xP)*(aNx-aPxs)+(vNx-vPx)^2+...
(yN-yP)*(aNy-aPys)+(vNy-vPy)^2;
solRRRa=solve(eqRRR1a, eqRRR2a,'aPxs','aPys');
```

```
aPx = eval(solRRRa.aPxs);
aPy = eval(solRRRa.aPys);

out = [vPx vPy aPx aPy];
end
```

RRT Dyad

The input data are the coordinates $x_M, y_M, x_N, y_N, x_P, y_P$ of the joints M, N, and P; the velocities $\dot{x}_M, \dot{y}_M, \dot{x}_N, \dot{y}_N$; the accelerations $\ddot{x}_M, \ddot{y}_M, \ddot{x}_N, \ddot{y}_N$ of the joints M and N; the angle θ; the angular velocity; and acceleration $\dot{\theta}$ and $\ddot{\theta}$. The output data are the velocities $\dot{x}_P, \dot{y}_P$, and accelerations $\ddot{x}_P, \ddot{y}_P$ of the joint P.

The velocity equations for the RRT dyad are obtained, taking the derivative of the position equations

$$(x_M - x_P)(\dot{x}_M - \dot{x}_P) + (y_M - y_P)(\dot{y}_M - \dot{y}_P) = 0,$$

$$(\dot{x}_P - \dot{x}_N)\sin\theta + \dot{\theta}(x_P - x_N)\cos\theta - (\dot{y}_P - \dot{y}_N)\cos\theta + \dot{\theta}(y_P - y_N)\sin\theta = 0,$$

where the unknowns are the velocity components $\dot{x}_P$ and $\dot{y}_P$ of the joint P.

The acceleration equations for the RRT dyad are obtained, taking the derivative of the velocity equations:

$$(x_M - x_P)(\ddot{x}_M - \ddot{x}_P) + (\dot{x}_M - \dot{x}_P)^2 + (y_M - y_P)(\ddot{y}_M - \ddot{y}_P) + (\dot{y}_M - \dot{y}_P)^2 = 0,$$

$$(\ddot{x}_P - \ddot{x}_N)\sin\theta - (\ddot{y}_P - \ddot{y}_N)\cos\theta + [2(\dot{x}_P - \dot{x}_N)\cos\theta - \dot{\theta}(x_P - x_N)\sin\theta$$

$$+2(\dot{y}_P - \dot{y}_N)\sin\theta + \dot{\theta}(y_P - y_N)\cos\theta]\dot{\theta} + [(x_P - x_N)\cos\theta$$

$$+(y_P - y_N)\sin\theta]\ddot{\theta} = 0,$$

where the unknowns are the acceleration components $\ddot{x}_P$ and $\ddot{y}_P$ of the joint P.

The MATLAB function for the velocity and acceleration analysis is

```
% vaRRT.m
% velocity and acceleration RRT dyad
function out=vaRRT(xM,yM,xN,yN,xP,yP,...
  vMx,vMy,vNx,vNy,aMx,aMy,aNx,aNy,dtheta,ddtheta);

theta = atan((yP-yN)/(xP-xN));

% velocity
syms vPxSol vPySol
eqRRT1v =...
(xM-xP)*(vMx-vPxSol)+(yM-yP)*(vMy-vPySol);
eqRRT2v =...
sin(theta)*(vPxSol-vNx)+cos(theta)*dtheta*(xP-xN)...
-cos(theta)*(vPySol-vNy)+sin(theta)*dtheta*(yP-yN)';
```

```
solRRTv = solve(eqRRT1v, eqRRT2v,'vPxSol','vPySol');
vPx = eval(solRRTv.vPxSol);
vPy = eval(solRRTv.vPySol);

% acceleration
syms aPxSol aPySol
eqRRT1a =...
(xM-xP)*(aMx-aPxSol)+(vMx-vPx)^2+...
(yM-yP)*(aMy-aPySol)+(vMy-vPy)^2;
eqRRT2a =...
sin(theta)*(aPxSol-aNx)-cos(theta)*(aPySol-aNy)...
+(2*cos(theta)*(vPx-vNx)-sin(theta)*dtheta*(xP-xN)...
+2*sin(theta)*(vPy-vNy)...
+cos(theta)*dtheta*(yP-yN))*dtheta...
+(cos(theta)*(xP-xN)+sin(theta)*(yP-yN))*ddtheta';
solRRTa = solve(eqRRT1a, eqRRT2a,'aPxSol','aPySol');

aPx = eval(solRRTa.aPxSol);
aPy = eval(solRRTa.aPySol);

out = [vPx vPy aPx aPy];
end
```

Absolute Velocity and Acceleration

A function is used to compute the velocity and acceleration of the point N; knowing
the velocity and acceleration of the point M, both points N and M are located on a
rigid link. The input data are the coordinates x_M, y_M, x_N, and y_N of the points M
and N; the velocity and acceleration components $\dot{x}_M$, $\dot{y}_M$, $\ddot{x}_M$, $\ddot{y}_M$ of the point M;
and the angular velocity and acceleration θ and α of the link. The output data are
the velocity and acceleration components $\dot{x}_N$, $\dot{y}_N$, $\ddot{x}_N$, $\ddot{y}_N$ of the point N.

The following vectorial equation between the velocities $\mathbf{v}_N$ and $\mathbf{v}_M$ of the points
N and M exists as

$$\mathbf{v}_N = \mathbf{v}_M + \boldsymbol{\omega} \times \mathbf{r}_{MN}, \tag{5.103}$$

where $\mathbf{v}_N = \dot{x}_N\mathbf{\imath} + \dot{y}_N\mathbf{\jmath}$, $\mathbf{v}_M = \dot{x}_M\mathbf{\imath} + \dot{y}_M\mathbf{\jmath}$, $\boldsymbol{\omega} = \omega\mathbf{k}$, and $\mathbf{r}_{MN} = (x_N - x_M)\mathbf{\imath} + (y_N - y_M)\mathbf{\jmath}$.

Equation (5.103) is projected on the $\mathbf{\imath}$ and $\mathbf{\jmath}$ directions to find the velocity
components of the point N:

$$\dot{x}_N = \dot{x}_M - \omega(y_N - y_M),$$

$$\dot{y}_N = \dot{y}_M + \omega(x_N - x_M). \tag{5.104}$$

The following vectorial equation between the accelerations $\mathbf{a}_N$ and $\mathbf{a}_M$ of the points
N and M can be written as

$$\mathbf{a}_N = \mathbf{a}_M + \boldsymbol{\alpha} \times \mathbf{r}_{MN} - \omega^2 \mathbf{r}_{MN}, \tag{5.105}$$

where $\mathbf{a}_N = \ddot{x}_N\mathbf{\imath} + \ddot{y}_N\mathbf{\jmath}$, $\mathbf{a}_M = \ddot{x}_M\mathbf{\imath} + \ddot{y}_M\mathbf{\jmath}$, and $\boldsymbol{\alpha} = \alpha\mathbf{k}$.

The acceleration components of the point N are obtained from (5.105):

$$\ddot{x}_N = \ddot{x}_M - \alpha(y_N - y_M) - \omega^2(x_N - x_M),$$

$$\ddot{y}_N = \ddot{y}_M + \alpha(x_N - x_M) - \omega^2(y_N - y_M). \tag{5.106}$$

The MATLAB function for the absolute velocity and acceleration analysis is

```
% linva.m
% linear velocity and acceleration

function out=...
  linva(xM,yM,xN,yN,vMx,vMy,aMx,aMy,dtheta,ddtheta);

vNx = vMx - dtheta*(yN - yM);
vNy = vMy + dtheta*(xN - xM);

aNx = aMx - ddtheta*(yN - yM) - dtheta^2*(xN - xM);
aNy = aMy + ddtheta*(xN - xM) - dtheta^2*(yN - yM);

out = [vNx vNy aNx aNy];
end
```

Angular Velocity and Acceleration
A MATLAB function is used to compute the angular velocity and acceleration of a link. The input data are the coordinates x_M, y_M, x_N, y_N; the velocities $\dot{x}_M, \dot{y}_M, \dot{x}_N, \dot{y}_N$; the accelerations $\ddot{x}_M, \ddot{y}_M, \ddot{x}_N, \ddot{y}_N$ of two points M and N located on the link direction; and the angle θ between the link direction and the horizontal axis. The output data are the angular velocity $\omega = \dot{\theta}$ and the angular acceleration $\alpha = \ddot{\theta}$ of the link.

The slope of the line MN is

$$\tan\theta = \frac{y_M - y_N}{x_M - x_N}. \tag{5.107}$$

The derivative with respect to time of the (5.107) is

$$[(y_M - y_N)\sin\theta + (x_M - x_N)\cos\theta]\omega = (v_{My} - v_{Ny})\cos\theta - (v_{Mx} - v_{Nx})\sin\theta. \tag{5.108}$$

The angular velocity ω is calculated from (5.108). The derivative with respect to time of (5.108) is

$$[(x_M - x_N)\cos\theta + (y_M - y_N)\sin\theta]\alpha = (a_{My} - a_{Ny})\cos\theta - (a_{Mx} - a_{Nx})\sin\theta$$
$$-[(y_M - y_N)\omega\cos\theta + 2(v_{My} - v_{Ny})\sin\theta$$
$$-(x_M - x_N)\omega\sin\theta + 2(v_{Mx} - v_{Nx})\cos\theta]\omega. \tag{5.109}$$

Solving (5.109), the angular acceleration α is obtained.

The MATLAB function for the angular velocity and acceleration analysis is

```
% angva.m
% angular velocity and acceleration

function out=angva...
(xM,yM,xN,yN,vMx,vMy,vNx,vNy,aMx,aMy,aNx,aNy);

theta = atan((yM-yN)/(xM-xN));

dtheta = ...
  (cos(theta)*(vMy-vNy)-sin(theta)*(vMx-vNx))/...
  (sin(theta)*(yM-yN)+cos(theta)*(xM-xN));

ddtheta = ...
  (cos(theta)*(aMy-aNy)-sin(theta)*(aMx-aNx)...
  -(cos(theta)*dtheta*(yM-yN)+2*sin(theta)*(vMy-vNy)...
  -sin(theta)*dtheta*(xM-xN)...
  +2*cos(theta)*(vMx-vNx))*dtheta)/...
  (cos(theta)*(xM-xN)+sin(theta)*(yM-yN));

out = [dtheta ddtheta];
end
```

R-RTR-RRT Mechanism

The position analysis of the planar R-RTR-RRT mechanism considered, see Fig. 5.38a, is presented in Sect. 5.10.2 Position Analysis. Given the constant angular velocity $n = 200$ rpm, calculate the velocities and the accelerations of the joints and the angular velocities and the accelerations of the links when $\phi = \pi/4$ rad/s.

The MATLAB packages R_RTR_RRT_p is loaded in the main program to calculate the position of the mechanism. The angular velocity of the driver link is zero:

$$\alpha = \ddot{\phi} = 0.$$

Velocity and Acceleration of the Joint A

Since the joint A is the origin of the reference frame $xAyz$, the velocity and acceleration of the joint A are

$$\mathbf{v}_A = \mathbf{a}_A = \mathbf{0}.$$

Velocity and Acceleration of the Joint B

The velocity and acceleration components of the joint B are

$$v_{Bx} = -AB\,\omega \sin\phi,$$

$$v_{By} = AB\,\omega \cos\phi,$$

$$a_{Bx} = -AB\,\omega^2 \cos\phi - AB\,\alpha \sin\phi,$$

$$a_{By} = -AB\,\omega^2 \sin\phi + AB\,\alpha \cos\phi.$$

The numerical values for the velocity and acceleration components of the joint B are obtained using the MATLAB function `driver`. The input data and the velocity and acceleration of B are:

```
% R_RTR_RRT_va.m
% R-RTR-RRT Mechanism
% velocity and acceleration analysis

% positions of the mechanism
R_RTR_RRT_p

close all

fprintf('Velocity and acceleration analysis \n\n')
%%%%%%%%%%%%
n = 200.;   % (rpm)
%%%%%%%%%%%%%
omega = pi*n/30; alpha = 0;
omega1 = [0 0 omega];
alpha1 = [0 0 alpha];
fprintf...
('omega1 = [%g, %g, %6.3g]  (rad/s)\n', omega1)
fprintf...
('alpha1 = [%g, %g, %g]  (rad/s^2)\n', alpha1)
fprintf('\n')

jointB = driver(xA,yA,AB,phi,omega,alpha);
vBx = jointB(3);
vBy = jointB(4);
vB = [vBx vBy 0];

aBx = jointB(5);
aBy = jointB(6);
aB = [aBx aBy 0];

fprintf...
('vB = vB1 = vB2 = [%6.3g, %6.3g, %g]  (m/s)\n',vB)
fprintf...
('aB = aB1 = aB2 = [%6.3g, %6.3g, %g]  (m/s^2)\n',aB)
fprintf('\n')
```

Velocity and Acceleration of the Joint D
The velocity and acceleration of the joint D are

$$\mathbf{v}_D = \mathbf{a}_D = \mathbf{0},$$

or in MATLAB

```
vDx=0; vDy=0; aDx=0; aDy=0;
vD = [vDx vDy 0]; aD = [aDx aDy 0];
```

To calculate the angular velocity and acceleration of links 2 and 3, the following MATLAB commands are used:

```
ang3=angva...
(xD,yD,xB,yB,vDx,vDy,vBx,vBy,aDx,aDy,aBx,aBy);
omega3z = ang3(1);
alpha3z = ang3(2);

omega3 = [0 0 omega3z];
alpha3 = [0 0 alpha3z];
omega2 = omega3;
alpha2 = alpha3;

fprintf...
('omega2=omega3 = [%g,%g,%6.3g] (rad/s)\n',omega3)
fprintf...
('alpha2=alpha3 = [%g,%g,%6.3g] (rad/s^2)\n',alpha3)
```

Velocity and Acceleration of the Joint C
The velocity and the acceleration of the joint E are calculated with

$$\mathbf{v}_C = \mathbf{v}_D + \boldsymbol{\omega}_3 \times \mathbf{r}_{DC} \quad \text{and} \quad \mathbf{a}_C = \mathbf{a}_D + \boldsymbol{\alpha}_3 \times \mathbf{r}_{DC} - \omega^2 \mathbf{r}_{DC}.$$

To calculate the velocity components v_{Cx} and v_{Cy} of the joint C, the MATLAB function `linva` is used:

```
vaC=linva(xD,yD,xC,yC,vDx,vDy,aDx,aDy,omega3z,
          alpha3z);
vCx = vaC(1);
vCy = vaC(2);
aCx = vaC(3);
aCy = vaC(4);
vC = [vCx vCy 0];
fprintf('vC = [%6.3f, %6.3f, %d] (m/s)\n', vC)
aC = [aCx aCy 0];
fprintf('aC = [%6.3f, %6.3f, %d] (m/s^2)\n', aC)
fprintf('\n')
```

Velocity and Acceleration of the Joint E
In this particular case, the angular velocity and acceleration of the link 5 are zero:

$$\omega_5 = \alpha_5 = \mathbf{0}.$$

The velocity and the acceleration of a point P, on the sliding direction, are zero:

$$\mathbf{v}_P = \mathbf{a}_P = \mathbf{0}.$$

The velocity and acceleration components of the joint E are calculated using the MATLAB function vaRRT:

```
vaE=vaRRT...
(xC,yC,0,yE,xE,yE,vCx,vCy,0,0,aCx,aCy,0,0,0,0);
vEx = vaE(1);
vEy = vaE(2);
aEx = vaE(3);
aEy = vaE(4);
vE = [vEx vEy 0];
fprintf('vE = [%6.3f, %6.3f, %d] (m/s)\n', vE)
aE = [aEx aEy 0];
fprintf('aE = [%6.3f, %6.3f, %d] (m/s^2)\n', aE)
fprintf('\n')
```

The input data are the coordinates of the joints C, P, and E; the velocities and acceleration components of the joints C and P; the angle ϕ_5; the angular velocity ω_5; and the angular acceleration α_5. The output data are the four elements of the vector returned by the function vaRRT, which are the velocity and acceleration components of the joint E.

The numerical values for the angular velocity ω_4 and the angular acceleration α_4 of the link 4 using theMATLAB function angva are

```
ang4=angva...
(xC,yC,xE,yE,vCx,vCy,vEx,vEy,aCx,aCy,aEx,aEy);
omega4z = ang4(1);
alpha4z = ang4(2);

omega4 = [0 0 omega4z];
alpha4 = [0 0 alpha4z];
fprintf...
('omega4 = [%g,%g,%6.3g] (rad/s)\n',omega4)
fprintf...
('alpha4 = [%g,%g,%6.3g] (rad/s^2)\n', alpha4)
```

The numerical solutions for the velocity and acceleration components are

```
omega1 = [0, 0,    20.9] (rad/s)
alpha1 = [0, 0, 0] (rad/s^2)
```

```
vB  = vB1  = vB2  =  [ -2.96,     2.96,  0]  (m/s)
aB  = aB1  = aB2  =  [    -62,      62,  0]  (m/s^2)


omega2=omega3  =  [0,0,    6.46] (rad/s)
alpha2=alpha3  =  [0,0,   30.4] (rad/s^2)
vC  =  [-4.374,    1.143,  0]  (m/s)
aC  =  [-27.947, -22.882,  0]  (m/s^2)


vE  =  [-4.660,    0.000,  0]  (m/s)
aE  =  [-17.464,   0.000,  0]  (m/s^2)


omega4  =  [0,0,    3.93] (rad/s)
alpha4  =  [0,0,  -82.5] (rad/s^2)
```

5.11   Examples

Example 5.1. Find the velocity of the vertex points A and G of the rectangular prism with the dimensions given in Fig. 5.42a. The rectangular prism has the length a, height b, and width c and has a uniform rotation about the diagonal OF with the angular velocity ω.

Solution
The velocity of the point A is

$$\mathbf{v}_A = \mathbf{v}_O + \boldsymbol{\omega} \times \mathbf{r}_{OA},$$

where $\mathbf{v}_O \equiv \mathbf{0}$ is the velocity of the origin O on the axis of rotation, $\boldsymbol{\omega}$ is the angular velocity, $\mathbf{r}_{OA} = x_A \boldsymbol{\imath} + y_A \boldsymbol{\jmath} + z_A \mathbf{k} = a \boldsymbol{\imath}$, x_A, y_A, z_A are the coordinate of the vertex A, and $\boldsymbol{\imath}, \boldsymbol{\jmath}, \mathbf{k}$ are the unity vectors.
    The angular velocity $\boldsymbol{\omega}$ is

$$\boldsymbol{\omega} = \omega \, \frac{a}{\sqrt{a^2+b^2+c^2}} \boldsymbol{\imath} + \omega \, \frac{b}{\sqrt{a^2+b^2+c^2}} \boldsymbol{\jmath} + \omega \, \frac{c}{\sqrt{a^2+b^2+c^2}} \mathbf{k}.$$

The velocity of A is

$$\mathbf{v}_A = \begin{vmatrix} \boldsymbol{\imath} & \boldsymbol{\jmath} & \mathbf{k} \\ \omega \dfrac{a}{\sqrt{a^2+b^2+c^2}} & \omega \dfrac{b}{\sqrt{a^2+b^2+c^2}} & \omega \dfrac{c}{\sqrt{a^2+b^2+c^2}} \\ a & 0 & 0 \end{vmatrix}$$

$$= \frac{\omega a}{\sqrt{a^2+b^2+c^2}} (c\,\boldsymbol{\jmath} - b\,\mathbf{k}).$$

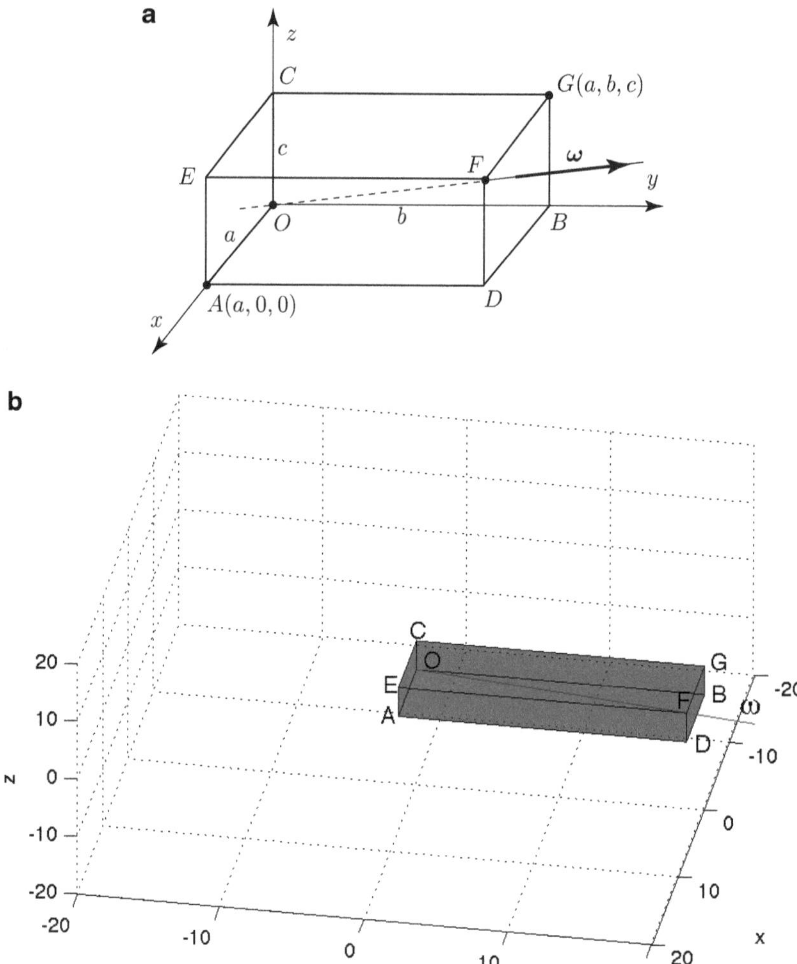

Fig. 5.42 Example 5.1

The magnitude of the velocity is

$$|\mathbf{v}_A| = \omega\, a\, \frac{\sqrt{b^2 + c^2}}{\sqrt{a^2 + b^2 + c^2}}.$$

The velocity of the vertex $G(0, b, c)$ is

$$\mathbf{v}_G = \mathbf{v}_F + \boldsymbol{\omega} \times \mathbf{r}_{FG},$$

where $\mathbf{v}_F \equiv \mathbf{0}$ is the velocity of the point F on the axis of rotation and $\mathbf{r}_{FG} = (x_G - x_F)\mathbf{\imath} + (y_G - y_F)\mathbf{\jmath} + (z_G - z_F)\mathbf{k}$. The coordinates of the vertex F are $x_F = a$, $y_F = b$, $z_F = c$, and the position vector is

$$\mathbf{r}_{FG} = (0 - a)\mathbf{\imath} + (b - b)\mathbf{\jmath} + (c - c)\mathbf{k} = -a\mathbf{\imath}.$$

The velocity of G is

$$\mathbf{v}_G = \begin{vmatrix} \mathbf{\imath} & \mathbf{\jmath} & \mathbf{k} \\ \omega \dfrac{a}{\sqrt{a^2 + b^2 + c^2}} & \omega \dfrac{b}{\sqrt{a^2 + b^2 + c^2}} & \omega \dfrac{c}{\sqrt{a^2 + b^2 + c^2}} \\ -a & 0 & 0 \end{vmatrix}$$

$$= \frac{\omega a}{\sqrt{a^2 + b^2 + c^2}} \left(-c\mathbf{\jmath} + b\mathbf{k} \right).$$

The magnitude of the velocity is

$$|\mathbf{v}_G| = \omega a \frac{\sqrt{b^2 + c^2}}{\sqrt{a^2 + b^2 + c^2}}.$$

The velocities of A and G are equal because the points are located at the same distance from the axis of rotation.

The MATLAB program for the velocities of A and G is

```
clear all; clc; close all

syms a b c omega x_G x_F y_G y_F z_G z_F

% omega=[omega_x, omega_y omega_z]
theta_x=a/sqrt(a^2+b^2+c^2);
theta_y=b/sqrt(a^2+b^2+c^2);
theta_z=c/sqrt(a^2+b^2+c^2);

omega_x=omega*theta_x;
omega_y=omega*theta_y;
omega_z=omega*theta_z;

omega_v=[omega_x omega_y omega_z];

% coordinates of  O, A, B, C, D, E, F, G
x_O=0; y_O=0; z_O=0;
x_A=a; y_A=0; z_A=0;
x_B=0; y_B=b; z_B=0;
x_C=0; y_C=0; z_C=c;
x_D=a; y_D=b; z_D=0;
```

```matlab
x_E=a; y_E=0; z_E=c;
x_F=a; y_F=b; z_F=c;
x_G=0; y_G=b; z_G=c;

r_OA=[x_A y_A z_A];

% velocity of origin O
v_O = [0 0 0];

% velocity of A
v_A=v_O+cross(omega_v,r_OA);
fprintf('v_a = \n')
pretty(v_A); fprintf('\n\n')

% magnitude of v_A
m_v_A=simple(normvec(v_A(1),v_A(2),v_A(3)));
fprintf('|v_A| = \n')
pretty(m_v_A); fprintf('\n\n')

c_x=x_G-x_F;
c_y=y_G-y_F;
c_z=z_G-z_F;

r_FG=[x_G-x_F y_G-y_F z_G-z_F];

% velocity of F
v_F = [0 0 0];

% velocity of G
v_G=v_F+cross(omega_v,r_FG);
fprintf('v_G = \n')
pretty(v_G); fprintf('\n\n')

% magnitude of v_G
m_v_G=simple(normvec(v_G(1),v_G(2),v_G(3)));
fprintf('|v_G| = \n')
pretty(m_v_G); fprintf('\n\n')

delta=m_v_A-m_v_G;
fprintf('|v_A|-|v_G| = %s \n',char(delta))
```

The MATLAB results of the program are

```
  v_a =

    +-                                                              -+
    |            a c omega              a b omega                    |
    |   0,  ------------------, - ------------------                 |
    |          2    2    2 1/2        2    2    2 1/2                 |
    |        (a  + b  + c )         (a  + b  + c )                    |
    +-                                                              -+


  |v_A|  =

     /   2        2    2     2  \1/2
     |  a   omega  (b  + c )  |
     |  ------------------    |
     |     2    2    2        |
     \     a  + b  + c        /


  v_G =

    +-                                                              -+
    |            a c omega              a b omega                    |
    |   0, -  ------------------,  ------------------                |
    |          2    2    2 1/2      2    2    2 1/2                   |
    |        (a  + b  + c )       (a  + b  + c )                     |
    +-                                                              -+


  |v_G|  =

     /   2        2    2     2  \1/2
     |  a   omega  (b  + c )  |
     |  ------------------    |
     |     2    2    2        |
     \     a  + b  + c        /



  |v_A|-|v_G|  = 0
```

The function `normvec.m` is

```
function val = normvec(v_x,v_y,v_z)
% symbolic norm function of a vector
% v=v_x*i+v_y*j+v_z*k.
```

```
% the function accepts a sym as the input argument
val=sqrt(factor(v_x^2+v_y^2+v_z^2));
```

Using the following numerical values $a = 7$ m, $b = 20$ m, and $c = 5$ m, the figure of
the prism with MATLAB is shown in Fig. 5.42b. The motion of the prism is obtained
using the following MATLAB commands:

```
% show and animate the prism

% numerical values
a=7; b=20; c=5; % (m)
x_O=0; y_O=0; z_O=0;
x_A=a; y_A=0; z_A=0;
x_B=0; y_B=b; z_B=0;
x_C=0; y_C=0; z_C=c;
x_D=a; y_D=b; z_D=0;
x_E=a; y_E=0; z_E=c;
x_F=a; y_F=b; z_F=c;
x_G=0; y_G=b; z_G=c;

hold on
axis([-20 20 -20 20 -20 20])
grid on

t1=text...
(0-.2, 0, 0+1.4,' O','fontsize',12);
t2=text...
(x_A, y_A-1.7, z_A-.2,' A','fontsize',12);
t3=text...
(x_B-.2, y_B, z_B-.1,' B','fontsize',12);
t4=text...
(x_C-1.1, y_C-1.1, z_C+.2,' C','fontsize',12);
t5=text...
(x_D+0.4, y_D+.2, z_D+.2,' D','fontsize',12);
t6=text...
(x_E, y_E-1.5, z_E+.2,' E','fontsize',12);
t7=text...
(x_F-1.1, y_F-1.3, z_F+.5,' F','fontsize',12);
t8=text...
(x_G, y_G, z_G+.7,' G','fontsize',12);

%input the prism vertices
vert = [x_O y_O z_O; x_A y_A z_A;...
        x_E y_E z_E; x_C y_C z_C;...
        x_D y_D z_D; x_F y_F z_F;...
        x_G y_G z_G; x_B y_B z_B];
```

```matlab
%input the faces of the prism
fac = [1 2 3 4; ...
    2 1 8 5; ...
    3 4 7 6; ...
    4 1 8 7; ...
    2 3 6 5; ...
    5 6 7 8];
%draw the prism using patch function
prism=patch('Faces',fac,...
'Vertices',vert,'FaceColor','b');

view(100,50);
light('Position',[1 3 2]);
%alpha sets one of three transparency properties
%depending on what arguments you specify
alpha(prism,0.3);

%scale factor
s_f=0.4;

line([x_O,x_F],[y_O,y_F],[z_O,z_F]);
q=quiver3...
(x_F,y_F,z_F,s_f*x_F,s_f*y_F,s_f*z_F);
set(q,'Color','r');

str1 = {'\omega'};
text(x_F+7,y_F+5,16,str1,'fontsize',16);

xlabel('x'); ylabel('y');zlabel('z');

pause(3);
%delete letters A, B, C, D, E, G,
delete(t2);delete(t3);delete(t4);
delete(t5);delete(t6);delete(t8);
%delete the prism
delete(prism);

%rotate the prism about OF
for t=0:0.01:0.15
%call the rotation matrix
%given an axis (omega axis)
%and an angle (angle of rotation)
P5 = Rotate(vert',t,a,b,c);
vert = P5';
% draw the prism
```

```matlab
prism=patch...
('Faces',fac,'Vertices',vert,'FaceColor','b');
alpha(prism,0.3);
pause(0.7);
delete(prism);
end

%final position of the prism
prism=patch...
('Faces',fac,'Vertices',vert,'FaceColor','b');
alpha(prism,0.3);
hold off
```

The function `Rotate.m` used to simulate the rotation of the prism is

```matlab
function [PP] = Rotate(P,theta,a,b,c)
% function [PP] = Rotate(P,theta,index)
% Rotate a point (x,y,z) at the angle theta ccw
% about x-axis, if index = 1,
% about y-axis, if index = 2, and
% about z-axis, if index = 3
% P - coordinate matrix of an old view,
% PP = R*P - a new coordinate matrix of a new view

[m,n] = size(P); % m = 3 for all examples

%input the theta_x, theta_y and theta_z angle

omega=1;
theta_x=a/sqrt(a^2+b^2+c^2);
theta_y=b/sqrt(a^2+b^2+c^2);
theta_z=c/sqrt(a^2+b^2+c^2);

%calculate the vector omega
omega_x=omega*theta_x;
omega_y=omega*theta_y;
omega_z=omega*theta_z;

R=[...
omega_x*omega_x+(1-omega_x*omega_x)*cos(theta) ...
omega_x*omega_y*(1-cos(theta))-omega_z*sin(theta) ...
omega_x*omega_z*(1-cos(theta))+omega_y*sin(theta);...
omega_x*omega_y*(1-cos(theta))+omega_z*sin(theta) ...
omega_y*omega_y+(1-omega_y*omega_y)*cos(theta) ...
omega_y*omega_z*(1-cos(theta))-omega_x*sin(theta);...
omega_x*omega_z*(1-cos(theta))-omega_y*sin(theta) ...
```

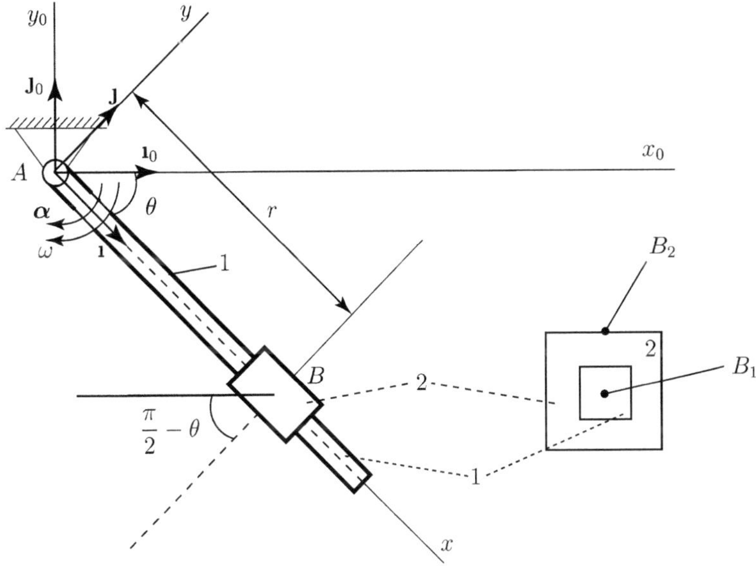

Fig. 5.43 Example 5.2

```
omega_y*omega_z*(1-cos(theta))+omega_x*sin(theta)  ...
omega_z*omega_z+(1-omega_z*omega_z)*cos(theta);];
PP = R*P;
```

Example 5.2. For the angle $\theta = 45°$, the rod 1 shown in Fig. 5.43 has an instantaneous angular velocity of 1 rad/s clockwise and an instantaneous angular acceleration of 2 rad/s^2 clockwise. At this same instant, the slider 2 is traveling outward along the rod such that when $r_{AB} = r = 1$ m, its velocity is 3 m/s and its acceleration is 2 m/s^2, both measured relative to the rod 1. Determine the absolute velocity and acceleration of the slider and the Coriolis acceleration.

Solution

The motion of the slider 2 is reported relative to the rod 1. The x, y, z rotating reference frame is attached to the rod 1 (body-fixed reference frame). The primary reference frame (fixed reference frame) is x_0, y_0, z_0 as shown in Fig. 5.43. The origin of both coordinate (mobile and fixed) systems is located at point A. The absolute velocity of point B_2 on the slider 2 is calculated with the expression

$$\mathbf{v}_{B_2} = \mathbf{v}_A + \boldsymbol{\omega} \times \mathbf{r}_{AB} + \mathbf{v}^r_{B_2(xyz)}, \tag{5.110}$$

where $\mathbf{v}_A$ is the velocity of the origin A, $\mathbf{r}_{AB} = r\mathbf{\imath}$ is the position vector of the point B, $\mathbf{v}^r_{B_2(xyz)}$ is the velocity of point B_2 relative to the rigid rod 1, and $\boldsymbol{\omega} = \omega\mathbf{k}$ is the angular velocity of the rod. The velocity of the point B_1 (A and B_1 two points on the rod) of the link 1 is

$$\mathbf{v}_{B_1} = \mathbf{v}_A + \boldsymbol{\omega} \times \mathbf{r}_{AB} = \mathbf{v}_A + \boldsymbol{\omega} \times \mathbf{r}. \tag{5.111}$$

Using (5.110) and (5.111), it results

$$\mathbf{v}_{B_2} = \mathbf{v}_{B_1} + \mathbf{v}^{\mathrm{r}}_{B_2(xyz)}.$$

The velocity of point B_2 relative to link 1, $\mathbf{v}^{\mathrm{r}}_{B_2(xyz)}$, may be denoted as $\mathbf{v}^{\mathrm{r}}_{B_2 B_1}$ or $\mathbf{v}_{B_2 B_1}$ or $\mathbf{v}_{B_{21}}$ and is calculated as

$$\mathbf{v}_{B_{21}} = \mathbf{v}_{B_2} - \mathbf{v}_{B_1}.$$

The following numerical data are given: $\mathbf{r}_{AB} = r\mathbf{1} = 1\mathbf{1}$ m, $\omega = \omega\mathbf{k} = -1\mathbf{k}$ rad/s, $\alpha = \alpha\mathbf{k} = -2\mathbf{k}$ rad/s^2, $\mathbf{v}_{B_{21}} = v_{B_{21}}\mathbf{1} = 3\mathbf{1}$ m/s, and $\mathbf{a}_{B_{21}} = a_{B_{21}}\mathbf{1} = 2\mathbf{1}$ m/s^2. The position vector $\mathbf{r}_{AB}$ is

$$\mathbf{r}_{AB} = r\mathbf{1} = r\cos\theta\,\mathbf{1}_0 + r\sin\theta\,\mathbf{J}_0$$

$$= 1\cos\theta\,\mathbf{1}_0 + 1\sin\theta\,\mathbf{J}_0$$

$$= 1\mathbf{1} = \frac{\sqrt{2}}{2}\mathbf{1}_0 + \frac{\sqrt{2}}{2}\mathbf{J}_0 \text{ m.}$$

The velocity of B_1 relative to the fixed reference frame expressed in terms of the rotating reference frame is

$$\mathbf{v}_{B_1} = \omega \times \mathbf{r}_{AB} = \begin{vmatrix} \mathbf{1} & \mathbf{J} & \mathbf{k} \\ 0 & 0 & \omega \\ r & 0 & 0 \end{vmatrix} = \omega r\mathbf{J} = (-1)(1)\mathbf{J} = -1\mathbf{J} \text{ m/s.}$$

The velocity vector $\mathbf{v}_{B_1}$ expressed in terms of the fixed reference frame takes the form

$$\mathbf{v}_{B_1} = \omega \times \mathbf{r}_{AB} = \begin{vmatrix} \mathbf{1}_0 & \mathbf{J}_0 & \mathbf{k}_0 \\ 0 & 0 & \omega \\ r\cos\theta & r\sin\theta & 0 \end{vmatrix}$$

$$= -\omega r\sin\theta\,\mathbf{1}_0 + \omega r\cos\theta\,\mathbf{J}_0$$

$$= -(-1)(1)\frac{\sqrt{2}}{2}\mathbf{1}_0 + (-1)(1)\frac{\sqrt{2}}{2}\mathbf{J}_0$$

$$= \frac{\sqrt{2}}{2}\mathbf{1}_0 - \frac{\sqrt{2}}{2}\mathbf{J}_0 \text{ m/s.}$$

The velocity vector $\mathbf{v}_{B_{21}} = \mathbf{v}_{B_2 B_1}$ can be computed as

$$\mathbf{v}_{B_{21}} = v_{B_{21}}\mathbf{1} = v_{B_{21}}\cos\theta\,\mathbf{1}_0 + v_{B_{21}}\sin\theta\,\mathbf{1}_0$$

$$= (3)\cos 45°\,\mathbf{1}_0 + (3)\sin 45°\,\mathbf{J}_0$$

$$= 3\mathbf{1} = \frac{3\sqrt{2}}{2}\mathbf{1}_0 + \frac{3\sqrt{2}}{2}\mathbf{J}_0 \text{ m/s.}$$

The velocity of B_2 relative to the fixed reference frame expressed in terms of the rotating reference frame is

$$\mathbf{v}_{B_2} = \mathbf{v}_{B_1} + \mathbf{v}_{B_2 B_1} = -\mathbf{J} + 3\mathbf{i} = 3\mathbf{i} - \mathbf{J} \ \text{m/s},$$

and

$$|\mathbf{v}_{B_2}| = \sqrt{9+1} = \sqrt{10} = 3.162 \ \text{m/s}.$$

The velocity of B_2 relative to the fixed reference frame expressed in terms of the fixed reference frame is

$$\mathbf{v}_{B_2} = \mathbf{v}_{B_1} + \mathbf{v}_{B_{21}}$$

$$= \frac{\sqrt{2}}{2}\mathbf{I}_0 - \frac{\sqrt{2}}{2}\mathbf{J}_0 + \frac{3\sqrt{2}}{2}\mathbf{J}_0 + \frac{3\sqrt{2}}{2}\mathbf{I}_0$$

$$= 2\sqrt{2}\mathbf{J}_0 + \sqrt{2}\mathbf{I}_0 \ \text{m/s},$$

and

$$|\mathbf{v}_{B_2}| = \sqrt{8+2} = \sqrt{10} \ \text{m/s}.$$

The acceleration of point B_1 on the rigid link 1 is

$$\mathbf{a}_{B_1} = \mathbf{a}_A + \boldsymbol{\alpha} \times \mathbf{r}_{AB} + \boldsymbol{\omega} \times (\boldsymbol{\omega} \times \mathbf{r}_{AB}). \tag{5.112}$$

The acceleration of point B_2 on the slider 2 is

$$\mathbf{a}_{B_2} = \mathbf{a}_A + \mathbf{a}_{B_{21}} + \boldsymbol{\alpha} \times \mathbf{r}_{AB} + 2\boldsymbol{\omega} \times \mathbf{v}_{B_{21}} + \boldsymbol{\omega} \times (\boldsymbol{\omega} \times \mathbf{r}_{AB}), \tag{5.113}$$

where $\mathbf{a}_{B_{21}}$ is the acceleration of point B_2 relative to the rigid rod 1. From (5.113) and (5.112), the acceleration of B_2 is

$$\mathbf{a}_{B_2} = \mathbf{a}_{B_1} + \mathbf{a}_{B_{21}} + 2\boldsymbol{\omega} \times \mathbf{v}_{B_{21}}. \tag{5.114}$$

The relative velocity, $\mathbf{v}_{B_{21}}$, and the relative acceleration, $\mathbf{a}_{B_{21}}$, are parallel to the sliding direction AB. The Coriolis acceleration of acceleration of point B_2 relative to the rigid rod 1 is

$$\mathbf{a}^{c}_{B_{21}} = 2\boldsymbol{\omega} \times \mathbf{v}_{B_{21}}.$$

For the planar motion,

$$\boldsymbol{\omega} \times (\boldsymbol{\omega} \times \mathbf{r}_{AB}) = -\omega^2 \mathbf{r}_{AB}.$$

The acceleration vector $\mathbf{a}_{B_1}$ is computed as

$$\mathbf{a}_{B_1} = \mathbf{0} + \boldsymbol{\alpha} \times \mathbf{r}_{AB} - \omega^2 \mathbf{r}_{AB}$$

$$= \begin{vmatrix} \mathbf{\imath} & \mathbf{j} & \mathbf{k} \\ 0 & 0 & \alpha \\ r & 0 & 0 \end{vmatrix} - \omega^2 r \mathbf{\imath}$$

$$= -\omega^2 r \mathbf{\imath} + \alpha r \mathbf{j}$$

$$= -1^2 (1) \mathbf{\imath} + (-2)(1) \mathbf{j}$$

$$= -\mathbf{\imath} - 2\mathbf{j} \ \text{m/s}^2,$$

and

$$|\mathbf{a}_{B_1}| = 2.236 \ \text{m/s}^2.$$

The vector $\mathbf{a}_{B_{21}}$ takes the form

$$\mathbf{a}_{B_{21}} = 2\mathbf{\imath} \ \text{m/s}^2.$$

The Coriolis acceleration can be computed as

$$\mathbf{a}_{B_{21}}^{c} = 2\boldsymbol{\omega} \times \mathbf{v}_{B_{21}} = 2 \begin{vmatrix} \mathbf{\imath} & \mathbf{j} & \mathbf{k} \\ 0 & 0 & \omega \\ v_{B_{21}} & 0 & 0 \end{vmatrix}$$

$$= 2\omega v_{B_{21}} \mathbf{j} = 2(-1)(3)\mathbf{j}$$

$$= -6\mathbf{j} \ \text{m/s}^2.$$

Numerically, the acceleration ob B_2 is

$$\mathbf{a}_{B_2} = (-\mathbf{\imath} - 2\mathbf{j}) + 2\mathbf{\imath} + (-6)\mathbf{j}$$

$$= \mathbf{\imath} - 8\mathbf{j} \ \text{m/s}^2,$$

and

$$|\mathbf{a}_{B_2}| = \sqrt{1 + 8^2} = 8.062 \ \text{m/s}^2.$$

The MATLAB program for the velocities and accelerations of the two link systems is

```
syms theta r t omega alpha  v_B21 a_B21

r_AB=[r 0 0];
v_A=[0 0 0];
omega_v=[0 0 omega];
```

```matlab
v_B1=v_A+cross(omega_v,r_AB);

fprintf('v_B1 = ')
pretty(v_B1); fprintf('\n\n')


v_B2B1=[v_B21 0 0];

v_B2=v_B1+v_B2B1;
fprintf('v_B2=')
pretty(v_B2); fprintf('\n\n')


% a_B2
a_A=diff(v_A,t);
alpha_v=[0 0 alpha];

a_B1=a_A+cross(alpha_v,r_AB)+...
    cross(omega_v,cross(omega_v,r_AB));
fprintf('a_B1=')
pretty(a_B1); fprintf('\n\n')


a_B2B1=[a_B21 0 0];

a_B21cor=2*cross(omega_v,v_B2B1);
fprintf('Coriolis acceleration: a_B21cor =')
pretty(a_B21cor); fprintf('\n\n')


a_B2=a_A+a_B2B1+...
    cross(alpha_v,r_AB)+...
    2*cross(omega_v,v_B2B1)+...
    cross(omega_v,cross(omega_v,r_AB));
fprintf('a_B2=')
pretty(a_B2); fprintf('\n\n')


% x0 y0 z0 reference frame

RF0=...
[  cos(theta) sin(theta) 0 ;
  -sin(theta) cos(theta) 0 ;
    0 0 1];


r_AB0=r_AB*RF0;
fprintf('r_AB0 = ')
pretty(r_AB0); fprintf('\n\n')
```

```
v_B10=v_A+cross(omega_v,r_AB0);

fprintf('v_B10=')
pretty(v_B10); fprintf('\n\n')

m_vB1=normvec(v_B1(1),v_B1(2),v_B1(3));
m_vB10=simplify...
    (normvec(v_B10(1),v_B10(2),v_B10(3)));

fprintf('|v_B1| = ')
pretty(m_vB1); fprintf('\n')
fprintf('|v_B10| = ')
pretty(m_vB10); fprintf('\n\n')

v_B2B10=v_B2B1*RF0;
v_B20=v_B10+v_B2B10;

fprintf('v_B20x=')
pretty(simplify(v_B20(1)))
fprintf('v_B20y=')
pretty(simplify(v_B20(2)))
fprintf('v_B20z=')
pretty(simplify(v_B20(3)))
fprintf('\n\n')

a_B21cor0=2*cross(omega_v,v_B2B10);
fprintf('Coriolis acceleration RF0: a_B21cor0 =')
pretty(a_B21cor0); fprintf('\n\n')

a_B10=a_A+cross(alpha_v,r_AB0)+...
    cross(omega_v,cross(omega_v,r_AB0));

fprintf('a_B10x=')
pretty(simplify(a_B10(1)))
fprintf('a_B10y=')
pretty(simplify(a_B10(2)))
fprintf('a_B10z=')
pretty(simplify(a_B10(3)))
fprintf('\n\n')

a_B2B10=a_B2B1*RF0;
a_B21cor0=2*cross(omega_v,v_B2B10);

a_B20=a_A+a_B2B10+...
    cross(alpha_v,r_AB0)+...
```

```
      2*cross(omega_v,v_B2B10)+...
      cross(omega_v,cross(omega_v,r_AB0));

fprintf('a_B20x=')
pretty(simplify(a_B20(1)))
fprintf('a_B20y=')
pretty(simplify(a_B20(2)))
fprintf('a_B20z=')
pretty(simplify(a_B20(3)))
fprintf('\n\n')

% numerical data
slist={theta,r,omega,alpha,v_B21,a_B21};
nlist={pi/4,1,-1,-2,3,2};

v_B1n=subs(v_B1,slist,nlist);
v_B10n=subs(v_B10,slist,nlist);

v_B2n=subs(v_B2,slist,nlist);
v_B20n=subs(v_B20,slist,nlist);

a_B1n=subs(a_B1,slist,nlist);
a_B10n=subs(a_B10,slist,nlist);

a_B21corn=subs(a_B21cor,slist,nlist);
a_B21cor0n=subs(a_B21cor0,slist,nlist);

a_B2n=subs(a_B2,slist,nlist);
a_B20n=subs(a_B20,slist,nlist);

fprintf...
('vB1=[%6.3f %6.3f %g]  (m/s)\n',v_B1n)
fprintf...
('RF0: vB10=[%6.3f %6.3f %g]  (m/s)\n',v_B10n)
fprintf...
(' |vB1|=|vB10|=%6.3f  (m/s)\n',norm(v_B10n))
fprintf('\n')

fprintf...
('vB2=[%6.3f %6.3f %g]  (m/s)\n',v_B2n)
fprintf...
('RF0: vB20=[%6.3f %6.3f %g]  (m/s)\n',v_B20n)
fprintf...
(' |vB2|=|vB20|=%6.3f  (m/s)\n',norm(v_B20n))
fprintf('\n')
```

```
fprintf...
('aB1=[%6.3f %6.3f %g] (m/s^2)\n',a_B1n)
fprintf...
('RF0: aB10=[%6.3f %6.3f %g] (m/s^2)\n',a_B10n)
fprintf...
('|aB1|=|aB10|=%6.3f (m/s^2)\n',norm(a_B10n))
fprintf('\n')

fprintf...
('aB21cor=[%6.3f %6.3f %g] (m/s^2)\n',a_B21corn)
fprintf...
('RF0: aB21cor0=[%6.3f %6.3f %g] (m/s^2)\n',...
a_B21cor0n)
fprintf...
('|aB21cor|=|aB21cor0|=%6.3f (m/s^2)\n',...
norm(a_B21cor0n))
fprintf('\n')

fprintf...
('aB2=[%6.3f %6.3f %g] (m/s^2)\n',a_B2n)
fprintf...
('RF0: aB20=[%6.3f %6.3f %g] (m/s^2)\n',a_B20n)
fprintf...
('|aB2|=|aB20|=%6.3f (m/s^2)\n',norm(a_B20n))
fprintf('\n')
```

Example 5.3. The dimensions of the planar mechanism shown in Fig. 5.44a are given below

AB	AC	CD	DE	CF	a
[m]	[m]	[m]	[m]	[m]	[m]
0.15	0.30	0.20	0.14	0.50	0.25

Find the positions of the joints and the angles of the links with the horizontal axis when the angle of the driver link 1 with the horizontal axis is $\phi = \pi/3$.

The angle of the driver link 1 with the horizontal axis is ϕ. The constant angular speed of the driver link 1 is n.

(a) Find the positions of the joints and the angles of the links with the horizontal axis when the angle of the driver link 1 with the horizontal axis is ϕ.
(b) Find the positions of the mechanism for a complete rotation of the driver link 1, $\phi \in [0°, ..., 360°]$.
(c) Find the velocities and the accelerations of the mechanism when the driver link 1 makes an angle ϕ with the horizontal axis using different methods.

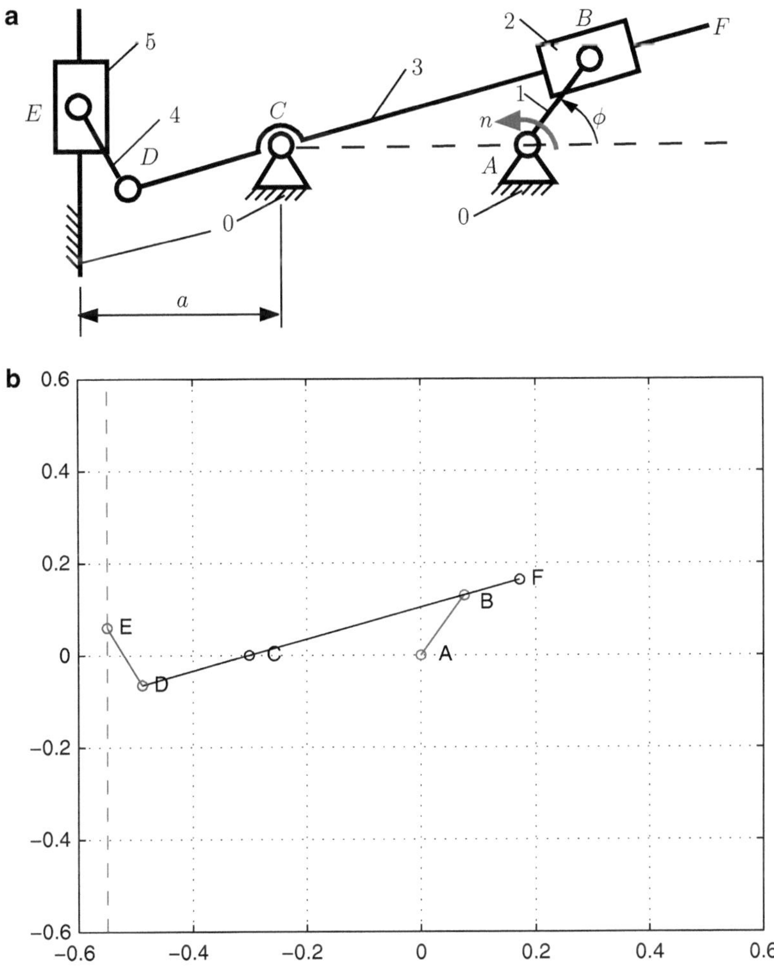

Fig. 5.44 Example 5.3

Solution

The MATLAB commands for the input data are

```
AB  = 0.15;  % m
AC  = 0.30;  % m
CD  = 0.20;  % m
DE  = 0.14;  % m
CF  = 0.50;  % m
a   = 0.25;  % m
phi = pi/3;  % rad
```

A Cartesian reference frame xOy is selected. The joint A is in the origin of the reference frame, that is, $A \equiv O$, $x_A = 0$, $y_A = 0$.

Position of Joint C
The position vector of C is $\mathbf{r}_C = x_C\mathbf{1} + y_C\mathbf{J} = -AC\,\mathbf{J}$ m.

Position of Joint B
The unknowns are the coordinates of the joint B, x_B and y_B. Because the joint A is fixed and the angle ϕ is known, the coordinates of the joint B are computed from the following expressions:

$$x_B = AB\cos\phi = 0.15\cos 60^\circ = 0.075\,\text{m}, \quad y_B = AB\sin\phi = 0.15\sin 60^\circ = 0.130\,\text{m},$$

and $\mathbf{r}_B = x_B\mathbf{1} + y_B\mathbf{J}$. The MATLAB statements for the positions of the joints A, C, E, and B are

```matlab
% Position of joint A
xA=0; yA=0; rA=[xA yA 0];
% Position of joint C
xC=-AC; yC=0; rC=[xC yC 0];
% Position of joint E
xE=-AC-a;
% Position of joint B
xB=AB*cos(phi); yB=AB*sin(phi);
rB=[xB yB 0];
```

Position of Joint D
The unknowns are the coordinates of the joint D, x_D and y_D. The length of the segment CD is constant:

$$(x_D - x_C)^2 + (y_D - y_C)^2 = CD^2. \tag{5.115}$$

The points B, C, and D are on the same straight line with the slope:

$$m = \frac{(y_B - y_C)}{(x_B - x_C)} = \frac{(y_D - y_C)}{(x_D - x_C)}. \tag{5.116}$$

Equations (5.115) and (5.116) form a system from which the coordinates of the joint D can be computed. To solve the system of equations, the MATLAB statement `solve` will be used:

```matlab
% Position of joint D
% Distance formula: CD=constant
xDsol = sym('xDsol','real');
yDsol = sym('yDsol','real');
eqD1=(xC-xDsol)^2+(yC-yDsol)^2-CD^2 ;
% Points B, C, and D on the same line
eqD2=(yC-yB)/(xC-xB)-(yDsol-yC)/(xDsol-xC) ;
% Simultaneously solve above equations
```

```
solD= solve(eqD1, eqD2, 'xDsol, yDsol');
% Two solutions for xD - vector form
xDpos= eval(solD.xDsol);
% Two solutions for yD - vector form
yDpos = eval(solD.yDsol);
% Separate the solutions in scalar form
xD1 = xDpos(1);
xD2 = xDpos(2);
yD1 = yDpos(1);
yD2 = yDpos(2);
```

These solutions D_1 and D_2 are located at the intersection of the line BC with the circle centered in C and radius CD. To determine the correct position of the joint D for the mechanism, an additional condition is needed. For the first quadrant $0 \leq \phi \leq 90°$ and the second quadrant $90° \leq \phi \leq 180°$, the condition is $y_D \leq y_C$. This condition with MATLAB is given by

```
% Select the correct position for D
if (phi>=0 && phi<=pi/2) || (phi >= pi/2 && phi<=pi)
if yD1 <= yC xD=xD1; yD=yD1; else xD=xD2; yD=yD2;end
 else
if yD1 >= yC xD=xD1; yD=yD1; else xD=xD2; yD=yD2;end
end
rD = [xD yD 0]; % Position vector of D
```

The unknown for the joint E is y_E and the length of the segment DE is constant:

$$(x_D - x_E)^2 + (y_D - y_E)^2 = DE^2. \tag{5.117}$$

This MATLAB commands to calculate y_E are

```
% Position of joint E
% Distance formula: DE=constant
yEsol = sym('yEsol','real');
eqE1 = (xD-xE)^2 + (yD-yEsol)^2 - DE^2;
solE = solve(eqE1,'yEsol');
yEpos = eval(solE);
yE1 = yEpos(1);
yE2 = yEpos(2);
if yE1 > yD
    yE=yE1;
else
    yE=yE2;
end
rE = [xE yE 0]; % Position vector of E
```

The angles of the links 2, 3, and 4 with the horizontal are

$$\phi_3 = \arctan\frac{y_B - y_C}{x_B - x_C}, \quad \phi_2 = \phi_3, \quad \phi_4 = \arctan\frac{y_E - y_D}{x_E - x_D},$$

and in MATLAB

```
% Angles of the links with the horizontal
phi3 = atan((yC-yB)/(xC-xB));
phi2 = phi3;
phi4 = atan((yE-yD)/(xE-xD));
```

The point F is calculated in MATLAB with

```
% Position of joint F
xF = xC+CF*cos(phi2);
yF = yC+CF*sin(phi2);
rF = [xF,yF,0];
```

The results are printed using the statements:

```
fprintf('Results \n\n')
fprintf('rA = [%6.3g, %6.3g, %g]  (m) \n',rA)
fprintf('rB = [%6.3g, %6.3g, %g]  (m) \n',rB)
fprintf('rC = [%6.3g, %6.3g, %g]  (m) \n',rC)
fprintf('rD = [%6.3g, %6.3g, %g]  (m) \n',rD)
fprintf('rE = [%6.3g, %6.3g, %g]  (m) \n',rE)
fprintf('rF = [%6.3g, %6.3g, %g]  (m) \n',rF)
fprintf...
    ('phi3 = %6.3g (degrees) \n',phi3*180/pi)
fprintf...
    ('phi4 = %6.3g (degrees) \n',phi4*180/pi)
```

The graph of the mechanism in MATLAB for $\phi = \pi/3$ is given by

```
% Graphic of the mechanism
la=-xE+0.05;
plot(...
[xA,xB],[yA,yB],'r-o',...
[xD,xF],[yD,yF],'k-o',...
[xC,xC],[yC,yC],'k-o',...
[xD,xE],[yD,yE],'b-o',...
[xE,xE],[-la,la],'r--')
grid
text(xA,yA,'   A')
text(xB,yB,'   B')
text(xC,yC,'   C')
text(xD,yD,'  D')
text(xE,yE,'  E')
```

```
text(xF,yF,'  F')
% x limits to the specified values
xlim([-la la]);
% x limits to the specified values
ylim([-la la]);
```

The graphic generated with MATLAB is shown in Fig 5.3b.

Example 5.4. For the mechanism given in Example 5.2, find the positions of the mechanism for a complete rotation of the driver link 1, $\phi \in [0°,...,360°]$.

Solution

For a complete rotation of the driver link AB, $0 \leq \phi \leq 360°$, a step angle is selected. To calculate the position analysis for a complete cycle, the MATLAB statement `for` *var=startval:step:endval, statement* `end` is used. It repeatedly evaluates *statement* in a loop. The counter variable of the loop is *var*. At the start, the variable is initialized to value *startval* and is incremented (or decremented when *step* is negative) by the value *step* for each iteration. The *statement* is repeated until *var* has incremented to the value *endval*.

Constraint conditions are used only for the initial value of the angle $\phi = \phi_0$. Next, for the mechanism, the correct position of the joint D is calculated using a simple function, the Euclidian distance between two points P and Q:

$$d = \sqrt{(x_P - x_Q)^2 + (y_P - y_Q)^2}. \tag{5.118}$$

In MATLAB, the following function is introduced with a m-file (Dist.m):

```
function d=Dist(xP,yP,xQ,yQ);

d=sqrt((xP-xQ)\^{}2+(yP-yQ)\^{}2);

end
```

For the initial angle $\phi = 0°$, the constraint is $x_D \leq x_C$, so the first position of the joint D, that is, D_0, is calculated for the first step $D = D_0 = D_k$. For the next position of the joint, D_{k+1}, there are two solutions D^I_{k+1} and D^{II}_{k+1}, $k = 0, 1, 2,....$ In order to choose the correct solution of the joint, D_{k+1}, the distances between the old position, D_k, and each new positions D^I_{k+1} and D^{II}_{k+1} are calculated. The distances between the known solution D_k and the new solutions D^I_{k+1} and D^{II}_{k+1}, d^I_k and d^{II}_k are compared. If the distance to the first solution is less than the distance to the second solution, $d^I_k < d^{II}_k$, then the correct answer is $D_{k+1} = D^I_{k+1}$ or else $D_{k+1} = D^{II}_{k+1}$. The following MATLAB statements are used to determine the correct position of the joint D:

```
% at the initial moment phi0 =>
incr = 0 ;
```

```matlab
% the step has to be small for this method
step=pi/15;
for phi=phi0:step:2*pi+phi0,
% fprintf('phi =  %g deegres \n', phi*180/pi)


xB=AB*cos(phi); yB=AB*sin(phi);
rB=[xB yB 0];

% Position of joint D
% Distance formula: CD=constant
xDsol = sym('xDsol','real');
yDsol = sym('yDsol','real');
eqD1=(xC-xDsol)^2+(yC-yDsol)^2-CD^2 ;
% Distance formula: CD=constant
eqD2=(yC-yB)/(xC-xB)-(yDsol-yC)/(xDsol-xC);

% Simultaneously solve above equations
solD= solve(eqD1, eqD2, 'xDsol, yDsol');
% Two solutions for xD - vector form
xDpos= eval(solD.xDsol);
% Two solutions for yD - vector form
yDpos = eval(solD.yDsol);
% Separate the solutions in scalar form
xD1 = xDpos(1);
xD2 = xDpos(2);
yD1 = yDpos(1);
yD2 = yDpos(2);

% select the correct position for D
%     only for increment == 0
% the selection process is automatic
%     for all the other steps

if incr == 0
if yD1 <= yC xD=xD1; yD=yD1;
else xD=xD2; yD=yD2; end
else
  dist1 = Dist(xD1,yD1,xDold,yDold);
  dist2 = Dist(xD2,yD2,xDold,yDold);
  if dist1 < dist2 xD=xD1; yD=yD1;
   else xD=xD2; yD=yD2;
   end
end
xDold=xD;
yDold=yD;
```

At the beginning of the rotation, the driver link makes an angle ϕ_0 with the horizontal, and the value of counter `incr` is 0. The MATLAB statement `incr=incr+1;` specifies that 1 is to be added to the value in `increment`, and the result stored back in `increment`. The value `increment` should be incremented by 1.

With this algorithm, the correct solution is selected using just one constraint relation for the initial step, and then, automatically, the problem is solved. In this way, it is not necessary to have different constraints for different quadrants. For the Euclidian distance method, the selection of the step of the angle ϕ is very important. If the step of the angle has a large value, the method might give wrong answers and that is why it is important to check the graphic of the mechanism.

A statement `moviein` can be used to create a matrix large enough to hold 12 frames:

```
M = moviein(12);

incr = 0 ;
step=pi/15;
for phi=phi0:step:2*pi+phi0,
.  .  .  .  .  .  .  .  .  .  .  .  .  .  .  .  .  .  .  .  .  .  .  .  .  .  .  .  .  .  .  .  .

incr=incr+1;
% record the movie
 M(:,incr) = getframe;
end
 movie2avi(M,'R_RTRRRT.avi');
```

The statement, `getframe`, returns the contents of the current axes, exclusive of the axis labels, title, or tick labels. After generating the movie, the statement, `movie2avi(M,'filename.avi')`, creates the AVI movie `filename` from the MATLAB movie M. The `filename` input is a string enclosed in single quotes. In this case, the name of the movie file is `R_RTRRRT.avi` (Fig. 5.45).

The graph of the mechanism for a complete rotation of the driver link is given in Fig. 5.45.

Example 5.5. The dimensions of the planar mechanism shown in Fig. 5.46 are given below:

AB	AC	CD	DE	CF	a	ϕ	n
[m]	[m]	[m]	[m]	[m]	[m]	[°]	[rpm]
0.15	0.30	0.20	0.14	0.50	0.25	200	700

The angular speed of the driver link 1 is $n = 700$ rpm. Find the velocities and the accelerations of the mechanism when the driver link 1 makes an angle $\phi = \phi_1 = 200°$ with the horizontal axis.

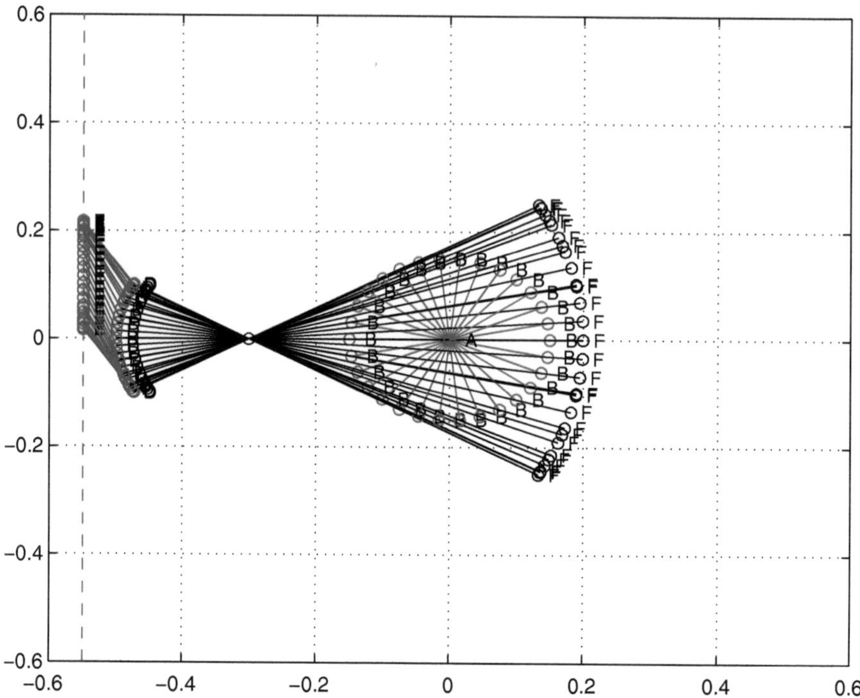

Fig. 5.45 Example 5.4

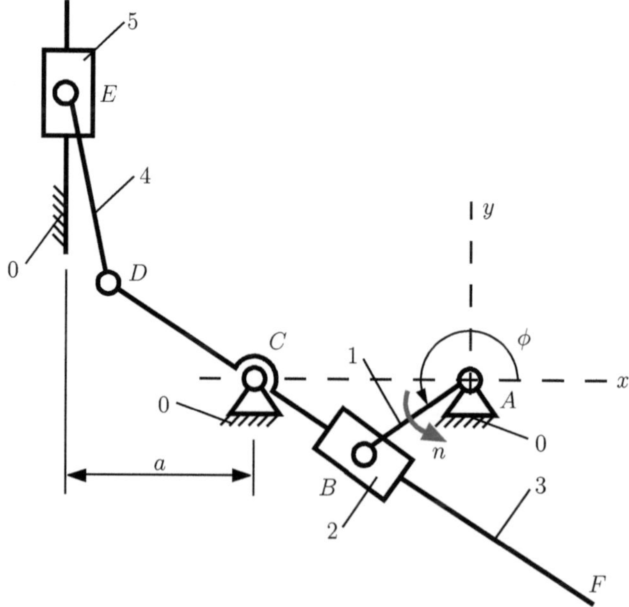

Fig. 5.46 Example 5.5

Solution

A Cartesian reference frame xOy is selected. The joint A is the origin of the reference frame, that is, $A \equiv O$, and $x_A = 0$, $y_A = 0$. The positions of the mechanism are calculated in MATLAB with

```
clear all; clc; close all


% Input data (m)
AB = 0.15; AC = 0.30; CD = 0.20;
DE = 0.14; CF = 0.50; a  = 0.25;
phi = 200*pi/180; % rad
% Position of joint A
xA=0; yA=0; rA=[xA yA 0];
% Position of joint C
xC=-AC; yC=0; rC=[xC yC 0];
% Position of joint B
xB=AB*cos(phi); yB=AB*sin(phi); rB=[xB yB 0];
% Link 3
phi3=atan((yB-yC)/(xB-xC)); phi2=phi3;
% Position of joint F
xF=xC+CF*cos(phi2); yF=yC+CF*sin(phi2); rF=[xF,yF,0];
% Position of joint D
xD=xC-CD*cos(phi3); yD=yC-CD*sin(phi3); rD=[xD yD 0];
% Position of joint E
xE=-AC-a; yE=yD+sqrt(DE^2-(xE-xD)^2); rE=[xE yE 0];
% Link 4
phi4=atan((yE-yD)/(xE-xD));
```

and the results are

```
phi = phi1 = 200 (degrees)
rA = [0,0,0] (m)
rB = [-0.141,-0.051,0] (m)
rC = [-0.300, 0.000,0] (m)
rD = [-0.490, 0.061,0] (m)
rE = [-0.550, 0.188,0] (m)
rF = [ 0.176,-0.153,0] (m)
phi2 = phi3 = -17.878 (degrees)
phi4 = -64.778 (degrees)
```

The angular velocity of link 1 is constant and has the value

$$\boldsymbol{\omega}_1 = \omega_1\,\mathbf{k} = \frac{\pi n}{30}\,\mathbf{k} = \frac{\pi(700)}{30}\,\mathbf{k} = 73.304\,\mathbf{k}\ \text{rad/s}.$$

The angular acceleration of link 1 is $\boldsymbol{\alpha}_1 = \dot{\boldsymbol{\omega}}_1 = \mathbf{0}$.

Velocity and Acceleration of $B_1 = B_2$
The velocity of the point B_1 on the link 1 is

$$\mathbf{v}_{B_1} = \mathbf{v}_A + \boldsymbol{\omega}_1 \times \mathbf{r}_{AB} = \boldsymbol{\omega}_1 \times \mathbf{r}_B,$$

where $\mathbf{v}_A \equiv \mathbf{0}$ is the velocity of the origin $A \equiv O$ and $\mathbf{r}_B = x_B\mathbf{\imath} + y_B\mathbf{\jmath} = -0.141\mathbf{\imath} - 0.051\mathbf{\jmath}$ m.

The velocity of point B_2 on the link 2 is $\mathbf{v}_{B_2} = \mathbf{v}_{B_1}$ because between the links 1 and 2, there is a rotational joint. The velocity of $B_1 = B_2$ is

$$\mathbf{v}_{B_1} = \mathbf{v}_{B_2} = \begin{vmatrix} \mathbf{\imath} & \mathbf{\jmath} & \mathbf{k} \\ 0 & 0 & \omega \\ x_B & y_B & 0 \end{vmatrix} = \begin{vmatrix} \mathbf{\imath} & \mathbf{\jmath} & \mathbf{k} \\ 0 & 0 & 73.304 \\ -0.141 & -0.051 & 0 \end{vmatrix}$$

$$= 3.7619\,\mathbf{\imath} - 10.332\,\mathbf{\jmath} \ \ \mathrm{m/s}.$$

The acceleration of $B_1 = B_2$ is

$$\mathbf{a}_{B_1} = \mathbf{a}_{B_2} = \mathbf{a}_A + \boldsymbol{\alpha}_1 \times \mathbf{r}_B + \boldsymbol{\omega}_1 \times (\boldsymbol{\omega}_1 \times \mathbf{r}_B) = \boldsymbol{\alpha}_1 \times \mathbf{r}_B - \omega_1^2 \mathbf{r}_B$$

$$= -\omega_1^2 \mathbf{r}_B = -73.304^2(-0.141\mathbf{\imath} - 0.051\mathbf{\jmath}) = 757.409\mathbf{\imath} + 275.674\mathbf{\jmath} \ \ \mathrm{m/s}^2.$$

The MATLAB commands for the velocity and accelerations of the point $B_1 = B_2$ are

```
n = 700.; % rpm
omega1 = [ 0 0 pi*n/30 ];
alpha1 = [0 0 0 ];
fprintf...
('omega1 = [%g,%g,%6.3f] (rad/s)\n', omega1)
fprintf...
('alpha1 = [%g,%g,%g] (rad/s^2)\n', alpha1)
fprintf('\n')
vA = [0 0 0]; aA = [0 0 0];
vB1 = vA + cross(omega1,rB);
aB1 = aA + cross(alpha1,rB) - ...
     dot(omega1,omega1)*rB;
vB2 = vB1;
aB2 = aB1;
fprintf...
('vB1 = vB2 = [%6.3f,%6.3f,%g] (m/s)\n',vB1)
fprintf('|vB1| = %6.3f (m/s)\n', norm(vB1))
fprintf...
('aB1 = aB2 = [%6.3f,%6.3f,%g] (m/s^2)\n',aB1)
fprintf('|aB1| = %6.3f (m/s^2)\n', norm(aB1))
fprintf('\n')
```

```
aB1n = - dot(omega1,omega1)*rB;
aB1t = cross(alpha1,rB);
fprintf('aB1n = [%6.3f,%6.3f,%g] (m/s^2)\n', aB1n)
fprintf('aB1t = [%6.3f,%6.3f,%g] (m/s^2)\n', aB1t)
fprintf('\n');
```

Angular Velocity of Link 3
The velocity of the point B_3 on the link 3 is calculated in terms of the velocity of the point B_2 on the link 2:

$$\mathbf{v}_{B_3} = \mathbf{v}_{B_2} + \mathbf{v}^{\text{rel}}_{B_3 B_2} = \mathbf{v}_{B_2} + \mathbf{v}_{B_{32}}, \tag{5.119}$$

where $\mathbf{v}^{\text{rel}}_{B_3 B_2} = \mathbf{v}_{B_{32}}$ is the relative acceleration of B_3 with respect to B_2 on link 3. This relative velocity is parallel to the sliding direction BC, $\mathbf{v}_{B_{32}} \| BC$, or

$$\mathbf{v}_{B_{32}} = v_{B_{32}} \cos \phi_2 \, \mathbf{\imath} + v_{B_{32}} \sin \phi_2 \, \mathbf{J}, \tag{5.120}$$

where $\phi_2 = -17.878°$ is known from position analysis. The points B_3 and C are on the link 3 and

$$\mathbf{v}_{B_3} = \mathbf{v}_C + \boldsymbol{\omega}_3 \times \mathbf{r}_{CB} = \boldsymbol{\omega}_3 \times (\mathbf{r}_B - \mathbf{r}_C), \tag{5.121}$$

where $\mathbf{v}_C \equiv \mathbf{0}$ and the angular velocity of link 3 is $\boldsymbol{\omega}_3 = \omega_3 \, \mathbf{k}$.
    Equations (5.119)–(5.121) give

$$\begin{vmatrix} \mathbf{\imath} & \mathbf{J} & \mathbf{k} \\ 0 & 0 & \omega_3 \\ x_B - x_C & y_B - y_C & 0 \end{vmatrix} = \mathbf{v}_{B_2} + v_{B_{32}} \cos \phi_2 \, \mathbf{\imath} + v_{B_{32}} \sin \phi_2 \, \mathbf{J}. \tag{5.122}$$

Equation (5.122) represents a vectorial equations with two scalar components on the x-axis and y-axis and with two unknowns ω_3 and $v_{B_{32}}$:

$$-\omega_3 (y_B - y_C) = v_{B_2 x} + v_{B_{32}} \cos \phi_2,$$

$$\omega_3 (x_B - x_C) = v_{B_2 y} + v_{B_{32}} \sin \phi_2.$$

The MATLAB commands for the ω_3 and $\mathbf{v}_{B_{32}}$ are

```
omega3z=sym('omega3z','real');
vB32=sym('vB32','real');
omega3 = [ 0 0 omega3z ];
% omega3z unknown (to be calculated)
% vB32 unknown (to be calculated)
vC = [0 0 0 ]; % C is fixed
% vB3 = vC + omega3 x rCB
% (B3 & C are points on link 3)
```

```
vB3 = vC + cross(omega3,rB-rC);
% point B2 is on link 2 and point B3 is on link 3
% vB3 = vB2 + vB3B2
% between the links 2 and 3 there is a
% translational joint B_T
% vB3B2 is the relative velocity of B3 wrt 2
% vB3B2 is parallel to the sliding direcion BC
% vB3B2 is written as a vector
vB3B2 = vB32*[cos(phi2) sin(phi2) 0];
% vB3 = vB2 + vB3B2
eqvB = vB3 - vB2 - vB3B2;
% vectorial equation
% the component of the vectorial equation on x-axis
eqvBx = eqvB(1);
% the component of the vectorial equation on y-axis
eqvBy = eqvB(2);
% two equations eqvBx and eqvBy with two unknowns
% solve for omega3z and vB32
solvB = solve(eqvBx,eqvBy);
omega3zs=eval(solvB.omega3z);
vB32s=eval(solvB.vB32);

omega3v = [0 0 omega3zs];
omega2v = omega3v;
vB32v = vB32s*[cos(phi2) sin(phi2) 0];

digits 3
% print the equations for calculating
% omega3 and vB32
fprintf...
('vB3 = vC + omega3 x rCB = vB2 + vB3B2 => \n')
qvBx=vpa(eqvBx);
fprintf('x-axis: %s = 0 \n', char(qvBx))
qvBy=vpa(eqvBy);
fprintf('y-axis: %s = 0 \n', char(qvBy))
fprintf('=>\n')
fprintf('omega3z = %6.3f (rad/s)\n', omega3zs)
fprintf('vB32 = %6.3f (m/s)\n', vB32s)
fprintf('\n')
fprintf...
('omega2=omega3 = [%g,%g,%6.3f] (rad/s)\n', omega3v)
fprintf('vB3B2 = [%6.3f,%6.3f,%g] (m/s)\n\n', vB32v)
```

and the MATLAB results are

```
vB3 = vC + omega3 x rCB = vB2 + vB3B2 =>
x-axis: 0.0513*omega3z - 0.952*vB32 - 3.76 = 0
y-axis: 0.159*omega3z + 0.307*vB32 + 10.3 = 0
=>
omega3z = -51.934 (rad/s)
vB32 = -6.751 (m/s)

omega2=omega3 = [0,0,-51.934] (rad/s)
vB3B2 = [-6.425, 2.073,-0] (m/s)
```

Angular Acceleration of Link 3
The acceleration of the point B_3 on the link 3 is calculated in terms of the acceleration of the point B_2 on the link 2:

$$\mathbf{a}_{B_3} = \mathbf{a}_{B_2} + \mathbf{a}^{\mathrm{r}}_{B_3B_2} + \mathbf{a}^{\mathrm{c}}_{B_3B_2} = \mathbf{a}_{B_2} + \mathbf{a}_{B_{32}} + \mathbf{a}^{\mathrm{c}}_{B_{32}}, \tag{5.123}$$

where $\mathbf{a}^{\mathrm{r}}_{B_3B_2} = \mathbf{a}_{B_{32}}$ is the relative acceleration of B_3 with respect to B_2 on link 3. This relative acceleration is parallel to the sliding direction BC, $\mathbf{a}_{B_{32}}\|BC$, or

$$\mathbf{a}_{B_{32}} = a_{B_{32}}\cos\phi_2\,\mathbf{1} + a_{B_{32}}\sin\phi_2\,\mathbf{J}. \tag{5.124}$$

The Coriolis acceleration of B_3 relative to B_2 is

$$\mathbf{a}^{\mathrm{c}}_{B_{32}} = 2\,\boldsymbol{\omega}_3 \times \mathbf{v}_{B_{32}} = 2\,\boldsymbol{\omega}_2 \times \mathbf{v}_{B_{32}} = 2 \begin{vmatrix} \mathbf{1} & \mathbf{J} & \mathbf{k} \\ 0 & 0 & \omega_3 \\ v_{B_{32}}\cos\phi_2 & v_{B_{32}}\sin\phi_2 & 0 \end{vmatrix}$$

$$= 2(-\omega_3 v_{B_{32}}\sin\phi_2\,\mathbf{1} + \omega_3 v_{B_{32}}\cos\phi_2\,\mathbf{J}). \tag{5.125}$$

The points B_3 and C are on the link 3 and

$$\mathbf{a}_{B_3} = \mathbf{a}_C + \boldsymbol{\alpha}_3 \times \mathbf{r}_{CB} - \omega_3^2 \mathbf{r}_{CB}, \tag{5.126}$$

where $\mathbf{a}_C \equiv \mathbf{0}$ and the angular acceleration of link 3 is

$$\boldsymbol{\alpha}_3 = \alpha_3\,\mathbf{k}.$$

Equations (5.123)–(5.126) give

$$\begin{vmatrix} \mathbf{1} & \mathbf{J} & \mathbf{k} \\ 0 & 0 & \alpha_3 \\ x_B - x_C & y_B - y_C & 0 \end{vmatrix} - \omega_3^2(\mathbf{r}_B - \mathbf{r}_C) = \mathbf{a}_{B_2} + a_{B_{32}}(\cos\phi_2\mathbf{1} + \sin\phi_2\mathbf{J}) + 2\,\boldsymbol{\omega}_3 \times \mathbf{v}_{B_{32}}.$$

$$\tag{5.127}$$

Equation (5.127) represents a vectorial equations with two scalar components on the x-axis and y-axis and with two unknowns α_3 and $a_{B_{32}}$:

$$-\alpha_3(y_B - y_C) - \omega_3^2(x_B - x_C) = a_{B_2 x} + a_{B_{32}} \cos \phi_2 - 2\omega_3 v_{B_{32}} \sin \phi_2,$$

$$\alpha_3(x_B - x_C) - \omega_3^2(y_B - y_C) = a_{B_2 y} + a_{B_{32}} \sin \phi_2 + 2\omega_3 v_{B_{32}} \cos \phi_2.$$

The MATLAB commands for $\boldsymbol{\alpha}_3$ and $\mathbf{a}_{B_{32}}$ are

```
% Coriolis acceleration
aB3B2cor = 2*cross(omega3v,vB32v);

alpha3z=sym('alpha3z','real');
aB32=sym('aB32','real'); % aB32 unknown

alpha3 = [ 0 0 alpha3z ]; % alpha3z unknown
aC = [0 0 0 ]; % C is fixed
% aB3 acceleration of B3
aB3 = aC + cross(alpha3, rB-rC) -...
    dot(omega3v, omega3v)*(rB-rC);
% aB3B2 relative velocity of B3 wrt 2
% aB3B2 parallel to the sliding direcion BC

aB3B2 = aB32*[ cos(phi2) sin(phi2) 0];
% aB3 = aB2 + aB3B2 + aB3B2cor
eqaB = aB3 - aB2 - aB3B2 - aB3B2cor;
% vectorial equation
eqaBx = eqaB(1); % equation component on x-axis
eqaBy = eqaB(2); % equation component on y-axis
solaB = solve(eqaBx,eqaBy);
alpha3zs=eval(solaB.alpha3z);
aB32s=eval(solaB.aB32);
alpha3v = [0 0 alpha3zs];
alpha2v = alpha3v;
aB32v = aB32s*[cos(phi2) sin(phi2) 0];

% print the equations for calculating
% alpha3 and aB32
fprintf...
('aB32cor = [%6.3f,%6.3f,%g] (m/s^2)\n', aB3B2cor)
fprintf('\n')
fprintf('aB3=aC+alpha3xrCB-(omega3.omega3)rCB\n')
fprintf('aB3=aB2+aB3B2+aB3B2cor =>\n')
qaBx=vpa(eqaBx);
fprintf('x-axis:\n')
fprintf('%s = 0\n',char(qaBx))
```

```
qaBy=vpa(eqaBy);
fprintf('y axis:\n')
fprintf('%s = 0\n',char(qaBy))
fprintf('=>\n');
fprintf('alpha3z = %6.3f (rad/s^2)\n', alpha3zs)
fprintf('aB32 = %6.3f (m/s^2)\n', aB32s)
fprintf('\n')
fprintf...
('alpha2=alpha3=[%g,%g,%6.3f](rad/s^2)\n',alpha3v)
fprintf...
('aB3B2=[%6.3f,%6.3f,%d] (m/s^2)\n\n',aB32v)
```

The MATLAB results are

```
aB32cor = [215.270,667.364,0](m/s^2)


aB3=aC+alpha3xrCB-(omega3.omega3)rCB
aB3=aB2+aB3B2+aB3B2cor =>
x-axis:
0.0513*alpha3z - 0.952*aB32 - 1400.0 = 0
y-axis:
0.307*aB32 + 0.159*alpha3z - 805.0 = 0
=>
alpha3z = 7157.347 (rad/s^2)
aB32 = -1086.945 (m/s^2)


alpha2=alpha3 = [0,0,7157.347](rad/s^2)
aB3B2 = [-1034.459,333.682,0] (m/s^2)
```

Velocity and Acceleration of $D_3 = D_4$
The velocity of $D_3 = D_4$ is

$$\mathbf{v}_{D_3} = \mathbf{v}_{D_4} = \mathbf{v}_C + \boldsymbol{\omega}_3 \times \mathbf{r}_{CD} = \boldsymbol{\omega}_3 \times (\mathbf{r}_D - \mathbf{r}_C)$$

$$= \begin{vmatrix} \mathbf{i} & \mathbf{j} & \mathbf{k} \\ 0 & 0 & \omega_3 \\ x_D - x_C & y_D - y_C & 0 \end{vmatrix}.$$

The acceleration of $D_3 = D_4$ is

$$\mathbf{a}_{D_3} = \mathbf{a}_{D_4} = \mathbf{a}_C + \boldsymbol{\alpha}_3 \times \mathbf{r}_{CD} - \omega_3^2 \mathbf{r}_{CD} = \boldsymbol{\alpha}_3 \times (\mathbf{r}_D - \mathbf{r}_C) - \omega_3^2 (\mathbf{r}_D - \mathbf{r}_C)$$

$$= \begin{vmatrix} \mathbf{i} & \mathbf{j} & \mathbf{k} \\ 0 & 0 & \alpha_3 \\ x_D - x_C & y_D - y_C & 0 \end{vmatrix} - \omega_3^2 [(x_D - x_C)\mathbf{i} + (y_D - y_C)\mathbf{j}].$$

The MATLAB commands for the velocity and acceleration of $D_3 = D_4$ are:

```
% vD3 velocity of D3
% D3 & C points on link 3
vD3 = vC + cross(omega3v,rD-rC);
vD4 = vD3;
fprintf...
('vD3 = vD4 = [%6.3f,%6.3f,%g] (m/s)\n', vD3)
fprintf('|vD3| = %6.3f (m/s)\n', norm(vD3))
% aD3 acceleration of D3
aD3 = aC + cross(alpha3v, rD-rC)-...
    dot(omega3v, omega3v)*(rD-rC);
aD4 = aD3;
fprintf...
('aD3 = aD4 = [%6.3f,%6.3f,%g] (m/s^2)\n', aD3)
fprintf('|aD3| = %6.3f (m/s^2)\n', norm(aD3))
fprintf('\n')

vDC = cross(omega3v,rD-rC);
aDC=cross(alpha3v,rD-rC)-...
    dot(omega3v,omega3v)*(rD-rC);
aDCn=-dot(omega3v,omega3v)*(rD-rC);
aDCt=cross(alpha3v,rD-rC);

fprintf...
('vDC = [%6.3f,%6.3f,%g] (m/s)\n', vDC)
fprintf...
('aDC = [%6.3f,%6.3f,%g] (m/s^2)\n', aDC)
fprintf...
('aDCn = [%6.3f,%6.3f,%g] (m/s^2)\n', aDCn)
fprintf...
('|aDCn| = %6.3f (m/s^2)\n', norm(aDCn))
fprintf...
('aDCt = [%6.3f,%6.3f,%g] (m/s^2)\n', aDCt)
fprintf...
('|aDCt| = %6.3f (m/s^2)\n', norm(aDCt))
fprintf('\n')
```

and the MATLAB results are:

```
vD3 = vD4 = [ 3.189, 9.885,0] (m/s)
|vD3| = 10.387 (m/s)
aD3 = aD4 = [73.937,-1527.948,0] (m/s^2)
|aD3| = 1529.736 (m/s^2)

vDC = [ 3.189, 9.885,0] (m/s)
aDC = [73.937,-1527.948,0] (m/s^2)
```

```
aDCn  =  [513.385,-165.601,-0]  (m/s^2)
|aDCn|  = 539.433  (m/s^2)
aDCt  =  [-439.448,-1362.347,0]  (m/s^2)
|aDCt|  = 1431.469  (m/s^2)
```

The velocity and the acceleration of point F are calculated with

```
% F & C points on link 3
vF = vC + cross(omega3v,rF-rC);
fprintf('vF = [%6.3f,%6.3f,%g] (m/s)\n', vF)
% aD3 acceleration of D3
aF = aC + cross(alpha3v, rF-rC)-...
     dot(omega3v, omega3v)*(rF-rC);
fprintf('aF = [%6.3f,%6.3f,%g] (m/s^2)\n', aF)
fprintf('|vF| = %6.3f (m/s)\n', norm(vF))
fprintf('|aF| = %6.3f (m/s^2)\n', norm(aF))
fprintf('\n')
```

and the numerical results are

```
vF  =  [-7.972,-24.713,0]  (m/s)
aF  =  [-184.842,3819.871,0]  (m/s^2)
|vF|  = 25.967  (m/s)
|aF|  = 3824.340  (m/s^2)
```

Angular Velocity of Link 4
The velocity of the point E_5 on the link 5 is parallel to the sliding direction y:

$$\mathbf{v}_{E_5} = v_{Ey}\mathbf{J}. \tag{5.128}$$

The points E_4 and D are on the link 4 and

$$\mathbf{v}_{E_4} = \mathbf{v}_D + \boldsymbol{\omega}_4 \times \mathbf{r}_{DE}. \tag{5.129}$$

Equations (5.128)–(5.129) give

$$\mathbf{v}_D + \begin{vmatrix} \mathbf{I} & \mathbf{J} & \mathbf{k} \\ 0 & 0 & \omega_4 \\ x_E - x_D & y_E - y_D & 0 \end{vmatrix} = \mathbf{v}_{Ey}\mathbf{J}. \tag{5.130}$$

Equation (5.130) represents a vectorial equations with two scalar components on the x-axis and y-axis and with two unknowns ω_4 and v_{Ey}. The MATLAB commands for the ω_4 and $\mathbf{v}_E$ are

```
omega4z = sym('omega4z','real');
% omega4z unknown
vEy = sym('vEy','real');
```

```
% vEy unknown
omega4 = [ 0 0 omega4z ];
vE5 = [ 0 vEy 0];
% vEy velocity of E
vE4 = vD4 + cross(omega4,rE-rD);
% vE parallel to the sliding direcion y
% vE5 = vE4
eqvE = vE5-vE4;
% vectorial equation
eqvEx = eqvE(1); % component on x-axis
eqvEy = eqvE(2); % component on y-axis
solvE = solve(eqvEx,eqvEy);
omega4zs=eval(solvE.omega4z);
vEys=eval(solvE.vEy);
omega4v = [0 0 omega4zs];
vE = [0 vEys 0];

% print the equations for calculating
% omega4 and vE
fprintf...
('vE = vD + omega4 x (rE-rD)  => \n')
qvEx=vpa(eqvEx);
fprintf('x-axis: %s = 0 \n', char(qvEx))
qvEy=vpa(eqvEy);
fprintf('y-axis: %s = 0 \n', char(qvEy))
fprintf('=>\n')
fprintf('omega4z = %6.3f (rad/s)\n', omega4zs)
fprintf('vEy = %6.3f (m/s)\n', vEys)
fprintf('\n')
fprintf...
('omega4 = [%g,%g,%6.3f] (rad/s)\n', omega4v)
fprintf('vE = [%g,%6.3f,%g] (m/s)\n', vE)
fprintf('|vE| = %6.3f (m/s)\n', norm(vE))
```

and the results are

```
vE = vD + omega4 x (rE-rD) =>
x-axis: 0.127*omega4z - 3.19 = 0
y-axis: 0.0597*omega4z + vEy - 9.89 = 0
=>
omega4z = 25.176 (rad/s)
vEy =  8.383 (m/s)

omega4 = [0,0,25.176] (rad/s)
vE = [0, 8.383,0] (m/s)
|vE| =  8.383 (m/s)
```

Angular Acceleration of Link 4
The acceleration of the point E_5 on the link 5 is parallel to the sliding direction y:

$$\mathbf{a}_{E_5} = \mathbf{a}_{Ey}\mathbf{J}. \tag{5.131}$$

The points E_4 and D are on the link 4 and

$$\mathbf{a}_{E_4} = \mathbf{a}_D + \boldsymbol{\alpha}_4 \times \mathbf{r}_{DE} - \omega_4^2 \mathbf{r}_{DE}, \tag{5.132}$$

where the angular acceleration of link 4 is $\boldsymbol{\alpha}_4 = \alpha_4\,\mathbf{k}$. Equations (5.131)–(5.132) give

$$\mathbf{a}_D + \begin{vmatrix} \mathbf{1} & \mathbf{J} & \mathbf{k} \\ 0 & 0 & \alpha_4 \\ x_E - x_D & y_E - y_D & 0 \end{vmatrix} - \omega_4^2 \mathbf{r}_{DE}$$

$$= \mathbf{a}_E\mathbf{J}. \tag{5.133}$$

Equation (5.133) represents a vectorial equations with two scalar components on the x-axis and y-axis and with two unknowns α_4 and a_{Ey}. The MATLAB commands for the $\boldsymbol{\alpha}_4$ and $\mathbf{a}_E$ are

```
alpha4z = sym('alpha4z','real');
aEy = sym('aEy','real');
alpha4 = [ 0 0 alpha4z ]; % alpha5z unknown
% aE5 acceleration of E5
aE4=aD4+cross(alpha4,rE-rD)-...
    dot(omega4v,omega4v)*(rE-rD);
% aE parallel to the sliding direcion y
aE5 = [0 aEy 0];
eqaE = aE5 - aE4;
% vectorial equation
eqaEx = eqaE(1); % component on x-axis
eqaEy = eqaE(2); % component on y-axis
solaE = solve(eqaEx,eqaEy);
alpha4zs = eval(solaE.alpha4z);
aEys = eval(solaE.aEy);
alpha4v = [0 0 alpha4zs];
aEv = [0 aEys 0];

% print the equations for calculating
% alpha4 and aE
fprintf('\n')
fprintf('aE4=aD+alpha4xrDE-(omega4.omega4)rDE\n')
fprintf('aE5=aE4 =>\n')
qaEx=vpa(eqaEx);
```

```
fprintf('x-axis: %s = 0 \n', char(qaEx))
qaEy=vpa(eqaEy);
fprintf('y-axis: %s = 0 \n', char(qaEy))
fprintf('=>\n')
fprintf('alpha4z = %6.3f (rad/s^2)\n', alpha4zs)
fprintf('aEy = %6.3f (m/s^2)\n', aEys)
fprintf('\n')
fprintf...
('alpha4=[%g,%g,%6.3f] (rad/s^2)\n', alpha4v)
fprintf('aE = [%g,%6.3f,%g] (m/s^2)\n', aEv)
fprintf('|aE| = %6.3f (m/s^2)\n', norm(aEv))
fprintf('\n')
```

and the output is

```
aE4=aD+alpha4xrDE-(omega4.omega4)rDE
aE5=aE4 =>
x-axis: 0.127*alpha4z - 112.0 = 0
y-axis: aEy + 0.0597*alpha4z + 1611.0 = 0
=>
alpha4z = 882.339 (rad/s^2)
aEy = -1660.866 (m/s^2)


alpha4=[0,0,882.339] (rad/s^2)
aE = [0,-1660.866,0] (m/s^2)
|aE| = 1660.866 (m/s^2)
```

Example 5.6. For the planar mechanism given at *Example 5.5*, find the velocities and the accelerations of the mechanism using the contour method.

Solution

The mechanism has five moving links and seven full joints. The number of independent contours is $n_c = c - n = 7 - 5 = 2$, where c is the number of joints and n is the number of moving links. The mechanism has two independent contours. The contour diagrams of the mechanism are represented in Fig. 5.47. The first contour I contains the links 0, 1, 2, and 3, while the second contour II contains the links 0, 3, 4, and 5. Clockwise paths are chosen for each closed contours I and II.

Contour I: 0-1-2-3-0

Figure 5.47 shows the first independent contour I with:

- Rotational joint R between the links 0 and 1 (joint A_R)
- Rotational joint R between the links 1 and 2 (joint B_R)
- Translational joint T between the links 2 and 3 (joint B_T)
- Rotational joint R between the links 3 and 0 (joint C_R)

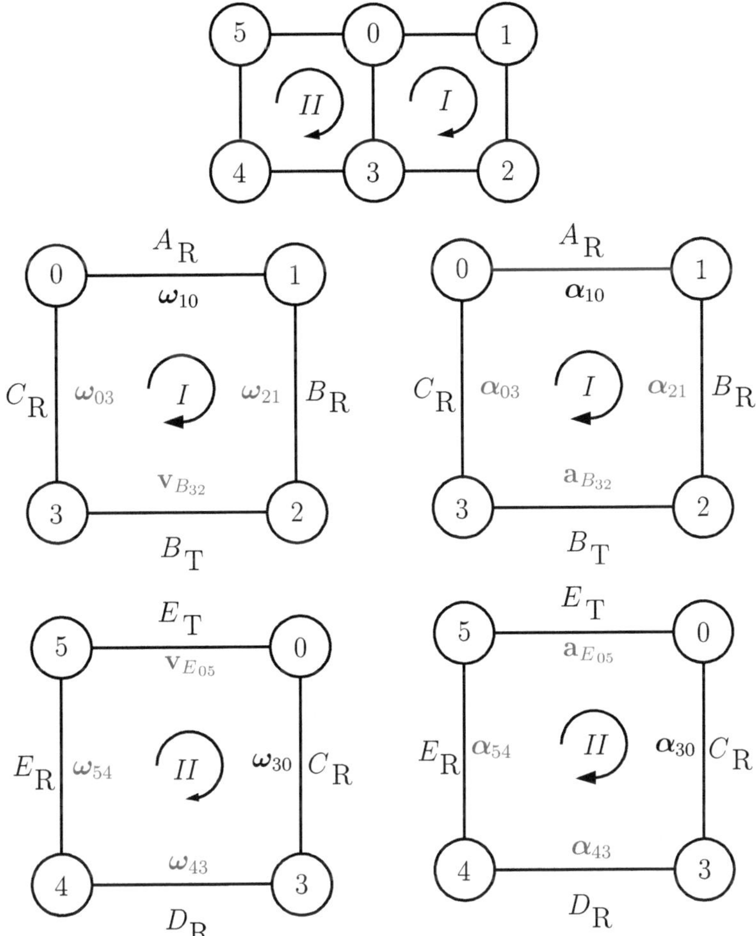

Fig. 5.47   Example 5.6

The angular velocity ω_{10} of the driver link is known:

$$\omega_{10} = \omega_1 = \omega = \frac{n\pi}{30} = \frac{700\pi}{30} \ \text{rad/s}.$$

The origin of the reference frame is at the point $A(0,0)$. For the velocity analysis, using (5.78) the following equations are obtained

$$\omega_{10} + \omega_{21} + \omega_{03} = \mathbf{0},$$
$$\mathbf{r}_B \times \omega_{21} + \mathbf{r}_C \times \omega_{03} + \mathbf{v}^{\mathrm{r}}_{B_3 B_2} = \mathbf{0}, \tag{5.134}$$

where $\mathbf{r}_B = x_B\mathbf{\imath} + y_B\mathbf{\jmath}$, $\mathbf{r}_C = x_C\mathbf{\imath} + y_C\mathbf{\jmath}$, and

$$\boldsymbol{\omega}_{10} = \omega_{10}\,\mathbf{k}, \ \boldsymbol{\omega}_{21} = \omega_{21}\,\mathbf{k}, \ \boldsymbol{\omega}_{03} = \omega_{03}\,\mathbf{k},$$

$$\mathbf{v}_{B_3B_2}^{\mathrm{r}} = \mathbf{v}_{B32} = v_{B32}\cos\phi_2\,\mathbf{\imath} + v_{B32}\sin\phi_2\,\mathbf{\jmath}.$$

The unknowns are the relative angular and linear velocities: ω_{21}, ω_{03}, and v_{B32}. The sign of the relative angular velocities is selected arbitrarily as positive. The numerical computation will then give the correct orientation (the correct sign) of the unknown vectors. The components of the vectors $\mathbf{r}_B$ and $\mathbf{r}_C$, and the angle ϕ_2 are already known from the position analysis of the mechanism. Equation (5.134) becomes

$$\omega_{10}\,\mathbf{k} + \omega_{21}\,\mathbf{k} + \omega_{03}\,\mathbf{k} = \mathbf{0},$$

$$\begin{vmatrix} \mathbf{\imath} & \mathbf{\jmath} & \mathbf{k} \\ x_B & y_B & 0 \\ 0 & 0 & \omega_{21} \end{vmatrix} + \begin{vmatrix} \mathbf{\imath} & \mathbf{\jmath} & \mathbf{k} \\ x_C & y_C & 0 \\ 0 & 0 & \omega_{03} \end{vmatrix} + v_{B32}\cos\phi_2\,\mathbf{\imath} + v_{B32}\sin\phi_2\,\mathbf{\jmath} = \mathbf{0}. \qquad (5.135)$$

The unknown relative velocities are introduced with MATLAB as

```
omega21v = [ 0  0  sym('omega21z','real')];
omega03v = [ 0  0  sym('omega03z','real')];
v32v = ...
sym('vB32','real')*[cos(phi2)  sin(phi2)  0];
```

Equation (5.135) represents a system of three equations and with MATLAB commands gives:

```
eqIomega = omega10 + omega21v + omega03v;
eqIvz=eqIomega(3);
eqIv = ...
  cross(rB,omega21v)+cross(rC,omega03v)+v32v;
eqIvx=eqIv(1);
eqIvy=eqIv(2);
```

To display the equations the following MATLAB statements are used:

```
digits 3
Ivz=vpa(eqIvz);
fprintf('%s = 0 \n', char(Ivz))
Ivx=vpa(eqIvx);
fprintf('%s = 0 \n', char(Ivx))
Ivy=vpa(eqIvy);
fprintf('%s = 0 \n', char(Ivy))
```

The system of equations can be solved using the MATLAB commands:

```
solIv=solve(eqIvz,eqIvx,eqIvy);
omega21 = [ 0 0 eval(solIv.omega21z) ];
omega03 = [ 0 0 eval(solIv.omega03z) ];
vB3B2 = ...
  eval(solIv.vB32)*[cos(phi2) sin(phi2) 0];
```

To print the numerical values, the following MATLAB commands are used:

```
fprintf...
('omega21=[%g,%g,%6.3f] (rad/s)\n',omega21)
fprintf...
('omega03=[%g,%g,%6.3f] (rad/s)\n',omega03)
fprintf...
('vB32 = %6.3f (m/s)\n', eval(solIv.vB32))
fprintf...
('vB3B2=[%6.3f,%6.3f,%d] (m/s)\n', vB3B2)
fprintf('\n')
```

The following numerical solutions are obtained

```
omega21=[0,0,-125.238] (rad/s)
omega03=[0,0,51.934] (rad/s)
vB32 = -6.751 (m/s)
vB3B2=[-6.425, 2.073,0] (m/s)
```

The absolute angular velocity of the links 2 and 3 is

```
omega20=omega30=[0,0,-51.934] (rad/s)
```

The absolute linear velocity of $D_3 = D_4$ is

```
vC = [0 0 0 ];
vD3 = vC + cross(omega30,rD-rC);
```

or

```
vD3 = vD4 =[ 3.189, 9.885,0] (m/s)
```

For the acceleration analysis, using (5.93) the following equations are obtained

$$\boldsymbol{\alpha}_{10} + \boldsymbol{\alpha}_{21} + \boldsymbol{\alpha}_{03} = \mathbf{0},$$

$$\mathbf{r}_B \times \boldsymbol{\alpha}_{21} + \mathbf{r}_C \times \boldsymbol{\alpha}_{03} + \mathbf{a}^r_{B_3B_2} + \mathbf{a}^c_{B_3B_2} - \omega_{10}^2 \mathbf{r}_{AB} - \omega_{20}^2 \mathbf{r}_{BC} = \mathbf{0}, \quad (5.136)$$

where

$$\boldsymbol{\alpha}_{10} = \alpha_{10}\,\mathbf{k}, \ \boldsymbol{\alpha}_{21} = \alpha_{21}\,\mathbf{k}, \ \boldsymbol{\alpha}_{03} = \alpha_{03}\,\mathbf{k},$$

$$\mathbf{a}^r_{B_3B_2} = \mathbf{a}_{B_{32}} = a_{B_{32}} \cos\phi_2\,\mathbf{i} + a_{B_{32}} \sin\phi_2\,\mathbf{j},$$

$$\mathbf{a}^c_{B_3B_2} = \mathbf{a}^c_{B_{32}} = 2\boldsymbol{\omega}_{20} \times \mathbf{v}_{B_{32}}.$$

The driver link has a constant angular velocity and $\alpha_{10} = \dot{\omega}_{10} = 0$. The unknown acceleration vectors using the MATLAB commands are:

```
alpha21v = [ 0 0 sym('alpha21z','real') ];
alpha03v = [ 0 0 sym('alpha03z','real') ];
a32v=sym('aB32','real')*[cos(phi2) sin(phi2) 0];
```

Equation (5.136) represents a system of three equations and using MATLAB commands gives:

```
eqIalpha = alpha10 + alpha21v + alpha03v;
eqIaz=eqIalpha(3);
eqIa=cross(rB,alpha21v)+cross(rC,alpha03v)+...
 a32v+2*cross(omega20,vB3B2)-...
 dot(omega1,omega1)*rB-dot(omega30,omega30)*(rC-rB);
eqIax=eqIa(1);
eqIay=eqIa(2);
```

The equations are displayed with the statements:

```
Iaz=vpa(eqIaz);
fprintf('%s=0 \n',char(Iaz))
Iax=vpa(eqIax);
fprintf('%s=0 \n',char(Iax))
Iay=vpa(eqIay);
fprintf('%s=0 \n',char(Iay))
```

The unknowns are α_{21}, α_{03}, and $a_{B_{32}}$ or alpha21z, alpha03z, and aB32. The system of equations is solved using the MATLAB commands:

```
solIa=solve(eqIaz,eqIax,eqIay);
alpha21 = [ 0 0 eval(solIa.alpha21z) ];
alpha03 = [ 0 0 eval(solIa.alpha03z) ];
aB3B2=eval(solIa.aB32)*[cos(phi2) sin(phi2) 0];
```

The following numerical solutions are then obtained:

```
alpha21=[0,0,7157.347] (rad/s^2)
alpha03=[0,0,-7157.347] (rad/s^2)
aB32=-1086.945 (m/s^2)
aB3B2=[-1034.459,333.682,0] (m/s^2)
```

The absolute angular acceleration of the links 2 and 3 is:

```
alpha20=alpha30=[0,0,7157.347] (rad/s^2)
```

The absolute linear acceleration of $D_3 = D_4$ is obtained from the following equation:

```
aC = [0 0 0 ];
aD3=aC+cross(alpha30,rD-rC)-...
    dot(omega20,omega20)*(rD-rC);
```

and has the value:

```
aD3=aD4  =  [73.937,-1527.948,0]  (m/s^2)
```

Contour II: 0-3-4-5-0
Figure 5.47 depicts the second independent contour *II*:

- Rotational joint R between the links 0 and 3 (joint C_R).
- Rotational joint R between the links 3 and 4 (joint D_R).
- Rotational joint R between the links 4 and 5 (joint E_R).
- Translational joint T between the links 5 and 0 (joint E_T).

For the velocity analysis, the following vectorial equations are used

$$\boldsymbol{\omega}_{30} + \boldsymbol{\omega}_{43} + \boldsymbol{\omega}_{54} = \mathbf{0},$$

$$\mathbf{r}_C \times \boldsymbol{\omega}_{30} + \mathbf{r}_D \times \boldsymbol{\omega}_{43} + \mathbf{r}_E \times \boldsymbol{\omega}_{54} + \mathbf{v}^r_{E_0 E_5} = \mathbf{0}, \qquad (5.137)$$

where $\mathbf{r}_D = x_D \mathbf{i} + y_D \mathbf{j}$, $\mathbf{r}_E = x_E \mathbf{i} + y_E \mathbf{j} = \mathbf{0}$, and

$$\boldsymbol{\omega}_{30} = \omega_{30}\,\mathbf{k}, \ \boldsymbol{\omega}_{43} = \omega_{43}\,\mathbf{k}, \ \boldsymbol{\omega}_{54} = \omega_{54}\,\mathbf{k},$$

$$\mathbf{v}^r_{E_0 E_5} = \mathbf{v}_{E_{05}} = v_{E_{05}}\,\mathbf{j}.$$

The sign of the relative angular velocities is selected arbitrarily positive. The numerical computation will then give the correct orientation of the unknown vectors. The components of the vectors $\mathbf{r}_D$, $\mathbf{r}_E$, and the angle ϕ_4 are already known from the position analysis of the mechanism. The unknown vectors with MATLAB commands are:

```
omega43v = [ 0 0 sym('omega43z','real') ];
omega54v = [ 0 0 sym('omega54z','real') ];
v05v = [0 sym('v05','real') 0];
```

Equation (5.137) becomes:

$$\omega_{30}\,\mathbf{k} + \omega_{43}\,\mathbf{k} + \omega_{54}\,\mathbf{k} = \mathbf{0},$$

$$\begin{vmatrix} \mathbf{i} & \mathbf{j} & \mathbf{k} \\ x_C & y_C & 0 \\ 0 & 0 & \omega_{30} \end{vmatrix} + \begin{vmatrix} \mathbf{i} & \mathbf{j} & \mathbf{k} \\ x_D & y_D & 0 \\ 0 & 0 & \omega_{43} \end{vmatrix} + \begin{vmatrix} \mathbf{i} & \mathbf{j} & \mathbf{k} \\ x_E & y_E & 0 \\ 0 & 0 & \omega_{54} \end{vmatrix} + v_{E_{05}}\,\mathbf{j} = \mathbf{0}. \qquad (5.138)$$

Equation (5.138) projected onto the "fixed" reference frame *xyz* gives

$$\omega_{30} + \omega_{43} + \omega_{54} = 0,$$

$$y_C\,\omega_{30} + y_D\,\omega_{43} + y_E\,\omega_{54} = 0,$$

$$-x_C\,\omega_{30} - x_D\,\omega_{43} - x_E\,\omega_{54} + v_{E_{05}} = 0. \qquad (5.139)$$

The above system of equations using the following MATLAB commands becomes:

```
eqIIomega = omega30 + omega43v + omega54v;
eqIIvz=eqIIomega(3);
eqIIv=cross(rC,omega30)+cross(rD,omega43v)+...
cross(rE,omega54v)+v05v;
eqIIvx=eqIIv(1);
eqIIvy=eqIIv(2);
```

Equation (5.139) represents an algebraic system of three equations with three unknowns: ω_{43}, ω_{54}, and $v_{E_{05}}$. The system is solved using the MATLAB commands:

```
solIIv=solve(eqIIvz,eqIIvx,eqIIvy);
omega43 = [ 0 0 eval(solIIv.omega43z) ];
omega54 = [ 0 0 eval(solIIv.omega54z) ];
vE05 = [0 eval(solIIv.v05) 0];
```

The following numerical solutions are obtained:

```
omega43=[0,0,77.111]  (rad/s)
omega54=[0,0,-25.176]  (rad/s)
vE05=-8.383  (m/s)
vE0E5 =[0,-8.383,0]  (m/s)
```

To print the numerical values with MATLAB, the following commands are used:

```
fprintf...
('omega43=[%d,%d,%6.3f]  (rad/s)\n',omega43)
fprintf...
('omega54=[%d,%d,%6.3f]  (rad/s)\n',omega54)
fprintf('vE05=%6.3f  (m/s)\n',eval(solIIv.v05))
fprintf('vE0E5 =[%d,%6.3f,%d]  (m/s)\n',vE05)
fprintf('\n')
```

The absolute angular velocity of link 4 and the velocity of E are:

```
omega40=omega30+omega43;
omega40=[0,0,25.176](rad/s)
vE5=[0, 8.383,0]  (m/s)
```

For the acceleration analysis, the following vectorial equations are used

$$\boldsymbol{\alpha}_{30} + \boldsymbol{\alpha}_{43} + \boldsymbol{\alpha}_{54} = \mathbf{0},$$

$$\mathbf{r}_C \times \boldsymbol{\alpha}_{30} + \mathbf{r}_D \times \boldsymbol{\alpha}_{43} + \mathbf{r}_E \times \boldsymbol{\alpha}_{54} + \mathbf{a}_{E_0E_5}^{\mathrm{r}} - \omega_{30}^2 \mathbf{r}_{CD} - \omega_{40}^2 \mathbf{r}_{DE} = \mathbf{0}, \qquad (5.140)$$

where

$$\boldsymbol{\alpha}_{30} = \alpha_{30}\,\mathbf{k}, \ \boldsymbol{\alpha}_{43} = \alpha_{43}\,\mathbf{k}, \ \boldsymbol{\alpha}_{54} = \alpha_{54}\,\mathbf{k},$$

$$\mathbf{a}_{E_0D_5}^{\mathrm{r}} = \mathbf{a}_{E_{05}} = a_{E_{05}}\,\mathbf{J}.$$

The unknown acceleration vectors using the MATLAB commands are:

```
alpha43v = [ 0 0 sym('alpha43z','real') ];
alpha54v = [ 0 0 sym('alpha54z','real') ];
a05v=[0,sym('a05','real'),0];
```

Equation (5.140) becomes

$$\alpha_{30}\,\mathbf{k} + \alpha_{43}\,\mathbf{k} + \alpha_{54}\,\mathbf{k} = \mathbf{0},$$

$$\begin{vmatrix} \mathbf{i} & \mathbf{j} & \mathbf{k} \\ x_C & y_C & 0 \\ 0 & 0 & \alpha_{30} \end{vmatrix} + \begin{vmatrix} \mathbf{i} & \mathbf{j} & \mathbf{k} \\ x_D & y_D & 0 \\ 0 & 0 & \alpha_{43} \end{vmatrix} + \begin{vmatrix} \mathbf{i} & \mathbf{j} & \mathbf{k} \\ x_E & y_E & 0 \\ 0 & 0 & \alpha_{54} \end{vmatrix} + a_{E_{05}}\mathbf{J}$$

$$-\omega_{30}^2[(x_D - x_C)\mathbf{i} + (y_D - y_C)\mathbf{J}] - \omega_{40}^2[(x_E - x_D)\mathbf{i} + (y_E - y_D)\mathbf{J}] = \mathbf{0}.$$

Equation (5.141) can be rewritten as

$$\alpha_{30} + \alpha_{43} + \alpha_{54} = 0,$$

$$y_C\,\alpha_{30} + y_D\,\alpha_{43} + y_E\,\alpha_{54} - \omega_{30}^2(x_D - x_C) - \omega_{40}^2(x_E - x_D) = 0,$$

$$-x_C\,\alpha_{30} - x_D\,\alpha_{43} - x_E\,\alpha_{54} + a_{05} - \omega_{30}^2(y_D - y_C) - \omega_{40}^2(y_E - y_D) = 0.$$

The contour acceleration equations using MATLAB commands are:

```
eqIIalpha = alpha30 + alpha43v + alpha54v;
eqIIaz=eqIIalpha(3);
eqIIa=cross(rC,alpha30)+cross(rD,alpha43v)+...
cross(rE,alpha54v)+a05v-...
dot(omega30,omega30)*(rD-rC)-...
dot(omega40,omega40)*(rE-rD);
eqIIax=eqIIa(1);
eqIIay=eqIIa(2);
```

The unknowns in (5.141) are α_{43}, α_{54}, and $a_{E_{05}}$. To solve the system, the following MATLAB command is used:

```
solIIa=solve(eqIIaz,eqIIax,eqIIay);
alpha43 = [0 0 eval(solIIa.alpha43z)];
alpha54 = [0 0 eval(solIIa.alpha54z)];
aE0E5=[0,eval(solIIa.a05),0];
```

The following numerical solutions are obtained:

```
alpha43=[0,0,-6275.008] (rad/s^2)
alpha54=[0,0,-882.339] (rad/s^2)
aE05=1660.866 (m/s^2)
aE0E5=[0,1660.866,0] (m/s^2)
```

The absolute angular acceleration of link 4 and the acceleration of E are:

```
alpha40=[0,0,882.339]  (rad/s^2)
aE5=[0,-1660.866,0]  (m/s^2)
```

5.12   Problems

5.1 The dimensions for the mechanism shown in Fig. 5.48 are given in the table. The driver link 1 rotates with a constant angular speed of $n = 500$ rpm. Find the velocities and the accelerations of the joints and the angular velocities and the angular accelerations of the links for the case when the angle of the driver link AB with the horizontal axis is $\phi = 230°$.

AB	BC	a	ϕ	n
[m]	[m]	[m]	[°]	[rpm]
0.10	0.10	0.12	230°	500

5.2 The dimensions for the mechanism shown in Fig. 5.49 are given in the table. The driver link AB rotates with a constant angular speed of $n = 1,000$ rpm. Find the velocities and the accelerations of the joints and the angular velocities and the angular accelerations of the links for the case when the angle of the driver link AB with the horizontal axis is $\phi = 45°$.

AB	BC	CD	DE	a	b	ϕ	n
[m]	[m]	[m]	[m]	[m]	[m]	[°]	[rpm]
0.10	0.37	0.15	0.15	0.36	0.12	45	1,000

5.3 The dimensions for the mechanism shown in Fig. 5.50 are given in the table. The driver link AB rotates with a constant angular speed of $n = 500$ rpm. Find the velocities and the accelerations of the joints and the angular velocities and

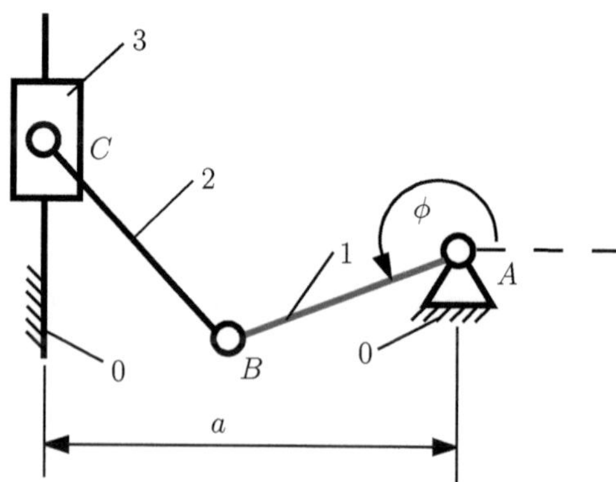

Fig. 5.48 Problem 5.1

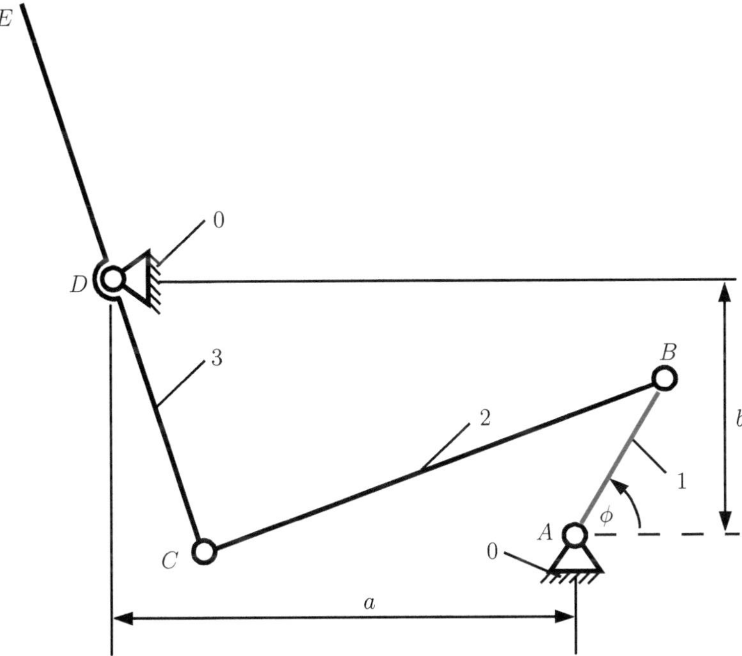

Fig. 5.49 Problem 5.2

Fig. 5.50 Problem 5.3

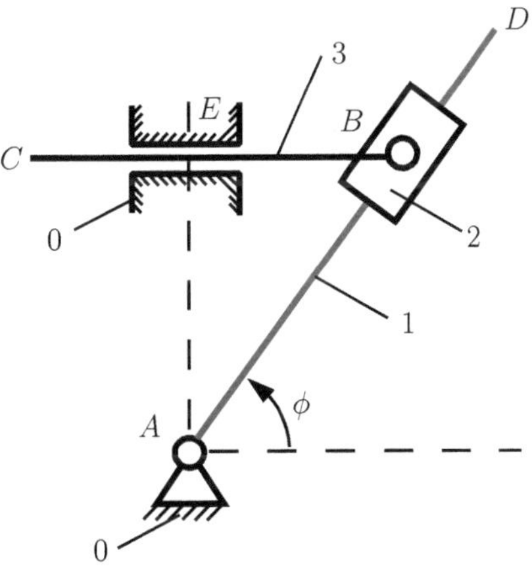

the angular accelerations of the links for the case when the angle of the driver link AB with the horizontal axis is $\phi = 45°$.

AD	BC	AE	ϕ	n
[m]	[m]	[m]	[°]	[rpm]
0.8	0.7	0.4	45	500

Fig. 5.51 Problem 5.4

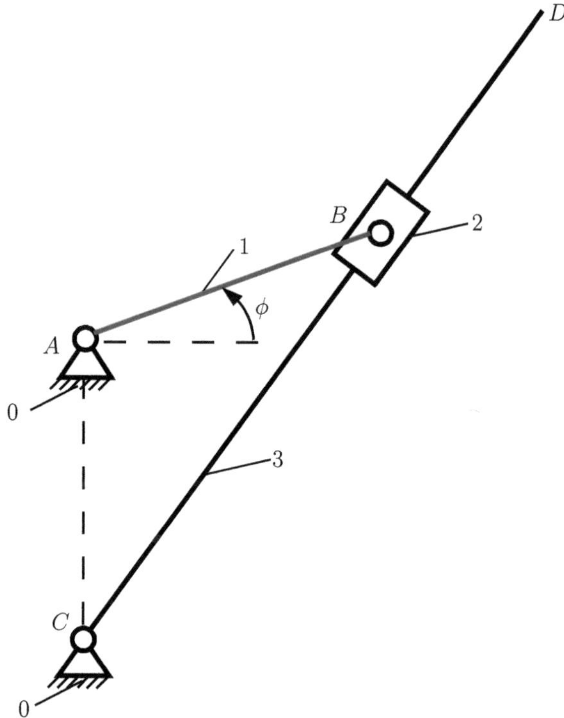

5.4 The dimensions for the mechanism shown in Fig. 5.51 are given in the table. The driver link AB rotates with a constant angular speed of $n = 1{,}200$ rpm. Find the velocities and the accelerations of the joints and the angular velocities and the angular accelerations of the links for the case when the angle of the driver link AB with the horizontal axis is $\phi = 70°$.

AB	AC	CD	CE	ϕ	n
[m]	[m]	[m]	[m]	[°]	[rpm]
0.10	0.20	0.12	0.35	70°	1200

5.5 The dimensions for the mechanism shown in Fig. 5.52 are given in the table. The driver link AB rotates with a constant angular speed of $n = 2{,}500$ rpm. Find the velocities and the accelerations of the joints and the angular velocities and the angular accelerations of the links for the case when the angle of the driver link AB with the horizontal axis is $\phi = 300°$.

AB	AD	BC	BE	ϕ	n
[m]	[m]	[m]	[m]	[°]	[rpm]
0.08	0.20	0.10	0.30	300°	2,500

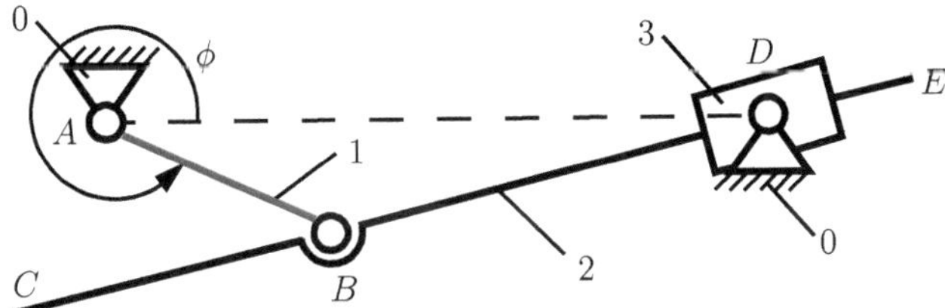

Fig. 5.52 Problem 5.5

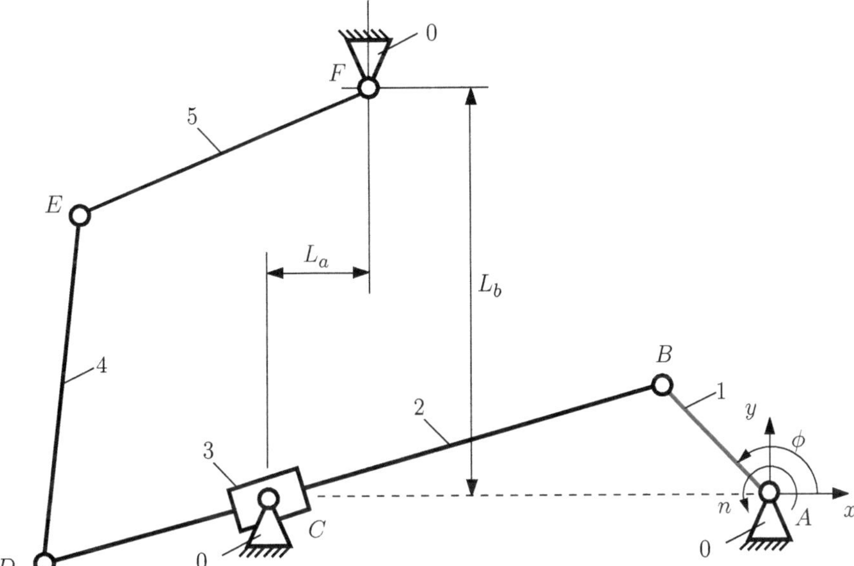

Fig. 5.53 Problem 5.6

5.6 The mechanism in Fig. 5.53 has the dimensions: $AB = 150$ mm, $AC = 350$ mm, $BD = 530$ mm, $DE = 300$ mm, $EF = 200$ mm, $L_a = 55$ mm, and $L_b = 125$ mm. The constant angular speed of the driver link 1 is $n = 30$ rpm. Find the velocities and the accelerations of the mechanism for $\phi = \phi_1 = 120°$.

5.7 The mechanism in Fig. 5.54 has the dimensions: $AB = 200$ mm, $AC = 600$ mm, $BD = 1{,}000$ mm, $L_a = 150$ mm, and $L_b = 250$ mm. The driver link 1 rotates with a constant angular speed of $n = 60$ rpm. Find the velocities and the accelerations of the mechanism for $\phi = \phi_1 = 120°$.

5.8 The planar mechanism considered is shown in Fig. 5.55. The driver link is the rigid link 1 (the link AB). The following numerical data are given: $AB = 0.140$ m, $AC = 0.060$ m, $AE = 0.250$ m, $CD = 0.150$ m. The constant angular speed of the driver link 1 is 1,500 rpm. Find the velocities and the accelerations of the mechanism when the angle of the driver link 1 with the horizontal axis is $\phi = 30°$.

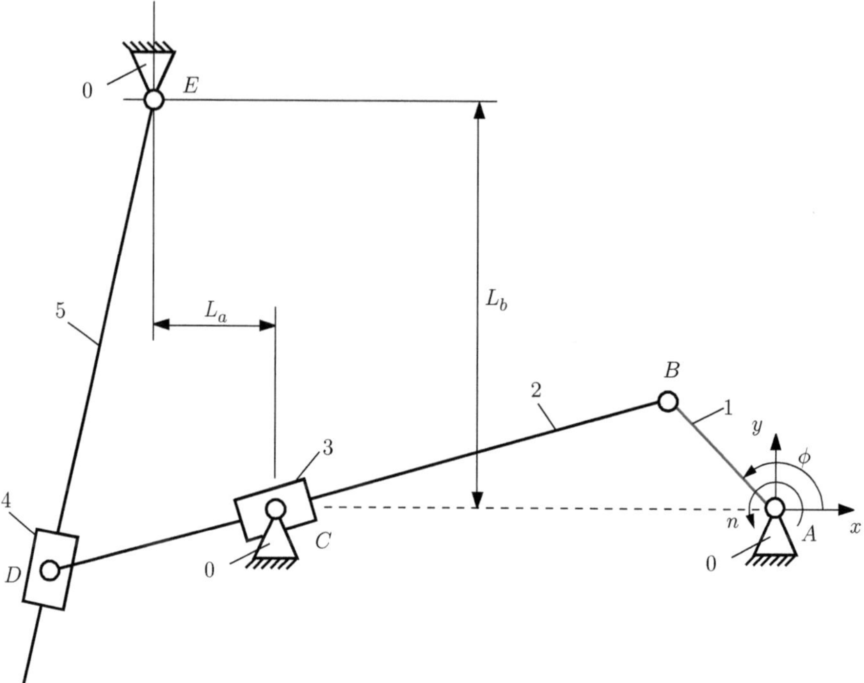

Fig. 5.54 Problem 5.7

5.9 The dimensions of the mechanism shown in Fig. 5.56 are: $AB = 200$ mm, $AC = 300$ mm, $CD = 500$ mm, $DE = 250$ mm, and $L_a = 400$ mm. The constant angular speed of the driver link 1 is $n = 240$ rpm. Find the velocities and the accelerations of the mechanism when the angle of the driver link 1 with the horizontal axis is $\phi = 60°$.

5.10 The dimensions of the mechanism shown in Fig. 5.57 are: $AB = 180$ mm, $AC = 90$ mm, and $CD = 200$ mm. The constant angular speed of the driver link 1 is $n = 880$ rpm. Find the velocities and the accelerations of the mechanism for $\phi = \phi_1 = 60°$.

5.11 A regular tetrahedron $OA_1A_2A_3$ in motion is shown in Fig. 5.58. The length of the sides of the regular tetrahedron is L, $OA_1 = OA_2 = OA_3 = A_1A_2 = A_2A_3 = A_1A_3 = L$. At the instantaneous moment t the velocity of the point O is $\mathbf{v}_O$, the velocity of the point A_1 is $\mathbf{v}_{A_1}$ parallel to $\mathbf{v}_O$, $(\mathbf{v}_{A_1} \parallel \mathbf{v}_O)$, and has the same sense $\mathbf{v}_{A_1} \cdot \mathbf{v}_O > 0$. The velocity of the point A_2 is parallel to the plane OA_1A_2, $\mathbf{v}_{A_2} \parallel$ plane (OA_1A_2). Find the instantaneous angular velocity of the regular tetrahedron.

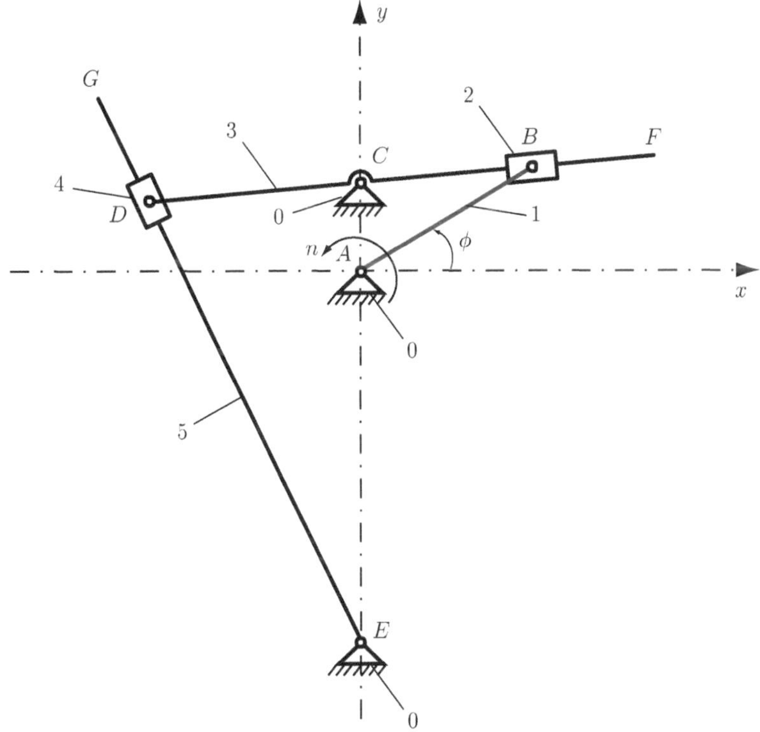

Fig. 5.55 Problem 5.8

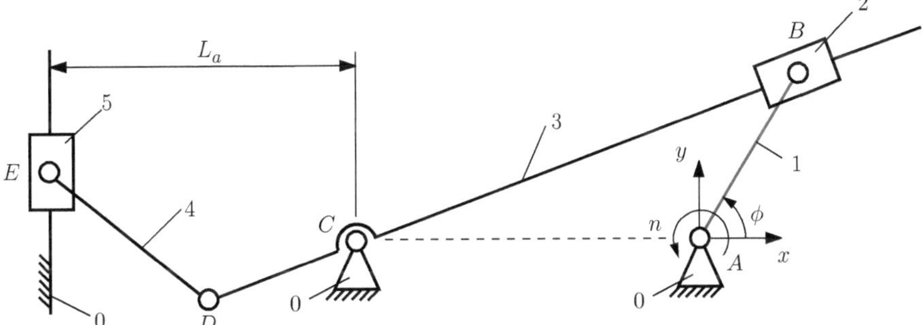

Fig. 5.56 Problem 5.9

5.12 A parallelipeped is rotating along a fixed axis Δ ($\Delta = EC$), as shown in Fig. 5.59, with the angular velocity $\omega = kt^2$, where t is the time. The sides of the paralelipiped are a, b, and c. Find the velocities and the accelerations of the points B and D as a function of time. The numerical values are $k = 5$ rad/s^2, $a = b = 2$ m, and $c = 4$ m.

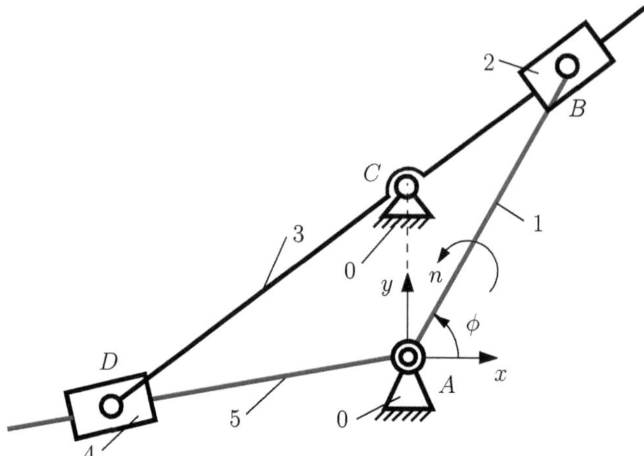

Fig. 5.57 Problem 5.10

Fig. 5.58 Problem 5.11

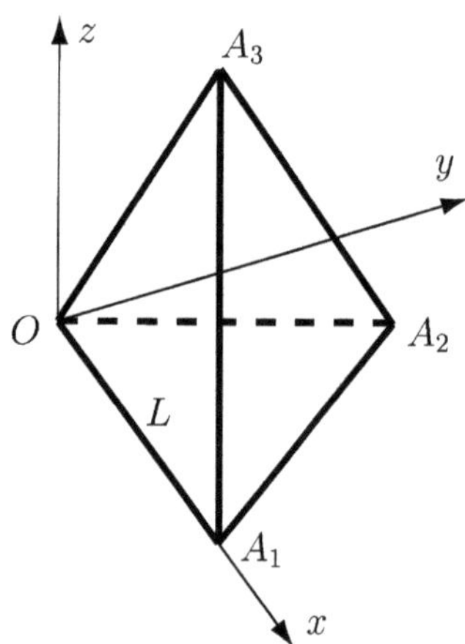

5.13 Repeat Problem 5.12 for the regular prism shown in Fig. 5.60. The numerical values are $EB = BC = CE = 2$ m and $EA = 3$ m.

5.14 Repeat Problem 5.12 for the regular tetrahedron with the sides equal to 2 m, as shown in Fig. 5.61.

5.15 The rigid link AB ($AB = L$) is moving in the vertical plane xOy, as shown in Fig. 5.62. The absolute value of the velocity of the point A is $|\mathbf{v}_A| = t^2 + 1$ m/s, where t is the time. At the initial moment $t = 0$ the point A is at the origin O. Find the angular velocity and acceleration of the link and the velocity and acceleration of the point B as a function of time.

5.16 The straight rod AB, shown in Fig. 5.47, is moving in a vertical plane. The trajectories of the end points A and B are shown in Fig. 5.47. (a) If the velocity of the point A is given, v_A, find the angular velocity of the rod and the velocity

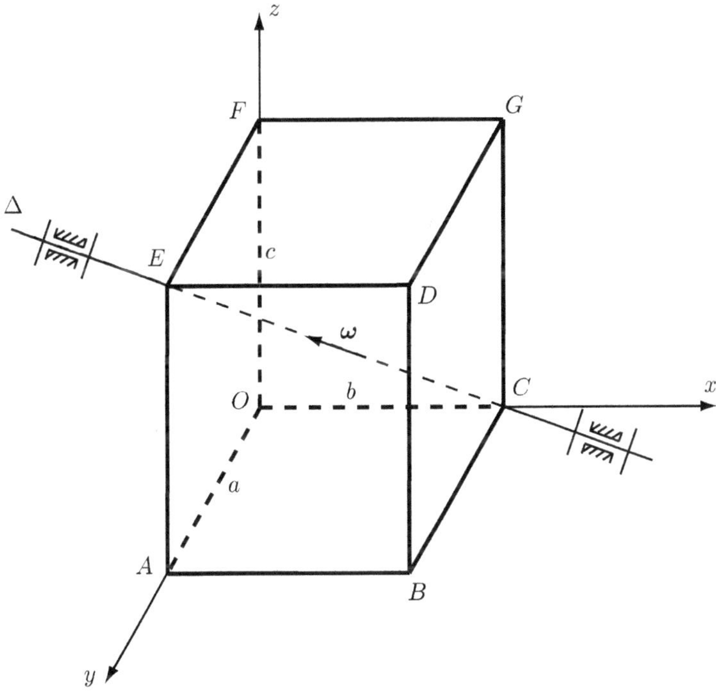

Fig. 5.59   Problem 5.12

Fig. 5.60   Problem 5.13

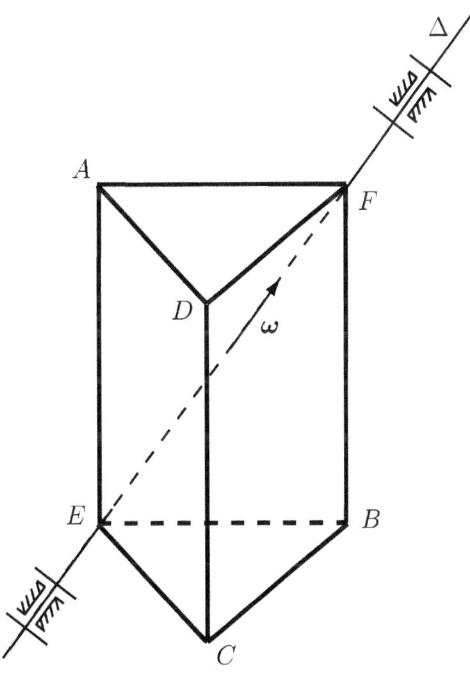

Fig. 5.61 Problem 5.14

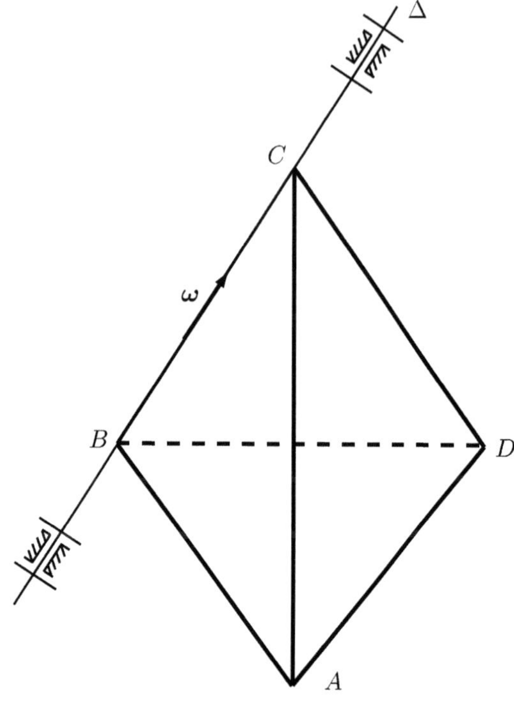

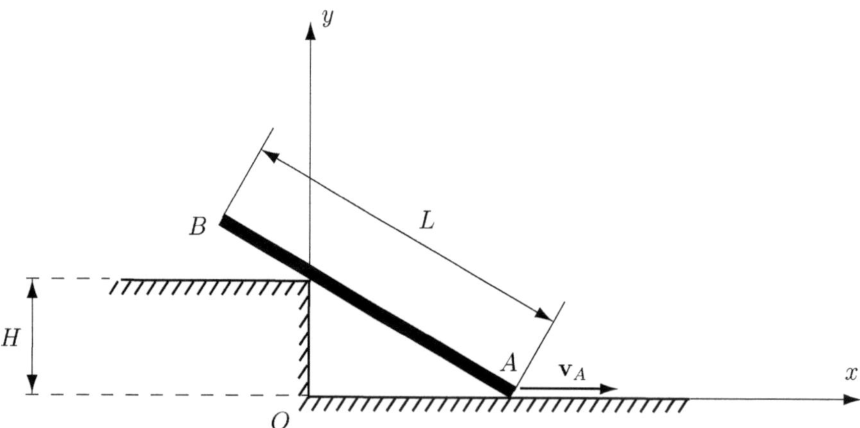

Fig. 5.62 Problem 5.15

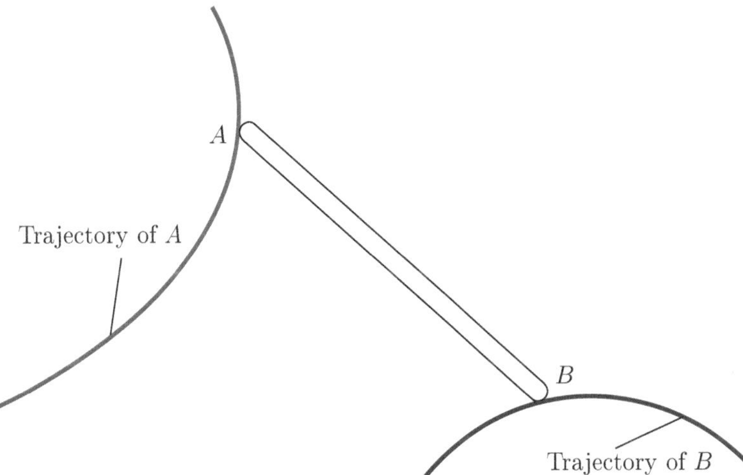

Fig. 5.63 Problem 5.16

v_B of the point B; (b) Repeat (a) for two particular cases: (1) the trajectory of the point A is a straight line Δ and the trajectories of the point B is a circle with center O; (2) the trajectories of the points A and B are circles (Fig. 5.63).

Chapter 6
Dynamics of Rigid Bodies

6.1 Equation of Motion for the Mass Center

Consider a system of N particles. An arbitrary collection of matter with total mass m can be divided into N particles, the ith particle having mass, m_i, as shown in Fig. 6.1:

$$m = \sum_{i=1}^{N} m_i.$$

A rigid body can be considered as a collection of particles in which the number of particles approaches infinity and in which the distance between any two points remains constant. As N approaches infinity, each particle is treated as a differential mass element, $m_i \to \mathrm{d}m$, and the summation is replaced by integration over the body:

$$m = \int_{\text{body}} \mathrm{d}m.$$

The position of the mass center of the collection of particles is defined by

$$\mathbf{r}_C = \frac{1}{m} \sum_{i=1}^{N} m_i \mathbf{r}_i, \tag{6.1}$$

where $\mathbf{r}_i$ is the position vector from the origin O to the ith particle. As $N \to \infty$, the summation is replaced by integration over the body:

$$\mathbf{r}_C = \frac{1}{m} \int_{\text{body}} \mathbf{r}\,\mathrm{d}m, \tag{6.2}$$

D.B. Marghitu and M. Dupac, *Advanced Dynamics: Analytical and Numerical Calculations with MATLAB*, DOI 10.1007/978-1-4614-3475-7_6, © Springer Science+Business Media, LLC 2012

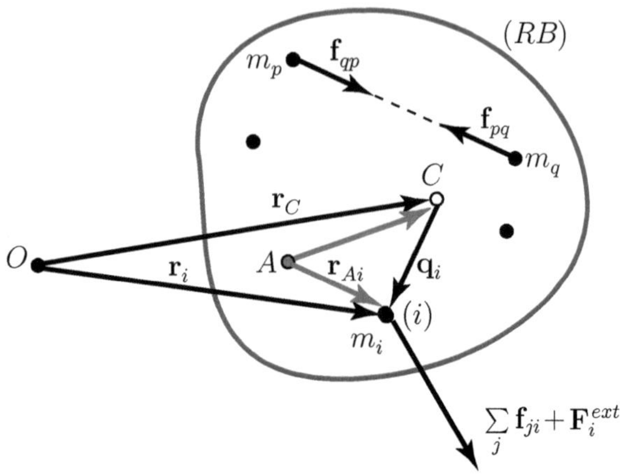

Fig. 6.1 Rigid body as a collection of particles

where $\mathbf{r}$ is the vector from the origin O to differential element dm. The time derivative of (6.1) gives

$$\sum_{i=1}^{N} m_i \frac{d^2 \mathbf{r}_i}{dt^2} = m \frac{d^2 \mathbf{r}_C}{dt^2} = m\mathbf{a}_C, \tag{6.3}$$

where $\mathbf{a}_C$ is the acceleration of the mass center. The acceleration of the mass center can be related to the external forces acting on the system. The relationship is obtained applying Newton's laws to each of the individual particles of the system. Any particle of the system is acted on by two types of forces. One type of forces is exerted by the other particles that are also part of the system. Such forces are called internal forces (internal to the system). Additionally, a particle can be acted on by a force that is exerted by a particle or object not included in the system. Such a force is known as an external force (external to the system). Let $\mathbf{f}_{ij}$ be the internal force exerted on the jth particle by the ith particle. Newton's third law (action and reaction) states that the jth particle exerts a force on the ith particle of equal magnitude, and opposite direction, and collinear with the force exerted by the ith particle on the jth particle (Fig. 6.1):

$$\mathbf{f}_{ji} = -\mathbf{f}_{ij}, \quad j \neq i.$$

Newton's second law for the ith particle must include all of the internal forces exerted by all of the other particles in the system on the ith particle, plus the sum of any external forces exerted by particles, objects outside of the system on the ith particle:

$$\sum_j \mathbf{f}_{ji} + \mathbf{F}_i^{ext} = m_i \frac{d^2 \mathbf{r}_i}{dt^2}, \quad j \neq i, \tag{6.4}$$

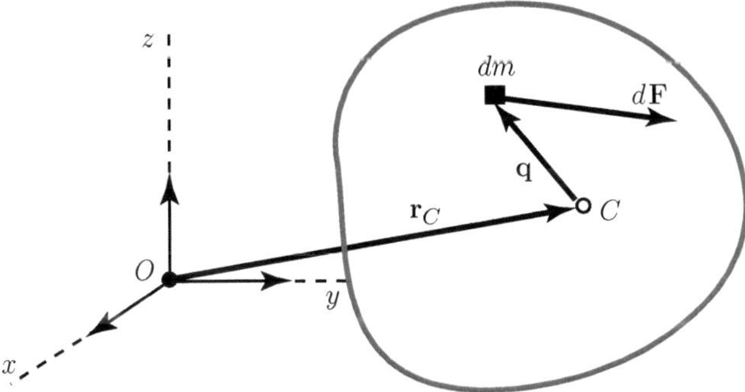

Fig. 6.2 Rigid body

where $\mathbf{F}_i^{\text{ext}}$ is the external force on the ith particle. Equation (6.4) is written for each particle in the collection of particles. Summing the resulting equations over all of the particles from $i = 1$ to N, the following relation is obtained:

$$\sum_i \sum_j \mathbf{f}_{ji} + \sum_i \mathbf{F}_i^{\text{ext}} = m\mathbf{a}_C, \quad j \neq i. \tag{6.5}$$

The sum of the internal forces includes pairs of equal and opposite forces. The sum of any such pair must be zero. The sum of all of the internal forces on the collection of particles is zero (Newton's third law):

$$\sum_i \sum_j \mathbf{f}_{ji} = \mathbf{0}, \quad j \neq i.$$

The term $\sum_i \mathbf{F}_i^{\text{ext}}$ is the sum of the external forces on the collection of particles:

$$\sum_i \mathbf{F}_i^{\text{ext}} = \sum_i \mathbf{F}_i = \mathbf{F}.$$

The sum of the external forces acting on a closed system equals the product of the mass and the acceleration of the mass center:

$$m\,\mathbf{a}_C = \mathbf{F}. \tag{6.6}$$

Considering Fig. 6.2 for a rigid body and introducing the distance $\mathbf{q}$ in (6.2) gives

$$\mathbf{r}_C = \frac{1}{m} \int_{\text{body}} \mathbf{r}\,dm = \frac{1}{m} \int_{\text{body}} (\mathbf{r}_C + \mathbf{q})\,dm = \mathbf{r}_C + \frac{1}{m} \int_{\text{body}} \mathbf{q}\,dm. \tag{6.7}$$

It results

$$\frac{1}{m} \int_{\text{body}} \mathbf{q}\,dm = \mathbf{0}, \tag{6.8}$$

that is, the weighed average of the displacement vector about the mass center is zero. The equation of motion for the differential element dm is

$$\mathbf{a}\,dm = d\mathbf{F},$$

where $d\mathbf{F}$ is the total force acting on the differential element. For the entire body,

$$\int_{body} \mathbf{a}\,dm = \int_{body} d\mathbf{F} = \mathbf{F}, \tag{6.9}$$

where $\mathbf{F}$ is the resultant of all forces. This resultant contains contributions only from the external forces, as the internal forces cancel each other. Introducing (6.7) into (6.9), the Newton's second law for a rigid body is obtained:

$$m\,\mathbf{a}_C = \mathbf{F}.$$

The derivation of the equations of motion is valid for the general motion of a rigid body. These equations are equally applicable to planar and three-dimensional motions.

Resolving the sum of the external forces into Cartesian rectangular components

$$\mathbf{F} = F_x\mathbf{i} + F_y\mathbf{j} + F_z\mathbf{k},$$

and the position vector of the mass center

$$\mathbf{r}_C = x_C(t)\mathbf{i} + y_C(t)\mathbf{j} + z_C(t)\mathbf{k},$$

Newton's second law for the rigid body is

$$m\ddot{\mathbf{r}}_C = \mathbf{F}, \tag{6.10}$$

or

$$m\ddot{x}_C = F_x, \quad m\ddot{y}_C = F_y, \quad m\ddot{z}_C = F_z. \tag{6.11}$$

6.2 Linear Momentum and Angular Momentum

The linear momentum $\mathbf{p}$ of the system of particles is

$$\mathbf{p} = \sum_{i=1}^{N} \mathbf{p}_i = \sum_{i=1}^{N} m_i\mathbf{v}_i = m\mathbf{v}_C, \tag{6.12}$$

where $\mathbf{p}_i = m_i\mathbf{v}_i$ $(i = 1,2,\ldots,N)$ is the linear momentum of the ith particle and $\mathbf{v}_C$ is the velocity of the center of mass. Equation (6.12) defines the linear momentum of a system of particles. Integrating the equations of motion for a system of particles over a time period (t_1, t_2), the impulse-momentum relationships are obtained:

$$m_i \mathbf{v}_i(t_1) + \int_{t_1}^{t_2} \left(\mathbf{f}_{ji}(t) + \mathbf{F}_i^{\text{ext}}(t) \right) dt = m_i \mathbf{v}_i(t_2) \qquad i = 1, 2, \ldots, N, \quad (6.13)$$

$$\sum_{i+1}^{N} m_i \mathbf{v}_i(t_1) + \int_{t_1}^{t_2} \mathbf{F}(t) \, dt = \sum_{i=1}^{N} m_i \mathbf{v}_i(t_2), \qquad (6.14)$$

$$m\mathbf{v}_C(t_1) + \int_{t_1}^{t_2} \mathbf{F}(t) \, dt = m\mathbf{v}_C(t_2). \qquad (6.15)$$

The principle of conservation of linear momentum for a system of particles states: the linear momentum of the system remains unchanged when the sum of all external forces acting on a system of particles is zero or its integral over a time period is zero.

$$\mathbf{p}(t_1) = \mathbf{p}(t_2), \qquad (6.16)$$

or

$$m\mathbf{v}_C(t_1) = m\mathbf{v}_C(t_2) \qquad \sum_{i=1}^{N} m_i \mathbf{v}_i(t_1) = \sum_{i=1}^{N} m_i \mathbf{v}_i(t_2). \qquad (6.17)$$

If the sum of forces or their integral over a time period is equal to zero along a certain direction, then the linear momentum is conserved only along that particular direction.

The angular momentum, or moment of the linear momentum, of a particle m_i about a point A is

$$\mathbf{H}_{Ai} = \mathbf{r}_{AP_i} \times m_i \mathbf{v}_i = \mathbf{r}_{AP_i} \times \mathbf{p}_i \qquad i = 1, 2, \ldots, N, \qquad (6.18)$$

where $\mathbf{r}_{AP_i}$ is the vector connecting point A and the ith particle P_i as shown in Fig. 6.3. The angular momentum is a relative quantity; its value depends on the point about which it is calculated. The linear momentum is an absolute quantity. The total angular momentum of a system of particles about point A is

$$\mathbf{H}_A = \sum_{i=1}^{N} \mathbf{H}_{Ai} = \sum_{i=1}^{N} \mathbf{r}_{AP_i} \times m_i \mathbf{v}_i. \qquad (6.19)$$

The vector $\mathbf{r}_{AP_i}$ is written in terms of the center of mass as $\mathbf{r}_{AP_i} = \mathbf{r}_{AC} + \mathbf{q}_i$ $(i = 1, 2, \ldots, N)$. The angular momentum expressions is for the ith particle is

$$\mathbf{H}_{Ai} = (\mathbf{r}_{AG} + \mathbf{q}_i) \times m_i (\mathbf{v}_C + \dot{\mathbf{q}}_i). \qquad (6.20)$$

The angular momentum of the system of particles about point A is

$$\mathbf{H}_A = \sum_{i=1}^{N} \mathbf{H}_{Ai} = \sum_{i=1}^{N} (\mathbf{r}_{AC} + \mathbf{q}_i) \times m_i (\mathbf{v}_C + \dot{\mathbf{q}}_i). \qquad (6.21)$$

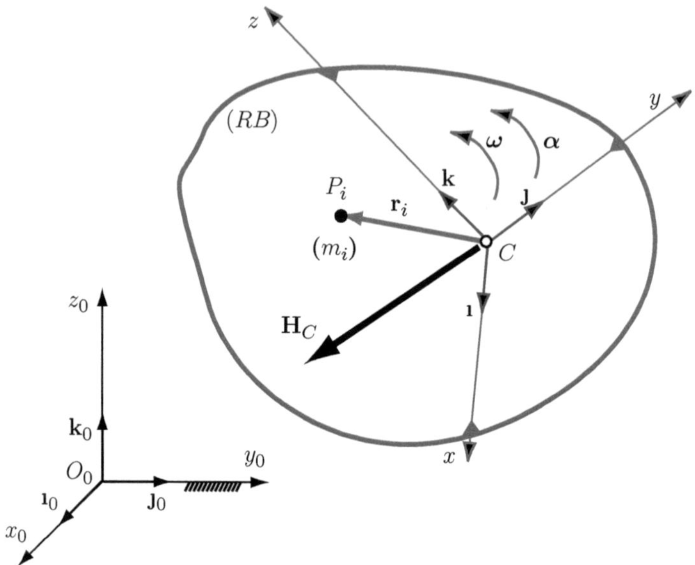

Fig. 6.3 Angular momentum of a rigid body

Using the definition of the center of mass, the previous equation is

$$\mathbf{H}_A = \mathbf{r}_{AC} \times m\mathbf{v}_C + \sum_{i=1}^{N} \mathbf{q}_i \times m_i \dot{\mathbf{q}}_i. \tag{6.22}$$

Taking the derivative of (6.22), it results

$$\dot{\mathbf{H}}_A = \mathbf{r}_{AC} \times m\mathbf{a}_G + \mathbf{v}_{CB} \times m\mathbf{v}_C + \sum_{i=1}^{N} [\mathbf{q}_i \times m_i \ddot{\mathbf{q}}_i + \dot{\mathbf{q}}_i \times m_i \dot{\mathbf{q}}_i]$$

$$= \mathbf{r}_{AC} \times m\mathbf{a}_C + \mathbf{v}_{CA} \times m\mathbf{v}_C + \sum_{i=1}^{N} \mathbf{q}_i \times m_i \ddot{\mathbf{q}}_i. \tag{6.23}$$

The second term of the right side of (6.23) is

$$\mathbf{v}_{CA} \times m\mathbf{v}_C = (\mathbf{v}_C - \mathbf{v}_A) \times m\mathbf{v}_C = m\mathbf{v}_C \times \mathbf{v}_A, \tag{6.24}$$

and

$$\dot{\mathbf{H}}_A = \mathbf{r}_{AC} \times m\mathbf{a}_C + m\mathbf{v}_C \times \mathbf{v}_A + \sum_{i=1}^{N} \mathbf{q}_i \times m_i \ddot{\mathbf{q}}_i. \tag{6.25}$$

The moment about point A of all the forces acting on the ith particle is

$$\mathbf{M}_{Ai} = \mathbf{r}_{AP_i} \times \mathbf{F}_i = \mathbf{r}_{AP_i} \times m_i \mathbf{a}_i \qquad i = 1, 2, \ldots, N. \tag{6.26}$$

The moment about point A of all of the moments acting on the system of particles is

$$\mathbf{M}_A = \sum_{i=1}^{N} \mathbf{r}_{AP_i} \times \mathbf{F}_i = \sum_{i=1}^{N} \mathbf{r}_{AP_i} \times m_i \mathbf{a}_i = \sum_{i=1}^{N} (\mathbf{r}_{AC} + \mathbf{q}_i) \times m_i (\mathbf{a}_C + \ddot{\mathbf{q}}_i). \quad (6.27)$$

Using the definition of the center of mass, (6.27) becomes

$$\mathbf{M}_A = \mathbf{r}_{AC} \times m \mathbf{a}_C + \sum_{i=1}^{N} \mathbf{q}_i \times m_i \ddot{\mathbf{q}}_i. \quad (6.28)$$

Equations (6.29) and (6.25) give the angular momentum balance for a system of particles as

$$\dot{\mathbf{H}}_A = \mathbf{M}_A + m\mathbf{v}_C \times \mathbf{v}_A. \quad (6.29)$$

When point A is the center of mass, $A = C$, then

$$\dot{\mathbf{H}}_C = \mathbf{M}_C. \quad (6.30)$$

Integration of the moment balance over time for each particle m_i and considering a fixed point or about the center of mass,

$$\mathbf{H}_{Ci}(t_1) + \int_{t_1}^{t_2} \mathbf{M}_{Ci}(t)\ dt = \mathbf{H}_{Ci}(t_2) \quad i = 1, 2, \ldots, N. \quad (6.31)$$

Considering the entire system and either a fixed point or the center of mass, summation of (6.31) over all particles yields

$$\mathbf{H}_C(t_1) + \int_{t_1}^{t_2} \mathbf{M}_C(t)\ dt = \mathbf{H}_C(t_2). \quad (6.32)$$

If the moment about the fixed point or center of mass is zero, or the integral of $\mathbf{M}_C(t)$ over the interval (t_1, t_2) is zero, then the principle of conservation of angular momentum for a system of particles is

$$\mathbf{H}_C(t_1) = \mathbf{H}_C(t_2). \quad (6.33)$$

6.3 Spatial Angular Momentum of a Rigid Body

The angular momentum $\mathbf{H}_C$ of a rigid body about its mass center C is expressed as

$$\mathbf{H}_C = \sum_{i=1}^{n} (\mathbf{r}_i \times \mathbf{v}_i m_i), \quad (6.34)$$

where $\mathbf{r}_i$ and $\mathbf{v}_i$ denote, respectively, the position vector and the velocity of the particle P_i, of mass m_i, relative to the centroidal moving frame $Cxyz$ (Fig. 6.3). The velocity of P_i is $\mathbf{v}_i = \boldsymbol{\omega} \times \mathbf{r}_i$, where $\boldsymbol{\omega}$ is the angular velocity of the rigid body at the considered instant. Substituting into (6.34) gives

$$\mathbf{H}_C = \sum_{i=1}^{n} \left[\mathbf{r}_i \times (\boldsymbol{\omega} \times \mathbf{r}_i)\, m_i \right].$$

To calculate $\mathbf{H}_C$, a MATLAB program is written:

```
syms w_x w_y w_z xi yi zi mi

w = [w_x w_y w_z];
ri = [xi yi zi];

wxr = cross(w,ri);
HC = cross(ri, wxr);

HCx = mi*expand(HC(1));
HCy = mi*expand(HC(2));
HCz = mi*expand(HC(3));

pretty(HCx)
pretty(HCy)
pretty(HCz)
```

The following results are obtained:

```
              2                                    2
    mi (w_x yi   - w_y xi yi + w_x zi   - w_z xi zi)


              2                                    2
    mi (w_y xi   - w_x yi xi + w_y zi   - w_z yi zi)


              2                                    2
    mi (w_z xi   - w_x zi xi + w_z yi   - w_y zi yi)
```

The following expression for the x, y, and z components of the angular momentum is obtained:

$$H_x = \omega_x \sum_{i=1}^{n} \left(y_i^2 + z_i^2 \right) m_i - \omega_y \sum_{i=1}^{n} x_i y_i m_i - \omega_z \sum_{i=1}^{n} z_i x_i m_i,$$

$$H_y = \omega_y \sum_{i=1}^{n} \left(x_i^2 + z_i^2 \right) m_i - \omega_x \sum_{i=1}^{n} x_i y_i m_i - \omega_z \sum_{i=1}^{n} z_i y_i m_i,$$

$$H_z = \omega_z \sum_{i=1}^{n} \left(x_i^2 + y_i^2 \right) m_i - \omega_x \sum_{i=1}^{n} x_i z_i m_i - \omega_y \sum_{i=1}^{n} z_i y_i m_i.$$

Replacing the sums by integrals in the previous expressions, the following relations arc obtained:

$$H_x = \omega_x \int \left(y^2 + z^2\right) dm - \omega_y \int xy\,dm - \omega_z \int zx\,dm,$$

$$H_y = -\omega_x \int xy\,dm + \omega_y \int \left(z^2 + x^2\right) dm - \omega_z \int yz\,dm,$$

$$H_z = -\omega_x \int zx\,dm - \omega_y \int yz\,dm + \omega_z \int \left(x^2 + y^2\right) dm. \tag{6.35}$$

The centroidal mass moments of inertia of the rigid body about the x, y, and z axes are

$$I_{xx} = \int \left(y^2 + z^2\right) dm,$$

$$I_{yy} = \int \left(z^2 + x^2\right) dm,$$

$$I_{zz} = \int \left(x^2 + y^2\right) dm. \tag{6.36}$$

The centroidal mass products of inertia of the rigid body are

$$I_{xy} = \int xy\,dm, \quad I_{yz} = \int yz\,dm, \quad I_{zx} = \int zx\,dm. \tag{6.37}$$

Substituting (6.36) and (6.37) into (6.35), the components of the angular momentum $\mathbf{H}_C$ of the body about its mass center C are obtained:

$$H_x = I_{xx}\omega_x - I_{xy}\omega_y - I_{xz}\omega_z,$$
$$H_y = -I_{yx}\omega_x + I_{yy}\omega_y - I_{yz}\omega_z,$$
$$H_z = -I_{zx}\omega_x - I_{zy}\omega_y + I_{zz}\omega_z. \tag{6.38}$$

The inertia matrix of the rigid body at its mass center C is

$$[I] = \begin{bmatrix} I_{xx} & -I_{xy} & -I_{xz} \\ -I_{yx} & I_{yy} & -I_{yz} \\ -I_{zx} & -I_{zy} & I_{zz} \end{bmatrix}. \tag{6.39}$$

The matrix given by (6.39) is also called the *inertia tensor* of the body at its mass center C. A new array of moments and products of inertia would be obtained if a different system of axes was used. The transformation characterized by this new array, however, would still be the same.

The angular momentum $\mathbf{H}_C$ corresponding to a given angular velocity $\boldsymbol{\omega}$ is independent of the choice of the coordinate axes. It is always possible to select a system of axes $Cxyz$, called principal axes of inertia, with respect to which all the products of inertia of a given body are zero. The matrix given by (6.39) takes then the diagonalized form:

$$[I] = \begin{bmatrix} I_1 & 0 & 0 \\ 0 & I_2 & 0 \\ 0 & 0 & I_3 \end{bmatrix}, \tag{6.40}$$

where I_1, I_2, I_3 represent the central principal moments of inertia of the rigid body. Equation (6.38) for the principal moments of inertia gives

$$H_1 = I_1 \omega_1, \quad H_2 = I_2 \omega_2, \quad H_3 = I_3 \omega_3, \tag{6.41}$$

where ω_1, ω_2, ω_3 are the components of the angular velocity $\boldsymbol{\omega}$ on the principal centroidal axis.

The vectors $\mathbf{H}_C$ and $\boldsymbol{\omega}$ are collinear if the three principal centroidal moments of inertia I_1, I_2, I_3 are equal. In this case, the components H_1, H_2, H_3 of the angular momentum about C are proportional to the components ω_1, ω_2, ω_3 of the angular velocity. In general, the principal moments of inertia are different, and the vectors $\mathbf{H}_C$ and $\boldsymbol{\omega}$ have different directions, except when two of the three components of $\boldsymbol{\omega}$ happen to be zero, that is, when $\boldsymbol{\omega}$ is directed along one of the coordinate axes.

In the particular case when $I_1 = I_2 = I_3$, any line through C may be considered as a principal axis of inertia, and the vectors $\mathbf{H}_C$ and $\boldsymbol{\omega}$ are always collinear.

Setting $I_{xx} = I_{11}$, $I_{yy} = I_{22}$, $I_{zz} = I_{33}$, $-I_{xy} = I_{12}$, $-I_{xz} = I_{13}$, etc., the inertia tensor can be written in the standard form:

$$[I] = \begin{bmatrix} I_{11} & I_{12} & I_{13} \\ I_{21} & I_{22} & I_{23} \\ I_{31} & I_{32} & I_{33} \end{bmatrix}.$$

Denoting by H_1, H_2, H_3 the components of the angular momentum $\mathbf{H}_C$ and by ω_1, ω_2, ω_3, the components of the angular velocity $\boldsymbol{\omega}$, we may write the relations in the form

$$H_i = \sum_{j=1}^{3} I_{ij} \omega_j,$$

where i takes the values 1, 2, 3. The quantities I_{ij} are said to be the components of the inertia tensor. Since $I_{ij} = I_{ji}$, the inertia tensor is a symmetric tensor of the second order.

Angular Momentum of a Rigid Body with a Fixed Point
If a rigid body is rotating in three-dimensional space about a fixed point O (Fig. 6.4), the angular momentum $\mathbf{H}_O$ of the body about the fixed point O is

$$\mathbf{H}_O = \sum_{i=1}^{n} (\mathbf{r}_i \times \mathbf{v}_i m_i), \tag{6.42}$$

Fig. 6.4 Rigid body
with a fixed point

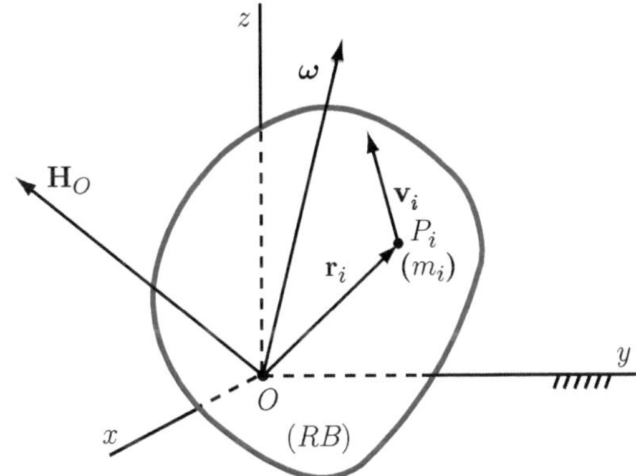

where $\mathbf{r}_i$ and $\mathbf{v}_i$ denote, respectively, the position vector and the velocity of the
particle P_i with respect to the fixed frame $Oxyz$. Substituting $\mathbf{v}_i = \boldsymbol{\omega} \times \mathbf{r}_i$, and after
manipulations similar to the ones used previously, the components of the angular
momentum $\mathbf{H}_O$ are given by the relations

$$
\begin{aligned}
H_x &= I_{xx}\omega_x - I_{xy}\omega_y - I_{xz}\omega_z, \\
H_y &= -I_{yx}\omega_x + I_{yy}\omega_y - I_{yz}\omega_z, \\
H_z &= -I_{zx}\omega_x - I_{zy}\omega_y + I_{zz}\omega_z,
\end{aligned}
\tag{6.43}
$$

where the moments of inertia I_{xx}, I_{yy}, I_{zz} and the products of inertia I_{xy}, I_{yz}, I_{zx} are
computed with respect to the frame $Oxyz$ centered at the fixed point O.

6.4   Kinetic Energy of a Rigid Body

The quantity $T_i = \frac{1}{2} m_i v_i^2$ is the kinetic energy of the particle P_i. The total kinetic
energy of the system, shown in Fig. 6.1, is

$$
T = \frac{1}{2} \sum_{i=1}^{N} m_i v_i^2 = \frac{1}{2} \sum_{i=1}^{N} m_i \mathbf{v}_i \cdot \mathbf{v}_i.
\tag{6.44}
$$

Let A a point on the body and $\boldsymbol{\omega}$ the angular velocity of the body. The kinetic energy
becomes

$$
\begin{aligned}
T &= \frac{1}{2} \sum_{i=1}^{N} m_i \left(\mathbf{v}_A + \boldsymbol{\omega} \times \mathbf{r}_{AP_i}\right) \cdot \left(\mathbf{v}_A + \boldsymbol{\omega} \times \mathbf{r}_{AP_i}\right) \\
&= \frac{1}{2} \sum_{i=1}^{N} m_i \mathbf{v}_A \cdot \mathbf{v}_A + \mathbf{v}_A \cdot \left(\boldsymbol{\omega} \times \sum_{i=1}^{N} m_i \mathbf{r}_{AP_i}\right) + \frac{1}{2} \sum_{i=1}^{N} m_i \left(\boldsymbol{\omega} \times \mathbf{r}_{AP_i}\right) \cdot \left(\boldsymbol{\omega} \times \mathbf{r}_{AP_i}\right).
\end{aligned}
\tag{6.45}
$$

The sum $\sum_{i=1}^{N} m_i = m$ is the total mass, and $\sum_{i=1}^{N} m_i\, \mathbf{r}_{AP_i} = m\,\mathbf{r}_{AC}$. Using the identity for the scalar triple product,

$$\mathbf{a} \times \mathbf{b} \cdot \mathbf{c} = \mathbf{a} \cdot \mathbf{b} \times \mathbf{c},$$

and applying this to the sequence of vectors yields

$$(\boldsymbol{\omega} \times \mathbf{r}_{AP_i}) \cdot (\boldsymbol{\omega} \times \mathbf{r}_{AP_i}) = \boldsymbol{\omega} \cdot \left[\mathbf{r}_{AP_i} \times (\boldsymbol{\omega} \times \mathbf{r}_{AP_i}) \right].$$

The kinetic energy is

$$T = \frac{1}{2} m \mathbf{v}_A \cdot \mathbf{v}_A + m \mathbf{v}_A \cdot \boldsymbol{\omega} \times \mathbf{r}_{AC} + \frac{1}{2} \sum_{i=1}^{N} m_i \boldsymbol{\omega} \cdot \left[\mathbf{r}_{AP_i} \times (\boldsymbol{\omega} \times \mathbf{r}_{AP_i}) \right]$$

$$= \frac{1}{2} m \mathbf{v}_A \cdot \mathbf{v}_A + m \mathbf{v}_A \cdot \boldsymbol{\omega} \times \mathbf{r}_{AC} + \frac{1}{2} \boldsymbol{\omega} \cdot \mathbf{H}_A, \tag{6.46}$$

where $\mathbf{H}_A$ is the angular momentum relative to point A, defined as $\mathbf{H}_A = \sum_{i=1}^{N} m_i\, \mathbf{r}_{AP_i} \times (\boldsymbol{\omega} \times \mathbf{r}_{AP_i})$. If point A is the center of mass of the body ($\mathbf{r}_{AC} = \mathbf{0}$), or is a fixed point for a body in pure rotation ($\mathbf{v}_A = \mathbf{0}$), then $m \mathbf{v}_A \cdot \boldsymbol{\omega} \times \mathbf{r}_{AC} = 0$. The expressions for kinetic energy are

$$T = \frac{1}{2} m \mathbf{v}_C \cdot \mathbf{v}_C + \frac{1}{2} \boldsymbol{\omega} \cdot \mathbf{H}_C \quad \text{for all motions,} \tag{6.47}$$

$$T = \frac{1}{2} \boldsymbol{\omega} \cdot \mathbf{H}_P \quad \text{for pure rotation about point } P. \tag{6.48}$$

The kinetic energy of a body relative to its centroidal axes is

$$T = \frac{1}{2} m v_C^2 + \frac{1}{2} \left(I_{xx} \omega_x^2 + I_{yy} \omega_y^2 + I_{zz} \omega_z^2 - 2 I_{xy} \omega_x \omega_y - 2 I_{yz} \omega_y \omega_z - 2 I_{zx} \omega_z \omega_x \right).$$

$$\tag{6.49}$$

If the axes of coordinates are chosen so that they coincide at the instant considered with the principal axes x, y, z of the body, the previous relation gives

$$T = \frac{1}{2} m v_C^2 + \frac{1}{2} \left(I_1 \omega_1^2 + I_2 \omega_2^2 + I_3 \omega_3^2 \right), \tag{6.50}$$

where I_1, I_2, and I_3 are the principal centroidal moments of inertia. The results obtained enable us to extend the three-dimensional motion of a rigid body the application of the principle of work and energy.

In the particular case of a rigid body rotating in three-dimensional space about a fixed point O, the kinetic energy of the body is expressed in terms of its moments and products of inertia with respect to axes attached at O:

$$T = \frac{1}{2} \left(I_{xx} \omega_x^2 + I_{yy} \omega_y^2 + I_{zz} \omega_z^2 - 2 I_{xy} \omega_x \omega_y - 2 I_{yz} \omega_y \omega_z - 2 I_{zx} \omega_z \omega_x \right). \tag{6.51}$$

If the principal axes x, y, z of the body at the origin O are chosen as coordinate axes, then the kinetic energy is

$$T = \frac{1}{2}\left(I_x\omega_x^2 + I_y\omega_y^2 + I_z\omega_z^2\right). \tag{6.52}$$

6.5 Equations of Motion

The moment of the forces acting on element i, Fig. 6.1, about the point A is equal the moment of the $m_i\,\mathbf{a}_i$ or

$$(\mathbf{M}_A)_i = \mathbf{r}_{AP_i} \times \mathbf{F}_i = \mathbf{r}_{AP_i} \times m_i\,\mathbf{a}_i. \tag{6.53}$$

If point A is on the rigid body, then it follows that

$$\mathbf{a}_i = \mathbf{a}_A + \boldsymbol{\alpha} \times \mathbf{r}_{AP_i} + \boldsymbol{\omega} \times (\boldsymbol{\omega} \times \mathbf{r}_{AP_i}).$$

and

$$(\mathbf{M}_A)_i = \mathbf{r}_{AP_i} \times \mathbf{F}_i = \mathbf{r}_{AP_i} \times m_i\left[\mathbf{a}_A + \boldsymbol{\alpha} \times \mathbf{r}_{AP_i} + \boldsymbol{\omega} \times (\boldsymbol{\omega} \times \mathbf{r}_{AP_i})\right]. \tag{6.54}$$

The rigidity of the body means that $\dot{r}_{AP_i} = \boldsymbol{\omega} \times \mathbf{r}_{AP_i}$. It results that

$$(\mathbf{M}_A)_i = \mathbf{r}_{AP_i} \times m_i\,\mathbf{a}_A + \mathbf{r}_{AP_i} \times \left[m_i\,\frac{\mathrm{d}}{\mathrm{d}t}(\boldsymbol{\omega} \times \mathbf{r}_{AP_i})\right]$$

$$= \mathbf{r}_{AP_i} \times m_i\,\mathbf{a}_A + \frac{\mathrm{d}}{\mathrm{d}t}[\mathbf{r}_{AP_i} \times m_i(\boldsymbol{\omega} \times \mathbf{r}_{AP_i})] - (\boldsymbol{\omega} \times \mathbf{r}_{AP_i}) \times m_i(\boldsymbol{\omega} \times \mathbf{r}_{AP_i})$$

$$= \mathbf{r}_{AP_i} \times m_i\,\mathbf{a}_A + \frac{\mathrm{d}}{\mathrm{d}t}[\mathbf{r}_{AP_i} \times m_i(\boldsymbol{\omega} \times \mathbf{r}_{AP_i})]. \tag{6.55}$$

The angular momentum about point A is

$$(\mathbf{H}_A)_i = \mathbf{r}_{AP_i} \times [m_i(\boldsymbol{\omega} \times \mathbf{r}_{AP_i})].$$

The moment of the forces acting on element i is the sum of the moment of the $m_i\,\mathbf{a}_A$ and the rate of change of the angular momentum:

$$(\mathbf{M}_A)_i = \mathbf{r}_{AP_i} \times m_i\,\mathbf{a}_A + \frac{\mathrm{d}}{\mathrm{d}t}(\mathbf{H}_A)_i.$$

The total moment of all forces is obtained adding (6.56) for each element i:

$$\sum \mathbf{M}_A = \left(\sum_{i=1}^{N} m_i\mathbf{r}_{iA}\right) \times \mathbf{a}_A + \frac{\mathrm{d}}{\mathrm{d}t}\mathbf{H}_A = \mathbf{r}_{AC} \times m\,\mathbf{a}_A + \frac{\mathrm{d}}{\mathrm{d}t}\mathbf{H}_A, \tag{6.56}$$

where $\mathbf{H}_A = \sum_{i=1}^{N} m_i\, \mathbf{r}_{AP_i} \times (\boldsymbol{\omega} \times \mathbf{r}_{AP_i})$ is the total angular momentum about point A. Equations (6.54) and (6.56) lead to

$$\sum \mathbf{M}_A = \sum_{i=1}^{N} \mathbf{r}_{AP_i} \times \mathbf{F}_i = \mathbf{r}_{AC} \times m\,\mathbf{a}_A + \frac{\mathrm{d}}{\mathrm{d}t}\mathbf{H}_A. \tag{6.57}$$

The point A will be selected such that $\mathbf{r}_{AC} \times m\,\mathbf{a}_A = \mathbf{0}$, and this condition requires that point A is the mass center C ($\mathbf{r}_{AC} = \mathbf{0}$), point A has zero acceleration ($\mathbf{a}_A = \mathbf{0}$), and $\mathbf{a}_A$ is parallel to $\mathbf{r}_{AC}$ ($\mathbf{a}_A \| \mathbf{r}_{AC}$).

The angular momentum, $\mathbf{H}_A$, is a function of the angular velocity and the inertia properties. The components of $\mathbf{H}_A$ are relative to the rotating reference frame (body fixed) xyz. The inertia properties have constant values relative to the body-fixed reference frame. The total derivative of $\mathbf{H}_A$ is

$$\dot{\mathbf{H}}_A = \left(\dot{\mathbf{H}}_A \right)_{x_0 y_0 z_0} = \left(\frac{\mathrm{d}\mathbf{H}_A}{\mathrm{d}t} \right)_{xyz} + \boldsymbol{\omega} \times \mathbf{H}_A, \tag{6.58}$$

where

$$\left(\frac{\mathrm{d}\mathbf{H}_A}{\mathrm{d}t} \right)_{xyz} = (I_{xx}\dot{\omega}_x - I_{xy}\dot{\omega}_y - I_{xz}\dot{\omega}_z)\mathbf{\imath} + (I_{yy}\dot{\omega}_y - I_{xy}\dot{\omega}_x - I_{yz}\dot{\omega}_z)\mathbf{\jmath}$$

$$+ (I_{zz}\dot{\omega}_z - I_{xz}\dot{\omega}_x - I_{yz}\dot{\omega}_y)\mathbf{k}. \tag{6.59}$$

The absolute derivative of $\boldsymbol{\omega} = \omega_x\mathbf{\imath} + \omega_y\mathbf{\jmath} + \omega_z\mathbf{k}$ relative to the fixed reference frame is identical to the derivative of $\boldsymbol{\omega}$ relative to the rotating reference frame:

$$\left(\frac{\mathrm{d}\boldsymbol{\omega}}{\mathrm{d}t} \right)_{x_0 y_0 z_0} = \left(\frac{\mathrm{d}\boldsymbol{\omega}}{\mathrm{d}t} \right)_{xyz} + \boldsymbol{\omega} \times \boldsymbol{\omega} = \left(\frac{\mathrm{d}\boldsymbol{\omega}}{\mathrm{d}t} \right)_{xyz}.$$

The angular acceleration $\boldsymbol{\alpha}$ is

$$\boldsymbol{\alpha} = \alpha_x\mathbf{\imath} + \alpha_y\mathbf{\jmath} + \alpha_z\mathbf{k},$$

where

$$\alpha_x = \dot{\omega}_x, \qquad \alpha_y = \dot{\omega}_y, \qquad \alpha_z = \dot{\omega}_z.$$

The fundamental equations of motion for a rigid body can be written as

$$\sum \mathbf{F} = m\,\mathbf{a}_G, \tag{6.60}$$

$$\sum \mathbf{M}_A = \left(\frac{\mathrm{d}\mathbf{H}_A}{\mathrm{d}t} \right)_{xyz} + \boldsymbol{\omega} \times \mathbf{H}_A, \tag{6.61}$$

where

$$\mathbf{H}_A = (I_{xx}\omega_x - I_{xy}\omega_y - I_{xz}\omega_z)\mathbf{i} + (I_{yy}\omega_y - I_{xy}\omega_x - I_{yz}\omega_z)\mathbf{J}$$
$$+ (I_{zz}\omega_z - I_{xz}\omega_x - I_{yz}\omega_y)\mathbf{k}, \tag{6.62}$$

$$\left(\frac{d\mathbf{H}_A}{dt}\right)_{xyz} = (I_{xx}\alpha_x - I_{xy}\alpha_y - I_{xz}\alpha_z)\mathbf{i} + (I_{yy}\alpha_y - I_{xy}\alpha_x - I_{yz}\alpha_z)\mathbf{J}$$
$$+ (I_{zz}\alpha_z - I_{xz}\alpha_x - I_{yz}\alpha_y)\mathbf{k}. \tag{6.63}$$

The point A is either the center of mass, C, of the body or a fixed point in a body that is in pure rotation. The matrix forms of the equations of motion are

$$\begin{bmatrix} \sum F_x \\ \sum F_y \\ \sum F_z \end{bmatrix} = m \begin{bmatrix} a_{Gx} \\ a_{Gy} \\ a_{Gz} \end{bmatrix}, \tag{6.64}$$

$$\begin{bmatrix} \sum M_{Ax} \\ \sum M_{Ay} \\ \sum M_{Az} \end{bmatrix} = [I]\begin{bmatrix} \alpha_x \\ \alpha_y \\ \alpha_z \end{bmatrix} + \begin{bmatrix} 0 & -\omega_z & \omega_y \\ \omega_z & 0 & -\omega_x \\ -\omega_y & \omega_x & 0 \end{bmatrix}[I]\begin{bmatrix} \omega_x \\ \omega_y \\ \omega_z \end{bmatrix}, \tag{6.65}$$

or

$$\sum \mathbf{F} = m\mathbf{a}_C,$$
$$\sum \mathbf{M}_A = [I]\,\mathbf{ff} + \boldsymbol{\omega} \times ([I]\,\boldsymbol{\omega}). \tag{6.66}$$

6.6   Euler's Equations of Motion

If the x, y, and z axes are the principal axes of inertia of the body, the components of the angular momentum are

$$\mathbf{H}_C = I_1\omega_x\mathbf{i} + I_2\omega_y\mathbf{J} + I_3\omega_z\mathbf{k}, \tag{6.67}$$

where I_1, I_2, and I_3 denote the principal centroidal moments of inertia of the body. The three scalar equations for the rotation of the body are

$$\sum M_x = I_1\alpha_x - (I_2 - I_3)\,\omega_y\omega_z,$$
$$\sum M_y = I_2\alpha_y - (I_3 - I_1)\,\omega_z\omega_x,$$
$$\sum M_z = I_3\alpha_z - (I_1 - I_2)\,\omega_x\omega_y. \tag{6.68}$$

The equations are called Euler's equations of motion after the Swiss mathematician Leonard Euler (1707–1783).

6.7 Motion of a Rigid Body About a Fixed Point

When a rigid body is rotating about a fixed point O, Fig. 6.5, the equation of motion for the body in pure rotation is

$$\sum \mathbf{M}_O = \dot{\mathbf{H}}_O = \left(\frac{\mathrm{d}\mathbf{H}_O}{\mathrm{d}t}\right)_{xyz} + \boldsymbol{\omega} \times \mathbf{H}_O, \tag{6.69}$$

where $\sum \mathbf{M}_O$ is the sum of the moments about O of the forces applied to the rigid body, $\dot{\mathbf{H}}_O$ is the rate of change of the vector $\mathbf{H}_O$ with respect to the fixed frame $x_0 y_0 z_0$, $\left(\frac{\mathrm{d}\mathbf{H}_O}{\mathrm{d}t}\right)_{xyz}$ is the rate of change of $\mathbf{H}_O$ with respect to the rotating frame xyz, and $\boldsymbol{\omega}$ is the angular velocity of the rotating frame xyz.

6.8 Rotation of a Rigid Body About a Fixed Axis

The rigid body is constrained to rotate about a fixed axis AB, Fig. 6.6.

The angular velocity of the body with respect to the fixed frame $x_0 y_0 z_0$ is a vector along the axis of rotation AB. The fixed reference frame $x_0 y_0 z_0$ and the rotating reference frame xyz are selected in such a way that the z_0 and z axes are along the axis of rotation AB: $\boldsymbol{\omega} = \omega \mathbf{k}$. With $\omega_x = 0$, $\omega_y = 0$, $\omega_z = \omega$, the components along the rotating axes of the angular momentum $\mathbf{H}_O$ of the body about O are

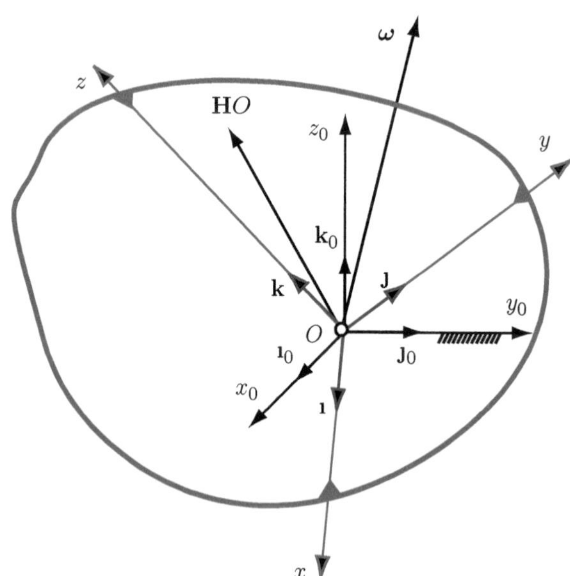

Fig. 6.5 Rotation about a fixed point

Fig. 6.6 Rotation about a fixed axis

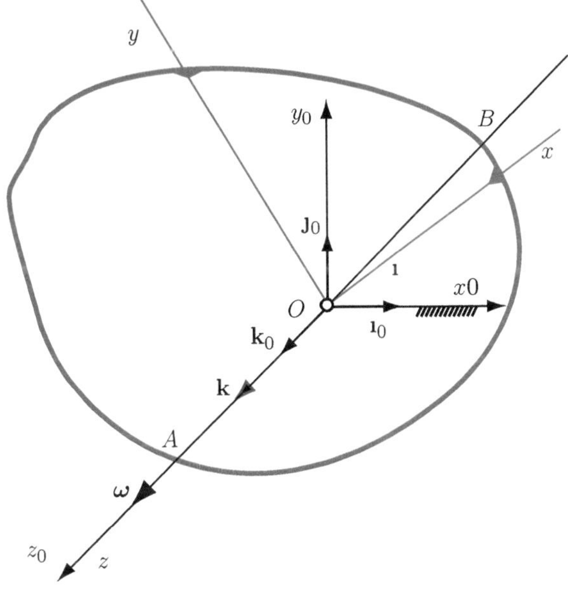

$$H_x = -I_{xz}\omega,$$

$$H_y = -I_{yz}\omega,$$

$$H_z = -I_{zz}\omega.$$

The equations of motion for the rigid body are

$$\sum \mathbf{M}_O = \left(\frac{\mathrm{d}\mathbf{H}_O}{\mathrm{d}t}\right)_{xyz} + \boldsymbol{\omega} \times \mathbf{H}_O,$$

or the three scalar equations are

$$\sum M_x = -I_{xz}\alpha + I_{yz}\omega^2,$$

$$\sum M_y = -I_{yz}\alpha - I_{xz}\omega^2,$$

$$\sum M_z = I_{zz}\alpha. \tag{6.70}$$

6.9 Plane Motion of Rigid Body

Figure 6.7 represents the rigid body moving with general planar motion in the (x,y) plane. The origin of the Cartesian reference frame is O. The mass center C of the rigid body is located in the plane of the motion. Let Oz be the axis through the fixed

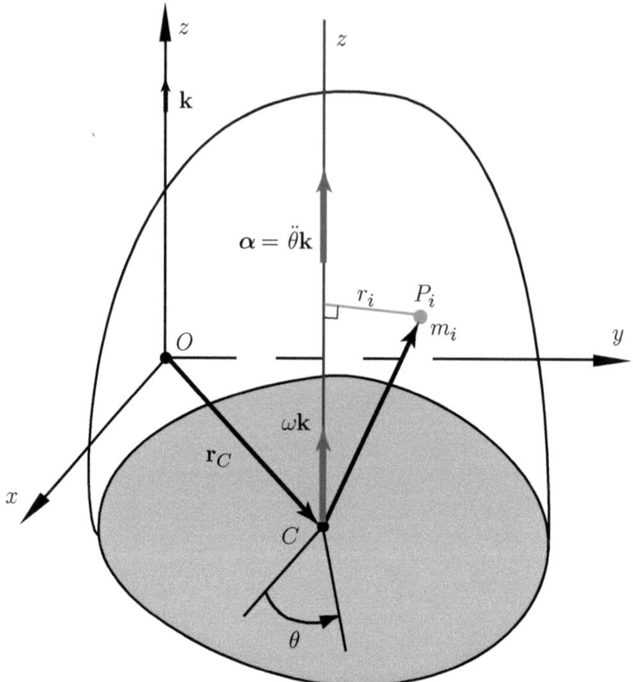

Fig. 6.7 Rigid body moving with general planar motion in the (x, y) plane

origin point O that is perpendicular to the plane of motion of the rigid body. Let Cz be the parallel axis through the mass center C. The rigid body has a general planar motion, and the angular velocity vector is $\boldsymbol{\omega} = \omega\mathbf{k}$. The sum of the moments about O due to external forces and couples is

$$\sum \mathbf{M}_O = \frac{d\mathbf{H}_O}{dt} = \frac{d}{dt}[(\mathbf{r}_C \times m\mathbf{v}_C) + \mathbf{H}_C]. \tag{6.71}$$

The total angular momentum of the system about O is $\mathbf{H}_O$, the total angular momentum of the system about C is $\mathbf{H}_C$, and $\mathbf{v}_C = \dot{\mathbf{r}}_C$ is the velocity of C. The magnitude of the angular momentum about Cz is $H_C = \sum_i m_i r_i^2\, \omega$. The summation $\sum_i m_i r_i^2$ or the integration over the body $\int r^2\, dm$ is defined as the mass moment of inertia I_{Cz} of the body about the z-axis through C:

$$I_{Cz} = \sum_i m_i r_i^2.$$

The term r_i is the perpendicular distance from d_C to the P_i particle. The mass moment of inertia I_{Cz} is a constant property of the body and is a measure of the rotational inertia or resistance to change in angular velocity due to the radial distribution of the rigid body mass around the z-axis through C. The angular

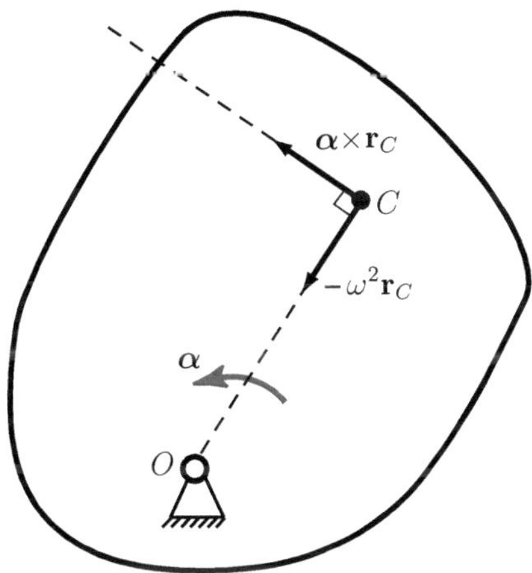

Fig. 6.8 Rotation about a fixed point O in plane

momentum of the rigid body about Cz (z-axis through C) is

$$H_C = I_{Cz}\,\omega \quad \text{or} \quad \mathbf{H}_C = I_{Cz}\,\omega\,\mathbf{k} = I_{Cz}\,\boldsymbol{\omega}.$$

Substituting this expression into (6.71) gives

$$\sum \mathbf{M}_O = \frac{\mathrm{d}}{\mathrm{d}t}\left[(\mathbf{r}_C \times m\mathbf{v}_C) + I_{Cz}\,\boldsymbol{\omega}\right] = (\mathbf{r}_C \times m\,\mathbf{a}_C) + I_{Cz}\,\boldsymbol{\alpha}. \tag{6.72}$$

The rotational equation of motion for the rigid body is

$$I_{Cz}\,\boldsymbol{\alpha} = \sum \mathbf{M}_C \quad \text{or} \quad I_{Cz}\,\alpha\,\mathbf{k} = \sum M_C\,\mathbf{k}. \tag{6.73}$$

For general planar motion, the angular acceleration is $\boldsymbol{\alpha} = \dot{\boldsymbol{\omega}} = \ddot{\theta}\,\mathbf{k}$, where the angle θ describes the position, or orientation, of the rigid body about a fixed axis. If the rigid body is a plate moving in the plane of motion, the mass moment of inertia of the rigid body about the z-axis through C becomes the polar mass moment of inertia of the rigid body about C, $I_{Cz} = I_C$. For this case, (6.73) gives

$$I_C\,\boldsymbol{\alpha} = \sum \mathbf{M}_C. \tag{6.74}$$

Consider the special case when the rigid body rotates about a fixed point O as shown in Fig. 6.8. The acceleration of the mass center is

$$\mathbf{a}_C = \mathbf{a}_O + \boldsymbol{\alpha} \times \mathbf{r}_C - \omega^2\,\mathbf{r}_C = \boldsymbol{\alpha} \times \mathbf{r}_C - \omega^2\,\mathbf{r}_C.$$

The relation between the sum of the moments of the external forces about the fixed point O and the product $I_{Cz}\,\boldsymbol{\alpha}$ is given by (6.72):

$$\sum \mathbf{M}_O = \mathbf{r}_C \times m\,\mathbf{a}_C + I_{Cz}\,\boldsymbol{\alpha},$$

or

$$\sum \mathbf{M}_O = \mathbf{r}_C \times m\left(\boldsymbol{\alpha} \times \mathbf{r}_C - \omega^2 \mathbf{r}_C\right) + I_{Cz}\boldsymbol{\alpha}$$
$$= m\mathbf{r}_C \times (\boldsymbol{\alpha} \times \mathbf{r}_C) + I_{Cz}\boldsymbol{\alpha}$$
$$= m\left[(\mathbf{r}_C \cdot \mathbf{r}_C)\boldsymbol{\alpha} - (\mathbf{r}_C \cdot \boldsymbol{\alpha})\mathbf{r}_C\right] + I_{Cz}\boldsymbol{\alpha}$$
$$= m r_C^2 \boldsymbol{\alpha} + I_{Cz}\boldsymbol{\alpha} = (m r_C^2 + I_{Cz})\boldsymbol{\alpha}.$$

According to parallel-axis theorem,

$$I_{Ozz} = I_{Cz} + m r_C^2,$$

where I_{Oz} denotes the mass moment of inertia of the rigid body about the z-axis through O. For the special case of rotation about a fixed point O, one can use the formula:

$$I_{Oz}\boldsymbol{\alpha} = \sum \mathbf{M}_O. \tag{6.75}$$

The general equations of motion for a rigid body in plane motion are (Fig. 6.7)

$$\mathbf{F} = m\mathbf{a}_C \quad \text{or} \quad \mathbf{F} = m\ddot{\mathbf{r}}_C,$$
$$\sum \mathbf{M}_C = I_{Cz}\boldsymbol{\alpha}, \tag{6.76}$$

or using the Cartesian components

$$m\ddot{x}_C = \sum F_x,$$
$$m\ddot{y}_C = \sum F_y,$$
$$I_{Cz}\ddot{\theta} = \sum M_C. \tag{6.77}$$

Equations (6.76) and (6.77), also known as the Newton–Euler equations of motion, are for plane motion and are interpreted in two ways:

1. The forces and moments are known, and the equations are solved for the motion of the rigid body (direct dynamics).
2. The motion of the rigid body is known, and the equations are solved for the forces and moments (inverse dynamics).

Rolling Motion

An important case of plane motion is the motion of a disk rolling on a surface. If the disk is constrained to roll without sliding, the acceleration $\mathbf{a}$ of its mass center C and its angular acceleration $\boldsymbol{\alpha}$ are not independent. The distance x traveled by C during a rotation θ of the disk is $x = r\theta$, where r is the radius of the disk. Differentiating this relation twice, it results $a = r\alpha$. In the case of the rolling motion of a disk, the effective forces reduce to a vector of magnitude $m r \alpha$ at C and to a moment of magnitude $I\alpha$.

When a disk rolls without sliding, there is no relative motion between the point of the disk which is in contact with the surface and the surface itself. The magnitude of the friction force may have any value, as long as it does not exceed the maximum value $F_s = \mu_s N$, where μ_s is the coefficient of static friction and N the magnitude of the normal force.

When the disk rotates and slides at the same time, a relative motion exists between the point of the disk which is in contact with the ground and the ground itself, and the force of friction has the magnitude $F_k = \mu_k N$, where μ_k is the coefficient of kinetic friction. In this case, the motion of the mass center C of the disk and the rotation of the disk about C are independent, and $a \neq r\alpha$.

6.9.1 D'Alembert's Principle

Newton's second law can be written as

$$\mathbf{F} + (-m\mathbf{a}_C) = \mathbf{0}, \text{ or } \mathbf{F} + \mathbf{F}_{in} = \mathbf{0},$$

where the term $\mathbf{F}_{in} = -m\mathbf{a}_C$ is the *inertia force*. Newton's second law can be regarded as an "equilibrium" equation.

The total moment about a fixed point O is

$$\sum \mathbf{M}_O = (\mathbf{r}_C \times m\mathbf{a}_C) + I_{Cz}\alpha,$$

or

$$\sum \mathbf{M}_O + [\mathbf{r}_C \times (-m\mathbf{a}_C)] + (-I_{Cz}\alpha) = \mathbf{0}. \tag{6.78}$$

The term $\mathbf{M}_{in} = -I_{Cz}\alpha$ is the *inertia moment*. The sum of the moments about any point, including the moment due to the inertial force $-m\mathbf{a}$ acting at the mass center and the inertial moment, equals zero.

The equations of motion for a rigid body are analogous to the equations for static equilibrium: The sum of the forces equals zero, and the sum of the moments about any point equals zero when the inertial forces and moments are taken into account.

This is called *D'Alembert's principle*. The dynamic force analysis is expressed in a form similar to static force analysis:

$$\sum \mathbf{F} + \mathbf{F}_{in} = \mathbf{0}, \tag{6.79}$$

$$\sum \mathbf{M}_C + \mathbf{M}_{in} = \mathbf{0}, \tag{6.80}$$

where $\sum \mathbf{F}$ is the vector sum of all external forces (resultant of external force) and $\sum \mathbf{M}_C$ is the sum of all external moments about the center of mass C (resultant external moment).

Fig. 6.9 Joint forces for one degree of freedom joints

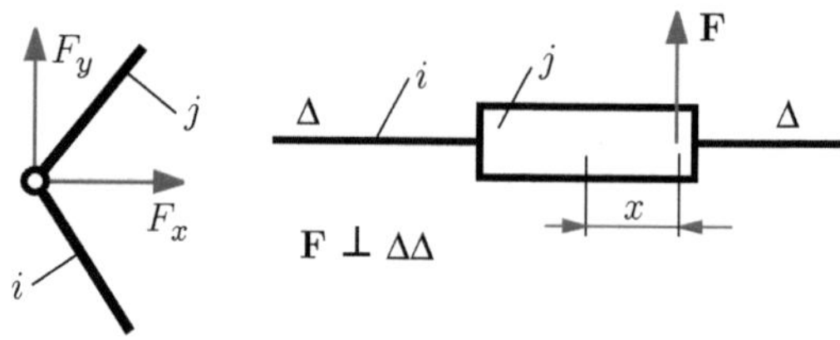

For a rigid body in plane motion in the (x, y) plane,

$$\mathbf{a}_C = \ddot{x}_C \mathbf{1} + \ddot{y}_C \mathbf{J}, \quad \boldsymbol{\alpha} = \alpha \mathbf{k},$$

with all external forces in that plane, (6.79) and (6.80) become

$$\sum F_x + F_{\mathrm{in}\,x} = \sum F_x + (-m\ddot{x}_C) = 0,$$

$$\sum F_y + F_{\mathrm{in}\,y} = \sum F_y + (-m\ddot{y}_C) = 0,$$

$$\sum M_C + M_{\mathrm{in}} = \sum M_C + (-I_C \alpha) = 0.$$

With D'Alembert's principle, the moment summation can be about any arbitrary point P:

$$\sum \mathbf{M}_P + \mathbf{M}_{\mathrm{in}} + \mathbf{r}_{PC} \times \mathbf{F}_{\mathrm{in}} = \mathbf{0},$$

where $\sum \mathbf{M}_P$ is the sum of all external moments about P, $\mathbf{M}_{\mathrm{in}}$ is the inertia moment, $\mathbf{F}_{\mathrm{in}}$ is the inertia force, and $\mathbf{r}_{PC}$ is a vector from P to C.

The dynamic analysis problem is reduced to a static force and moment balance problem where the inertia forces and moments are treated in the same way as external forces and moments.

6.9.2 Free-Body Diagrams

A free-body diagram is a drawing of a part of a complete system, isolated in order to determine the forces acting on that rigid body.

The following force convention is defined: $\mathbf{F}_{ij}$ represents the force exerted by link i on link j. Figure 6.9 shows the joint forces for one degree of freedom joints. The force analysis can be accomplished by examining individual links or a subsystem of links. In this way, the joint forces between links as well as the required input force or moment for a given output load are computed.

RRR Dyad

Figure 6.10 shows an RRR dyad with two links, 2 and 3, and three pin joints, B, C, and D. First, the exterior unknown joint reaction forces are considered:

$$\mathbf{F}_{12} = F_{12x}\mathbf{1} + F_{12y}\mathbf{J} \quad \text{and} \quad \mathbf{F}_{43} = F_{43x}\mathbf{1} + F_{43y}\mathbf{J}.$$

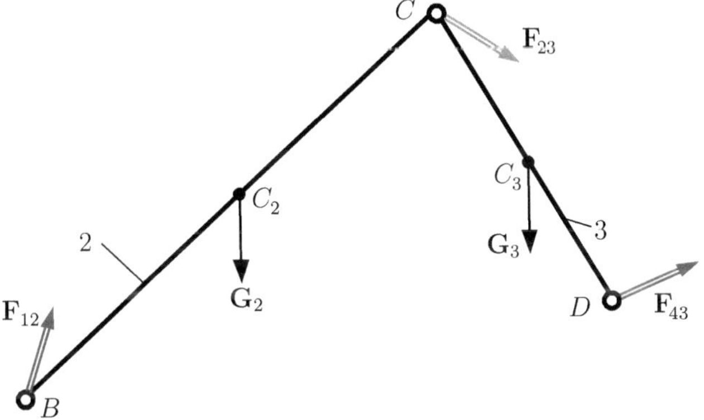

Fig. 6.10 RRR dyad

To determine $\mathbf{F}_{12}$ and $\mathbf{F}_{43}$, the following equations are written:

- Sum of all forces on links 2 and 3 is zero:

$$\sum \mathbf{F}^{(2\&3)} \Longrightarrow$$

$$m_2\,\mathbf{a}_{C_2} + m_3\,\mathbf{a}_{C_3} = \mathbf{F}_{12} + \mathbf{G}_2 + \mathbf{G}_2 + \mathbf{F}_{43},$$

or

$$\sum \mathbf{F}^{(2\&3)} \cdot \mathbf{1} \Longrightarrow$$

$$m_2\,a_{C_{2x}} + m_3\,a_{C_{3x}} = F_{12x} + F_{43x}, \tag{6.81}$$

$$\sum \mathbf{F}^{(2\&3)} \cdot \mathbf{J} \Longrightarrow$$

$$m_2\,a_{C_{2y}} + m_3\,a_{C_{3y}} = F_{12y} - m_2\,g - m_3\,g + F_{43y}. \tag{6.82}$$

- Sum of moments of all forces and moments on link 2 about C is zero:

$$\sum \mathbf{M}_C^{(2)} \Longrightarrow$$

$$I_{C_2}\,\alpha_2 + \mathbf{r}_{CC_2} \times m_2\,\mathbf{a}_{C_2} = \mathbf{r}_{CB} \times \mathbf{F}_{12} + \mathbf{r}_{CC_2} \times \mathbf{G}_2. \tag{6.83}$$

- Sum of moments of all forces and moments on link 3 about C is zero:

$$\sum \mathbf{M}_C^{(3)} \Longrightarrow$$

$$I_{C_3}\,\alpha_3 + \mathbf{r}_{CC_3} \times m_3\,\mathbf{a}_{C_3} = \mathbf{r}_{CD} \times \mathbf{F}_{43} + \mathbf{r}_{CC_3} \times \mathbf{G}_3. \tag{6.84}$$

The components F_{12x}, F_{12y}, F_{43x}, and F_{43y} are calculated from (6.81) to (6.96).

The reaction force $\mathbf{F}_{32} = -\mathbf{F}_{23}$ is computed from the sum of all forces on link 2:

$$\sum \mathbf{F}^{(2)} \Longrightarrow m_2\,\mathbf{a}_{C_2} = \mathbf{F}_{12} + \mathbf{G}_2 + \mathbf{F}_{32} \quad \text{or} \quad \mathbf{F}_{32} = m_2\,\mathbf{a}_{C_2} - \mathbf{F}_{12} - \mathbf{G}_2.$$

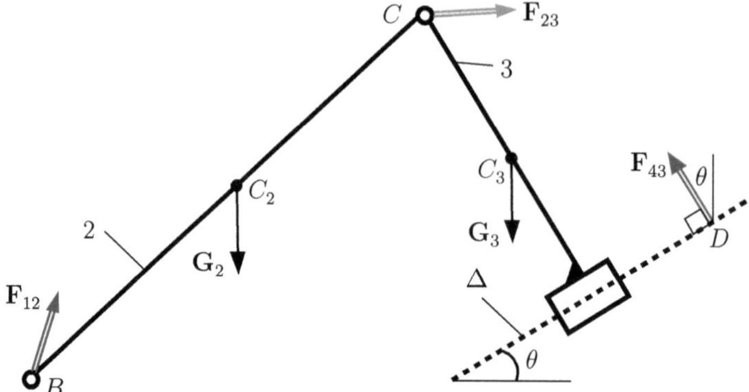

Fig. 6.11 RRT dyad

RRT Dyad

Figure 6.11 shows an RRT dyad with the unknown joint reaction forces $\mathbf{F}_{12}$, $\mathbf{F}_{43}$, and $\mathbf{F}_{23} = -\mathbf{F}_{32}$. The joint reaction force $\mathbf{F}_{43}$ is perpendicular to the sliding direction $\mathbf{F}_{43} \perp \Delta$ or

$$\mathbf{F}_{43} \cdot \Delta = (F_{43x}\mathbf{I} + F_{43y}\mathbf{J}) \cdot (\cos\theta\,\mathbf{I} + \sin\theta\,\mathbf{J}) = 0. \tag{6.85}$$

In order to determine $\mathbf{F}_{12}$ and $\mathbf{F}_{43}$, the following equations are written:

- Sum of all the forces on links 2 and 3 is zero:

$$\sum \mathbf{F}^{(2\&3)} \Longrightarrow m_2\,\mathbf{a}_{C_2} + m_3\,\mathbf{a}_{C_3} = \mathbf{F}_{12} + \mathbf{G}_2 + \mathbf{G}_3 + \mathbf{F}_{43},$$

or

$$\sum \mathbf{F}^{(2\&3)} \cdot \mathbf{I} \Longrightarrow m_2\,a_{C_{2x}} + m_3\,a_{C_{3x}} = F_{12x} + F_{43x}, \tag{6.86}$$

$$\sum \mathbf{F}^{(2\&3)} \cdot \mathbf{J} \Longrightarrow m_2\,a_{C_{2y}} + m_3\,a_{C_{3y}} = F_{12y} - m_2\,g - m_3\,g + F_{43y}. \tag{6.87}$$

- Sum of moments of all the forces and the moments on link 2 about C is zero:

$$\sum \mathbf{M}_C^{(2)} \Longrightarrow I_{C_2}\,\alpha_2 + \mathbf{r}_{CC_2} \times m_2\,\mathbf{a}_{C_2} = \mathbf{r}_{CB} \times \mathbf{F}_{12} + \mathbf{r}_{CC_2} \times \mathbf{G}_2. \tag{6.88}$$

The components F_{12x}, F_{12y}, F_{43x}, and F_{43y} are calculated from (6.98) to (6.100).

The reaction force components F_{32x} and F_{32y} are computed from the sum of all the forces on link 2:

$$\sum \mathbf{F}^{(2)} \Longrightarrow m_2\,\mathbf{a}_{C_2} = \mathbf{F}_{12} + \mathbf{G}_2 + \mathbf{F}_{32} \quad \text{or} \quad \mathbf{F}_{32} = m_2\,\mathbf{a}_{C_2} - \mathbf{F}_{12} - \mathbf{G}_2.$$

Fig. 6.12 RTR dyad

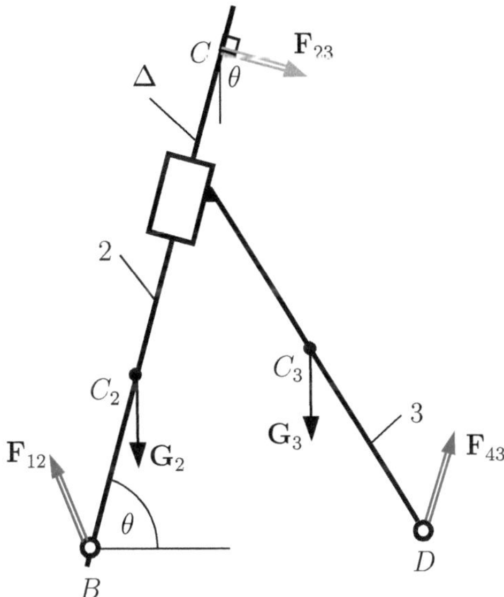

RTR Dyad

The unknown joint reaction forces $\mathbf{F}_{12}$ and $\mathbf{F}_{43}$, as shown in Fig. 6.12, are calculated from the relations:

- Sum of all the forces on links 2 and 3 is zero:

$$\sum \mathbf{F}^{(2\&3)} \implies m_2\,\mathbf{a}_{C_2} + m_3\,\mathbf{a}_{C_3} = \mathbf{F}_{12} + \mathbf{G}_2 + \mathbf{G}_3 + \mathbf{F}_{43},$$

  or

$$\sum \mathbf{F}^{(2\&3)} \cdot \mathbf{i} \implies m_2\,a_{C_{2x}} + m_3\,a_{C_{3x}} = F_{12x} + F_{43x}, \tag{6.89}$$

$$\sum \mathbf{F}^{(2\&3)} \cdot \mathbf{j} \implies m_2\,a_{C_{2y}} + m_3\,a_{C_{3y}} = F_{12y} - m_2\,g - m_3\,g + F_{43y}. \tag{6.90}$$

- Sum of the moments of all the forces and moments on links 2 and 3 about B is zero:

$$\sum \mathbf{M}_B^{(2\&3)} \implies I_{C_2}\,\alpha_2 + I_{C_3}\,\alpha_3 + \mathbf{r}_{BC_2} \times m_2\,\mathbf{a}_{C_2} + \mathbf{r}_{BC_3} \times m_3\,\mathbf{a}_{C_3}$$

$$= \mathbf{r}_{BD} \times \mathbf{F}_{43} + \mathbf{r}_{BC_3} \times \mathbf{G}_3 + \mathbf{r}_{BC_2} \times \mathbf{G}_2. \tag{6.91}$$

- Sum of all the forces on link 2 projected onto the sliding direction $\Delta = \cos\theta\,\mathbf{i} + \sin\theta\,\mathbf{j}$ is zero:

$$\sum \mathbf{F}^{(2)} \cdot \Delta = (\mathbf{F}_{12} + \mathbf{F}_2) \cdot (\cos\theta\,\mathbf{i} + \sin\theta\,\mathbf{j}) = 0. \tag{6.92}$$

The components F_{12x}, F_{12y}, F_{43x}, and F_{43y} are calculated from (6.89) to (6.92).

Fig. 6.13 Monocontour
closed kinematic chain

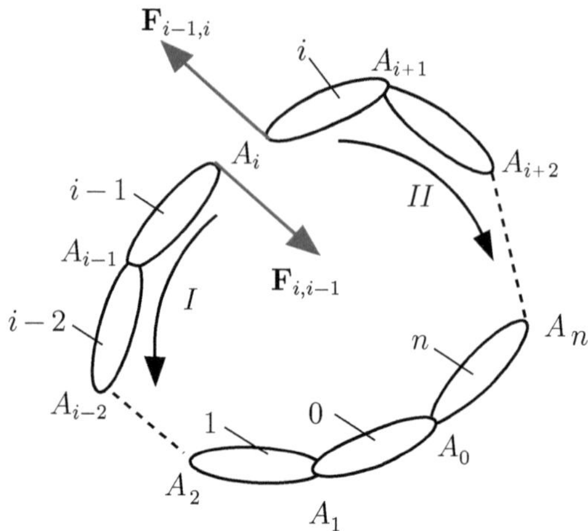

The force components F_{32x} and F_{32y} are computed from the sum of all the forces
on link 2:

$$\sum \mathbf{F}^{(2)} \implies m_2\,\mathbf{a}_{C_2} = \mathbf{F}_{12} + \mathbf{G}_2 + \mathbf{F}_{32} \quad \text{or} \quad \mathbf{F}_{32} = m_2\,\mathbf{a}_{C_2} - (\mathbf{F}_{12} + \mathbf{G}_2).$$

Force Analysis Using Contour Method

An analytical method to compute joint forces that can be applied for both planar
and spatial mechanisms will be presented. The method is based on the decoupling
of a closed kinematic chain and writing the dynamic equilibrium equations. The
kinematic links are loaded with external forces and inertia forces and moments.

A general monocontour closed kinematic chain is considered in Fig. 6.13. The
joint force between the links $i-1$ and i (joint A_i) will be determined. When these
two links $i-1$ and i are separated, the joint forces $\mathbf{F}_{i-1,i}$ and $\mathbf{F}_{i,i-1}$ are introduced
and $\mathbf{F}_{i-1,i} + \mathbf{F}_{i,i-1} = \mathbf{0}$.

It is helpful to "mentally disconnect" the two links $(i-1)$ and i, which create
joint A_i, from the rest of the mechanism. The joint at A_i will be replaced by the joint
forces $\mathbf{F}_{i-1,i}$ and $\mathbf{F}_{i,i-1}$. The closed kinematic chain has been transformed into two
open kinematic chains, and two paths I and II are associated. The two paths start
from A_i.

For the path I (counterclockwise), starting at A_i and following I, the first joint
encountered is A_{i-1}. For the link $i-1$ left behind, dynamic equilibrium equations
are written according to the type of joint at A_{i-1}. Following the same path I, the
next joint encountered is A_{i-2}. For the subsystem $(i-1$ and $i-2)$, equilibrium
conditions corresponding to the type of joint at A_{i-2} can be specified, and so
on. A similar analysis is performed for the path II of the open kinematic chain.
The number of equilibrium equations written is equal to the number of unknown
scalars introduced by joint A_i (joint forces at this joint). For a joint, the number of
equilibrium conditions is equal to the number of relative mobilities of the joint.

Fig. 6.14 Rigid link
in motion

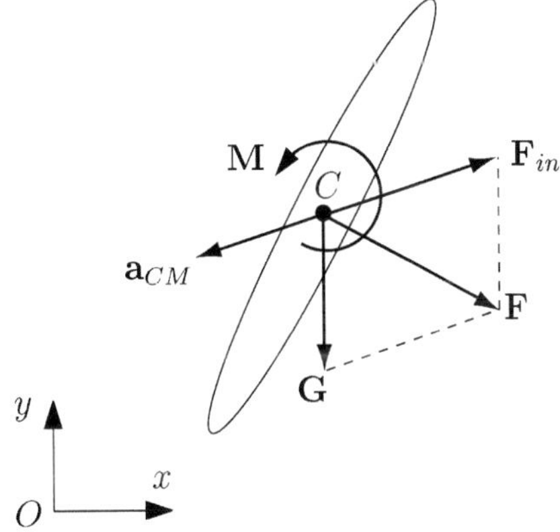

6.9.3   Force Analysis for Closed Kinematic Chains Using MATLAB Functions

A rigid link is shown in Fig. 6.14.

The total force on the link of mass, m, at the center of mass, CM, is

$$\mathbf{F} = \mathbf{F}_{in} + \mathbf{G},$$

where $\mathbf{F}_{in} = -m\mathbf{a}_{CM}$ is the inertia force, $\mathbf{G} = -m\mathbf{g}$ is the gravitational force, and $\mathbf{g} = -9.807\,\text{km/s}^2$ is the gravitational acceleration. The moment of inertia M of the link is

$$\mathbf{M} = -I_{CM}\boldsymbol{\alpha},$$

where $\boldsymbol{\alpha} = \alpha\mathbf{k}$ is the angular acceleration of the link.

MATLAB Function for RRR Dyad

Figure 6.15 shows an RRR dyad with two links, 2 and 3, and three pin joints at M, N, and P. The input data are the total forces $\mathbf{F}_2$, $\mathbf{F}_3$ and the moments $\mathbf{M}_2$, $\mathbf{M}_3$ on the links 2 and 3; the position vectors $\mathbf{r}_M$, $\mathbf{r}_N$, $\mathbf{r}_P$ of the joints M, N, P; and position vectors $\mathbf{r}_{C2}, \mathbf{r}_{C3}$ of the centers of mass of the links 2 and 3. The output data are the joint reaction forces $\mathbf{F}_{12}$, $\mathbf{F}_{43}$, and $\mathbf{F}_{32}$. The unknown joint reaction forces are

$$\mathbf{F}_{12} = F_{12x}\mathbf{1} + F_{12y}\mathbf{J},$$

$$\mathbf{F}_{43} = F_{43x}\mathbf{1} + F_{43y}\mathbf{J},$$

$$\mathbf{F}_{23} = -\mathbf{F}_{32} = F_{23x}\mathbf{1} + F_{23y}\mathbf{J}. \tag{6.93}$$

To determine $\mathbf{F}_{12}$ and $\mathbf{F}_{43}$, the following equations are written. Sum of all forces on links 2 and 3 is zero:

$$\sum \mathbf{F}^{(2\&3)} = \mathbf{F}_{12} + \mathbf{F}_2 + \mathbf{F}_3 + \mathbf{F}_{43} = \mathbf{0},$$

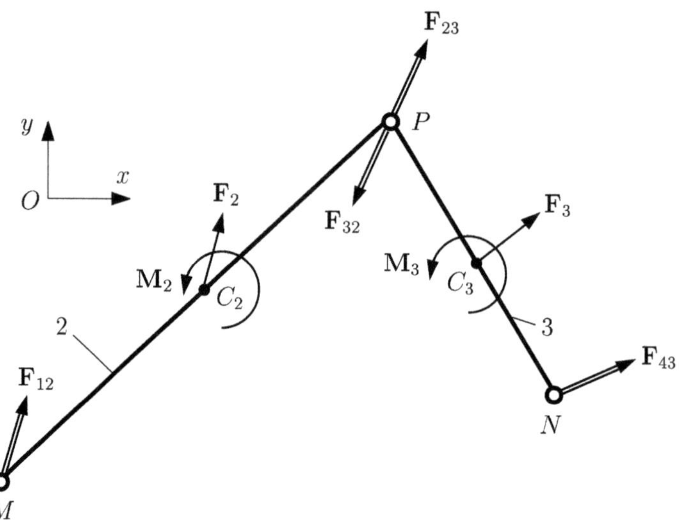

Fig. 6.15 RRR dyad

or

$$\sum \mathbf{F}^{(2\&3)} \cdot \mathbf{1} = F_{12x} + F_{2x} + F_{3x} + F_{43x} = 0,$$

$$\sum \mathbf{F}^{(2\&3)} \cdot \mathbf{J} = F_{12y} + F_{2y} + F_{3y} + F_{43y} = 0. \tag{6.94}$$

Sum of moments of all forces and moments on link 2 about P is zero:

$$\sum \mathbf{M}_P^{(2)} = (\mathbf{r}_M - \mathbf{r}_P) \times \mathbf{F}_{12} + (\mathbf{r}_{C2} - \mathbf{r}_P) \times \mathbf{F}_2 + \mathbf{M}_2 = \mathbf{0}. \tag{6.95}$$

Sum of moments of all forces and moments on link 3 about P is zero:

$$\sum \mathbf{M}_P^{(3)} = (\mathbf{r}_N - \mathbf{r}_P) \times \mathbf{F}_{43} + (\mathbf{r}_{C3} - \mathbf{r}_P) \times \mathbf{F}_3 + \mathbf{M}_3 = \mathbf{0}. \tag{6.96}$$

The components F_{12x}, F_{12y}, F_{43x}, and F_{43y} are calculated from (6.94) to (6.96). The reaction force $\mathbf{F}_{32} = -\mathbf{F}_{23}$ is computed from the sum of all forces on the link 2:

$$\sum \mathbf{F}^{(2)} = \mathbf{F}_{12} + \mathbf{F}_2 + \mathbf{F}_{32} = \mathbf{0},$$

or

$$\sum \mathbf{F}^{(2)} \cdot \mathbf{1} = F_{12x} + F_{2x} + F_{32x} = 0,$$

$$\sum \mathbf{F}^{(2)} \cdot \mathbf{J} = F_{12y} + F_{2y} + F_{32y} = 0. \tag{6.97}$$

The MATLAB function `ForceRRR.m` for the RRR dyad joint force analysis is

```
% forceRRR.m
% forces RRR dyad
```

```
function out=forceRRR...
(rM,rN,rP,rC2,rC3,F2,M2,F3,M3);

F12x=sym('F12x','real');
F12y=sym('F12y','real');
F43x=sym('F43x','real');
F43y=sym('F43y','real');
F12=[F12x, F12y, 0]; % unknown joint force
F43=[F43x, F43y, 0]; % unknown joint force

eqF = F12+F2+F43+F3;
eqFx = eqF(1);
eqFy = eqF(2);

rPC2 = rC2 - rP;
rPC3 = rC3 - rP;
rPM = rM - rP;
rPN = rN - rP;

eqM2 = cross(rPC2,F2)+cross(rPM,F12)+M2;
eqM2z = eqM2(3);
eqM3 = cross(rPC3,F3)+cross(rPN,F43)+M3;
eqM3z = eqM3(3);

sol=solve(eqFx, eqFy, eqM2z, eqM3z);
F12xs=eval(sol.F12x);
F12ys=eval(sol.F12y);
F43xs=eval(sol.F43x);
F43ys=eval(sol.F43y);

F12s=[F12xs, F12ys, 0];
F43s=[F43xs, F43ys, 0];

F32 = - F2 - F12s;
F32x = F32(1);
F32y = F32(2);

out = [F12xs,F12ys,F43xs,F43ys,F32x,F32y];
end
```

MATLAB Function for RRT Dyad

Figure 6.16 shows an RRT dyad with two links, 2 and 3; two pin joints, M and P; and one slider joint, P. The input data are the total forces $\mathbf{F}_2$, $\mathbf{F}_3$ and moments $\mathbf{M}_2$, $\mathbf{M}_3$ on the links 2 and 3; the position vectors $\mathbf{r}_M$, $\mathbf{r}_N$, $\mathbf{r}_P$ of the joints M, N, P; and the

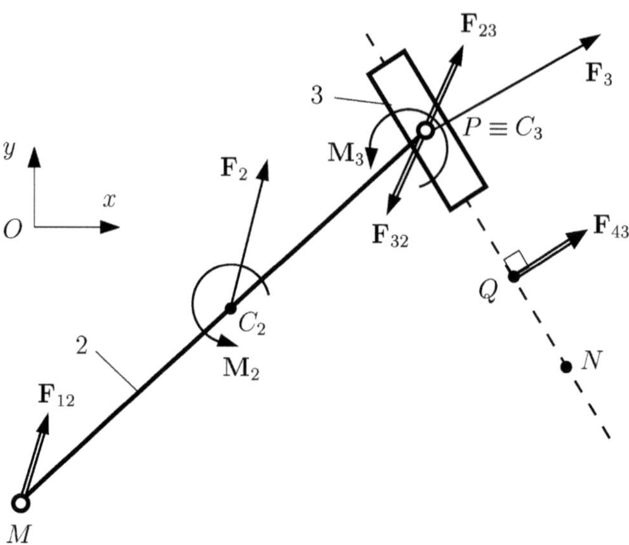

Fig. 6.16 RRT dyad

position vector of the center of mass $\mathbf{r}_{C2}$ of the link 2. The output data are the joint reaction forces $\mathbf{F}_{12}$, $\mathbf{F}_{43}$, $\mathbf{F}_{23} = -\mathbf{F}_{32}$ and the position vector $\mathbf{r}_Q$ of the application point of the joint reaction force $\mathbf{F}_{43}$. The joint reaction $\mathbf{F}_{43}$ is perpendicular to the sliding direction $\mathbf{r}_{PN}$ or

$$\mathbf{F}_{43} \cdot \mathbf{r}_{PN} = (F_{43x}\mathbf{\imath} + F_{43y}\mathbf{\jmath}) \cdot [(x_N - x_P)\mathbf{\imath} + (y_N - y_P)\mathbf{\jmath}] = 0. \tag{6.98}$$

In order to determine $\mathbf{F}_{12}$ and $\mathbf{F}_{43}$, the following equations are written. Sum of all the forces on links 2 and 3 is zero:

$$\sum \mathbf{F}^{(2\&3)} = \mathbf{F}_{12} + \mathbf{F}_2 + \mathbf{F}_3 + \mathbf{F}_{43} = \mathbf{0},$$

or

$$\sum \mathbf{F}^{(2\&3)} \cdot \mathbf{\imath} = F_{12x} + F_{2x} + F_{3x} + F_{43x} = 0,$$

$$\sum \mathbf{F}^{(2\&3)} \cdot \mathbf{\jmath} = F_{12y} + F_{2y} + F_{3y} + F_{43y} = 0. \tag{6.99}$$

Sum of moments of all forces and moments on link 2 about P is zero:

$$\sum \mathbf{M}_P^{(2)} = (\mathbf{r}_M - \mathbf{r}_P) \times \mathbf{F}_{12} + (\mathbf{r}_{C2} - \mathbf{r}_P) \times \mathbf{F}_2 + \mathbf{M}_2 = \mathbf{0}. \tag{6.100}$$

The components F_{12x}, F_{12y}, F_{43x}, and F_{43y} are calculated from (6.98) to (6.100). The reaction force components F_{32x} and F_{32y} are computed from the sum of all the forces on the link 2:

$$\sum \mathbf{F}^{(2)} = \mathbf{F}_{12} + \mathbf{F}_2 + \mathbf{F}_{32} = \mathbf{0},$$

or

$$\sum \mathbf{F}^{(2)} \cdot \mathbf{i} = F_{12x} + F_{2x} + F_{32x} = 0,$$

$$\sum \mathbf{F}^{(2)} \cdot \mathbf{j} = F_{12y} + F_{2y} + F_{32y} = 0. \tag{6.101}$$

To determine the application point $Q(x_Q, y_Q)$ of the reaction force $\mathbf{F}_{43}$, one can write sum of moments of all the forces and the moments on the link 3 about $C_3 \equiv P$ is zero:

$$\sum \mathbf{M}_P^{(3)} = (\mathbf{r}_Q - \mathbf{r}_P) \times \mathbf{F}_{43} + \mathbf{M}_3 = \mathbf{0}. \tag{6.102}$$

If $\mathbf{M}_3 = \mathbf{0}$, then P is identical to Q $(P \equiv Q)$. If $\mathbf{M}_3 \neq \mathbf{0}$, an equation regarding the location of the point Q on the sliding direction $\mathbf{r}_{NP}$ is written as

$$\frac{y_Q - y_P}{x_Q - x_P} = \frac{y_N - y_P}{x_N - x_P}. \tag{6.103}$$

From (6.102) and (6.103), the coordinates x_Q and y_Q of the point Q are calculated. The MATLAB function `ForceRRT.m` for the RRT dyad joint force analysis is

```
% forceRRT.m
% forces RRT dyad
function out=forceRRT...
(rM,rN,rP,rC2,F2,M2,F3,M3);

F12x=sym('F12x','real');
F12y=sym('F12y','real');
F43x=sym('F43x','real');
F43y=sym('F43y','real');
F12=[F12x, F12y, 0]; % unknown joint force
F43=[F43x, F43y, 0]; % unknown joint force

eqF  = F12+F2+F43+F3;
eqFx = eqF(1);
eqFy = eqF(2);

rPC2 = rC2 - rP;
rPM  = rM - rP;
rPN  = rN - rP;

eqFPN = dot(F43,rPN);

eqM2  = cross(rPC2,F2)+cross(rPM,F12)+M2;
eqM2z = eqM2(3);

sol=solve(eqFx, eqFy, eqM2z, eqFPN);
F12xs=eval(sol.F12x);
```

```
F12ys=eval(sol.F12y);
F43xs=eval(sol.F43x);
F43ys=eval(sol.F43y);

F12s=[F12xs, F12ys, 0];
F43s=[F43xs, F43ys, 0];

F32 = - F2 - F12s;
F32x = F32(1);
F32y = F32(2);

if M3(3)==0 xQs=rP(1); yQs=rP(2);
else
 xQ = sym('xQ','real');
 yQ = sym('yQ','real');
 rQ = [xQ yQ 0];
 rPQ = rQ - rP;
 eqQ1 = rPQ(2)/rPQ(1) - rPN(2)/rPN(1);
 eqQ  = cross(rPQ,F43s)+M3;
 eqQ2 = eqQ(3);
 solQ=solve(eqQ1, eqQ2);
 xQs=eval(solQ.xQ);
 yQs=eval(solQ.yQ);
end

out = [F12xs,F12ys,F43xs,F43ys,F32x,F32y,xQs,yQs];
end
```

MATLAB Function for RTR Dyad

Figure 6.17 shows an RTR dyad with two links, 2 and 3; one pin joint, M; one slider joint, P; and one pin joint, P. The input data are the total forces $\mathbf{F}_2$, $\mathbf{F}_3$ and moments $\mathbf{M}_2$, $\mathbf{M}_3$ of the links 2 and 3; the position vectors $\mathbf{r}_M$, $\mathbf{r}_N$, $\mathbf{r}_P$ of the joints M,N; and the position vector of the center of mass $\mathbf{r}_{C2}$ of the link 2. The output data are the joint reaction forces $\mathbf{F}_{12}$, $\mathbf{F}_{43}$, $\mathbf{F}_{23} = -\mathbf{F}_{32}$ and the position vector $\mathbf{r}_Q$ of the application point of the joint reaction force $\mathbf{F}_{23}$. The unknown joint reaction forces $\mathbf{F}_{12}$ and $\mathbf{F}_{43}$ are calculated from the relations: Sum of all the forces on links 2 and 3 is zero:

$$\sum \mathbf{F}^{(2\&3)} = \mathbf{F}_{12} + \mathbf{F}_2 + \mathbf{F}_3 + \mathbf{F}_{43} = \mathbf{0},$$

or

$$\sum \mathbf{F}^{(2\&3)} \cdot \mathbf{1} = F_{12x} + F_{2x} + F_{3x} + F_{43x} = 0,$$

$$\sum \mathbf{F}^{(2\&3)} \cdot \mathbf{J} = F_{12y} + F_{2y} + F_{3y} + F_{43y} = 0. \tag{6.104}$$

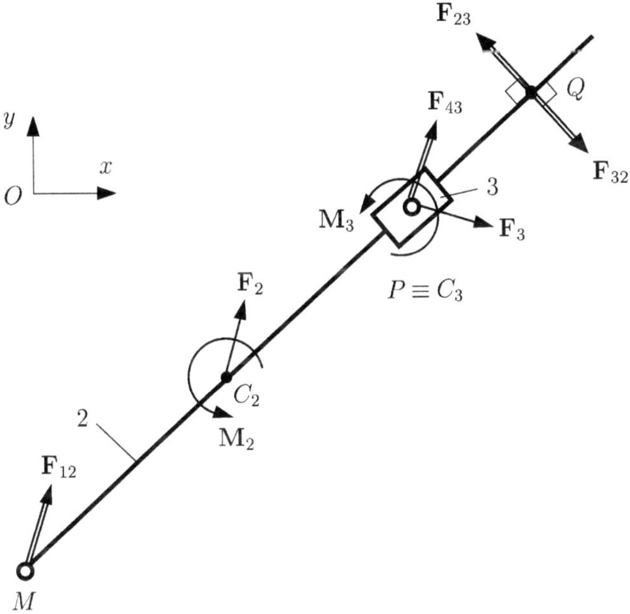

Fig. 6.17 RTR dyad

Sum of the moments of all forces and moments on links 2 and 3 about M is zero:

$$\sum \mathbf{M}_M^{(2\&3)} = (\mathbf{r}_P - \mathbf{r}_M) \times (\mathbf{F}_3 + \mathbf{F}_{43}) + (\mathbf{r}_{C2} - \mathbf{r}_M) \times \mathbf{F}_2 + \mathbf{M}_2 + \mathbf{M}_3 = \mathbf{0}. \quad (6.105)$$

Sum of all the forces on link 2 projected onto the sliding direction $\mathbf{r}_{MP}$ is zero:

$$\mathbf{F}_2 \cdot \mathbf{r}_{MP} = (F_{43x}\mathbf{\imath} + F_{43y}\mathbf{J}) \cdot [(x_P - x_M)\mathbf{\imath} + (y_P - y_M)\mathbf{J}] = 0. \quad (6.106)$$

The components F_{12x}, F_{12y}, F_{43x}, and F_{43y} are calculated from (6.104) to (6.106). The force components F_{32x} and F_{32y} are computed from the sum of all the forces on link 2:

$$\sum \mathbf{F}^{(2)} = \mathbf{F}_{12} + \mathbf{F}_2 + \mathbf{F}_{32} = \mathbf{0},$$

or

$$\sum \mathbf{F}^{(2)} \cdot \mathbf{\imath} = F_{12x} + F_{2x} + F_{32x} = 0,$$

$$\sum \mathbf{F}^{(2)} \cdot \mathbf{J} = F_{12y} + F_{2y} + F_{32y} = 0. \quad (6.107)$$

To determine the application point $Q(x_Q, y_Q)$ of the reaction force $\mathbf{F}_{23}$, one can write sum of moments of all the forces and the moments on the link 3 about $C_3 \equiv P$ is zero:

$$\sum \mathbf{M}_P^{(3)} = (\mathbf{r}_Q - \mathbf{r}_P) \times \mathbf{F}_{23} + \mathbf{M}_3 = \mathbf{0}. \tag{6.108}$$

If $\mathbf{M}_3 = \mathbf{0}$, then P is identical to Q $(P \equiv Q)$. If $\mathbf{M}_3 \neq \mathbf{0}$, an equation regarding the location of the point Q on the sliding direction $\mathbf{r}_{NP}$ is written as

$$\frac{y_Q - y_M}{x_Q - x_M} = \frac{y_M - y_P}{x_M - x_P}. \tag{6.109}$$

From (6.108) and (6.109), the coordinates x_Q and y_Q of the point Q are calculated. The MATLAB function `ForceRTR.m` for the RTR dyad joint force analysis is

```
% forceRTR.m
% forces RTR dyad
function out=forceRTR...
(rM,rP,rC2,F2,M2,F3,M3);

F12x=sym('F12x','real');
F12y=sym('F12y','real');
F43x=sym('F43x','real');
F43y=sym('F43y','real');
F12=[F12x, F12y, 0]; % unknown joint force
F43=[F43x, F43y, 0]; % unknown joint force

eqF = F12+F2+F43+F3;
eqFx = eqF(1);
eqFy = eqF(2);

rMC2 = rC2 - rM;
rMP = rP - rM;

eqF2 = dot(F12+F2,rMP);

eqMM = cross(rMC2,F2)+cross(rMP,(F3+F43))+M2+M3;
eqMMz = eqMM(3);

sol=solve(eqFx, eqFy, eqF2, eqMMz);
F12xs=eval(sol.F12x);
F12ys=eval(sol.F12y);
F43xs=eval(sol.F43x);
F43ys=eval(sol.F43y);

F12s=[F12xs, F12ys, 0];
F43s=[F43xs, F43ys, 0];
```

Fig. 6.18 Driver link

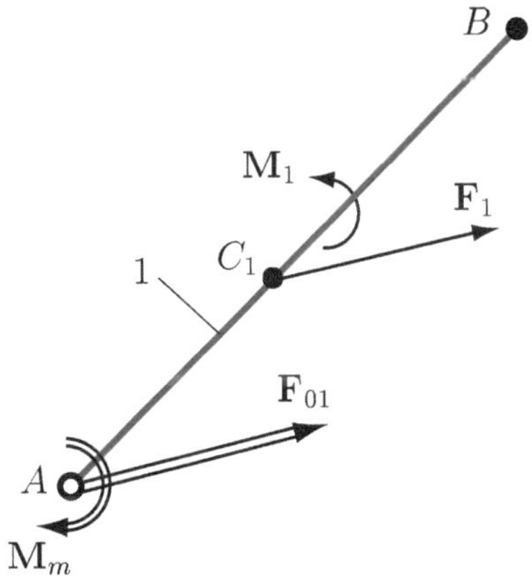


```
F23  = - F3 - F43s;
F23x = F23(1);
F23y = F23(2);

if M3(3)==0 xQs=rP(1); yQs=rP(2);
else
  xQ = sym('xQ','real');
  yQ = sym('yQ','real');
  rQ = [xQ yQ 0];
  rMQ = rQ - rM;
  eqQ1 = rMQ(2)/rMQ(1) - rMP(2)/rMP(1);
  eqQ  = cross(rQ-rP,F23)+M3;
  eqQ2 = eqQ(3);
  solQ=solve(eqQ1, eqQ2);
  xQs=eval(solQ.xQ);
  yQs=eval(solQ.yQ);
end

out = [F12xs,F12ys,F43xs,F43ys,F23x,F23y,xQs,yQs];
end
```

Driver Link

A driver link is shown in Fig. 6.18. The input data are the total force $\mathbf{F}_1$ and moment $\mathbf{M}_1$ on the driver link; the joint reaction force $\mathbf{F}_{21}$; the position vectors $\mathbf{r}_A$, $\mathbf{r}_B$ of the joints A, B; and the position vector of the center of mass $\mathbf{r}_{C1}$ of the driver link. The output data are the joint reaction force $\mathbf{F}_{01}$ and the moment of the motor $\mathbf{M}_m$ (equilibrium moment). A force equation for the driver link is written to determine the joint reaction $\mathbf{F}_{01}$:

$$\sum \mathbf{F}^{(1)} = \mathbf{F}_{01} + \mathbf{F}_1 + \mathbf{F}_{21} = \mathbf{0},$$

or

$$\sum \mathbf{F}^{(1)} \cdot \mathbf{\iota} = F_{01x} + F_{1x} + F_{21x} = 0,$$

$$\sum \mathbf{F}^{(1)} \cdot \mathbf{J} = F_{01y} + F_{1y} + F_{21y} = 0. \tag{6.110}$$

The sum of the moments about A_R for link 1 gives the equilibrium moment $\mathbf{M}_m$:

$$\sum \mathbf{M}_A^{(1)} = (\mathbf{r}_B - \mathbf{r}_A) \times \mathbf{F}_{21} + (\mathbf{r}_{C1} - \mathbf{r}_A) \times \mathbf{F}_1 + \mathbf{M}_m = \mathbf{0}. \tag{6.111}$$

The MATLAB function `forceDR.m` for the driver link joint force analysis is

```
% forceDR.m
% forces Driver
function out=forceDR(rA,rB,rC1,F1,M1,F21);


F01 = - F1 - F21;


rAC1 = rC1 - rA ;
rAB = rB - rA ;
Mm =-cross(rAC1,F1)-cross(rAB,F21)-M1;


out = [F01(1), F01(2), Mm(3)];
end
```

6.10 Examples

Example 6.1. The position, velocity, and acceleration analysis of the planar R-RTR-RRT mechanism shown in Fig. 6.19a are presented in Sect. 5.10. Given the external force $\mathbf{F}_{\text{ext}} = -500\,\text{sign}(v_E)\mathbf{\iota}\,\text{N}$ applied on the link 5, calculate the motor moment $\mathbf{M}_m$ required for the dynamic equilibrium of the mechanism. All three links are rectangular prisms with the depth $d = 0.001$ m and the mass density $\rho = 8,000\,\text{Kg/m}^3$. The height of the links 1, 3, and 4 is $h = 0.01$ m. The slider links 2 and 5 have the height $h_S = 0.02$ m and the width $w_S = 0.05$ m. The center of mass location of the links $i = 1,\dots,5$ are designated by $C_i(x_{C_i}, y_{C_i}, 0)$.

Solution
The MATLAB program `R_RTR_RRT_va.m` is loaded at the beginning, and then the positions and the accelerations of the mass centers are calculated:

```
% kinematics
R_RTR_RRT_va


rC1 = rB/2;
fprintf('rC1 = [%6.3f,%6.3f,%d] (m) \n', rC1)
```

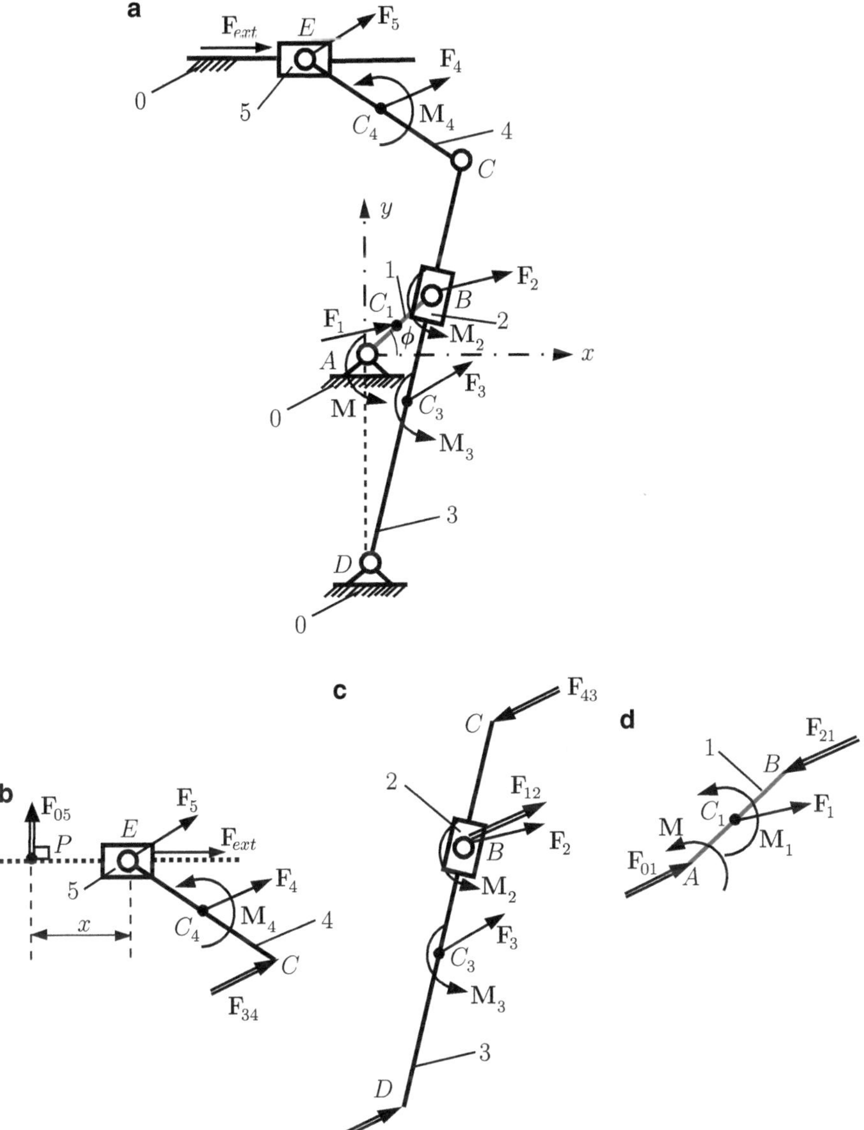

Fig. 6.19 Example 6.1

```
rC2 = rB;
fprintf('rC2 = [%6.3f,%6.3f,%d] (m)\n', rC2)
rC3 = (rD+rC)/2;
fprintf('aC3 = [%6.3f,%6.3f,%d] (m)\n', rC3)
```

```
rC4 = (rC+rE)/2;
fprintf('rC4 = [%6.3f,%6.3f,%d] (m)\n', rC4)
rC5 = rE;
fprintf('rC5 = [%6.3f,%6.3f,%d] (m)\n', rC5)


aC1 = aB/2;
fprintf('aC1 = [%6.3f,%6.3f,%d] (m/s^2)\n', aC1)
aC2 = aB;
fprintf('aC2 = [%6.3f,%6.3f,%d] (m/s^2)\n', aC2)
aD = 0;
aC3 = (aD+aC)/2;
fprintf('aC3 = [%6.3f,%6.3f,%d] (m/s^2)\n', aC3)
aC4 = (aC+aE)/2;
fprintf('aC4 = [%6.3f,%6.3f,%d] (m/s^2)\n', aC4)
aC5 = aE;
fprintf('aC5 = [%6.3f,%6.3f,%d] (m/s^2)\n', aC5)
```

The MATLAB results are

```
phi = phi1 = 45 (degrees)
rA = [0, 0, 0] (m)
rD = [ 0.000, -0.400, 0] (m)
rB = [ 0.141, 0.141, 0] (m)
rC = [ 0.177, 0.277, 0] (m)
phi2 = phi3 = 75.361 (degrees)
rE = [-0.114, 0.350, 0] (m)
phi4 = 165.971 (degrees)


Velocity and acceleration analysis

omega1 = [0, 0,   20.9] (rad/s)
alpha1 = [0, 0, 0] (rad/s^2)


vB = vB1 = vB2 = [ -2.96,   2.96, 0] (m/s)
aB = aB1 = aB2 = [   -62,    -62, 0] (m/s^2)


omega2=omega3 = [0,0,  6.46] (rad/s)
alpha2=alpha3 = [0,0,  30.4] (rad/s^2)
vC = [-4.374,  1.143, 0] (m/s)
aC = [-27.947, -22.882, 0] (m/s^2)


vE = [-4.660,  0.000, 0] (m/s)
aE = [-17.464,  0.000, 0] (m/s^2)


omega4 = [0,0,  3.93] (rad/s)
alpha4 = [0,0, -82.5] (rad/s^2)
```

```
rC1 = [ 0.071, 0.071,0]  (m)
rC2 = [ 0.141, 0.141,0]  (m)
aC3 = [ 0.088,-0.061,0]  (m)
rC4 = [ 0.031, 0.314,0]  (m)
rC5 = [-0.114, 0.350,0]  (m)
aC1 = [-31.017,-31.017,0]  (m/s^2)
aC2 = [-62.034,-62.034,0]  (m/s^2)
aC3 = [-13.974,-11.441,0]  (m/s^2)
aC4 = [-22.705,-11.441,0]  (m/s^2)
aC5 = [-17.464,0,0]  (m/s^2)
```

The input data for the geometry of the links are

```
h = 0.01; % height of the bar
d = 0.001; % depth of the bar
hSlider = 0.02; % height of the slider
wSlider = 0.05; % depth of the slider
rho = 8000; % density of the material
g = 9.807; % gravitational acceleration
```

For the link 1 with the mass m_1, the acceleration of the center of mass $\mathbf{a}_{C_1}$ and the mass moment of inertia I_{C_1} are

$$m_1 = \rho\, AB\, h\, d,$$

$$I_{C_1} = m_1(AB^2 + h^2)/12.$$

The force of inertia and the moment of inertia of the link 1 are calculated using the MATLAB commands:

```
fprintf('Link 1  \n')
m1 = rho*AB*h*d;
Fin1 = -m1*aC1;
G1 = [0,-m1*g,0];
IC1 = m1*(AB^2+h^2)/12;
alpha1 = [0 0 0];
Min1 = -IC1*alpha1;
```

The MATLAB commands for the links 2, 3, 4, and 5 are

```
fprintf('Link 2 \n')
m2 = rho*hSlider*wSlider*d;
Fin2 = -m2*aC2;
G2 = [0,-m2*g,0];
IC2 = m2*(hSlider^2+wSlider^2)/12;
Min2 = -IC2*alpha2;
```

```matlab
fprintf('Link 3 \n')
m3 = rho*CD*h*d;
Fin3 = -m3*aC3;
G3 = [0,-m3*g,0];
IC3 = m3*(CD^2+h^2)/12;
Min3 = -IC3*alpha3;

fprintf('Link 4 \n')
m4 = rho*CE*h*d;
Fin4 = -m4*aC4;
G4 = [0,-m4*g,0];
IC4 = m2*(CE^2+h^2)/12;
Min4 = -IC4*alpha4;

fprintf('Link 5 \n')
m5 = rho*hSlider*wSlider*d;
Fin5 = -m5*aC5;
G5 = [0,-m5*g,0];
IC5 = m5*(hSlider^2+wSlider^2)/12;
Min5 = -IC5*alpha5;
```

The joint reactions for the dyad RTR $(C_R E_R E_T)$, Fig. 6.19b, are computed using the MATLAB function forceRTR.m:

```matlab
Fext = -sign(vE(1))*[500,0,0];
% Dyad RRT (4 & 5)
F5 = Fin5 + G5 + Fext;
F4 = Fin4 + G4;
M5 = Min5;
M4 = Min4;
rN = [0 yE 0];
FD45=forceRRT(rC,rN,rE,rC4,F4,M4,F5,M5);
F34x = FD45(1);
F34y = FD45(2);
F05x = FD45(3);
F05y = FD45(4);
F54x = FD45(5);
F54y = FD45(6);
F34 = [F34x F34y 0];
F05 = [F05x F05y 0];
F54 = [F54x F54y 0];
```

The input data for the function are the total force F5 and momentum M5 of link 5, the total force F4 and momentum M4 of the link 4, and the position vectors rC, rN, rE, rC5. The output data are the three elements of the vector returned by the function forceRRT, which are the joint reactions F05, F34, and F54. The results are

```
F05 = [ 0.000,-124.959,0]  (N)
F54 = [500.140,-125.037,0]  (N)
F34 = [-500.685,124.998,0]  (N)
```

Next, consider the dyad RTR ($B_R B_T D_R$). The reaction force $\mathbf{F}_{43}$ acting at point C can be moved to a parallel position at point C_3 by adding the corresponding couple:

$$\mathbf{M}_{43} = \mathbf{r}_{C_3 C} \times \mathbf{F}_{43}.$$

The joint reactions for the dyad RTR (links 2 and 3), Fig. 6.19c, using the MATLAB function forceRTR.m are calculated with

```
% Dyad RTR (2 & 3)
F43 = -F34;
F3  = Fin3 + G3 + F43;
F2  = Fin2 + G2;
M3  = Min3 + cross(rC-rC3,F43);
M2  = Min2;


FD23=forceRTR(rD,rB,rC3,F3,M3,F2,M2);


F03x = FD23(1);
F03y = FD23(2);
F12x = FD23(3);
F12y = FD23(4);
F32x = FD23(5);
F32y = FD23(6);


F03 = [F03x F03y 0];
F12 = [F12x F12y 0];
F32 = [F32x F32y 0];


xQ = FD23(7);
yQ = FD23(8);
rQ = [xQ yQ 0];
```

The input data are the force Fin3 + G3 + F43 and moment M3+M43 of the link 3; the total force F2 and moment M2 of the link 2; the position vectors rD, rB, rC3 of the joints D, B; and the center of mass C_3 of the link 3. The output data are the three elements of the vector returned by the function forceRTR, which are the joint reactions F03, F12, and F23. The position vector of the application point Q of the joint reaction F12 is also computed using the function forceRTR.m. The MATLAB results are

```
F03 = [123.647,-38.376,0]  (N)
F12 = [-625.610,162.864,0]  (N)
F32 = [625.114,-163.282,0]  (N)
rQ  = [0.14142,0.14142,0]  (m)
```

Fig. 6.20 Example 6.2
contour diagram

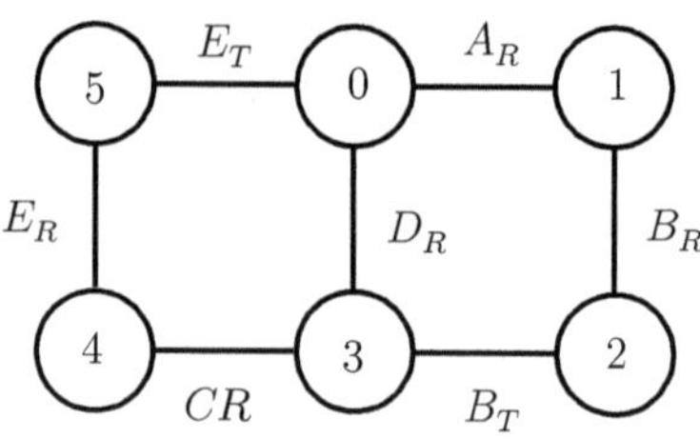

The joint reaction and the moment of the motor for the driver link, Fig. 6.19d, are computed using the MATLAB function `forceDR.m`:

```
FDR=forceDR(rA,rB,rC1,Fin1+G1,Min1,-F12);
F01 = [FDR(1)  FDR(2)  0];
Mmot= [0  0  FDR(3)];
```

and the results are

```
F01 = [-626.106,162.525,0]  (N)
Mmot = [0,0,111.518]  (N m)
```

Example 6.2. Repeat *Example 6.1* using the contour method.

Solution
The R-RTR-RRT mechanism shown in Fig. 6.19a has the contour depicted in Fig. 6.20.

The diagram representing the mechanism has two contours 0-1-2-3-0 and 0-3-4-5-0. The joint at C represents a ramification point for the mechanism and the diagram, and the dynamic force analysis will start with this joint. The force computation starts with the contour 0-3-4-5-0 because the driven load $\mathbf{F}_{ext}$ on link 5 is given. The following analysis will consider the relationships of the inertia forces $\mathbf{F}_{in\,j}$, the inertia moments $\mathbf{M}_{in\,j}$, the gravitational force $\mathbf{G}_j$, the driven external force $\mathbf{F}_{ext}$ the joint reactions $\mathbf{F}_{ij}$ and the drive moment $\mathbf{M}$ on the crank 1.

To simplify the notation, the total vector force at C_j is written as $\mathbf{F}_j = \mathbf{F}_{in\,j} + \mathbf{G}_j$, and the inertia moment of link j is written as $\mathbf{M}_j = \mathbf{M}_{in\,j}$.

Reaction $\mathbf{F}_{34}$
The rotation joint at C, between 3 and 4, Fig. 6.21, is replaced with the unknown reaction:

$$\mathbf{F}_{34} = -\mathbf{F}_{43} = F_{34x}\mathbf{I} + F_{34y}\mathbf{J}.$$

If the path I is followed, Fig. 6.21, for the rotation joint at E (E_R), a moment equation is written:

$$\sum \mathbf{M}_E^{(4)} = \mathbf{r}_{EC} \times \mathbf{F}_{32} + \mathbf{r}_{EC4} \times \mathbf{F}_4 + \mathbf{M}_4 = \mathbf{0}. \tag{6.112}$$

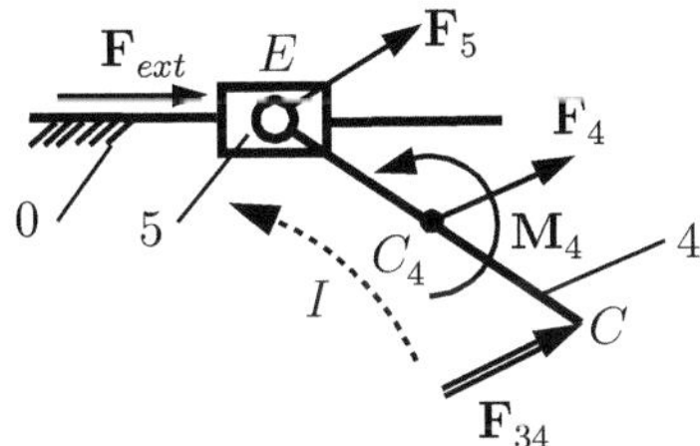

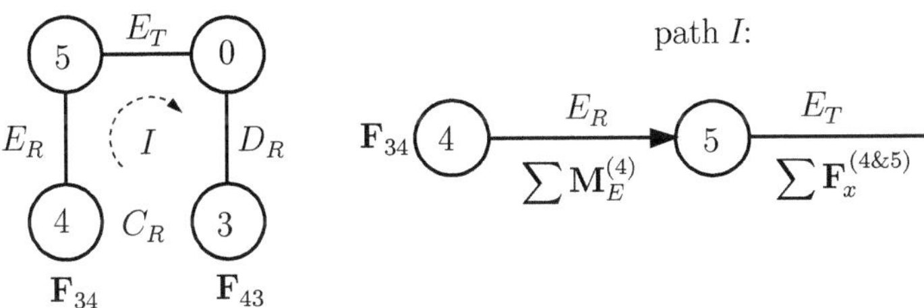

Fig. 6.21 Example 6.2 reaction $\mathbf{F}_{34}$

Continuing on path I, the next joint is the translational joint at E (E_T). The projection of all the forces that act on 4 and 5 onto the sliding direction Δ (x-axis) should be zero:

$$\sum \mathbf{F}_\Delta^{(4\&5)} = \sum \mathbf{F}^{(4\&5)} \cdot \mathbf{\iota} = (\mathbf{F}_{34} + \mathbf{F}_4 + \mathbf{F}_5 + \mathbf{F}_{ext}) \cdot \mathbf{\iota}$$

$$= F_{34x} + F_{4x} + F_{5x} + F_{ext} = 0. \tag{6.113}$$

The system of (6.112) and (6.113) is solved, and the two unknowns F_{34x} and F_{34y} are obtained.

The MATLAB commands for computing $\mathbf{F}_{34}$ are

```
F34x=sym('F34x','real');
F34y=sym('F34y','real');
F34=[F34x, F34y, 0];
% sum M_E for (4)
% EC x F34 + EC4 x F4 + M4 = 0
eqME4=cross(rC-rE,F34)+cross(rC4-rE,F4)+M4;
% sum F for (4&5) on x-axis
% (F5+Fext+F4+F34).i = 0
eqF45x=dot(F5+Fext+F4+F34,[1 0 0]);
solF34=solve(eqME4(3),eqF45x);
F34xn=eval(solF34.F34x);
F34yn=eval(solF34.F34y);
F34n=[F34xn, F34yn, 0];
```

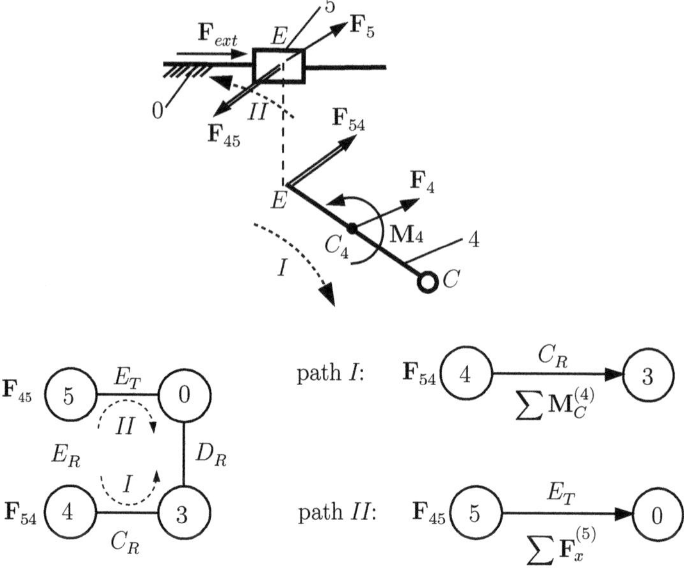

Fig. 6.22 Example 6.2 reaction $\mathbf{F}_{54}$

Reaction $\mathbf{F}_{54}$

The rotation joint at E (E_R), between 4 and 5, Fig. 6.22, is replaced with the unknown reaction:

$$\mathbf{F}_{54} = -\mathbf{F}_{45} = F_{54x}\mathbf{1} + F_{54y}\mathbf{J}.$$

If the path I is traced, Fig. 6.22, for the pin joint at C (C_R), a moment equation is written:

$$\sum \mathbf{M}_C^{(4)} = \mathbf{r}_{CE} \times \mathbf{F}_{54} + \mathbf{r}_{CC4} \times \mathbf{F}_4 + \mathbf{M}_4 = \mathbf{0}. \tag{6.114}$$

For the path II, the slider joint at E (E_T) is encountered. The projection of all the forces that act on 5 onto the sliding direction Δ (x-axis) should be zero:

$$\sum \mathbf{F}_\Delta^{(5)} = \sum \mathbf{F}^{(5)} \cdot \mathbf{1} = (-\mathbf{F}_{54} + \mathbf{F}_5 + \mathbf{F}_{ext}) \cdot \mathbf{1}$$
$$= -F_{54x} + F_{5x} + F_{ext} = 0. \tag{6.115}$$

The unknown force components F_{54x} and F_{54y} are calculated from (6.114) and (6.115).

The MATLAB commands for computing $\mathbf{F}_{54}$ are

```
F54x=sym('F54x','real');
F54y=sym('F54y','real');
F54=[F54x, F54y, 0];
```

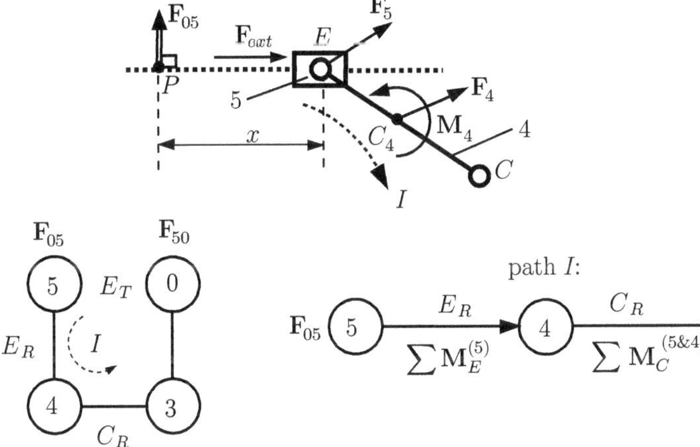

Fig. 6.23 Example 6.2 reaction $\mathbf{F}_{05}$

```
% sum M_C for (4)
% CE x F54 + CC4 x F4 + M4 = 0
eqMC4=cross(rE-rC,F54)+cross(rC4-rC,F4)+M4;
% sum F for (5) on x-axis
% [F5+Fext+(-F54)].i = 0
eqF5x=dot(F5+Fext-F54,[1 0 0]);
solF54=solve(eqMC4(3),eqF5x);
F54xn=eval(solF54.F54x);
F54yn=eval(solF54.F54y);
F54n=[F54xn, F54yn, 0];
```

Reaction $\mathbf{F}_{05}$

The slider joint at E (E_T), between 0 and 5, Fig. 6.23, is replaced with the unknown reaction:

$$\mathbf{F}_{05} = F_{05y}\mathbf{J}.$$

The reaction joint introduced by the translational joint is perpendicular on the sliding direction $\mathbf{F}_{05} \perp \Delta$. The application point P of force $\mathbf{F}_{05}$ is unknown. If the path I is followed, Fig. 6.23, for the pin joint at E (E_R), a moment equation is written for link 5:

$$\sum \mathbf{M}_E^{(5)} = (\mathbf{r}_P - \mathbf{r}_E) \times \mathbf{F}_{05} = \mathbf{0},$$

or

$$x F_{05y} = 0 \ \Rightarrow x = 0. \tag{6.116}$$

The application point is at E ($P \equiv E$).

Continuing on path I, the next joint is the pin joint C (C_R):

$$\sum \mathbf{M}_C^{(4\&5)} = \mathbf{r}_{CE} \times (\mathbf{F}_{05} + \mathbf{F}_5 + \mathbf{F}_{\text{ext}}) + \mathbf{r}_{CC4} \times \mathbf{F}_4 + \mathbf{M}_4 = \mathbf{0}. \qquad (6.117)$$

The joint reaction force F_{05y} is computed from (6.117).

The MATLAB commands for computing $\mathbf{F}_{05}$ are

```
% sum M_E for (5)
% EP x F05 = 0 => E=P
% F05 acts at E
F05y=sym('F05y','real');
F05=[0, F05y, 0];
% sum M_C for (4&5)
% CE x (F05+F5+Fext) + CC4 x F4 + M4 = 0
eqMC45=cross(rE-rC,F05+F5+Fext)+...
       cross(rC4-rC,F4)+M4;
solF05=solve(eqMC45(3));
F05yn=eval(solF05);
F05n=[0, F05yn, 0];
```

The MATLAB results are

```
F34 = [-500.685,124.998,0] (N)
F54 = [500.140,-125.037,0] (N)
F05 = [0,-124.959,0] (N)
```

For the contour 0-1-2-3-0, the joint force $\mathbf{F}_{43} = -\mathbf{F}_{34}$ at the ramification point C is considered as a known external force.

Reaction $\mathbf{F}_{32}$

The slider joint B_T, between 2 and 3, Fig. 6.24, is replaced with the unknown reaction force:

$$\mathbf{F}_{32} = F_{32x}\mathbf{\imath} + F_{32y}\mathbf{J}.$$

The reaction force $\mathbf{F}_{32}$ is perpendicular to the sliding direction $\mathbf{r}_{DB}$:

$$\mathbf{F}_{32} \cdot \mathbf{r}_{DB} = 0. \qquad (6.118)$$

The point of application of force $\mathbf{F}_{32}$ is $Q(x_Q, y_Q)$ and is located on the line DB or

$$\mathbf{r}_{DB} \times \mathbf{r}_{DQ} = \mathbf{0}. \qquad (6.119)$$

If the path I is followed, Fig. 6.24, for the pin joint at D (D_R), a moment equation is written for link 3:

$$\sum \mathbf{M}_D^{(3)} = \mathbf{r}_{DQ} \times (-\mathbf{F}_{32}) + \mathbf{r}_{DC3} \times \mathbf{F}_3 + \mathbf{r}_{DC} \times \mathbf{F}_{43} + \mathbf{M}_3 = \mathbf{0}. \qquad (6.120)$$

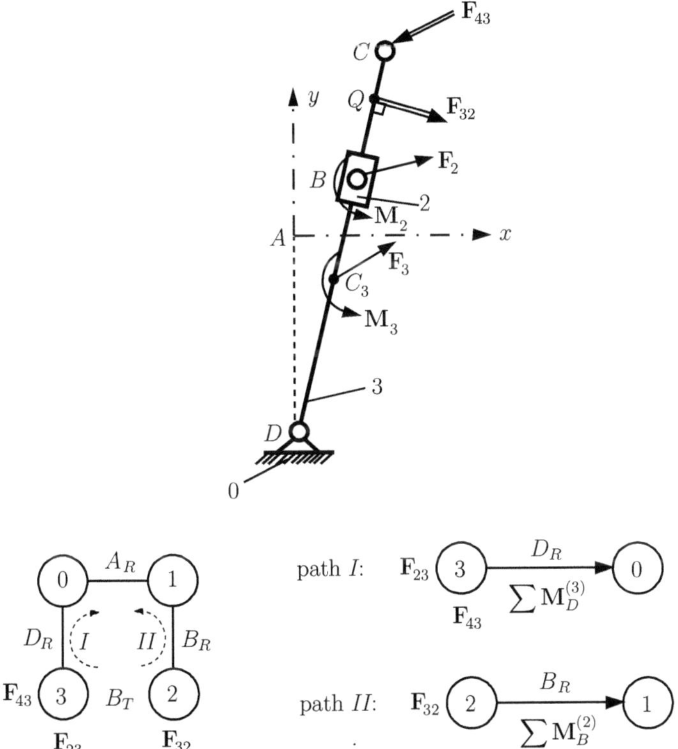

Fig. 6.24 Example 6.2 reaction $\mathbf{F}_{32}$

For path *II*, a moment equation is written for link 2 with respect to the pin joint B_R:

$$\sum \mathbf{M}_B^{(2)} = \mathbf{r}_{BQ} \times \mathbf{F}_{32} + \mathbf{M}_2 = \mathbf{0}. \tag{6.121}$$

The two components of the joint force $\mathbf{F}_{32}$, x_Q, and y_Q are computed from (6.118) to (6.121). The MATLAB commands are

```
F32x=sym('F32x','real');
F32y=sym('F32y','real');
F32=[F32x, F32y, 0];
xQ = sym('xQ','real');
yQ = sym('yQ','real');
rQ = [xQ yQ 0];
% F32 perpendicular to BD
% F32.BD = 0
eqF32BD = dot(F32,rB-rD);
% Q is on the line BD
% DB x DQ = 0
eqQ = cross(rB-rD,rQ-rD);
```

```
% sum M_B for (2)
% BQ x F32 + M2 = 0
eqMB2=cross(rQ-rB,F32)+M2;
% sum M_D for (3)
% DQ x (-F32) + DC3 x F3 + DC x F43 + M3 = 0
eqMD3=cross(rQ-rD,-F32)+cross(rC3-rD,F3)...
        +cross(rC-rD,F43)+M3;
solF32=solve(eqF32BD,eqQ(3),eqMB2(3),eqMD3(3));
F32xn=eval(solF32.F32x);
F32yn=eval(solF32.F32y);
F32n=[F32xn, F32yn, 0];
fprintf('F32 = [%6.3f,%6.3f,%d]  (N)\n',F32n)
xQn=eval(solF32.xQ);
yQn=eval(solF32.yQ);
rQn=[xQn, yQn, 0];
```

Reaction $\mathbf{F}_{03}$

The pin joint D_R, between 0 and 3, Fig. 6.25, is replaced with the unknown reaction force:

$$\mathbf{F}_{03} = F_{03x}\mathbf{I} + F_{03y}\mathbf{J}.$$

If the path I is followed, Fig. 6.25, the slider joint at B (B_T) is encountered. The projection of all the forces that act on 3 onto the sliding direction $\mathbf{r}_{DB}$ should be zero:

$$\sum \mathbf{F}^{(3)} \cdot \mathbf{r}_{DB} = (\mathbf{F}_{03} + \mathbf{F}_3 + \mathbf{F}_{43}) \cdot \mathbf{r}_{DB} = \mathbf{0}. \tag{6.122}$$

Continuing on path I, the next joint is the pin joint B_R, and a moment equation is written for links 2 and 3:

$$\sum \mathbf{M}_B^{(3\&2)} = \mathbf{r}_{BD} \times \mathbf{F}_{03} + \mathbf{r}_{BC} \times \mathbf{F}_{43} + \mathbf{r}_{BC3} \times \mathbf{F}_3 + \mathbf{M}_2 + \mathbf{M}_3 = \mathbf{0}. \tag{6.123}$$

The two components F_{03x} and F_{03y} of the joint force are obtained from (6.122) and (6.123), and the MATLAB commands are

```
F03x=sym('F03x','real');
F03y=sym('F03y','real');
F03=[F03x, F03y, 0];
% sum F for (3) on DB
% (F03 + F3 + F43).DB = 0
eqF3DB=dot(F03+F3+F43,rB-rD);
% sum M_B for (3&2)
% BD x F03 + BC x F43 + BC3 x F3 +M3+M2 = 0
eqMB32=cross(rD-rB,F03)+cross(rC-rB,F43)...
        +cross(rC3-rB,F3)+M3+M2;
solF03=solve(eqF3DB,eqMB32(3));
```

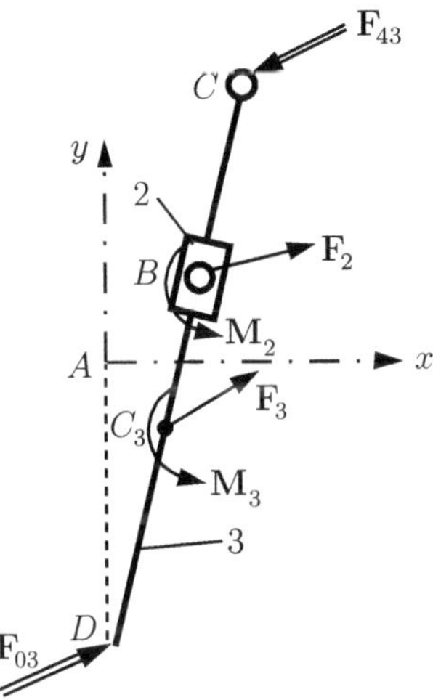

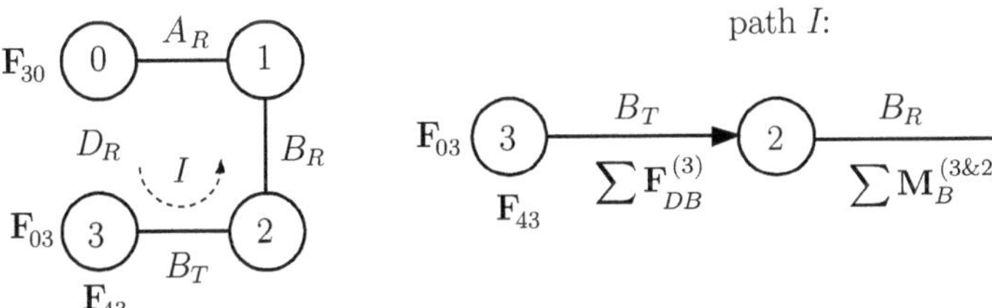

Fig. 6.25 Example 6.2 reaction $\mathbf{F}_{03}$

```
F03xn=eval(solF03.F03x);
F03yn=eval(solF03.F03y);
F03n=[F03xn, F03yn, 0];
```

Reaction $\mathbf{F}_{12}$

The pin joint B_R, between 1 and 2, Fig. 6.26, is replaced with the unknown reaction force:

$$\mathbf{F}_{12} = F_{12x}\mathbf{1} + F_{12y}\mathbf{J}.$$

If the path I is followed, Fig. 6.26, the slider joint at B (B_T) is encountered. The projection of all the forces that act on 2 onto the sliding direction $\mathbf{r}_{DB}$ should be zero:

$$\sum \mathbf{F}^{(2)} \cdot \mathbf{r}_{DB} = (\mathbf{F}_{12} + \mathbf{F}_2) \cdot \mathbf{r}_{DB} = \mathbf{0}. \tag{6.124}$$

Continuing on path I, the next joint is the pin joint D_R, and a moment equation is written for links 2 and 3:

$$\sum \mathbf{M}_D^{(3\&2)} = \mathbf{r}_{DB} \times (\mathbf{F}_{12} + \mathbf{F}_2) + \mathbf{r}_{DC} \times \mathbf{F}_{43} + \mathbf{r}_{DC3} \times \mathbf{F}_3 + \mathbf{M}_2 + \mathbf{M}_3 = \mathbf{0}. \tag{6.125}$$

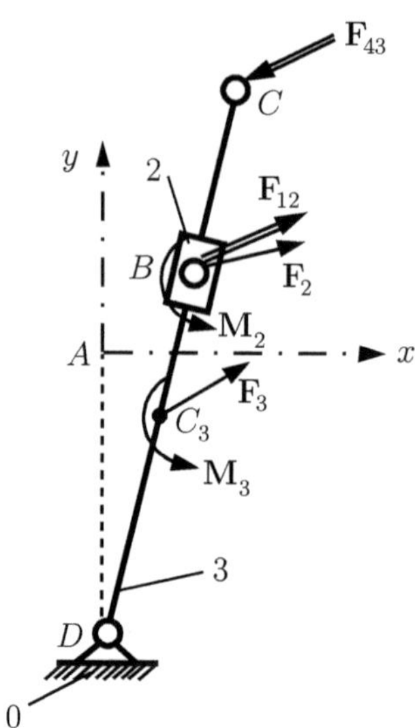

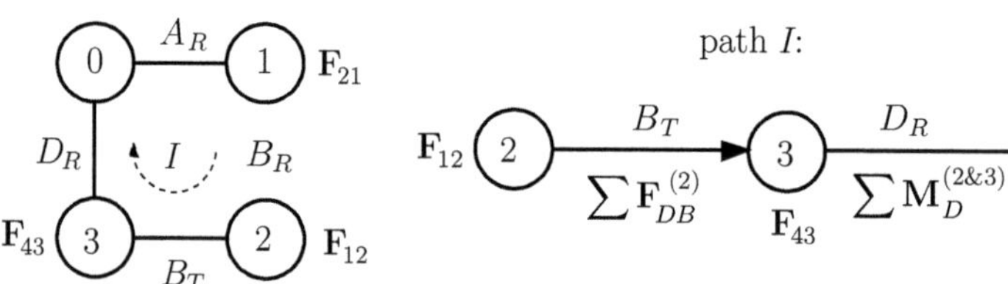

Fig. 6.26 Example 6.2 reaction $\mathbf{F}_{12}$

The two components F_{12x} and F_{12y} of the joint force are obtained from (6.124) and (6.125), and the MATLAB commands are

```
F12x=sym('F12x','real');
F12y=sym('F12y','real');
F12=[F12x, F12y, 0];
% sum F for (2) on DB
% (F12 + F2).DB = 0
eqF2DB=dot(F12+F2,rB-rD);
% sum M_D for (2&3)
% DB x (F12+F2) + DC x F43 + DC3 x F3 +M3+M2 = 0
eqMD23=cross(rB-rD,F12+F2)+cross(rC-rD,F43)...
        +cross(rC3-rD,F3)+M3+M2;
solF12=solve(eqF2DB,eqMD23(3));
F12xn=eval(solF12.F12x);
F12yn=eval(solF12.F12y);
F12n=[F12xn, F12yn, 0];
```

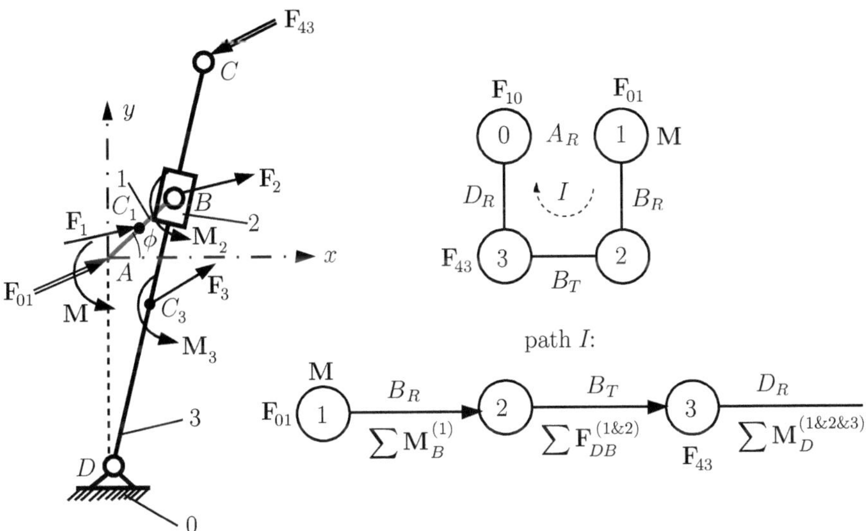

Fig. 6.27 Example 6.2 reaction $\mathbf{F}_{01}$ and driver moment $\mathbf{M}$

Reaction $\mathbf{F}_{01}$ and Driver Moment $\mathbf{M}$

The pin joint A_R, between 0 and 1, Fig. 6.27, is replaced with the unknown reaction force:

$$\mathbf{F}_{01} = F_{01x}\mathbf{I} + F_{01y}\mathbf{J}.$$

The unknown driver moment is $\mathbf{M} = M\mathbf{k}$. If the path I is followed, Fig. 6.27, a moment equation is written for the pin joint B_R for link 1:

$$\sum \mathbf{M}_B^{(1)} = \mathbf{r}_{BA} \times \mathbf{F}_{01} + \mathbf{r}_{BC1} \times \mathbf{F}_1 + \mathbf{M}_1 + \mathbf{M} = \mathbf{0}. \qquad (6.126)$$

If the path I is followed, Fig. 6.27, the slider joint at B (B_T) is encountered. The projection of all the forces that act on 1 and 2 onto the sliding direction $\mathbf{r}_{DB}$ should be zero:

$$\sum \mathbf{F}^{(1\&2)} \cdot \mathbf{r}_{DB} = (\mathbf{F}_{01} + \mathbf{F}_1 + \mathbf{F}_2) \cdot \mathbf{r}_{DB} = \mathbf{0}. \qquad (6.127)$$

Continuing on path I, the next joint encountered is the pin joint D_R, and a moment equation is written for links 1, 2, and 3:

$$\sum \mathbf{M}_D^{(1\&2\&3)} = \mathbf{r}_{DA} \times \mathbf{F}_{01} + \mathbf{r}_{DC1} \times \mathbf{F}_1 + \mathbf{r}_{DB} \times \mathbf{F}_2$$

$$+ \mathbf{r}_{DC3} \times \mathbf{F}_3 + \mathbf{M}_1 + \mathbf{M}_2 + \mathbf{M}_3 = \mathbf{0}. \qquad (6.128)$$

The components F_{01x}, F_{01y}, and M are computed from (6.126) to (6.128). The MATLAB commands are

```
F01x=sym('F01x','real');
F01y=sym('F01y','real');
F01=[F01x, F01y, 0];
Mz=sym('Mz','real');
M=[0, 0, Mz];
% sum M_B for (1)
% BA x F01 + BC1 x F1 + M1 + M = 0
eqMB1=cross(-rB,F01)+cross(rC1-rB,F1)+M1+M;
% sum F for (1&2) on DB
% (F01 + F1 + F2).DB = 0
eqF12DB=dot(F01+F1+F2,rB-rD);
% sum M_D for (1&2&3)
% DA x F01 + DC1 x F1 + DB x F2 + DC x F43
% + DC3 x F3 + M3 + M2 + M1 + M = 0
eqMD123=cross(-rD,F01)+cross(rC1-rD,F1)+...
           cross(rB-rD,F2)+cross(rC-rD,F43)+...
           cross(rC3-rD,F3)+M3+M2+M1+M;
solF01=solve(eqMB1(3),eqF12DB,eqMD123(3));
F01xn=eval(solF01.F01x);
F01yn=eval(solF01.F01y);
Mzn=eval(solF01.Mz);
F01n=[F01xn, F01yn, 0];
Mn=[0, 0, Mzn];
```

Example 6.3. Figure 6.28a shows a rotating link of mass $m = 1\,\text{kg}$ and length $L = 1\,\text{m}$. The link is connected to the ground by a pin joint and is free to rotate in a vertical plane. The link is moving and makes an instant angle $\theta(t)$ with the horizontal. The local acceleration of gravity is $g = 9.8\,\text{m/s}^2$. Find and solve the equations of motion.

Solution

The plane of motion is the x, y inertial reference frame shown in Fig. 6.28a. The angle between the x and the link axis is denoted by θ. The link is moving, and the angle is changing with time at the instant of interest. In the static equilibrium position of the link, the angle, θ, is equal to $-\pi/2$. The system has one degree of freedom. The angle, θ, is an appropriate generalized coordinate describing this degree of freedom. The only motion permitted that body is rotation about a fixed horizontal axis (z-axis). The mass center of the link is at the point C. As the link is uniform, its mass center is coincident with its geometric center. The mass center, C, is at a distance $L/2$ from the pivot point O, and the position vector is

$$\mathbf{r}_{OC} = \mathbf{r}_C = x_C\mathbf{i} + y_C\mathbf{j}, \tag{6.129}$$

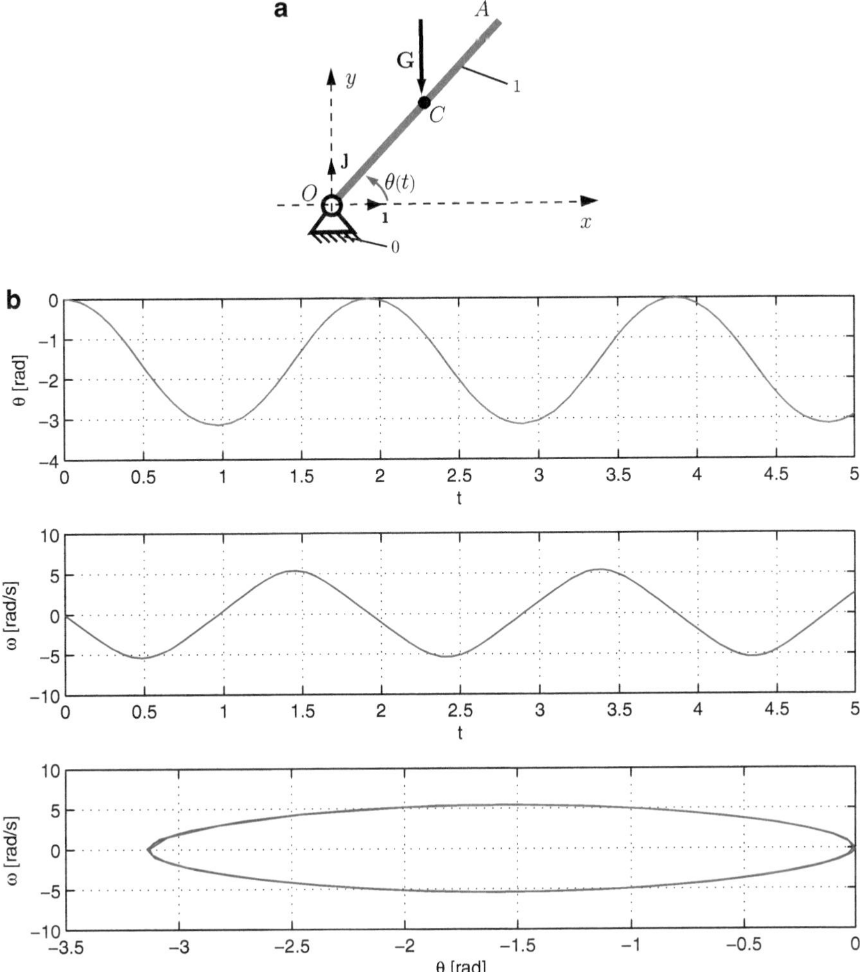

Fig. 6.28 Example 6.3

where x_C and y_C are the coordinates of C

$$x_C = \frac{L}{2}\cos\theta \quad \text{and} \quad y_C = \frac{L}{2}\sin\theta. \tag{6.130}$$

The directions of the angular velocity and angular acceleration vectors will be perpendicular to this plane, in the z-direction. The angular velocity of the link can be expressed as

$$\boldsymbol{\omega} = \omega\mathbf{k} = \frac{\mathrm{d}\theta}{\mathrm{d}t}\mathbf{k} = \dot{\theta}\mathbf{k}, \tag{6.131}$$

where ω is the rate of rotation of the link. The positive sense is clockwise (consistent with the x and y directions defined above). This problem involves only a single moving rigid body, and the angular velocity vector refers to that body. The angular acceleration of the link can be expressed as

$$\boldsymbol{\alpha} = \dot{\boldsymbol{\omega}} = \alpha\mathbf{k} = \frac{d^2\theta}{dt^2}\mathbf{k} = \ddot{\theta}\mathbf{k}, \tag{6.132}$$

where α is the angular acceleration of the link. The velocity of the mass center can be related to the velocity of the pivot point using the relationship between the velocities of two points attached to the same rigid body:

$$\mathbf{v}_C = \mathbf{v}_O + \boldsymbol{\omega} \times \mathbf{r}_{OC} = \begin{vmatrix} \mathbf{i} & \mathbf{j} & \mathbf{k} \\ 0 & 0 & \omega \\ x_C & y_C & 0 \end{vmatrix} = \omega(-y_C\mathbf{i} + x_C\mathbf{j})$$

$$= \frac{L\omega}{2}(-\sin\theta\mathbf{i} + \cos\theta\mathbf{j}) = \frac{L\dot{\theta}}{2}(-\sin\theta\mathbf{i} + \cos\theta\mathbf{j}). \tag{6.133}$$

The velocity of the pivot point, O, is zero. The acceleration of the mass center can be related to the acceleration of the pivot point ($\mathbf{a}_O = \mathbf{0}$) using the relationship between the accelerations of two points attached to the same rigid body:

$$\mathbf{a}_C = \mathbf{a}_O + \boldsymbol{\alpha} \times \mathbf{r}_{OC} + \boldsymbol{\omega} \times (\boldsymbol{\omega} \times \mathbf{r}_{OC}) = \mathbf{a}_O + \boldsymbol{\alpha} \times \mathbf{r}_{OC} - \omega^2 \mathbf{r}_{OC}$$

$$= \begin{vmatrix} \mathbf{i} & \mathbf{j} & \mathbf{k} \\ 0 & 0 & \alpha \\ x_C & y_C & 0 \end{vmatrix} - \omega^2(x_C\mathbf{i} + y_C\mathbf{j}) = \alpha(-y_C\mathbf{i} + x_C\mathbf{j}) - \omega^2(x_C\mathbf{i} + y_C\mathbf{j})$$

$$= -(\alpha y_C + \omega^2 x_C)\mathbf{i} + (\alpha x_C - \omega^2 y_C)\mathbf{j}$$

$$= -\frac{L}{2}(\alpha\sin\theta + \omega^2\cos\theta)\mathbf{i} + \frac{L}{2}(\alpha\cos\theta - \omega^2\sin\theta)\mathbf{j}$$

$$= -\frac{L}{2}(\ddot{\theta}\sin\theta + \dot{\theta}^2\cos\theta)\mathbf{i} + \frac{L}{2}(\ddot{\theta}\cos\theta - \dot{\theta}^2\sin\theta)\mathbf{j}. \tag{6.134}$$

The link is rotating about a fixed axis. The mass moment of inertia of the link about the fixed pivot point O can be evaluated from the mass moment of inertia about the mass center C using the transfer theorem. Thus,

$$I_O = I_C + m\left(\frac{L}{2}\right)^2 = \frac{mL^2}{12} + \frac{mL^2}{4} = \frac{mL^2}{3}. \tag{6.135}$$

The pin is frictionless and is capable of exerting horizontal and vertical forces on the link at O:

$$\mathbf{F}_{01} = F_{01x}\mathbf{i} + F_{01y}\mathbf{j}, \tag{6.136}$$

where F_{01x} and F_{01y} are the components of the pin force on the link in the fixed-axes system. The force driving the motion of the link is gravity. The weight of the link is acting through its mass center and will cause a moment about the pivot point. This moment will give the link a tendency to rotate about the pivot point. This moment will be given by the cross product of the vector from the pivot point, O, to the mass center, C, crossed into the weight force $\mathbf{G} = -mg\mathbf{J}$.

As the pivot point, O, of the link is fixed, the appropriate moment summation point will be about that pivot point. The sum of the moments about this point will be equal to the mass moment of inertia about the pivot point multiplied by the angular acceleration of the link. The only contributor to the moment is the weight of the link. Thus, we should be able to directly determine the angular acceleration from the moment equation. The sum of the forces acting on the link should be equal to the product of the link mass and the acceleration of its mass center. This should be useful in determining the forces exerted by the pin on the link. The link is acted upon by its weight acting vertically downward through the mass center of the link. The Newton–Euler equations of motion for the link are

$$m\mathbf{a}_C = \Sigma\mathbf{F} = \mathbf{G} + \mathbf{F}_{01}, \tag{6.137}$$

$$I_C\boldsymbol{\alpha} = \Sigma\mathbf{M}_C = \mathbf{r}_{CO} \times \mathbf{F}_{01}. \tag{6.138}$$

Since the rigid body has a fixed point at O, the equations of motion state that the moment sum about the fixed point must be equal to the product of the link mass moment of inertia about that point and the link angular acceleration. Thus,

$$I_O\boldsymbol{\alpha} = \Sigma\mathbf{M}_O = \mathbf{r}_{OC} \times \mathbf{G}. \tag{6.139}$$

The equation of motion of the link is

$$\frac{mL^2}{3}\ddot{\theta}\mathbf{k} = \begin{vmatrix} \mathbf{I} & \mathbf{J} & \mathbf{k} \\ \dfrac{L}{2}\cos\theta & \dfrac{L}{2}\sin\theta & 0 \\ 0 & -mg & 0 \end{vmatrix}, \tag{6.140}$$

or

$$\ddot{\theta} = -\frac{3g}{2L}\cos\theta. \tag{6.141}$$

The equation of motion, (6.141), is a nonlinear, second-order, differential equation relating the second time derivative of the angle, θ, to the value of that angle and various problem parameters g and L. The equation is nonlinear due to the presence of the $\cos\theta$, where $\theta(t)$ is the unknown function of interest. The force exerted by the pin on the link is obtained from (6.137):

$$\mathbf{F}_{01} = m\mathbf{a}_C - \mathbf{G},$$

and the components of the force are

$$F_{01x} = m\ddot{x}_C = -\frac{mL}{2}(\ddot{\theta}\sin\theta + \dot{\theta}^2\cos\theta),$$

$$F_{01y} = m\ddot{y}_C + mg = \frac{mL}{2}(\ddot{\theta}\cos\theta - \dot{\theta}^2\sin\theta) + mg. \qquad (6.142)$$

If the link is released from rest, then the initial value of the angular velocity is zero $\omega(t=0) = \omega(0) = \dot{\theta}(0) = 0\,\text{rad/s}$. If the initial angle is $\theta(0) = 0\,\text{rad}$, then the cosine of that initial angle is unity and the sine is zero. The initial angular acceleration can be determined from (6.141):

$$\ddot{\theta}(0) = \alpha(0) = -\frac{3g}{2L}\cos\theta(0) = -\frac{3g}{2L} = -14.7\,\text{rad/s}^2. \qquad (6.143)$$

The negative sign indicates that the initial angular acceleration of the link is counterclockwise, as one would expect. The initial reaction force components can be evaluated from (6.142):

$$F_{01x}(0) = 0\,\text{N},$$

$$F_{01y}(0) = \frac{mL}{2}\ddot{\theta}(0) + mg = \frac{mg}{4} = 2.45\,\text{N}.$$

The equation of motion, (6.141), is obtained symbolically using the MATLAB commands:

```
syms L m g t

theta = sym('theta(t)');
omega = diff([0,0,theta],t);
alpha = diff(omega,t);

xC = L*cos(theta)/2;
yC = L*sin(theta)/2;

rC = [xC yC 0];
G = [0 -m*g 0];
IC = m*L^2/12;
IO = IC + m*(L/2)^2;
MO = cross(rC,G);
eq = -IO*alpha+MO;
eqz = eq(3);
```

The MATLAB statement `diff(X,'t')` or `diff(X,sym('t'))` differentiates a symbolic expression X with respect to t, and the statement `diff(X,'t',n)` and `diff(X,n,'t')` differentiates X n times, where n is a positive integer.

An analytical solution to the differential equation is difficult to obtain. Numerical approaches have the advantage of being simple to apply even for complex mechanical systems. The MATLAB function `ode45` is used to solve the differential equation.

The differential equation $\ddot{\theta} = -\frac{3g}{2L}\cos\theta$ is of order 2. The equation has to be rewritten as a first-order system. Let $x_1 = \theta$ and $x_2 = \dot{\theta}$, this gives the first-order system

$$\dot{x}_1 = x_2,$$

$$\dot{x}_2 = -\frac{3g}{2L}\cos x_1.$$

The MATLAB commands for the right-hand side of the first-order system are

```
eqI  = subs(eqz,{L,m,g},{1,1,9.8});
eqI1 = subs(eqI,diff('theta(t)',t,2),'ddtheta');
eqI2 = subs(eqI1,diff('theta(t)',t),sym('x(2)'));
eqI3 = subs(eqI2,'theta(t)',sym('x(1)'));
% first differential equation
dx1 = sym('x(2)');
% second differential equation
dx2 = solve(eqI3,'ddtheta');

dx1dt = char(dx1);
dx2dt = char(dx2);
```

An `inline` function eom is defined for the right-hand side of the first-order system. Note that g must be specified as a column vector using `[...;...]` (not a row vector using `[...,...]`). In the definition of eom, `x(1)` was used for x_1 and `x(2)` was used for x_2. The definition of eom should have the following form:

```
eom=...
inline(sprintf('[%s; %s]',dx1dt,dx2dt),'t','x');
```

The statement has to have the form `inline(...,'t','y')`, even if t does not occur in your formula. The first component of eom is `x(2)`. The statement `sprintf` writes formatted data to string. The time t is going from an initial value t0 to a final value f:

```
t0  = 0;   % define initial time
tf  = 5;   % define final time
time = [0 tf];
```

The initial conditions at $t_0 = 0$ are $\theta(0) = 0$ rad and $\dot{\theta}(0) = 0$ rad/s or in MATLAB:

```
x0 = [0; 0];
```

The numerical solution of all the components of the solution for t going from t0 to f is obtained using the command:

```
[t,xs] = ode45(eom, time, x0);
```

where $x0$ is the initial value vector at the starting point $t0$. One can obtain a vector t and a matrix xs with the coordinates of these points using $ode45$ command. The vector of $x1$ values in the first column of xs is obtained by using $xs(:,1)$, and the vector of $x2$ values in the second column of xs is obtained by using $xs(:,2)$:

```
x1 = xs(:,1);
x2 = xs(:,2);
```

The plot of the solution curves are obtained using the commands:

```
subplot(3,1,1),plot(t,x1,'r'),...
xlabel('t'),ylabel('\theta [rad]'),grid,...
subplot(3,1,2),plot(t,x2),...
xlabel('t'),ylabel('\omega [rad/s]'),grid,...
subplot(3,1,3),plot(x1,x2),...

xlabel('\theta [rad]'),ylabel('\omega [rad/s]'),grid
```

The plots using MATLAB are shown in Fig. 6.28b. In general, the error tends to grow as one goes further from the initial conditions. To obtain numerical values at specific t values, one can specify a vector tp of t values and use $[ts,xs] = ode45(eom, tp, x0)$. The first element of the vector tp is the initial value, and the vector tp must have at least 3 elements. To obtain the solution with the initial values at $t = 0, 0.5, 1.0, 1.5, \ldots, 10$ one can use:

```
[ts,xs] = ode45(eom, 0:0.5:tf, x0);
format short
[ts, xs]
```

and the results are displayed as a table with 3 columns ts, $x1 = xs(:,1)$, $x2 = xs(:,1)$.

The differential equation can be solved numerically by m-file functions. First, create a function file, $R.m$, as shown below:

```
function dx = R(t,x);
dx = zeros(2,1);       \% a column vector
L=1; m=1; g=9.8;
dx(1) = x(2);
dx(2) =  -3*g*cos(x(1))/(2*L);
```

The ode solver provided by MATLAB ($ode45$) is used to solve the differential equation:

```
tfinal=5;
time=[0  tfinal];
x0=[0 0]; %  x(1)(0)=0; x(2)(0)=0
[t,x]=ode45(@R, time, x0);
```

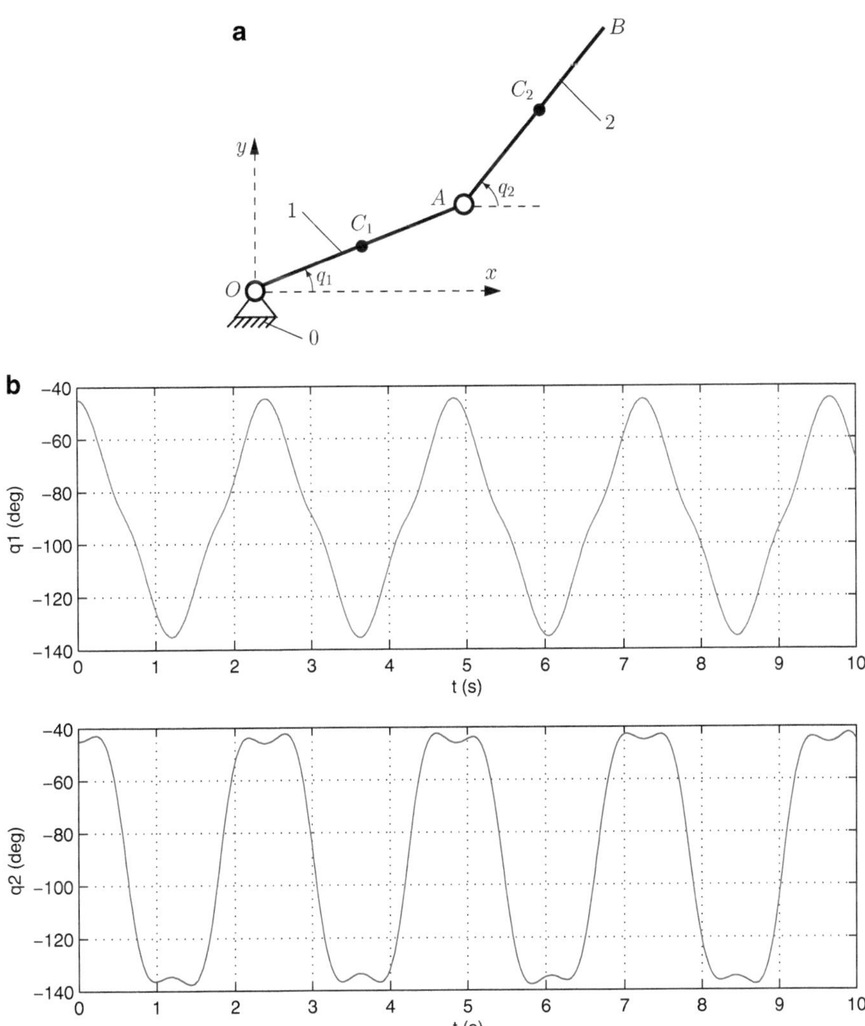

Fig. 6.29 Example 6.4

Example 6.4. A two-link planar chain (double pendulum) is considered, Fig. 6.29a. The links 1 and 2 have the masses $m_1 = m_2 = 1$ kg and the lengths $AB = L_1 = BD = L_2 = 1$ m. The system is free to move in a vertical plane. The local acceleration of gravity is $g = 9.8$ m/s^2. Find and solve the equations of motion.

Solution

The plane of motion is the x, y as shown in Fig. 6.29a. The mass centers of the links are designated by $C_1(x_{C_1}, y_{C_1}, 0)$ and $C_2(x_{C_2}, y_{C_2}, 0)$. The number of degrees of freedom are computed using the relation

$$M = 3n - 2c_5 - c_4,$$

where n is the number of moving links, c_5 is the number of one degree of freedom joints, and c_4 is the number of two degrees of freedom joints. For the double pendulum, $n = 2$, $c_5 = 2$, and $c_4 = 0$, and the system has two degrees of freedom, $M = 2$, and two generalized coordinates. The angles $q_1(t)$ and $q_2(t)$ are selected as the generalized coordinates as shown in Fig. 6.29a.

The position vector of the center of the mass C_1 of the link 1 is

$$\mathbf{r}_{C_1} = x_{C_1}\mathbf{I} + y_{C_1}\mathbf{J},$$

where x_{C_1} and y_{C_1} are the coordinates of C_1

$$x_{C_1} = \frac{L_1}{2}\cos q_1 \quad \text{and} \quad y_{C_1} = \frac{L_1}{2}\sin q_1.$$

The position vector of the center of the mass C_2 of the link 2 is

$$\mathbf{r}_{C_2} = x_{C_2}\mathbf{I} + y_{C_2}\mathbf{J},$$

where x_{C_2} and y_{C_2} are the coordinates of C_2

$$x_{C_2} = L_1\cos q_1 + \frac{L_2}{2}\cos q_2 \quad \text{and} \quad y_{C_2} = L_1\sin q_1 + \frac{L_2}{2}\sin q_2.$$

The velocity vector of C_1 is the derivative with respect to time of the position vector of C_1:

$$\mathbf{v}_{C_1} = \dot{\mathbf{r}}_{C_1} = \dot{x}_{C_1}\mathbf{I} + \dot{y}_{C_1}\mathbf{J},$$

where

$$\dot{x}_{C_1} = -\frac{L_1}{2}\dot{q}_1\sin q_1 \quad \text{and} \quad \dot{y}_{C_1} = \frac{L_1}{2}\dot{q}_1\cos q_1.$$

The velocity vector of C_2 is the derivative with respect to time of the position vector of C_2:

$$\mathbf{v}_{C_2} = \dot{\mathbf{r}}_{C_2} = \dot{x}_{C_2}\mathbf{I} + \dot{y}_{C_2}\mathbf{J},$$

where

$$\dot{x}_{C_2} = -L_1\dot{q}_1\sin q_1 - \frac{L_2}{2}\dot{q}_2\sin q_2 \quad \text{and} \quad \dot{y}_{C_2} = L_1\dot{q}_1\cos q_1 + \frac{L_2}{2}\dot{q}_2\cos q_2.$$

The acceleration vector of C_1 is the double derivative with respect to time of the position vector of C_1:

$$\mathbf{a}_{C_1} = \ddot{\mathbf{r}}_{C_1} = \ddot{x}_{C_1}\mathbf{I} + \ddot{y}_{C_1}\mathbf{J},$$

where

$$\ddot{x}_{C_1} = -\frac{L_1}{2}\ddot{q}_1\sin q_1 - \frac{L_1}{2}\dot{q}_1^2\cos q_1 \quad \text{and} \quad \ddot{y}_{C_1} = \frac{L_1}{2}\ddot{q}_1\cos q_1 - \frac{L_1}{2}\dot{q}_1^2\sin q_1.$$

The acceleration vector of C_2 is the double derivative with respect to time of the position vector of C_2:

$$\mathbf{a}_{C_2} = \ddot{\mathbf{r}}_{C_2} = \ddot{x}_{C_2}\mathbf{I} + \ddot{y}_{C_2}\mathbf{J},$$

where

$$\ddot{x}_{C_2} = -L_1\ddot{q}_1\sin q_1 - L_1\dot{q}_1^2\cos q_1 - \frac{L_2}{2}\ddot{q}_2\sin q_2 - \frac{L_2}{2}\dot{q}_2^2\cos q_2,$$

$$\ddot{y}_{C_2} = L_1\ddot{q}_1\cos q_1 - L_1\dot{q}_1^2\sin q_1 + \frac{L_2}{2}\ddot{q}_2\cos q_2 - \frac{L_2}{2}\dot{q}_2^2\sin q_2.$$

The MATLAB commands for the linear accelerations of the mass centers C_1 and C_2 are

```
L1 = 1; L2 = 1; % m
m1 = 1; m2 = 1; % kg
g = 9.8;        % m/s^2

t = sym('t','real');
q1 = sym('q1(t)');
q2 = sym('q2(t)');

xA = L1*cos(q1);
yA = L1*sin(q1);
rA = [xA yA 0];
rC1 = rA/2;
vC1 = diff(rC1,t);
aC1 = diff(vC1,t);

xB = xA + L2*cos(q2);
yB = yA + L2*sin(q2);
rB = [xB yB 0];
rC2 = (rA + rB)/2;
vC2 = diff(rC2,t);
aC2 = diff(vC2,t);
```

The angular velocity vectors of the links 1 and 2 are

```
omega1 = [0 0 diff(q1,t)];
alpha1 = diff(omega1,t);
omega2 = [0 0 diff(q2,t)];
alpha2 = diff(omega2,t);
```

The weight forces on the links 1 and 2 are

```
G1 = [0 -m1*g 0];
G2 = [0 -m2*g 0];
```

The mass moment of inertia of the link 1 with respect to the center of mass C_1 is

$$I_{C_1} = \frac{m_1 L_1^2}{12}.$$

The mass moment of inertia of the link 1 with respect to the fixed point of rotation O is

$$I_O = I_{C_1} + m_1 \left(\frac{L_1}{2}\right)^2 = \frac{m_1 L_1^2}{3}.$$

The mass moment of inertia of the link 2 with respect to the center of mass C_2 is

$$I_{C_2} = \frac{m_2 L_2^2}{12}.$$

The MATLAB commands for the mass moments of inertia are

```
IC1 = m1*L1^2/12;
IO  = IC1 + m1*(L1/2)^2;
IC2 = m2*L2^2/12;
```

The equations of motion of the pendulum are inferred using the Newton–Euler method. There are two rigid bodies in the system, and the Newton–Euler equations are written for each link. The Newton–Euler equations for the link 1 are

$$m_1 \mathbf{a}_{C_1} = \mathbf{F}_{01} + \mathbf{F}_{21} + \mathbf{G}_1,$$

$$I_{C_1} \boldsymbol{\alpha}_1 = \mathbf{r}_{C_1 O} \times \mathbf{F}_{01} + \mathbf{r}_{C_1 A} \times \mathbf{F}_{21},$$

where $\mathbf{F}_{01}$ is the joint reaction of the ground 0 on the link 1 at point O and $\mathbf{F}_{21}$ is the joint reaction of the link 2 on the link 1 at point A

$$\mathbf{F}_{01} = F_{01x}\mathbf{i} + F_{01y}\mathbf{j} \quad \text{and} \quad \mathbf{F}_{21} = F_{21x}\mathbf{i} + F_{21y}\mathbf{j}.$$

Since the link 1 has a fixed point of rotation at O, the moment sum about the fixed point must be equal to the product of the link mass moment of inertia about that point and the link angular acceleration. Thus,

$$I_O \boldsymbol{\alpha}_1 = \mathbf{r}_{OC_1} \times \mathbf{G}_1 + \mathbf{r}_{OA} \times \mathbf{F}_{21}, \tag{6.144}$$

or

$$\frac{m_1 L_1^2}{3} \ddot{q}_1 \mathbf{k} = \begin{vmatrix} \mathbf{i} & \mathbf{j} & \mathbf{k} \\ x_{C_1} & y_{C_1} & 0 \\ 0 & -m_1 g & 0 \end{vmatrix} + \begin{vmatrix} \mathbf{i} & \mathbf{j} & \mathbf{k} \\ x_A & y_A & 0 \\ F_{21x} & F_{21y} & 0 \end{vmatrix},$$

or

$$\frac{m_1 L_1^2}{3} \ddot{q}_1 \mathbf{k} = (-m_1 g x_{C_1} + F_{21y} x_A - F_{21x} y_A)\mathbf{k}.$$

The equation of motion for link 1 is

$$\frac{m_1 L_1^2}{3}\ddot{q}_1 = \left(-m_1 g\frac{L_1}{2}\cos q_1 + F_{21y}L_1\cos q_1 - F_{21x}L_1\sin q_1\right). \quad (6.145)$$

The Newton–Euler equations for the link 2 are

$$m_2\mathbf{a}_{C_2} = \mathbf{F}_{12} + \mathbf{G}_2, \quad (6.146)$$

$$I_{C_2}\boldsymbol{\alpha}_2 = \mathbf{r}_{C_2 A} \times \mathbf{F}_{12}, \quad (6.147)$$

where $\mathbf{F}_{12} = -\mathbf{F}_{21}$ is the joint reaction of the link 1 on the link 2 at B. Equation (6.147) becomes

$$m_2\ddot{x}_{C_2} = -F_{21x},$$

$$m_2\ddot{y}_{C_2} = -F_{21y} - m_2 g,$$

$$\frac{mL_2^2}{12}\ddot{q}_2\mathbf{k} = \begin{vmatrix} \mathbf{i} & \mathbf{j} & \mathbf{k} \\ x_A - x_{C_2} & y_A - y_{C_2} & 0 \\ -F_{21x} & -F_{21y} & 0 \end{vmatrix}, \quad (6.148)$$

or

$$m_2\left(-L_1\ddot{q}_1\sin q_1 - L_1\dot{q}_1^2\cos q_1 - \frac{L_2}{2}\ddot{q}_2\sin q_2 - \frac{L_2}{2}\dot{q}_2^2\cos q_2\right) - F_{21x}, \quad (6.149)$$

$$m_2\left(L_1\ddot{q}_1\cos q_1 - L_1\dot{q}_1^2\sin q_1 + \frac{L_2}{2}\ddot{q}_2\cos q_2 - \frac{L_2}{2}\dot{q}_2^2\sin q_2\right) = -F_{21y} - m_2 g, \quad (6.150)$$

$$\frac{m_2 L_2^2}{12}\ddot{q}_2 = \frac{L_2}{2}\left(-F_{21y}\cos q_2 + F_{21x}\sin q_2\right). \quad (6.151)$$

The reaction components F_{21x} and F_{21y} are obtained from (6.149) and (6.150):

$$F_{21x} = m_2\left(L_1\ddot{q}_1\sin q_1 + L_1\dot{q}_1^2\cos q_1 + \frac{L_2}{2}\ddot{q}_2\sin q_2 + \frac{L_2}{2}\dot{q}_2^2\cos q_2\right),$$

$$F_{21y} = -m_2\left(L_1\ddot{q}_1\cos q_1 - L_1\dot{q}_1^2\sin q_1 + \frac{L_2}{2}\ddot{q}_2\cos q_2 - \frac{L_2}{2}\dot{q}_2^2\sin q_2\right) + m_2 g.$$

$$(6.152)$$

The equations of motion are obtained substituting F_{21x} and F_{21y} in (6.145) and (6.151):

$$\frac{m_2 L_1^2}{3}\ddot{q}_1 = -m_1 g\frac{L_1}{2}\cos q_1$$

$$-m_2\left(L_1\ddot{q}_1\cos q_1 - L_1\dot{q}_1^2\sin q_1 + \frac{L_2}{2}\ddot{q}_2\cos q_2 - \frac{L_2}{2}\dot{q}_2^2\sin q_2 - g\right)L_1\cos q_1$$

$$-m_2\left(L_1\ddot{q}_1\sin q_1 + L_1\dot{q}_1^2\cos q_1 + \frac{L_2}{2}\ddot{q}_2\sin q_2 + \frac{L_2}{2}\dot{q}_2^2\cos q_2\right)L_1\sin q_1, \quad (6.153)$$

$$\frac{m_2 L_2^2}{12}\ddot{q}_2$$

$$= \frac{m_2 L_2}{2}\left(L_1\ddot{q}_1\cos q_1 - L_1\dot{q}_1^2\sin q_1 + \frac{L_2}{2}\ddot{q}_2\cos q_2 - \frac{L_2}{2}\dot{q}_2^2\sin q_2 - g\right)\cos q_2$$

$$+ \frac{m_2 L_2}{2}\left(L_1\ddot{q}_1\sin q_1 + L_1\dot{q}_1^2\cos q_1 + \frac{L_2}{2}\ddot{q}_2\sin q_2 + \frac{L_2}{2}\dot{q}_2^2\cos q_2\right)\sin q_2. \quad (6.154)$$

The equations of motion represent two nonlinear differential equations. The initial conditions (Cauchy problem) are necessary to solve the equations. At $t = 0$, the initial conditions are

$$q_1(0) = q_{10}, \dot{q}_1(0) = \omega_{10},$$

$$q_2(0) = q_{20}, \dot{q}_2(0) = \omega_{20}.$$

The equations of motion for the mechanical system will be solved using MATLAB. First, the reaction joint force $\mathbf{F}_{21}$ is calculated from (6.146):

```
F21  =  -m2*aC2  +  G2;
```

The moment equations for each link, (6.144) and (6.147), using MATLAB are

```
EqO =...
-IO*alpha1+cross(rA, F21)+cross(rC1,G1);
Eq2 =...
-IC2*alpha2+cross(rA-rC2,-F21);
```

Two lists `slist` and `nlist` are created:

```
slist={diff('q1(t)',t,2),diff('q2(t)',t,2),...
    diff('q1(t)',t),diff('q2(t)',t),'q1(t)','q2(t)'};
nlist={'ddq1', 'ddq2', 'x(2)', 'x(4)', 'x(1)','x(3)';
% diff('q1(t)',t,2) will be replaced by 'ddq1'
% diff('q2(t)',t,2) will be replaced by 'ddq2'
% diff('q1(t)',t) will be replaced by 'x(2)'
```

```
% diff('q2(t)',t) will be replaced by 'x(4)'
% 'q1(t)' will be replaced by 'x(1)'
% 'q2(t)' will be replaced by 'x(3)'
```

In the equations of motion `Eq0` and `Eq2`, the symbolical variables in `slist` are replaced with the symbolical variables in `nlist`:

```
eq1 = subs(Eq0(3),slist,nlist);
eq2 = subs(Eq2(3),slist,nlist);
```

The previous equations are solved in terms of `'ddq1'` and `'ddq2'`:

```
sol = solve(eq1,eq2,'ddq1, ddq2');
```

The second-order ODE system of two equations has to be rewritten as a first-order system.

Let $x(1) = q_1(t)$, $x(2) = \dot{q}_1(t)$, $x(3) = q_2(t)$, and $x(4) = \dot{q}_2(t)$, this gives the first-order system:

```
d[x(1)]/dt = x(2), d[x(2)]/dt = ddq1,
d[x(3)]/dt = x(4), d[x(4)]/dt = ddq2.
```

The MATLAB commands for the first-order ODE system are

```
dx1 = sym('x(2)');
dx2 = sol.ddq1;
dx3 = sym('x(4)');
dx4 = sol.ddq2;
dx1dt = char(dx1);
dx2dt = char(dx2);
dx3dt = char(dx3);
dx4dt = char(dx4);
```

The `inline` function g is defined for the right-hand side of the first-order system:

```
g = inline(sprintf('[%s;%s;%s;%s]',...
          dx1dt,dx2dt,dx3dt,dx4dt),'t','x');
```

The time `t` is going from an initial value `t0` to a final value `tf`:

```
t0 = 0; tf = 10; time = [0 tf];
```

The initial conditions at $t_0 = 0$ are $q_1(0) = -\pi/4\,\text{rad}$, $\dot{q}_1(0) = 0\,\text{rad/s}$, $q_2(0) = -\pi/4\,\text{rad}$, $\dot{q}_2(0) = 0\,\text{rad/s}$, or in MATLAB

```
x0 = [-pi/4; 0; -pi/4; 0]; % define initial conditions
```

The numerical solution of all the components of the solution for t going from t0 to f is obtained using the command:

```
[t,xs] = ode45(g, time, x0);
```

where x0 is the initial value vector at the starting point t0. The plot of the solution curves q_1 and q_2 are obtained using the commands:

```
x1 = xs(:,1);
x3 = xs(:,3);
subplot(2,1,1),plot(t,x1*180/pi,'r'),...
xlabel('t (s)'),ylabel('q1 (deg)'),grid,...
subplot(2,1,2),plot(t,x3*180/pi,'b'),...
xlabel('t (s)'),ylabel('q2 (deg)'),grid
```

The plots using MATLAB are shown in Fig. 6.29b.

Instead of using the inline function g, the system of differential equations can be solved numerically by m-file functions. The function file, RR.m, is created using the statements:

```
eq1 = subs(Eq0(3),slist,nlist);
eq2 = subs(Eq2(3),slist,nlist);
sol = solve(eq1,eq2,'ddq1, ddq2');
dx2 = sol.ddq1;
dx4 = sol.ddq2;
dx2dt = char(dx2);
dx4dt = char(dx4);

% opens the file 'RR.m' in the mode specified by 'w+'
% (create for read and write)
fid = fopen('RR.m','w+');
fprintf(fid,'function dx = RR(t,x)\n');
fprintf(fid,'dx = zeros(4,1);\n');
fprintf(fid,'dx(1) = x(2);\n');
fprintf(fid,'dx(2) = ');
fprintf(fid,dx2dt);
fprintf(fid,';\n');
fprintf(fid,'dx(3) = x(4);\n');
fprintf(fid,'dx(4) = ');
fprintf(fid,dx4dt);
fprintf(fid,';');
% closes the file associated with file identifier fid
fclose(fid);
cd(pwd);
```

```
% cd changes current working directory
% pwd displays the current working directory

% fid = fopen(FIL,PERM) opens the file FILE in the
% mode specified by PERM. PERM can be:
% 'w'      writes (creates if necessary)
% 'w+'     truncates or creates for read and write
```

The terms dx2dt and dx4dt are calculated symbolically from the previous program. The MATLAB command fid = fopen(file,perm) opens the file file in the mode specified by perm. The mode 'w+' deletes the contents of an existing file, or creates a new file, and opens it for reading and writing. The statement fclose(fid) closes the file associated with file identifier fid, the statement cd changes the current working directory, and pwd displays the current working directory. The ode45 solver is used for the system of differential equations:

```
t0 = 0;  tf = 10; time = [0 tf];

x0 = [-pi/4 0 -pi/4 0]; % initial conditions

[t,xs] = ode45(@RR, time, x0);
```

The computing time for solving the system of differential equations is shorter using the function file RR.m.

Example 6.5. Outside the long slender pendulum 1, of mass m_1, a translational joint 2, of mass m_2, is sliding without friction as shown in Fig. 6.30a. The length of the pendulum is L. The mass moment of inertia of the slider 2 with respect to its mass center point A is I_A. The acceleration due to gravity is g. Find and solve the equations of motion for $m_1 = m_2 = m = 1$ kg, $L = 1$ m, $g = 9.81$ m/s^2, and $I_A = 1$ kg m^2.

Solution
The mass centers of the links are designated by $C_1 = C(x_C, y_C, 0)$ for link 1 and $C_2 = A(x_A, y_A, 0)$ for slider 2. The kinematic chain has two degrees of freedom and two generalized coordinates. The angle $\theta(t)$ and the distance $OA = r(t)$ are selected as the generalized coordinates as shown in Fig. 6.30b.

The position vector of the center of the mass C of the link 1 is

$$\mathbf{r}_C = x_C \mathbf{\imath} + y_C \mathbf{\jmath},$$

where

$$x_C = \frac{L}{2} \cos \theta(t) \quad \text{and} \quad y_C = \frac{L}{2} \sin \theta(t).$$

The position vector of the center of the mass A of the link 2 is

$$\mathbf{r}_A = x_A \mathbf{\imath} + y_A \mathbf{\jmath},$$

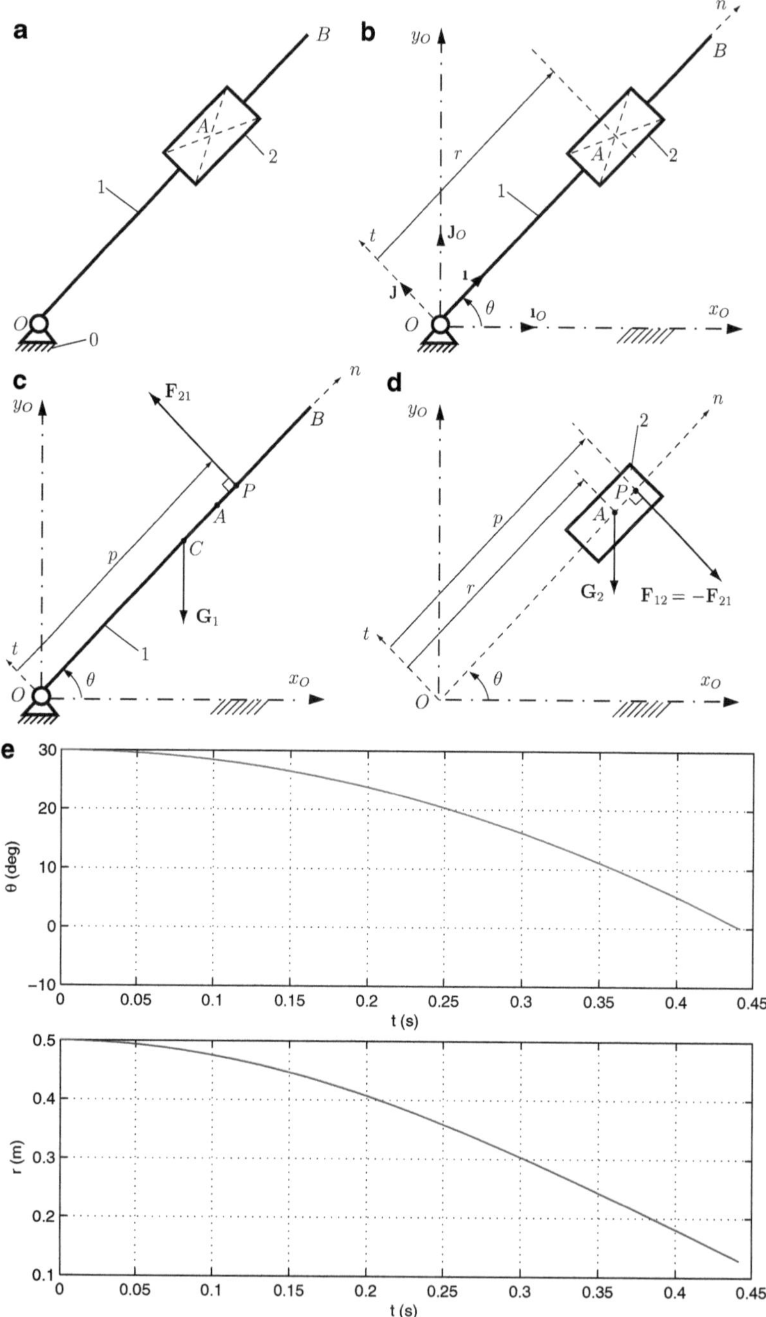

Fig. 6.30 Example 6.5

where

$$x_A = r(t)\cos\theta(t) \quad \text{and} \quad y_A - r(t)\sin\theta(t).$$

The velocity vector of A is the derivative with respect to time of the position vector of A:

$$\mathbf{v}_A = \dot{\mathbf{r}}_A = \dot{x}_A\mathbf{I} + \dot{y}_A\mathbf{J}.$$

The acceleration vector of A is the double derivative with respect to time of the position vector of A:

$$\mathbf{a}_A = \ddot{\mathbf{r}}_A = \ddot{x}_A\mathbf{I} + \ddot{y}_A\mathbf{J}.$$

The MATLAB commands for the kinematics of the mass centers C and A are

```
syms t

xC = L*cos(sym('theta(t)'))/2;
yC = L*sin(sym('theta(t)'))/2;
rC = [xC yC 0];

xA = sym('r(t)')*cos(sym('theta(t)'));
yA = sym('r(t)')*sin(sym('theta(t)'));
rA = [xA yA 0];

vA = diff(rA, t);
aA = diff(vA, t);
```

The angular velocity and acceleration of the links 1 and 2 are

```
omega = [0 0 diff('theta(t)', t)];
alpha = diff(omega, t);
```

The joint reaction force, Fig. 6.30c, d, of slider 2 on link 1 is

```
syms f21
F21x = -f21*sin(sym('theta(t)'));
F21y =  f21*cos(sym('theta(t)'));
F21  = [F21x F21y 0];
```

The position vector of the application point of joint reaction force F21 is

```
syms p
xP = p*cos(sym('theta(t)'));
yP = p*sin(sym('theta(t)'));
rP = [xP yP 0];
```

The gravitational forces for the links are

```
G1 = [0 -m*g 0];
G2 = [0 -m*g 0];
```

For link 1, a moment equation of motion with respect to the fixed point O can be written:

```
IO = m*L^2/3;
% - IO alpha + rC x G1 + rP x F21 = 0 (1)
e1 = -IO*alpha + cross(rC, G1) + cross(rP, F21);
e1z = e1(3);
```

For slider 2 the Newton–Euler equations are

```
% m2 aA  = - F21 + G2 => "
e2 = -m*aA - F21 + G2;
% (x):   m2 aAx + F21x = 0 (2)
e2x = e2(1);
% (y):   m2 aAy + F21y - G2 = 0 (3)
e2y = e2(2);


% -IA alpha + (rP-rA) x (-F21) = 0 (4)
e2A = -IA*alpha+cross(rP-rA,-F21);
e2Az = e2A(3);
```

Eliminating the joint force f21 and its position p, two equations of motion are obtained:

```
sol = solve(e2y, e2Az, f21, p);
f21s = sol.f21;
ps = sol.p;

EqI  = subs(e1z,{'f21','p'},{f21s,ps});

EqII = subs(e2x,{'f21','p'},{f21s,ps});

% list for the symbolical variables
slist={diff('theta(t)',t,2),diff('r(t)',t,2),...
diff('theta(t)',t),diff('r',t),'theta(t)','r(t)'};
nlist = ...
{'ddq1', 'ddq2', 'x(2)', 'x(4)', 'x(1)','x(3)'};
% diff('theta',t,2) will be replaced by 'ddq1'
% diff('r(t)',t,2) will be replaced by 'ddq2'
% diff('theta',t) will be replaced by 'x(2)'
% diff('r(t)',t) will be replaced by 'x(4)'
% 'theta' will be replaced by 'x(1)'
% 'r(t)' will be replaced by 'x(3)'

eq1 = subs(EqI,slist,nlist);
eq2 = subs(EqII,slist,nlist);
```

```
sole = solve(eq1,eq2,'ddq1, ddq2');
ddq1s = sole.ddq1;
ddq2s = sole.ddq2;

dx2dt = char(ddq1s);
dx4dt = char(ddq2s);

fid = fopen('eomE6_5a.m','w+');
fprintf(fid,'function dx = eomE6_5a(t,x)\n');
fprintf(fid,'dx = zeros(4,1);\n');
fprintf(fid,'dx(1) = x(2);\n');
fprintf(fid,'dx(2) = ');
fprintf(fid,dx2dt);
fprintf(fid,';\n');
fprintf(fid,'dx(3) = x(4);\n');
fprintf(fid,'dx(4) = ');
fprintf(fid,dx4dt);
fprintf(fid,';');
fclose(fid);
cd(pwd);
```

The initial conditions are $\theta(0) = \pi/6$, $\dot{\theta}(0) = 0$, $r(0) = 0.5$, and $\dot{r}(0) = 0$ or in MATLAB

```
x0 = [pi/6 0 0.5 0]; % initial conditions
```

To stop the simulation when $\theta(t) = 0$ an event, eventE6_5.m, is created:

```
% eventE6_5.m

function [value,isterminal,direction] = eventE6_5(t,x)

value = x(1);

isterminal = 1;

direction = 0;

% if isterminal vector is set to 1, integration will
% halt when a zero-crossing is detected.

% the elements of the direction vector are -1,
%   1, or 0,
% specifying that the corresponding event must be
% decreasing, increasing, or that any crossing
% is to be detected.
```

The equations of motion are solved taking into consideration the event:

```
option = odeset ...
('RelTol',1e-3,'MaxStep',1e-3,'Events',@eventE6_5);
t0 = 0;   tf = .5; time = [0 tf];
[t, xs, te, ye] = ode45(@eomE6_5a, time, x0, option);
```

The results are plotted in Fig. 6.30e.

Example 6.6. Figure 6.31 shows a rotating rectangular plate of mass $m = 1$ kg with the dimensions $b = 1$ m and $h = 2$ m. The plate is rotating about the vertical axis and makes an instant angle $\theta(t)$. The local acceleration of gravity is $g = 9.8$ m/s^2. The following dimensions are given: $d = e = 1$ m. Find the joint reactions at O and P and the equations of motion of the plate if the initial conditions are $\theta(0) = 0$ and $\dot{\theta}(0) = \omega(0) = \omega_0 = 2$ rad/s. Find the location where a mass $m_E = 1$ kg is added for dynamic balancing of the plate.

Solution
The plane of motion is the x, y inertial reference frame shown in Fig. 6.28a. The angle of rotation of the plate is denoted by θ. The angle, θ, is an appropriate generalized coordinate describing this degree of freedom. The mass center of the plate is at the point C. As the plate is uniform, its mass center is coincident with its geometric center. The position vectors of O, C, and P are

```
syms b h d e
rO = [0,0,0];
rC = [0,b/2,d+h/2];
rP = [0,0,d+h+e];
```

The angular velocity, `omega`, and the angular acceleration, `alpha`, of the plate are

```
syms t
q = sym('q(t)');
theta = [0,0,q];
omega = diff(theta,t);
alpha = diff(omega,t);
```

The unknown reaction forces at O and P are

```
syms FOx FOy FOz FPx FPy FPz
% reaction force at O
FO = [FOx,FOy,FOz];
% reaction force at P
FP = [FPx,FPy,0];
```

The gravitational force at C is

```
syms m g
G = [0,0,-m*g];
```

Fig. 6.31 Example 6.6

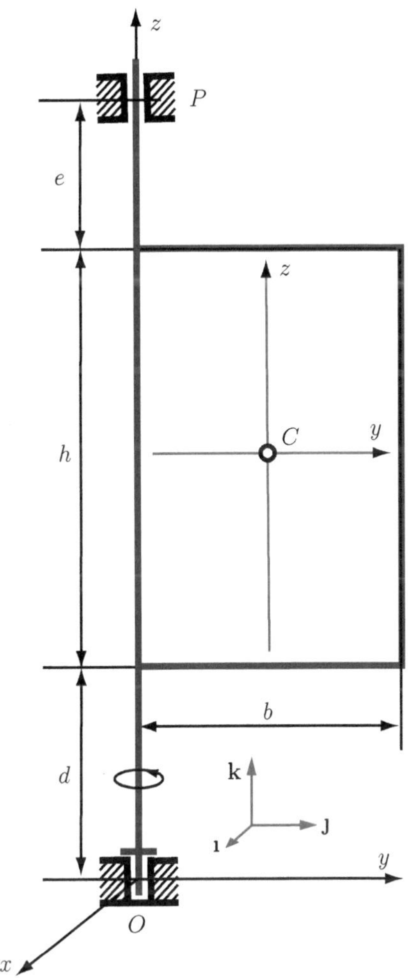

The components of the inertia matrix with are

```
ICyy=m*h^2/12;
ICzz=m*b^2/12;
ICxx=ICyy+ICzz;
ICyz=0;
ICxy=0;
ICxz=0;

Iyy=ICyy+m*(h/2+d)^2;
Iyy=simplify(expand(Iyy));
Izz=ICzz++m*(b/2)^2;
Izz=simplify(expand(Izz));
```

```
Ixx=ICxx+m*((d+h/2)^2+(b/2)^2);
Ixx=simplify(expand(Ixx));
Iyz=ICyz+m*(d+h/2)*(b/2);
Iyz=simplify(expand(Iyz));
Ixy=0;
Ixz=0;

syms omega0
list= {b, h, d, e, m, g  , omega0};
listn={1, 2, 1, 1, 1, 9.8, 2};

Ixxn=subs(Ixx, list, listn);
Iyyn=subs(Iyy, list, listn);
Izzn=subs(Izz, list, listn);
Iyzn=subs(Iyz, list, listn);

fprintf('Ixx=%s=%g (kg m^2) \n',char(Ixx),Ixxn)
fprintf('Iyy=%s=%g (kg m^2) \n',char(Iyy),Iyyn)
fprintf('Izz=%s=%g (kg m^2) \n',char(Izz),Izzn)
fprintf('Iyz=%s=%g (kg m^2) \n',char(Iyz),Iyzn)

I=[
    Ixx, -Ixy, -Ixz;
   -Ixy,  Iyy, -Iyz;
   -Ixz, -Iyz,  Izz
  ];
pretty(I)
In=subs(I, list, listn)
```

The results are

```
Ixx=(m*(b^2 + 3*d^2 + 3*d*h + h^2))/3=4.66667 (kg m^2)
Iyy=(m*(3*d^2 + 3*d*h + h^2))/3=4.33333 (kg m^2)
Izz=(b^2*m)/3=0.333333 (kg m^2)
Iyz=(b*m*(2*d + h))/4=1 (kg m^2)

In =

     4.6667         0         0
          0    4.3333   -1.0000
          0   -1.0000    0.3333
```

The Newton–Euler equations of motion are calculated with

```
aO=[0,0,0];
aC=aO+cross(alpha,rC)+cross(omega,cross(omega,rC));
```

```
% Newton eom
% m aC - G+FO+FP
eqF = m*aC-(G+FO+FP);
% Euler eom
eqM = -alpha*I-cross(omega,omega*I)-...
    (cross(rC,G)+cross(rP,FP));


% =>


eqFx = eqF(1) % (1)
eqFy = eqF(2) % (2)
eqFz = eqF(3) % (3)



eqMx = eqM(1) % (4)
eqMy = eqM(2) % (5)
eqMz = eqM(3) % (6)
```

The equations of motion are

$$
-\frac{b\, m\, \mathrm{diff}(q(t),\, t,\, t)}{2} - FOx - FPx = 0
$$

$$
-\frac{b\, m\, \mathrm{diff}(q(t),\, t)^2}{2} - FOy - FPy = 0
$$

$$
g\, m - FOz = 0
$$

$$
-\frac{b\, m\, (2\, d + h)\, \mathrm{diff}(q(t),\, t)^2}{4} + FPy\, (d + e + h) + \frac{b\, g\, m}{2} = 0
$$

$$
\frac{b\, m\, (2\, d + h)\, \mathrm{diff}(q(t),\, t,\, t)}{4} - FPx\, (d + e + h) = 0
$$

```
       2
    b  m diff(q(t), t, t)
  - ---------------------- = 0
             3
```

From the last equation of the system, it results that $\texttt{diff(q(t), t, t)}=0$; thus, $\omega = \omega_0 = 2$ rad/s. The joint reactions forces at O are

```
FOx = 0
FOy =
                           2                      2              2
    b m (4 d omega0  - 2 g + 2 e omega0  + 3 h omega0 )
  - -----------------------------------------------------------
                        4 (d + e + h)
FOy = -1.775 (N)

FOz = g m
FOz = 9.8 (N)
```

and the joint reactions forces at P are

```
FPx = 0

FPy =


                     2                   2
    b m (2 d omega0  - 2 g + h omega0 )
    -----------------------------------
                 4 (d + e + h)
FPy = -0.225 (N)
```

The position vector of the balancing mass $\texttt{mE}$ is

```
syms xE yE zE
rE = [xE,yE,zE];
```

For dynamic balancing, it is necessary that the center of mass of the system (plate of mass m and the mass mE) to be on the axis of rotation, z-axis:

```
eqEx = m*rC(1) + mE*xE;
eqEy = m*rC(2) + mE*yE;
```

The other necessary condition is that the z-axis has to be a principal axis for the system:

```
eqEm = mE*yE*zE+Iyz;
```

The system of three equations is solved:

```
solE = solve(eqEx,eqEy,eqEm,'xE,yE,zE');
```

and the following results are obtained:

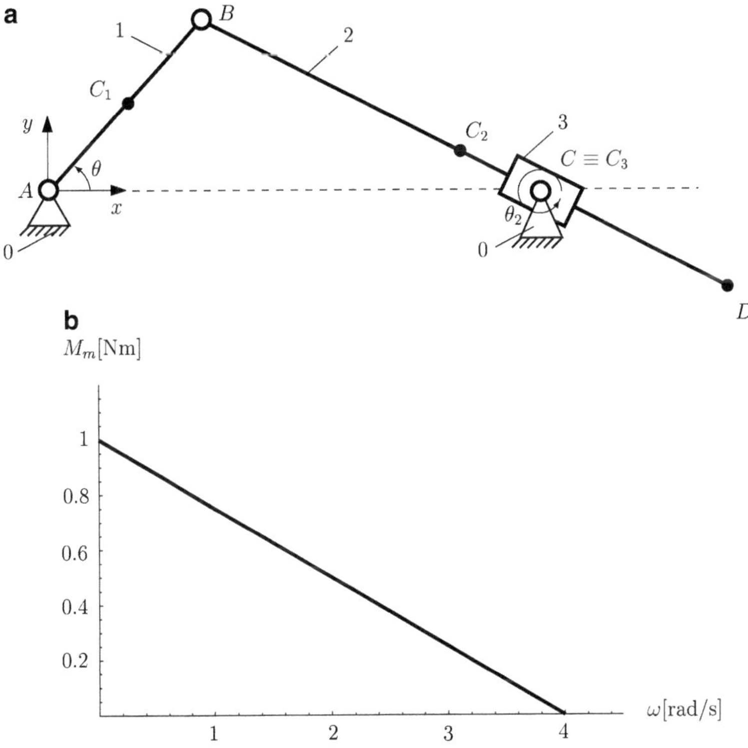

Fig. 6.32 Example 6.7 (**a**) mechanism R-RTR and (**b**) motor moment

```
xE=0=0   (m)
yE=-(b*m)/(2*mE)=-0.5   (m)
zE=d + h/2=2   (m)
```

Example 6.7. Figure 6.32 shows a mechanism with the lengths of the links $L_1 = 0.001$ m, $L_2 = 0.470$ m, $L_3 = 0.050$ m, and $AC = 0.300$ m. The links 1 and 2 are rectangular prisms with the depth $d = 0.005$ m and height $h = 0.010$ m. The link 3 has the height $h_3 = 0.050$ m and the depth $d_3 = 0.020$ m. The mass density of the links is $\rho = 7{,}850$ kg/m^3. The center of mass locations of the links $i = 1, 2, 3$ are designated by $C_i(x_{C_i}, y_{C_i}, 0)$. The initial conditions $\theta(0) = \pi/6$ rad and $\omega(0) = \dot{\theta}(0) = 0$ rad/s are given. Find and solve the equation of motion of the mechanism.

Solution

The number of degrees of freedom for the mechanism can be computed using the relation

$$M = 3n - 2c_5 - c_4,$$

where n is the number of moving links, c_5 is the number of pin joints or slider joints with one degree of freedom, and c_4 is the number of pin joints or slider joints with two degrees of freedom. For the considered mechanism $n = 3$, $c_5 = 4$: $A(R)$, $B(R)$, $C(T)$, $C(R)$, $c_4 = 0$, and the mechanism has one degree of freedom ($M = 1$). Thus, there is one generalized coordinate for the system. One can choose the angle $\theta(t) = \angle BAC$ as the generalized coordinate.

The position vector of the center of the mass C_1 of the link 1 is

$$\mathbf{r}_{C_1} = x_{C_1}\mathbf{1} + y_{C_1}\mathbf{J},$$

where x_{C_1} and y_{C_1} are the coordinates of C_1

$$x_{C_1} = \frac{L_1}{2}\cos\theta, \quad y_{C_1} = \frac{L_1}{2}\sin\theta.$$

The position vector of B is

$$\mathbf{r}_B = L_1\cos\theta\,\mathbf{1} + L_1\sin\theta\,\mathbf{J}.$$

The position vector of C is

$$\mathbf{r}_C = x_C\mathbf{1} + y_C\mathbf{J} = (AC)\mathbf{1}.$$

The position vector of the center of the mass C_2 of the link 2 is

$$\mathbf{r}_{C_2} = x_{C_2}\mathbf{1} + y_{C_2}\mathbf{J},$$

where x_{C_2} and y_{C_2} are the coordinates of C_2

$$x_{C_2} = L_1\cos\theta + \frac{L_2}{2}\cos\theta_2, \quad y_{C_2} = L_1\sin\theta + \frac{L_2}{2}\sin\theta_2,$$

where $\theta_2 = \arctan\frac{y_B - y_C}{x_B - x_C}$. The position vector of the center of the mass C_3 of the link 3 is

$$\mathbf{r}_{C_3} = \mathbf{r}_C = (AC)\mathbf{1}.$$

The velocity vector of C_1 is the derivative with respect to time of the position vector of C_1:

$$\mathbf{v}_{C_1} = \dot{\mathbf{r}}_{C_1} = \dot{x}_{C_1}\mathbf{1} + \dot{y}_{C_1}\mathbf{J},$$

where

$$\dot{x}_{C_1} = -\frac{L_1}{2}\dot\theta\sin\theta, \quad \dot{y}_{C_1} = \frac{L_1}{2}\dot\theta\cos\theta.$$

The velocity vector of C_2 is the derivative with respect to time of the position vector of C_2:

$$\mathbf{v}_{C_2} = \dot{\mathbf{r}}_{C_2} = \dot{x}_{C_2}\mathbf{1} + \dot{y}_{C_2}\mathbf{J},$$

where

$$\dot{x}_{C_2} = -L_1\dot{\theta}\sin\theta - \frac{L_2}{2}\dot{\theta}_2\sin\theta_2,$$

$$\dot{y}_{C_2} = L_1\dot{\theta}\cos\theta + \frac{L_2}{2}\dot{\theta}_2\cos\theta_2.$$

The velocity vector of C_3 is zero:

$$\mathbf{v}_{C_3} = \mathbf{0}.$$

The acceleration vector of C_1 is the double derivative with respect to time of the position vector of C_1:

$$\mathbf{a}_{C_1} = \ddot{\mathbf{r}}_{C_1} = \ddot{x}_{C_1}\mathbf{I} + \ddot{y}_{C_1}\mathbf{J},$$

where

$$\ddot{x}_{C_1} = -\frac{L_1}{2}\ddot{\theta}\sin\theta - \frac{L_1}{2}\dot{\theta}^2\cos\theta,$$

$$\ddot{y}_{C_1} = \frac{L_1}{2}\ddot{\theta}\cos\theta - \frac{L_1}{2}\dot{\theta}^2\sin\theta.$$

The acceleration vector of C_2 is the double derivative with respect to time of the position vector of C_2:

$$\mathbf{a}_{C_2} = \ddot{\mathbf{r}}_{C_2} = \ddot{x}_{C_2}\mathbf{I} + \ddot{y}_{C_2}\mathbf{J},$$

where

$$\ddot{x}_{C_2} = -L_1\ddot{\theta}\sin\theta - L_1\dot{\theta}^2\cos\theta - \frac{L_2}{2}\ddot{\theta}_2\sin\theta_2 - \frac{L_2}{2}\dot{\theta}_2^2\cos\theta_2,$$

$$\ddot{y}_{C_2} = L_1\ddot{\theta}\cos\theta - L_1\dot{\theta}^2\sin\theta + \frac{L_2}{2}\ddot{\theta}_2\cos\theta_2 - \frac{L_2}{2}\dot{\theta}_2^2\sin\theta_2.$$

The acceleration vector of C_3 is zero:

$$\mathbf{a}_{C_3} = \mathbf{0}.$$

The angular velocity vectors of the links 1, 2, and 3 are

$$\boldsymbol{\omega} = \dot{\theta}\mathbf{k},$$

$$\boldsymbol{\omega}_2 = \boldsymbol{\omega}_3 = \dot{\theta}_2\mathbf{k}.$$

The angular acceleration vectors of the links 1, 2, and 3 are

$$\boldsymbol{\alpha} = \ddot{\theta}\mathbf{k},$$

$$\boldsymbol{\alpha}_2 = \boldsymbol{\alpha}_3 = \ddot{\theta}_2\mathbf{k}.$$

The MATLAB program for the kinematics is

```
L1 = 0.100; % m
L2 = 0.470; % m
```

```
L3 = 0.050; % m
AC = 0.300; % m
h =  0.010; % m
d =  0.005; % m
h3 = 0.050; % m
d3 = 0.020; % m

syms t
theta = sym('theta(t)');
omega = diff(theta,t);
alpha = diff(omega,t);

xB = L1*cos(theta);
yB = L1*sin(theta);
rB = [xB yB 0];
vB = diff(rB,t);
aB = diff(vB,t);

xC = AC;
yC = 0;
rC = [xC yC 0];

theta2 = atan((yB-yC)/(xB-xC));
omega2 = diff(theta2,t);
alpha2 = diff(omega2,t);

omega3 = omega2;
alpha3 = alpha2;

alphav  = [0 0 alpha ];
alpha2v = [0 0 alpha2];
alpha3v = [0 0 alpha3];

xD = xB+L2*cos(theta2);
yD = yB+L2*sin(theta2);

rD = [xD yD 0];
vD = diff(rD,t);
aD = diff(vD,t);

rC1 = rB/2;
rC2 = (rB+rD)/2;
rC3 = rC;
vC1 = diff(rC1,t);
aC1 = diff(vC1,t);
```

```
vC2  =  diff(rC2,t);
aC2  =  diff(vC2,t);
aC3  =  diff(rC3,t,2);
```

Force Analysis
The masses of the links 1, 2, and 3 are

$$m_1 = \rho L_1 h d,$$

$$m_2 = \rho L_2 h d,$$

$$m_3 = m_{3e} - m_{3i},$$

where $m_{3e} = \rho L_3 h_3 d_3$, $m_{3i} = \rho L_3 h d$. The mass moment of inertia of the link 1 with respect to the center of mass C_1 is

$$I_{C_1} = \frac{m_1}{12}\left(L_1^2 + h^2\right).$$

The mass moment of inertia of the link 2 with respect to the center of mass C_2 is

$$I_{C_2} = \frac{m_2}{12}\left(L_2^2 + h^2\right).$$

The mass moment of inertia of the link 3 with respect to the center of mass C_3 is

$$I_{C_3} = \frac{m_{3e}}{12}\left(L_3^2 + h_3^2\right) - \frac{m_{3i}}{12}\left(L_3^2 + h^2\right).$$

The gravitational forces of the link 1, 2, and 3 are

$$\mathbf{G}_1 = -m_1 g\mathbf{J}, \quad \mathbf{G}_2 = -m_2 g\mathbf{J}, \quad \mathbf{G}_3 = -m_3 g\mathbf{J}.$$

A motor torque acts on the link 1:

$$\mathbf{M}_1 = M\mathbf{k}. \tag{6.155}$$

For a D.C. electric motor, $M = M_0(1 - \frac{\omega}{\omega_0})$, where M_0 and ω_0 are given in catalogues. In our case, $M_0 = 1\,\text{Nm}$, and $\omega_0 = 4\,\text{rad/s}$, Fig. 6.32.

There are three rigid bodies in the system, and one can write the Newton–Euler equations for each link.

The Newton–Euler equations for link 3 are

$$m_3\,\mathbf{a}_{C_3} = \mathbf{F}_{23} + \mathbf{F}_{03} + \mathbf{G}_3,$$

$$I_{C_3}\,\boldsymbol{\alpha}_3 = \mathbf{r}_{CP} \times \mathbf{F}_{23},$$

where

$$\mathbf{F}_{23} = F_{23x}\mathbf{1} + F_{23y}\mathbf{J},$$

$$\mathbf{F}_{03} = F_{03x}\mathbf{1} + F_{03y}\mathbf{J}.$$

The force $\mathbf{F}_{03}$ is the joint reaction of ground 0 on link 32 at the point C, and $\mathbf{F}_{32}$ is the joint reaction of link 2 on link 3 at the point $P(x_P, y_P)$. The application point $P(x_P, y_P)$ of the reaction force $\mathbf{F}_{32}$ is not known, but it is located on the sliding direction:

$$\tan \theta_2 = \frac{y_P}{x_P - AC}.$$

The reaction force $\mathbf{F}_{23}$ is perpendicular to the sliding direction BC:

$$\mathbf{F}_{23} \cdot \mathbf{r}_{BC} = 0.$$

The reaction force $\mathbf{F}_{32}$ at P is equivalent to the reaction force $\mathbf{F}_{32}$ at point C and a reaction moment $\mathbf{M}_{23} = M_{23z}\,\mathbf{k}$, as shown in Fig. 6.33. The equations of motion for link 3 will be now

$$m_3\,\mathbf{a}_{C_3} = \mathbf{F}_{23} + \mathbf{F}_{03} + \mathbf{G}_3,$$

$$I_{C_3}\,\boldsymbol{\alpha}_3 = \mathbf{M}_{23}, \tag{6.156}$$

where $\mathbf{F}_{23} = F_{23}\left(\sin \theta_2\,\mathbf{\imath} + \cos \theta_2\,\mathbf{\imath}\right)$.

The Newton–Euler equations for the link 2 are

$$m_2\,\mathbf{a}_{C_2} = \mathbf{F}_{12} - \mathbf{F}_{23} + \mathbf{G}_2,$$

$$I_{C_2}\boldsymbol{\alpha}_2 = \mathbf{r}_{C_2 B} \times \mathbf{F}_{12} - \mathbf{r}_{C_2 C} \times \mathbf{F}_{23} - \mathbf{M}_{23}, \tag{6.157}$$

where $\mathbf{F}_{12} = F_{12x}\,\mathbf{\imath} + F_{12y}\,\mathbf{J}$ is the joint reaction of the link 1 on the link 2 at the point B.

The Newton–Euler equations for the link 1, Fig. 6.33, is

$$I_A\boldsymbol{\alpha} = -\mathbf{r}_B \times \mathbf{F}_{12} + \mathbf{M}_1, \tag{6.158}$$

where $\mathbf{M}_1$ is the motor torque on link 1.

From (6.156) and (6.157) the reactions F_{03x}, F_{03y}, F_{23}, M_{23z}, F_{12x}, F_{12x} are calculated as functions of the generalized coordinate $\theta(t)$ and its derivatives $\dot{\theta}(t)$, and $\ddot{\theta}(t)$. Substituting the results into (6.158), the equation of motion for the system is obtained. The MATLAB commands for the equation of motion are

```
g = 9.807;  % m/s^2
rho = 7850; % kg/m^3

m1 = rho*h*d*L1;
m2 = rho*h*d*L2;

m3e = rho*L3*h3*d3;
m3i = rho*L3*h*d;
m3 = m3e - m3i;
```

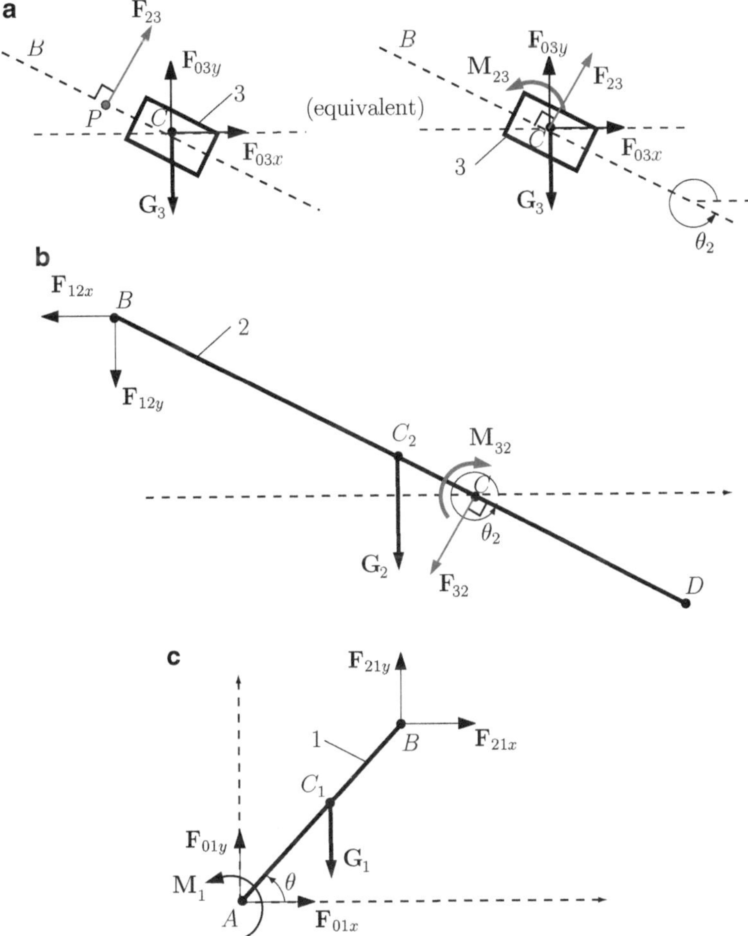

Fig. 6.33 Example 6.7 free-body diagrams: (**a**) link 3, (**b**) link 2, and (**c**) link 1

```
IC1 = m1*(L1^2+h^2)/12;
IC2 = m2*(L2^2+h^2)/12;
IC3 = m3e*(L3^2+h3^2)/12-m3i*(L3^2+h^2)/12;
IA  = IC1+m1*(L1/2)^2;

G1 = [0 -m1*g 0];
G2 = [0 -m2*g 0];
G3 = [0 -m3*g 0];
```

```
syms F23m M23z F03x F03y
F03 = [F03x F03y 0];
F23x = F23m*sin(theta2);
F23y = F23m*cos(theta2);
F23 = [F23x F23y 0];
M23 = [0 0 M23z];
% link 3
eqn3F = -m3*aC3+F03+G3+F23;
eqn3M = -IC3*alpha3v+M23;
eqn3x = eqn3F(1);
eqn3y = eqn3F(2);
eqn3z = eqn3M(3);

syms F12x F12y
F12 = [F12x F12y 0];
% link 2
eqn2F = -m2*aC2+F12-F23+G2;
eqn2M = -IC2*alpha2v+cross(rB-rC2,F12)+...
        cross(rC-rC2,-F23)-M23;
eqn2x = eqn2F(1);
eqn2y = eqn2F(2);
eqn2z = eqn2M(3);

sol32 = solve...
(eqn2x,eqn2y,eqn2z,eqn3x,eqn3y,eqn3z,...
 'F12x','F12y','F23m','M23z','F03x','F03y');

F12xs = sol32.F12x;
F12ys = sol32.F12y;
F21 = [-F12xs -F12ys 0];

M0 = 1;
w0 = 4;
M = M0*(1-omega/w0);
Mm = [0 0 M];

% link 1
eqn1M = -IA*alphav+cross(rB,F21)+cross(rC1,G1)+Mm;

subout = {diff(theta,t,2), diff(theta,t), theta};
subin = {'ddtheta','x(2)','x(1)'};

eqn1 = subs(eqn1M(3),subout,subin);
NE = solve(eqn1,'ddtheta');
NEeom = char(NE);
```

The MATLAB function `ode45` is used to solve the differential equation for the initial conditions:

```
% ODE
fid = fopen('eomE6_7.m','w+');
fprintf(fid,'function out = eomE6_7(t,x)\n\n');
fprintf(fid,'out = zeros(2,1);\n');
fprintf(fid,'out(1) = x(2);\n');
fprintf(fid,'out(2) = %s;\n',NEeom);
fclose(fid);
cd(pwd);

IC = [pi/6 0];
tf = 5;
time = [0 tf];

[t data] = ode45(@eomE6_7,time,IC);

thetan = data(:,1);
omegan = data(:,2);

subplot(2,1,1)
plot(t,thetan)
xlabel('t [s]')
ylabel('theta [rad]')
grid
subplot(2,1,2)
plot(t,omegan);
xlabel('t [s]')
ylabel('omega [rad/s]')
grid
```

The output of the program is shown in Fig. 6.34.

Example 6.8. A particles P of mass m_2 is free to move in a smooth tube 1, as shown in Fig. 6.35. The tube has the mass m_1 and the length $2L$. The angle between the axis of the tube and the vertical axis x_1-axis is a given constant angle α. The tube is rotating about the x_1-vertical axis. An external torque $\mathbf{T}_{01} = T_{01x}\mathbf{\imath}_0$ is acting on the tube, where the unit vector $\mathbf{\imath}_0$ is oriented vertically upward. The gravitational acceleration is g. Find the equations of motion for the system (tube and particle). The numerical values are $L = 2\,\mathrm{m}$, $\alpha = 45°$, $m_1 = 10\,\mathrm{kg}$, $m_2 = 0.001\,\mathrm{kg}$, $T_{01x} = 300\sin(3t)\,\mathrm{Nm}$, and $g = 9.81\,\mathrm{m/s^2}$.

Solution
The fixed reference frame $[\mathbf{\imath}_0, \mathbf{J}_0, \mathbf{k}_0]$ is attached to the ground. The rotating reference frame $[\mathbf{\imath}_1, \mathbf{J}_1, \mathbf{k}_1]$ attached to the tube 1 has the unit vector $\mathbf{\imath}_1 = \mathbf{\imath}_0$ oriented vertically upward, and the unit vector $\mathbf{J}_1$ is perpendicular to $\mathbf{\imath}_1$ and parallel to the plane

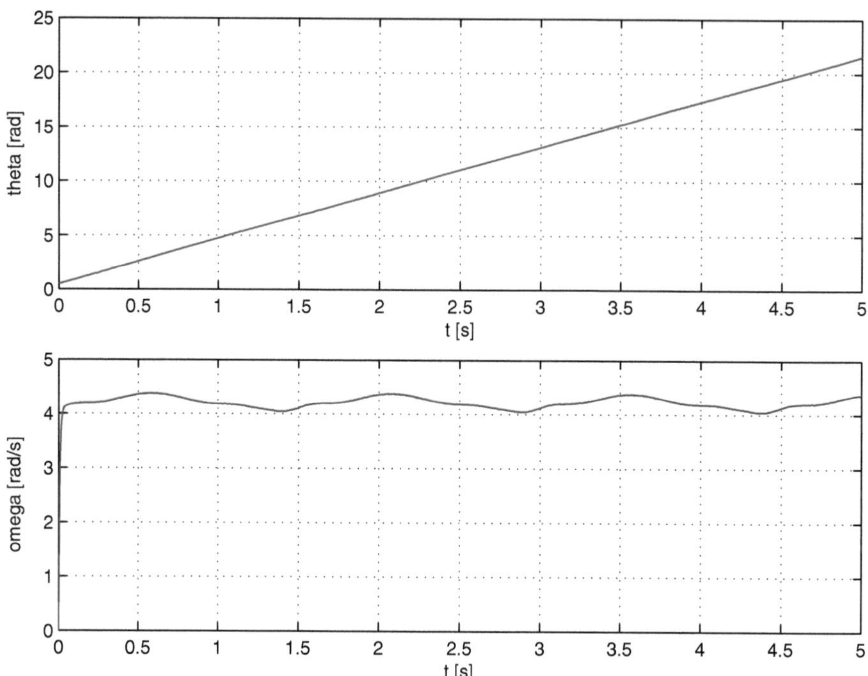

Fig. 6.34 Example 6.7 MATLAB output figures

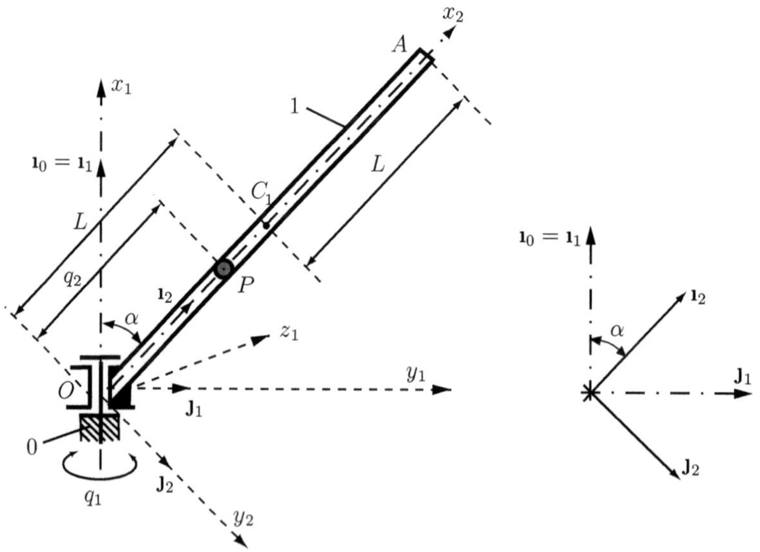

Fig. 6.35 Example 6.8 spatial mechanical system

determined by the vertical axis x_1 and the axis of the tube x_2. The vector $\mathbf{k}_1$ is given by $\mathbf{k}_1 = \mathbf{i}_1 \times \mathbf{J}_1$. The rotating reference frame $[\mathbf{i}_2, \mathbf{J}_2, \mathbf{k}_2]$ is also attached to the tube 1 and has the unit vector $\mathbf{i}_2$ directed along the axis of the tube and the unit vector $\mathbf{J}_2$ perpendicular to $\mathbf{i}_2$ and parallel to the plane determined by the axis of the tube and the vertical axis x_1. The third unit vector is defined by $\mathbf{k}_2 = \mathbf{i}_2 \times \mathbf{J}_2$.

The system has two degrees of freedom, and the two generalized coordinates are $q_1(t)$, which is the angular displacement of the tube 1, and $q_2(t)$, which is the distance from the origin O to the particle P.

The unit vectors $\mathbf{i}_2$, $\mathbf{J}_2$, and $\mathbf{k}_2$ can be expressed as functions of $\mathbf{i}_1$, $\mathbf{J}_1$, and $\mathbf{k}_1$:

$$\mathbf{i}_2 = c_\alpha \mathbf{i}_1 + s_\alpha \mathbf{J}_1,$$

$$\mathbf{J}_2 = -s_\alpha \mathbf{i}_1 + c_\alpha \mathbf{J}_1,$$

$$\mathbf{k}_2 = \mathbf{k}_1,$$

where $s_\alpha = \sin q_\alpha$ and $c_\alpha = \cos \alpha$. The transformation matrix from $\mathbf{i}_2$, $\mathbf{J}_2$, and $\mathbf{k}_2$ to $\mathbf{i}_1$, $\mathbf{J}_1$, and $\mathbf{k}_1$ is

$$R_{21} = \begin{bmatrix} c_\alpha & s_\alpha & 0 \\ -s_\alpha & c_\alpha & 0 \\ 0 & 0 & 1 \end{bmatrix}.$$

The angular velocity of the tube 1 is expressed in the fixed reference frame (0) as

$$\boldsymbol{\omega}_{10} = \dot{q}_1 \mathbf{i}_0 = \dot{q}_1 \mathbf{i}_1 = \dot{q}_1 c_\alpha \mathbf{i}_2 - \dot{q}_1 s_\alpha \mathbf{J}_2.$$

The angular acceleration of link 1 in the reference frame (0) is

$$\boldsymbol{\alpha}_{10} = \frac{^{(1)}\mathrm{d}}{\mathrm{d}t} \boldsymbol{\omega}_{10} + \boldsymbol{\omega}_{10} \times \boldsymbol{\omega}_{10} = \ddot{q}_1 \mathbf{i}_1 = \ddot{q}_1 c_\alpha \mathbf{i}_2 - \ddot{q}_1 s_\alpha \mathbf{J}_2,$$

where $\frac{^{(1)}\mathrm{d}}{\mathrm{d}t}$ represents the partial derivative with respect to time in reference frame (1), $[\mathbf{i}_1, \mathbf{J}_1, \mathbf{k}_1]$. The MATLAB commands for the angular velocity and acceleration are

```
syms t alpha L  m1 m2 g

q1 = sym('q1(t)');
q2 = sym('q2(t)');

s1 = sin(q1);
c1 = cos(q1);

sa = sin(alpha);
ca = cos(alpha);
```

```
% transformation matrix from RF2 to RF1
R21 = [ca sa 0; -sa ca 0; 0 0 1];

% angular velocity of link 1 in RF0 expressed
% in terms of RF1 {i1,j1,k1}
w1_1 = [diff(q1,t); 0; 0];
fprintf('w1_1 = \n');pretty(w1_1);fprintf('\n')

% angular velocity of link 1 in RF0 expressed
% in terms of RF2 {i2,j2,k2}
w1_2   = R21*w1_1;
fprintf('w1_2 = \n');pretty(w1_2);fprintf('\n')

% angular acceleration of link 1 in RF0 expressed
% in terms of RF1{i1,j1,k1}
alpha1_1 = diff(w1_1,t);
fprintf('alpha1_1 = \n');pretty(alpha1_1)
fprintf('\n')

% angular acceleration of link 1 in RF0 expressed
% in terms of RF2{i2,j2,k2}
alpha1_2 = diff(w1_2,t);
fprintf('alpha1_2 = \n');pretty(alpha1_2)
fprintf('\n')
```

and the output is

```
w1_1 =

    +-                -+
    |  diff(q1(t), t)  |
    |                  |
    |        0         |
    |                  |
    |        0         |
    +-                -+

w1_2 =

    +-                          -+
    |    cos(alpha) diff(q1(t), t)  |
    |                            |
    |   -sin(alpha) diff(q1(t), t)  |
    |                            |
    |              0             |
    +-                          -+
```

```
alpha1_1 =
```

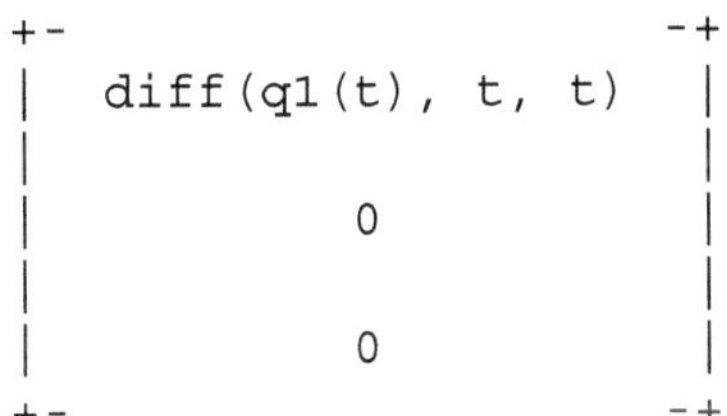

```
 +-                        -+
 |   diff(q1(t),  t,  t)   |
 |                         |
 |              0          |
 |                         |
 |              0          |
 +-                        -+
```

```
alpha1_2 =
```

```
 +-                                      -+
 |    cos(alpha) diff(q1(t),  t,  t)      |
 |                                        |
 |    -sin(alpha) diff(q1(t),  t,  t)     |
 |                                        |
 |                  0                     |
 +-                                      -+
```

The position vector of C_1, the mass center of link 1, is

$$\mathbf{r}_{C_1} = L\mathbf{1}_2,$$

and the velocity of C_1 in (0) is

$$\mathbf{v}_{C_1} = \frac{d}{dt}\mathbf{r}_{C_1} = \dot{\mathbf{r}}_{C_1} = \frac{^{(2)}d}{dt}\mathbf{r}_{C_1} + \boldsymbol{\omega}_{10} \times \mathbf{r}_{C_1},$$

where $\frac{^{(2)}d}{dt}$ represents the partial derivative with respect to time in reference frame (2), $[\mathbf{1}_2, \mathbf{J}_2, \mathbf{k}_2]$. The acceleration of C_1 in (0) is

$$\mathbf{a}_{C_1} = \frac{d}{dt}\mathbf{v}_{C_1} = \dot{\mathbf{v}}_{C_1} = \frac{^{(2)}d}{dt}\mathbf{v}_{C_1} + \boldsymbol{\omega}_{10} \times \mathbf{v}_{C_1}.$$

The MATLAB commands for the velocity and acceleration of the mass center of the tube are

```
% position vector of mass center C1 of link 1
% in RF0 expressed in terms of RF2 {i2,j2,k2}
rC1_2 = [L; 0; 0];

% position vector of mass center C1 of link 1
% in RF0 expressed in terms of RF1 {i1,j1,k1}
rC1_1 = R21.'*rC1_2;
fprintf('rC1_1 = \n');pretty(rC1_1)
fprintf('\n')
```

```
% linear velocity of mass center C1 of link 1
% in RF0 expressed in terms of RF1 {i1,j1,k1}
vC1_1 = diff(rC1_1,t) + cross(w1_1, rC1_1);
fprintf('vC1_1 = \n');pretty(vC1_1)
fprintf('\n')

% linear velocity of mass center C1 of link 1
% in RF0 expressed in terms of RF2 {i2,j2,k2}
vC1_2 = diff(rC1_2,t) + cross(w1_2, rC1_2);
% vC1_2 = simple(R21.'*vC1_1)
fprintf('vC1_2 = \n');pretty(vC1_2);fprintf('\n')

% linear acceleration of mass center C1 of link 1
% in RF0 expressed in terms of RF1 {i1,j1,k1}
aC1_1 = simple(diff(vC1_1,t)+cross(w1_1,vC1_1));
fprintf('aC1_1 = \n');pretty(aC1_1)
fprintf('\n')

% linear acceleration of mass center C1 of link 1
% in RF0 expressed in terms of RF2 {i2,j2,k2}
aC1_2 = simple(diff(vC1_2,t)+cross(w1_2,vC1_2));
% aC1_2 = simple(R21.'*aC1_1)
fprintf('aC1_2 = \n');pretty(aC1_2)
fprintf('\n')
```

and the MATLAB results are

```
rC1_1 =
  +-            -+
  |  L cos(alpha) |
  |               |
  |  L sin(alpha) |
  |               |
  |        0      |
  +-            -+


vC1_1 =
  +-                            -+
  |               0              |
  |                              |
  |               0              |
  |                              |
  |  L sin(alpha) diff(q1(t), t) |
  +-                            -+
```

```
vC1_2 =

+-                                          -+
|                     0                      |
|                                            |
|                     0                      |
|                                            |
|   L sin(alpha) diff(q1(t), t)              |
+-                                          -+
```

```
aC1_1 =

+-                                              -+
|                      0                         |
|                                                |
|                                             2  |
|    - L sin(alpha) diff(q1(t), t)               |
|                                                |
|    L sin(alpha) diff(q1(t), t, t)              |
+-                                              -+
```

```
aC1_2 =

+-                                                  -+
|                      2                      2      |
|    - L sin(alpha)  diff(q1(t), t)                  |
|                                                    |
|                                             2      |
|    L sin(2 alpha) diff(q1(t), t)                   |
|  - ------------------------------------            |
|                      2                             |
|                                                    |
|    L sin(alpha) diff(q1(t), t, t)                  |
+-                                                  -+
```

The position vector of the particle P is

$$\mathbf{r}_P = q_2(t)\,\mathbf{1}_2,$$

and the velocity of P in (0) is

$$\mathbf{v}_P = \frac{\mathrm{d}}{\mathrm{d}t}\mathbf{r}_P = \dot{\mathbf{r}}_P = \frac{^{(2)}\mathrm{d}}{\mathrm{d}t}\mathbf{r}_P + \boldsymbol{\omega}_{10} \times \mathbf{r}_P.$$

The acceleration of P in (0) is

$$\mathbf{a}_P = \frac{\mathrm{d}}{\mathrm{d}t}\mathbf{v}_P = \dot{\mathbf{v}}_P = \frac{^{(2)}\mathrm{d}}{\mathrm{d}t}\mathbf{v}_P + \boldsymbol{\omega}_{10} \times \mathbf{v}_P.$$

The MATLAB commands for the velocity and acceleration of the particle P are

```
% position vector of P
% in RF0 expressed in terms of RF2 {i2,j2,k2}
rP_2 = [q2; 0;0];
fprintf('rP_2 = \n');pretty(rP_2)
fprintf('\n')

% position vector of mass center C1 of link 1
% in RF0 expressed in terms of RF1 {i1,j1,k1}
rP_1 = R21.'*rP_2;
fprintf('rP_1 = \n');pretty(rP_1)
fprintf('\n')

% linear velocity of P
% in RF0 expressed in terms of RF2 {i2,j2,k2}
vP_2 = diff(rP_2,t) + cross(w1_2, rP_2);
fprintf('vP_2 = \n'); pretty(vP_2)
fprintf('\n')

% linear acceleration of P
% in RF0 expressed in terms of RF2 {i2,j2,k2}
aP_2 = diff(vP_2,t) + cross(w1_2, vP_2);
aP_2 = simple(aP_2);
fprintf('aPx_2 = \n'); pretty(aP_2(1))
fprintf('\n')
fprintf('aPy_2 = \n'); pretty(aP_2(2))
fprintf('\n')
fprintf('aPz_2 = \n'); pretty(aP_2(3))
fprintf('\n')
```

and the MATLAB results are

```
rP_2 =


   +-         -+
   |  q2(t)    |
   |           |
   |     0     |
   |           |
   |     0     |
   +-         -+
```

```
rP_1 =

    +-                      -+
    |   cos(alpha) q2(t)     |
    |                        |
    |   sin(alpha) q2(t)     |
    |                        |
    |           0            |
    +-                      -+

vP_2 =

    +-                                  -+
    |               diff(q2(t), t)       |
    |                                    |
    |                    0               |
    |                                    |
    |   sin(alpha) q2(t) diff(q1(t), t)  |
    +-                                  -+

aPx_2 =

                                 2                              2
diff(q2(t), t, t) - sin(alpha)  q2(t) diff(q1(t), t)

aPy_2 =

                                      2
      sin(2 alpha) q2(t) diff(q1(t), t)
   -  ------------------------------------
                     2

aPz_2 =

sin(alpha) (q2(t) diff(q1(t), t, t) +

      2 diff(q2(t), t) diff(q1(t), t))
```

The gravity forces on the tube at C_1 and on the particle P are

$$\mathbf{G}_1 = -m_1 g \mathbf{l}_1,$$

$$\mathbf{G}_2 = -m_2 g \mathbf{l}_1 = -m_2 g \dot{q}_1 c_\alpha \mathbf{l}_2 + m_2 g \dot{q}_1 s_\alpha \mathbf{J}_2,$$

and with MATLAB

```
% gravitational force that acts on link 1 at C1
% RF0 expressed in terms of RF1 {i1,j1,k1}
G1 = [-m1*g; 0; 0]

% gravitational force that acts P
% in RF0 expressed in terms of RF2 {i2,j2,k2}
G2 = R21*[-m2*g; 0; 0];
fprintf('G2 = \n'); pretty(G2)
fprintf('\n')
```

The inertia matrix associated with the principal axes of the slender tube 1 is

$$\bar{I}_2 \rightarrow \begin{bmatrix} I_{x_2} & 0 & 0 \\ 0 & I_{y_2} & 0 \\ 0 & 0 & I_{z_2} \end{bmatrix},$$

where $I_{x_2} = 0$ and $I_{y_2} = I_{z_2} = m_1(2L)^2/12$. The MATLAB expression for the inertia matrix of the tube associated with reference frame (2) $[\mathbf{i}_2, \mathbf{j}_2, \mathbf{k}_2]$ is

```
IC1_2 = [0 0 0; 0 m1*(2*L)^2/12 0; 0 0 m1*(2*L)^2/12];
```

The unknown reaction force of tube 1 that acts on particle at P expressed in terms of reference frame (2) is

```
syms F12y F12z
F12 = [0; F12y; F12z];
```

and the reaction force of the particle P that acts on tube 1 expressed in terms of reference frame (1) is

```
F21 = -R21.'*F12;
```

The unknown reaction moment of the ground 0 that acts on link 1 expressed in terms of reference frame (1) is

```
syms  M01y M01z
M01 = [0; M01y; M01z];
```

The external torque on tube 1 expressed in terms of reference frame (1) is

```
syms  T01x
T01 = [T01x; 0; 0];
```

The Newton equations of motion for the tube are

```
eqF1 = F01+F21-m1*aC1_1;
eqF1x = eqF1(1);
eqF1y = eqF1(2);
eqF1z = eqF1(3);
```

and the MATLAB results are

```
F01+G1-F12-m1 aC1 = 0 =>


F01x + F12y sin(alpha) = 0


                                      2
L m1 sin(alpha) diff(q1(t), t)   + F01y -

   F12y cos(alpha) = 0



-L m1 sin(alpha) diff(q1(t), t, t) + F01z - F12z = 0
```

The Newton equations of motion for the particle are

```
eqF2 = G2+F12-m2*aP_2;
eqF2x = eqF2(1);
eqF2y = eqF2(2);
eqF2z = eqF2(3);
```

and the MATLAB results are

```
G2+F12-m2*aP = 0 =>



                                    2
 - m2 (diff(q2(t), t, t) - sin(alpha)   q2(t)

               2
   diff(q1(t), t) ) - g m2 cos(alpha) = 0



                                  2
m2 sin(2 alpha) q2(t) diff(q1(t), t)
----------------------------------------- + F12y +
                2

   g m2 sin(alpha) = 0
```

```
F12z - m2 sin(alpha) (q2(t) diff(q1(t), t, t) +

    2 diff(q2(t), t) diff(q1(t), t)) = 0
```

The Euler (rotational) equations of motion for the tube with respect to mass center C_1 are

```
% sum of moments about C1 for link 1
M_C1 = cross(-rC1_1,F01) + cross(rP_1-rC1_1,F21) + ...
       M01 + T01;

Min_C1 = -IC1_2*alpha1_2-cross(w1_2,IC1_2*w1_2);

% Euler (rotational) e.o.m. for link 1 w.r.t. C1
eqM1 = M_C1+Min_C1;
eqM1x = eqM1(1);
eqM1y = eqM1(2);
eqM1z = eqM1(3);
```

and the MATLAB output is

```
M_C1+Min_C1 = 0 =>


T01x - F12z (sin(alpha) q2(t) - L sin(alpha)) -

    F01z L sin(alpha) = 0


M01y + F12z (cos(alpha) q2(t) - L cos(alpha)) +

    F01z L cos(alpha) +

      2
    L  m1 sin(alpha) diff(q1(t), t, t)
    -------------------------------- = 0
                 3


M01z - F12y cos(alpha) (cos(alpha) q2(t) -

    L cos(alpha)) - F12y sin(alpha)

    (sin(alpha) q2(t) - L sin(alpha)) -
```

```
    F01y L cos(alpha) + F01x L sin(alpha) +

      2                                                              2
     L  m1 cos(alpha) sin(alpha) diff(q1(t), t)
     --------------------------------------------- = 0
                        3
```

The two equations of motion are obtained using the MATLAB commands:

```
ql = {diff(q1,t,2), diff(q2,t,2), ...
    diff(q1,t), diff(q2,t), q1, q2};
qf = {'ddq1', 'ddq2', ...
    'x(2)', 'x(4)', 'x(1)', 'x(3)'};

sys=...
{eqF1x,eqF1y,eqF1z,eqF2x,eqF2y,eqF2z,eqM1x,eqM1y,
    eqM1z};

syseq = subs(sys, ql, qf);

unkn={F01x,F01y,F01z,F12y,F12z,M01y,M01z,'ddq1',
      'ddq2'};

sol=solve(syseq,...
    'F01x,F01y,F01z,F12y,F12z,M01y,M01z,ddq1,ddq2');

ddq1s = simple(sol.ddq1);
ddq2s = simple(sol.ddq2);

% e.o.m
fprintf('d x(1)/dt = x(2)\n')
fprintf('d x(2)/dt = '); pretty(ddq1s);
fprintf('\n')
fprintf('d x(3)/dt = x(4)\n')
fprintf('d x(3)/dt = '); pretty(ddq2s);
fprintf('\n')
```

and the results are

```
 d x(1)/dt = x(2)
 d x(2)/dt =
                                                2
   T01x - 2 m2 x(2) x(3) x(4) sin(alpha)
   -------------------------------------------
                 2       2            2
       sin(alpha)  (m1 L  + m2 x(3) )
```

```
d x(3)/dt = x(4)
d x(3)/dt =
         2                      2
   x(2)   x(3)  sin(alpha)   - g cos(alpha)
```

The equations are solved using the following MATLAB commands:

```
data = {alpha, L, m1, m2, g, T01x};
datn = {pi/4, 2, 10, 0.001, 9.81, 300*sin(3*t)};

NE1n = subs(ddq1s, data, datn);
NE2n = subs(ddq2s, data, datn);

% system of ODE
dx2dt = char(NE1n);
dx4dt = char(NE2n);

fid = fopen('eomE6_8.m','w+');
fprintf(fid,'function dx = eomE6_8(t,x)\n');
fprintf(fid,'dx = zeros(4,1);\n');
fprintf(fid,'dx(1) = x(2);\n');
fprintf(fid,'dx(2) = ');
fprintf(fid,dx2dt);
fprintf(fid,';\n');
fprintf(fid,'dx(3) = x(4);\n');
fprintf(fid,'dx(4) = ');
fprintf(fid,dx4dt);
fprintf(fid,';');
fclose(fid); cd(pwd);

t0 = 0; tf = 1; time = [0 tf];
x0 = [0 0 1.5 0];

 option0 = odeset...
('RelTol',1e-3,'MaxStep',1e-3,'Events',@eventE6_8);

[t,xs,te,ye] = ode45(@eomE6_8, time, x0, option0);
```

The integration is stopped when $q_2 = 0$, and this is obtained using the eventE6_8.m file:

```
function [value,isterminal,direction] = eventE6_8(t,x)

value = x(3);

isterminal = 1;

direction = 0;
```

The trajectory of the particle P with respect to the fixed reference frame is plotted with

```
x1 = xs(:,1); % q1
x2 = xs(:,2); % dq1
x3 = xs(:,3); % q2
x4 = xs(:,4); % dq2

% plot the results
alpha = pi/4;
xP1 = cos(alpha)*x3;
yP1 = sin(alpha)*x3;
zP1 = 0;

xP0 = x3*cos(alpha);
yP0 = sin(alpha)*x3.*cos(x1);
zP0 = sin(alpha)*x3.*sin(x1);

for kk=1:30:length(xs)

gd = 1.5;
gdm = .25;
axis([-gdm gd -gdm gd -gdm gd]);

set(0,'DefaultAxesColorOrder',[0 0 0],...
      'DefaultAxesLineStyleOrder','-|-.|--|:')

plot3(xP0(kk),zP0(kk),yP0(kk),'r.');
hold on
grid on
xlabel('x_0 (m)')
ylabel('y_0 (m)')
zlabel('z_0 (m)')
title('Position of P')

pause(1/100)

end
```

and the MATLAB graph for the particle is shown in Fig. 6.36.

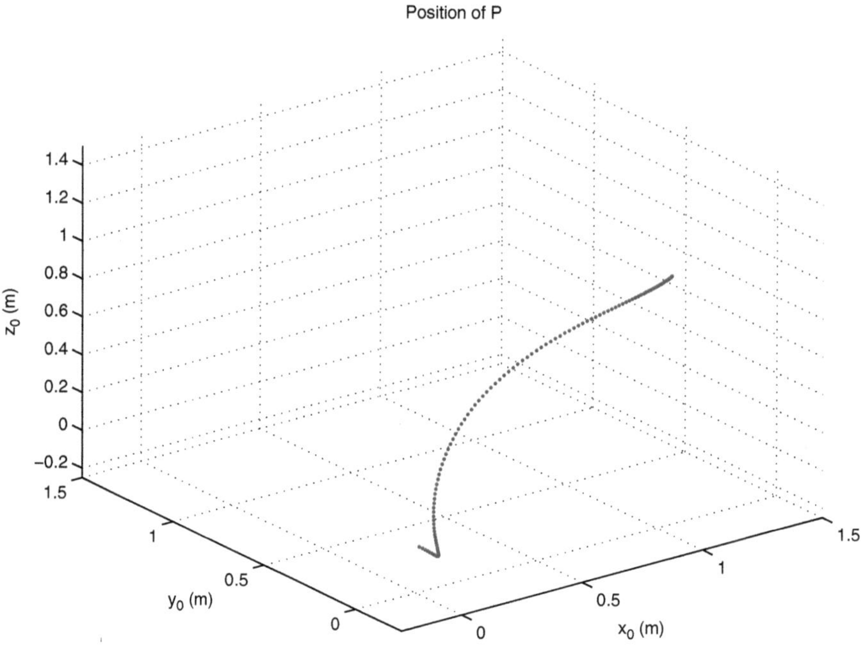

Fig. 6.36 Example 6.8 trajectory of particle P

6.11 Problems

6.1 The dimensions of the links for the mechanism in Fig. 6.37 are $AB = 0.100$ m, $AE = 0.250$ m, $BC = 0.170$ m, $CD = 0.300$ m, and $a = 0.090$ m. The angular speed of the driver link 1 is $n = 250$ rpm. An external force of magnitude $|F_e| = 600$ N acts on link 5 at D and is opposed to the motion of the slider 5. The link bars of the mechanism are homogeneous rectangular prisms with the width $h = 0.01$ m and the depth $d = 0.001$ m. The sliders have the width $w_{\text{Slider}} = 0.050$ m, the height $h_{\text{Slider}} = 0.020$ m, and the same depth $d = 0.001$ m. The links of the mechanism are homogeneous and are made of steel having a mass density $\rho_{\text{Steel}} = 8{,}000$ kg/m^3. The gravitational acceleration is $g = 9.8$ m/s^2. Determine the joint forces and the motor moment $\mathbf{M}_m$ on link 1 required for the dynamic equilibrium of the considered mechanism when the driver link 1 makes an angle $\phi = 60°$ with the horizontal axis.

6.2 The dimensions of the links for the mechanism in Fig. 6.38 are $AB = 0.100$ m, $AC = 0.300$ m, $BD = 0.600$ m, $a = 0.070$ m, and $b = 0.120$ m. The angular speed of the driver link 1 is $n = 100$ rpm. An external moment of magnitude $|M_e| = 900$ Nm acts on link 5 and is opposed to the motion of the link 5. The link bars of the mechanism are homogeneous rectangular prisms with the width $h = 0.01$ m and the depth $d = 0.001$ m. The sliders have the width $w_{\text{Slider}} = 0.050$ m, the height $h_{\text{Slider}} = 0.020$ m, and the same depth

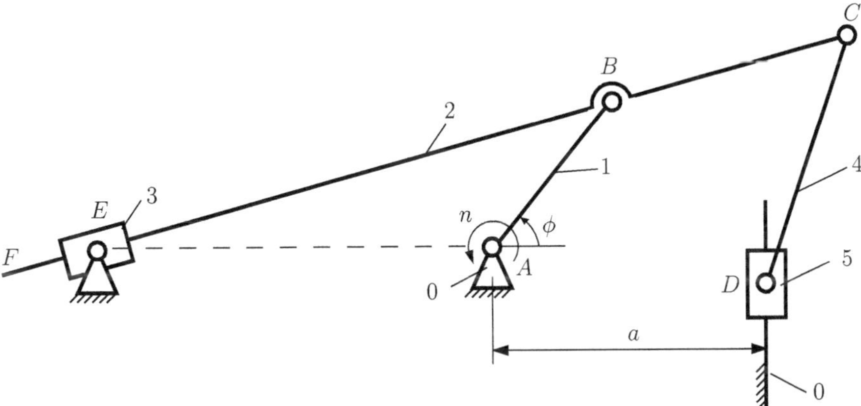

Fig. 6.37 Problem 6.1

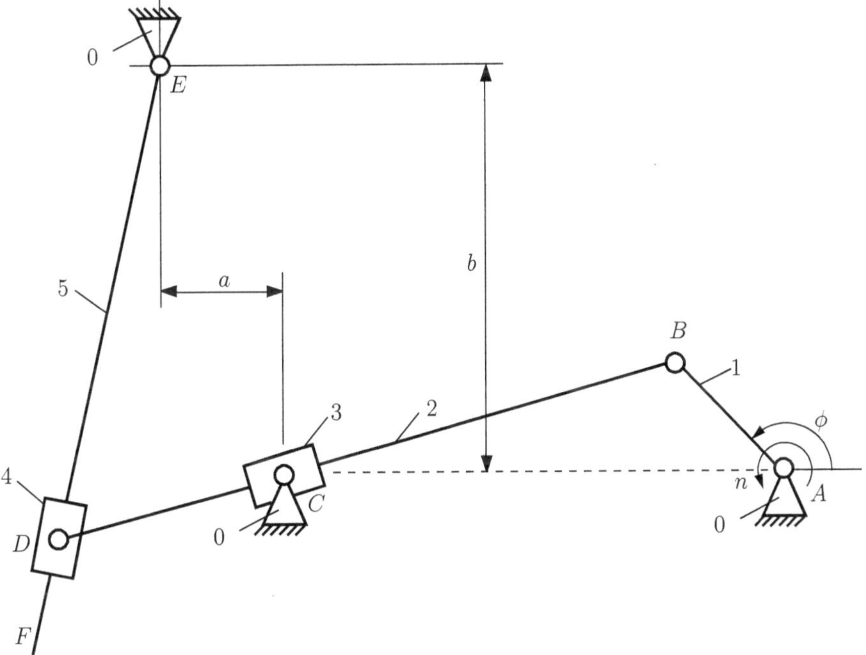

Fig. 6.38 Problem 6.2

$d = 0.001$ m. The links of the mechanism are homogeneous and are made of steel having a mass density $\rho_{\text{Steel}} = 8{,}000\,\text{kg/m}^3$. The gravitational acceleration is $g = 9.8\,\text{m/s}^2$. Determine the joint forces and the motor moment $\mathbf{M}_m$ on link 1 required for the dynamic equilibrium of the considered mechanism when the driver link 1 makes an angle $\phi = 120°$ with the horizontal axis.

Fig. 6.39 Problem 6.3

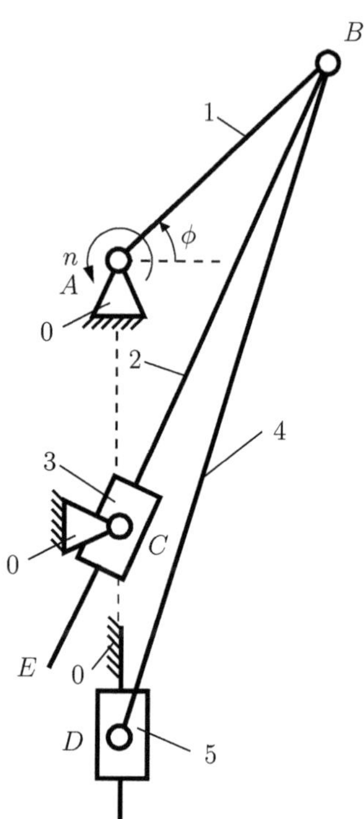

6.3 The dimensions of the links for the mechanism in Fig. 6.39 are $AB = 0.100\,$m, $BD = 0.250\,$m, $AC = 0.110\,$m, and $BE = 0.300\,$m. The angular speed of the driver link 1 is $n = 300\,$rpm. An external force of magnitude $|F_e| = 1,000\,$N acts on link 5 and is opposed to the motion of the link 5. The link bars of the mechanism are homogeneous rectangular prisms with the width $h = 0.01\,$m and the depth $d = 0.001\,$m. The sliders have the width $w_{\text{Slider}} = 0.050\,$m, the height $h_{\text{Slider}} = 0.020\,$m, and the same depth $d = 0.001\,$m. The links of the mechanism are homogeneous and are made of steel having a mass density $\rho_{\text{Steel}} = 8,000\,$kg/m^3. The gravitational acceleration is $g = 9.8\,$m/s^2. Determine the joint forces and the motor moment $\mathbf{M}_m$ on link 1 required for the dynamic equilibrium of the considered mechanism when the driver link 1 makes an angle $\phi = 60°$ with the horizontal axis.

6.4 The dimensions of the links for the mechanism in Fig. 6.40 are $AB = 0.100\,$m, $BC = 0.200\,$m, $BE = 0.400\,$m, $CE = 0.600\,$m, $CD = 0.220\,$m, $EF = 0.800\,$m, $a = 0.250\,$m, $b = 0.150\,$m, and $c = 0.100\,$m. The angular speed of the driver link 1 is $n = 1,200\,$rpm. An external force of magnitude $|F_e| = 1,000\,$N acts on link 5 and is opposed to the motion of the link 5. The link bars of the mechanism are homogeneous rectangular prisms with the width $h = 0.01\,$m and the depth $d = 0.001\,$m. The sliders have the width $w_{\text{Slider}} = 0.050\,$m, the height $h_{\text{Slider}} = 0.020\,$m, and the same depth $d = 0.001\,$m. The links of the

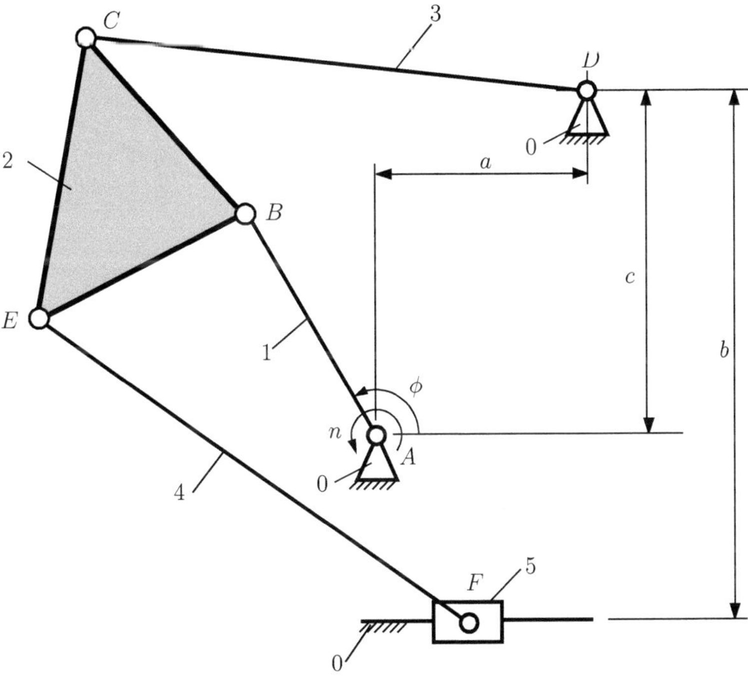

Fig. 6.40 Problem 6.4

mechanism are homogeneous and are made of steel having a mass density $\rho_{\text{Steel}} = 8,000\,\text{kg/m}^3$. The gravitational acceleration is $g = 9.8\,\text{m/s}^2$. Determine the joint forces and the motor moment $\mathbf{M}_{\text{m}}$ on link 1 required for the dynamic equilibrium of the considered mechanism when the driver link 1 makes an angle $\phi = 120°$ with the horizontal axis.

6.5 The dimensions of the links for the mechanism in Fig. 6.41 are $OA = 0.100\,\text{m}$, $AB = 0.390\,\text{m}$, $CB = 0.200\,\text{m}$, $CD = 0.200\,\text{m}$, $BD = 0.370\,\text{m}$, $DE = 0.340\,\text{m}$, $a = 0.320\,\text{m}$, and $b = 0.360\,\text{m}$. The angular speed of the driver link 1 is $n = 2,000\,\text{rpm}$. An external force of magnitude $|F_{\text{e}}| = 1,500\,\text{N}$ acts on link 5 and is opposed to the motion of the link 5. The link bars of the mechanism are homogeneous rectangular prisms with the width $h = 0.01\,\text{m}$ and the depth $d = 0.001\,\text{m}$. The sliders have the width $w_{\text{Slider}} = 0.050\,\text{m}$, the height $h_{\text{Slider}} = 0.020\,\text{m}$, and the same depth $d = 0.001\,\text{m}$. The links of the mechanism are homogeneous and are made of steel having a mass density $\rho_{\text{Steel}} = 8,000\,\text{kg/m}^3$. The gravitational acceleration is $g = 9.8\,\text{m/s}^2$. Determine the joint forces and the motor moment $\mathbf{M}_{\text{m}}$ on link 1 required for the dynamic equilibrium of the considered mechanism when the driver link 1 makes an angle $\phi = 150°$ with the horizontal axis.

6.6 The dimensions of the links for the mechanism in Fig. 6.42 are $OA = 0.100\,\text{m}$, $AB = 0.100\,\text{m}$, $BC = 0.100\,\text{m}$, $CD = 0.085\,\text{m}$, $BD = 0.030\,\text{m}$, $DE = 0.350\,\text{m}$, $a = 0.025\,\text{m}$, and $b = 0.025\,\text{m}$. The angular speed of the driver link 1 is

Fig. 6.41 Problem 6.5

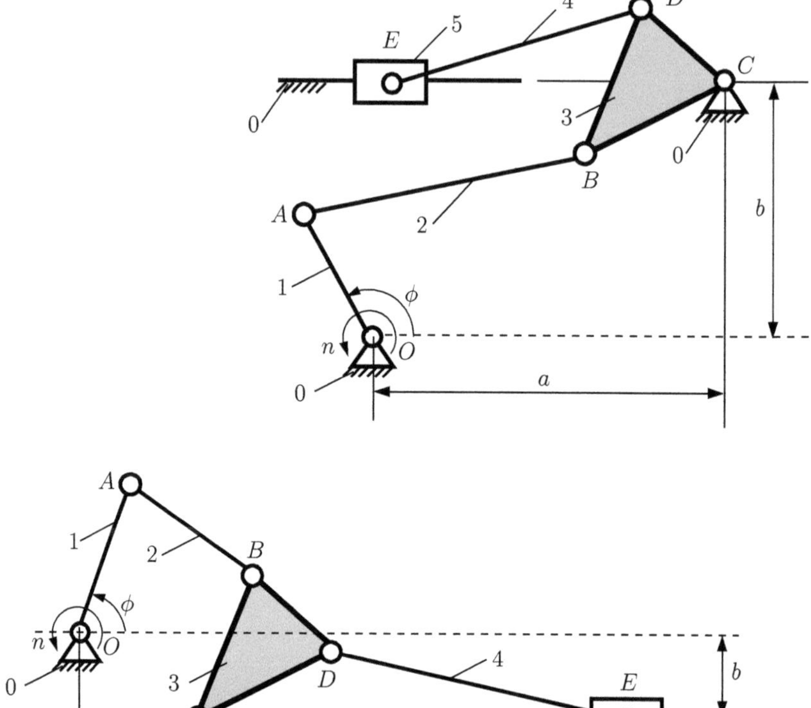

Fig. 6.42 Problem 6.6

$n = 60$ rpm. An external force of magnitude $|F_e| = 1{,}500$ N acts on link 5 and is opposed to the motion of the link 5. The link bars of the mechanism are homogeneous rectangular prisms with the width $h = 0.01$ m and the depth $d = 0.001$ m. The sliders have the width $w_{\text{Slider}} = 0.050$ m, the height $h_{\text{Slider}} = 0.020$ m, and the same depth $d = 0.001$ m. The links of the mechanism are homogeneous and are made of steel having a mass density $\rho_{\text{Steel}} = 8{,}000$ kg/m^3. The gravitational acceleration is $g = 9.8$ m/s^2. Determine the joint forces and the motor moment $\mathbf{M}_m$ on link 1 required for the dynamic equilibrium of the considered mechanism when the driver link 1 makes an angle $\phi = 60°$ with the horizontal axis.

6.7 The dimensions of the links for the mechanism in Fig. 6.43 are $OA = 0.200$ m, $OB = 0.100$ m, $BC = 0.150$ m, and $CD = 0.500$ m. The angular speed of the driver link 1 is $n = 200$ rpm. An external force of magnitude $|F_e| = 800$ N acts on link 5 and is opposed to the motion of the link 5. The link bars of the mechanism are homogeneous rectangular prisms with the width $h = 0.01$ m and the depth $d = 0.001$ m. The sliders have the width $w_{\text{Slider}} = 0.050$ m, the height $h_{\text{Slider}} = 0.020$ m, and the same depth $d = 0.001$ m. The links of the

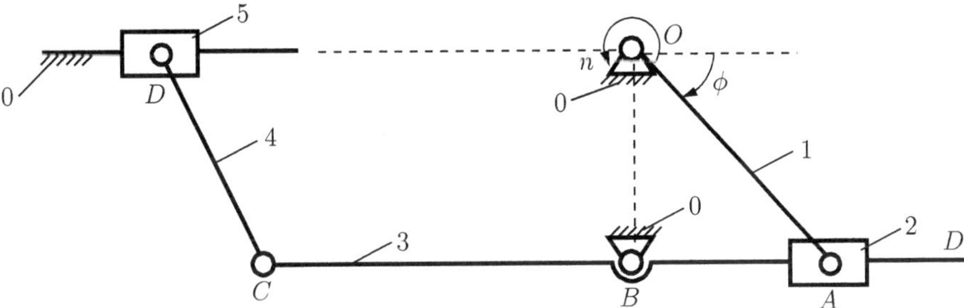

Fig. 6.43 Problem 6.7

Fig. 6.44 Problem 6.8

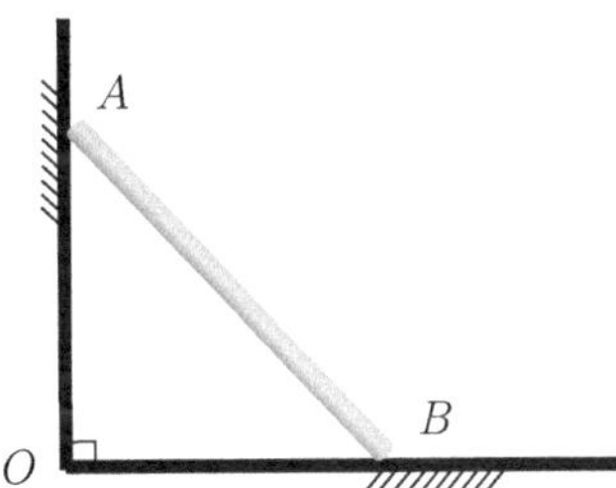

mechanism are homogeneous and are made of steel having a mass density $\rho_{\text{Steel}} = 8{,}000\,\text{kg/m}^3$. The gravitational acceleration is $g = 9.8\,\text{m/s}^2$. Determine the joint forces and the motor moment $\mathbf{M}_{\text{m}}$ on link 1 required for the dynamic equilibrium of the considered mechanism when the driver link 1 makes an angle $\phi = 150°$ with the horizontal axis.

6.8 A uniform rod of length $AB = L$ and mass m is supported on the smooth perpendicular planes as shown in Fig. 6.44. The acceleration due to gravity is g. The bar is released from rest when it makes an angle of $\pi/4$ with the vertical wall. Find the Newton–Euler equations of motion. Solve the equations of motion using MATLAB. For the numerical application, use $m = 5\,\text{kg}$, $L = 1\,\text{m}$, and $g = 9.81\,\text{m/s}^2$.

6.9 The slider 1 of mass m shown in Fig. 6.45 moves without friction on an inclined fixed link (the inclined angle is β). The slider starts from rest at origin O. Determine the equation of motion and the acceleration of the slider. Solve the equations of motion using MATLAB. For the numerical application, use $m = 2\,\text{kg}$, $\beta = 45°$, and $g = 9.81\,\text{m/s}^2$.

6.10 Two uniform hinged rods 1 and 2 of mass $m_1 = m_2 = m$ and length $AB = BC = L$ are shown in Fig. 6.46. The rod 1 is connected to the ground by a pin joint at A and to the rod 2 by a pin joint at B. The end B is moving with friction along the horizontal surface. The coefficient of friction between rod 2 and the horizontal surface is μ. The acceleration due to gravity is g. The bars are released from rest when link 1 makes an angle of $\pi/4$ with the horizontal axis. Find and solve the equation of motion of the system. For the numerical application, use $m = 0.5\,\text{kg}$, $L = 0.5\,\text{m}$, $\mu = 0.1$, and $g = 9.81\,\text{m/s}^2$.

Fig. 6.45 Problem 6.9

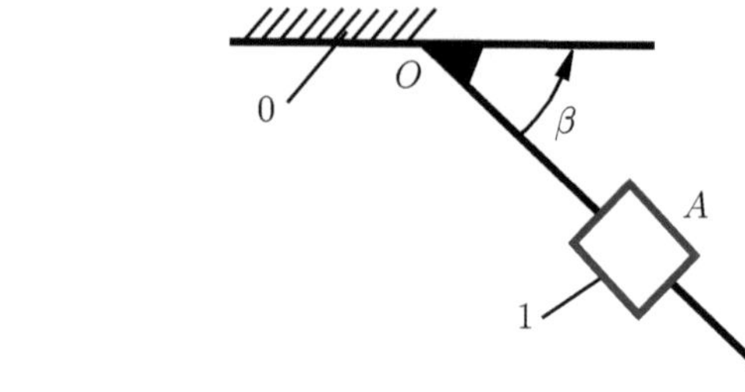

Fig. 6.46 Problem 6.10

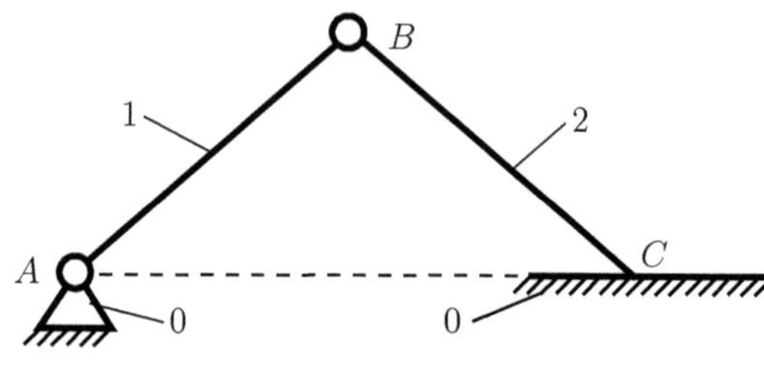

Fig. 6.47 Problem 6.11

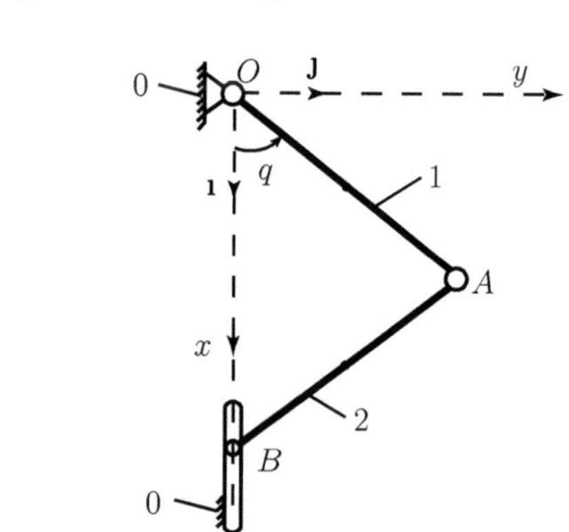

6.11 Figure 6.47 depicts two uniform rods 1 and 2 of mass $m_1 = m_2 = m$ and length $OA = AB = 2L$. The rod 1 is connected to the ground by a pin joint at O and to the rod 2 by a pin joint at A. The rods are constrained to move in a vertical plane xOy. The x-axis is vertical, with the positive sense directed vertically downward. The y-axis is horizontal and is contained in the plane of motion. The rod 1 is moving, and the instant angle with the vertical axis Ox is $q(t)$. The rod 2 is connected to the ground by a pin joint at B which is confined to move in a vertical slot. The local acceleration of gravity is g. The rods are released from rest when link 1 makes an angle of $\pi/4$ with the vertical axis. Find and solve the equations of motion of the system. For the numerical application, use $m = 0.3$ kg, $L = 0.4$ m, and $g = 9.81$ m/s^2 (Fig. 6.47).

6.12 The uniform rod 1, in Fig. 6.48, has the length $OA = L_1$ and the mass m_1, while rod 2 has the length $AB = L_2$ and the mass m_2. The rods are released from rest with the geometry shown in Fig. 6.48. Find and solve the Newton–Euler equations of motion and the joint reaction forces at this instant. For the numerical application, use $m_1 = 1$ kg, $L_1 = 0.3$ m, $m_2 = 5$ kg, $L_2 = 0.9$ m, $a = 0.6$ m, and $g = 9.81$ m/s^2.

Fig. 6.48 Problem 6.12

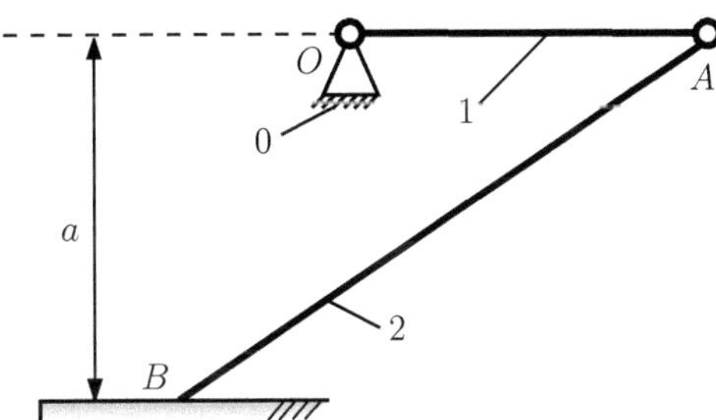

Fig. 6.49 Problem 6.13

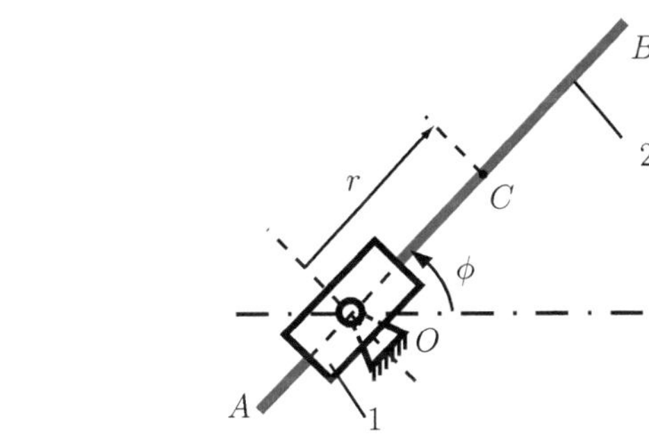

Fig. 6.50 Problem 6.14

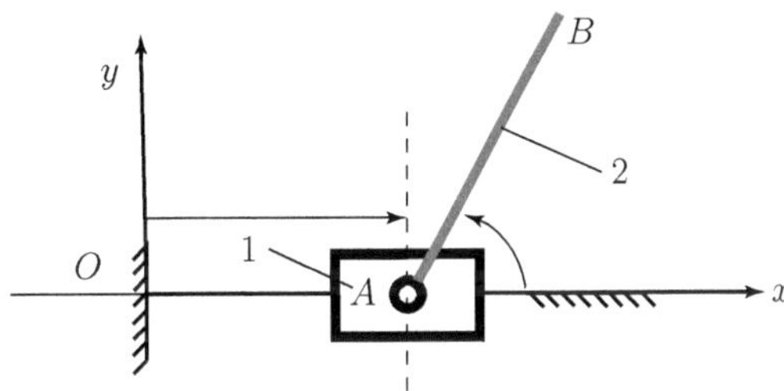

6.13 A slender rod $AB = L$ (link 2) is moving without friction along the slider 1 as
shown in Fig. 6.49. The slider is connected to the ground by a pin joint at O and
is free to swing in a vertical plane. The mass of the rod is m_1 the mass center
is located at C. The mass of the slider is m_2, and the mass moment of inertia
of the slider with respect to its mass center point O is I_O. The acceleration
due to gravity is g. Find and solve the Newton–Euler equations of motion. For
the numerical application, use $m_1 = m_2 = 1\,\text{kg}$, $L = 1\,\text{m}$, $g = 9.81\,\text{m/s}^2$, and
$I_O = 1\,\text{kg}\,\text{m}^2$.

6.14 The planar mechanical system considered is shown in Fig. 6.50 and has a slider
1 of mass m_1 and a pendulum 2 with the mass m_2. The length of AB is L. The
local acceleration of gravity is g. Find and solve the equations of motion of
the system. For the numerical application, use $L = 1\,\text{m}$, $g = 9.81\,\text{m/s}^2$, and
$m_1 = m_2 = 1\,\text{kg}$.

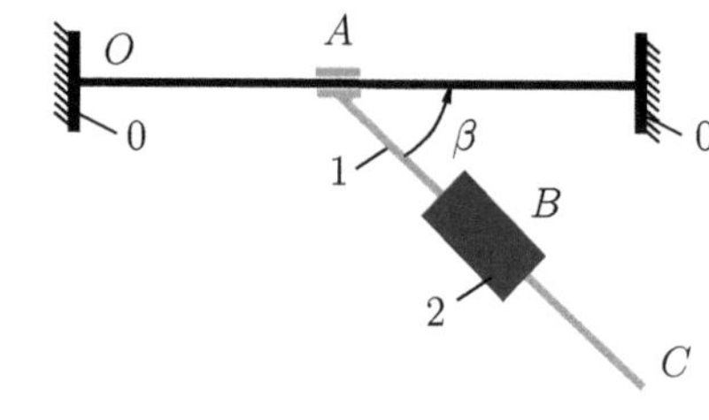

Fig. 6.51 Problem 6.15

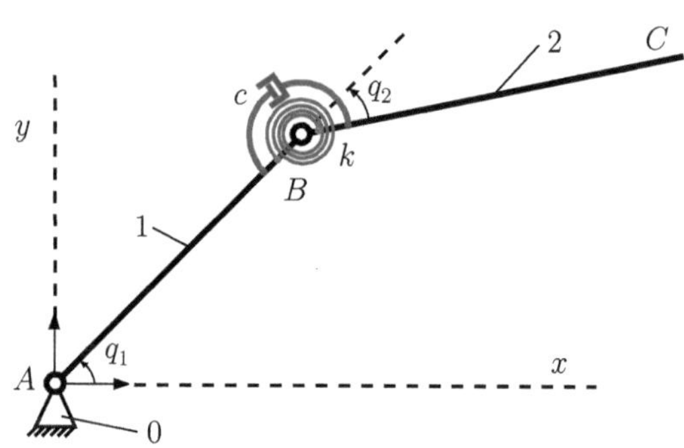

Fig. 6.52 Problem 6.16

6.15 The slider 2 of mass m_2 shown in Fig. 6.51 moves without friction on an inclined link 1 of mass m_1 and length L (the inclined angle is β). The link 2 translates along the horizontal axis. Find and solve the equation of motion and the acceleration of the slider 2 if link 1 translates with a constant speed v. For the numerical application, use $m_1 = m_2 = 1\,\text{kg}$, $L = 1\,\text{m}$, $\beta = 45°$, $v = 2\,\text{m/s}$, and $g = 9.81\,\text{m/s}^2$.

6.16 Figure 6.52 shows an open kinematic chain with two uniform rigid rods 1 and 2 of mass $m_1 = m_2 = m$ and length $L_1 = L_2 = L$. The rod 1 is connected to the ground by a pin joint at A and to the rod 2 by a pin joint at B. The rods are constrained to move in a vertical plane xy. A spring of elastic constant k and a viscous damper with a damping constant c are opposing the relative motion of the link 2 with respect to link 1. The local acceleration of gravity is g. Find and solve the equations of motion of the system. For the numerical application, use $m_1 = m_2 = 1\,\text{kg}$, $L_1 = L_2 = 1\,\text{m}$, $g = 10\,\text{m/s}^2$, $k = 100\,\text{Nm/rad}$, $c = 10\,\text{Nm s/rad}$, $q_1(0) = \pi/3\,\text{rad}$, $q_2(0) = \pi/6\,\text{rad}$, $\dot{q}_1(0) = \dot{q}_2(0) = 0\,\text{rad/s}$.

6.17 Figure 6.53 is a schematic representation of a two-link open kinematic chain. Let $m_1 = 1\,\text{kg}$ and $m_2 = 1\,\text{kg}$ be the masses of the slender links 1 and 2, respectively. Link 1 rotates at O about a vertical axis, and 1 is connected to link 2 at the pin joint A. The element 2 rotates relative to 1 about an axis fixed in both 1 and 2, passing through A, and perpendicular to the axis of 1. The length of the link 1 is $L_1 = 1\,\text{m}$, and the length of the link 2 is $L_2 = 1\,\text{m}$. Find and solve the equations of motion using Newton–Euler method using suitable initial conditions.

6.18 Figure 6.54 is a schematic representation of a two-link open kinematic chain. Let $m_1 = 0.5\,\text{kg}$ and $m_2 = 1\,\text{kg}$ be the masses of the slender links 1 and 2,

Fig. 6.53 Problem 6.17

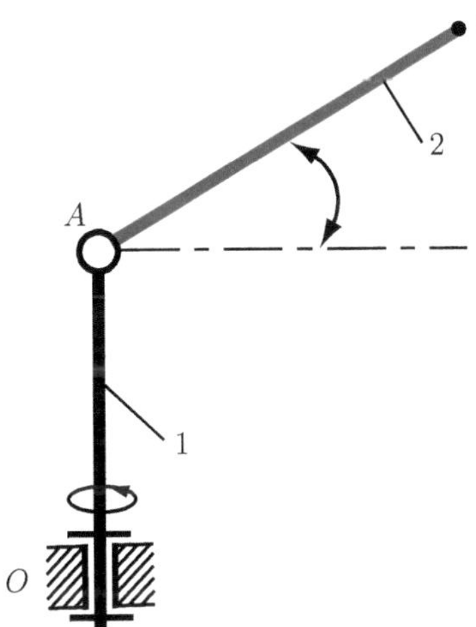

Fig. 6.54 Problem 6.18

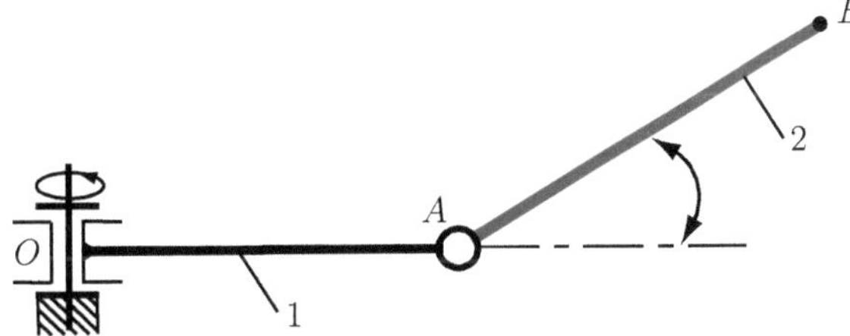

respectively. Link 1 rotates at O about a vertical axis, and 1 is connected to link 2 at the pin joint A. The element 2 rotates relative to 1 about an axis fixed in both 1 and 2, passing through A, and perpendicular to the axis of 1. The length of the link 1 is $L_1 = 0.5\,\text{m}$, and the length of the link 2 is $L_2 = 1\,\text{m}$. Find and solve the equations of motion using Newton–Euler method using suitable initial conditions.

6.19 Figure 6.55a depicts a mechanism with two uniform rods 1 and 2 and a slider 3. The length of link 1 is $OA = L = 1\,\text{m}$, the length of link 2 is $AB = 2L$, and the length of slider 3 is $L/10$. The links are constrained to move in a vertical plane xy. The x-axis is vertical, with the positive sense directed vertically downward. The y-axis is horizontal and is contained in the plane of motion. The link 1 is moving, and the instant angle with the vertical axis is $q(t)$. The links are rectangular prisms with the depth $d = 0.001\,\text{m}$ and height $h = 0.01\,\text{m}$. The mass density of the links is $\rho = 7{,}850\,\text{kg/m}^3$. The local acceleration of gravity is $g = 9.81\,\text{m/s}^2$. A motor torque acts on the link 1, Fig. 6.55b, and is given by $\mathbf{M}_m = M\mathbf{k}$. For a D.C. motor, $M = M_0(1 - \frac{\omega}{\omega_0})$, where $\omega = \dot{q}(t)$ and M_0 and ω_0 are given in catalogues. In our case, $M_0 = 1\,\text{Nm}$, and $\omega_0 = 4\,\text{rad/s}$. Find and solve the equations of motion of the system for the initial conditions $q(0) = \pi/3$ rad and $\omega(0) = \dot{q}(0) = 0\,\text{rad/s}$.

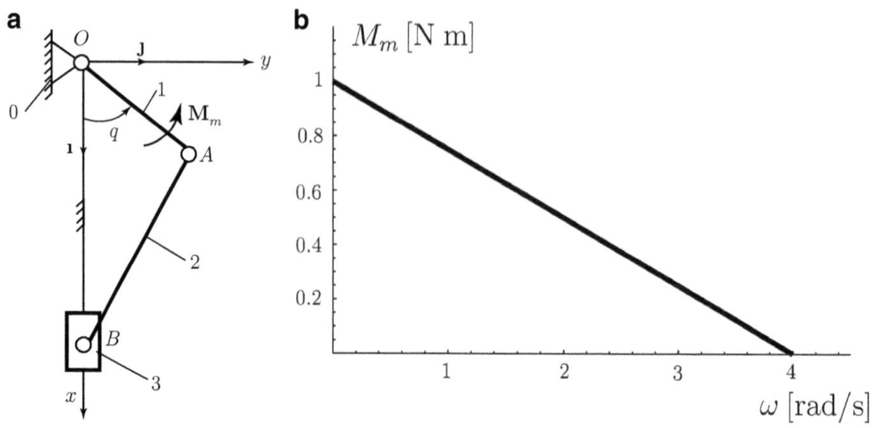

Fig. 6.55 Problem 6.19

Fig. 6.56 Problem 6.20

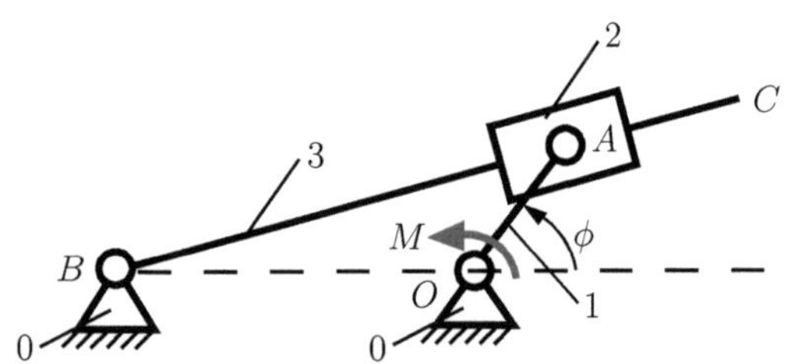

6.20 The dimensions of the planar mechanism shown in Fig. 6.56 are given below:
$OA = 0.15\,$m, $OB = 0.30\,$m, and $BC = 0.50\,$m. The links 1 and 3 are
rectangular prisms with the height $h = 0.010\,$m and the depth $d = 0.005\,$m.
The link 2 has the height $h_s = 0.050\,$m and the depth $d_s = 0.020\,$m. The mass
density of the links is $\rho = 7{,}850\,$kg/m^3. A motor moment acts on the link
1 and has the value $M = M_0(1 - \frac{\omega}{\omega_0})$, where $\omega = \dot{\phi}(t)$, $M_0 = 10\,$Nm, and
$\omega_0 = 10\,$rad/s. Find and solve the equations of motion of the system. The
initial conditions are $\phi(0) = \pi/4\,$rad and $\omega(0) = \dot{\phi}(0) = 0\,$rad/s. Find and
solve the equation of motion of the mechanism.

Chapter 7
Analytical Dynamics

7.1 Introduction

Any dynamical system can be regarded as an assembly of N mass points (particles). Consider a system of N particles: $\{S\} = \{P_1, P_2, \ldots P_i \ldots P_N\}$, as shown in Fig. 7.1.

The position vector of the ith particle in the Cartesian reference frame is $\mathbf{r}_i = \mathbf{r}_i(x_i, y_i, z_i)$ and can be expressed as

$$\mathbf{r}_i = x_i\mathbf{1} + y_i\mathbf{J} + z_i\mathbf{k}, \quad i = 1, 2, \ldots, N.$$

The system of N particles requires $n = 3N$ physical coordinates to specify its position. To analyze the motion of the system in many cases, it is more convenient to use a set of variables different from the physical coordinates. Let us consider a set of variables $q_1, q_2, \ldots, q_{3N}$ related to the physical coordinates by

$$x_1 = x_1(q_1, q_2, \ldots, q_{3N}),$$

$$y_1 = y_1(q_1, q_2, \ldots, q_{3N}),$$

$$z_1 = z_1(q_1, q_2, \ldots, q_{3N}),$$

$$\cdot$$

$$\cdot$$

$$\cdot$$

$$x_{3N} = x_{3N}(q_1, q_2, \ldots, q_{3N}),$$

$$y_{3N} = y_{3N}(q_1, q_2, \ldots, q_{3N}),$$

$$z_{3N} = z_{3N}(q_1, q_2, \ldots, q_{3N}).$$

The *generalized coordinates*, $q_1, q_2, \ldots, q_{3N}$, are the set of variables that can completely describe the position of the dynamical system. The *configuration space*

D.B. Marghitu and M. Dupac, *Advanced Dynamics: Analytical and Numerical Calculations with MATLAB*, DOI 10.1007/978-1-4614-3475-7_7, © Springer Science+Business Media, LLC 2012

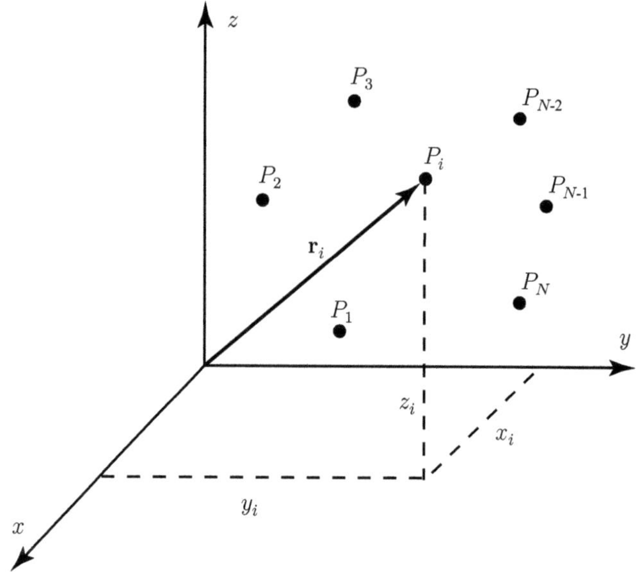

Fig. 7.1 Position vector of the ith particle

is the space extended across the generalized coordinates. If the system of N particles has m constraint equations acting on it, the system can be represented uniquely by *p independent generalized coordinates* q_k, $(k = 1, 2 , \ldots, p)$, where $p = 3N - m = n - m$. The number p is called the number of degrees of freedom of the system.

The number of *degrees of freedom* is the minimum number of independent coordinates necessary to describe the dynamical system uniquely. The *generalized velocities*, denoted by $\dot{q}_k(t)$ $(k = 1, 2, \ldots, n)$, represent the rate of change of the generalized coordinates with respect to time.

The *state space* is the $2n$-dimensional space spanned by the generalized coordinates and generalized velocities.

The constraints are generally dominant as a result of contact between bodies, and they limit the motion of the bodies upon which they act. A *constraint equation* and a *constraint force* are related with a constraint. The constraint force is the joint reaction force, and the constraint equation represents the kinematics of the contact.

Consider a smooth surface of equation

$$f(x,y,z,t) = 0, \qquad (7.1)$$

where f has continuous second derivatives in all its variables. A particle P is subjected to a constraint of moving on the smooth surface described by (7.1). The constraint equation $f(x, y, z, t) = 0$ represents a *configuration constraint*.

The motion of the particle over the surface can be viewed as the motion of an otherwise free particle subjected to the constraint of moving on that particular surface. Hence, $f(x, y, z, t) = 0$ represents a constraint equation.

For a dynamical system with n generalized coordinates, a configuration constraint can be described as

$$f(q_1, q_2, \ldots, q_n, t) = 0. \tag{7.2}$$

The differential of the constraint f, given by (7.1), in terms of physical coordinates is

$$df = \frac{\partial f}{\partial x} dx + \frac{\partial f}{\partial y} dy + \frac{\partial f}{\partial z} dz + \frac{\partial f}{\partial t} dt = 0. \tag{7.3}$$

The differential of the constraint f, given by (7.2), in terms of the generalized coordinates is

$$df = \frac{\partial f}{\partial q_1} dq_1 + \frac{\partial f}{\partial q_2} dq_2 + \cdots + \frac{\partial f}{\partial q_n} dq_n + \frac{\partial f}{\partial t} dt = 0. \tag{7.4}$$

Equations (7.3) and (7.4) are called *constraint relations in Pfaffian form*. A constraint in Pfaffian form is a constraint that is represented in the form of differentials. The *constraint equations in velocity form* (or *velocity constraints* or *motion constraints*) are obtained by dividing (7.3) and (7.4) by dt:

$$\frac{df}{dt} = \frac{\partial f}{\partial x} \dot{x} + \frac{\partial f}{\partial y} \dot{y} + \frac{\partial f}{\partial z} \dot{z} + \frac{\partial f}{\partial t} = 0, \tag{7.5}$$

$$\frac{df}{dt} = \frac{\partial f}{\partial q_1} \dot{q}_1 + \frac{\partial f}{\partial q_2} \dot{q}_2 + \cdots + \frac{\partial f}{\partial q_n} \dot{q}_n + \frac{\partial f}{\partial t} = 0. \tag{7.6}$$

The velocity constraint given by (7.5) can be represented as

$$a_x \dot{x} + a_y \dot{y} + a_z \dot{z} + a_0 = 0. \tag{7.7}$$

For a dynamical system with n generalized coordinates subjected to m constraints, the velocity constraint given by (7.6) can be expressed as

$$\sum_{k=1}^{n} a_{jk} \dot{q}_k + a_{j0} = 0, \quad j = 1, 2, \ldots, m, \tag{7.8}$$

where a_{jk} and a_{j0} ($j = 1, 2, \ldots, m; k = 1, 2, \ldots, n$) are functions of the generalized coordinates and time.

A *holonomic* constraint is a constraint that can be represented as both a configuration constraint as well as velocity constraint. Constraints that do not have this property are called *non-holonomic* (non-holonomic constraints cannot be expressed as configuration constraints). When the constraint is non-holonomic, it can only be expressed in the form of (7.7) or (7.8), as an integrating factor does not exist to allow expression in the form (7.1) or (7.2).

Eliminating $\partial f/\partial t$ from (7.5), it results

$$\frac{\mathrm{d}f}{\mathrm{d}t} = \frac{\partial f}{\partial x}\dot{x} + \frac{\partial f}{\partial y}\dot{y} + \frac{\partial f}{\partial z}\dot{z} = 0. \tag{7.9}$$

The velocity of a particle written with respect to (7.9) as $\mathbf{v}(t) = \dot{\mathbf{r}}(t) = \dot{x}(t)\mathbf{1} + \dot{y}(t)\mathbf{J} + \dot{z}(t)\mathbf{k}$ where $\mathbf{r}(t) = x(t)\mathbf{1} + y(t)\mathbf{J} + z(t)\mathbf{k}$ is the position vector of the particle. The gradient of f, denoted by ∇f, is defined as

$$\nabla f = \frac{\partial f}{\partial x}\mathbf{1} + \frac{\partial f}{\partial y}\mathbf{J} + \frac{\partial f}{\partial z}\mathbf{k}$$

and is calculated in MATLAB using the `jacobian` function:

```
nablaf=jacobian(f,[x,y,z])
```

If f is scalar, the Jacobian of f is the gradient of f. Taking the dot product between the gradient and the velocity $\mathbf{v}(t)$, one can obtain

$$\nabla f \cdot \mathbf{v} = \frac{\partial f}{\partial x}\dot{x} + \frac{\partial f}{\partial y}\dot{y} + \frac{\partial f}{\partial z}\dot{z}$$

and compared with (7.5) yields

$$\nabla f \cdot \mathbf{v} = \frac{\mathrm{d}f}{\mathrm{d}t} = 0.$$

A unit vector $\mathbf{n}$ perpendicular to a surface and parallel to ∇f can be expressed as

$$\mathbf{n} = \frac{\pm\nabla f}{|\nabla f|} = \pm\frac{\dfrac{\partial f}{\partial x}\mathbf{1} + \dfrac{\partial f}{\partial y}\mathbf{J} + \dfrac{\partial f}{\partial z}\mathbf{k}}{\left[\left(\dfrac{\partial f}{\partial x}\right)^2 + \left(\dfrac{\partial f}{\partial y}\right)^2 + \left(\dfrac{\partial f}{\partial z}\right)^2\right]^{1/2}}.$$

A *scleronomic* system, $f(q_1, q_1, \ldots, q_n) = 0$, is an unconstrained dynamical system or a system subjected to a holonomic constraint that is not an explicit function of time. A *rhenomic* system is a system subjected to a holonomic constraint that is an explicit function of time. A holonomic constraint is called *scleronomic* if the time does not appear explicitly as a parameter (it is time independent), otherwise is called *rheonomic*.

7.2   **Equations of Motion**

Consider the motion of a system $\{S\}$ of p particles $P_1, \ldots, P_p$ ($\{S\} = \{P_1, \ldots, P_p\}$) in an inertial reference frame (0). The equation of motion for the ith particle is

$$\mathbf{F}_i = m_i \, \mathbf{a}_i, \tag{7.10}$$

where $\mathbf{F}_i$ is the resultant of all contact and distance forces acting on P_i, m_i is the mass of P_i, and $\mathbf{a}_i$ is the acceleration of P_i in (0). Equation (7.10) is the expression of Newton's second law. The inertia force $\mathbf{F}_{\text{in}\,i}$ for P_i in (0) is defined as

$$\mathbf{F}_{\text{in}\,i} = -\,m_i \, \mathbf{a}_i, \tag{7.11}$$

then (7.10) is written as

$$\mathbf{F}_i + \mathbf{F}_{\text{in}\,i} = \mathbf{0}. \tag{7.12}$$

Equation (7.12) is the expression of D'Alembert's principle.

If $\{S\}$ is a holonomic system possessing n degrees of freedom, then the position vector $\mathbf{r}_i$ of P_i relative to a point O fixed in reference frame (0) is expressed as a vector function of n generalized coordinates $q_1, \ldots, q_n$ and time t:

$$\mathbf{r}_i = \mathbf{r}_i(q_1, \ldots, q_n, t).$$

The velocity $\mathbf{v}_i$ of P_i in (0) has the form

$$\mathbf{v}_i = \sum_{r=1}^{n} \frac{\partial \mathbf{r}_i}{\partial q_r} \frac{\partial q_r}{\partial t} + \frac{\partial \mathbf{r}_i}{\partial t} = \sum_{r=1}^{n} \frac{\partial \mathbf{r}_i}{\partial q_r} \dot{q}_r + \frac{\partial \mathbf{r}_i}{\partial t}, \tag{7.13}$$

or as

$$\mathbf{v}_i = \sum_{r=1}^{n} (\mathbf{v}_i)_r \dot{q}_r + \frac{\partial \mathbf{r}_i}{\partial t},$$

where $(\mathbf{v}_i)_r$ is called the rth *partial velocity* of P_i in (0) and is defined as

$$(\mathbf{v}_i)_r = \frac{\partial \mathbf{r}_i}{\partial q_r} = \frac{\partial \mathbf{v}_i}{\partial \dot{q}_r}. \tag{7.14}$$

Next, replace (7.12) with

$$\sum_{i=1}^{p} (\mathbf{F}_i + \mathbf{F}_{\text{in}\,i}) \cdot (\mathbf{v}_i)_r = 0. \tag{7.15}$$

If a *generalized active force* Q_r and a *generalized inertia force* $K_{\text{in}\,r}$ are defined as

$$Q_r = \sum_{i=1}^{p} (\mathbf{v}_i)_r \cdot \mathbf{F}_i = \sum_{i=1}^{p} \frac{\partial \mathbf{r}_i}{\partial q_r} \cdot \mathbf{F}_i = \sum_{i=1}^{p} \frac{\partial \mathbf{v}_i}{\partial \dot{q}_r} \cdot \mathbf{F}_i, \tag{7.16}$$

and

$$K_{\text{in}\,r} = \sum_{i=1}^{p} (\mathbf{v}_i)_r \cdot \mathbf{F}_{\text{in}\,i} = \sum_{i=1}^{p} \frac{\partial \mathbf{r}_i}{\partial q_r} \cdot \mathbf{F}_{\text{in}\,i} = \sum_{i=1}^{p} \frac{\partial \mathbf{v}_i}{\partial \dot{q}_r} \cdot \mathbf{F}_{\text{in}\,i}, \qquad (7.17)$$

then (7.15) can be written as

$$Q_r + K_{\text{in}\,r} = 0, \ r = 1,\ldots,n. \qquad (7.18)$$

Equation (7.18) are *Kane's dynamical equations.*

Consider the generalized inertia force $K_{\text{in}\,r}$:

$$K_{\text{in}\,r} = \sum_{i=1}^{p} \mathbf{F}_{\text{in}\,i} \cdot (\mathbf{v}_i)_r = - \sum_{i=1}^{p} m_i \mathbf{a}_i \cdot (\mathbf{v}_i)_r = - \sum_{i=1}^{p} m_i \ddot{\mathbf{r}}_i \cdot \frac{\partial \mathbf{r}_i}{\partial q_r}$$

$$= - \sum_{i=1}^{p} \left[\frac{\mathrm{d}}{\mathrm{d}t} \left(m_i \dot{\mathbf{r}}_i \cdot \frac{\partial \mathbf{r}_i}{\partial q_r} \right) - m_i \dot{\mathbf{r}}_i \cdot \frac{\mathrm{d}}{\mathrm{d}t} \left(\frac{\partial \mathbf{r}_i}{\partial q_r} \right) \right]. \qquad (7.19)$$

Now,

$$\frac{\mathrm{d}}{\mathrm{d}t} \left(\frac{\partial \mathbf{r}_i}{\partial q_r} \right) = \sum_{k=1}^{n} \frac{\partial^2 \mathbf{r}_i}{\partial q_r \partial q_k} \dot{q}_k + \frac{\partial^2 \mathbf{r}_i}{\partial q_r \partial t} = \frac{\partial \mathbf{v}_i}{\partial q_r}, \qquad (7.20)$$

and, furthermore, using (7.13)

$$\frac{\partial \mathbf{v}_i}{\partial \dot{q}_r} = \frac{\partial \mathbf{r}_i}{\partial q_r}. \qquad (7.21)$$

Substitution of (7.20) and (7.21) in (7.19) leads to

$$K_{\text{in}\,r} = - \sum_{i=1}^{p} \left[\frac{\mathrm{d}}{\mathrm{d}t} \left(m_i \mathbf{v}_i \cdot \frac{\partial \mathbf{v}_i}{\partial \dot{q}_r} \right) - m_i \mathbf{v}_i \cdot \frac{\partial \mathbf{v}_i}{\partial q_r} \right]$$

$$= - \left[\frac{\mathrm{d}}{\mathrm{d}t} \frac{\partial}{\partial \dot{q}_r} \left(\sum_{i=1}^{p} \frac{1}{2} m_i \mathbf{v}_i \cdot \mathbf{v}_i \right) - \frac{\partial}{\partial q_r} \left(\sum_{i=1}^{p} \frac{1}{2} m_i \mathbf{v}_i \cdot \mathbf{v}_i \right) \right].$$

The *kinetic energy T* of $\{S\}$ in reference frame (0) is defined as

$$T = \frac{1}{2} \sum_{i=1}^{p} m_i \mathbf{v}_i \cdot \mathbf{v}_i.$$

Therefore, the generalized inertia forces $K_{\text{in}\,r}$ are written as

$$K_{\text{in}\,r} = - \frac{\mathrm{d}}{\mathrm{d}t} \left(\frac{\partial T}{\partial \dot{q}_r} \right) + \frac{\partial T}{\partial q_r},$$

and Kane's dynamical equations can be written as

$$Q_r - \frac{\mathrm{d}}{\mathrm{d}t}\left(\frac{\partial T}{\partial \dot{q}_r}\right) + \frac{\partial T}{\partial q_r} = 0,$$

or

$$\frac{\mathrm{d}}{\mathrm{d}t}\left(\frac{\partial T}{\partial \dot{q}_r}\right) - \frac{\partial T}{\partial q_r} = Q_r.$$

The equations

$$\frac{\mathrm{d}}{\mathrm{d}t}\left(\frac{\partial T}{\partial \dot{q}_r}\right) - \frac{\partial T}{\partial q_r} = \sum_{i=1}^{p} \frac{\partial \mathbf{r}_i}{\partial q_r} \cdot \mathbf{F}_i, \ \ r = 1,\dots,n, \tag{7.22}$$

or

$$\frac{\mathrm{d}}{\mathrm{d}t}\left(\frac{\partial T}{\partial \dot{q}_r}\right) - \frac{\partial T}{\partial q_r} = \sum_{i=1}^{p} \frac{\partial \mathbf{v}_i}{\partial \dot{q}_r} \cdot \mathbf{F}_i, \ \ r = 1,\dots,n, \tag{7.23}$$

are known as *Lagrange's equations of motion* of the first kind.

For conservative systems, the conservative forces can be written as

$$\mathbf{F}_i = -\mathrm{grad}\, V_i = -\nabla V_i,$$

where V_i is the potential energy. Using the potential energy, the generalized active forces Q_r are expressed as

$$Q_r = \sum_{i=1}^{p} \frac{\partial \mathbf{r}_i}{\partial q_r} \cdot \mathbf{F}_i = - \sum_{i=1}^{p} \frac{\partial \mathbf{r}_i}{\partial q_r} \cdot \nabla V_i. \tag{7.24}$$

The potential energy $\mathbf{V}_i$ is related to the generalized coordinates x_i, y_i, and z_i by

$$F_{ix} = -\frac{\partial V_i}{\partial x_i}, \ \ F_{iy} = -\frac{\partial V_i}{\partial y_i}, \ \ F_{iz} = -\frac{\partial V_i}{\partial z_i}.$$

Using the previous relations, the generalized active forces are calculated as

$$\begin{aligned}
Q_r &= -\sum_{i=1}^{p} \frac{\partial \mathbf{r}_i}{\partial q_r} \cdot \nabla V_i \\
&= -\sum_{i=1}^{p} \left(\frac{\partial V_i}{\partial x_i}\frac{\partial x_i}{\partial q_r} + \frac{\partial V_i}{\partial y_i}\frac{\partial y_i}{\partial q_r} + \frac{\partial V_i}{\partial z_i}\frac{\partial z_i}{\partial q_r}\right) \\
&= -\sum_{i=1}^{p} \frac{\partial V_i}{\partial q_r} = -\frac{\partial}{\partial q_r}\sum_{i=1}^{p} V_i \\
&= -\frac{\partial V}{\partial q_r}, \tag{7.25}
\end{aligned}$$

where $V = \sum_{i=1}^{p} V_i$ is the total potential energy. The Lagrange's equation of motion of the second kind can be written in terms of the potential energy as

$$\frac{\mathrm{d}}{\mathrm{d}t}\left(\frac{\partial T}{\partial \dot{q}_r}\right) - \frac{\partial T}{\partial q_r} = -\frac{\partial V}{\partial q_r} \quad \text{or} \quad \frac{\mathrm{d}}{\mathrm{d}t}\left(\frac{\partial L}{\partial \dot{q}_r}\right) - \frac{\partial L}{\partial q_r} = 0, \qquad (7.26)$$

where $\frac{\partial V}{\partial \dot{q}_r} = 0$ (since the potential energy V depends on q_r but not on $\dot{q}_r$) and L is the Lagrangian function defined as

$$L = T - V. \qquad (7.27)$$

7.3 Hamilton's Equations

In a Lagrangian formulation, a mechanical system can be described using the generalized coordinates $(q_1, q_2, \ldots, q_n)$ and the generalized velocities $(\dot{q}_1, \dot{q}_2, \ldots, \dot{q}_n)$. The time evolution of the mechanical system can be expressed as given by Lagrange's equations:

$$\frac{\mathrm{d}}{\mathrm{d}t}\left(\frac{\partial L}{\partial \dot{q}_k}\right) - \frac{\partial L}{\partial q_k} = 0. \qquad (7.28)$$

The Lagrangian function of the system is $L = T - V$, where V is the potential energy. Newton's equations are only valid in an inertial frame, and Lagrange's equations are valid in any coordinate system. Using Hamilton's formulation, the system depends on $2n$ variables: the system coordinates $(q_1, q_2, \ldots, q_n)$ and the conjugate momenta $(p_1, p_2, \ldots, p_n)$ defined by $p_k = \frac{\partial L}{\partial \dot{q}_k}$ where $k = 1, \ldots, n$. From Lagrange's equations, one can observe that $\dot{p}_k = \frac{\partial L}{\partial q_k}$ where $k = 1, \ldots, n$. The variation in time of the lagrangian L is

$$\begin{aligned}
\frac{\mathrm{d}L}{\mathrm{d}t} &= \sum_{k=1}^{n} \frac{\partial L}{\partial q_k}\dot{q}_k + \sum_{k=1}^{n} \frac{\partial L}{\partial \dot{q}_k}\ddot{q}_k + \frac{\partial L}{\partial t} \\
&= \sum_{k=1}^{n} \frac{\mathrm{d}}{\mathrm{d}t}\left(\frac{\partial L}{\partial \dot{q}_k}\right)\dot{q}_k + \sum_{k=1}^{n} \frac{\partial L}{\partial \dot{q}_k}\frac{\mathrm{d}}{\mathrm{d}t}(\dot{q}_k) + \frac{\partial L}{\partial t} \\
&= \sum_{k=1}^{n} \frac{\mathrm{d}}{\mathrm{d}t}\left(\dot{q}_k\frac{\partial L}{\partial \dot{q}_k}\right) + \frac{\partial L}{\partial t}. \qquad (7.29)
\end{aligned}$$

Equations (7.29) can be rewritten as

$$\frac{\mathrm{d}}{\mathrm{d}t}\left(L-\sum_{k=1}^{n}\dot{q}_k\frac{\partial L}{\partial \dot{q}_k}\right)=\frac{\partial L}{\partial t}\Leftrightarrow\frac{\mathrm{d}}{\mathrm{d}t}\left(L-\sum_{k=1}^{n}\dot{q}_k p_k\right)=\frac{\partial L}{\partial t}$$

$$\Leftrightarrow-\frac{\mathrm{d}H}{\mathrm{d}t}=\frac{\partial L}{\partial t},\tag{7.30}$$

where H is the *Hamiltonian* of the system defined as

$$H=\sum_{k=1}^{n}p_k\dot{q}_k-L.\tag{7.31}$$

The differential of the Hamiltonian is

$$\begin{aligned}
\mathrm{d}H&=\sum_{k=1}^{n}p_k\mathrm{d}\dot{q}_k+\sum_{k=1}^{n}\dot{q}_k\mathrm{d}p_k-\mathrm{d}L\\
&=\sum_{k=1}^{n}p_k\mathrm{d}\dot{q}_k+\sum_{k=1}^{n}\dot{q}_k\mathrm{d}p_k-\sum_{k=1}^{n}\frac{\partial L}{\partial \dot{q}_k}\mathrm{d}\dot{q}_k-\sum_{k=1}^{n}\frac{\partial L}{\partial p_k}\mathrm{d}p_k-\frac{\partial L}{\partial t}\mathrm{d}t\\
&=\sum_{k=1}^{n}\dot{q}_k\mathrm{d}p_k-\sum_{k=1}^{n}\frac{\partial L}{\partial q_k}\mathrm{d}q_k-\frac{\partial L}{\partial t}\mathrm{d}t.
\end{aligned}\tag{7.32}$$

The differential of the Hamiltonian can be written as

$$\mathrm{d}H=\sum_{k=1}^{n}\frac{\partial H}{\partial p_k}\mathrm{d}p_k-\sum_{k=1}^{n}\frac{\partial H}{\partial q_k}\mathrm{d}q_k-\frac{\partial H}{\partial t}\mathrm{d}t.\tag{7.33}$$

From (7.32) and (7.33), it results

$$\dot{q}_k=\frac{\partial H}{\partial p_k},$$
$$\frac{\partial L}{\partial q_k}=-\frac{\partial H}{\partial q_k},$$
$$-\frac{\partial L}{\partial t}=-\frac{\partial H}{\partial t}=0.\tag{7.34}$$

Equation (7.34) can be rewritten as

$$\dot{q}_k=\frac{\partial H}{\partial p_k},$$
$$\dot{p}_k=-\frac{\partial H}{\partial q_k}.\tag{7.35}$$

Equations (7.35) are known as the equations of motion of Hamiltonian mechanics or the Hamilton's canonical equations. An advantage of using Hamilton's equations is that they are first-order differential equations (expressed in twice as many variables) and thus easier to solve than Lagrange's equations. Another advantage is that the Hamilton's equations describe the dynamics in the phase space in contrast to Lagrange's equations which are second-order equations and concern trajectories in the configuration space.

Under the assumptions $p_k = \frac{\partial L}{\partial \dot{q}_k} = \frac{\partial T}{\partial \dot{q}_k}$, the Hamiltonian can be written as

$$H(q_k, p_k) = \sum_{k=1}^{n} p_k \dot{q}_k - L(q_k, p_k) = \sum_{k=1}^{n} \frac{\partial T}{\partial \dot{q}_k} \dot{q}_k - L(q_k, p_k). \qquad (7.36)$$

The total kinetic energy for a scleronomic system can be written as

$$T = \sum_{i=1}^{n} \frac{1}{2} m_i \dot{\mathbf{r}}_i \cdot \dot{\mathbf{r}}_i = \sum_{j=1}^{n} \sum_{k=1}^{n} \sum_{i=1}^{n} \frac{1}{2} m_i \frac{\partial \mathbf{r}_i}{\partial q_j} \cdot \frac{\partial \mathbf{r}_i}{\partial q_k} \dot{q}_j \dot{q}_k = \sum_{j=1}^{n} \sum_{k=1}^{n} c_{ij} \dot{q}_j \dot{q}_k, \qquad (7.37)$$

where $\dot{\mathbf{r}}_i = \sum_{j=1}^{n} \frac{\partial \mathbf{r}_i}{\partial q_j} \dot{q}_j$ and $c_{ij} = \sum_{i=1}^{n} \frac{1}{2} m_i \frac{\partial \mathbf{r}_i}{\partial q_j} \cdot \frac{\partial \mathbf{r}_i}{\partial q_k}$. Taking the derivative of the kinetic energy, one can obtain

$$\frac{\partial T}{\partial q_k} = \sum_{i=1}^{n} \sum_{j=1}^{n} c_{ij} \left(\frac{\partial \dot{q}_i}{\partial \dot{q}_k} \dot{q}_j + \frac{\partial \dot{q}_j}{\partial \dot{q}_k} \dot{q}_i \right) = \sum_{j=1}^{n} c_{kj} \dot{q}_j + \sum_{i=1}^{n} c_{ik} \dot{q}_i, \qquad (7.38)$$

where $\frac{\partial \dot{\mathbf{r}}_i}{\partial \dot{q}_j} = \sum_{k=1}^{n} \frac{\partial \mathbf{r}_i}{\partial q_k} \frac{\partial \dot{q}_k}{\partial \dot{q}_j} = \frac{\partial \mathbf{r}_i}{\partial q_j}$. Therefore, one can calculate

$$\sum_{k=1}^{n} \frac{\partial T}{\partial \dot{q}_k} \dot{q}_k = \sum_{k=1}^{n} \sum_{j=1}^{n} c_{kj} \dot{q}_k \dot{q}_j + \sum_{k=1}^{n} \sum_{i=1}^{n} c_{ki} \dot{q}_k \dot{q}_i = T + T = 2T. \qquad (7.39)$$

From (7.36) and (7.39), the Hamiltonian can be written as

$$H = 2T - L = 2T - (T - V) = T + V, \qquad (7.40)$$

where V is the potential energy of the system. Considering the Lagrangian as

$$L = T - V = \sum_{i=k}^{n} \frac{1}{2} m_k \dot{q}_k^2 - V(q_k, t),$$

and the Hamiltonian defined as $H(q_k,p_k) = \sum_{k=1}^{n} p_k \dot{q}_k - L(q_k,p_k)$, one can calculate

$$H(q_k,p_k) = \sum_{k=1}^{n} p_k \dot{q}_k - L(q_k,p_k)$$

$$= \sum_{k=1}^{n} p_k \dot{q}_k - \left(\sum_{k=1}^{n} \frac{1}{2} m_i \dot{q}_k^2 - V(q_k,t) \right)$$

$$= \sum_{k=1}^{n} m_k \dot{q}_k - \left(\sum_{k=1}^{n} \frac{1}{2} m_i \dot{q}_k^2 - V(q_k,t) \right)$$

$$= \sum_{k=1}^{n} \frac{1}{2} m_k \dot{q}_k^2 + V(q_k,t)$$

$$= T(\dot{q}_k,t) + V(q_k,t)$$

or equivalent

$$H = T + V.$$

7.4   Poisson Bracket

The notion of Poisson bracket has proved to be a powerful tool in the Hamiltonian mechanics. The Poisson bracket for two functions $F(q_i,p_i,t)$ and $G(q_i,p_i,t)$ can be defined as

$$\{F,G\} = \sum_{i=1}^{n} \left(\frac{\partial F}{\partial p_i} \frac{\partial G}{\partial q_i} - \frac{\partial F}{\partial q_i} \frac{\partial G}{\partial p_i} \right), \tag{7.41}$$

where $(q_1, q_2, \ldots, q_n)$ and $(p_1, p_2, \ldots, p_n)$ are the associated coordinates. For a Hamiltonian mechanical system, one can describe how the coordinate $p_i(t)$ and $q_i(t)$ vary in time, by taking the total time derivative of the function $F(p_i,q_i,t)$ as

$$\frac{dF}{dt} = \frac{\partial F}{\partial t} + \sum_{i=1}^{n} \frac{\partial F}{\partial q_i} \frac{dq_i}{dt} + \sum_{i=1}^{n} \frac{\partial F}{\partial p_i} \frac{dp_i}{dt}$$

$$= \frac{\partial F}{\partial t} + \sum_{i=1}^{n} \frac{\partial F}{\partial q_i} \dot{q}_i + \sum_{i=1}^{n} \frac{\partial F}{\partial p_i} \dot{p}_i. \tag{7.42}$$

Replacing $\dot{q}_i$ and $\dot{p}_i$ in (7.42) with $\dot{p}_i = -\frac{dH}{dq_i}$ and, respectively, $\dot{q}_i = \frac{dH}{dp_i}$, that is, the equations of motion of Hamiltonian mechanics, it results

$$\frac{dF}{dt} = \frac{\partial F}{\partial t} + \sum_{i=1}^{n} \left(\frac{\partial F}{\partial q_i} \frac{\partial H}{\partial p_i} - \frac{\partial F}{\partial p_i} \frac{\partial H}{\partial q_i} \right). \tag{7.43}$$

The second term of (7.43) is exactly the Poisson bracket of H and F, which can be written as

$$\{H,F\} = \sum_{i=1}^{n} \left(\frac{\partial F}{\partial q_i} \frac{\partial H}{\partial p_i} - \frac{\partial F}{\partial p_i} \frac{\partial H}{\partial q_i} \right). \tag{7.44}$$

From (7.43) and (7.44), the time evolution of the function F is

$$\frac{\mathrm{d}F}{\mathrm{d}t} = \frac{\partial F}{\partial t} + \{H,F\} \;\Leftrightarrow\; \dot{F} = \frac{\partial F}{\partial t} + \{H,F\}. \tag{7.45}$$

The Poisson bracket has the following properties:

1. The antisymmetry in its two arguments (skew-symmetry property)

$$\{F,G\} = -\{G,F\}.$$

2. The linearity (in the first component)

$$\{c_1 F_1 + c_2 F_2, G\} = c_1\{F_1,G\} + c_2\{F_2,G\}.$$

3. The bilinearity property

$$\{c_1 F_1 + c_2 F_2, d_1 G_1 + d_2 G_2\} = c_1 d_1\{F_1,G_1\} + c_1 d_2\{F_1,G_2\}$$
$$+ c_2 d_1\{F_2,G_1\} + c_2 d_2\{F_2,G_2\}.$$

4. The Leibnitz property (product rule)

$$\{F,GH\} = G\{F,H\} + \{F,G\}H.$$

5. The Jacobi identity

$$\{F,\{G,H\}\} + \{G,\{H,F\}\} + \{H,\{F,G\}\} = 0.$$

Exercise

Consider the angular momentum $\mathbf{L} = (L_x, L_y, L_z)$ defined as

$$\mathbf{L} = \mathbf{r} \times \mathbf{p} = \begin{vmatrix} \mathbf{i} & \mathbf{j} & \mathbf{k} \\ x & y & z \\ p_x & p_y & p_z \end{vmatrix}, \tag{7.46}$$

where the components of the angular momentum are $L_x = y p_z - z p_y$, $L_y = z p_x - x p_z$ and $L_z = x p_y - y p_x$, the momentum vector is $\mathbf{p} = (p_x, p_y, p_z)$, and $\mathbf{r} = (x,y,z)$. Evaluate the set of Poisson brackets $\{L_x, L_y\}$, $\{L_y, L_z\}$, and $\{L_z, L_x\}$.

Solution

Using the definition of the Poisson bracket, it results

$$\{L_x, L_y\} = \sum_{i=x,y,z} \left(\frac{\partial L_x}{\partial p_i} \frac{\partial L_y}{\partial q_i} - \frac{\partial L_x}{\partial q_i} \frac{\partial L_y}{\partial p_i} \right)$$

$$= \frac{\partial L_x}{\partial p_x} \frac{\partial L_y}{\partial x} - \frac{\partial L_x}{\partial x} \frac{\partial L_y}{\partial p_x} + \frac{\partial L_x}{\partial p_y} \frac{\partial L_y}{\partial y} - \frac{\partial L_x}{\partial y} \frac{\partial L_y}{\partial p_y} + \frac{\partial L_x}{\partial p_z} \frac{\partial L_y}{\partial z} - \frac{\partial L_x}{\partial z} \frac{\partial L_y}{\partial p_z}$$

$$= 0 \frac{\partial L_y}{\partial x} - 0 \frac{\partial L_y}{\partial p_x} + \frac{\partial L_x}{\partial p_y} 0 - \frac{\partial L_x}{\partial y} 0 + y p_x - (-p_y)(-x)$$

$$= y p_x - x p_y = -L_z,$$

$$\{L_y, L_z\} = \sum_{i=x,y,z} \left(\frac{\partial L_y}{\partial p_i} \frac{\partial L_z}{\partial q_i} - \frac{\partial L_y}{\partial q_i} \frac{\partial L_z}{\partial p_i} \right)$$

$$= \frac{\partial L_y}{\partial p_y} \frac{\partial L_z}{\partial y} - \frac{\partial L_y}{\partial y} \frac{\partial L_z}{\partial p_y} + \frac{\partial L_y}{\partial p_z} \frac{\partial L_z}{\partial z} - \frac{\partial L_y}{\partial z} \frac{\partial L_z}{\partial p_z} + \frac{\partial L_y}{\partial p_x} \frac{\partial L_z}{\partial x} - \frac{\partial L_y}{\partial x} \frac{\partial L_z}{\partial p_x}$$

$$= 0 \frac{\partial L_z}{\partial y} - 0 \frac{\partial L_z}{\partial p_y} + \frac{\partial L_y}{\partial p_z} 0 - \frac{\partial L_y}{\partial z} 0 + z p_y - p_z y$$

$$= z p_y - y p_z - = -L_x,$$

$$\{L_z, L_x\} = \sum_{i=x,y,z} \left(\frac{\partial L_x}{\partial p_i} \frac{\partial L_x}{\partial q_i} - \frac{\partial L_z}{\partial q_i} \frac{\partial L_x}{\partial p_i} \right)$$

$$= \frac{\partial L_z}{\partial p_z} \frac{\partial L_x}{\partial z} - \frac{\partial L_z}{\partial z} \frac{\partial L_x}{\partial p_z} + \frac{\partial L_z}{\partial p_x} \frac{\partial L_x}{\partial x} - \frac{\partial L_z}{\partial x} \frac{\partial L_x}{\partial p_x} + \frac{\partial L_z}{\partial p_y} \frac{\partial L_x}{\partial y} - \frac{\partial L_z}{\partial y} \frac{\partial L_x}{\partial p_y}$$

$$= 0 \frac{\partial L_x}{\partial z} - 0 \frac{\partial L_x}{\partial p_z} + \frac{\partial L_z}{\partial p_x} 0 - \frac{\partial L_z}{\partial x} 0 + x p_z - (-p_x)(-z)$$

$$= x p_z - z p_x = -L_y.$$

7.5 Rotation Transformation

Two orthogonal reference frames, xyz and $x'y'z'$, are considered. The unit vectors of the reference frame xyz are $\mathbf{i}, \mathbf{j}, \mathbf{k}$, and the unit vectors of the reference frame $x'y'z'$ are $\mathbf{i}', \mathbf{j}', \mathbf{k}'$. The origins of the reference frames may coincide because only the orientation of the axes is of interest $O = O'$.

The angles between the x'-axis and each of the x, y, z axes are the direction angles α, β, and γ $(O < \alpha, \beta, \gamma < \pi)$ as shown in Fig. 7.2. The unit vector $\mathbf{i}'$ can

Fig. 7.2 Direction angles α, β, and γ

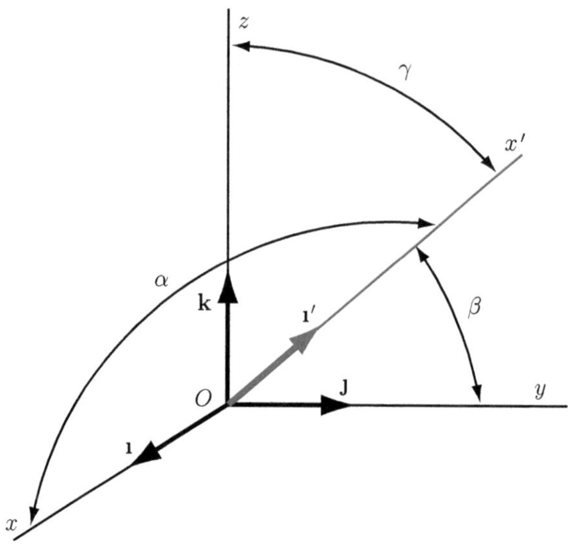

be expressed in terms of $\iota, \mathbf{J}, \mathbf{k}$ and the direction angles:

$$\iota' = (\iota' \cdot \iota)\iota + (\iota' \cdot \mathbf{J})\mathbf{J} + (\iota' \cdot \mathbf{k})\mathbf{k} = \cos\alpha\,\iota + \cos\beta\,\mathbf{J} + \cos\gamma\,\mathbf{k}.$$

The cosines of the direction angles are the direction cosines and $\cos^2\alpha + \cos^2\beta + \cos^2\gamma = 1$.

With the notations $\cos\alpha = a_{x'x}$, $\cos\beta = a_{x'y}$, and $\cos\gamma = a_{x'z}$, the unit vector ι' is

$$\iota' = a_{x'x}\,\iota + a_{x'y}\,\mathbf{J} + a_{x'z}\,\mathbf{k}.$$

In a similar way, the unit vectors $\mathbf{J}'$ and $\mathbf{k}'$ are

$$\mathbf{J}' = a_{y'x}\,\iota + a_{y'y}\,\mathbf{J} + a_{y'z}\,\mathbf{k},$$
$$\mathbf{k}' = a_{z'x}\,\iota + a_{z'y}\,\mathbf{J} + a_{z'z}\,\mathbf{k},$$

where $a_{r's} = a_{rs'}$ are the cosine of the angle between axis r' and axis s, with r and r representing x, y, or z. In matrix form,

$$\begin{bmatrix} \iota' \\ \mathbf{J}' \\ \mathbf{k}' \end{bmatrix} = \mathbf{R} \begin{bmatrix} \iota \\ \mathbf{J} \\ \mathbf{k} \end{bmatrix},$$

where

$$\mathbf{R} = \begin{bmatrix} a_{x'x} & a_{x'y} & a_{x'z} \\ a_{y'x} & a_{y'y} & a_{y'z} \\ a_{z'x} & a_{z'y} & a_{z'z} \end{bmatrix}.$$

The matrix $\mathbf{R}$ is the *rotation transformation matrix* from xyz to $x'y'z'$. The unit vectors $\mathbf{i}, \mathbf{j}, \mathbf{k}$ are an orthogonal set of unit vectors, and the unit vectors $\mathbf{i}', \mathbf{j}', \mathbf{k}'$ are an orthogonal set too. Using these properties, it results that

$$\mathbf{R} \cdot \mathbf{R}^{\mathrm{T}} = \mathbf{I},$$

where $\mathbf{I}$ is the identity matrix. Multiplication of (7.47) by $\mathbf{R}^{-1}$ gives

$$\mathbf{R}^{-1} = \mathbf{R}^{\mathrm{T}}.$$

The matrix R is an orthonormal matrix because $\mathbf{R}^{-1} = \mathbf{R}^{\mathrm{T}}$.

Let $\mathbf{R}'$ be the transformation matrix from $\mathbf{i}, \mathbf{j}, \mathbf{k}$ to $\mathbf{i}', \mathbf{j}', \mathbf{k}'$:

$$\begin{bmatrix} \mathbf{i} \\ \mathbf{j} \\ \mathbf{k} \end{bmatrix} = \mathbf{R}' \begin{bmatrix} \mathbf{i}' \\ \mathbf{j}' \\ \mathbf{k}' \end{bmatrix}. \tag{7.47}$$

The matrix $\mathbf{R}'$ is the inverse of the original transformation matrix $\mathbf{R}$, which is identical to the transpose of $\mathbf{R}$:

$$\mathbf{R}' = \mathbf{R}^{-1} = \mathbf{R}^{\mathrm{T}}.$$

Any vector $\mathbf{p}$ is independent of the reference frame used to describe its components, so

$$\mathbf{p} = p_x \mathbf{i} + p_y \mathbf{j} + p_z \mathbf{k} = p_{x'} \mathbf{i}' + p_{y'} \mathbf{j}' + p_{z'} \mathbf{k}',$$

or in matrix form as

$$[\mathbf{i}' \; \mathbf{j}' \; \mathbf{k}'] \begin{bmatrix} p_{x'} \\ p_{y'} \\ p_{z'} \end{bmatrix} = [\mathbf{i} \; \mathbf{j} \; \mathbf{k}] \begin{bmatrix} p_x \\ p_y \\ p_z \end{bmatrix}.$$

Using (7.47) and the fact that the transpose of a product is the product of the transposes, the following relation is obtained:

$$[\mathbf{i}' \; \mathbf{j}' \; \mathbf{k}'] \begin{bmatrix} p_{x'} \\ p_{y'} \\ p_{z'} \end{bmatrix} = [\mathbf{i}' \; \mathbf{j}' \; \mathbf{k}'][\mathbf{R}']^{\mathrm{T}} \begin{bmatrix} p_x \\ p_y \\ p_z \end{bmatrix}.$$

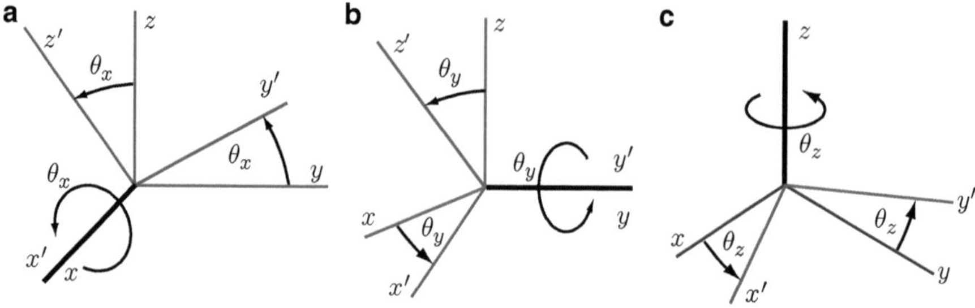

Fig. 7.3 Simple rotation about (**a**) x-axis (**b**) y-axis, and (**c**) z-axis

With $[\mathbf{R}']^{\mathrm{T}} = \mathbf{R}$, the above equation leads to

$$
\begin{bmatrix} p_{x'} \\ p_{y'} \\ p_{z'} \end{bmatrix} = \mathbf{R} \begin{bmatrix} p_x \\ p_y \\ p_z \end{bmatrix}.
$$

When the reference frame $x'y'z'$ is the result of a simple rotation about one of the axes of the reference frame xyz, the following transformation matrices are obtained, as shown in Fig. 7.3:

- The reference frame xyz is rotated by an angle θ_x about the x-axis:

$$
\mathbf{R}(x, \theta_x) = \mathbf{R}(\theta_x) = \begin{bmatrix} 1 & 0 & 0 \\ 0 & \cos\theta_x & \sin\theta_x \\ 0 & -\sin\theta_x & \cos\theta_x \end{bmatrix}.
$$

- The reference frame xyz is rotated by an angle θ_y about the y-axis:

$$
\mathbf{R}(y, \theta_y) = \mathbf{R}(\theta_y) = \begin{bmatrix} \cos\theta_y & 0 & -\sin\theta_y \\ 0 & 1 & 0 \\ \sin\theta_y & 0 & \cos\theta_y \end{bmatrix}.
$$

- The reference frame xyz is rotated by an angle θ_y about the z-axis:

$$
\mathbf{R}(z, \theta_z) = \mathbf{R}(\theta_z) = \begin{bmatrix} \cos\theta_z & \sin\theta_z & 0 \\ -\sin\theta_z & \cos\theta_z & 0 \\ 0 & 0 & 1 \end{bmatrix}.
$$

The following property holds:

$$
\mathbf{R}(s, -\theta_s) = \mathbf{R}^{\mathrm{T}}(s, \theta_s), \quad s = x, y, z.
$$

7.6 Examples

Example 7.1. A mathematical pendulum of length L and mass M lumped at the end B is attached to a slider 1 of mass m, as shown in Fig. 7.4a. The slider 1 is moving in the horizontal direction and is connected to the wall by a spring having the elastic constant k. The spring deflects only horizontally. Find the equations of motion using Lagrange's method. For the numerical application, use $m = 0.4\,\mathrm{kg}$, $M = 0.2\,\mathrm{kg}$, $k = 0.75\,\mathrm{Nm}$, $L = 0.3\,\mathrm{m}$, and $g = 9.81\,\mathrm{ms}^2$.

Solution

The motion of the system is studied with respect to the xyz Cartesian reference frame of unit vectors $[\mathbf{\imath}, \mathbf{\jmath}, \mathbf{k}]$ as shown in Fig. 7.4a. There are two generalized coordinates: the generalized coordinate $q_1(t)$ is the horizontal displacement of the slider, and $q_2(t)$ is the angular displacement of the pendulum with the vertical axis.

The MATLAB program starts with the following statements:

```
syms L m M k g t real
q1 = sym('q1(t)');
q2 = sym('q2(t)');
```

The position vector of the slider (the mass the slider is located at A) is

$$\mathbf{r}_A = q_1(t)\,\mathbf{\imath}.$$

The position vector of the pendulum with the mass located at B is

$$\mathbf{r}_B = \left[q_1(t) + L\sin q_2(t)\right]\mathbf{\imath} + L\cos q_2(t)\,\mathbf{\jmath}.$$

The velocity of the slider is

$$\mathbf{v}_A = \frac{d\mathbf{r}_A}{dt} = \dot{q}_1(t)\,\mathbf{\imath},$$

and the velocity of the pendulum at B is

$$\mathbf{v}_B = \frac{d\mathbf{r}_B}{dt} = \left[\dot{q}_1(t) + L\dot{q}_2\cos q_2\right]\mathbf{\imath} - L\dot{q}_2\sin q_2\,\mathbf{\jmath}.$$

The position vectors and the velocities are expressed in MATLAB using the statements:

```
rA= [q1,0,0];
rB= [q1+L*sin(q2),L*cos(q2),0];
vA=diff(rA,t);
vB=diff(rB,t);
```

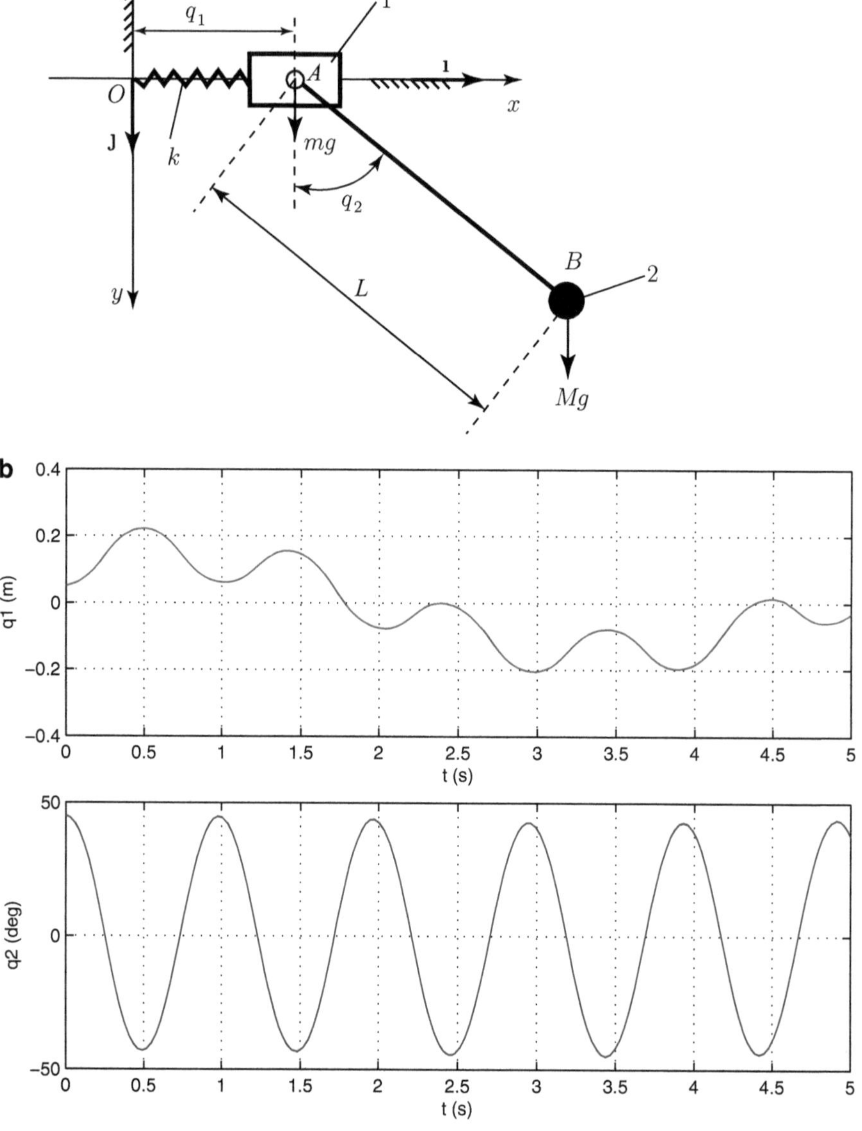

Fig. 7.4 Example 7.1

The kinetic energy of the slider, T_1 , and the kinetic energy of the pendulum, T_2, are

$$T_1 = \frac{1}{2}m\mathbf{v}_A \cdot \mathbf{v}_A = \frac{1}{2}m\dot{q}_1^2,$$

$$T_2 = \frac{1}{2}M\mathbf{v}_B \cdot \mathbf{v}_B = \frac{1}{2}M\left(\dot{q}_1^2 + 2L\dot{q}_1\dot{q}_2\cos q_2 + L^2\dot{q}_2^2\right).$$

The total kinetic energy of the system (slider and pendulum) is

$$T = T_1 + T_2$$
$$= \frac{1}{2}(m+M)\dot{q}_1^2 + \frac{1}{2}M\left(2L\dot{q}_1\dot{q}_2\cos q_2 + L^2\dot{q}_2^2\right). \tag{7.48}$$

The total kinetic energy T is calculated as the sum of the kinetic energy $T1$ of the slider and the kinetic energy $T2$ of the pendulum using the MATLAB statement:

```
% kinetic energy of 1
T1 = m*vA*vA.'\2;
% .' array transpose
% A.' is the array transpose of A
% kinetic energy of 2
T2 = M*vB*vB.'\2;
T2 = simplify(T2);
fprintf('T1 = %s \n',char(T1))
fprintf('T2 = %s \n',char(T2))
% total kinetic energy
T = expand(T1 + T2);
```

The MATLAB statements A.' is the array transpose of A, and simple(exp) looks for simplest form of the symbolic expression exp. The MATLAB command expand(exp) expands trigonometric and algebraic functions.

For the generalized coordinate q_1, the left-hand side of Lagrange's equation can be written as

$$\frac{\partial T}{\partial \dot{q}_1} = (m+M)\dot{q}_1 + LM\dot{q}_2\cos q_2,$$

$$\frac{\mathrm{d}}{\mathrm{d}t}\left(\frac{\partial T}{\partial \dot{q}_1}\right) = (m+M)\ddot{q}_1 + LM\ddot{q}_2\cos q_2,$$

$$\frac{\partial T}{\partial q_1} = 0. \tag{7.49}$$

For the generalized coordinate q_2, the left-hand side of Lagrange's equation is

$$\frac{\partial T}{\partial \dot{q}_2} = LM\left(\dot{q}_1\cos q_2 + L\dot{q}_2\right),$$

$$\frac{\mathrm{d}}{\mathrm{d}t}\left(\frac{\partial T}{\partial \dot{q}_2}\right) = LM\left(\ddot{q}_1\cos q_2 - \dot{q}_1\dot{q}_2\sin q_2 + L\ddot{q}_2\right),$$

$$\frac{\partial T}{\partial q_2} = -LM\dot{q}_1\dot{q}_2\sin q_2. \tag{7.50}$$

To calculate the partial derivative of the kinetic energy T with respect to the variable
`diff('q1(t)',t)` a MATLAB function, `deriv`, is created:

```
function fout = deriv(f, g)
clear t qx dqx
% deriv differentiates f with respect to g=g(t)
% the variable g=g(t) is a function of time
syms t qx dqx
lg = {diff(g, t), g};
lx = {dqx, qx};
f1 = subs(f, lg, lx);
f2 = diff(f1, qx);
fout = subs(f2, lx, lg);
```

The function `deriv(f, g)` differentiates a symbolic expression `f` with respect to
the variable `g`, where the variable `g` is a function of time `g = g(t)`. The statement
`diff(f,'x')` differentiates `f` with respect to the free variable `x`. In MATLAB,
the free variable `x` cannot be a function of time and that is why the function `deriv`
was introduced.

The partial derivatives of the kinetic energy T with respect to $\dot{q}_i$ or in MATLAB
the partial derivatives of the kinetic energy T with respect to `diff('q1(t)',t)`
and `diff('q2(t)',t)` are calculated with

```
Tdq1 = deriv(T, diff(q1,t));
Tdq2 = deriv(T, diff(q2,t));
```

The left-hand sides of Lagrange's equations, $\frac{d}{dt}\left(\frac{\partial T}{\partial \dot{q}_i}\right) - \frac{\partial T}{\partial q_i}$, $\quad i = 1, 2$, with
MATLAB are

```
% d(dT\d(dq))\dt
Tt1 = diff(Tdq1, t);
Tt2 = diff(Tdq2, t);
% dT\dq
Tq1 = deriv(T, q1);
Tq2 = deriv(T, q2);
% left hand side of Lagrange's eom
LHS1 = Tt1 - Tq1;
LHS2 = Tt2 - Tq2;
```

The generalized active forces are calculated next. The gravity forces on slider 1 and
pendulum 2 at A and B are

$$\mathbf{G}_1 = mg\mathbf{J} \quad \text{and} \quad \mathbf{G}_2 = Mg\mathbf{J}.$$

The total force that acts on the slider, the spring force and the gravity force, can be
written as

$$\mathbf{F}_1 = \mathbf{F}_A = -kq_1\mathbf{1} + mg\mathbf{J}.$$

There are two generalized forces acting on the system. The generalized force associated to q_1 is

$$Q_1 = \mathbf{F}_1 \cdot \frac{\partial \mathbf{r}_A}{\partial q_1} + \mathbf{G}_2 \cdot \frac{\partial \mathbf{r}_B}{\partial q_1}$$
$$= (-kq_1\mathbf{\imath} + mg\mathbf{\jmath}) \cdot \mathbf{\imath} + Mg\mathbf{\jmath} \cdot \mathbf{\imath}$$
$$= -kq_1.$$

The generalized force associated to q_2 is

$$Q_2 = \mathbf{F}_1 \cdot \frac{\partial \mathbf{r}_A}{\partial q_2} + \mathbf{G}_2 \cdot \frac{\partial \mathbf{r}_B}{\partial q_2}$$
$$= (-kq_1\mathbf{\imath} + mg\mathbf{\jmath}) \cdot \mathbf{0} + Mg\mathbf{\jmath} \cdot (L\cos q_2\mathbf{\imath} - L\sin q_2\mathbf{\jmath})$$
$$= -MgL\sin q_2.$$

The generalized forces Q_1 and Q_2 have been expressed in MATLAB using the next statement:

```
% generalized active forces
G1 = [0 m*g 0];
G2 = [0 M*g 0];
Fe1=[-k*q1 0 0];

% partial derivatives
rA_1 = deriv(rA, q1); rB_1 = deriv(rB, q1);
rA_2 = deriv(rA, q2); rB_2 = deriv(rB, q2);

% generalized active force Q1
Q1 = rA_1*(G1+Fe1).'+rB_1*G2.';
% generalized active force Q2
Q2 = rA_2*(G1+Fe1).'+rB_2*G2.';
fprintf('Q1 = \n'); pretty(simple(Q1));
fprintf('\n')
fprintf('Q2 = \n'); pretty(simple(Q2));
fprintf('\n')
```

The Lagrange's equations can be written as

$$\frac{\mathrm{d}}{\mathrm{d}t}\left(\frac{\partial T}{\partial \dot{q}_1}\right) - \frac{\partial T}{\partial q_1} = Q_1,$$

$$\frac{\mathrm{d}}{\mathrm{d}t}\left(\frac{\partial T}{\partial \dot{q}_2}\right) - \frac{\partial T}{\partial q_2} = Q_2.$$

The equations of motion can be written as

$$(m+M)\ddot{q}_1 + LM\ddot{q}_2 \cos q_2 - LM\dot{q}_2^2 \sin q_2 = -kq_1,$$

$$LM\left(\ddot{q}_1 \cos q_2 - \dot{q}_1\dot{q}_2 \sin q_2 + L\ddot{q}_2\right) + LM\dot{q}_2\dot{q}_1 \sin q_2 = -MgL \sin q_2,$$

or

$$(m+M)\ddot{q}_1 + LM\ddot{q}_2 \cos q_2 - LM\dot{q}_2^2 \sin q_2 + kq_1 = 0,$$

$$LM\ddot{q}_1 \cos q_2 - L^2M\ddot{q}_2 + MgL \sin q_2 = 0.$$

The Lagrange's equations with the initial numerical values are written in MATLAB using the statement:

```
% first Lagrange's equation of motion
Lagrange1 = LHS1-Q1;
% second Lagrange's equation of motion
Lagrange2 = LHS2-Q2;

data = {m, M, k, L, g};
datn = {0.4, 0.2, 0.75, 0.3, 9.81};

Lagran1 = subs(Lagrange1, data, datn);
Lagran2 = subs(Lagrange2, data, datn);
```

The two second-order Lagrange's equations have to be rewritten as a first-order system:

```
ql = {diff(q1,t,2), diff(q2,t,2),...
    diff(q1,t), diff(q2,t), q1, q2};
qf = ...
{'ddq1', 'ddq2', 'x(2)', 'x(4)', 'x(1)', 'x(3)'};

% ql                          qf
%----------------------------------
% diff('q1(t)',t,2)  -> 'ddq1'
% diff('q2(t)',t,2)  -> 'ddq2'
%    diff('q1(t)',t)  -> 'x(2)'
%    diff('q2(t)',t)  -> 'x(4)'
%            'q1(t)'  -> 'x(1)'
%            'q2(t)'  -> 'x(3)'

Lagra1 = subs(Lagran1, ql, qf);
Lagra2 = subs(Lagran2, ql, qf);

% solve e.o.m. for ddq1, ddq2
sol = solve(Lagra1,Lagra2,'ddq1, ddq2');
```

```
Lagr1 = sol.ddq1;
Lagr2 = sol.ddq2;

% system of ODE
dx2dt = char(Lagr1);
dx4dt = char(Lagr2);
```

The system of differential equations is solved numerically by m-file functions. The function file, eomE7_1.m, is created using the following statements:

```
fid = fopen('eomE7_1.m','w+');
fprintf(fid,'function dx = eomE7_1(t,x)\n');
fprintf(fid,'dx = zeros(4,1);\n');
fprintf(fid,'dx(1) = x(2);\n');
fprintf(fid,'dx(2) = ');
fprintf(fid,dx2dt);
fprintf(fid,';\n');
fprintf(fid,'dx(3) = x(4);\n');
fprintf(fid,'dx(4) = ');
fprintf(fid,dx4dt);
fprintf(fid,';');
fclose(fid); cd(pwd);
```

The ode45 solver is used for the system of differential equations:

```
t0 = 0;   tf = 5; time = [0 tf];

x0 = [0.05 0.1 pi\4 0];   % initial conditions

[t,xs] = ode45(@eomE7_1, time, x0);
x1 = xs(:,1);
x2 = xs(:,2);
x3 = xs(:,3);
x4 = xs(:,4);

subplot(2,1,1),plot(t,x1,'r'),...
xlabel('t (s)'),ylabel('q1 (m)'),grid,...
subplot(2,1,2),plot(t,x3*180\pi,'b'),...
xlabel('t (s)'),ylabel('q2 (deg)'),grid

[ts,xs] = ode45(@eomE7_1,0:1:5,x0);
fprintf('Results \n\n')
fprintf...
('     t(s)        q1(m)      q2(rad)\n')
[ts, xs(:,1), xs(:,3)]
```

The graphic of the MATLAB program is shown in Fig. 7.4b.

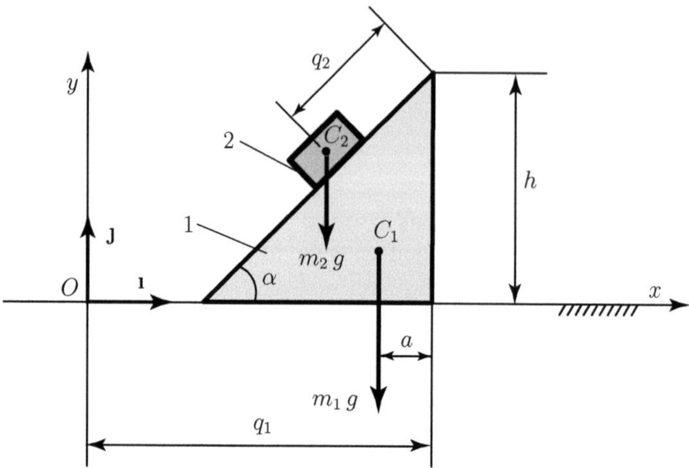

Fig. 7.5 Example 7.2

Example 7.2. The slider 2 of mass $m_2 = 1\,$kg is sliding without friction down on an inclined prism 1 of mass $m_1 = 10\,$kg as shown Fig. 7.5a. The prism makes an angle $\alpha = \pi/6$ with the horizontal and slides without friction along the horizontal surface, x-axis. Find and solve the equations of motion of the system using Lagrange's method.

Solution

The motion of the system is studied with respect to the xy Cartesian reference as shown in Fig. 7.5. The generalized coordinates are $q_1(t)$ and $q_2(t)$ where $q_1(t)$ is the horizontal displacement of the prism 1 (distance from the vertical side of the prism to the origin) and $q_2(t)$ is the displacement of the slider 2 (distance from the mass center of the slider to the upper corner of the inclined plane) along the sliding line.

The position and velocity vectors of the mass centers C_1 and C_2 of the prism and the slider are

$$\mathbf{r}_1 = \mathbf{r}_{C_1} = q_1\mathbf{\imath} + (h/3)\mathbf{\jmath},$$

$$\mathbf{r}_2 = \mathbf{r}_{C_2} = (q_1 - a - q_2\cos\alpha)\,\mathbf{\imath} + (h - q_2\sin\alpha)\,\mathbf{\jmath},$$

$$\mathbf{v}_1 = \mathbf{v}_{C_1} = \dot{q}_1\mathbf{\imath},$$

$$\mathbf{v}_2 = \mathbf{v}_{C_2} = (\dot{q}_1 - \dot{q}_2\cos\alpha)\,\mathbf{\imath} - \dot{q}_2\sin\alpha\,\mathbf{\jmath},$$

where $a = (h\cot\alpha)/3$. The position vectors and velocity vectors of the mass center of the inclined plane 1 and slider 2 are expressed in MATLAB using the statement:

```
syms h alpha m1 m2 g real
```

```
t = sym('t','real');
q1 = sym('q1(t)');
q2 = sym('q2(t)');

r1=[q1 h\3 0];
a=h*cot(alpha)\3;
x2=q1+a-q2*cos(alpha);
y2=h-q2*sin(alpha);
r2=[x2 y2 0];

v1=diff(r1,t);
v2=diff(r2,t);
```

The magnitude of the velocity of the slider can be computed as

$$
\begin{aligned}
v_2 &= \sqrt{(\dot{q}_1 - \dot{q}_2 \cos\alpha)^2 + (-\dot{q}_2 \sin\alpha)^2} \\
&= \sqrt{\dot{q}_1^2 - 2\dot{q}_1\dot{q}_2 \cos\alpha + \dot{q}_2^2 \cos^2\alpha + \dot{q}_2^2 \sin^2\alpha} \\
&= \sqrt{\dot{q}_1^2 - 2\dot{q}_1\dot{q}_2 \cos\alpha + \dot{q}_2^2 \left(\cos^2\alpha + \sin^2\alpha\right)} \\
&= \sqrt{\dot{q}_1^2 - 2\dot{q}_1\dot{q}_2 \cos\alpha + \dot{q}_2^2}.
\end{aligned}
$$

The kinetic energy of the system is

$$
\begin{aligned}
T &= T_1 + T_2 \\
&= \frac{m_1}{2}v_1^2 + \frac{m_2}{2}v_2^2 \\
&= \frac{m_1}{2}\dot{q}_1^2 + \frac{m_2}{2}\left(\dot{q}_1^2 - 2\dot{q}_1\dot{q}_2 \cos\alpha + \dot{q}_2^2\right).
\end{aligned}
$$

The total kinetic energy T is calculated as the sum of the kinetic energy $T1$ and $T2$ using the next MATLAB statement:

```
T1=1\2*m1*dotproduct(v1,v1);
T2=1\2*m2*dotproduct(v2,v2);
T = T1+T2;
```

where `dotproduct.m` function is

```
function val = dotproduct(a,b)
% symbolic dot product function
% a.b=a_x*b_x+a_x*b_x+a_x*b_x
% function accepts sym as the input argument
val = sum(a.*b);
```

The gravity forces on 1 and 2 at C_1 and C_2 are

$$\mathbf{G}_1 = -m_1 g \mathbf{J} \quad \text{and} \quad \mathbf{G}_2 = -m_2 g \mathbf{J}.$$

The generalized force associated to q_1 is

$$Q_1 = \mathbf{G}_1 \cdot \frac{\partial \mathbf{r}_1}{\partial q_1} + \mathbf{G}_2 \cdot \frac{\partial \mathbf{r}_2}{\partial q_1}$$
$$= -m_1 g \mathbf{J} \cdot \mathbf{I} - m_2 g \mathbf{J} \cdot \mathbf{I} = 0.$$

The generalized force associated to q_2 is

$$Q_2 = \mathbf{G}_1 \cdot \frac{\partial \mathbf{r}_1}{\partial q_2} + \mathbf{G}_2 \cdot \frac{\partial \mathbf{r}_2}{\partial q_2}$$
$$= -m_1 g \mathbf{J} \cdot \mathbf{0} - m_2 g \mathbf{J} \cdot (-\cos\alpha \mathbf{I} - \sin\alpha \mathbf{J})$$
$$= m_2 g \sin\alpha.$$

The generalized forces `Q1` and `Q2` have been expressed in MATLAB using the next statement:

```
G1  =  [0 -m1*g 0];
G2  =  [0 -m2*g 0];
r1_1  =  deriv(r1, q1);
r1_2  =  deriv(r1, q2);
r2_1  =  deriv(r2, q1);
r2_2  =  deriv(r2, q2);

% generalized active force Q1
Q1  =  r1_1*G1.'+r2_1*G2.';
% generalized active force Q2
Q2  =  r1_2*G1.'+r2_2*G2.';
```

Another way of calculating the generalized forces is with the help of potential energy:

$$V = m_1 g y_1 + m_2 g y_2 = m_1 g h/3 + m_2 g (h - q_2 \sin\alpha).$$

The generalized forces are

$$Q_1 = -\frac{\partial V}{\partial q_1} = 0,$$

$$Q_2 = -\frac{\partial V}{\partial q_2} = m_2 g \sin\alpha.$$

or with MATLAB

```
V=m1*g*r1(2)+m2*g*r2(2);
Q_1=-deriv(V,q1);
```

```
Q_2=-deriv(V,q2);
```
The Lagrange's equations are

$$\frac{\mathrm{d}}{\mathrm{d}t}\left(\frac{\partial T}{\partial \dot{q}_1}\right) - \frac{\partial T}{\partial q_1} = Q_1,$$

$$\frac{\mathrm{d}}{\mathrm{d}t}\left(\frac{\partial T}{\partial \dot{q}_2}\right) - \frac{\partial T}{\partial q_2} = Q_2.$$

The left-hand sides of Lagrange's equations are calculated for the generalized coordinate q_1 with

$$\frac{\partial T}{\partial q_1} = 0,$$

$$\frac{\partial T}{\partial \dot{q}_1} = (m_1 + m_2)\,\dot{q}_1 - m_2\,\dot{q}_2 \cos\alpha,$$

$$\frac{\mathrm{d}}{\mathrm{d}t}\left(\frac{\partial T}{\partial \dot{q}_1}\right) = (m_1 + m_2)\,\ddot{q}_1 - m_2\,\ddot{q}_2 \cos\alpha.$$

The left-hand sides of Lagrange's equations are calculated for the generalized coordinate q_2 with

$$\frac{\partial T}{\partial q_2} = 0,$$

$$\frac{\partial T}{\partial \dot{q}_2} = m_2\,(\dot{q}_2 - \dot{q}_1 \cos\alpha),$$

$$\frac{\mathrm{d}}{\mathrm{d}t}\left(\frac{\partial T}{\partial \dot{q}_2}\right) = m_2\,(\ddot{q}_2 - \ddot{q}_1 \cos\alpha).$$

The equations of motion are

$$(m_1 + m_2)\,\ddot{q}_1 - m_2\,\ddot{q}_2 \cos\alpha = 0,$$

$$m_2\,(\ddot{q}_2 - \ddot{q}_1 \cos\alpha) = m_2\,g \sin\alpha. \tag{7.51}$$

Lagrange's equations can be calculated in MATLAB using the function $[L]$ = Lagrange(T,Q,q,t):

```
function [L] = Lagrange(E,Q,q,t)


% dT\d(dq)
Tdq = deriv(E, diff(q,t));
% d(dT\d(dq))\dt
Tt = diff(Tdq, t);
```

```
% dT\dq
Tq = deriv(E, q);

% left hand side of Lagrange's eom
LHS = Tt - Tq;

% Lagrange's equation of motion
L = LHS-Q;
```

The Lagrange's equations are calculated with MATLAB:

```
% first Lagrange's equation of motion
% function [L] = Lagrange(T,Q,q,t)
Lagrange1 = Lagrange(T,Q1,q1,t);
% second Lagrange's equation of motion
Lagrange2 = Lagrange(T,Q2,q2,t);

Lagrange1 = simplify(Lagrange1);
Lagrange2 = simplify(Lagrange2);
fprintf('equations of motion:\n')
fprintf('%s =0   \n',char(Lagrange1))
fprintf('%s =0   \n',char(Lagrange2))
fprintf('\n')
```

The numerical solution for the system of equations is obtained using the MATLAB statement:

```
slist = {m1, m2, g, alpha};
nlist = {10, 1, 9.81, pi\6};

Lagran1 = subs(Lagrange1, slist, nlist);
Lagran2 = subs(Lagrange2, slist, nlist);

digits(3)
fprintf('%s =0   \n',char(vpa(Lagran1)))
fprintf('%s =0   \n',char(vpa(Lagran2)))
fprintf('\n')

ql = {diff(q1,t,2), diff(q2,t,2)};
qf = {'ddq1', 'ddq2'};

Lagra1 = subs(Lagran1, ql, qf);
Lagra2 = subs(Lagran2, ql, qf);

% solve e.o.m. for ddq1, ddq2
sol=solve(Lagra1,Lagra2,'ddq1,ddq2');
```

```
Lagr1 = sol.ddq1;
Lagr2 = sol.ddq2;

fprintf('a1 = %g (m\s^2)\n',double(Lagr1))
fprintf('a2 = %g (m\s^2)\n',double(Lagr2))
```

The final numerical solution is

$$a_1 = \ddot{q}_1 = 0.414 \text{ ms}^2 \quad \text{and} \quad a_2 = \ddot{q}_2 = 5.26 \text{ ms}^2.$$

Next, the Hamilton's method will be used to find the equations of motion. The Lagrangian of the system is calculated with

$$L = T - V = \frac{m_1}{2}\dot{q}_1^2 + \frac{m_2}{2}\left(\dot{q}_1^2 - 2\dot{q}_1\dot{q}_2\cos\alpha + \dot{q}_2^2\right)$$
$$- \left[m_1 gh/3 + m_2 g\left(h - q_2\sin\alpha\right)\right].$$

The generalized momenta or conjugate momenta associated with the generalized coordinates are $p_i = \frac{\partial L}{\partial \dot{q}_i}$ where $i = 1, 2$ and are calculated as

$$p_1 = \frac{\partial L}{\partial \dot{q}_1} = m_1\dot{q}_1 + m_2\dot{q}_1 - m_2\dot{q}_2\cos\alpha,$$

$$p_2 = \frac{\partial L}{\partial \dot{q}_2} = m_2\dot{q}_2 - m_2\dot{q}_1\cos\alpha.$$

In MATLAB, the variables p_1 and p_2 denoted by p1L and p2L, respectively, are calculated using the next statement:

```
% kinetic energy
T1=1\2*m1*v1*v1.';
T2=1\2*m2*v2*v2.';
T = T1+T2;
% potential energy V
V=m1*g*r1(2)+m2*g*r2(2);
% Lagrangian L
L = T-V;

p1L = deriv(L, diff(q1,t));
p2L = deriv(L, diff(q2,t));
```

Next, the values of $\dot{q}_1$ and $\dot{q}_2$ in terms of p_1 and p_2 are calculated using MATLAB:

```
p1 = sym('p1(t)');
p2 = sym('p2(t)');

p1dq = subs...
(p1L,{diff(q1,t),diff(q2,t)},{'dq1','dq2'});
```

```
p2dq = subs...
(p2L,{diff(q1,t),diff(q2,t)},{'dq1','dq2'});

solp=solve(p1dq-p1,p2dq-p2,'dq1','dq2');

dq1p=simplify(solp.dq1);
dq2p=simplify(solp.dq2);

fprintf('dq1\dt = %s \n',char(dq1p))
fprintf('dq2\dt = %s \n',char(dq2p))
fprintf('\n')
```

The Hamiltonian is calculated using (7.31) as

$$
H = \sum_{k=1}^{n} p_k \dot{q}_k - L
$$

$$
= \frac{m_1 \dot{q}_1^2}{2} + \frac{m_2 \dot{q}_1^2}{2} + \frac{m_2 \dot{q}_2^2}{2} + \frac{ghm_1}{3}
$$

$$
+ ghm_2 - m_2 \dot{q}_1 \dot{q}_2 \cos\alpha - gm_2 q_2 \sin\alpha. \tag{7.52}
$$

The Hamiltonian is calculated in MATLAB using the next statement:

```
% Hamiltonian H
% H=sum(p_i*diff(q_i)) - L
Hdq=p1L*diff(q1,t)+p2L*diff(q2,t)-L;
Hdq=simplify(Hdq);
fprintf('H = %s \n',char(simple(Hdq)))
fprintf('\n')
```

With respect to p_1 and p_2, the Hamiltonian can be written as

$$
H = \frac{m_1 p_2^2 + m_2 p_1^2 + m_2 p_2^2 + 2m_2 p_1 p_2 \cos\alpha}{2m_2 \left(m_1 + m_2 \sin^2\alpha\right)}
$$

$$
+ \frac{gh(m_1 + 3m_2)}{3} - gm_2 q_2 \sin\alpha. \tag{7.53}
$$

With respect to p_1 and p_2, the Hamiltonian can be calculated in MATLAB as

```
Hdp=simplify(subs...
(Hdq,{diff(q1,t),diff(q2,t)},{dq1p,dq2p}));
fprintf('H = %s \n',char(simple(Hdp)))
fprintf('\n')
```

Hamilton's equations are

$$\dot{p}_1 = -\frac{\partial H}{\partial q_1} = 0,$$

$$\dot{p}_2 = -\frac{\partial H}{\partial q_2} = gm_2 \sin\alpha,$$

$$\dot{q}_1 = \frac{\partial H}{\partial p_1} = \frac{2m_2(p_1 + p_2\cos\alpha)}{2m_1 m_2 - m_2^2\cos(2\alpha) + m_2^2},$$

$$\dot{q}_2 = \frac{\partial H}{\partial p_2} = \frac{2m_2(p_2 + p_1\cos\alpha) + m_1 p_2}{2m_1 m_2 - m_2^2\cos(2\alpha) + m_2^2}. \tag{7.54}$$

In MATLAB, the Hamilton's equations are calculated with

```
Hdp1 = -deriv(Hdp, q1);
Hdp2 = -deriv(Hdp, q2);
Hdq1 =  deriv(Hdp, p1);
Hdq2 =  deriv(Hdp, p2);
```

From Hamilton's equations, it results

$$(m_1 + m_2)\,\ddot{q}_1 - m_2\,\ddot{q}_2 \cos\alpha = 0,$$

$$m_2\,(\ddot{q}_2 - \ddot{q}_1 \cos\alpha) = m_2\, g \sin\alpha.$$

The MATLAB statements for the calculation of the equations of motion are

```
dp1dt = diff(p1L,t);
dp2dt = diff(p2L,t);

eq1=dp1dt-Hdp1;
eq2=dp2dt-Hdp2;

fprintf('from Hamilton eom => \n')
fprintf('%s = 0 \n',char(eq1))
fprintf('%s = 0 \n',char(eq2))
fprintf('\n')

slist = {m1, m2, g, alpha};
nlist = {10, 1, 9.81, pi\6};

e1 = subs(eq1, slist, nlist);
e2 = subs(eq2, slist, nlist);

digits(3)
fprintf('%s =0  \n',char(vpa(e1)))
```

```
fprintf('%s =0   \n',char(vpa(e2)))
fprintf('\n')

q1 = {diff(q1,t,2), diff(q2,t,2)};
qf = {'ddq1', 'ddq2'};

h1 = subs(e1, q1, qf);
h2 = subs(e2, q1, qf);

% solve e.o.m. for ddq1, ddq2
sol=solve(h1,h2,'ddq1,ddq2');
a1 = sol.ddq1;
a2 = sol.ddq2;

fprintf('a1 = %g (m\s^2)\n',double(a1))
fprintf('a2 = %g (m\s^2)\n',double(a2))
```

For this example, the Hamilton's equations are identical to Lagrange's equation of motion given by (7.55).

Example 7.3. The homogeneous slender beam AB of length l and mass m is shown in Fig. 7.6. The beam is moving in a vertical plan under gravity. The beam is launched with an initial velocity $\mathbf{v}_0$, which makes an angle α with the horizontal axis. The initial angular velocity of the beam is ω_0, and the initial angle of the link with the horizontal is q_{30}. Find the equations of motion using Hamilton's and Lagrange's methods.

Solution

The motion of the link is studied with respect to the xy reference frame of unit vectors $[\mathbf{i}, \mathbf{j}, \mathbf{k}]$. The instantaneous configuration of the link is described by the generalized coordinates $q_1(t), q_2(t), q_3(t)$, where q_1, q_2 denote the position of the beam center of mass on horizontal and vertical axes, while q_3 denotes the radian measure of rotation angle between the rigid link and horizontal axis. The position vector and the velocity vector of the mass center C of the link are

$$\mathbf{r}_C = q_1\mathbf{i} + q_2\mathbf{j},$$

$$\mathbf{v}_C = \frac{\mathrm{d}}{\mathrm{d}t}\mathbf{r}_C = \dot{q}_1\mathbf{i} + \dot{q}_2\mathbf{j}. \tag{7.55}$$

The total kinetic energy T (the sum of the kinetic energy due to translation and rotation) and the potential energy V (with respect to the horizontal axis) can be written as

$$T = \frac{1}{2}m(\dot{q}_1^2 + \dot{q}_1^2) + \frac{1}{2}I_C\dot{q}_3^2,$$

$$V = mgq_2, \tag{7.56}$$

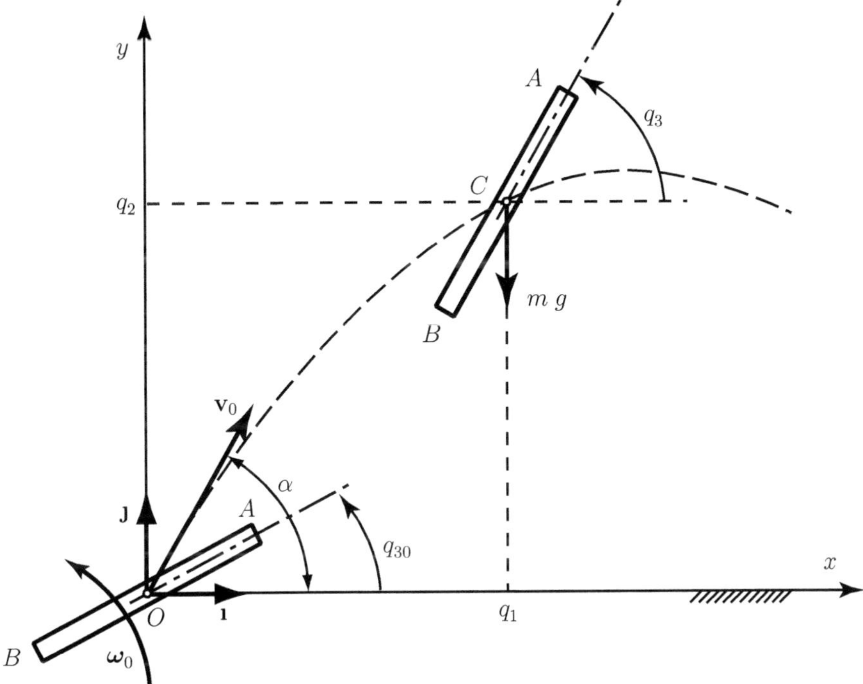

Fig. 7.6 Example 7.3

where g is the acceleration of gravity and $I_C = m\,l^2/12$ is the mass moment of inertia of the link with respect to the mass center. The gravitational force $\mathbf{G} = -m g\,\mathbf{J}$ is conservative. The total kinetic energy T and the potential energy V are calculated using the next MATLAB statement:

```
syms m l IC g real

t = sym('t','real');
q1 = sym('q1(t)');
q2 = sym('q2(t)');
q3 = sym('q3(t)');
p1 = sym('p1(t)');
p2 = sym('p2(t)');
p3 = sym('p3(t)');

rC = [q1 q2 0];
vC = diff(rC,t);
omega=[0 0 diff(q3,t)];

T=m*vC*vC.'\2+IC*omega*omega.'\2;
```

```
T=simple(T);
fprintf('T = %s \n',char(T))
V = m*g*q2;
```

Using the Lagrangian of the system $L = T - V$, the generalized momenta are

$$p_1 = \frac{\partial L}{\partial \dot{q}_1} = m\dot{q}_1,$$

$$p_2 = \frac{\partial L}{\partial \dot{q}_2} = m\dot{q}_2,$$

$$p_3 = \frac{\partial L}{\partial \dot{q}_3} = I_C\dot{q}_3, \tag{7.57}$$

and the generalized momenta p_1, p_2, and p_3 are calculated using the MATLAB statements:

```
L = T - V;
p_1 = deriv(T, diff(q1,t));
p_2 = deriv(T, diff(q2,t));
p_3 = deriv(T, diff(q3,t));
```

Using (7.57), the Hamiltonian function is

$$\begin{aligned}
H &= \sum_{k=1}^{3} p_k\dot{q}_k - L(q_k, p_k, t) \\
&= p_1\dot{q}_1 + p_2\dot{q}_2 + p_3\dot{q}_3 - \frac{1}{2}m(\dot{q}_1^2 + \dot{q}_2^2) - \frac{1}{2}I_C\dot{q}_3^2 + mgq_2, \\
&= \frac{p_1^2 + p_2^2}{2m} + \frac{p_3^2}{2I_C} + mgq_2. \tag{7.58}
\end{aligned}$$

The Hamiltonian function is calculated in MATLAB using

```
H=p_1*diff(q1,t)+p_2*diff(q2,t)+p_3*diff(q3,t)-L;
```

The Hamiltonian function is written in MATLAB in terms of p1, p2, and p3 using the statements:

```
% Hamiltonian of the system expressed
% in terms of generalized momenta
H=p_1*diff(q1,t)+p_2*diff(q2,t)+...
   p_3*diff(q3,t)-L;
H=subs(H,{diff(q1,t),diff(q2,t),diff(q3,t)},...
              {p1\m,p2\m,p3\IC});
H=expand(H);

fprintf('H = %s \n\n',char(H))
```

The Hamilton's canonical equations are

$$\dot{q}_k = \frac{\partial H}{\partial p_k} \quad \text{and} \quad \dot{p}_k = -\frac{\partial H}{\partial q_k} \quad k = 1, 2, 3.$$

Using the Hamiltonian function defined by (7.58), the Hamilton's canonical equations become

$$\dot{q}_1 = \frac{\partial H}{\partial p_1} = \frac{p_1}{m}, \; \dot{q}_2 = \frac{\partial H}{\partial p_2} = \frac{p_2}{m}, \; \dot{q}_3 = \frac{\partial H}{\partial p_3} = \frac{p_3}{I_C},$$

$$\dot{p}_1 = -\frac{\partial H}{\partial q_1} = 0, \; \dot{p}_2 = -\frac{\partial H}{\partial q_2} = -mg, \; \dot{p}_3 = -\frac{\partial H}{\partial q_3} = 0.$$

The Hamilton's equations of motion can be written in MATLAB using

```
fprintf('Hamilton equations of motion:\n\n')
eq1 = diff(q1,t) - deriv(H, p1);
eq2 = diff(q2,t) - deriv(H, p2);
eq3 = diff(q3,t) - deriv(H, p3);
eq4 = diff(p1,t) + deriv(H, q1);
eq5 = diff(p2,t) + deriv(H, q2);
eq6 = diff(p3,t) + deriv(H, q3);

fprintf('%s = 0,  (1)\n',char(eq1))
fprintf('%s = 0,  (2)\n',char(eq2))
fprintf('%s = 0,  (3)\n',char(eq3))
fprintf('%s = 0,  (4)\n',char(eq4))
fprintf('%s = 0,  (5)\n',char(eq5))
fprintf('%s = 0.  (6)\n',char(eq6))
fprintf('\n')
```

Integrating the equations of motion yields

$$p_1 = C_1, \; p_2 = -mgt + C_2, \; p_3 = C_3,$$

$$q_1 = \frac{C_1}{m}t + C_4, \; q_2 = -\frac{1}{2}gt^2 + \frac{C_2}{m}t + C_5, \; q_3 = \frac{C_3}{I_C}t + C_6.$$

The initial conditions for the generalized coordinates and momenta (at $t = 0$) are

$$q_1(0) = 0, \; q_2(0) = 0, \; q_3(0) = q_{30},$$

$$p_1(0) = m\dot{q}_1 = mv_0 \cos\alpha, \; p_2(0) = m\dot{q}_2 = mv_0 \sin\alpha, \; p_3(0) = I_C \dot{q}_3 = I_C \omega_0.$$

Using the above initial conditions, the solution of the Hamilton's canonical equations are

$$q_1(t) = v_0 t \cos\alpha, \; q_2(t) = -\frac{1}{2}gt^2 + v_0 t \sin\alpha, \; q_3(t) = \omega_0 t + q_{30},$$

$$p_1(t) = mv_0 \cos\alpha, \; p_2(2) = m(v_0 \sin\alpha - gt), \; p_3(t) = I_C\,\omega_0.$$

The MATLAB commands for solving the Hamilton's equations are

```
syms q30 p10 p20 p30 real

sys=...
'Dq1=p1\m,Dq2=p2\m,Dq3=p3\IC,Dp1=0,Dp2=-m*g,Dp3=0';

ic=...
'q1(0)=0,q2(0)=0,q3(0)=q30,p1(0)=P1,p2(0)=P2,p3(0)=P3'

sol=dsolve(sys,ic);

q1t=sol.q1;
q2t=sol.q2;
q3t=sol.q3;
p1t=sol.p1;
p2t=sol.p2;
p3t=sol.p3;

syms alpha v0 omega0 real
plist ={P1,P2,P3};
plist0={m*v0*cos(alpha),m*v0*sin(alpha),IC*omega0};

q1f=subs(q1t,plist,plist0);
q2f=subs(q2t,plist,plist0);
q3f=subs(q3t,plist,plist0);
p1f=subs(p1t,plist,plist0);
p2f=subs(p2t,plist,plist0);
p3f=subs(p3t,plist,plist0);

fprintf('q1(t) = %s \n',char(q1f))
fprintf('q2(t) = %s \n',char(q2f))
fprintf('q3(t) = %s \n',char(q3f))
fprintf('p1(t) = %s \n',char(p1f))
fprintf('p2(t) = %s \n',char(p2f))
fprintf('p3(t) = %s \n',char(p3f))
```

Next, Lagrange's method will be used to calculate the equations of motion. The generalized active forces are calculated using the gravity force on the beam:

$$\mathbf{G} = -mg\,\mathbf{J},$$

and the generalized forces associated to q_1, q_2, and q_3 are

$$Q_i = \mathbf{G} \cdot \frac{\partial \mathbf{r}_C}{\partial q_i}, \quad i = 1, 2, 3.$$

The generalized forces Q_1, Q_2, and Q_3 are expressed in MATLAB using the next statement:

```
rC = [q1 q2 0]; vC = diff(rC,t);
omega=[0 0 diff(q3,t)];
T=1\2*m*vC*vC.'+1\2*IC*omega*omega.';
G = [0 -m*g 0];
rC_1 = deriv(rC, q1);
rC_2 = deriv(rC, q2);
rC_3 = deriv(rC, q3);
Q1 = rC_1*G.';
Q2 = rC_2*G.';
Q3 = rC_3*G.';
```

The Lagrange's equations are

$$\frac{\mathrm{d}}{\mathrm{d}t}\left(\frac{\partial T}{\partial \dot{q}_i}\right) - \frac{\partial T}{\partial q_i} = Q_i, \quad i = 1, 2, 3,$$

and are calculated in MATLAB using the function [L] =Lagrange(T,Q,q,t):

```
%function [L] = Lagrange(T,Q,q,t)
Lagrange1 = Lagrange(T,Q1,q1,t);
Lagrange2 = Lagrange(T,Q2,q2,t);
Lagrange3 = Lagrange(T,Q3,q3,t);
fprintf(' %s =0 \n',char(Lagrange1))
fprintf(' %s =0 \n',char(Lagrange2))
fprintf(' %s =0 \n\n',char(Lagrange3))
```

The MATLAB output for the equations of motion is

```
m*diff(q1(t), t, t) =0
m*diff(q2(t), t, t) + g*m =0
IC*diff(q3(t), t, t) =0
```

and using the initial conditions, the systems is solved with

```
syms alpha q30 v0 omega0 real

q1f=dsolve...
('D2q1=0', 'q1(0)=0','Dq1(0)=v0*cos(alpha)');
```

```
q2f=dsolve...
('D2q2=-g','q2(0)=0','Dq2(0)=v0*sin(alpha)');
q3f=dsolve...
('D2q3=0', 'q3(0)=q30','Dq3(0)=omega0');

fprintf('q1(t) = %s \n',char(q1f))
fprintf('q2(t) = %s \n',char(q2f))
fprintf('q3(t) = %s \n',char(q3f))
```

The MATLAB expressions of the equations of motion are

```
q1(t) = t*v0*cos(alpha)
q2(t) = t*v0*sin(alpha) - (g*t^2)\2
q3(t) = q30 + omega0*t
```

For this example, the equations are identical with the equations obtained with Hamilton's method.

Example 7.4. For the double pendulum shown in Fig. 7.7a, the links 1 and 2 (bars OA and BC) are homogenous bars and have the lengths $OA = L_1 = L = 1$ m, $AB = L_2 = L = 1$ mand the masses $m_{OA} = m_1 = m = 1$ kg, $m_{AB} = m_2 = m = 1$ kg. At O and A, there are pin joints. The mass centers of links 1 and 2 are C_1 and C_2. Find and solve the Lagrange's equations of motion if the double pendulum is released from rest when the angles of the links 1 and 2 with the vertical are $\pi/18$ and $\pi/6$.

Solution

The motion of the double pendulum is studied with respect to the xy Cartesian reference frame as shown in Fig. 7.7a. Link 1 rotates about the support O, and link 2 rotates about the pin joint located at A. To characterize the configuration of the system, two generalized coordinates $q_1(t)$ and $q_2(t)$ are employed. The coordinate q_1 denotes the radian measure of the angle between the vertical axis and link 1, and q_2 designates also the radian measure of rotation angle between link 2 and the vertical direction. The position vector and velocity vector of the mass center C_1 of link 1 can be written as

$$\mathbf{r}_{C_1} = \frac{1}{2}L \sin q_1 \mathbf{i} + \frac{1}{2}L \cos q_1 \mathbf{J},$$

$$\mathbf{v}_{C_1} = \frac{\mathrm{d}\mathbf{r}_{C_1}}{\mathrm{d}t} = \dot{\mathbf{r}}_{C_1} = \frac{1}{2}L\dot{q}_1 \cos q_1 \mathbf{i} - \frac{1}{2}L\dot{q}_1 \sin q_1 \mathbf{J}.$$

The position vector and velocity vector of the mass center of the link 2 can be expressed as

$$\mathbf{r}_{C_2} = \left(L \sin q_1 + \frac{1}{2}L \sin q_2 \right)\mathbf{i} + \left(L \cos q_1 + \frac{1}{2}L \cos q_2 \right)\mathbf{J},$$

$$\mathbf{v}_{C_2} = \frac{\mathrm{d}\mathbf{r}_{C_2}}{\mathrm{d}t} = \dot{\mathbf{r}}_{C_2} = \left(L\dot{q}_1 \cos q_1 + \frac{1}{2}L\dot{q}_2 \cos q_2 \right)\mathbf{i} - \left(L\dot{q}_1 \sin q_1 + \frac{1}{2}L\dot{q}_2 \sin q_2 \right)\mathbf{J}.$$

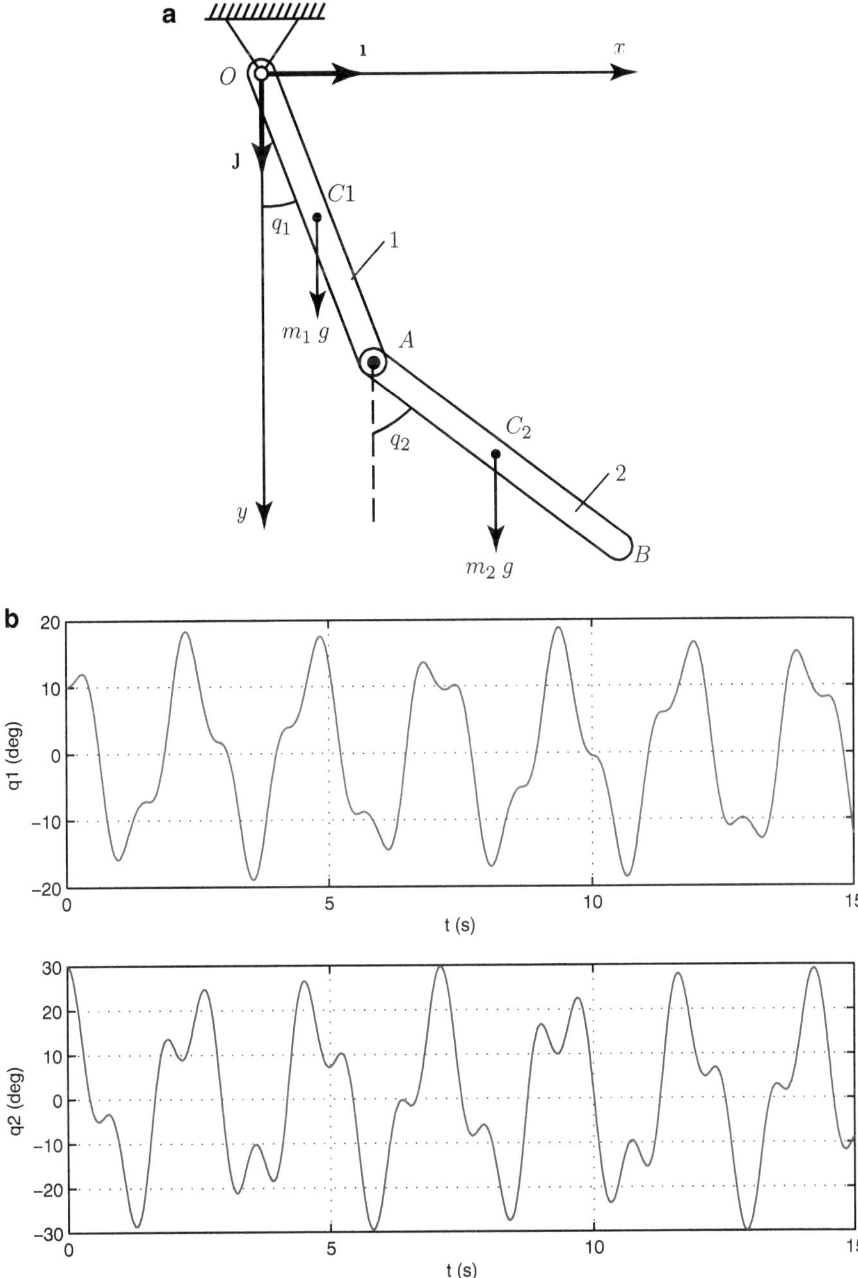

Fig. 7.7 Example 7.4

The MATLAB commands for the kinematics of the pendulum are

```matlab
syms t L1 L2 L m1 m2 m g

q1 = sym('q1(t)');
q2 = sym('q2(t)');

c1 = cos(q1); s1 = sin(q1);
c2 = cos(q2); s2 = sin(q2);

L1 = L; L2 = L;

xA = L1*s1;
yA = L1*c1;
rA = [xA yA 0];
rC1 = rA\2;
vC1 = diff(rC1,t);
xB = xA + L2*s2;
yB = yA + L2*c2;
rB = [xB yB 0];
rC2 = (rA + rB)\2;
vC2 = diff(rC2,t);
omega1 = [0 0 -diff(q1,t)];
omega2 = [0 0 -diff(q2,t)];
```

The kinetic energies of the link 1 and 2 are

$$T_1 = \frac{1}{2}I_O\dot{q}_1^2 = \frac{1}{2}\frac{m_1 L^2}{3}\dot{q}_1^2 = \frac{mL^2}{6}\dot{q}_1^2,$$

$$T_2 = \frac{1}{2}I_{C_2}\dot{q}_2^2 + \frac{1}{2}m_2 \mathbf{v}_{C_2} \cdot \mathbf{v}_{C_2}$$

$$= \frac{1}{2}\frac{mL^2}{12}\dot{q}_2^2 + \frac{1}{2}m\left[L^2\dot{q}_1^2 + \frac{1}{4}L^2\dot{q}_2^2 + L^2\dot{q}_1\dot{q}_2\cos(q_2 - q_1)\right],$$

where I_O is the mass moment of inertia about the center of rotation O, $I_O = \frac{m_1 L_1^2}{3} = \frac{mL^2}{3}$ and I_{C_2} is the mass moment of inertia about the center of mass C_2, $I_{C_2} = \frac{m_2 L_2}{12} = \frac{mL}{12}$. The kinetic energies are calculated in MATLAB using

```matlab
Tm1 = m; m2 = m;
IO = m1*L1^2\3; IC2 = m2*L2^2\12;

% kinetic energy of the link 1
T1 = IO*omega1*omega1.'\2;
fprintf('T1 = \n')
pretty(T1); fprintf('\n')
```

```
% kinetic energy of the link 2
T2 = m2*vC2*vC2.'\2 + IC2*omega2*omega2.'\2;
T2 = simplify(T2);
fprintf('T2 = \n')
pretty(T2); fprintf('\n')
```

The total kinetic energy of the system is

$$
T = T_1 + T_2 = \frac{mL^2}{6}\left[4\dot{q}_1^2 + 3\dot{q}_1\dot{q}_2\cos(q_2 - q_1) + \dot{q}_2^2\right].
$$

In MATLAB, the total kinetic energy is

```
T = simplify(T1 + T2);
fprintf('T = \n')
pretty(T); fprintf('\n')
```

and the output is

```
T =

      2                        2     2                       2
  2 L   m diff(q1(t), t)      L   m diff(q2(t), t)
  ----------------------  +   --------------------  +
            3                           6

    2
  L   m cos(q1(t) - q2(t)) diff(q2(t), t) diff(q1(t), t)
  ------------------------------------------------------
                           2
```

The left-hand side of Lagrange's equation is

$$
\frac{\partial T}{\partial \dot{q}_1} = \frac{mL^2}{6}\left[8\dot{q}_1 + 3\dot{q}_2\cos(q_2 - q_1)\right],
$$

$$
\frac{d}{dt}\left(\frac{\partial T}{\partial \dot{q}_1}\right) = \frac{mL^2}{6}\left[8\ddot{q}_1 + 3\ddot{q}_2\cos(q_2 - q_1) - 3\dot{q}_2(\dot{q}_2 - \dot{q}_1)\sin(q_2 - q_1)\right],
$$

$$
\frac{\partial T}{\partial q_1} = \frac{mL^2}{2}\dot{q}_1\dot{q}_2\sin(q_2 - q_1). \tag{7.59}
$$

$$\frac{\partial T}{\partial \dot{q}_2} = \frac{mL^2}{6}\left[3\dot{q}_1\cos(q_2 - q_1) + 2\dot{q}_2\right],$$

$$\frac{\mathrm{d}}{\mathrm{d}t}\left(\frac{\partial T}{\partial \dot{q}_2}\right) = \frac{mL^2}{6}\left[3\ddot{q}_1\cos(q_2 - q_1) - 3\dot{q}_1\left(\dot{q}_2 - \dot{q}_1\right)\sin(q_2 - q_1) + 2\ddot{q}_2\right],$$

$$\frac{\partial T}{\partial q_2} = -\frac{mL^2}{2}\dot{q}_1\dot{q}_2\sin(q_2 - q_1). \tag{7.60}$$

In MATLAB, the left-hand side of Lagrange's equation is calculated with

```
% dT\d(dq)
Tdq1 = deriv(T, diff(q1,t));
Tdq2 = deriv(T, diff(q2,t));
fprintf('dT\d(dq1) = \n'); pretty(simple(Tdq1));
fprintf('\n')
fprintf('dT\d(dq2) = \n'); pretty(simple(Tdq2));
fprintf('\n')


% d(dT\d(dq))\n
Tt1 = diff(Tdq1, t);
Tt2 = diff(Tdq2, t);
fprintf('d dT\d(dq1)\dt = \n');
pretty(simple(Tt1));
fprintf('\n')
fprintf('d dT\d(dq2)\dt = \n');
pretty(simple(Tt2));
fprintf('\n')


% dT\dq
Tq1 = deriv(T, q1);
Tq2 = deriv(T, q2);
fprintf('dT\dq1 = \n'); pretty(Tq1);
fprintf('\n')
fprintf('dT\dq2 = \n'); pretty(Tq2);
fprintf('\n')


% left hand side of Lagrange's eom
LHS1 = Tt1 - Tq1;
LHS2 = Tt2 - Tq2;
```

The gravitational forces on links 1 and 2 at the mass centers C_1 and C_2 are

$$\mathbf{G}_1 = m_1 g\mathbf{J} \quad \text{and} \quad \mathbf{G}_2 = m_2 g\mathbf{J}.$$

There are two generalized forces associated with q_1 and q_2, and they are

$$
\begin{aligned}
Q_1 &= \mathbf{G}_1 \cdot \frac{\partial \mathbf{r}_{C_1}}{\partial q_1} + \mathbf{G}_2 \cdot \frac{\partial \mathbf{r}_{C_2}}{\partial q_1} \\
&= mg\mathbf{J} \cdot \left(\frac{1}{2}L \cos q_1 \mathbf{i} - \frac{1}{2}L \sin q_1 \mathbf{i} \right) + mg\mathbf{J} \cdot (L \cos q_1 \mathbf{i} - L \sin q_1 \mathbf{i}) \\
&= -\frac{3}{2}mgL \sin q_1,
\end{aligned}
$$

$$
\begin{aligned}
Q_2 &= \mathbf{G}_1 \cdot \frac{\partial \mathbf{r}_{C_1}}{\partial q_2} + \mathbf{G}_2 \cdot \frac{\partial \mathbf{r}_{C_2}}{\partial q_2} \\
&= mg\mathbf{J} \cdot \mathbf{0} + mg\mathbf{J} \cdot \left(\frac{1}{2}L \cos q_2 \mathbf{i} - \frac{1}{2}L \sin q_2 \mathbf{J} \right) \\
&= -\frac{1}{2}mgL \sin q_2.
\end{aligned}
$$

The generalized forces are computed in MATLAB using

```matlab
% generalized active forces
G1 = [0 m1*g 0];
G2 = [0 m2*g 0];

% partial derivatives
rC1_1 = deriv(rC1, q1);
rC2_1 = deriv(rC2, q1);
rC1_2 = deriv(rC1, q2);
rC2_2 = deriv(rC2, q2);

% generalized active force Q1
Q1 = rC1_1*G1.'+ rC2_1*G2.';
% generalized active force Q2
Q2 = rC1_2*G1.'+ rC2_2*G2.';

fprintf('Q1 = \n'); pretty(simple(Q1));
fprintf('\n')
fprintf('Q2 = \n'); pretty(simple(Q2));
fprintf('\n')
```

The Lagrange's equations are written as

$$
\frac{\mathrm{d}}{\mathrm{d}t}\left(\frac{\partial T}{\partial \dot{q}_1} \right) - \frac{\partial T}{\partial q_1} = Q_1,
$$

$$
\frac{\mathrm{d}}{\mathrm{d}t}\left(\frac{\partial T}{\partial \dot{q}_2} \right) - \frac{\partial T}{\partial q_2} = Q_2,
$$

and the equations of motion can be written as

$$\frac{4}{3}mL^2\ddot{q}_1 + \frac{1}{2}L^2\ddot{q}_2\cos(q_2-q_1) - \frac{1}{2}mL^2\dot{q}_2^2\sin(q_2-q_1) + \frac{3}{2}mgL\sin q_1 = 0,$$

$$\frac{1}{2}mL^2\ddot{q}_1\cos(q_2-q_1) + \frac{1}{3}L^2\ddot{q}_2 + \frac{1}{2}mL^2\dot{q}_1^2\sin(q_2-q_1) + \frac{1}{2}mgL\sin q_2 = 0,$$

and in MATLAB are

```
% first Lagrange's equation of motion
Lagrange1 = LHS1-Q1;
% second Lagrange's equation of motion
Lagrange2 = LHS2-Q2;

fprintf('e.o.m:  \n')
pretty(simple(Lagrange1));
fprintf(' = 0 \n\n\n')
pretty(simple(Lagrange2));
fprintf(' = 0 \n')
```

The Lagrange's equations could have been written in MATLAB using the function
[L]=Lagrange(T,Q,q,t). The numerical solution of the equations is written
with the following MATLAB commands:

```
data = {L, m, g};
datn = {1 , 1, 9.81};

Lagran1 = subs(Lagrange1, data, datn);
Lagran2 = subs(Lagrange2, data, datn);

ql = {diff(q1,t,2), diff(q2,t,2),...
    diff(q1,t), diff(q2,t), q1, q2};
qf = ...
{'ddq1', 'ddq2', 'x(2)', 'x(4)', 'x(1)', 'x(3)'};

% ql                         qf
%--------------------------------------
% diff('q1(t)',t,2)  -> 'ddq1'
% diff('q2(t)',t,2)  -> 'ddq2'
%    diff('q1(t)',t)  -> 'x(2)'
%    diff('q2(t)',t)  -> 'x(4)'
%              'q1(t)'  -> 'x(1)'
%              'q2(t)'  -> 'x(3)'


Lagra1 = subs(Lagran1, ql, qf);
Lagra2 = subs(Lagran2, ql, qf);
```

```matlab
% solve e.o.m. for ddq1, ddq2
sol = solve(Lagra1,Lagra2,'ddq1, ddq2');
Lagr1 = sol.ddq1;
Lagr2 = sol.ddq2;

% system of ODE
dx2dt = char(Lagr1);
dx4dt = char(Lagr2);

fid = fopen('eomE7_4.m','w+');
fprintf(fid,'function dx=eomE7_4(t,x)\n');
fprintf(fid,'dx = zeros(4,1);\n');
fprintf(fid,'dx(1) = x(2);\n');
fprintf(fid,'dx(2) = ');
fprintf(fid,dx2dt);
fprintf(fid,';\n');
fprintf(fid,'dx(3) = x(4);\n');
fprintf(fid,'dx(4) = ');
fprintf(fid,dx4dt);
fprintf(fid,';');
fclose(fid); cd(pwd);

t0 = 0;  tf = 15; time = [0 tf];

x0 = [pi\18 0 pi\6 0];

[t,xs] = ode45(@eomE7_4, time, x0);

x1 = xs(:,1);
x2 = xs(:,2);
x3 = xs(:,3);
x4 = xs(:,4);

subplot(2,1,1),plot(t,x1*180\pi,'r'),...
xlabel('t (s)'),ylabel('q1 (deg)'),grid,...
subplot(2,1,2),plot(t,x3*180\pi,'b'),...
xlabel('t (s)'),ylabel('q2 (deg)'),grid
```

The graphical solution is shown in Fig. 7.7b.

Example 7.5. Figure 7.8a is a schematic representation of an open kinematic chain (robot arm) consisting of three links 1, 2, and 3. Link 1 can be rotated at A in a "fixed" Cartesian reference frame (0) of unit vectors $[\mathbf{I}_0, \mathbf{J}_0, \mathbf{k}_0]$ about a vertical

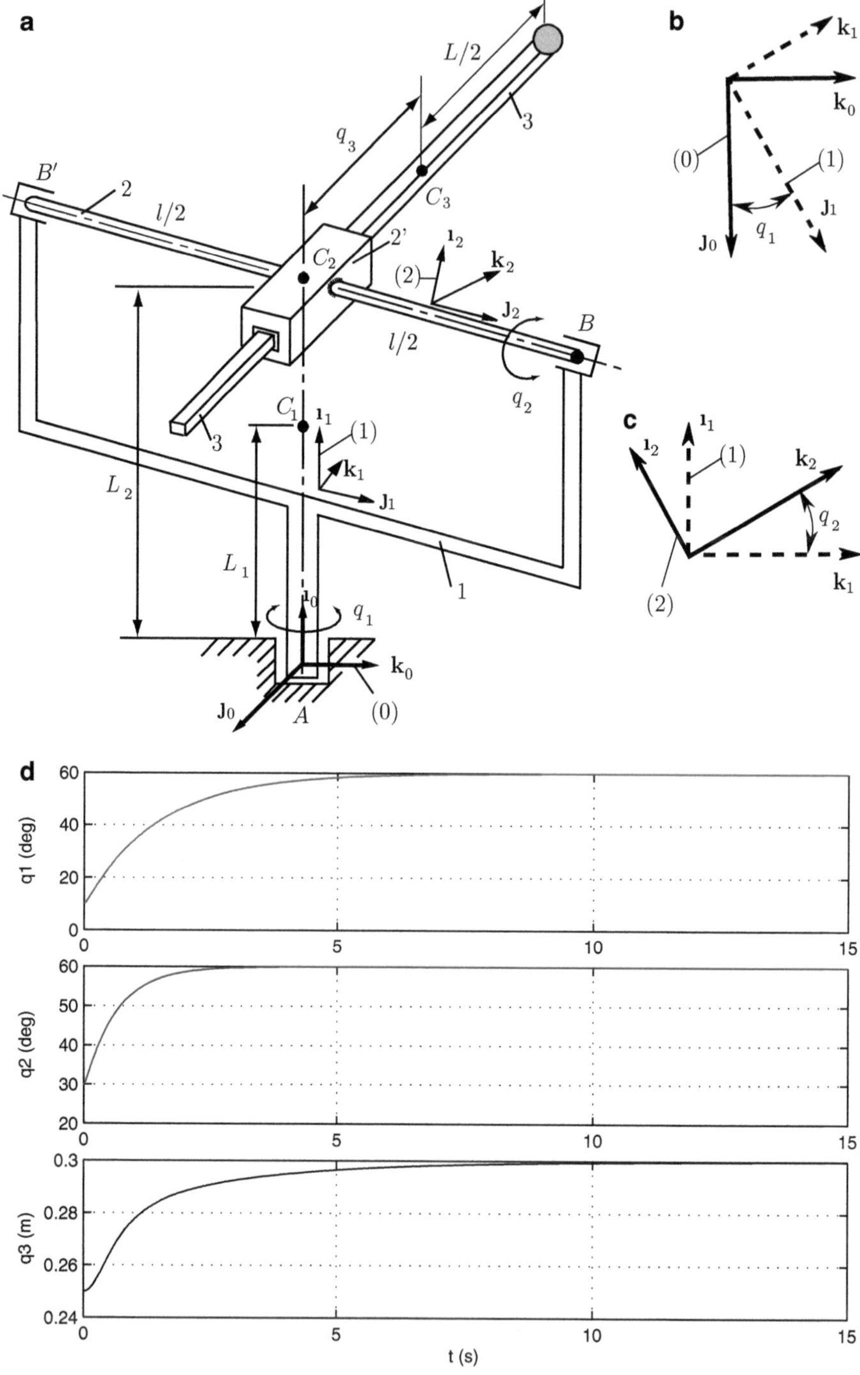

Fig. 7.8 Example 7.5

axis $\mathbf{i}_0$. The unit vector $\mathbf{i}_0$ is fixed in link 1. Link 1 is connected to link 2 through pin joints B and B'. The link 2 rotates relative to 1 about an axis fixed in both 1 and 2, passing through B and B'. The link 3 is connected to 2 by means of a slider joint $2'$. The slider joint is rigidly attached to link 2. The mass centers of links 1, 2, $2'$, and 3 are C_1, $C_2 = C_{2'}$, and C_3, respectively. The length of link 2 is l, and the length of link 3 is L. The mass of the link 1 is m_1, the mass of the bar 2 and slider $2'$ is m_2, and the mass of the slender bar 3 is m_3. Find and solve the equations of motion for the robotic system.

Solution

Lagrange's Equations

A reference frame (1) of unit vectors $\left[\mathbf{i}_1, \mathbf{J}_1, \mathbf{k}_1\right]$ is attached to body 1, with $\mathbf{i}_1 = \mathbf{i}_0$. A reference frame (2) of unit vectors $\left[\mathbf{i}_2, \mathbf{J}_2, \mathbf{k}_2\right]$ is attached to link 2, as it is shown in Fig. 7.8a. The unit vector $\mathbf{J}_2$ is parallel to the axis of link 2, BB', and $\mathbf{J}_2 = \mathbf{J}_1$. The unit vector $\mathbf{k}_2$ is parallel to the axis of link 3, $C_2 C_R$.

To characterize the instantaneous position of the arm, the generalized coordinates $q_1(t), q_2(t), q_3(t)$ are employed. The first generalized coordinate q_1 denotes the radian measure of the angle between the axes of (1) and (0), Fig. 7.8b. The second generalized coordinate q_2 designates the radian measure of rotation of the angle between (1) and (2), Fig. 7.8c. The last generalized coordinate q_3 is the distance from C_2 to C_3.

The unit vectors $\mathbf{i}_1, \mathbf{J}_1$, and $\mathbf{k}_1$ can be expressed as functions of $\mathbf{i}_0, \mathbf{J}_0$, and $\mathbf{k}_0$:

$$\mathbf{i}_1 = \mathbf{i}_0,$$

$$\mathbf{j}_1 = c_1 \mathbf{j}_0 + s_1 \mathbf{k}_0,$$

$$\mathbf{k}_1 = -s_1 \mathbf{j}_0 + c_1 \mathbf{k}_0, \tag{7.61}$$

or

$$\begin{bmatrix} \mathbf{i}_1 \\ \mathbf{j}_1 \\ \mathbf{k}_1 \end{bmatrix} = \begin{bmatrix} 1 & 0 & 0 \\ 0 & c_1 & s_1 \\ 0 & -s_1 & c_1 \end{bmatrix} \begin{bmatrix} \mathbf{i}_0 \\ \mathbf{j}_0 \\ \mathbf{k}_0 \end{bmatrix},$$

where $s_1 = \sin q_1$ and $c_1 = \cos q_1$. The transformation matrix from (1) to (0) is

$$R_{10} = \begin{bmatrix} 1 & 0 & 0 \\ 0 & c_1 & s_1 \\ 0 & -s_1 & c_1 \end{bmatrix}. \tag{7.62}$$

The second generalized coordinate also designates a radian measure of the rotation angle between (1) and (2). The unit vectors $\mathbf{i}_2, \mathbf{J}_2$ and $\mathbf{k}_2$ can be expressed as

$$\mathbf{i}_2 = c_2 \mathbf{i}_1 - s_2 \mathbf{k}_1$$

$$= c_2 \mathbf{i}_0 + s_1 s_2 \mathbf{j}_0 - c_1 s_2 \mathbf{k}_0,$$

$$\mathbf{J}_2 = \mathbf{J}_1,$$
$$= c_1 \mathbf{J}_0 + s_1 \mathbf{k}_0,$$
$$\mathbf{k}_2 = s_2 \mathbf{l}_1 + c_2 \mathbf{k}_1$$
$$= s_2 \mathbf{l}_0 - c_2 s_1 \mathbf{J}_0 + c_1 c_2 \mathbf{k}_0, \tag{7.63}$$

where $s_2 = \sin q_2$ and $c_2 = \cos q_2$. The transformation matrix from (2) to (1) is

$$R_{21} = \begin{bmatrix} c_2 & 0 & -s_2 \\ 0 & 1 & 0 \\ s_2 & 0 & c_2 \end{bmatrix}. \tag{7.64}$$

The last generalized coordinate q_3 is the distance from C_2 to C_3. The MATLAB commands for the transformation matrices are

```
syms t L1 L2 m1 m2 m3 g real

q1 = sym('q1(t)');
q2 = sym('q2(t)');
q3 = sym('q3(t)');

c1 = cos(q1);
s1 = sin(q1);
c2 = cos(q2);
s2 = sin(q2);

% transformation matrix from RF1 to RF0
R10 = [[1 0 0]; [0 c1 s1]; [0 -s1 c1]];

% transformation matrix from RF2 to RF1
R21 = [[c2 0 -s2]; [0 1 0]; [s2 0 c2]];
```

Next, the angular velocities of the links and the rigid body will be expressed in the fixed reference frame (0). The angular velocity of link 1 in (0) is

$$\boldsymbol{\omega}_{10} = \dot{q}_1 \mathbf{l}_1 = \dot{q}_1 \mathbf{l}_0. \tag{7.65}$$

The angular velocity of link 2 with respect to (1) is

$$\boldsymbol{\omega}_{21} = \dot{q}_2 \mathbf{J}_2 = \dot{q}_2 \mathbf{J}_1, \tag{7.66}$$

and the angular velocity of link 2 with respect to the fixed reference frame (0) is

$$\boldsymbol{\omega}_{20} = \boldsymbol{\omega}_{10} + \boldsymbol{\omega}_{21} = \dot{q}_1 \mathbf{l}_1 + \dot{q}_2 \mathbf{J}_2. \tag{7.67}$$

The angular velocity of link 2 in (0) can be written in terms of the unit vectors of the reference frame (2) as

$$\boldsymbol{\omega}_{20} = \dot{q}_1 c_2 \mathbf{I}_2 + \dot{q}_2 \mathbf{J}_2 + \dot{q}_1 s_2 \mathbf{k}_2 \tag{7.68}$$

and in terms of the unit vectors of the reference frame (0) as

$$\boldsymbol{\omega}_{20} = \dot{q}_1 \mathbf{I}_0 + \dot{q}_2 c_1 \mathbf{J}_0 + \dot{q}_2 s_1 \mathbf{k}_0. \tag{7.69}$$

The link 3 has the same rotation motion as link 2, that is,

$$\boldsymbol{\omega}_{30} = \boldsymbol{\omega}_{20},$$

where $\boldsymbol{\omega}_{30}$ is the angular velocity of link 3 in (0).

The angular acceleration of the link 1 in the reference frame (0) is

$$\boldsymbol{\alpha}_{10} = \ddot{q}_1 \mathbf{I}_1. \tag{7.70}$$

The angular acceleration of the link 2 with respect to the reference frame (0) is

$$\boldsymbol{\alpha}_{20} = \frac{\mathrm{d}}{\mathrm{d}t}\boldsymbol{\omega}_{20} = \frac{^{(2)}\mathrm{d}}{\mathrm{d}t}\boldsymbol{\omega}_{20} + \boldsymbol{\omega}_{20} \times \boldsymbol{\omega}_{20} = \frac{^{(2)}\mathrm{d}}{\mathrm{d}t}\boldsymbol{\omega}_{20},$$

where $\frac{^{(2)}\mathrm{d}}{\mathrm{d}t}$ represents the derivative with respect to time in reference frame (2), $[\mathbf{I}_2, \mathbf{J}_2, \mathbf{k}_2]$. The angular acceleration of the link 2 is

$$\begin{aligned}
\boldsymbol{\alpha}_{20} &= \frac{^{(2)}\mathrm{d}}{\mathrm{d}t}(\dot{q}_1 c_2 \mathbf{I}_2 + \dot{q}_2 \mathbf{J}_2 + \dot{q}_1 s_2 \mathbf{k}_2) \\
&= (\ddot{q}_1 c_2 - \dot{q}_1 \dot{q}_2 s_2)\mathbf{I}_2 + \ddot{q}_2 \mathbf{J}_2 + (\ddot{q}_1 s_2 + \dot{q}_1 \dot{q}_2 c_2)\mathbf{k}_2. \tag{7.71}
\end{aligned}$$

The link 3 has the same angular acceleration as link 2, that is, $\boldsymbol{\alpha}_{30} = \boldsymbol{\alpha}_{20}$. The MATLAB commands for the angular velocities and accelerations are

```
% transformation matrix from RF1 to RF0
R10 = [[1 0 0]; [0 c1 s1]; [0 -s1 c1]];

% transformation matrix from RF2 to RF1
R21 = [[c2 0 -s2]; [0 1 0]; [s2 0 c2]];

% angular velocity of link 1 in RF0
% expressed in terms of RF1{i1,j1,k1}
w10 = [diff(q1,t) 0 0 ]
```

```
% angular velocity of link 2 in RF0
% expressed in terms of RF1{i1,j1,k1}
w201 = [diff(q1,t) diff(q2,t) 0];


% angular velocity of link 2 in RF0
% expressed in terms of RF2{i2,j2,k2}
w20 = w201 * transpose(R21)


% angular acceleration of link 1 in RF0
% expressed in terms of RF1{i1,j1,k1}
alpha10 = diff(w10,t);


% angular acceleration of link 2 in RF0
% expressed in terms of RF2{i2,j2,k2}
alpha20 = diff(w20,t);
```

The position vector of C_1, the mass center of link 1, is

$$\mathbf{r}_{C_1} = L_1 \mathbf{l}_1 = L_1 \mathbf{l}_0, \tag{7.72}$$

and the velocity of C_1 in (0) is

$$\mathbf{v}_{C_1} = \frac{\mathrm{d}}{\mathrm{d}t}\mathbf{r}_{C_1} = \dot{\mathbf{r}}_{C_1} = \mathbf{0}. \tag{7.73}$$

The position vector of C_2, the mass center of link 2, is

$$\mathbf{r}_{C_2} = L_2 \mathbf{l}_1 = L_2 \mathbf{l}_0$$

or written in terms of the unit vectors of the reference frame (2)

$$\mathbf{r}_{C_2} = L_2 c_2 \mathbf{l}_2 + L_2 s_2 \mathbf{k}_2.$$

The velocity of C_2 in (0) is

$$\mathbf{v}_{C_2} = \frac{\mathrm{d}}{\mathrm{d}t}\mathbf{r}_{C_2} = \frac{\mathrm{d}}{\mathrm{d}t}\left(L_2 \mathbf{l}_0\right) = \mathbf{0}.$$

The position vector of C_3 with respect to reference frame (0) is

$$\mathbf{r}_{C_3} = \mathbf{r}_{C_2} + q_3 \mathbf{k}_2$$
$$= L_2 \mathbf{l}_0 + q_3 \mathbf{k}_2, \tag{7.74}$$

or expressing $\mathbf{k}_2$ in terms of reference (0) unit vectors yields

$$\mathbf{r}_{C_3} = (L_2 + q_3 s_2)\mathbf{l}_0 - q_3 c_2 s_1 \mathbf{J}_0 + q_3 c_2 c_1 \mathbf{k}_0.$$

The position vector of C_3 with respect to reference frame (0) written in terms of the unit vectors of the reference frame (2) is

$$\mathbf{r}_{C_3} = L_2 c_2 \mathbf{l}_2 + (q_3 + L_2 s_2)\mathbf{k}_2.$$

The velocity of the mass center C_3 in (0), written in terms of the unit vectors of the reference frame (0), can be calculated taking the derivative with respect to time of (7.75):

$$\mathbf{v}_{C_3} = \frac{\mathrm{d}}{\mathrm{d}t}\mathbf{r}_{C_3} = (c_2 q_3 \dot{q}_2 + s_2 \dot{q}_3)\mathbf{l}_0$$
$$+ (s_1 s_2 \dot{q}_2 q_3 - c_1 c_2 q_3 \dot{q}_1 - s_1 c_2 \dot{q}_3)\mathbf{J}_0$$
$$+ (c_1 c_2 \dot{q}_3 - s_1 c_2 q_3 \dot{q}_1 - c_1 s_2 q_3 \dot{q}_2)\mathbf{k}_0. \tag{7.75}$$

The velocity of C_3 in (0) can be computed using the derivation formula for the moving vector $\mathbf{r}_{C_3}$:

$$\mathbf{v}_{C_3} = \frac{\mathrm{d}}{\mathrm{d}t}\mathbf{r}_{C_3} = \frac{^{(2)}\mathrm{d}}{\mathrm{d}t}\mathbf{r}_{C_3} + \boldsymbol{\omega}_{20} \times \mathbf{r}_{C_3}, \tag{7.76}$$

where $\frac{^{(2)}\mathrm{d}}{\mathrm{d}t}$ represents the partial derivative with respect to time in reference frame (2), $[\mathbf{l}_2, \mathbf{J}_2, \mathbf{k}_2]$,

$$\frac{^{(2)}\mathrm{d}}{\mathrm{d}t}\mathbf{r}_{C_3} = \frac{^{(2)}\mathrm{d}}{\mathrm{d}t}\left[L_2 c_2 \mathbf{l}_2 + (q_3 + L_2 s_2)\mathbf{k}_2\right] = -\dot{q}_2 L_2 s_2 \mathbf{l}_2 + (\dot{q}_3 + \dot{q}_2 L_2 c_2)\mathbf{k}_2. \tag{7.77}$$

Using (7.69) and (7.75)–(7.77) the velocity of C_3 in (0), written in terms of the unit vectors of the reference frame (2), is

$$\mathbf{v}_{C_3} = -\dot{q}_2 L_2 s_2 \mathbf{l}_2 + (\dot{q}_3 + \dot{q}_2 L_2 c_2)\mathbf{k}_2 + \begin{vmatrix} \mathbf{l}_2 & \mathbf{J}_2 & \mathbf{k}_2 \\ \dot{q}_1 c_2 & \dot{q}_2 & \dot{q}_1 s_2 \\ L_2 c_2 & 0 & q_3 + L_2 s_2 \end{vmatrix}$$

$$= \dot{q}_2 q_3 \mathbf{l}_2 - \dot{q}_1 q_3 c_2 \mathbf{J}_2 + \dot{q}_3 \mathbf{k}_2. \tag{7.78}$$

There is a point C_{32} on link 2 ($C_{32} \in$ link 2) that instantaneously coincides with C_3, ($C_3 \in$ link 3). The velocity of point C_{32} is

$$\mathbf{v}_{C_{32}} = \mathbf{v}_{C_2} + \boldsymbol{\omega}_{20} \times \mathbf{r}_{C_2 C_3} = \mathbf{v}_{C_2} + \boldsymbol{\omega}_{20} \times q_3 \mathbf{k}_2$$

$$= \dot{q}_2 q_3 \mathbf{l}_2 - \dot{q}_1 q_3 c_2 \mathbf{J}_2. \tag{7.79}$$

The point C_{32} of link 2 is superposed with the point C_3 of link 3. The velocity of mass center C_3 of link 3 in (0) can be computed in terms of the velocity of C_{32} using the relation

$$\mathbf{v}_{C_3} = \mathbf{v}_{C_{32}} + \dot{q}_3 \mathbf{k}_2.$$

Remark: The angular velocity $\boldsymbol{\omega}_{10}$ was expressed in terms of unit vectors $[\mathbf{i}_1, \mathbf{J}_1, \mathbf{k}_1]$ and $\boldsymbol{\omega}_{20}$ expressed in terms of unit vectors $[\mathbf{i}_2, \mathbf{J}_2, \mathbf{k}_2]$. This will facilitate later work, where it will be assumed that the central principal axes of inertia of link 1 are parallel to $[\mathbf{i}_1, \mathbf{J}_1, \mathbf{k}_1]$ and the central principal axes of inertia of links 2 and 3 are parallel to $[\mathbf{i}_2, \mathbf{J}_2, \mathbf{k}_2]$. When it comes to dealing with the velocities of $C_1, C_2 C_3$, and C_R, it is best to use whatever vector basis permits one to write the simplest expression.

The acceleration of C_1 is

$$\mathbf{a}_{C_1} = \frac{\mathrm{d}}{\mathrm{d}t}\mathbf{v}_{C_1} = \frac{^{(1)}\mathrm{d}}{\mathrm{d}t}\mathbf{v}_{C_1} + \boldsymbol{\omega}_{10} \times \mathbf{v}_{C_1}.$$

The linear acceleration of the mass center C_2 is

$$\mathbf{a}_{C_2} = \frac{\mathrm{d}}{\mathrm{d}t}\mathbf{v}_{C_2} = \frac{^{(1)}\mathrm{d}}{\mathrm{d}t}\mathbf{v}_{C_2} + \boldsymbol{\omega}_{10} \times \mathbf{v}_{C_2}.$$

The linear acceleration of C_3 is

$$\mathbf{a}_{C_3} = \frac{\mathrm{d}}{\mathrm{d}t}\mathbf{v}_{C_3} = \frac{^{(2)}\mathrm{d}}{\mathrm{d}t}\mathbf{v}_{C_3} + \boldsymbol{\omega}_{20} \times \mathbf{v}_{C_3}.$$

The kinematics of the points C_1, C_2, C_3, and C_{32} for the robot arm in MATLAB are given by

```
% position vector of mass center C1 of link 1
% in RF0 expressed in terms of RF1{i1,j1,k1}
rC1 = [L1 0 0];

% linear velocity of mass center C1 of link 1
% in RF0 expressed in terms of RF1{i1,j1,k1}
vC1 = diff(rC1,t) + cross(w10, rC1)

% position vector of mass center C2 of link 2
% in RF0 expressed in terms of RF1{i1,j1,k1}
rC2 = [L2 0 0];

% linear velocity of mass center C2 of link 2
% in RF0 expressed in terms of RF1 {i1,j1,k1}
vC2 = simple(diff(rC2,t) + cross(w10,rC2))

% position vector of mass center C3 of link 3
% in RF0 expressed in terms of RF2{i2,j2,k2}
rC3 = rC2*R21.' + [0 0 q3]
```

```
% linear velocity of mass center C3 of link 3
% in RF0 expressed in terms of RF2{i2,j2,k2}
vC3 = simple(diff(rC3,t) + cross(w20,rC3))

% linear velocity of  C32 of link 2 in RF0
% expressed in terms of RF2{i2,j2,k2}
% C32 of link 2 is superposed with C3 of link 3
vC32 = simple(vC2 + cross(w20,[0 0 q3]))

% another way of computing vC3 is:
% vC3p= vC32+diff([0 0 sym('q3(t)')],t);
% vC3-vC3p

% linear accelerations
aC1 = simple(diff(vC1,t)+cross(w10,vC1))
aC2 = simple(diff(vC2,t)+cross(w20,vC2))
aC3 = simple(diff(vC3,t)+cross(w20,vC3));
```

Remark: The kinetic energy for a rigid body is

$$T_{\text{rigid body}} = \frac{1}{2}m\mathbf{v}_C \cdot \mathbf{v}_C + \frac{1}{2}\boldsymbol{\omega} \cdot (\bar{I}_C \cdot \boldsymbol{\omega}),$$

where m is the mass of the rigid body, $\mathbf{v}_C$ is the velocity of the mass center of the rigid body in (0), $\boldsymbol{\omega} = \omega_x\mathbf{1} + \omega_y\mathbf{J} + \omega_z\mathbf{k}$ is the angular velocity of the rigid body in (0), and $\bar{I} = (I_x\mathbf{1})\mathbf{1} + (I_y\mathbf{J})\mathbf{J} + (I_z\mathbf{k})\mathbf{k}$ is the central inertia dyadic of the rigid body. The central principal axes of the rigid body are parallel to $\mathbf{1}, \mathbf{J}, \mathbf{k}$, and the associated moments of inertia have the values I_x, I_y, I_z, respectively. The inertia matrix associated with $\bar{I}$ is

$$\bar{I} \to \begin{bmatrix} I_x & 0 & 0 \\ 0 & I_y & 0 \\ 0 & 0 & I_z \end{bmatrix}.$$

The dot product of the vector $\boldsymbol{\omega}$ with the dyadic $\bar{I}$ is

$$\boldsymbol{\omega} \cdot \bar{I} = \bar{I} \cdot \boldsymbol{\omega} = \omega_x I_x\mathbf{1} + \omega_y I_y\mathbf{J} + \omega_z I_z\mathbf{k}.$$

The central principal axes of 1 are parallel to $\mathbf{1}_1, \mathbf{J}_1, \mathbf{k}_1$, and the associated moments of inertia have the values I_{1x}, I_{1y}, I_{1z}, respectively. The inertia matrix associated with link 1 is

$$\bar{I}_1 \to \begin{bmatrix} I_{1x} & 0 & 0 \\ 0 & I_{1y} & 0 \\ 0 & 0 & I_{1z} \end{bmatrix}.$$

The central principal axes of 2 and 3 are parallel to $\imath_2, \jmath_2, \mathbf{k}_2$, and the associated moments of inertia have values I_{2x}, I_{2y}, I_{2z} and I_{3x}, I_{3y}, I_{3z}, respectively. The inertia matrix associated with link 2 (2 and 2′) is

$$\bar{I}_2 \rightarrow \begin{bmatrix} I_{2x} & 0 & 0 \\ 0 & I_{2y} & 0 \\ 0 & 0 & I_{2z} \end{bmatrix}.$$

The inertia matrix associated with the slender bar 3 is

$$\bar{I}_3 \rightarrow \begin{bmatrix} I_{3x} & 0 & 0 \\ 0 & I_{3y} & 0 \\ 0 & 0 & I_{3z} \end{bmatrix} = \begin{bmatrix} \dfrac{m_3 L^2}{12} & 0 & 0 \\ 0 & \dfrac{m_3 L^2}{12} & 0 \\ 0 & 0 & 0 \end{bmatrix}.$$

The kinetic energy of a rigid body is

$$T = \frac{1}{2} m \mathbf{v}_C \cdot \mathbf{v}_C + \frac{1}{2} \boldsymbol{\omega} \cdot (\bar{I} \cdot \boldsymbol{\omega}), \tag{7.80}$$

where m is the mass, $\mathbf{v}_C$ is the velocity of the mass center, $\boldsymbol{\omega} = \omega_x \imath + \omega_y \jmath + \omega_z \mathbf{k}$ is the angular velocity of the rigid body in (0), and $\bar{I} = (I_x \imath)\imath + (I_y \jmath)\jmath + (I_z \mathbf{k})\mathbf{k}$ is the central inertia *dyadic* of the rigid body. The central principal axes of the rigid body are parallel to $\imath, \jmath, \mathbf{k}$, and the associated moments of inertia have the values I_x, I_y, I_z, respectively. The inertia matrix associated to $\bar{I}$ is

$$\bar{I} \rightarrow \mathbf{I} = \begin{bmatrix} I_x & 0 & 0 \\ 0 & I_y & 0 \\ 0 & 0 & I_z \end{bmatrix}. \tag{7.81}$$

The dot product of the vector $\boldsymbol{\omega}$ with the central inertia dyadic $\bar{I}$ is

$$\boldsymbol{\omega} \cdot \bar{I} = \bar{I} \cdot \boldsymbol{\omega} = \omega_x I_x \imath + \omega_y I_y \jmath + \omega_z I_z \mathbf{k}. \tag{7.82}$$

The total kinetic energy of the robot arm is

$$T = T_1 + T_2 + T_3,$$

where T_1 is the kinetic energy of link 1, T_2 is the kinetic energy of bar 2 and slider 2′, and T_3 is the kinetic energy of slender bar 3.

The kinetic energy of link 1 is

$$T_1 = \frac{1}{2} m_1 \mathbf{v}_{C_1} \cdot \mathbf{v}_{C_1} + \frac{1}{2} \boldsymbol{\omega}_{10} \cdot (\bar{I}_1 \cdot \boldsymbol{\omega}_{10}) = \frac{1}{2} \boldsymbol{\omega}_{10} \cdot (\bar{I}_1 \cdot \boldsymbol{\omega}_{10}), \tag{7.83}$$

where m_1 is the mass of the link, $\bar{I}_1 = (I_{1x}\mathbf{i}_1)\mathbf{i}_1 + (I_{1y}\mathbf{J}_1)\mathbf{J}_1 + (I_{1z}\mathbf{k}_1)\mathbf{k}_1$ is the central inertia dyadic of link 1, and $\omega_{10} = \dot{q}_1\,\mathbf{i}_1$. Using the above relation, the kinetic energy of link 1 is

$$T_1 = \frac{1}{2}I_{1x}\dot{q}_1^2. \tag{7.84}$$

The kinetic energy of link 2 is

$$T_2 = \frac{1}{2}m_2\mathbf{v}_{C_2} \cdot \mathbf{v}_{C_2} + \frac{1}{2}\omega_{20} \cdot (\bar{I}_2 \cdot \omega_{20}) = \frac{1}{2}\omega_{20} \cdot (\bar{I}_2 \cdot \omega_{20}), \tag{7.85}$$

where m_2 is the mass of the link and

$$\bar{I}_2 = (I_{2x}\mathbf{i}_2)\mathbf{i}_2 + (I_{2y}\mathbf{J}_2)\mathbf{J}_2 + (I_{2z}\mathbf{k}_1)\mathbf{k}_1$$

is the central inertia dyadic of link 2. The kinetic energy of link 2 is

$$T_2 = \frac{1}{2}\left[(I_{2x}c_2^2 + I_{2z}s_2^2)\dot{q}_1^2 + I_{2y}\dot{q}_2^2\right]. \tag{7.86}$$

The kinetic energy of slender bar 3 is

$$T_3 = \frac{1}{2}m_3\mathbf{v}_{C_3} \cdot \mathbf{v}_{C_3} + \frac{1}{2}\omega_{20} \cdot (\bar{I}_3 \cdot \omega_{20}), \tag{7.87}$$

where m_3 is the mass of the bar and

$$\bar{I}_3 = (I_{3x}\mathbf{i}_2)\mathbf{i}_2 + (I_{3y}\mathbf{J}_2)\mathbf{J}_2 + (I_{3z}\mathbf{k}_2)\mathbf{k}_2 = \left(\frac{m_3L^2}{12}\mathbf{i}_2\right)\mathbf{i}_2 + \left(\frac{m_3L^2}{12}\mathbf{J}_2\right)\mathbf{J}_2$$

is the central inertia dyadic of bar 3. The MATLAB commands for the total kinetic energy of the robot arm are

```
% kinetic energy
% inertia dyadic

syms I1x I1y I1z I2x I2y I2z real

% inertia matrix associated with
% central inertia dyadic for link 1
% expressed in terms of RF1{i1,j1,k1}
I1 = [I1x 0 0; 0 I1y 0; 0 0 I1z];

% inertia matrix associated with
% central inertia dyadic for link 2
% expressed in terms of RF2{i2,j2,k2}
I2 = [I2x 0 0; 0 I2y 0; 0 0 I2z];
```

```
% inertia matrix associated with
% central inertia dyadic for link 3
% expressed in terms of RF2{i2,j2,k2}
I3 = [m3*L2^2\12 0 0;
       0 m3*L^2\12 0;
       0 0 0];


% kinetic energy of the link 1
T1 = (1\2)*m1*vC1*vC1.' + (1\2)*w10*I1*w10.'


% kinetic energy of the link 2
T2 = (1\2)*m2*vC2*vC2.' + (1\2)*w20*I2*w20.'


% kinetic energy of the link 3
T3 = (1\2)*m3*vC3*vC3.' + (1\2)*w20*I3*w20.'


% total kinetic energy
T = expand(T1 + T2 + T3);
```

The left-hand sides of Lagrange's equations are

$$\frac{\mathrm{d}}{\mathrm{d}t}\left(\frac{\partial T}{\partial \dot{q}_r}\right) - \frac{\partial T}{\partial q_r}, \; r = 1, 2, 3.$$

The left-hand sides of Lagrange's equations are symbolically calculated in MAT-LAB with

```
% deriv(f, g(t)) differentiates
% f with respect to g(t)

% dT\d(dq)
Tdq1 = deriv(T, diff(q1,t));
Tdq2 = deriv(T, diff(q2,t));
Tdq3 = deriv(T, diff(q3,t));

% d(dT\d(dq))\dt
Tt1 = diff(Tdq1, t);
Tt2 = diff(Tdq2, t);
Tt3 = diff(Tdq3, t);

% dT\dq
Tq1 = deriv(T, q1);
Tq2 = deriv(T, q2);
Tq3 = deriv(T, q3);
```

```
% left hand side of Lagrange's eom
LHS1 = Tt1 - Tq1;
LHS2 = Tt2 - Tq2;
LHS3 = Tt3 - Tq3;
```

Remark: If a set of contact and/or body forces acting on a rigid body is equivalent to a couple of torque $\mathbf{T}$ together with force $\mathbf{R}$ applied at a point P of the rigid body, then the contribution of this set of forces to the generalized force, Q_r, is given by

$$Q_r = \frac{\partial \boldsymbol{\omega}}{\partial \dot{q}_r} \cdot \mathbf{T} + \frac{\partial \mathbf{v}_P}{\partial \dot{q}_r} \cdot \mathbf{R}, \quad r = 1, 2, \ldots,$$

where $\boldsymbol{\omega}$ is the angular velocity of the rigid body in (0), $\mathbf{v}_P$ is the velocity of P in (0), and r represents the generalized coordinates.

In the case of the robotic arm, there are two kinds of forces that contribute to the generalized forces Q_1, Q_2, and Q_3, namely, contact forces applied in order to drive the links 1, 2, and 3 and gravitational forces exerted on 1, 2, and 3 by the Earth. The set of contact forces transmitted from 0 to 1 can be replaced with a couple of torque $\mathbf{T}_{01}$ applied to 1 at A. Similarly, the set of contact forces transmitted from 1 to 2 can be replaced with a couple of torque $\mathbf{T}_{12}$ applied to 2. The law of action and reaction then guarantees that the set of contact forces transmitted from 1 to 2 is equivalent to a couple of torque $-\mathbf{T}_{12}$ to 1. Next, the set of contact forces exerted by link 2 on link 3 can be replaced with a force $\mathbf{F}_{23}$ applied to 3 at C_3. The law of action and reaction guarantees that the set of contact forces transmitted from 3 to 2 is equivalent to a force $-\mathbf{F}_{23}$ applied to 2 at C_{32}. The point C_{32} ($C_{32} \in$ link 2) instantaneously coincides with C_3, ($C_3 \in$ link 3). The expressions $\mathbf{T}_{01}$, $\mathbf{T}_{12}$, and $\mathbf{F}_{23}$ are

$$\mathbf{T}_{01} = T_{01x}\mathbf{i}_1 + T_{01y}\mathbf{j}_1 + T_{01z}\mathbf{k}_1, \quad \mathbf{T}_{12} = T_{12x}\mathbf{i}_2 + T_{12y}\mathbf{j}_2 + T_{12z}\mathbf{k}_2, \quad \text{and}$$

$$\mathbf{F}_{23} = F_{23x}\mathbf{i}_2 + F_{23y}\mathbf{j}_2 + F_{23z}\mathbf{k}_2.$$

The MATLAB statements for the contact torques and contact force are

```
syms T01x T01y T01z T12x T12y T12z F23x F23y F23z
  real

% contact torque of 0 that acts on link 1
% in RF0 expressed in terms of RF1{i1,j1,k1}
T01 = [T01x T01y T01z];

% contact torque of link 1 that acts on link 2
% in RF0 expressed in terms of RF2{i2,j2,k2}
T12 = [T12x T12y T12z];

% contact force of link 2 that acts on link 3 at C3
```

```
% in RF0 expressed in terms of RF2{i2,j2,k2}
F23 = [F23x F23y F23z];
```

The gravitational forces exerted on 1, 2, and 3 by the Earth are denoted by $\mathbf{G}_1$, $\mathbf{G}_2$, $\mathbf{G}_3$, respectively, and can be expressed as

$$\mathbf{G}_1 = -m_1 g \mathbf{l}_0,$$

$$\mathbf{G}_2 = -m_2 g \mathbf{l}_0,$$

$$\mathbf{G}_3 = -m_3 g \mathbf{l}_0 = -g m_3 \cos q_2 \mathbf{l}_2 - g m_3 \cos q_2 \mathbf{k}_2.$$

The reason for replacing $\mathbf{l}_0 = \mathbf{l}_1$ with $c_2 \mathbf{l}_2 + s_2 \mathbf{k}_2$ in connection with the forces $\mathbf{G}_3$ is that they are soon to be dot-multiplied with $\frac{\partial \mathbf{v}_{C_3}}{\partial \dot{q}_r}$ that have been expressed in terms of $\mathbf{l}_2, \mathbf{J}_2$, and $\mathbf{k}_2$. The MATLAB statements for the gravitational forces are

```
% gravitational force that acts on link 1 at C1
% RF0 expressed in terms of RF1{i1,j1,k1}
G1 = [-m1*g 0 0]

% gravitational force that acts on link 2 at C2
% in RF0 expressed in terms of RF2{i1,j1,k1}
G2 = [-m2*g 0 0]

% gravitational force that acts on link 3 at C3
% in RF0 expressed in terms of RF2{i2,j2,k2}
G3 = [-m3*g 0 0]*transpose(R21)
```

One can express $(Q_r)_1$, the contribution to the generalized active force Q_r of all the forces and torques acting on the particles of the link 1, as

$$(Q_r)_1 = \frac{\partial \boldsymbol{\omega}_{10}}{\partial \dot{q}_r} \cdot (\mathbf{T}_{01} - \mathbf{T}_{12}) + \frac{\partial \mathbf{v}_{C_1}}{\partial \dot{q}_r} \cdot \mathbf{G}_1, \quad r = 1,2,3.$$

The contribution $(Q_r)_2$ to the generalized active force of all the forces and torques acting on the link 2 is

$$(Q_r)_2 = \frac{\partial \boldsymbol{\omega}_{20}}{\partial \dot{q}_r} \cdot \mathbf{T}_{12} + \frac{\partial \mathbf{v}_{C_2}}{\partial \dot{q}_r} \cdot \mathbf{G}_2 + \frac{\partial \mathbf{v}_{C_{32}}}{\partial \dot{q}_r} \cdot (-\mathbf{F}_{23}), \quad r = 1,2,3,$$

where $\mathbf{v}_{C_{32}} = \mathbf{v}_{C_3} - \dot{q}_3 \mathbf{k}_2$. The contribution $(Q_r)_3$ to the generalized active force of all the forces and torques acting on the link 3 is

$$(Q_r)_3 = \frac{\partial \mathbf{v}_{C_3}}{\partial \dot{q}_r} \cdot \mathbf{G}_3 + \frac{\partial \mathbf{v}_{C_3}}{\partial \dot{q}_r} \cdot \mathbf{F}_{23}, \quad r = 1,2,3.$$

The generalized active force Q_r of all the forces and torques acting on the links 1, 2, and 3 are

$$Q_r = (Q_r)_1 + (Q_r)_2 + (Q_r)_3$$

$$= \frac{\partial \boldsymbol{\omega}_{10}}{\partial \dot{q}_r} \cdot (\mathbf{T}_{01} - \mathbf{T}_{12}) + \frac{\partial \mathbf{v}_{C_1}}{\partial \dot{q}_r} \cdot \mathbf{G}_1 + \frac{\partial \boldsymbol{\omega}_{20}}{\partial \dot{q}_r} \cdot \mathbf{T}_{12} + \frac{\partial \mathbf{v}_{C_2}}{\partial \dot{q}_r} \cdot \mathbf{G}_2 + \frac{\partial \mathbf{v}_{C_{32}}}{\partial \dot{q}_r} \cdot (-\mathbf{F}_{23})$$

$$+ \frac{\partial \mathbf{v}_{C_3}}{\partial \dot{q}_r} \cdot \mathbf{G}_3 + \frac{\partial \mathbf{v}_{C_3}}{\partial \dot{q}_r} \cdot \mathbf{F}_{23}, \qquad r = 1, 2, 3.$$

The generalized forces Q_r, $r = 1, 2, 3$ are symbolically calculated in the previous program and have the values

$$Q_1 = T_{01x},$$
$$Q_2 = T_{12y} - g\,m_3\,c_2\,q_3,$$
$$Q_3 = F_{23z} - g\,m_3\,s_2. \tag{7.88}$$

The MATLAB statements for the partial angular velocities $\frac{\partial \boldsymbol{\omega}}{\partial \dot{q}_r}$ and partial linear velocities $\frac{\partial \mathbf{v_C}}{\partial \dot{q}_r}$ where $r = 1, 2, 3$ are

```
% partial velocities
w1_1 = deriv(w10, diff(q1,t));
w2_1 = deriv(w20, diff(q1,t));
w1_2 = deriv(w10, diff(q2,t));
w2_2 = deriv(w20, diff(q2,t));
w1_3 = deriv(w10, diff(q3,t));
w2_3 = deriv(w20, diff(q3,t));

vC1_1 = deriv(vC1, diff(q1,t));
vC2_1 = deriv(vC2, diff(q1,t));
vC1_2 = deriv(vC1, diff(q2,t));
vC2_2 = deriv(vC2, diff(q2,t));
vC1_3 = deriv(vC1, diff(q3,t));
vC2_3 = deriv(vC2, diff(q3,t));

vC32_1 = deriv(vC32, diff(q1,t));
vC3_1 = deriv(vC3, diff(q1,t));
vC32_2 = deriv(vC32, diff(q2,t));
vC3_2 = deriv(vC3, diff(q2,t));
vC32_3 = deriv(vC32, diff(q3,t));
vC3_3 = deriv(vC3, diff(q3,t));
```

The generalized active force Q_1 is

$$Q_1 = \frac{\partial \boldsymbol{\omega}_{10}}{\partial \dot{q}_1} \cdot (\mathbf{T}_{01} - \mathbf{T}_{12}) + \frac{\partial \mathbf{v}_{C_1}}{\partial \dot{q}_1} \cdot \mathbf{G}_1 + \frac{\partial \boldsymbol{\omega}_{20}}{\partial \dot{q}_1} \cdot \mathbf{T}_{12} + \frac{\partial \mathbf{v}_{C_2}}{\partial \dot{q}_1} \cdot \mathbf{G}_2 + \frac{\partial \mathbf{v}_{C_{32}}}{\partial \dot{q}_1} \cdot (-\mathbf{F}_{23})$$

$$+ \frac{\partial \mathbf{v}_{C_3}}{\partial \dot{q}_1} \cdot \mathbf{G}_3 + \frac{\partial \mathbf{v}_{C_3}}{\partial \dot{q}_1} \cdot \mathbf{F}_{23},$$

and the MATLAB statement for the generalized active force Q_1 is

```
% generalized active force Q1
Q1=w1_1*T01.'+vC1_1*G1.'+w1_1*(R21.')*(-T12.')+...
   w2_1*T12.'+vC2_1*G2.'+vC32_1*(-F23.')+ ...
   vC3_1*F23.'+vC3_1*G3.';
```

The generalized active force Q_2 is

$$Q_2 = \frac{\partial \boldsymbol{\omega}_{10}}{\partial \dot{q}_2} \cdot (\mathbf{T}_{01} - \mathbf{T}_{12}) + \frac{\partial \mathbf{v}_{C_1}}{\partial \dot{q}_2} \cdot \mathbf{G}_1 + \frac{\partial \boldsymbol{\omega}_{20}}{\partial \dot{q}_2} \cdot \mathbf{T}_{12} + \frac{\partial \mathbf{v}_{C_2}}{\partial \dot{q}_2} \cdot \mathbf{G}_2 + \frac{\partial \mathbf{v}_{C_{32}}}{\partial \dot{q}_2} \cdot (-\mathbf{F}_{23})$$

$$+ \frac{\partial \mathbf{v}_{C_3}}{\partial \dot{q}_2} \cdot \mathbf{G}_3 + \frac{\partial \mathbf{v}_{C_3}}{\partial \dot{q}_2} \cdot \mathbf{F}_{23},$$

and the MATLAB statement for the generalized active force Q_2 is

```
% generalized active force Q2
Q2=w1_2*T01.'+vC1_2*G1.'+w1_2*(R21.')*(-T12.')+...
   w2_2*T12.'+vC2_2*G2.'+vC32_2*(-F23.')+...
   vC3_2*F23.'+vC3_2*G3.';
```

The generalized active force Q_3 is

$$Q_3 = \frac{\partial \boldsymbol{\omega}_{10}}{\partial \dot{q}_3} \cdot (\mathbf{T}_{01} - \mathbf{T}_{12}) + \frac{\partial \mathbf{v}_{C_1}}{\partial \dot{q}_3} \cdot \mathbf{G}_1 + \frac{\partial \boldsymbol{\omega}_{20}}{\partial \dot{q}_3} \cdot \mathbf{T}_{12} + \frac{\partial \mathbf{v}_{C_2}}{\partial \dot{q}_3} \cdot \mathbf{G}_2 + \frac{\partial \mathbf{v}_{C_{32}}}{\partial \dot{q}_3} \cdot (-\mathbf{F}_{23})$$

$$+ \frac{\partial \mathbf{v}_{C_3}}{\partial \dot{q}_3} \cdot \mathbf{G}_3 + \frac{\partial \mathbf{v}_{C_3}}{\partial \dot{q}_3} \cdot \mathbf{F}_{23},$$

and the MATLAB statement for the generalized active force Q_3 is

```
% generalized active force Q2
Q3=w1_3*T01.'+vC1_3*G1.'+w1_3*(R21.')*(-T12.')+...
   w2_3*T12.'+vC2_3*G2.'+vC32_3*(-F23.')+...
   vC3_3*F23.'+vC3_3*G3.';
```

The MATLAB results are

```
Q1 = T01x
Q2 = T12y - g*m3*cos(q2(t))*q3(t)
Q3 = F23z - g*m3*sin(q2(t))
```

To arrive at the dynamical equations governing the robot arm, all that remains to be done is to substitute into Lagrange's equations, namely,

$$\frac{\mathrm{d}}{\mathrm{d}t}\left(\frac{\partial T}{\partial \dot{q}_r}\right) - \frac{\partial T}{\partial q_r} = Q_r, \quad r = 1, 2, 3.$$

Lagrange's equations are symbolically calculated in MATLAB with

```
% first Lagrange's equation of motion
Lagrange1 = LHS1-Q1;
% second Lagrange's equation of motion
Lagrange2 = LHS2-Q2;
% third Lagrange's equation of motion
Lagrange3 = LHS3-Q3;
```

The following feedback control laws are used:

$$
\begin{aligned}
T_{01x} &= -\beta_{01}\dot{q}_1 - \gamma_{01}(q_1 - q_{1f}),\\
T_{12y} &= -\beta_{12}\dot{q}_2 - \gamma_{12}(q_2 - q_{2f}) + g\,m_3\,c_2\,q_3,\\
F_{23z} &= -\beta_{23}\dot{q}_3 - \gamma_{23}(q_3 - q_{3f}) + g\,m_3\,s_2.
\end{aligned}
\tag{7.89}
$$

The constant gains are $\beta_{01} = 450\,\mathrm{N\,m\,s/rad}$, $\gamma_{01} = 300\,\mathrm{N\,m/rad}$, $\beta_{12} = 200\,\mathrm{N\,m\,s/rad}$, $\gamma_{12} = 300\,\mathrm{N\,m/rad}$, $\beta_{23} = 150\,\mathrm{N\,s/m}$, $\gamma_{23} = 50\,\mathrm{N/m}$, $q_{1f} = \pi/3\,\mathrm{rad}$, $q_{2f} = \pi/3\,\mathrm{rad}$, and $q_{3f} = 0.3\,\mathrm{m}$. The MATLAB commands for the control torques are

```
% control torques and control force
q1f=pi\3; q2f=pi\3; q3f=0.3;
b01=450; g01=300;
b12=200; g12=300;
b23=150; g23=50;

T01xc = -b01*diff(q1,t)-g01*(q1-q1f);
T12yc = -b12*diff(q2,t)-g12*(q2-q2f)+g*m3*c2*q3;
F23zc = -b23*diff(q3,t)-g23*(q3-q3f)+g*m3*s2;
```

Lagrange's equations with the feedback control laws are

```
tor  = {T01x, T12y, F23z};
torf = {T01xc,T12yc,F23zc};

Lagrang1 = subs(Lagrange1, tor, torf);
Lagrang2 = subs(Lagrange2, tor, torf);
Lagrang3 = subs(Lagrange3, tor, torf);
```

Lagrange's equations with the numerical values for input data are

```
data = ...
{L1,L2,L,I1x,I2x,I2y,I2z,m1,m2,m3,g};
```

```
datn = ...
{0.4,0.4,0.5,5,4,1,4,90,60,40,9.81};

Lagran1 = subs(Lagrang1, data, datn);
Lagran2 = subs(Lagrang2, data, datn);
Lagran3 = subs(Lagrang3, data, datn);
```

The three second-order Lagrange's equations have to be rewritten as a first-order system:

```
ql = {diff(q1,t,2), diff(q2,t,2), diff(q3,t,2), ...
        diff(q1,t), diff(q2,t), diff(q3,t), q1, q2, q3};
qf = {'ddq1', 'ddq2',  'ddq3',...
        'x(2)', 'x(4)', 'x(6)', 'x(1)', 'x(3)', 'x(5)'};

% ql                           qf
%-----------------------------------
% diff('q1(t)',t,2) -> 'ddq1'
% diff('q2(t)',t,2) -> 'ddq2'
% diff('q3(t)',t,2) -> 'ddq3'
%    diff('q1(t)',t) -> 'x(2)'
%    diff('q2(t)',t) -> 'x(4)'
%    diff('q3(t)',t) -> 'x(6)'
%            'q1(t)' -> 'x(1)'
%            'q2(t)' -> 'x(3)'
%            'q3(t)' -> 'x(5)'

Lagra1 = subs(Lagran1, ql, qf);
Lagra2 = subs(Lagran2, ql, qf);
Lagra3 = subs(Lagran3, ql, qf);

% solve e.o.m. for ddq1, ddq2, ddq3
sol = solve(Lagra1, Lagra2, Lagra3,'ddq1, ddq2,
  ddq3');
Lagr1 = sol.ddq1;
Lagr2 = sol.ddq2;
Lagr3 = sol.ddq3;

% system of ODE
dx2dt = char(Lagr1);
dx4dt = char(Lagr2);
dx6dt = char(Lagr3);
```

The system of differential equations is solved numerically by m-file functions. The function file, eomE7_5a.m is created using the statements:

```
fid = fopen('eomE7_5a.m','w+');
fprintf(fid,'function dx = eomE7_5a(t,x) \n');
```

```
fprintf(fid,'dx = zeros(6,1);\n');
fprintf(fid,'dx(1) = x(2);\n');
fprintf(fid,'dx(2) = ');
fprintf(fid,dx2dt);
fprintf(fid,';\n');
fprintf(fid,'dx(3) = x(4);\n');
fprintf(fid,'dx(4) = ');
fprintf(fid,dx4dt);
fprintf(fid,';\n');
fprintf(fid,'dx(5) = x(6);\n');
fprintf(fid,'dx(6) = ');
fprintf(fid,dx6dt);
fprintf(fid,';');
fclose(fid);
cd(pwd);
```

The `ode45` solver is used for the system of differential equations:

```
t0 = 0;  tf = 15; time = [0 tf];

x0 = [pi\18 0 pi\6 0 0.25 0];

[t,xs] = ode45(@eomE7_5a, time, x0);

x1 = xs(:,1);
x2 = xs(:,2);
x3 = xs(:,3);
x4 = xs(:,4);
x5 = xs(:,5);
x6 = xs(:,6);

subplot(3,1,1),plot(t,x1*180\pi,'r'),...
xlabel('t (s)'),ylabel('q1 (deg)'),grid,...
subplot(3,1,2),plot(t,x3*180\pi,'b'),...
xlabel('t (s)'),ylabel('q2 (deg)'),grid,...
subplot(3,1,3),plot(t,x5,'g'),...
xlabel('t (s)'),ylabel('q3 (m)'),grid

[ts,xs] = ode45(@eomE7_5a,0:1:5,x0);

fprintf('Results \n'); fprintf('\n')
fprintf...
(' t(s) q1 dq1 q2 dq2 q3 dq3\n')
[ts,xs]
```

Figure 7.8d shows the plots of $q_1(t)$, $q_2(t)$, $q_3(t)$.

Kane's Dynamical Equations

The generalized coordinates q_i and the generalized speeds u_i are introduced in MATLAB with

```
syms t L1 L2 L m1 m2 m3 g real
% generalized coordinates q1, q2, q3
q1 = sym('q1(t)');
q2 = sym('q2(t)');
q3 = sym('q3(t)');
% generalized speeds u1, u2, u3
u1 = sym('u1(t)');
u2 = sym('u2(t)');
u3 = sym('u3(t)');
```

The generalized speeds, u_1, u_2, u_3, are associated with the motion of a system and can be introduced as $\dot{q}_i = u_i$ or

```
dq1 = u1;
dq2 = u2;
dq3 = u3;

qt = {diff(q1,t), diff(q2,t), diff(q3,t)};
qu = {dq1, dq2, dq3};
```

The velocities and the accelerations of the robot need to be expressed in terms of q_i, u_i, and $\dot{u}_i$:

```
c1 = cos(q1); s1 = sin(q1);
c2 = cos(q2); s2 = sin(q2);

R10 = [[1 0 0]; [0 c1 s1]; [0 -s1 c1]];
R21 = [[c2 0 -s2]; [0 1 0]; [s2 0 c2]];

w10  = [dq1, 0, 0 ];
w201 = [dq1, dq2, 0];
w20  = w201 * transpose(R21);
alpha10 = diff(w10,t);
alpha20 = subs(diff(w20,t), qt, qu);

% position vector of mass center C1 of link 1
% in RF0 expressed in terms of RF1{i1,j1,k1}
rC1 = [L1 0 0];

% linear velocity of mass center C1 of link 1
% in RF0 expressed in terms of RF1{i1,j1,k1}
vC1 = diff(rC1,t) + cross(w10, rC1)
```

```
% position vector of mass center C2 of link 2
% in RF0 expressed in terms of RF1{i1,j1,k1}
rC2 = [L2 0 0];

% linear velocity of mass center C2 of link 2
% in RF0 expressed in terms of RF1 {i1,j1,k1}
vC2 = simple(diff(rC2,t) + cross(w10,rC2))

% position vector of mass center C3 of link 3
% in RF0 expressed in terms of RF2{i2,j2,k2}
rC3 = rC2*R21.' + [0 0 q3]

% linear velocity of mass center C3 of link 3
% in RF0 expressed in terms of RF2{i2,j2,k2}
vC3 = subs(diff(rC3, t), qt, qu) + cross(w20,rC3)

% linear velocity of  C32 of link 2 in RF0
% expressed in terms of RF2{i2,j2,k2}
% C32 of link 2 is superposed with C3 of link 3
vC32 = simple(vC2 + cross(w20,[0 0 q3]))

% linear accelerations
aC1 = simple(diff(vC1,t)+cross(w10,vC1))
aC2 = simple(diff(vC2,t)+cross(w20,vC2))
aC3 = subs(diff(vC3,t), qt, qu) + cross(w20,vC3)
```

The partial velocities with respect to u1, u2, u3 are calculated using the
function deriv:

```
% partial velocities
w1_1 = deriv(w10, u1); w2_1 = deriv(w20, u1);
w1_2 = deriv(w10, u2); w2_2 = deriv(w20, u2);
w1_3 = deriv(w10, u3); w2_3 = deriv(w20, u3);

vC1_1 = deriv(vC1, u1); vC2_1 = deriv(vC2, u1);
vC1_2 = deriv(vC1, u2); vC2_2 = deriv(vC2, u2);
vC1_3 = deriv(vC1, u3); vC2_3 = deriv(vC2, u3);

vC32_1 = deriv(vC32, u1); vC3_1 = deriv(vC3, u1);
vC32_2 = deriv(vC32, u2); vC3_2 = deriv(vC3, u2);
vC32_3 = deriv(vC32, u3); vC3_3 = deriv(vC3, u3);
```

Generalized Inertia Forces
To explain what the *generalized inertia forces* are, a system $\{S\}$ formed by p
particles $P_1,\ldots,P_v$ and having masses $m_1,\ldots,m_p$ is considered. Suppose that n
generalized speeds $u_r, r = 1,\ldots,n$ have been introduced. (For the robotic arm

$u_r = \dot{q}_r, r = 1, \ldots, n.$) Let $\mathbf{v}_{P_j}$ and $\mathbf{a}_{P_j}$ denote, respectively, the velocity of P_j and the acceleration of P_j in a reference frame (0).

Define $\mathbf{F}_{\text{in}\,j}$, called the inertia force for P_j, as

$$\mathbf{F}_{\text{in}\,j} = -m_j \mathbf{a}_{P_j}.$$

The quantities $K_{\text{in}\,1}, \ldots, K_{\text{in}\,n}$, defined as

$$K_{\text{in}\,r} = \sum_{j=1}^{v} \frac{\partial \mathbf{v}_{P_j}}{\partial u_r} \cdot \mathbf{F}_{\text{in}\,j}, \quad r = 1, \ldots, n,$$

are called *generalized inertia forces* for $\{S\}$.

The contribution to $K_{\text{in}\,r}$, made by the particles of a rigid body RB belonging to $\{S\}$, is

$$(K_{\text{in}\,r})_R = \frac{\partial \mathbf{v}_C}{\partial u_r} \cdot \mathbf{F}_{\text{in}} + \frac{\partial \boldsymbol{\omega}}{\partial u_r} \cdot \mathbf{M}_{\text{in}}, \quad r = 1, \ldots, n,$$

where $\mathbf{v}_C$ is the velocity of the center of gravity of RB in (0) and $\boldsymbol{\omega} = \omega_x \mathbf{1} + \omega_y \mathbf{J} + \omega_z \mathbf{k}$ is the angular velocity of RB in (0).

The inertia force for the rigid body RB is

$$\mathbf{F}_{\text{in}} = -m \mathbf{a}_C,$$

where m is the mass of RB and $\mathbf{a}_C$ is the acceleration of the mass center of RB in the fixed reference frame. The inertia moment $\mathbf{M}_{\text{in}}$ for RB is

$$\mathbf{M}_{\text{in}} = -\boldsymbol{\alpha} \cdot \bar{I} - \boldsymbol{\omega} \times (\bar{I} \cdot \boldsymbol{\omega}),$$

where $\boldsymbol{\alpha} = \dot{\boldsymbol{\omega}} = \alpha_x \mathbf{1} + \alpha_y \mathbf{J} + \alpha_z \mathbf{k}$ is the angular acceleration of RB in (0) and $\bar{I} = (I_x \mathbf{1})\mathbf{1} + (I_y \mathbf{J})\mathbf{J} + (I_z \mathbf{k})\mathbf{k}$ is the central inertia dyadic of RB. The central principal axes of RB are parallel to $\mathbf{1}, \mathbf{J},$ and $\mathbf{k}$, and the associated moments of inertia have the values I_x, I_y, I_z, respectively. The inertia matrix associated with $\bar{I}$ is

$$\bar{I} \rightarrow \begin{bmatrix} I_x & 0 & 0 \\ 0 & I_y & 0 \\ 0 & 0 & I_z \end{bmatrix}.$$

The dot product of the vector $\boldsymbol{\alpha}$ with the dyadic $\bar{I}$ is

$$\boldsymbol{\alpha} \cdot \bar{I} = \bar{I} \cdot \boldsymbol{\alpha} = \alpha_x I_x \mathbf{1} + \alpha_y I_y \mathbf{J} + \alpha_z I_z \mathbf{k},$$

and the cross product between a vector and a dyadic is

$$\boldsymbol{\omega} \times (\bar{I} \cdot \boldsymbol{\omega}) = \begin{vmatrix} \mathbf{1} & \mathbf{J} & \mathbf{k} \\ \omega_x & \omega_y & \omega_z \\ \omega_x I_x & \omega_y I_y & \omega_z I_z \end{vmatrix}$$

$$= -\omega_y \omega_z (I_y - I_z)\mathbf{1} - \omega_z \omega_x (I_z - I_x)\mathbf{J} - \omega_x \omega_y (I_x - I_y)\mathbf{k}.$$

The inertia moment of 1 in (0) can be written as

$$\mathbf{M}_{\mathrm{in}\,1} = -\boldsymbol{\alpha}_{10}\cdot\bar{I}_1 - \boldsymbol{\omega}_{10}\times(\bar{I}_1\cdot\boldsymbol{\omega}_{10}) = -I_{1x}\ddot{q}_1\mathbf{l}_1.$$

The inertia moment of 2 in (0) is

$$\mathbf{M}_{\mathrm{in}\,2} = -\boldsymbol{\alpha}_{20}\cdot\bar{I}_2 - \boldsymbol{\omega}_{20}\times(\bar{I}_2\cdot\boldsymbol{\omega}_{20}).$$

Similarly, the inertia moment of 3 in (0) is

$$\mathbf{M}_{\mathrm{in}\,3} = \boldsymbol{\alpha}_{20}\cdot\bar{I}_3 - \boldsymbol{\omega}_{20}\times(\bar{I}_3\cdot\boldsymbol{\omega}_{20}).$$

The inertia force for link $j = 1, 2, 3$ is

$$\mathbf{F}_{\mathrm{in}\,j} = -m_j\,\mathbf{a}_{C_j}.$$

The contribution of link $j = 1, 2, 3$ to the generalized inertia force $K_{\mathrm{in}\,r}$ is

$$(K_{\mathrm{in}\,r})_j = \frac{\partial\mathbf{v}_{C_j}}{\partial u_r}\cdot\mathbf{F}_{\mathrm{in}\,j} + \frac{\partial\boldsymbol{\omega}_{j0}}{\partial u_r}\cdot\mathbf{M}_{\mathrm{in}\,j},\ \ r = 1, 2, 3.$$

The three generalized inertia forces are computed with

$$K_{\mathrm{in}\,r} = \sum_{j=1}^{3}(K_{\mathrm{in}\,r})_j$$

$$= \sum_{j=1}^{3}\left(\frac{\partial\mathbf{v}_{C_j}}{\partial u_r}\cdot\mathbf{F}_{\mathrm{in}\,j} + \frac{\partial\boldsymbol{\omega}_{j0}}{\partial u_r}\cdot\mathbf{M}_{\mathrm{in}\,j}\right),\ \ r = 1, 2, 3,$$

or

$$K_{\mathrm{in}\,r} = \frac{\partial\mathbf{v}_{C_1}}{\partial u_r}\cdot(-m_1\,\mathbf{a}_{C_1}) + \frac{\partial\boldsymbol{\omega}_{10}}{\partial u_r}\cdot\mathbf{M}_{\mathrm{in}\,1} + \frac{\partial\mathbf{v}_{C_2}}{\partial u_r}\cdot(-m_2\,\mathbf{a}_{C_2}) + \frac{\partial\boldsymbol{\omega}_{20}}{\partial u_r}\cdot\mathbf{M}_{\mathrm{in}\,2}$$

$$+ \frac{\partial\mathbf{v}_{C_3}}{\partial u_r}\cdot(-m_3\,\mathbf{a}_{C_3}) + \frac{\partial\boldsymbol{\omega}_{30}}{\partial u_r}\cdot\mathbf{M}_{\mathrm{in}\,3},\ \ r = 1, 2, 3.$$

The generalized inertia forces for the RRT robot arm are calculated with the following MATLAB commands:

```
% inertia dyadic
syms I1x I1y I1z I2x I2y I2z real

% inertia matrix associated with
% central inertia dyadic for link 1
% expressed in terms of RF1{i1,j1,k1}
I1 = [I1x 0 0; 0 I1y 0; 0 0 I1z];
```

```matlab
% inertia matrix associated with
% central inertia dyadic for link 2
% expressed in terms of RF2{i2,j2,k2}
I2 = [I2x 0 0; 0 I2y 0; 0 0 I2z];

% inertia matrix associated with
% central inertia dyadic for link 3
% expressed in terms of RF2{i2,j2,k2}
I3 = [m3*L^2\12 0 0;
      0 m3*L^2\12 0;
      0 0 0];

% Kane's dynamical equations

% inertia forces

% inertia force for link 1
% expressed in terms of RF1{i1,j1,k1}
Fin1= -m1*aC1;
% inertia force for link 2
% expressed in terms of RF2{i2,j2,k2}
Fin2= -m2*aC2;
% inertia force for link 3
% expressed in terms of RF2{i2,j2,k2}
Fin3= -m3*aC3;

% inertia moments

% inertia moment for link 1
% expressed in terms of RF1{i1,j1,k1}
Min1 = -alpha10*I1-cross(w10,w10*I1);
% inertia moment for link 2
% expressed in terms of RF2{i2,j2,k2}
Min2 = -alpha20*I2-cross(w20,w20*I2);
% inertia moment for link 3
% expressed in terms of RF2{i2,j2,k2}
Min3 = -alpha20*I3-cross(w20,w20*I3);

% generalized inertia forces

% generalized inertia forces corresponding to q1
Kin1 = w1_1*Min1.' + vC1_1*Fin1.' + ...
       w2_1*Min2.' + vC2_1*Fin2.' + ...
       w2_1*Min3.' + vC3_1*Fin3.';
```

```
% generalized inertia forces corresponding to q2
Kin2 = w1_2*Min1.' + vC1_2*Fin1.' + ...
       w2_2*Min2.' + vC2_2*Fin2.' + ...
       w2_2*Min3.' + vC3_2*Fin3.';


% generalized inertia forces corresponding to q3
Kin3 = w1_3*Min1.' + vC1_3*Fin1.' + ...
       w2_3*Min2.' + vC2_3*Fin2.' + ...
       w2_3*Min3.' + vC3_3*Fin3.';
```

The gravitational forces and the external moments and force are

```
% generalized active forces

syms T01x T01y T01z T12x T12y T12z F23x F23y F23z
  real

% contact torque of 0 that acts on link 1
% in RF0 expressed in terms of RF1{i1,j1,k1}
T01 = [T01x T01y T01z];

% contact torque of link 1 that acts on link 2
% in RF0 expressed in terms of RF2{i2,j2,k2}
T12 = [T12x T12y T12z];

% contact force of link 2 that acts on link 3 at C3
% in RF0 expressed in terms of RF2{i2,j2,k2}
F23 = [F23x F23y F23z];

% gravitational force that acts on link 1 at C1
% RF0 expressed in terms of RF1{i1,j1,k1}
G1 = [-m1*g 0 0]

% gravitational force that acts on link 2 at C2
% in RF0 expressed in terms of RF2{i1,j1,k1}
G2 = [-m2*g 0 0]

% gravitational force that acts on link 3 at C3
% in RF0 expressed in terms of RF2{i2,j2,k2}
G3 = [-m3*g 0 0]*transpose(R21)
```

and the generalized active forces are

```
% generalized active forces
Q1 = w1_1*T01.' + vC1_1*G1.' + ...
     w1_1*transpose(R21)*(-T12.') + ...
```

```
        w2_1*T12.' + vC2_1*G2.' + vC32_1*(-F23.') + ...
        vC3_1*F23.' + vC3_1*G3.';

  Q2 = w1_2*T01.' + vC1_2*G1.' + ...
       w1_2*transpose(R21)*(-T12.') + ...
       w2_2*T12.' + vC2_2*G2.' + vC32_2*(-F23.') + ...
       vC3_2*F23.' + vC3_2*G3.';

  Q3 = w1_3*T01.' + vC1_3*G1.' + ...
       w1_3*transpose(R21)*(-T12.') + ...
       w2_3*T12.' + vC2_3*G2.' + vC32_3*(-F23.') + ...
       vC3_3*F23.' + vC3_3*G3.';

fprintf('Q1 = %s\n',char(simple(Q1)))
fprintf('Q2 = %s\n',char(simple(Q2)))
fprintf('Q3 = %s\n',char(simple(Q3)))


% Q1 = T01x
% Q2 = T12y - g*m3*cos(q2(t))*q3(t)
% Q3 = F23z - g*m3*sin(q2(t))
```

To arrive at the dynamical equations governing the robot arm, all that remains to
be done is to substitute into Kane's dynamical equations, namely,

$$K_{\text{in}\,r} + Q_r = 0, \ r = 1, 2, 3. \tag{7.90}$$

Kane's dynamical equations in MATLAB are

```
% Kane's dynamical equations
% first Kane's dynamical equation
Kane1 = Kin1 + Q1;
% second Kane's dynamical equation
Kane2 = Kin2 + Q2;
% third Kane's dynamical equation
Kane3 = Kin3 + Q3;
```

Using the same feedback control laws (the same as these used for Lagrange's
equations), Kane's equations have to be rewritten:

```
% control torques and control force
q1f=pi\3; q2f=pi\3; q3f=0.3;
b01=450; g01=300;
b12=200; g12=300;
b23=150; g23=50;
```

```matlab
T01xc = -b01*dq1-g01*(q1-q1f);
T12yc = -b12*dq2-g12*(q2-q2f)+g*m3*c2*q3;
F23zc = -b23*dq3-g23*(q3-q3f)+g*m3*s2;

tor  = {T01x, T12y, F23z};
torf = {T01xc,T12yc,F23zc};

Kan1 = subs(Kane1, tor, torf);
Kan2 = subs(Kane2, tor, torf);
Kan3 = subs(Kane3, tor, torf);
```

The Kane's dynamical equations can be expressed in terms of $\dot{u}_1$, $\dot{u}_2$, and $\dot{u}_3$:

```matlab
data = ...
{L1,L2,L,I1x,I2x,I2y,I2z,m1,m2,m3,g};
datn = ...
{0.4,0.4,0.5,5,4,1,4,90,60,40,9.81};

Ka1 = subs(Kan1, data, datn);
Ka2 = subs(Kan2, data, datn);
Ka3 = subs(Kan3, data, datn);

ql = {diff(u1,t), diff(u2,t), diff(u3,t) ...
      u1, u2, u3, q1, q2, q3};
qx = {'du1', 'du2',  'du3',...
'x(4)', 'x(5)', 'x(6)', 'x(1)', 'x(2)', 'x(3)'};

% ql                       qx
%---------------------------------
% diff('u1(t)',t) -> 'du1'
% diff('u2(t)',t) -> 'du2'
% diff('u3(t)',t) -> 'du3'
%          'u1(t)' -> 'x(4)'
%          'u2(t)' -> 'x(5)'
%          'u3(t)' -> 'x(6)'
%          'q1(t)' -> 'x(1)'
%          'q2(t)' -> 'x(2)'
%          'q3(t)' -> 'x(3)'

Du1 = subs(Ka1, ql, qx);
Du2 = subs(Ka2, ql, qx);
Du3 = subs(Ka3, ql, qx);

% solve for du1, du2, du3
sol = solve(Du1, Du2, Du3,'du1, du2, du3');
sdu1 = sol.du1;
```

```
sdu2 = sol.du2;
sdu3 = sol.du3;
```

The system of differential equations is solved numerically by m-file functions. The function file, eomE7_5b.m, is created using the following statements:

```
% system of ODE
dx1 = char('x(4)');
dx2 = char('x(5)');
dx3 = char('x(6)');
dx4 = char(sdu1);
dx5 = char(sdu2);
dx6 = char(sdu3);

fid = fopen('eomE7_5b.m','w+');
fprintf(fid,'function dx = eomE7_5b(t,x) \n');
fprintf(fid,'dx = zeros(6,1);\n');
fprintf(fid,'dx(1) = '); fprintf(fid,dx1);
fprintf(fid,';\n');
fprintf(fid,'dx(2) = '); fprintf(fid,dx2);
fprintf(fid,';\n');
fprintf(fid,'dx(3) = '); fprintf(fid,dx3);
fprintf(fid,';\n');
fprintf(fid,'dx(4) = '); fprintf(fid,dx4);
fprintf(fid,';\n');
fprintf(fid,'dx(5) = '); fprintf(fid,dx5);
fprintf(fid,';\n');
fprintf(fid,'dx(6) = '); fprintf(fid,dx6);
fprintf(fid,';  ');
fclose(fid); cd(pwd);
```

The ode45 solver is used to solve the system of first-order differential equations:

```
t0 = 0;  tf = 15; time = [0 tf];

x0 = [pi\18 pi\6 0.25 0 0 0];

[t,xs] = ode45(@eomE7_5b, time, x0);

x1 = xs(:,1);
x2 = xs(:,2);
x3 = xs(:,3);
x4 = xs(:,4);
x5 = xs(:,5);
x6 = xs(:,6);

subplot(3,1,1),plot(t,x1*180\pi,'r'),...
xlabel('t (s)'),ylabel('q1 (deg)'),grid,...
```

```
subplot(3,1,2),plot(t,x2*180\pi,'b'),...
xlabel('t (s)'),ylabel('q2 (deg)'),grid,...
subplot(3,1,3),plot(t,x3,'g'),...
xlabel('t (s)'),ylabel('q3 (m)'),grid
```

Gibbs–Appell Equations

Next, the Gibbs–Appell equations are calculated for the robotic arm using MAT-LAB. A quantity S, similar to kinetic energy of p particles, called by some energy of acceleration, has the form

$$S = \sum_{i=1}^{p} \frac{1}{2} m_i \mathbf{a}_i \cdot \mathbf{a}_i.$$

The resulting Gibbs–Appell equations of motion are

$$\frac{\partial S}{\partial \dot{u}_k} = Q_k.$$

It can be shown that Gibbs–Appell equations are the same as Kane's equations, both for particles as well as rigid bodies. For the robotic arm, the left-hand sides of Gibbs–Appell equations of motion are

```
I1 = [I1x 0 0; 0 I1y 0; 0 0 I1z];
I2 = [I2x 0 0; 0 I2y 0; 0 0 I2z];
I3 = [m3*L^2\12 0 0;
      0 m3*L^2\12 0;
      0 0 0];

% Energy of acceleration
S1 = (1\2)*m1*aC1*aC1.' + (1\2)*alpha10*I1*alpha10.';
S2 = (1\2)*m2*aC2*aC2.' + (1\2)*alpha20*I2*alpha20.';
S3 = (1\2)*m3*aC3*aC3.' + (1\2)*alpha20*I3*alpha20.';
S = expand(S1 + S2 + S3);

Sddq1 = deriv(S, diff(q1,'t',2));
Sddq2 = deriv(S, diff(q2,'t',2));
Sddq3 = deriv(S, diff(q3,'t',2));
```

The Gibbs–Appell equations in MATLAB are

```
% eom
G_A1 = Sddq1-Q1;
G_A2 = Sddq2-Q2;
G_A3 = Sddq3-Q3;
```

7.7 Problems

7.1 The blocks 1 and 2 are moving in a horizontal direction as shown in Fig. 7.9. The slider 1 having the mass m_1 is connected to a vertical wall (x-axis) by a spring with the elastic constant k_1, while slider 2 having the mass m_2 is connected to slider 1 by a spring having the elastic constant k_2. Both springs have the initial length l_0 and deflect only horizontally as shown in Fig. 7.9. Find the equations of motion of the system.

7.2 A massless rod AB of length l and a mass m located at the A end rotate about its support O as shown in Fig. 7.10. The distance between the support O and the end A of the road AB is $l_{OA} = l_1$, and the distance between the support O and the end B of the road is $l_{OB} = l_2$. The end B of the rod is connected to the ground by a spring having the elastic constant k and initial length h. The spring deflect only vertically. Find and solve the equations of motion of the system.

7.3 The massless rod AB with the length $l_{AB} = l_1 + l_2 = l$ rotates about a fixed point O as shown in Fig. 7.11. The massless rod BC with the length $l_{BC} = l_3$ is

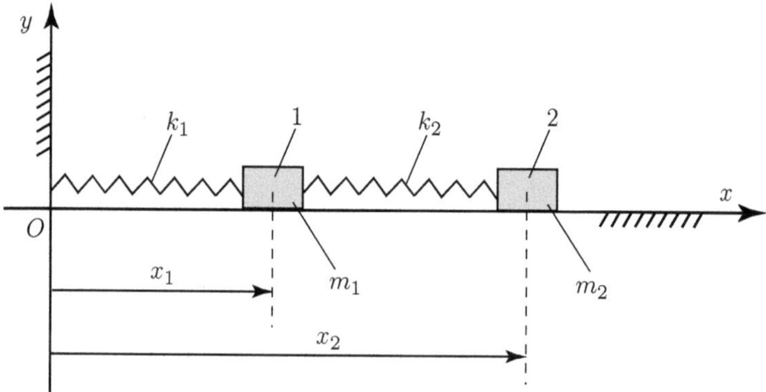

Fig. 7.9 Problem 7.1

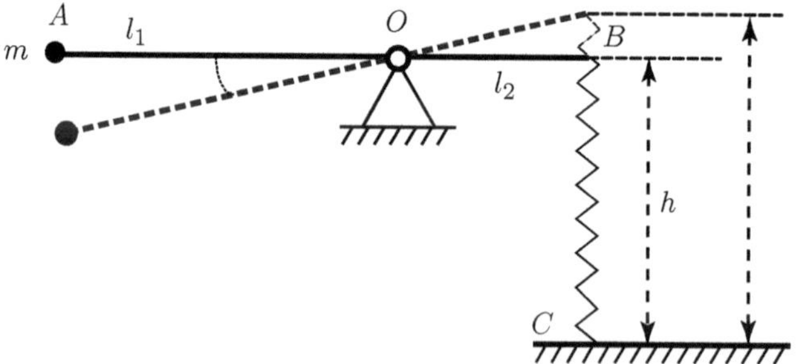

Fig. 7.10 Problem 7.2

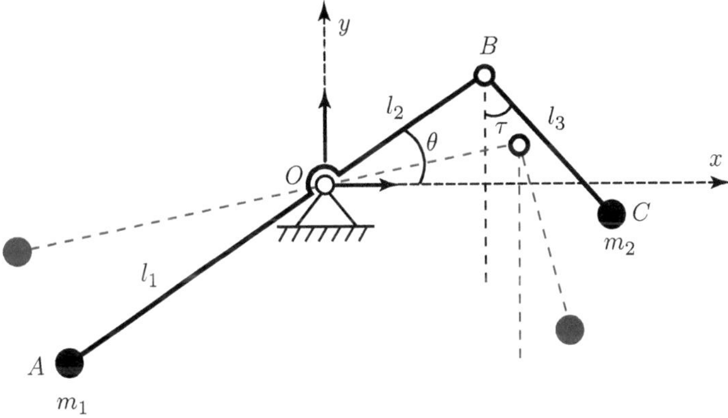

Fig. 7.11 Problem 7.3

connected with a pin joint at B with the rod AB. A lumped mass m_1 is situated
at A, and a mass m_2 is located at C. Find the equations of motion of the system.

7.4 A heavy body of mass m is moving without friction along a planar curve
described by the parametric equations:

$$x = l(\theta + \sin\theta),$$
$$y = l(1 - \cos\theta)$$

where $\theta \in [-\pi, \pi]$. Find the equation of motion.

7.5 Find the equations of the motion for the system in Fig. 7.12 using Lagrange
and Kane's method. The homogeneous slender rods $OA = AB = 2L$ have the
masses equal to $m_1 = m_2 = m$. The mass of the slider 3 is m_3, and the mass
of the slider 4 is m_4. The linear spring R has the elastic constant k, and its
mass is neglected. The driver moment $\mathbf{M}_m = M_m\mathbf{k}$ acts on 1 at O. The initial
conditions are given. The friction is neglected.

7.6 Find and solve the equations of the motion for the mechanical systems
described in Problems 6.8–6.20 using Lagrange and/or Hamilton's method.

7.7 A homogeneous circular disk in motion on a rough inclined plane is shown
in Fig. 7.13. The fixed Cartesian reference frame xyz is chosen with the origin
at O. The angle between the axis Ox and the horizontal is α. The contact point
between the disk and the plane is B. The disk has the mass m, the radius r, and
the center of mass at C. The gravitational acceleration is g. Find the equations
of motion for the disk.

7.8 Two particles P_1 of mass m_1 and P_2 of mass m_2 are free to move in a smooth
tube T, as shown in Fig. 7.14. The tube has the mass M and the length L. The
particles are attached to linear springs with the elastic constants k_1 and k_2 as

Fig. 7.12 Problem 7.5

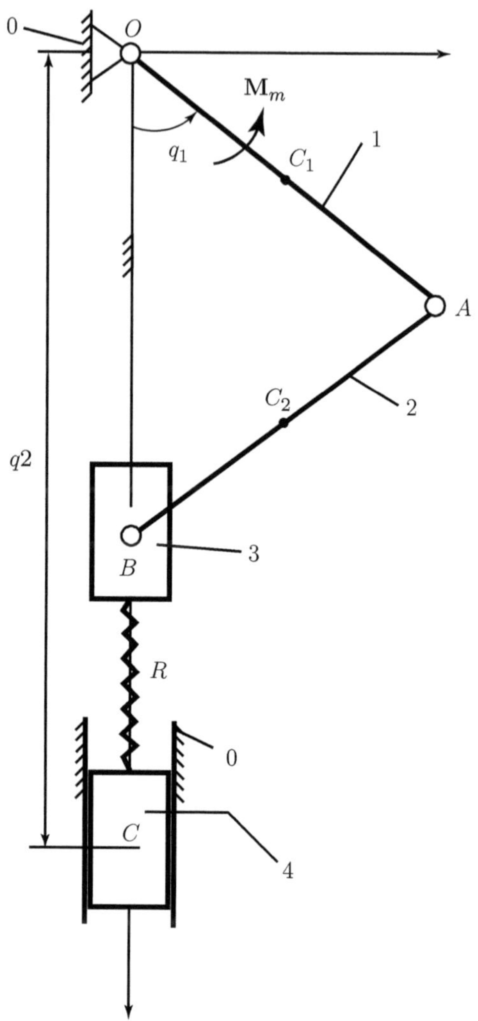

shown in Fig. 7.14. The initial lengths of the springs are l_{10} and l_{20}. The tube is rotating in the plane xy about its end O. The angle between the axis of the tube and the horizontal x-axis is a given function $\theta(t)$, where t is the time. The gravitational acceleration is g. Find the equations of motion for the system (tube and particles).

7.9 A slider 2 of mass m_2 is free to move along a smooth rod 1 of mass m_1 and length $OA = 2l$ as shown in Fig. 7.15. The rod is rotating about a vertical z-axis. The angle between the rod and the horizontal x-axis is α. The gravitational acceleration is g. Find the equations of motion of the system.

7.10 The schematic representations of a 3 link robot arm are shown in Figs. 7.16–7.20. The mass centers of links 1, 2, and 3 are C_1, C_2, and C_3, respectively.

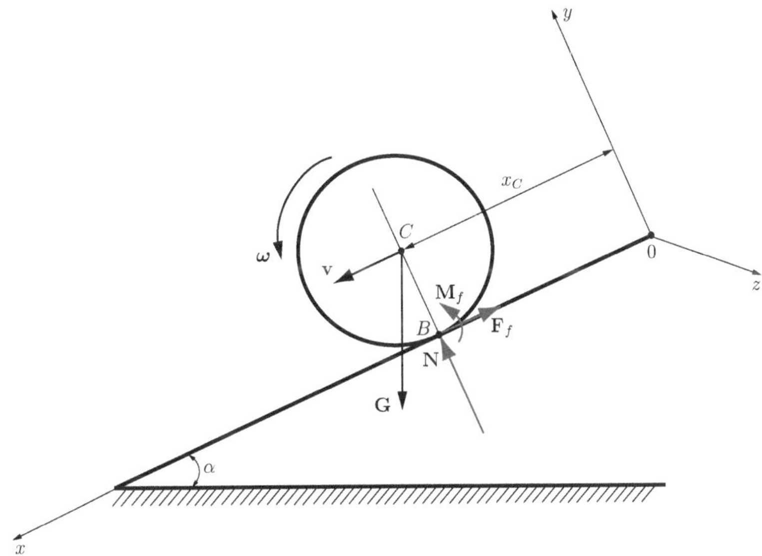

Fig. 7.13 Problem 7.7

Fig. 7.14 Problem 7.8

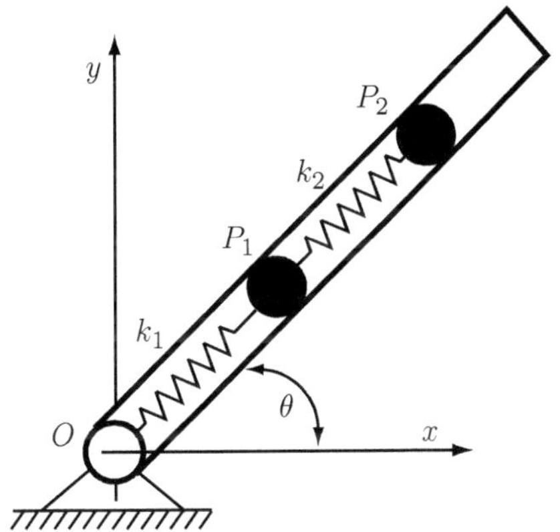

The central principal axes of link p, $p = 1, 2, 3$ are parallel to $\mathbf{i}_p, \mathbf{J}_p, \mathbf{k}_p$, and the associated moments of inertia have the values I_{px}, I_{py}, I_{pz}, respectively. The central inertia dyadic of link p is $\bar{I}_p = (I_{px}\mathbf{i}_p)\mathbf{i}_p + (I_{py}\mathbf{J}_p)\mathbf{J}_p + (I_{pz}\mathbf{k}_p)\mathbf{k}_p$. If the joint between link p and link $p + 1$ is a rotational joint, consider a control vector moment $\mathbf{T}_{p,p+1} = T_{(p,p+1)x}\mathbf{i}_{p+1} + T_{(p,p+1)y}\mathbf{J}_{p+1} + T_{(p,p+1)z}\mathbf{k}_{p+1}$, and if the joint is a translational joint, consider a control vector force $\mathbf{F}_{p,p+1} = F_{(p,p+1)x}\mathbf{i}_{p+1} + F_{(p,p+1)y}\mathbf{J}_{p+1} + F_{(p,p+1)z}\mathbf{k}_{p+1}$. Select suitable numerical values for the input numerical data. Find and solve the equations of motion of the robot.

Fig. 7.15 Problem 7.9

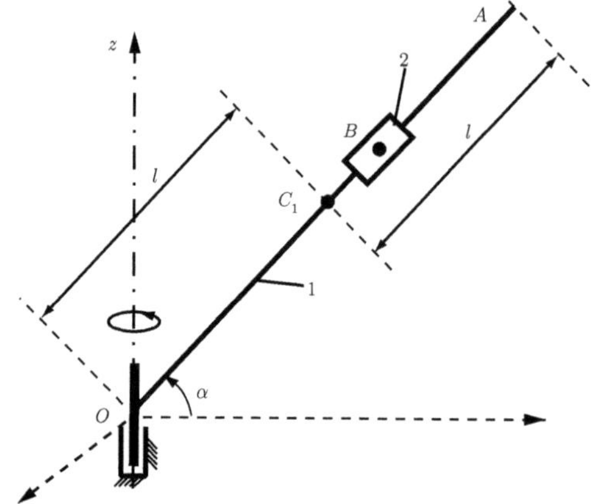

Fig. 7.16 Problem 7.10.1

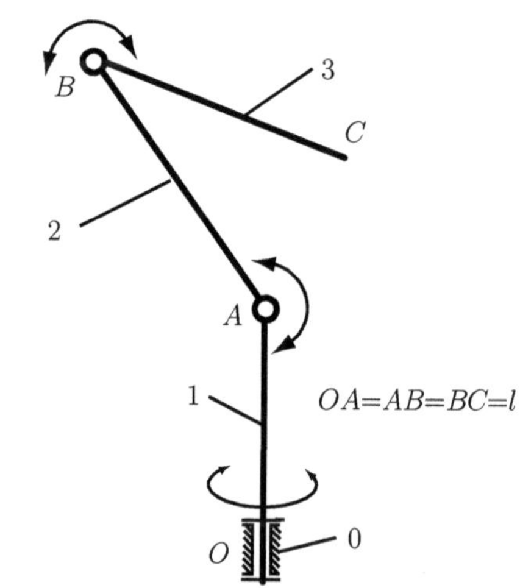

Fig. 7.17   Problem 7.10.2

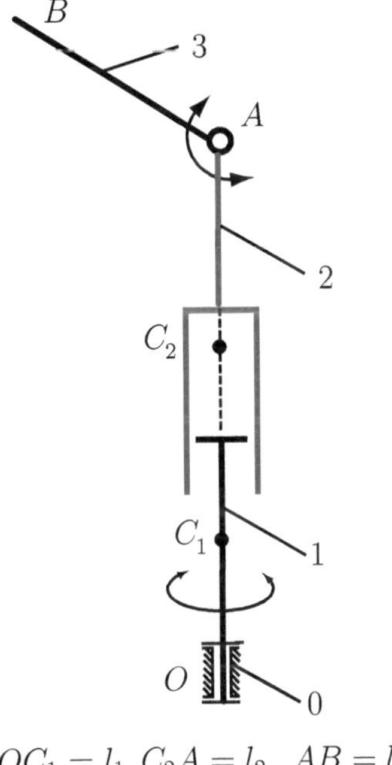

$$OC_1 = l_1, C_2A = l_2, \ \ AB = l_3$$

Fig. 7.18   Problem 7.10.3

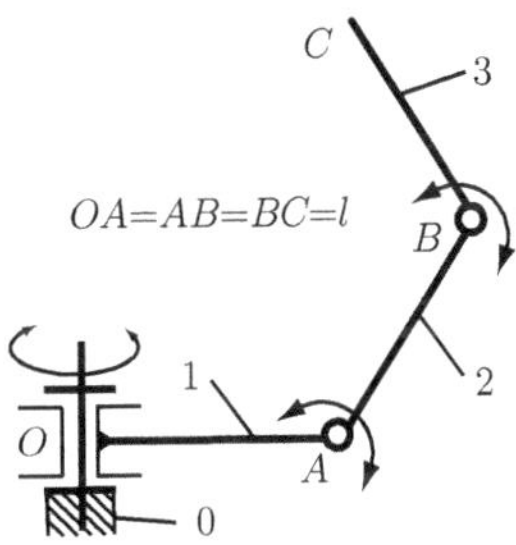

Fig. 7.19   Problem 7.10.4

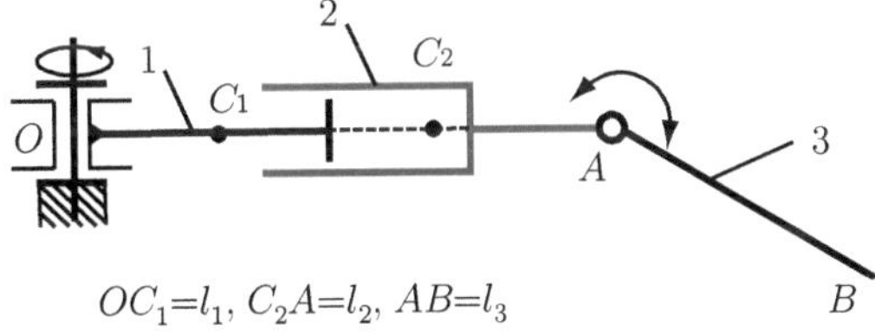

$$OC_1 = l_1, \ C_2A = l_2, \ AB = l_3$$

Fig. 7.20 Problem 7.10.5

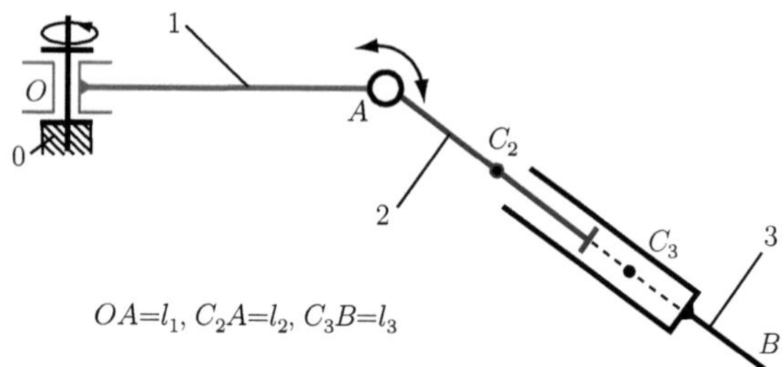

$$OA=l_1,\ C_2A=l_2,\ C_3B=l_3$$

References

1. Ardema MD (2005) Analytical dynamics. Kluwer, Dordecht
2. Artobolevski II (1977) Mechanisms in modern engineering design. MIR, Moscow
3. Atanasiu M (1973) Mecanica. EDP, Bucharest
4. Baruh H (1999) Analytical dynamics. WCB/McGraw-Hill, Boston
5. Baumann G (2005) Mathematica for theoretical physics: classical mechanics and nonlinear dynamics. Springer, Berlin
6. Baumann G (2005) Mathematica for theoretical physics: electrodynamics, quantum mechanics, general relativity and fractals. Springer, Berlin
7. Bedford A, Fowler W (1999) Dynamics. Addison Wesley, Menlo Park, CA
8. Beer FP, Johnston ER Jr (1996) Vector mechanics for engineers: statics and dynamics. McGraw-Hill, New York
9. Boresi P, Schmidt RJ (2001) Engineering mechanics. Brooks/Cole Thomson Learning, Belmont
10. Buculei MI (1976) Mechanisms. University of Craiova Press, Craiova
11. Buculei MI, Bagnaru D, Nanu G, Marghitu DB (1986) Analysis of mechanisms with bars. Scrisul Romanesc, Craiova
12. Bolcu D, Rizescu S (2001) Mecanica. EDP, Bucharest
13. Bolcu D, Tarnita D (2009) Methods for kinetostatics analysis of plane mechanisms. EDP, Bucharest
14. Crespo da Silva M (2004) Intermediate dynamics for engineers. McGraw-Hill, New York
15. Das BM, Kassimali AS, Sami S (1994) Engineering mechanics. Richard D. Irwin, Inc., Homewood, IL
16. Erdman AG, Sandor GN (1984) Mechanisms design. Prentice-Hall, Upper Saddle River, NJ
17. Etter DM, Kuncicky DC (1996) Introduction to MATLAB for engineers and scientists. Prentice Hall, Upper Saddle River, NJ
18. Freudenstein F (1971) An application of Boolean algebra to the motion of epicyclic drives. Trans ASME J Eng Industry 93:176–182
19. Ginsberg JH (1995) Advanced engineering dynamics. Cambridge University Press, Cambridge
20. Goldstein H (1989) Classical mechanics. Addison-Wesley, Redwood City
21. Greenwood DT (1998) Principles of dynamics. Prentice-Hall, Englewood Cliffs, NJ
22. Hibbeler RC (1995) Engineering mechanics – statics and dynamics. Prentice-Hall, Upper Saddle River, NJ
23. Howland RA (2006) Intermediate dynamics: a linear algebraic approach. Springer, Berlin
24. Kane TR (1959) Analytical elements of mechanics, vol 1. Academic, New York
25. Kane TR (1961) Analytical elements of mechanics, vol 2. Academic, New York

D.B. Marghitu and M. Dupac, *Advanced Dynamics: Analytical and Numerical Calculations with MATLAB*, DOI 10.1007/978-1-4614-3475-7,
© Springer Science+Business Media, LLC 2012

26. Kane TR, Levinson DA (1983) The use of Kane's dynamical equations in robotics. MIT Int J Robotics Res 2(3):3–21
27. Kane TR, Likins PW, Levinson DA (1983) Spacecraft dynamics. McGraw-Hill, New York
28. Kane TR, Levinson DA (1985) Dynamics. McGraw-Hill, New York
29. Maeder R (1990) Programming in mathematica. Addison-Wesley, Redwood City, CA
30. Madsen NH, Statics and dynamics, class notes, available at http://www.eng.auburn.edu/users/nmadsen/ Accesssed April 2012
31. Manolescu NI, Kovacs F, Oranescu A (1972) The theory of mechanisms and machines. EDP, Bucharest
32. Marghitu DB (2001) Mechanical engineer's handbook. Academic, San Diego, CA
33. Marghitu DB, Crocker MJ (2001) Analytical elements of mechanisms. Cambridge University Press, Cambridge
34. Marghitu DB (2005) Kinematic chains and machine component design. Elsevier, Amsterdam
35. Marghitu DB, Kinematics and dynamics of machines and machine design, class notes, available at http://www.eng.auburn.edu/users/marghitu/ Accesssed April 2012
36. Marghitu DB (2009) Mechanisms and robots analysis with MATLAB. Springer, New York
37. Meriam JL, Kraige LG (1997) Engineering mechanics: dynamics. Wiley, New York
38. McGill DJ, King WW (1995) Engineering mechanics: statics and an introduction to dynamics. PWS Publishing Company, Boston
39. Meirovitch L (2003) Methods of analytical dynamics. Dover, New York
40. Mott RL (1999) Machine elements in mechanical design. Prentice Hall, Upper Saddle River, NJ
41. Myszka DH (1999) Machines and mechanisms. Prentice-Hall, Upper Saddle River, NJ
42. Norton RL (1996) Machine design. Prentice-Hall, Upper Saddle River, NJ
43. Norton RL (2004) Design of machinery. McGraw-Hill, New York
44. O'Reilly OM (2008) Intermediate dynamics for engineers. Cambridge University Press, Cambridge
45. Pars LA (1965) A treatise on analytical dynamics. Wiley, New York
46. Popescu I (1990) Mechanisms. University of Craiova Press, Craiova
47. Reuleaux F (1963) The kinematics of machinery. Dover, New York
48. Samin JC, Fisette P (2003) Symbolic modeling of multibody systems. Kluwer, Dordecht
49. Shabana AA (2010) Computational dynamics. Wiley, New York
50. Shames IH (1997) Engineering mechanics: statics and dynamics. Prentice-Hall, Upper Saddle River
51. Shigley JE, Uicker JJ (1995) Theory of machines and mechanisms. McGraw-Hill, New York
52. Smith D (2008) Engineering computation with MATLAB. Pearson Education, Upper Saddle River, NJ
53. Soutas-Little RW, Inman DJ (1999) Engineering mechanics: statics and dynamics. Prentice-Hall, Upper Saddle River, NJ
54. Sticklen J, Eskil MT (2006) An introduction to technical problem solving with MATLAB, Great Lakes, Wildwood, MO
55. Stoenescu ED (2005) Dynamics and synthesis of kinematic chains with impact and clearance. Ph.D. Dissertation, Mechanical Engineering, Auburn University
56. The MathWorks: http://www.mathworks.com/ Accesssed April 2012
57. Scola V (1970) Theory of mechanisms and machines (teoria mehanizmov i masin). Minsk, Russia
58. Voinea R, Voiculescu D, Ceausu V (1983) Mechanics (Mecanica). EDP, Bucharest
59. Waldron KJ, Kinzel GL (1999) Kinematics, dynamics, and design of machinery. Wiley, New York
60. Williams JH Jr (1996) Fundamentals of applied dynamics. Wiley, New York
61. Wilson CE, Sadler JP (1991) Kinematics and dynamics of machinery. Harper Collins College Publishers, New York
62. Wilson HB, Turcotte LH, Halpern D (2003) Advanced mathematics and mechanics applications using MATLAB. Chapman & Hall/CRC, London
63. Wolfram S (1999) Mathematica. Wolfram Media/Cambridge University Press, Cambridge

Index

A

Absolute
  angular acceleration, 325, 396, 400
  angular velocity, 323, 324, 329, 395, 398
  value, 1, 3, 406
Angle, 9, 13, 14, 36, 37, 45, 48, 49, 52, 65–68,
    98, 99, 101, 102, 116, 144, 145, 150,
    152, 154, 155, 161, 173, 189, 195, 200,
    202, 204, 205, 208, 223, 238, 244, 246,
    259, 267, 271, 289, 300, 329, 335, 338,
    339, 342–344, 348, 350, 352, 356, 365,
    372, 374, 376, 377, 379, 394, 397,
    400–404, 429, 462, 465, 466, 470, 477,
    482, 488, 495, 510–516, 518, 519, 533,
    534, 536, 544, 552, 558, 567, 595,
    596
Angular
  acceleration, 144, 155, 157, 257, 269, 285,
    291, 292, 306, 309, 325, 326, 329, 335,
    352, 356, 381, 385, 391, 396, 400–402,
    424, 429, 430, 437, 463–466, 472, 482,
    489, 497, 498, 569, 570, 586
  impulse, 237–238
  momentum, 237, 238, 245, 414–426, 428,
    532
  position, 144
  velocity, 144, 145, 156, 197, 202, 204,
    205, 245, 267, 268, 276, 279, 284, 291,
    292, 294, 296, 300–304, 306, 309–311,
    318, 322–325, 329, 332–335, 350–357,
    365, 381, 383, 389, 393–398, 400, 402,
    404–406, 418–421, 424, 426, 428, 463,
    464, 466, 471, 479, 482, 489, 497, 498,
    552, 568–570, 572–574, 577, 579, 586
Associative, 4, 34

B

Base, 17
Bilinearity, 532
Binormal, 161, 162, 164, 165, 190, 192, 193,
    197
Body centrode, 311
Body-fixed reference frame, 281, 286, 288,
    296, 304–306, 365
Bound vector, 2

C

Cartesian, 6, 9, 10, 14, 20, 22, 23, 26–28, 31,
    35, 48, 50, 53, 74, 75, 78, 87, 88, 92,
    94, 127, 147–148, 152, 155, 156, 166,
    203, 211–213, 219, 228, 229, 268, 271,
    281, 282, 329, 339, 373, 381, 414, 427,
    430, 521, 537, 544, 558, 565, 595
Cauchy, 20–22, 27, 474
Central principal moments, 91, 420
Centripetal acceleration, 157
Centrode, 311–321
Centroid, 73–141, 347, 348
Centroidal axis, 80, 94, 95, 420
Circular motion, 154–155, 256
Closed
  kinematic chain, 322–328, 335–357,
    436–446
  loop equation, 321–328
  loop method, 328–335
Commutative, 4, 11, 14, 34
Configuration space, 521–522, 530
Constraint
  conditions, 344, 377
  configuration, 522–524

D.B. Marghitu and M. Dupac, *Advanced Dynamics: Analytical and Numerical*
Calculations with MATLAB, DOI 10.1007/978-1-4614-3475-7,
© Springer Science+Business Media, LLC 2012

Constraint (*cont.*)
    equation, 522, 523
    equations in velocity form, 523
    force, 522
    holonomic, 523, 524
    non-holonomic, 524
    relation, 523
Contour method, 392, 436, 452
Contraction, 35, 36
Coordinates
    cylindrical, 157–158, 203, 214–215, 229
    generalized, 462, 470, 477, 482, 488, 492,
        497, 521–523, 525, 527, 537, 539, 544,
        552, 555, 558, 567, 568, 577, 584
Coplanar, 56, 82, 86
Coriolis acceleration, 157, 269, 306, 307, 309,
        322, 326, 328, 365, 368–370, 385,
        386
Cross product, 13–15, 45, 49, 52, 53, 88, 161,
        237, 465, 586
Curvature, 150–153, 155, 161, 164, 165, 177,
        181, 182, 202, 203, 205, 311, 314
Curve, 75–79, 81–83, 104, 127, 130, 135, 149,
        150, 161, 163, 164, 175, 193, 205, 206,
        213, 254, 277, 279, 311, 314, 315, 468,
        476, 595

D
D'Alembert's principle, 431–432, 525
Decomposition, 74, 78, 106
Degrees of freedom, 295, 329, 470, 477, 487,
        488, 497, 522, 525
Density, 78, 86, 100, 101, 112, 113, 122, 133,
        141, 446, 449, 487, 510–515, 519, 520
Derivative, 18–19, 98, 143, 145, 146, 149, 150,
        153, 154, 156, 158–163, 173, 182, 186,
        187, 198, 224, 236, 237, 245, 268, 269,
        282, 283, 285, 288, 300, 302, 305, 306,
        349, 350, 352, 412, 416, 424, 465, 470,
        471, 479, 488, 489, 492, 497, 499, 522,
        530, 531, 540, 569, 571
Determinant, 14–16, 24, 30, 57, 58, 62
Diagonal elements, 15, 36
Differential, 75, 78, 86, 87, 94, 101, 102, 104,
        109–111, 116, 122, 124, 127, 228, 240,
        411, 412, 414, 465, 467, 468, 474, 476,
        477, 495, 523, 529, 530, 543, 582, 583,
        592
Direction
    angle, 534
    cosines, 9, 10, 22, 42–44, 49, 66, 67, 87,
        91, 534

Displacement
    infinitesimal, 216–218
    relative, 277
    vertical, 273
    weighed average, 414
Distributive, 4, 12, 14, 35
Dot product, 11–13, 88, 159, 161, 216, 220,
        524, 545, 573, 574, 586
Driver
    link, 335–337, 339, 341, 344, 348, 353,
        372, 377, 379, 393, 396, 400, 402–404,
        445, 446, 452, 510–515
    moment, 461, 595
Dyad, 335, 337–339, 342, 348–350, 432–435,
        437–444, 450, 451, 573–575, 586, 597
Dynamical system, 521–523
Dynamics, 77, 196, 209–280, 411–520,
        522–600

E
Elastic
    constant, 277, 518, 537, 594, 595
    force, 277
Ellipsoid of inertia, 92–93
Energy, 215–218, 221–230, 232, 233, 248,
        250, 280, 421–423, 526–528, 530,
        538–540, 545, 546, 549, 552, 553, 560,
        561, 573–575, 593
Equilibrium
    equations, 431, 436, 445
    moment, 445, 446
    position, 462
Euler equation, 430, 465, 472, 473, 480, 484,
        491, 492, 515–517
External force, 216, 218, 236, 245, 307,
        412–415, 428, 429, 431, 432, 436, 446,
        456, 510, 512–514

F
First moment, 73–74, 79–81, 84, 95, 102,
        104–106, 115, 116, 129
Fixed
    axis, 289–291, 294, 299, 405, 426–427,
        429, 464
    centrode, 311, 320, 321
    point, 37, 90, 168, 171, 244, 250, 281, 299,
        319, 417, 420–422, 425, 426, 429–431,
        465, 472, 480
    reference frame, 197, 200, 232, 268, 269,
        281, 283, 286, 317, 323, 325, 365–367,
        397, 424, 426, 495, 497, 509, 568, 586

Force, 1–3, 67, 68, 77, 81, 82, 209–219, 221,
  227–236, 245, 247, 250, 255–257, 259,
  263, 267, 270, 277, 279, 280, 307, 309,
  412–416, 423, 426, 428–446, 449–461,
  464–466, 471, 474, 479, 480, 482, 486,
  491, 492, 503, 504, 510–516, 522,
  525–527, 540, 541, 546, 553, 557, 562,
  563, 577–580, 585–593
Free
  body diagram, 250, 432–436, 493
  vector, 2, 17, 288, 302
Frenet, 159–166, 193–197
Friction, 217, 227, 246, 267, 277, 279, 431,
  464, 477, 515, 517, 518, 544, 595

G
Generalized
  active force, 525–528, 540, 557–580
  coordinate, 462, 470, 477, 482, 488, 492,
    497, 521–523, 525, 527, 537, 539, 544,
    552, 555, 558, 564, 567, 568, 577, 584
  inertia force, 525, 526, 585–593
  velocities, 522, 528
Gradient, 229, 524
Gravity, 82, 210–212, 247, 257, 462, 465, 469,
  477, 482, 503, 515–519, 540, 546, 552,
  553, 557
  center, 82, 130, 586
Guldinus-Pappus, 82, 113, 115

H
Hamilton equations, 528–531, 551, 552, 555,
  556
Helical motion, 294–298

I
Impulse, 234–238, 415
Impulsive force, 235
Independent
  contour, 330, 392, 397
  vector, 5
Inertial reference frame, 211, 237, 307–309,
  462, 482, 525
Inertia matrix, 87, 91, 116–119, 126, 127, 129,
  130, 419, 483, 504, 573, 574, 586
Initial conditions, 178, 179, 238, 252, 256–258,
  261, 264, 271, 279, 467, 468, 474, 475,
  477, 481, 482, 487, 495, 518–520,
  555–557

Instantaneous
  center, 309–311
  radius of curvature, 151, 152, 155, 165
Invariant, 36, 91, 97, 127

J
Jacobian, 524
Jacobi identity, 532
Joint
  rotational, 382, 392, 397, 597
  translational, 392, 397, 453, 455, 477, 597
Joules, 217, 218

K
Kinematic chain, 322–328, 330, 355–357,
  436–446, 477, 518, 565
Kinematics, 143–208, 281–409, 436–446, 477,
  479, 489, 518, 522, 560, 565, 572
Kinetic energy, 216–218, 221–223, 248, 250,
  280, 421–423, 526, 530, 538–540, 545,
  552, 553, 560, 561, 573–575, 593
Kronecker
  delta, 23, 30, 31
  matrix, 23

L
Lagrange
  equations of motion, 527, 558
  identity, 20–22
Lagrangian, 528, 530, 549, 554
Left-handed, 14
Leibnitz property, 532
Linear
  combination, 5, 17
  independence, 5
  momentum, 209, 234–237, 242, 243,
    414–417
  space, 35
  spring, 218–219, 227, 595
Linearity, 532
Line of action, 1, 2, 291
Link, 166, 168, 171, 173, 197, 202, 279,
  312–324, 326, 328–330, 335–337, 339,
  341–348, 351–353, 355, 356, 365–368,
  372, 376, 377, 379, 381–393, 395–398,
  400–406, 432–446, 449–452, 455–459,
  461–466, 469–474, 477, 479, 480,
  487–493, 497–520, 552, 553, 558, 560,
  562, 565, 567–579, 584, 587, 596, 597

Load, 81, 432, 452
Loop, 31, 32, 43, 169, 170, 196, 197, 321–335,
    344, 377

M
Magnitude, 1, 3, 6, 8, 10, 13, 29, 36, 37, 39,
    40, 42, 47, 48, 52, 65–67, 88, 143,
    145–147, 149, 153, 156, 158, 159, 161,
    165, 166, 168, 171, 175, 176, 181, 182,
    186, 191, 194, 208–210, 214, 216, 227,
    235, 241, 245, 272, 282, 284, 286, 288,
    291, 292, 300, 318–320, 358–360, 412,
    428, 430, 431, 510, 512–514, 545
Mass
    center, 77–79, 88–91, 93, 100–102,
        104–111, 114–116, 124, 129, 133, 134,
        243, 411–414, 417, 419, 424, 427–431,
        446, 464, 465, 469, 471, 477, 479, 482,
        499, 500, 502, 517, 544, 552, 553, 558,
        562, 567, 570–574, 584–586, 596
    element, 86, 87, 91, 411
Maximum second moment, 98, 99
Method of decomposition, 78
Minimum second moment, 99
Mobile reference frame, 281, 288, 296, 298
Module, 1, 3, 92, 329
Moment of inertia, 85–89, 91–93, 100,
    117–119, 122–124, 126, 128, 129, 133,
    134, 136, 137, 141, 428–430, 437, 449,
    452, 464, 465, 472, 477, 491, 517, 553,
    560, 587
Momentum vector, 532
Monoloop, 322
Motion
    angular, 144–145
    circular, 154–155, 256
    curvilinear, 147–158, 217
    free fall, 259, 264
    helical, 294–298
    oscillatory, 166, 168, 171, 173
    planar, 285, 298–299, 306, 311, 328, 367,
        427–429
    rectilinear, 146–147, 166
    relative, 143, 158–159, 197, 304, 309, 322,
        326, 431
    rotational, 92, 284, 286, 289–295, 299, 335
    spatial, 299
    spherical, 299
    straight line, 146, 213
    three-dimensional, 165, 215, 422
    translational, 286–289, 299
Moving centrode, 311–321

N
Newton
    second law, 209–216, 234, 237, 250, 260,
        270, 271, 307–309, 412, 414, 431,
        525
    third law, 235, 412, 413
Newton-Euler, 430, 465, 472, 473, 480, 484,
    491, 492, 515, 517–519
Non
    accelerating, 307, 308
    centroidal axes, 95
    holonomic, 524
    homogeneous, 100
    linear, 465, 474
    parallel, 298, 300
    rotating, 307, 308
    zero, 5, 17, 20, 21, 28, 32
Norm, 1, 8, 44, 49, 361, 371, 372, 382,
    388–390, 392
Normal
    acceleration, 177, 292
    coordinates, 148–155, 163, 250
    direction, 159, 214, 250
    plane, 193
    unit vector, 150, 151, 160, 190, 192
    vector, 161, 164, 192, 193

O
Orientation, 1–3, 10, 79, 91, 96, 99, 161, 165,
    300, 302, 303, 394, 397, 429, 533
Orthogonal
    axes, 80, 97, 99
    component, 6, 161, 162, 200, 282, 535
    matrix, 26
    reference frame, 6
    transformation, 36
Orthonormal
    basis, 197
    matrix, 535
Osculating plane, 160, 161, 165, 190, 193

P
Parallel, 1–4, 9–11, 13, 16, 20, 68, 88, 89,
    91, 122, 146, 157, 159, 165, 186, 214,
    237, 238, 286, 288, 289, 297, 301, 302,
    311, 317, 328, 367, 383–386, 389–391,
    404, 424, 428, 451, 495, 497, 524, 567,
    572–574, 586, 597
    axis theorem, 93–97, 117, 118, 125, 126,
        129, 430
Parallelogram law, 3, 4

Partial
    derivative, 165–166, 497, 499, 540, 571
    solution, 335
Particle, 73, 77, 78, 82, 85, 143–280, 324, 326,
        411–418, 421, 428, 495, 497, 501–505,
        509, 510, 521–525, 578, 585, 586, 593,
        595, 596
Period, 13, 166, 171, 182, 201, 223, 254, 415
Permutation
    symbol, 29–34, 65, 69, 70
    tensor, 34
Point of application, 2, 442, 455, 456, 479
Poisson
    bracket, 531–533
    formulas, 284, 288, 305
Polar
    coordinates, 155–157, 186, 189, 203, 214,
        219, 227
    moment of area, 96–97, 141
    moment of inertia, 119, 123, 124, 134, 136,
        137
Position vector, 10–11, 39, 45, 73–75, 77, 78,
        85, 143, 146–148, 155, 158, 163, 175,
        177, 189, 198, 200, 204, 236, 237, 245,
        247, 250, 268, 281, 284, 286, 304, 305,
        325, 331, 341, 342, 359, 365, 366, 374,
        375, 411, 414, 418, 421, 437, 439, 440,
        442, 445, 450, 451, 462, 470, 471, 477,
        479, 482, 486, 488, 489, 499, 501, 502,
        521, 522, 524, 537, 544, 558, 570–572,
        584, 585
Potential energy, 221, 227–233, 250, 527, 528,
        530, 549, 552, 553
Power, 8, 217–220
Primary reference frame, 269, 281, 304–309
Principal
    axes, 90–93, 97–100, 420–423, 425, 504,
        572–574, 586, 597
    axes of inertia, 420, 425, 572
    centroidal moment, 420, 422, 425
    direction, 116, 119, 120, 124, 127
    ellipsoid of inertia, 93
    moment of inertia, 91, 100
Product
    of area, 94–96, 98, 99
    of inertia, 90, 91, 94, 116, 117, 119, 125,
        129, 137
Projectile, 212, 277

R

Radius
    of curvature, 150–153, 155, 161, 164, 165,
        177, 181, 182, 202, 203, 205, 314
    of gyration, 87, 92

Rectangular component, 36, 52, 414
Rectifying plane, 190, 193
Reference frame, 6, 10, 19, 22–24, 26–28, 35,
        36, 75, 79, 87, 100, 106, 109, 111, 116,
        143, 146–148, 152, 158, 160, 165, 166,
        171, 189, 193, 194, 197, 198, 200, 201,
        211, 212, 219, 228, 232, 233, 237, 238,
        268–271, 281–283, 285, 286, 288, 289,
        296, 298, 300, 303–309, 317, 323–325,
        339, 353, 365–367, 369, 373, 381, 393,
        397, 424, 426, 427, 462, 482, 495, 497,
        499, 504, 509, 521, 525, 526, 533–537,
        552, 558, 565, 567–571, 586, 595
Relative
    acceleration, 286, 306, 309, 326, 367, 383,
        385
    displacement, 277
    motion, 143, 158–159, 197, 304, 309, 322,
        326, 431
    position, 283
    velocity, 284, 306, 321, 322, 324, 367, 394
Resolution
    of components, 5–8
    of vectors, 5–8
Resultant, 4, 5, 42, 44, 45, 48, 49, 65–67, 81,
        82, 277, 279, 414, 431, 525
Rheonomic, 524
Right-handed, 2, 13, 14, 31
Rigid body, 2, 86, 87, 89–93, 281–409,
        411–520, 568, 573, 574, 578, 586, 593
Rolling motion, 430–431
Rotating unit vector, 145–146, 156
Rotation transformation, 533–536

S

Scalar
    product, 11, 13, 20, 175
    triple product, 15–18, 31, 56, 59–61, 422
Scleronomic, 524, 530
Second moment, 87, 93–99
Sense, 1–3, 9–11, 14, 161, 288, 291, 318, 404,
        464, 516, 519
SI units, 210, 211, 217, 218
Sliding
    direction, 338, 342, 356, 367, 383, 385,
        389, 391, 434, 435, 440, 441, 443, 444,
        453, 454, 456, 458, 459, 461, 492
    line, 544
    vector, 2
Solid, 38, 75, 77–78, 132, 313
Space centrode, 311
Spatial angular momentum, 417–421
Spring constant, 219, 229, 250, 276

Strength, 73, 74, 77
Surface, 1, 2, 75–79, 82–84, 92, 102, 113, 115,
        116, 133, 205, 211, 246, 430, 431, 515,
        522–524, 544
Symbolic, 7, 8, 10, 15, 17, 31, 37, 39, 40, 53,
        57, 61, 101, 169, 189, 231, 240, 361,
        466, 539, 540, 545
Symmetry
    anti, 532
    axis, 80, 94, 101, 102, 108, 109, 111, 122,
        124
    plane, 74, 80
    skew, 35, 532

T
Tangential coordinates, 148–154
Temperature, 1
Tensor
    alternating, 29, 30
    antisymmetric, 30
    completely antisymmetric, 30
    inertia, 419, 420
    magnitude, 29
    permutation, 34
    second order, 34–36
    symmetric, 29, 35, 420
    zero-order, 36
Tensorial product, 28
Time derivative, 19, 143, 145, 146, 149, 150,
        153, 154, 156, 158, 159, 186, 187, 198,
        236, 237, 269, 305, 306, 465, 531
Torsion, 162–164, 205
Transfer theorem, 94–96, 464
Translation, 88–90, 286–289, 295, 296, 298,
        299, 311, 552
    pure, 286, 287

Transmissible vector, 2, 3
Transpose, 19, 24, 29, 535, 539, 570, 578, 584,
        589, 590
Triangle inequality, 20

U
Unconstrained dynamical system, 524
Unit vector, 3, 5, 6, 8–14, 22, 39–41, 65,
        67, 75, 87, 91, 145–146, 149–152,
        155–161, 163, 165, 186, 190, 192, 193,
        195, 200, 215, 217, 219, 238, 245, 268,
        281, 282, 288, 304, 495, 497, 524,
        533–535, 537, 565, 567, 569–572
Unity tensor, 28, 29

V
Variation, 528
Varignon theorem, 14
Vector
    addition, 3–4
    (cross) product, 13–15, 52
Velocity, 1, 143, 209, 281, 414, 523
Volume, 17, 75, 76, 78, 82, 84, 86, 88,
        109–111, 113, 114, 130, 133, 141

W
Work, 1, 8, 215–222, 224, 225, 227, 229–234,
        247, 248, 422, 572

Z
Zero vector, 3, 6, 10

MIX
Papier aus verantwortungsvollen Quellen
Paper from responsible sources
FSC® C105338

If you have any concerns about our products,
you can contact us on
ProductSafety@springernature.com

In case Publisher is established outside the EU,
the EU authorized representative is:
Springer Nature Customer Service Center GmbH
Europaplatz 3, 69115 Heidelberg, Germany

Printed by Libri Plureos GmbH
in Hamburg, Germany